The Receptors

Volume 26

Series Editor

Giuseppe Di Giovanni
Department of Physiology & Biochemistry Faculty of Medicine and Surgery,
University of Malta, Malta.

The Receptors book Series, founded in the 1980's, is a broad-based and wellrespected series on all aspects of receptor neurophysiology. The series presents published volumes that comprehensively review neural receptors for a specific hormone or neurotransmitter by invited leading specialists. Particular attention is paid to in-depth studies of receptors' role in health and neuropathological processes. Recent volumes in the series cover chemical, physical, modeling, biological, pharmacological, anatomical aspects and drug discovery regarding different receptors. All books in this series have, with a rigorous editing, a strong reference value and provide essential up-to-date resources for neuroscience researchers, lecturers, students and pharmaceutical research.

More information about this series at
http://www.springer.com/series/7668

Robin A.J. Lester
Editor

Nicotinic Receptors

Humana Press

Editor
Robin A.J. Lester
Department of Neurobiology
University of Alabama at Birmingham
Birmingham, AL, USA

ISBN 978-1-4939-1166-0 ISBN 978-1-4939-1167-7 (eBook)
DOI 10.1007/978-1-4939-1167-7
Springer New York Heidelberg Dordrecht London

Library of Congress Control Number: 2014947538

Printed on acid-free paper

Humana Press is a brand of Springer
Springer is part of Springer Science+Business Media (www.springer.com)

Foreword

Few molecules in the nervous system have been studied as intensively by so many scientists as the family of nicotinic acetylcholine receptors (nAChRs). From the first description of the "receptive substance" by Langley in 1905 to ongoing development of medications for smoking cessation, pain, cognitive dysfunction, and other neurological or psychiatric illnesses, studies of nAChRs have served as models for exploration of receptors in the nervous system for more than 100 years.

nAChRs were the first receptors to be reconstituted into a lipid bilayer, the first channels to be recorded in a patch clamp preparation, and the first neurotransmitter receptors to be cloned. The presence of nAChRs at the neuromuscular junction and the high conservation of these receptors from *Caenorhabditis elegans* to Torpedo electricus up to *Homo sapiens* has made them a model for biophysical and structure-function studies. The ubiquity of the nAChRs in the brain, peripheral nervous system, and non-neuronal tissues has allowed studies to be performed on complex functions in areas as diverse as homeostasis, motor control, mood, reward, and cognition. The community of nicotinians (scientists studying nAChRs) includes structural biologists, biophysicists, biochemists, cell biologists, physiologists, anatomists, pharmacologists, behavioral scientists, radiologists, clinicians, and more. The nicotinians provide an excellent example of how data obtained at one level of complexity can provide insights into many other levels of biological inquiry. The study of nAChRs provides the ultimate potential for translation of very basic science to studies of therapeutic relevance for human patients.

A unique bioassay for the function of nAChRs in a complex system comes from the addiction to tobacco smoking in humans that was imported from the Americas to Europe by Sir Walter Raleigh in the late sixteenth century. Many clues to the behavioral consequences of nAChR function and dysfunction come from human subjects who report the effects of the nicotine in tobacco on their subjective experience. An understanding of the role of nAChRs in nicotine reinforcement and withdrawal, as well as clues about the genetic basis for susceptibility to addiction, has come from studies of these receptors in human smokers. A fundamental understanding of the biophysical properties of nAChRs has been an important tool in medication development for smoking cessation, and this, in turn, has led the way to development of potential therapeutics to treat other indications as diverse as myasthenia gravis to schizophrenia.

The multiple levels at which studies of nAChR structure, localization, and function have enhanced our understanding of fundamental biological systems are covered in this volume. A historical perspective from Dr. Robin A.J. Lester sets the stage for understanding how pioneering studies of these receptors have paved the way for studies of neurotransmitter receptors of many classes. Several chapters provide an understanding of the nAChR family at the molecular level. Dr. Steen Pedersen discusses structure-function relationships in nAChR gating, Dr. Paul Gardner describes the determinants of transcriptional regulation of the receptor family, and Dr. Michael Marks and Dr. Sharon Grady review the presynaptic function of nAChRs in neurotransmitter release from synaptosomes. Another set of chapters provides a framework for understanding how the electrophysiological properties of these receptors can alter the function of different brain circuits. Dr. John Dani reviews the homomeric α7 nAChR subtype and the effects of nAChR signaling during development are covered by Dr. Darwin Berg. Chapters on the distribution and function of nAChRs by Dr. Jerry Yakel, the presynaptic regulation of network activity by nAChRs from Dr. Lorna Role, and the slow nicotinic responses seen in several brain areas covered by Dr. Rory McQuiston complete the reviews at the electrophysiological level.

At the systems level, reviews of the effects of nAChRs in the autonomic ganglia by Dr. Peter Sargent and the spinal cord by Dr. Philippe Ascher show how nAChRs play essential roles in the physiology of critical neurobiological systems that carry out essential homeostatic functions. At the behavioral level, a review of the role of nAChRs in learning and memory by Dr. Tom Gould provides a larger context in which to understand how the modulation of neuronal excitability and brain networks by this receptor family can alter complex responses to the environment.

Numerous levels of nAChR function are critical to understanding smoking behavior, including the role of nAChRs in reward and withdrawal reviewed by Dr. Andrew Tapper and Dr. Mariella De Debiasi, respectively; the genetics of human nAChR variants affecting tobacco addiction discussed by Dr. Jerry Stitzel, Dr. Laura Bierut, and Dr. Inez Ibanez-Tallon; and the interactions with other neurotransmitter systems, such as the serotonin system, outlined by Dr. Giuseppe Di Giovanni. It is fitting that the volume closes with a number of reviews outlining how dysfunction of various nAChRs can contribute to human illness, such as the neurodegeneration in Alzheimer's and Parkinson's disease covered by Dr. Kelly Dineley and the cognitive dysfunction in schizophrenia summarized by Dr. Sherry Leonard, and how nAChRs may be therapeutic targets for treatment of human disorders from pain, discussed by Dr. M. Imad Damaj, and other CNS disorders, reviewed by Dr. Stephen Arneric and Dr. Mani Sher.

This is a rich volume that ties together the historical context for studies of nAChRs to current-day problems in systems neurobiology and human disease that can be approached only because of the fundamental molecular studies that have provided information on the structure, function, and anatomy of the nAChR family. This book provides optimism about how far the nicotinic field has advanced, and provides guideposts for where we need continued focus to move this knowledge forward to solve fundamental neurobiological problems that are critical to human health.

Marina Picciotto
marina.picciotto@yale.edu
http://psychiatry.yale.edu/people/marina_picciotto-1.profile

Preface

Not so long ago, there was nothing known about receptors, neurotransmitters, and synapses. This was all to change, beginning around the middle of the nineteenth century, with the investigations of Claude Bernard into the mechanisms underlying drug-induced muscle paralysis, which together with a desire to understand autonomic transmission has led to fundamental insights into synaptic function (reflected by Nobel Prizes awarded in Physiology and Medicine), much of it derived from the cholinergic-nicotinic system. About a hundred years ago, several scientists, in particular the anatomists, Santiago Ramon Y Cajal and Camillo Golgi (1906), together with the physiologists, Charles Sherrington and Edgar Adrian (1932), convinced the scientific community that the basic building blocks of the nervous system were individual neurons that communicated with each other via synapses. Soon after, Henry Dale and Otto Loewi (1936) provided clarity in the "soups and sparks" communication conundrum by identifying acetylcholine as one of the first synaptic neurotransmitters. Next, Julius Axelrod, Ulf von Euler, and Bernard Katz (1970) demonstrated that neurotransmitters were stored in vesicles in the presynaptic terminal, and that chemical transmission was initiated by the influx of calcium ions. While these pioneering studies spurred the field forwards, the postsynaptic nicotinic acetylcholine receptor remained elusive, and it was not until the last quarter of the twentieth century, when Erwin Neher and Bert Sakmann (1991) had sufficiently refined existing techniques, were researchers able to observe the activation of single-nicotinic receptors by the neurotransmitter acetylcholine. By the end of the twentieth century, it had been firmly established that reliable neuromuscular synaptic transmission occurred as a result of the random combination of presynaptically released acetylcholine molecules with postsynaptic nicotinic acetylcholine receptors. We now know more about this receptor than any other and are starting to see how the binding of transmitter/drug initiates the structural twists and turns that open, close, and desensitize the channel. A comprehensive understanding of how nicotinic receptors function, at the molecular level, seems at last to be just over the horizon.

I view nicotinic receptors as one of the heroes of a multi-plot adventure story. By demystifying the spiritual usage of the drug nicotine, we have defined synaptic transmission, from its beginnings at the neuromuscular junction to its seemingly more cryptic deployment in the central nervous system. In doing so we have made major inroads into understanding how cholinergic-nicotinic circuitry contributes to fundamental aspects of sensation and movement, in addition to more complex brain states and behaviors including motivation and reward, learning and memory, and our nebulous conception of consciousness. Along the way we have had to dig deeper into the genetic basis that controls the expression and distribution of this family of receptors, their regulatory importance during development, and their interactions with other neurotransmitter systems, in particular dopamine and serotonin. In turn we have gained valuable insights into mechanisms of neurologic and psychiatric disease, and as a consequence, potential drug intervention strategies are emerging. Moreover, the discrete localization of receptor subtypes throughout the nervous system makes them particularly attractive drug targets if we want to restrict and tweak their activity within specific brain regions.

While nicotinic receptors are the locks under discussion in this book, nicotine remains one of the major keys used to access brain function. As such, addiction to nicotine must be a central theme, not only due to its societal impact, affecting more than 20% of the world's population, but also because it ties together genes, proteins, synapses, circuitry, and behavior, and continues to provide much motivation to understand nicotinic receptors and the brain. Recently by delving into the flip side of reward, and gaining an understanding of the mechanisms of aversion and withdrawal, we have unmasked additional regions of the brain that contribute to the devastation produced, not only by nicotine, but possibly by all drugs of abuse. Nicotine is a somewhat unique drug, a double-edged sword, neither an upper nor a downer, in some ways a mood stabilizer, which not only helps to explain why it is so addictive through self-medication, but also provides an explanation for the involvement of nicotinic receptors in so many psychological states, and as a consequence so many psychiatric disorders. Nicotine addiction, through gene linkage studies that correlate smoking behavior with specific DNA mutations, has more recently reopened the debate into nature versus nurture, and thus presents us with opportunities for tackling this disease on multiple levels.

It is true that good science is inspired and moved forward by rigorous and honest competition, but efficient progress requires the unselfish sharing of ideas and collaborative research. The nicotinic receptor "family" has provided me a relative late arrival to the field, a nurturing environment, in which to develop my own ideas about the role of these receptors in the brain. I feel honored to have been asked to assemble this collection of chapters, which I hope as a compilation reflects the breadth of the field, not only as it stands now, but also its growth towards the future. It is obviously an impossible task to invite everyone to contribute to this volume, although one thing is clear: none of these chapters would have been possible without substantial research from all who work on nicotinic receptors. My personal bias will be apparent in the selection of topics, which I have organized around a synaptic theme, but which I hope successfully brings together genes, molecules, and circuitry in order to explain behavior and disease.

I have tried to include all parts of the nicotinic receptor story in the book.

Birmingham, AL, USA Robin A.J. Lester

Contents

Contributors

Jessica L. Ables, M.D., Ph.D. Laboratory of Molecular Biology, The Rockefeller University, New York, NY, USA

Stephen P. Arneric, Ph.D. Pain and Migraine Drug Hunting Team, Neuroscience Discovery, Lilly Research Centre, Eli Lilly and Company, Indianapolis, IN, USA

Philippe Ascher, Ph.D. Laboratoire de Physiologie cérébrale, Université Paris Descartes, Paris, France

Deniz Bagdas, Ph.D. Experimental Animals Breeding and Research Center, Faculty of Medicine, Uludag University, Bursa, Turkey

Darwin K. Berg, Ph.D. Neurobiology Section, Division of Biological Sciences, University of California, San Diego, La Jolla, San Diego, CA 92093, USA

Mariella De Biasi, Ph.D. Department of Psychiatry, Perelman School of Medicine, University of Pennsylvania, Philadelphia, PA, USA

Laura J. Bierut, M.D. Department of Psychiatry, Washington University School of Medicine, St. Louis, MO, USA

Stephanie Chambers Department of Medical Sciences, Exeter Medical School, St. Luke's Campus, Exeter, Devon, UK

M. Imad Damaj, Ph.D. Department of Pharmacology and Toxicology, Virginia Commonwealth University, School of Medicine, Richmond, VA, USA

John A. Dani, Ph.D. Department of Neuroscience, Perelman School of Medicine, University of Pennsylvania, Philadelphia, PA, USA

Dang Q. Dao, Ph.D. Department Neuroscience, Baylor College of Medicine, Houston, TX, USA

Philippe De Deurwaerdère, Ph.D. Centre National de la Recherche Scientifique, Institut des Maladies Neurodégénératives (UMR CNRS 5293), Universitè de Bordeaux, Bordeaux Cedex, France

Boris Lamotte d'Incamps, Ph.D. Centre de Neurophysique, Physiologie, Pathologie, Université Paris Descartes, Paris, France

Kelly T. Dineley, B.A., M.S. Ph.D. Department of Neurology, University of Texas Medical Branch, Galveston, TX, USA

Catarina C. Fernandes, Ph.D. Neurobiology Section, Division of Biological Sciences, University of California, San Diego, La Jolla, San Diego, CA 92093, USA

Pamela Flood, M.D. Department of Anesthesia and Perioperative Care, University of California at San Francisco, San Francisco, CA, USA

Kelen Freitas Department of Pharmacology and Toxicology, Virginia Commonwealth University, School of Medicine, Richmond, VA, USA

Paul D. Gardner, Ph.D. Department of Psychiatry, Brudnick Neuropsychiatric Research Institute, University of Massachusetts Medical School, Worcester, MA, USA

Giuseppe Di Giovanni, Ph.D. Faculty of Medicine, Department of Physiology and Biochemistry, Malta University, Msida, Malta

Thomas J. Gould, Ph.D. Department of Psychology and Neuroscience, Temple University, Philadelphia, PA, USA

Sharon R. Grady, Ph.D. Institute for Behavioral Genetics, University of Colorado, Boulder, CO, USA

Linzy M. Hendrickson, Ph.D. Department of Psychiatry, Brudnick Neuropsychiatric Research Institute, University of Massachusetts Medical School, Worcester, MA, USA

Ines Ibañez-Tallon, Ph.D. Laboratory of Molecular Biology, The Rockefeller University, New York, NY, USA

Sherry Leonard, Ph.D. Department of Psychiatry, University of Colorado, Anschutz Medical Campus, Aurora, CO, USA

Robin A.J. Lester, Ph.D. Department of Neurobiology, University of Alabama at Birmingham, Birmingham, AL, USA

Gretchen Y. López-Hernández, Ph.D. Department of Neurobiology and Behavior, Center for Brain and Spinal Cord Research, Stony Brook University, Stony Brook, NY, USA

Adrian F. Lozada, Ph.D. Neurobiology Section, Division of Biological Sciences, University of California, San Diego, La Jolla, San Diego, CA 92093, USA

Michael J. Marks, B.S., M.S., Ph.D. Department of Psychology and Neuroscience, Institute for Behavioral Genetics, University of Colorado, Boulder, CO, USA

Richard Martindale, B.S. University of Alabama at Birmingham, Birmingham, AL, USA

Tristan D. McClure-Begley, Ph.D. Institute for Behavioral Genetics, University of Colorado, Boulder, CO, USA

A. Rory McQuiston, Ph.D. Department of Anatomy and Neurobiology, Virginia Commonwealth University, Richmond, VA, USA

Heidi C. O'Neill, Ph.D. Institute for Behavioral Genetics, University of Colorado, Boulder, CO, USA

Michael Paolini, M.D. Department of Neuroscience, Baylor College of Medicine, Houston, TX, USA

Lucy Partridge Department of Bioscience, Cardiff School of Biosciences, South Glamorgan, Wales, UK

Steen E. Pedersen Ph.D. Department of Molecular Physiology and Biophysics, Baylor College of Medicine, Houston, TX, USA

Massimo Pierucci, B.Sc. Faculty of Medicine, Department of Physiology and Biochemistry, Malta University, Msida, Malta

Lorna W. Role, Ph.D. Department of Neurobiology and Behavior, Center for Brain and Spinal Cord Research, Stony Brook University, Stony Brook, NY, USA

Ramiro Salas, Ph.D. Department of Psychiatry, Baylor College of Medicine, Houston, TX, USA

Peter B. Sargent, Ph.D. Department of Cell and Tissue Biology, University of California at San Francisco, San Francisco, CA, USA

Michael D. Scofield, Ph.D. Department of Neurosciences, Medical University of South Carolina, Charleston, SC, USA

Emanuele Sher, M.D., Ph.D. Neuroscience Discovery, Lilly Research Centre, Eli Lilly and Company, Windlesham, Surrey, UK

Jerry A. Stitzel, Ph.D. Institute for Behavioral Genetics, University of Colorado, Boulder, CO, USA

David A. Talmage, Ph.D. Department of Neurobiology and Behavior, Center for Brain and Spinal Cord Research, Stony Brook University, Stony Brook, NY, USA

Andrew R. Tapper, Ph.D. Department of Psychiatry, Brudnick Neuropsychiatric Research Institute, University of Massachusetts Medical School, Worcester, MA, USA

Jerrel L. Yakel, Ph.D. Department of Neurobiology, National Institute of Environmental Health Sciences, Research Triangle Park, NC, USA

Cristian A. Zambrano, Ph.D. Institute for Behavioral Genetics, University of Colorado, Boulder, CO, USA

Chongbo Zhong, M.D., Ph.D. Department of Neurobiology and Behavior, Center for Brain and Spinal Cord Research, Stony Brook University, Stony Brook, NY, USA

Chapter 1
On the Discovery of the Nicotinic Acetylcholine Receptor Channel

Richard Martindale and Robin A.J. Lester

Abstract The discovery and characterization of the nicotinic acetylcholine receptor (nAChR) is in essence the story of receptor pharmacology in general; arguably one of the greatest journeys in neuroscience, spanning more than 150 years. From its beginnings as the site of action of the poison, curare, and the psychotropic drug, nicotine, to its high-resolution structure, it has touched every subfield of biology. It has shaped how transmitter–receptor interactions are analyzed quantitatively, along the way introducing the scientific community to many novel receptor concepts and kinetic mechanisms, in addition to methods, techniques, and/or their refinement, particularly for understanding single channel behavior. Important to note, is that our knowledge of fast synaptic transmission would not be the same without analysis of nAChRs at the neuromuscular junction. Although determination of nAChR function has benefited from parallel discoveries on other proteins, it can be reasonably argued that all ligand-gated ion channels have their roots somewhere in this receptor. We highlight some of the chronological steps in the discovery and characterization of the receptor, together with some of the key players.

Keywords Nicotine • Curare • Bungarotoxin • Hill • Langley • Katz

1 Before the Receptor: A Tale of Two Alkaloids

The discovery of the nicotinic acetylcholine receptor (nAChR) is pure pharmacology; curiosity about how native botanical compounds exert their profound effects on the human body and mind. At the outset, it was, and continues to be, the story of two poisons from the Americas, an agonist, nicotine, and its antagonist curare, and how their mystery became unraveled.

R. Martindale, B.S.
University of Alabama at Birmingham, Birmingham, AL 35233, USA
e-mail: rmmarti4@uab.edu

R.A.J. Lester, Ph.D. (✉)
Department of Neurobiology, University of Alabama at Birmingham,
1825 University Boulevard, SHEL 1006, Birmingham, AL 35294, USA
e-mail: nicotine@uab.edu

R.A.J. Lester (ed.), *Nicotinic Receptors*, The Receptors 26,
DOI 10.1007/978-1-4939-1167-7_1

Nicotine: Human interest in tobacco has endured for centuries. The early American Indians revered the plant. Its psychotropic properties, which they exploited for spiritual, social, medicinal, and recreational purposes, clearly indicated an interaction with the central nervous system (CNS). Soon after Christopher Columbus discovered the Americas, and with it, tobacco, European explorers plundered and bartered for the valuable crop and introduced it to more civilized territories for profit where it would be cultivated on nonnative soil [1]. While it was widely heralded as a praiseworthy plant, salubrious and medicinal in nature, there were others who were much more skeptical of the new foreign wonder. Amongst them, Sir Francis Bacon and King James I, who discuss the observably unpleasant consequences of tobacco use along with its addictive potential [2]:

> …many in this kingdom have had such a continual use of taking this unsavory smoke, as now they are not able to forbear the same no more than an old drunkard can abide to be long sober without falling into an incurable weakness and evil constitution… a custom loathsome to the eye, hateful to the nose, harmful to the brain, dangerous to the lungs, and in the black stinking fume thereof nearest resembling the horrible stygian smoke of the pit that is bottomless.

The intriguing debate as to whether tobacco was of a toxic or curative nature [3, 4], in addition to the curiosity that would surround any botanical novelty from the Americas, led ultimately to the extraction and purification of nicotine. After many centuries of attempts using ever advancing chemical technologies resulted only in partial isolation of nicotine, the credit of discovery would instead come to rest upon the minds of two young students at the University of Heidelberg [5].

Curare: Around the time when the toxicity of nicotine was becoming increasingly apparent, another poisonous alkaloid had become the subject of intense scientific scrutiny. Curare, a poison derived from the plant *chondrodendron tomentosum*, was commonly used by native Indians for hunting, by applying the plant poison to the tips of their arrows to paralyze and kill wild game [6]. Humboldt [7] describes how an extract of curare was prepared by Orinoco River Indians, in a rather sophisticated "chemical laboratory," through a process of boiling and filtration. Even prior to the full discovery of its mechanism of action in the motor system (Bernard and Langley, see below) centuries later, it is clear that the Indians knew something about the action of curare on the body because it was safe to eat the toxin-poisoned meat.

2 Determination of the Site of Action of Curare (1840–1900)

Although the nicotinic acetylcholine receptor (nAChR) eventually became named for its agonist, nicotine (Dale differentiated between the actions of muscarine and nicotine-like compounds [8]), as often is the case in pharmacology, it was the physiological investigation of the antagonist, curare, that allowed scientists to hone in on the anatomical location of the receptor, but not without difficulty. Claude Bernard (Fig. 1.1), possibly because of his high profile and other significant contributions to science, was given much credit for this discovery, but others were involved.

Fig. 1.1 Tombstone of Claude Bernard. Père Lachaise Cemetery, Paris, France

After being exposed to the use of curare in hunting, Bernard experimented with a curare-tipped arrow by inserting it into the subcutaneous tissue of a rabbit's thigh, observing that, although the rabbit was paralyzed within minutes and eventually died, its heart function was curiously unaffected. Such observations complemented those of Brodie [9], who showed that, in addition, as long as respiration is maintained, muscle paralysis does not lead to death. In his attempts to pinpoint where on the motor axis curare acted, Bernard convincingly demonstrated that sensory nerves were unaffected (see below), but perhaps the observation that most seemed to hold him to the view that curare poisoned the motor nerve rather than the muscle it innervated, was that curare did not prevent direct electrical excitation of the muscle (see [10]).

In one set of experiments, Bernard placed a ligature around the waist of a frog in order to restrict the spread of curare. When stimuli were applied to the anterior curarized area, no movement occurred as the muscles were paralyzed; however, reflexive movements were detectable in the non-curarized posterior half of the body in response to sensory stimuli applied in the (poisoned) anterior half, clearly indicating that sensory function remained intact despite the presence of curare.

Having noted that curare acted solely on the motor pathway, Bernard designed "the watch glass" experiment, which involved a motor nerve, the skeletal muscle it innervated, and a bath of curare. He observed that when the muscle was submerged in the curare bath, and the nerve was electrically stimulated while lying outside of the curare bath, the muscle did not contract; however, when the muscle was situated outside of the bath and the nerve within the curare bath was electrically stimulated, this time the muscle contracted. Along with more refined experiments, Bernard deduced that curare acted upon the nerve at the point in which it terminated in the muscle [10, 11].

In the 1860s, Alfred Vulpian, after repeating Bernard's experiments and consideration of the opinions of other physiologists, publicly articulated, but not without vacillation, the idea that curare was acting at some junctional zone between the motor nerve and its muscle. His conclusions did not appear to specify a structure upon which curare acted, but spoke in terms of disrupted communication between nerve and muscle [11].

3 The Receptor Entity (1905)

By the turn of the twentieth century, the physiological synapse had been defined [12], but for the concept of chemical transmission to gel, an additional structure was required to explain the actions of chemicals, and to formally separate the nerve and the effector tissue. To attain this goal, evidence drawn from many preparations including, autonomic ganglia (especially the postganglionic sympathetic system), in addition to the nerve-striated muscle system, was required.

British physiologist John Newport Langley is credited with introducing the world to the word receptor, using the term "receptive substance" for drug action [13], and his influence has been documented in detail elsewhere [10, 14]. Initially contemplating the early experiments of Hirschmann, who, similar to Bernard, suggested that nicotine exerted its physiological effects on the postganglionic tissue via an action on nerve endings, Langley and Dickerson [15], showed that this could not be the case. Thus, they observed that, while application of nicotine blocked the effects of sympathetic stimulation "below" (i.e., after preganglionic stimulation) the superior cervical ganglion (SCG), stimulation "above" (i.e., of postganglionic fibers) was without effect, and correctly concluded that the paralysis occurred at the postganglionic nerve cells in the SCG. In more definitive follow up studies, preganglionic fibers were severed and allowed to fully degenerate, eliminating all possibility that the alkaloid was interfering with electrical activity in the presynaptic nerves. Moreover, rather than being a pure paralytic (an effect that is now attributable to desensitization, see below), like curare, its agonist, or initial excitatory effects were also described [16]:

> The application of warm 1 p.c. nicotine to a sympathetic ganglion in the cat produces effects like those produced by brief stimulation of its pre-ganglionic fibres.

As a side note, and along a similar theme, Langley, inadvertently [17], and with regards to the sympathetic nerves, may have introduced the neurotransmitter concept of "identity of action" prior to the observations of Elliot:

> In many cases the effects produced by the extract and by electrical stimulation of the sympathetic nerve correspond exactly.

It would be many decades later before the local application of acetylcholine (ACh) could be shown to precisely mimic the effects of a single quantum of transmitter released from the presynaptic motor neuron (see [18]). Prior to the conclusions that ACh was the transmitter at the NMJ [19], early ideas on chemical transmission arose out of experiments and speculation on the action of adrenal extract on the sympathetic nervous system, in particular those performed by Elliot (see [10]). These studies, along with scholarly discussions between the two colleagues, were invaluable to Langley's conceptualization of the "accessory substance" for nicotinic compounds. He states [13]:

> Elliott considers that the substance which is stimulated by adrenalin is not an intrinsic part of the muscle, but is developed from the muscle in consequence of its union with a sympathetic fibre. This substance is formed at the junction of the nerve and muscle, and establishes continuity between them, hence he calls it after Brodie and Dixon the myo-neural junction.

However, formalization of the idea that there was a structural interaction between biological substances and an auxiliary substance originated in the form of Emil Fischer's "lock and key" hypothesis in the late 1800s, as he imaged the interaction between enzyme and substrate. A few years later, Paul Erlich advanced his side-chain theory of cells, endowing them immunological specificity when binding to circulating products (see [20]). But it was not until after Langley's seminal work [13] that Ehrlich conformed to the opinion that drugs could indeed interact chemically with the side chains of certain receptors. Erlich is, however, credited with the term "chemoreceptors" (see [10, 20]).

Along the path to elucidating the "receptive substance," Langley made some important and necessary observations regarding drug antagonism. Interestingly, in his earlier work, he had proposed the idea of interaction between certain muscarinic compounds, although a location on the nerve or tissue was not specified at that time [21]. Subsequently, he was able to extend this concept to include nicotinic compounds. When added in combination with one another, it became evident that the muscular contraction induced by nicotine was antagonized by the administration of curare, and vice versa. Langley surmised that these findings suggested a similar or shared mechanism of action for the two poisonous alkaloids, as he wrote [13]:

> From these experiments it is I think clear that curari and nicotine are antagonistic in large as well as in small amounts and that the extent is proportional to the relative amounts of the two poisons… the mutual antagonism can only mean that the two poisons act on the same protoplasmic substance or substances.

Knowing, as Bernard did, that curare did not block direct electrical stimulation of muscle, Langley utilized his previous experimental approach—denervation—in order to remove any doubt left over from the work of Bernard and Vulpian concerning a presynaptic site of action of curare (and/or nicotine) on skeletal muscle. He concluded [13]:

> …that nicotine and curari do not act on the axon-endings but on the muscle itself. Further, since both nicotine and curare prevent nervous impulses from affecting the contractile substance, but do not prevent the muscle from contracting on direct stimulation, I conclude that the poisons do not act directly on the contractile substance, but on other substances in the muscle which may be called **receptive substances**.

The nicotinic acetylcholine receptor (nAChR) had arrived. Although Langley expounded on his receptor theory in later studies (see [14]), his 1905 manuscript [13], can be considered especially influential, and contained several important deductions with implications extending beyond the nature of the "receptive substance." With much credit to Elliott (see [10]), he generalized his observations to all synapses [13]:

> So we may suppose that in all cells two constituents at least are to be distinguished, a chief substance [effector organ], which is concerned with the chief function of the cell as contraction and secretion, and receptive substances [receptors] which are acted upon by chemical bodies and in certain cases by nervous stimuli. The receptive substance affects or is capable of affecting the metabolism of the chief substance.

4 Quantitative Analysis of Receptor Function (1900–1950)

Agonism: With little idea of the physical nature of the "receptive substance," A.V. Hill, under the guidance of Langley, began the journey that would provide a comprehensive account of the interactions (rates and equilibrium concentrations) between drugs and the nAChR (for a thorough historical review see Colquhoun [22]). Hill started from Langley's premise that a "reversible chemical combination" between receptor and drug was responsible for the contraction and relaxation of the muscle. In his initial experiments, Hill [23] carefully measured the onset and offset time courses of the muscle response to varying concentrations of nicotine, showing that they could be fitted by exponential functions. From this result, he considered the rate limiting processes for the contraction and relaxation; either diffusion of nicotine into the tissue or binding of nicotine to the "receptive substance." He eliminated diffusion of nicotine, based on the weak temperature dependence of the rates (but see [24] for a discussion), and settled on the chemical interaction between nicotine, N, and some component of the muscle, A, represented by the reversible bimolecular reaction:

$$A + N \rightleftharpoons NA$$

That being the case he demonstrated that, the 'height of the [muscle] contraction', y, was directly proportional, at any given time, to the amount of NA, itself governed by the law of mass action (where the rate of contraction, but not relaxation, will depend on $[N]$), and writing at the time an early version of the Hill equation (amended for simplicity):

$$y = \frac{N}{k' + kN}$$

where, k and k' are the velocity (rate) constants governing muscle contraction and relaxation, respectively. Despite the fact that the response (muscle contraction) is somewhat downstream from the amount of nicotine-receptor complex, a problem that to this day continues to plague determination of both rate and affinity constants for agonists (see below and Colquhoun, [22, 24–26]), Hill essentially provided the first equation to fit dose–response curves [23].

Further consideration of disparate findings regarding the size and number of 0_2 sites on haemoglobin "aggregates" led to the general equation (again amended) for the combination of multiple molecules to a "receptor" [27], now known as the Hill equation (or Langmuir isotherm, see Colquhoun [22] for discussion):

$$y = \frac{Kx^n}{1 + Kx^n}$$

The infamous Hill coefficient, n, often written as n_{H} was at that time interpreted by Hill as *just* a constant (as was the "affinity constant," K, for that matter), although it has received considerable attention since with respect to the number of molecules needed for receptor activation (see below). Alfred Clark [28] has been credited with the first representation of nAChR data as a Hill plot (albeit using the Langmuir isotherm; see [22]). He estimated a K value of 40 μM for ACh, very close to that derived from more contemporary single channel data (Colquhoun and Ogden, 1988), and concluded that, because n varied around one, it was probably unity (but see below). Clark did, however, show the use of the concentration–response method for comparing agonist potency across different tissues, and thus, albeit indirectly, implicating different types of receptor [28]. Antagonists were later used to provide the first definitive separation of ganglionic (neuronal) and muscle nAChRs (see [29]).

Antagonism: Although Hill analyzed the 'antagonistic' action of curare [23], formal descriptions of competitive drug interactions were not complete until many years later. As Colquhoun [22] discusses, unlike agonists (see the binding-gating problem below), it is relatively straightforward to estimate the dissociation constant (K_{d}; binding affinity) of pure competitive antagonists from the equipotent responses of agonists in the presence and absence of the blocker (dose-ratio). This experimental approach was first recognized and applied by Schild (see [22]), although it was not until later that the method was used to estimate the affinity of tubocurarine at the neuromuscular junction [30].

5 More Complete Receptor Mechanisms (1950–)

Affinity and efficacy: Our picture of ligand-gated ion channel gating in general owes a great deal to the work performed on the muscle type nicotinic acetylcholine receptor, and, in particular, to the development of specific and testable receptor mechanisms [22]. As discussed above, analysis of agonists is complicated primarily because of downstream events following the binding step. Stephenson [31] states concisely the nature of the problem in his paper "A modification of receptor theory":

> Clark, however, went a step further than this and, by taking y as the response, used equation [Hill equation/Langmuir isotherm] to relate the response of the tissue to the concentration of drug. Thus there was in Clark's treatment the implicit assumption that the percentage of receptors occupied is equal to the percentage response of the tissue.

Stephenson [31] then hypothesizes that the activity [potency] of agonists:

> "is the product of their affinity and their efficacy", with efficacy defined as a "drugs varying capacity to initiate a response."

After consideration of the actions of partial agonists/antagonists, in particular choline, at the nAChR, del Castillo and Katz [32] restate this argument in terms of a mechanism that defines separate "binding" and "gating" steps (i.e., in the case of partial ligands, whereas, at high enough drug concentrations, all the receptors are bound, *SR*, only some make it to the active state, *SR**):

$$S + R \overset{\text{binding}}{\rightleftharpoons} SR \overset{\text{gating}}{\rightleftharpoons} SR^*$$

Although still not always appreciated, the affinity of an agonist, measured from either physiological or radiolabeled ligand studies, is a combination of both of these steps, and does not reflect the affinity as it would for an antagonist (which has zero efficacy). This "binding-gating" problem was solved with the introduction of single channel recording, when the two steps in the mechanism could be fully resolved (see [25, 26]). But first came the observation of additional distinct conformational states of the nAChR.

Desensitization. It was further realized by Katz that a full description of the interaction between ACh and the nAChR required consideration of the long-lasting slowly accessed "refractory" period after prolonged agonist exposure [33]. Receptor desensitization is the reason that nicotine was seen to produce neuromuscular and ganglionic block described earlier (e.g., [15]. Again wanting to propose a specific "kinetic" mechanism to explain these results, as before they first considered linear schemes, but in order to satisfy such schemes [32, 33], where SA is the active state and SB is refractory to agonist (desensitized):

$$S + A \overset{\text{fast}}{\rightleftharpoons} SA \overset{\text{slow}}{\rightleftharpoons} SB^*$$

> …the onset of desensitization must always be faster than the recovery.

Otherwise receptors, according to this model, would not accumulate in the desensitized conformation (*SB*). They go on to note that:

> These predictions are clearly at variance with what has been consistently observed. Half-desensitization was found to develop at a rate about equal to, or lower than, that of the subsequent recovery.

This forced Katz and Thesleff [33] to conclude that there must be another path out of the desensitized state, and led to a model in which the desensitized conformations could in theory pre-exist (i.e., in the absence of agonist) in thermodynamic equilibrium with the resting state (see [34]). Thus, the first 'cyclical' model for the nAChR emerged:

$$\begin{array}{ccc} S + A & \rightleftharpoons & SA \\ \upharpoonleft\downharpoonright & & \upharpoonleft\downharpoonright \\ S + B & \rightleftharpoons & SB \end{array}$$

Activation requires multiple molecules of agonist: In their desensitization paper, Katz and Thesleff [33] also comment on the nonlinear (or sigmoidal) start to the dose–response relationship for ACh at the neuromuscular junction. After eliminating possible complications due to action at cholinesterases, one explanation was that the:

> S-shaped relation could be the result of a reaction in which two (or more) drug molecules become attached to the receptor molecule.

It was subsequently demonstrated that agonist (log-log) concentration–response curves show a limiting slope close to 2, consistent with the suggestion that two molecules of agonist were required for channel activation [35–37]. These data are in agreement with biochemical observations on the number of agonist binding sites (see below).

Allosteric models: If a multi agonist mechanism is valid, then as Stephenson [31] discusses the value of n_H, from a Hill plot will likely not be constrained to unity. It is influenced by not only the number of binding sites/agonist molecules but also their potential binding interdependency along with agonist efficacy. Thus, there would appear to be cooperativity in the binding and activation of receptors. This brings us back to hemoglobin and to the allosteric model of conformational changes in proteins. Colquhoun [22] sums up the pioneering contribution of Wyman:

> …the binding of oxygen to hemoglobin could be explained very economically if it were supposed that the two different conformations of hemoglobin (already known then) had different affinities for oxygen, so binding of oxygen would shift the conformational equilibrium towards the high-affinity form. This would explain the observed 'cooperativity' of oxygen binding without having to postulate an interaction between different binding sites.

The pseudosymmetrical multi-subunit/agonist nAChR was recognized as a possible candidate for the allosteric mode of operation [38]. With little knowledge of the exact make-up of the receptor, Karlin [39] first applied the allosteric model to explain the observed cooperativity at nAChRs. Consistent with this view of gating, Jackson [40] showed that receptors could open spontaneously, although rarely, in the absence of agonist (i.e., like hemoglobin, both the open and closed states pre-existed), and the binding of one (and more so two) molecules of agonist would stabilize (more of) the higher affinity open conformation (for a comprehensive treatment of this idea see [41]).

6 Single nAChR Transitions and Synaptic Transmission (1970–)

The application of noise and single channel analysis in the 1970s not only began to define individual nAChR properties but also answered fundamental questions about the nature of synaptic transmission. The [rapid nature of the] time course of the end-plate potential (EPP), measured through intracellular electrodes, had been described many years before, and was known to be influenced by the passive properties of the membrane [42]. Importantly, because curare and prostigmine (esterase inhibitor) altered the amplitude of, in this case, the unitary miniature end-plate potential (mEPP) in a graded fashion, it could be concluded that the mEPP resulted from the combination of 1000s molecules of ACh with numerous nAChRs on the muscle ([43]; see also [18]). But exactly how a brief pulse of transmitter, limited by enzymatic hydrolysis and diffusion [44], combined with nAChRs to produce a fast-activating and exponentially decaying synaptic response was unclear. A first step was to remove the contamination due to the membrane resistance and capacitance, by isolating the conductance change solely due to synaptically activated nAChRs. Using extracellular electrodes positioned close to the postsynaptic membrane and near to the quantal event, the recorded synaptic "current" was observed to be much faster/briefer than the mEPP [43]. Subsequently, Takeuchi and Takeuchi [45] applied the newly developed voltage-clamp technique to obtain the first intracellular synaptic end-plate currents. The next step was to obtain information about the behavior of individual nAChRs, in particular their "active" lifetime, when exposed to ACh, and then relate these properties to the synaptic currents [24].

As Colquhoun [26] notes, Bernard Katz was the first to apply the term "channel" to nAChRs. Katz and Miledi [46] used the technique of noise-analysis to estimate the properties of single nAChR channels. The single channel 'lifetime' was on the order of 1 ms, and the conductance in the pS range. A few years later, these results were verified when the first single channel measurements were made using the new patch-clamp technique [47]. But, in the midst of these discoveries, on both sides of the Atlantic, the relationship between the behavior of the resolved single channel events and the synaptic response became much clearer. It was reported that the decay of the end-plate currents and single nAChR channel "lifetimes" (from noise analysis) had the same voltage- and time-dependency, thereby linking the two mechanistically [48]. In their studies, Katz and Miledi [49] where able also to separate the action of ACh at nAChRs from its effects on esterase:

> Prostigmine [cholinesterase inhibitor] has little or no effect on the duration of the molecular 'gating action', [single channel using noise-analysis] while it greatly prolongs the quantal [synaptic response] conductance change.

They concluded that:

> After inhibition of ACh hydrolysis, the removal of the transmitter from the synapse is generally too slow to be accounted for by free diffusion. It is suggested that diffusion is delayed by binding to post-synaptic receptors. This is consistent with the finding that receptor blockage by curare or α-bungarotoxin shortens as well as reduces quantal transmitter action.

Thus, because of the dimensions of the synaptic cleft, esterase activity was required to prevent the rebinding of ACh to nAChRs, which otherwise would have led to prolongation of the synaptic event. After the refinement of single channel recording [50] it became possible to explain the time course of the synaptic event in terms of the probabilistic nature of the unbinding, opening and closing transitions of single nAChRs (for review see [24]). The conclusion that, under normal conditions, the end plate channels opened once only, (because channel opening was initially thought to be slow [51]), and after closing, the transmitter dissociated, was *not* strictly true, rather it was the lifetime of the transmitter-receptor complex which determined the duration of the synaptic response. Colquhoun and Ogden [52] conclude with a sentence that adequately sums up the optimization of neuromuscular transmission:

> These attributes are exactly what is needed for an efficient fast neurotransmitter. ACh molecules will bind rapidly because the transmitter concentration is high in the synaptic cleft, most channels will open very soon after binding (large opening rate constant, β) and ACh will dissociate rapidly after a few openings have occurred in quick succession.

7 The Receptor Protein (1970–)

More than half a century passed before the "receptive substance" [13] was realized as the protein structure we know now as the nAChR. Even the renowned British physiologist Sir Henry Dale, albeit with respect to the adrenergic system, remained skeptical of the whole idea of receptors as a target for competitive drug interactions (see [53]). Changeux [54] remarks:

> But the Langley receptor substance had remained a mysterious entity, and the concept itself was heavily criticized by Sir Henry Dale, an established authority in the field.

Katz, although embracing the concept, could only refer to the elementary events underlying receptor–drug interactions as the 'shot effect' due to the transient opening of 'ionic gates' [50, 55]. But by then biochemical analysis was beginning to shed light on the physical nature of receptors. Largely due to the abundant source of nAChRs in the electric organs of fish, plus the means to extract them, it soon became possible to identify the esterase and the nAChR as separate entities. In particular, Changeux [54] remembers:

> One spring afternoon in 1970, Chen-Yuan Lee, a Taiwanese pharmacologist, unexpectedly came into my laboratory. He informed me of his work on a snake venom toxin, α-bungarotoxin, which he had isolated and purified, and which, according to him, did not interact with acetylcholinesterase but almost irreversibly blocked the neuromuscular junction of higher vertebrates at the postsynaptic level.

This allowed for the unequivocal tagging of receptors and showed that the binding site competition between α-bungarotoxin (α-BTX) and quaternary ammonium agonists was preserved after extraction of the nAChR [56]. The receptor was purified from *Torpedo* and separated into 4 subunits, with the α subunit, identified as containing, at least part of, the ligand binding site [57]. The amino acid sequences of the subunits are homologous, with the α subunit occurring twice in the $\alpha_2\beta\gamma\delta$ pentameric structure [58].

8 Receptor Cloning and Structure (1980–)

The cloning of the first subunit [59] led to an explosion in the molecular biology of nAChRs and solidified the importance of both muscle and neuronal receptors. The subtypes of these receptors are discussed elsewhere in this volume. With knowledge of the gene and amino acid sequences structural analysis was begun in earnest. Here, we will provide just a few highlights. Hydropathy plots indicate that each subunit contributes four transmembrane domains with the N and C terminals both extracellular [60]. Once the basic structure was known then major goals became assigning functions, e.g., the pore, the gate, and the binding sites, to specific parts of the receptor, with the ultimate challenge being able to see the whole process of channel activation from a structural point of view (see 61). Using a chimeric approach involving substitution of parts of the δ subunit and noting the effects on single channel conductance, Imoto et al. [62] argued that the second transmembrane domain (M2) contributes to the pore. This finding was substantiated by photolabeling nAChRs with the channel blocker, chlorpromazine, producing a pattern consistent with an α-helix in the M2 region [63]. Rings of negatively charged amino acids likely contribute to the selectivity and permeability of the channel [26, 61, 64], and the bulk of a M2 residue near to the cytoplasmic face of the pore has profound effects on the channel conductance, and is implicated in the selectivity filter [65]. The nature of the gate, controlling the closed to open transition, has remained a controversial and unresolved issue since the outset of structural studies. Pioneering work based on EM images of the nAChR, proposed that a ring of leucines near the midpoint of the M2 domain may introduce a kink in the α-helix—and provide a putative gate for the channel [66]. Consistent with this notion, mutation of these amino acids to the less bulky serine residues appears to destabilize the closed state of the channel, such that the nAChR is much more readily opened by agonist [67, 68]. Conversely, using a cysteine mutation approach to probe the closed and open states of the pore, Akabas et al. [69] suggest a location for the gate that is more cytoplasmic than the 9'leucines. The gate remains elusive. With respect to other functional domains, crystallization of the ACh binding protein [70], has allowed conformational changes at the ligand binding site to be resolved [71]. The chemical nature of the side chain interaction between ACh and its binding site has also been identified [72]. Along with higher resolution EM reconstructions of the nAChR it has become possible to provide speculative answers to questions such as [73]:

> How do the rotational movements in the ligand-binding domain communicate through the M2 helices to open the channel?

More recent crystal structures of prokaryotic pentameric receptor channels have allowed further insight into channel gating [74]:

> Assuming that GLIC and ELIC structures faithfully represent the open and closed forms of pLGICs [pentameric ligand-gated ion channels] embedded in a phospholipid bilayer, insight into the opening mechanism can be provided by analysing their rearrangements,

These authors infer that the rotational changes in the extracellular domain result in tilting movements of the TM2 and TM3 alpha helices and opening of the pore

[74]. Currently, the picture of nAChR channel (or the cys-loop pentameric family) gating is still incomplete (for more insight see Yakel [75, 76]):

> In conclusion, the exact nature of the structure of the cys-loop ligand-gated ion channel subunits and the movements observed during and after ligand binding, gating and desensitization are still unknown. Nevertheless, a general hypothesis has emerged that indicates agonist binding induces closure of the C-loop [in the binding site], which is conveyed to the M2 pore region [through a coupling domain], resulting in channel opening. Thus, the transduction pathway involves many regions of the channel.

Moreover, antagonists, such as α-BTX trap the C-loop in an open conformation [77], preventing gating of the channel.

Ultimately, one hopes that it will be possible to comprehensively map the functional transitions onto the nAChR structure, and fully realize the molecular nature of the binding-gating steps (see [78]). As the single channel analysis of nAChR binding-gating behavior by Grosman et al. [79] indicate the structural changes should be consistent with the allosteric model:

> …the conformational change proceeds in a wave-like manner with the low-to-high affinity change at the transmitter binding sites proceeding the complete opening of the pore.

Likewise, structural rearrangements across multiple domains of the protein have been recognized in the process of nAChR desensitization [80]. The reader is referred to Chap. 2 for further discussions of the structural-functional operation of the receptor.

9 Summary

In the span of 150 years, the nicotinic acetylcholine receptor has gone from being the nebulous site of action of the poison curare to arguably the most thoroughly characterized ligand-gated ion channel in the nervous system. But, although we can now almost completely visualize it at the atomic level, and have identified many other family members, we are only beginning to unravel its physiological role in the brain, its contribution to disease, and its usefulness as a therapeutic target.

References

1. Dani JA, Balfour DJ. Historical and current perspective on tobacco use and nicotine addiction. Trends Neurosci. 2011;34:383–92.
2. James I King of England. A Counter-Blaste to Tobacco. Goldsmid E, editor. Dodo Press; 1604.
3. Stewart GG. A history of the medicinal use of tobacco 1492–1860. Med Hist. 1967;11:228–68.
4. Charlton A. Medicinal uses of tobacco in history. J R Soc Med. 2004;97:292–6.
5. Posselt W, Reimann L. Chemische Untersuchungen des Tabaks und Darstellung des eigenhumlichen wirksamen Principes dieser Pflanze. Geigers Magazin der Pharmazie. 1828;24:138–61.
6. Cheney RH. Geographic and taxonomic distribution of american plant arrow poisons. Am J Bot. 1931;18:136–45.

7. Humboldt A. von. Personal Narrative of Travels to the Equinoctial Regions of America, During the Year 1799–1804. Http://www.gutenberg.org/ebooks/7014
8. Dale HH. The action of certain esters and ethers of choline, and their relation to muscarine. J Pharmacol Exp Ther. 1914;6:147–90.
9. Brodie BC. Experiments and observations on the different modes in which death is produced by certain vegetable poisons. Phil Trans Roy Soc Lond. 1811;101:178–207.
10. Bennett MR. The concept of transmitter receptors: 100 years on. Neuropharmacology. 2000;39:523–46.
11. Cousin MT. History of anaesthesia: who discovered the neuromuscular junction? The opposing views of Claude Bernard and Alfred Vulpian. Eur J Anaesthesiol. 2013;30:1–4.
12. Sherrington CS. In: A textbook of physiology 7th ed. Pt 3. Foster M, Sherrington CS, editors. 915–1000. London: Macmillan; 1897.
13. Langley JN. On the reaction of cells and of nerve-endings to certain poisons, chiefly as regards the reaction of striated muscle to nicotine and to curari. J Physiol. 1905;33:374–413.
14. Maehle AH. "Receptive substances": John Newport Langley (1852–1925) and his path to a receptor theory of drug action. Med Hist. 2004;48:153–74.
15. Langley JN, Dickenson WL. On the local paralysis of peripheral ganglia and on the connection of different classes of nerve fibres with them. Proc R Soc Lond. 1889;46:423–31.
16. Langley JN. On the stimulation and paralysis of nerve-cells and of nerve-endings: Part I. J Physiol. 1901;27:224–36.
17. Langley JN. Observations on the physiological action of extracts of the supra-renal bodies. J Physiol. 1901;27:237–56.
18. Kuffler SW, Yoshikami D. The number of transmitter molecules in a quantum: an estimate from iontophoretic application of acetylcholine at the neuromuscular synapse. J Physiol. 1975;251:465–82.
19. Dale HH, Feldberg W, Vogt M. Release of acetylcholine at voluntary motor nerve endings. J Physiol. 1936;86:353–80.
20. Bosch F, Rosich L. The contributions of Paul Ehrlich to pharmacology: a tribute on the occasion of the centenary of his Nobel Prize. Pharmacology. 2008;82:171–9.
21. Langley JN. On the physiology of the salivary secretion: Part II. On the mutual antagonism of atropin and pilocarpin, having especial reference to their relations in the Sub-maxillary gland of the Cat. J Physiol. 1878;1:339–69.
22. Colquhoun D. The quantitative analysis of drug-receptor interactions: a short history. Trends Pharmacol Sci. 2006;27:149–57.
23. Hill AV. The mode of action of nicotine and curari determined by the form of the contraction curve and the method of temperature coefficients. J Physiol. 1909;39:361–73.
24. Colquhoun D. How fast do drugs work? Trends Pharmacol Sci. 1981;8:212–7.
25. Colquhoun D. Binding, gating, affinity and efficacy: the interpretation of structure-activity relationships for agonists and of the effects of mutating receptors. Br J Pharmacol. 1998;125:924–47.
26. Colquhoun D. Agonist-activated ion channels. Br J Pharmacol. 2006;147 Suppl 1:S17–26.
27. Hill AV. The possible effects of the aggregation of the molecules of haemoglobin on its dissociation curve. J Physiol. 1910;40:IV–VII.
28. Clark AJ. The reaction between acetyl choline and muscle cells. J Physiol. 1926;61:530–46.
29. Brown DA. The history of the neuronal nicotinic receptors. In: Neuronal Nicotinic Receptors. F. Clementi et al., editor. Berlin Heidelberg: Springer-Verlag; 2000. p 3–11.
30. Jenkinson DH. The antagonism between tubocurarine and substances which depolarize the motor end-plate. J Physiol. 1960;152:309–24.
31. Stephenson RP. A modification of receptor theory. Br J Pharmacol Chemother. 1956;11:379–93.
32. Del Casillo J, Katz B. Interaction at end-plate receptors between different choline derivatives. Proc R Soc Lond B Biol Sci. 1957;146:369–81.
33. Katz B, Thesleff S. A study of the desensitization produced by acetylcholine at the motor end-plate. J Physiol. 1957;138:63–80.

34. Quick MW, Lester RAJ. Desensitization of neuronal nicotinic receptors. J Neurobiol. 2002;53: 457–78.
35. Adams PR. An analysis of the dose-response curve at voltage-clamped frog-endplates. Pflugers Arch. 1975;360:145–53.
36. Lester HA, Changeux JP, Sheridan RE. Conductance increases produced by bath application of cholinergic agonists to Electrophorus electroplaques. J Gen Physiol. 1975;65(6):797–816.
37. Dionne VE, Steinbach JH, Stevens CF. An analysis of the dose-response relationship at voltage-clamped frog neuromuscular junctions. J Physiol. 1978;281:421–44.
38. Changeux JP, Edelstein SJ. Allosteric receptors after 30 years. Neuron. 1998;21:959–80.
39. Karlin A. On the application of "a plausible model" of allosteric proteins to the receptor for acetylcholine. J Theor Biol. 1967;16:306–20.
40. Jackson MB. Kinetics of unliganded acetylcholine receptor channel gating. Biophys J. 1986;49:663–72.
41. Auerbach A. Thinking in cycles: MWC is a good model for acetylcholine receptor-channels. J Physiol. 2012;590:93–8.
42. Eccles J, Katz B, Kuffler S. Nature of the "end plate potential" in curarized muscle. J Neurophysiol. 1941;4:362–87.
43. Fatt P, Katz B. Spontaneous subthreshold activity at motor nerve endings. J Physiol. 1952;117:109–28.
44. Eccles JC, Jaeger JC. The relationship between the mode of operation and the dimensions of the junctional regions at synapses and motor end-organs. Proc R Soc Lond B Biol Sci. 1958;148:38–56.
45. Takeuchi A, Takeuchi N. Active phase of frog's end-plate potential. J Neurophysiol. 1959;22: 395–411.
46. Katz B, Miledi R. Membrane noise produced by acetylcholine. Nature. 1970;226:962–3.
47. Neher E, Sakmann B. Single-channel currents recorded from membrane of denervated frog muscle fibres. Nature. 1976;260:799–802.
48. Anderson CR, Stevens CF. Voltage clamp analysis of acetylcholine produced end-plate current fluctuations at frog neuromuscular junction. J Physiol. 1973;235:655–91.
49. Katz B, Miledi R. The binding of acetylcholine to receptors and its removal from the synaptic cleft. J Physiol. 1973;231:549–74.
50. Colquhoun D, Sakmann B. Fast events in single-channel currents activated by acetylcholine and its analogues at the frog muscle end-plate. J Physiol. 1985;369:501–57.
51. Magleby KL, Stevens CF. A quantitative description of end-plate currents. J Physiol. 1972; 223:173–97.
52. Colquhoun D, Ogden DC. Activation of ion channels in the frog end-plate by high concentrations of acetylcholine. J Physiol. 1988;395:131–59.
53. Schild HO. Dale and the development of pharmacology. Lecture given at Sir Henry Dale Centennial Symposium, Cambridge, 17-19 September 1975. Br J Pharmacol. 1975;1976(56):3–7.
54. Changeux J-P. Allosteric receptors: from electric organ to cognition. Ann Rev Pharmacol Toxicol. 2010;50:1–38.
55. Katz B, Miledi R. The statistical nature of the acetycholine potential and its molecular components. J Physiol. 1972;224:665–99.
56. Changeux JP, Kasai M, Lee CY. Use of a snake venom toxin to characterize the cholinergic receptor protein. Proc Natl Acad Sci U S A. 1970;67:1241–7.
57. Weill CL, McNamee MG, Karlin A. Affinity-labeling of purified acetylcholine receptor from Torpedo californica. Biochem Biophys Res Commun. 1974;61:997–1003.
58. Raftery MA, Hunkapiller MW, Strader CD, Hood LE. Acetylcholine receptor: complex of homologous subunits. Science. 1980;208:1454–356.
59. Noda M, Takahashi H, Tanabe T, Toyosato M, Furutani Y, Hirose T, Asai M, Inayama S, Miyati T, Numa S. Primary structure of alpha-subunit precursor of Torpedo californica acetylcholine receptor deduced from cDNA sequence. Nature. 1982;299:793–7.
60. Claudio T, Ballivet M, Patrick J, Heinemann S. Nucleotide and deduced amino acid sequences of Torpedo californica acetylcholine receptor gamma subunit. Proc Natl Acad Sci U S A. 1983;80:1111–5.

61. Unwin N. Neurotransmitter action: opening of ligand-gated ion channels. Cell. 1993; 72(Suppl):31–41.
62. Imoto K, Methfessel C, Sakmann B, Mishina M, Mori Y, Konno T, Fukuda K, Kurasaki M, Bujo H, Fujita Y, Numa S. Location of a delta-subunit region determining ion transport through the acetylcholine receptor channel. Nature. 1986;324:670–4.
63. Revah F, Galzi JL, Giraudat J, Haumont PY, Lederer F, Changeux JP. The noncompetitive blocker [3H]chlorpromazine labels three amino acids of the acetylcholine receptor gamma subunit: implications for the alpha-helical organization of regions MII and for the structure of the ion channel. Proc Natl Acad Sci U S A. 1990;87:4675–9.
64. Imoto K, Busch C, Sakmann B, Mishina M, Konno T, Nakai J, Bujo H, Mori Y, Fukuda K, Numa S. Rings of negatively charged amino acids as determinants of acetylcholine receptor channel conductance. Nature. 1988;335:645–8.
65. Villarroel A, Sakmann B. Threonine in the selectivity filter of the acetylcholine receptor channel. Biophys J. 1992;62:196–205.
66. Unwin N. Acetylcholine receptor channel imaged in the open state. Nature. 1995;373:37–43.
67. Revah F, Bertrand D, Galzi JL, Devillers-Thiéry A, Mulle C, Hussy N, Bertrand S, Ballivet M, Changeux JP. Mutations in the channel domain alter desensitization of a neuronal nicotinic receptor. Nature. 1991;353:846–9.
68. Labarca C, Nowak MW, Zhang H, Tang L, Deshpande P, Lester HA. Channel gating governed symmetrically by conserved leucine residues in the M2 domain of nicotinic receptors. Nature. 1995;376:514–6.
69. Akabas MH, Kaufmann C, Archdeacon P, Karlin A. Identification of acetylcholine receptor channel-lining residues in the entire M2 segment of the alpha subunit. Neuron. 1994;13: 919–27.
70. Brejc K, Van Dijk WJ, Klaassen RV, Schuurmans M, Van Der Oost J, Smit AB, Sixma TK. Crystal structure of an ACh-binding protein reveals the ligand-binding domain of nicotinic receptors. Nature. 2001;411:269–76.
71. Gao F, Mer G, Tonelli M, Hansen SB, Burghardt TP, Taylor P, Sine SM. Solution NMR of acetylcholine binding protein reveals agonist-mediated conformational change of the C-loop. Mol Pharmacol. 2006;70:1230–5.
72. Zhong W, Gallivan JP, Zhang Y, Li L, Lester HA, Dougherty DA. From ab initio quantum mechanics to molecular neurobiology: a cation-pi binding site in the nicotinic receptor. Proc Natl Acad Sci U S A. 1998;95:12088–93.
73. Unwin N. Refined structure of the nicotinic acetylcholine receptor at 4A resolution. J Mol Biol. 2005;346:967–89.
74. Bocquet N, Nury H, Baaden M, Le Poupon C, Changeux JP, Delarue M, Corringer PJ. X-ray structure of a pentameric ligand-gated ion channel in an apparently open conformation. Nature. 2009;457:111–4.
75. Yakel JL. Gating of nicotinic ACh receptors: latest insights into ligand binding and function. J Physiol. 2010;588:597–602.
76. Yakel J. Advances and hold-ups in the study of structure, function and regulation of Cys-loop ligand-gated ion channels and receptors. J Physiol. 2010;588:555–6.
77. Huang S, Li SX, Bren N, Cheng K, Gomoto R, Chen L, Sine SM. Complex between α-bungarotoxin and an α7 nicotinic receptor ligand-binding domain chimaera. Biochem J. 2013;454:303–10.
78. Unwin N. Structure and action of the nicotinic acetylcholine receptor explored by electron microscopy. FEBS Lett. 2003;555:91–5.
79. Grosman C, Zhou M, Auerbach A. Mapping the conformational wave of acetylcholine receptor channel gating. Nature. 2000;2000(403):773–6.
80. Zhang J, Xue F, Liu Y, Yang H, Wang X. The structural mechanism of the cys-loop receptor desensitization. Mol Neurobiol. 2013;48(1):97–108.

Chapter 2
Molecular Structure, Gating, and Regulation

Steen E. Pedersen

Abstract Gating of ion channels is the opening response of the channel to a stimulus and its subsequent desensitization or inactivation. For nicotinic acetylcholine receptors, the stimulus is the binding of the neurotransmitter acetylcholine to an extracellular domain followed by a sequence of conformational changes resulting in rapid channel opening. In most channels a persistent stimulus invokes a second gating event driving the channel into an inactive, desensitized state that may or may not contribute to the net response. A full understanding of gating requires a correlation of structural changes with the kinetics of channel opening and desensitization; an understanding of how these changes result in rapid, large changes in ion flux through the channel; and how they are terminated. In this article the current structural changes and the current understanding of nicotinic receptor channel kinetics are reviewed and correlated. The analysis necessarily draws on inferences from the larger family of ligand-gated ion channels and related proteins. The focus will be predominantly on the opening event but will also include consideration of desensitization.

Keywords Nicotinic acetylcholine receptor • Gating • Ligand binding • Conformational regulation • Linear free energy analysis • Allosteric ligand

Abbreviations

AChBP	Acetylcholine Binding Protein
ELIC	*Erwinia chrysanthemi* ligand-gated ion channel
GLIC	*Gloeobacter violaceus* ligand-gated ion channel
Glu-Cl	*C. elegans* glutamate-gated chloride channel
LFER	Linear free energy relationship
LGIC	Pentameric ligand-gated ion channel

S.E. Pedersen, Ph.D. (✉)
Department of Molecular Physiology and Biophysics, Baylor College of Medicine, One Baylor Plaza, Houston, TX 77030, USA
e-mail: pedersen@bcm.edu

R.A.J. Lester (ed.), *Nicotinic Receptors*, The Receptors 26,
DOI 10.1007/978-1-4939-1167-7_2

MWC	Monod-Wyman-Changeux cooperativity model
nAChR	Nicotinic acetylcholine receptor
TID	3-(trifluoromethyl)-3-(m-iodophenyl)diazirine
TM	Transmembrane domain

1 Introduction

The family of pentameric ligand-gated ion channels constitutes a class of fast synaptic neurotransmitter receptors (Fig. 2.1). This family includes the nicotinic receptors, the $GABA_A$ receptors, the glycine receptors, and the 5-hydroxy-tryptamine receptors. The focus in this review will be on understanding gating of nicotinic acetylcholine receptors (nAChR), which are themselves divided into two classes: muscle-type and neuronal nAChR. The muscle-type receptors comprise 4 distinct subunits arranged pentamerically ($\alpha_2\beta\gamma\delta$ for the embryonic form; $\alpha_2\beta\varepsilon\delta$ for the adult form). The neuronal forms comprise various mixtures subunit subtypes designated α2 through α10 and β2 to β4. The distribution and putative function of these receptors has been reviewed extensively [1] and will be addressed in detail elsewhere in this volume. Notably, subunits can mix as heteropentamers, e.g., $\alpha4_2\beta2_3$, or for some subunits, homo-pentamers, notably $\alpha7_5$. However, structural and mechanistic information is derived from across the family; the receptors are all structurally similar, and, in broad strokes, likely undergo quite similar mechanisms of gating.

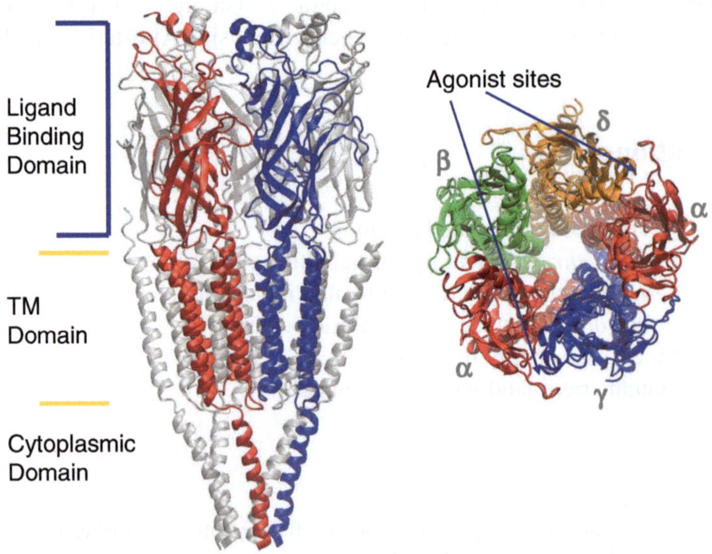

Fig. 2.1 Structure of the Torpedo nAChR. The *left* figure shows the *Torpedo* nAChR in cross-section. The *right* figure shows the receptor from the extracellular side [20]. The *arrows* indicate the location of the ligand-binding sites at the α–γ and α–δ subunit interfaces

$$R \rightleftharpoons A_2R \rightleftharpoons A_2O \rightleftharpoons A_2D$$

Fig. 2.2 A minimal model of nAChR activation. **R** represents the resting, or closed, conformation of the nAChR. **O** refers to the open, active conformation, and **D** the desensitized state. **A** refers to agonist or acetylcholine binding, occurring in two independent steps

The pentameric ligand-gated ion channels, as with most channels, have at least two gating actions: opening and desensitization. Both gating steps are ultimately initiated by the binding of neurotransmitter or ligand to binding sites. The binding first triggers rapid opening followed by desensitization; desensitization is entry into a nonconducting state that exists in the persistent presence of neurotransmitter. In practice, to analyze binding and single channel data, these are often described as chemical reaction steps in kinetic and thermodynamic schemes (see Fig. 2.2 for a simple example). Under ideal conditions, the opening step, or activation, can be isolated to a single elementary reaction, amenable to detailed kinetic and thermodynamic analysis. In contrast, desensitization, which is the equivalent of inactivation in voltage gated channels, appears to be a complex, multistep process, less amenable to a simple description; less is understood about this process.

High-resolution structural information on ligand-gated ion channels and some of their bacterial congeners has been forthcoming at an increasing rate and has shed a great deal of light on the possible mechanisms for opening and desensitization. These data are reviewed in the first section; for a more detailed review of structure-based analysis of the conformational changes, also see the excellent exposition by daCosta and Baenziger [2]. Nonetheless, a clear and comprehensive picture of gating at the level of the channel is still lacking. The second section considers the functional and indirect structural data that informs gating in the context of the high-resolution structural data.

2 Atomic-Resolution Structural Analysis of Gating

2.1 The Structural Basis of Gating at the Ligand-Binding Sites

As seen in Fig. 2.1, the family of pentameric ligand-gated ion channels constitutes pseudo-symmetric proteins with homologous or identical subunits surrounding a central pore that provides the ion conducting pathway. The ligand-binding event that triggers gating occurs in the extracellular (ligand-binding) domain. Structural studies on the homologous Acetylcholine Binding Protein (AChBP; [3–5]) and more recently on an α7 chimera [6] have provided a clear description of the structural changes that accompany binding within the ligand-binding domain.

As shown in Fig. 2.3 binding of agonists induces a pronounced change in the loop C region, an area of the protein that had already been identified as critical for ligand binding and characteristic of the α-subunits in the nicotinic receptor family. The idea that gating is triggered by a general inward movement of loop C was

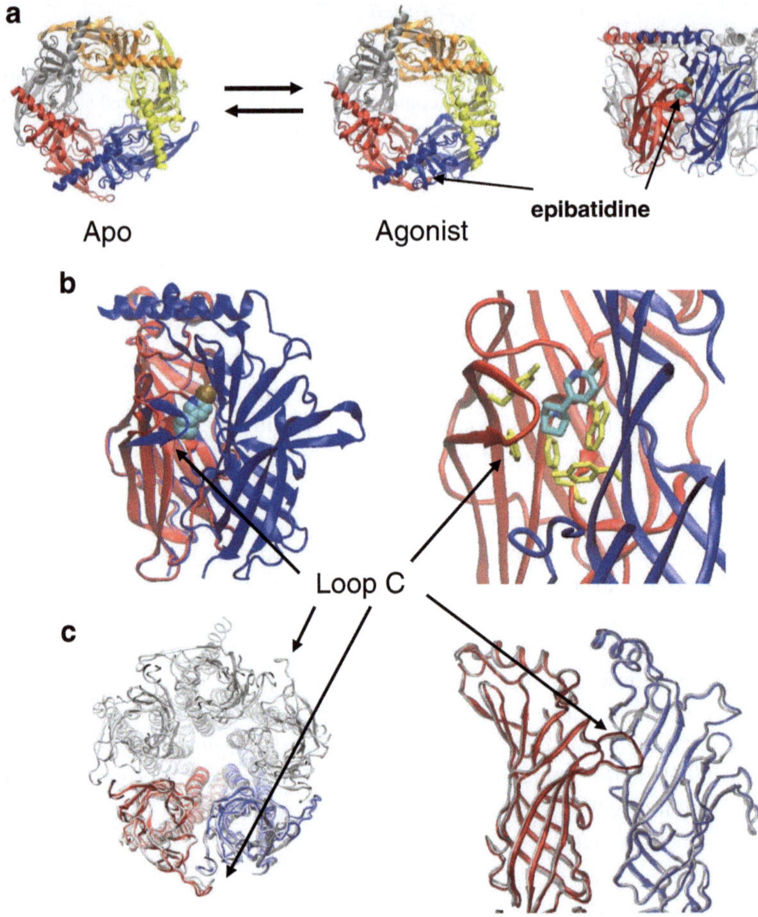

Fig. 2.3 Acetylcholine Binding Protein structures define agonist binding motions. Panel **a**, *left*, top views (extracellular side) of AChBP shown in its unliganded, apo form (pdb 2BYN) and bound to the agonist epibatidine (pdb 2BYP; [5]. Panel **a**, *right*, side view of the AChBP with epibatidine bound. Panel **b**, *left*, Superposition of individual subunit structures in the bound and unbound states: *red*—unbound; *blue*, bound. Panel **b**, *right*, detail of bound epibatidine surrounded by the aromatic residues known to stabilize cation binding. Panel **c**, Gating-changes in the nAChR Ligand-Binding Domain of the *Torpedo* nAChR by comparison of closed and open structures. *Left* side shows the open conformation in gray and the closed conformation colored for the α- and γ-subunits only. Panel **c**, *right*, shows details of the structural changes ascribed to opening

confirmed when antagonist binding was observed to interfere with loop closure [5]. This is consistent with the generally larger size of antagonists (e.g. conotoxin and *d*-tubocurarine) whereby a simple model of steric hindrance of loop closure prevents the normal loop C closure [5].

The details of further structural changes remain somewhat ambiguous. There are only modest changes in the overall β-sheet structure twist of the AChBPs between the two conformations (Fig. 2.3). The advantage of these high-resolution structures as models for gating changes is mitigated by the absence of the conformational restrictions imposed by the presence of the transmembrane, channel domain and therefore, the relevant changes that communicate binding to the transmembrane domain may not be observed in AChBPs.

Data from full channels include recent electron microscopic structural investigations on the *Torpedo* nAChR. Changes in the extra-cellular domain seen by comparison of the electron-microscopic structures from Unwin's lab in the closed and open states also show only subtle changes (See Fig. 2.3c). The loops C are extended, similar to the unbound structure of the AChBP and the loop C changes upon opening are small relative to those seen in the AChBPs, but may give some indication of the general direction of those changes [7]. X-ray crystal data from the bacterial homologues ELIC [8] and GLIC [9], as well as several structures of the nematode Glu-Cl channel, which is also in the LGIC family [10] all show the loops C in a conformation similar to the agonist-bound AChBP structures. The structural data on loops C in the Glu-Cl channel is likely constrained by interaction with the cocrystallized FAB fragments [10]; whether gating in the bacterial channels involves changes in the extracellular domain is not known.

2.2 *Binding Site Inequality*

The two binding sites are nonequivalent. This is apparent from the structural organization of the heteropentamer in the muscle-type nAChRs (Fig. 2.1) in that elements from both the α-subunits contribute to binding, as well as elements of the adjoining subunits, the γ- and δ-subunits for the embryonic form of the muscle receptor. The site differences were first determined through *d*-tubocurarine binding and subsequent photoaffinity labeling studies with this compound identified the α-γ site and having higher affinity for *d*-tubocurarine and, through subsequent studies, lower affinity for agonist (acetylcholine and Dansyl-C6-Choline) in the resting state [11–13]. A number of chimeric and photoaffinity labeling studies identified residues responsible for the distinct agonist affinities in the γ-, δ-, and ε-subunits [14, 15], as well as for the antagonists d-tubocurarine, α-conotoxin, and α-bungarotoxin [16–19]. The affinity differences between the sites contribute substantially to shaping the dose-response curves in the embryonic $\alpha_2\beta\gamma\delta$ nAChRs; whereas in the adult $\alpha_2\beta\varepsilon\delta$ form, the affinities are more or less equal. The studies identifying the sources of site-selectivity have been critical in identifying the key elements of the binding sites before atomic resolution structures became available. For obvious reasons, these differences do not apply to the homo-pentameric nAChRs, such as the α7 nAChR. The consequences of site-selectivity will be discussed further under the functional aspects of gating.

A key question, not fully answered by the model studies on AChBP, is how the observed structural changes communicate the information to the transmembrane domain to affect opening of the pore. This question has been the topic of intense investigation through site-directed mutagenesis (see below).

2.3 *The Structural Basis of Gating at the Channel*

Substantial data have emerged showing a variety of LGICs with distinct structures in the channel region. High-resolution data from prokaryotic channels has shown clear, distinct structures from the two channels, ELIC [8] and GLIC [9]. These have been interpreted as reflecting closed and open states, respectively and the changes in the transmembrane region appear consistent with this interpretation (see Fig. 2.4). The nematode Glu-Cl channel also appears to be in an open conformation. However, two conformations, resting and open, determined on the same channel, the *Torpedo* nAChR, show only subtle differences and it is unclear whether those modest structural changes are sufficient to account for gating. But they may indicate the general shape and direction of the gating motion [7].

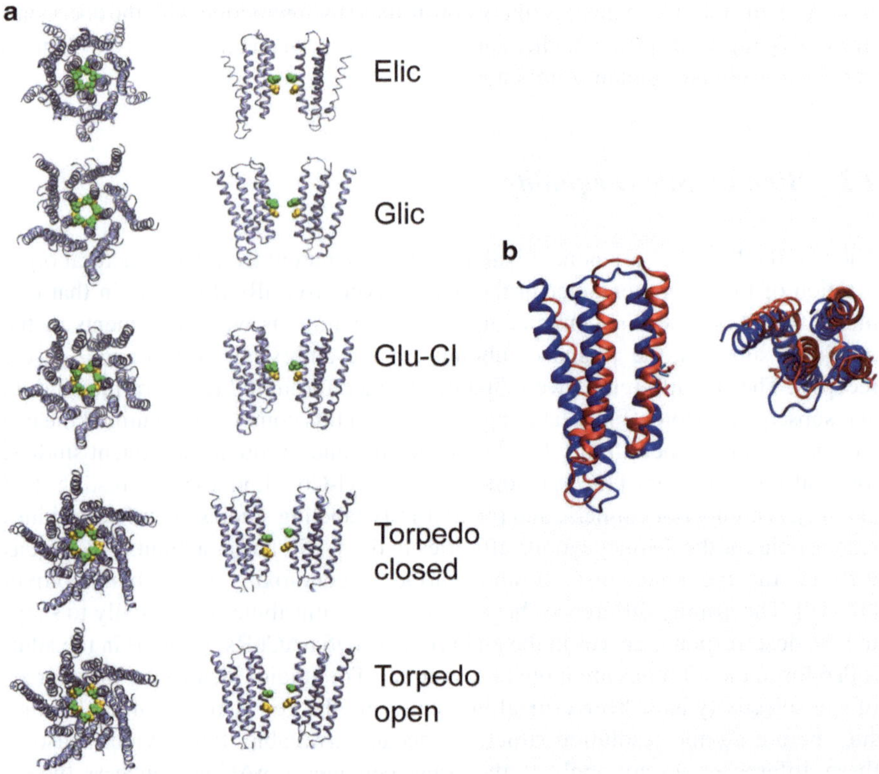

Fig. 2.4 Structures of the TM domain in open and closed conformations. Panel **a**, the left hand side shows the pentameric structure of the TM domain from the extracellular side. The right hand side shows the membrane cross section showing only two of the subunits, for clarity. Shown are: the ELIC channel (nominally closed, pdb 2VL0); the GLIC channel (pdb 3EHZ); the Glu-Cl channel, picrotoxin bound (nominally open, pdb 3RI5); *Torpedo* nAChR closed (pdb 4AQ5); *Torpedo* nAChR open (pdb 4AQ9). Panel **b**, side and top views of a single TM domain ELIC (*red*) and GLIC (*blue*) channels aligned. M1 was omitted in the side view for clarity

An initial EM structure of the *Torpedo* nAChR [20] showed a relatively accessible transmembrane (TM) domain that narrowed to a 6 Å diameter pore that was nonetheless apparently a closed structure (Fig. 2.1). The TM domain consists of four transmembrane α-helices from each subunit, named as M1 through M4. The M2-helices constitute the pore lining, and the M4-helix is the most exposed to the lipid milieu. This assessment was based on the apparent outward Loop C binding site configuration, indicating a conformation unbound by ligand and the general hydrophobicity of the pore (see Fig. 2.4). However the high-resolution structure of the prokaryotic channel ELIC revealed a few years later, showed a dramatically distinct TM-domain structure with much more closely apposed, straight and nearly parallel M2 α-helices [8]. This appeared more consistent with a generally closed structure where the close apposition of hydrophobic residues near the extracellular end of the channel clearly precludes ion permeation. Shortly thereafter, an acid-activated prokaryotic channel (GLIC) was crystallized, in low pH and the resultant structure indicated an apparent open-channel conformation [9, 21, 22]. The pore-lining M2 α-helices are splayed apart at the extracellular end and have sufficient space for ion permeation near the cytoplasmic end. The M2–M3 helices of ELIC and GLIC overlay nicely after an inward rotation of the helices from the splayed GLIC helices to those of ELIC. This suggests a simple outward and twist motion of the TM-helix bundle helices as a trigger for opening (see Fig. 2.4b).

This putative, open state was fairly consistent with the structures of the nematode Glu-Cl channel, that was crystallized in the presence of the endogenous agonist, glutamate, as well as in the presence of an open-channel blocker, picrotoxin, and in the presence of an allosteric activator [10]. These structures together argue for the Glu-Cl TM-domain being captured in an open state. Recent, higher-resolution structures of the GLIC channel reveal the presence of a hydrated sodium ion in what appears to be part of the selectivity filter of the channel. Further computer simulations on the ion pathway argue strongly for an open-channel configuration [23]. To further test whether the GLIC conformation reflected an open state, rather than an alternative closed state, cysteine crosslinks were engineered into GLIC. The cross-linked form demonstrated the capacity of GLIC to adopt a closed-like conformation, similar in structure to that of ELIC, as determined by crystallography [24], and shown to be nonconducting.

The apparent contradiction between these structures and the quite subtle changes found for the *Torpedo* nAChR are not readily reconciled. The *Torpedo* structure shows quite modest changes upon agonist exposure, both at the agonist sites and the level of the TM domain [7]. The TM domain in the putative closed state is closer to the open configuration as defined by the GLIC and Glu-Cl structures. However, the ligand-binding domain appears to be in an unliganded conformation. Given the close coupling between binding and opening and desensitization, it is unclear whether the gating movement in the *Torpedo* nAChR is fundamentally subtler or whether the inability to determine the absolute conformation in crystal structures simply confounds interpretation.

2.4 *Structural Basis of Coupling Ligand Binding to Channel Gating*

The ligand-binding and transmembrane domains are distinct structures, the former being predominantly β-sheet and the latter largely α-helical. The interface is where communication between the ligand-binding event and the channel must take place. The primary features of the interface thought to mediate this are loops intertwining from the ligand-binding domain and the TM domain. The Cys-loop (β6–β7 loop), which contains the disulfide bond conserved among the eukaryotic LGIC subunits, and the β1–β2 loop (See Fig. 2.5) surround the M2–M3 linker on the TM domain (the loop linking the M2 and M3 TM α-helices). Mutational analysis early on identified the M2–M3 linker as a critical aspect of communication between the extracellular domain and the TM domain [25, 26]. In addition, there is a direct connection between the β9–β10 sheet (which includes the Loop C structure that binds agonist), and the first transmembrane domain, that seems likely to exist for communication between ligand-binding and the TM-domain. The motions triggered by ligand binding likely communicate these changes to the TM domain through movements in these linkers. There is substantial mutational evidence (see below) for communication between these loops, but the conclusions from the structures is less enlightening about motions that could effect the large changes seen among the putative closed and open channel structures.

Comparison of putative open (GLIC and Glu-Cl) or closed (ELIC) structures do not show obvious changes in this region that lead to a clear hypothesis of communication. The M2–M3 linker is longer for these channels than in the *Torpedo* nAChR. The recent comparative structures of the *Torpedo* nAChR channel show very small changes in the motion of the linkers, as well as only small motions in the rest of the protein, upon agonist binding. The smaller linker in the nAChR primarily engages the Cys-loop (β6–β7 loop) and the β1–β2 loop, which surround the highly conserved proline residue whereas in the prokaryotic channels and in Glu-Cl the M2–M3 linker has one helix coil unwound and interacts with the adjacent subunit as well (Fig. 2.5).

In summary, it seems apparent and logical that tension generated by ligand binding likely affects the β9–β10 loop, which connects directly to the TM domain at M1,

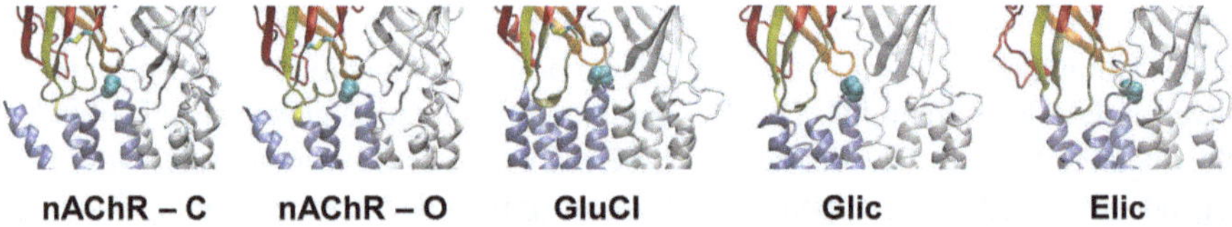

Fig. 2.5 Structures of the Domain interface and the coupling region. The M2–M3 linker is shown with the conserved proline in van der Waals dimensions. The β10 sheet is shown in yellow, the Cys-loop in tan, and the β1–β2 loop in orange. The pdb accession codes for the structures are nAChR closed (nAChR-**C**), 4AQ5; nAChR open (nAChR-**O**), 4AQ9; Glu-Cl, 3RI5; GLIC, 3EHZ; ELIC, 2VL0

and likely involves motions of the Cys-loop (β6–β7 loop) and the β1–β2 loop acting on the M2–M3 linker. While it is clear from the structural analysis and from mutational analysis that these regions are critical for communication between the extracellular domain and the transmembrane domain, nonetheless, a clear description of how ligand binding alters the conformation in this region and communicates the movement to the channel, is still lacking. This is partly because the prokaryotic channels do not evince the same extracellular ligand binding as the eukaryotic channels, possibly due to crystal constraints, and to the modest structural changes seen in the careful comparisons of the closed and activated *Torpedo* nAChR [7].

2.5 Desensitization

An added layer of complexity in our understanding of the structural basis of gating is introduced when we consider desensitization for LGICs [27]. The *Torpedo* nAChR is well-known to desensitize readily under the influence of a number of local anesthetics, detergents, and other agents, as well as in the prolonged presence of conventional agonists such as acetylcholine [28–32]. However, ambiguities in the functional state of channels where we have high-resolution structures make it unclear whether any of those structures fully define a distinct functional state. Is it possible that the ELIC structure or the nAChR structure represents a desensitized state rather than a simple closed configuration? This appears unlikely considering several general observations. Photo-affinity labeling studies have identified residues that change their exposure, or reactivity, or both when changing from closed to open to desensitized; channel blockers that preferentially bind the resting state, such as tetracaine [33] and 3-(trifluoromethyl)-3-(m-iodophenyl)diazirine (TID) are typically smaller, while the desensitized channel can accommodate substantially larger ligands such as meproadifen mustard, ethidium, and crystal violet. Large channel blockers display slow kinetics of binding and egress from the desensitized channel, indicating restricted access in the desensitized state [32, 34]. Yet, structural features consistent with restriction of ligand access to the channel pore have not been observed in any of the crystal structures, so it is possible that none of the known structures are representative of the desensitized state.

2.6 Summary of Structural Data

High-resolution structural data have provided a wealth of information on the basic structure of the channel. It provides a clear picture of changes in the ligand-binding domain, and also about possible changes that may constitute gating within the TM domain. The actual location of such conceptual features as the gate, the element that restricts flow in the closed state, and the selectivity filter, are less clear, given the distinctions between the nAChR structures and the bacterial LGICs. Other types of indirect structural information, such as accessibility studies and photo-affinity labeling, along with functional analysis further informs the structure and location of these features as discussed below.

3 Functional Analysis of Gating

Gating of nAChRs has been analyzed extensively using electrophysiology, ion flux assays, ligand-binding assays, and single channel recordings. These have been combined with such techniques as photoaffinity labeling and cysteine accessibility assays to probe the structures and structural changes at various sites on the receptor. These techniques have their distinct advantages and taken collectively present a reasonably cohesive, though incomplete, picture of gating in nicotinic receptors. The combination of single amino acid substitution, single channel gating measurements, and thermodynamically-based analysis is proving a powerful technique that provides substantial functional information on the role of individual amino acids in gating. It also provides insight into the order of events during the gating transition.

nAChR activation typically requires high micromolar concentrations of acetylcholine, which are substantially higher than the low nanomolar concentrations determined to bind fully at equilibrium [28, 35, 36]. nAChRs were closed at conditions of equilibrium binding, thereby describing the nonconducting desensitized state as having high affinity for acetylcholine [27]. These observations essentially set the stage for the current model of nAChR-activation and desensitization (see Fig. 2.2). Although a number of details embellish this model, as discussed below, this model captures the main, essential features of nAChR activation as we currently understand it.

One goal of functional studies is to determine the number of stable and intermediate states of the nAChRs so as to arrive at a complete energetic and structural description. The atomic-resolution structural information lags behind due to the inherent difficulty of crystallization. Therefore, functional studies are critical to understanding the basic route that activation and desensitization can occur and thereby inform basic structural constraints. The current questions include whether there are intermediate states between ligand binding and channel opening and between opening and full desensitization.

3.1 Energetics and Linear Free Energy Analysis of Gating

The energetics of gating can be examined through the effect of mutational changes on the overall gating equilibrium between the closed and open states (Figs. 2.2 and 2.6). This informs the amino acids that are critical in the gating motion or can affect it allosterically. A refinement of this analysis that provides a distinct element of information is Linear Free Energy Analysis (LFER; or Φ-value analysis). This requires an assumption that the closed-to-open state transition can be described as

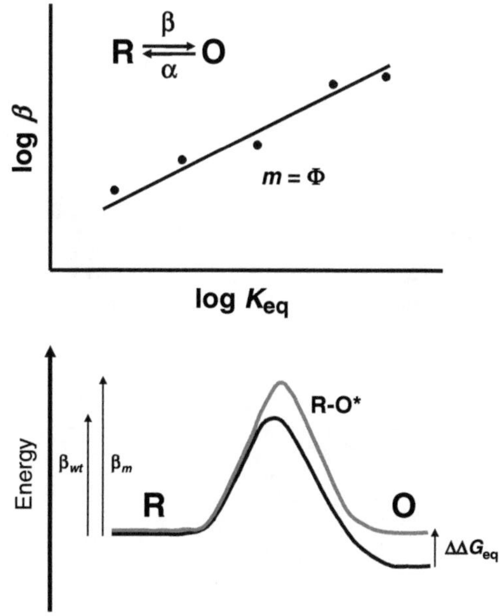

Fig. 2.6 Linear Free Energy Analysis. *Top*, the gating transition is confined to being described as a simple elementary reaction between **R** and **O** states with rate constants β (forward) and α (backward). The ratio β/α describes the equilibrium constant, K_{eq}. A plot of log β versus log K_{eq} from mutations or other perturbations at a discrete site is described by the slope Φ, which is a descriptor of how close the transition state (**R-O***) is to the open state structure, ranging from zero to one. *Bottom*, a schematic of the transition state diagram for *wt* and mutant (*m*) nAChR between the open and closed states. The diagrams for the two nAChRs are arbitrarily aligned at the resting state energy, showing changes in the equilibrium constant free energies ($ΔΔG_{eq}$). The *arrows* at the left correspond to the activation energies associated with the forward rate constants β and are proportional to log β

an elementary chemical transition and examines the gating transition state at any amino acid where a mutation perturbs gating. Based on perturbation of the closed-to-open equilibrium, the principles of the analysis are derived from basic thermodynamics and reaction kinetics. Grosman et al. [37] described its application to the nAChR: the technique requires single channel analysis of opening and closing rates (β and α, respectively). These also determine the overall equilibrium between the two states (see Fig. 2.6). Upon perturbation of the equilibrium through mutation or using various agonists, a log plot of β *versus* the equilibrium constant provides a slope, a Φ-value. The higher the slope for the forward rate constant the greater the transition state resembles the open state. This interpretation reflects the lack of change in the rate constant from open to closed, suggesting little perturbation between the open state and the transition state (Fig. 2.6b). For plots of the closing constant, α, the reverse argument holds true, high slopes reflect low Φ values and reflect a transition state that more closely approximates the closed state. By carrying out Φ-value analysis on the binding site and on other amino acids thought to be

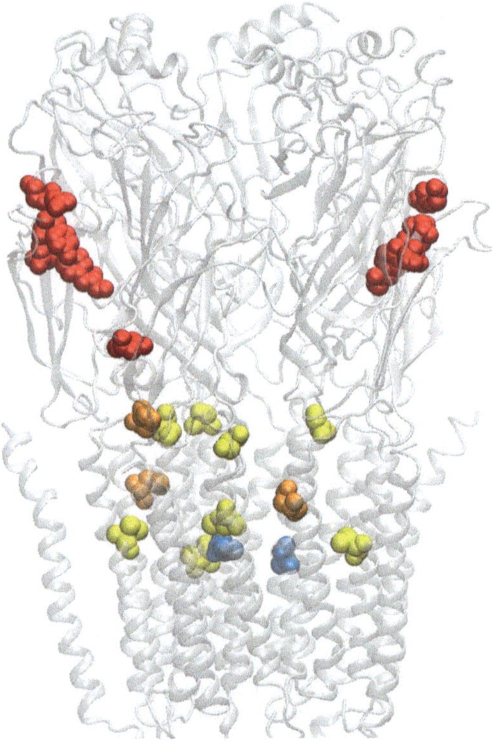

Fig. 2.7 A structural illustration of Φ-value analysis. The structure illustrates a few amino acids from the various zones of high, intermediate and low Φ-values. The residues are colored according to Φ-value ranges with the data taken from Grosman and Auerbach [37]: *Red* indicates a range of 0.9–1; *yellow*, 0.7–0.9; *orange*, 0.4–0.7; *blue*, 0–0.4. Subsequent work has fleshed out these results to show there are four distinct regions of similar Φ-values. These are not restricted to distinct domains but have some structural overlap. In general, amino acids closer to the binding site have higher Φ-values and those at the gating region and channel, lower Φ-values. Higher values indicate that the transition state structure is more like the open state; lower values more like the closed state. The four regions may indicate parts of the structure that move coordinately during the transition from closed to open

involved in gating, Auerbach and colleagues mapped the transition state patterns of the gating movement throughout the mouse muscle nAChR [38].

The results of this analysis reveals a pattern where the binding site more closely resembles the open state and parts of the channel more closely resemble the closed state during the transition state [39] (see Fig. 2.7 for a limited sample of residues tested). This appears to make sense as ligand-binding drives channel opening through an induced-fit mechanism [40] that is energetically driven by cation–pi interactions between acetylcholine and the aromatic side chains of the binding site [41–43]. The binding provides the ultimate strain that drives opening, so it is expected that this should more closely resemble the final, high-affinity open state. Likewise, it is not unexpected that the channel and the physical gate would more closely approximate the closed state. The full data set paints more detail on this simple picture of strain communicated from the binding site to the gate.

The Φ-values reflect the overall change of the transition state relative to the starting and end states. Thus, amino acids with similar Φ-values like move in a tightly-coupled manner. That is, by identifying the regions where the Φ-values changes, there will be greater flexibility in transmission of the strain from the binding site to the gates. This can give insight into the motions of the nAChR as it undergoes channel opening. Using this approach Auerbach's group [38] identified several areas that move in relative synchrony. These begin with a cluster in the ligand-binding site where Φ-values are close to one. Lower values occur in the other parts of the ligand-binding domain, including the region where linkage to the transmembrane domain occurs, and in the transmembrane domain. By identifying four distinct regions with similar Φ-values, the data argue for regions moving in blocks during the transition.

3.2 *Where Is the Actual Gate?*

Gating was initially proposed to lie near the conserved 9′-Leu that appeared congruent with a bend in the M2 helix [44, 45]. Consistent with this hypothesis is the profound effect of mutagenesis of this residue on the gating equilibrium [20, 44, 46, 47]. The structural studies on the nAChR are generally consistent with Substituted Cysteine Accessibility Measurements (SCAM) from Karlin's group, which showed the closed and open states to have a number of residues in M2 accessible in both states, but with residues deeper in the channel, near the intracellular end of M2, being accessible only in the open state [48]. This identified the cytoplasmic region of M2 as the effective gating structure. Modeling indicates that even with a relatively large-diameter pore in the closed state, change in hydrophobicity can affect conductivity dramatically. Thus, gating can be described as a local change in hydrophobicity due to modest changes in pore size and side-chain exposure. This can affect the ability a partly hydrated ion to permeate the pore [49]. This hydrophobicity change is likely accomplished either by twisting the M2 helices such as to change the exposure from hydrophobic to hydrophilic residues, or, as has been argued by Cymes et al. based on proton modification, a modest increase in the diameter at this region [50]. However, the model of gating evinced by the differences between ELIC and GLIC structures shows more dramatic changes, and gating may reflect larger overall changes in structure (see Fig. 2.4).

3.3 *The Flip Side of Activation*

Colquhoun's group initially proposed the existence of an intermediate state of activation based on experiments on the glycine receptor [51, 52]. This scheme is illustrated in Fig. 2.8 (for nAChRs) and shows a proposed state required to account for

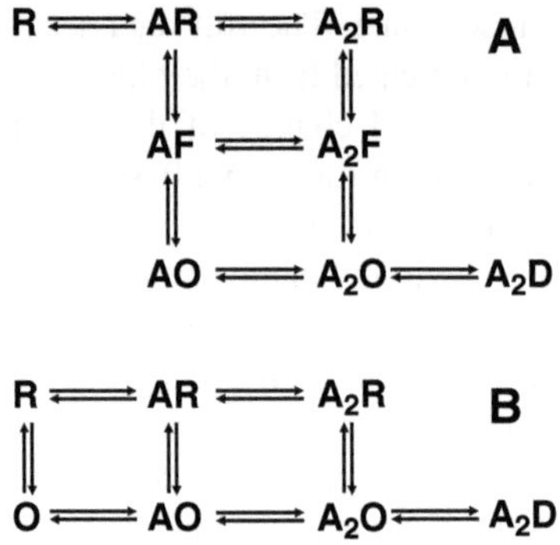

Fig. 2.8 Models for nAChR activation. Panel **a**, the model includes a pre-open, global flip state, with intermediate affinities for agonist. This state is hypothesized to be an intermediate transient state from which opening occurs. There is no direct transition from the initial, low-affinity binding state, **R**, to the open state, **O**. Based on the description by Lape et al.[53], the **F** (flip) state is include in models to adequately account for brief closing events in the presence of partial agonists. Panel **b**, a classic MWC-type model for opening [43, 77]

data from single channel measurements of short closing times, which are not accounted for by simpler models, but that also occur in the same time domain as open-channel block. It was further argued that this mechanism applies also to the nAChR on the basis of experiments using the partial agonists choline and tetramethyl ammonium [52, 53] on human muscle nAChR. The addition of this intermediate affinity, flip state (see Fig. 2.8), closed conformation may explain these observations, although it has been argued that it is unnecessary to account for agonist gating in the mouse muscle-type nicotinic receptor [43, 54] and that a typical MWC model accounts adequately for activation (Fig. 2.8b). Nevertheless, the addition of a new state to the repertoire of intermediate and quasi-stable states may require further structural characterization for a complete understanding of the movements of the LGICs as they activate.

3.4 Desensitization as a Gating Phenomenon

Desensitization is characterized by high agonist affinity and by a nonconducting channel and occurs spontaneously upon prolonged exposure to agonist or tetanic stimulation of a muscle endplate [27]. In the muscle-type nAChR, desensitization occurs relatively slowly, typically taking greater than 50 ms [28, 29, 35, 55]. This makes it unlikely to have a physiological role during typical moderate endplate stimulation but may be important during tetany or under pathological conditions. Some neuronal nAChRs desensitize substantially faster, particularly the $\alpha7$ homopentameric subtype. In those cases it is likely that desensitization plays a role in terminating the response to acetylcholine [56, 57]. Desensitization is less-well understood relative to our understanding of opening, despite its initial thorough characterization on the *Torpedo* nAChR [28, 29]. This partly reflects the difficulty of characterizing a nonconducting state that exits from the open state (Figs. 2.1 and 2.8b). Surprisingly, as it is the mostly stable bound state, it is yet unclear if any of the structures discussed above reflect a desensitized state rather than a closed state.

As noted by Katz and Thesleff [27], the desensitized state is the thermodynamically most stable state and should reflect the highest agonist-affinity. However, thermodynamic cycle analysis and measurements of ligand binding and open probabilities show that the open state also has high affinities for acetylcholine [58] that do not appear to differ appreciably from measured affinities of acetylcholine for the desensitized state [59].

This is a critical observation because it indicates that there is likely little, if any change in the ligand-binding affinity, and perhaps, little to no change in structure of the ligand-binding domain (acetylcholine cannot induce a structural change to a conformation with similar or lower affinity). Thus, structural changes in the progression from the open to desensitized state may be confined to the transmembrane domain. How can the desensitized state be more stable and have the same agonist affinity as the open state? This can be true if the unliganded desensitized state is also more stable than the open state. And this is generally true—the unliganded open state is quite rare (10^{-6} or so). But the unliganded resting-desensitized equilibrium is more modest (0.05 in Torpedo; a bit lower in mouse muscle). That implies a strong, residual unliganded open to desensitized equilibrium (~10^4; see also Fig. 2.9), which will be similar for the open to desensitized transition. This conclusion has important implications for any drug-discovery efforts that hope to differentiate between activation and desensitization of the nAChR, as might be useful for treatment of nicotine addiction. Drugs targeted to the agonist sites that activate may inevitably lead to desensitization and reinforce any addiction mediated by desensitization. Therefore allosteric activators may be better choices for such future therapeutics.

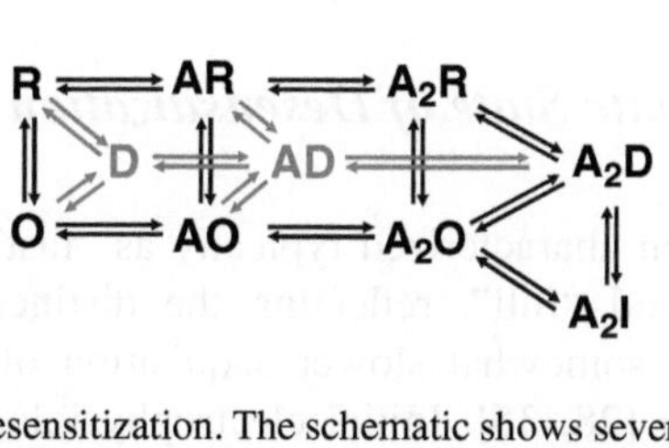

Fig. 2.9 MWC Model for desensitization. The schematic shows several possible routes for desensitization. The predominant route is undoubtedly through doubly-liganded binding to opening to the desensitized state. Some nAChRs, such at the *Torpedo* muscle type, have a substantial population in the preexisting desensitized state (in *gray*), which bind agonist with intrinsically high affinity. Desensitization from the open state like proceeds through one or more intermediate, structurally distinct states (**I**)

Desensitization involves rendering the channel nonconductive. In principle, it is possible that the channel reverts to the resting, or closed, state, while the ligand-binding domain remains in a high-affinity conformation, effectively uncoupling binding from channel gating. However, a substantial body of evidence indicates that the channel occupies a structurally distinct state when desensitized. This evidence comes from photoaffinity labeling studies, from cysteine accessibility studies, and from kinetic studies of ligand binding by channel blockers.

Photoaffinity labeling of the nAChR, using compounds such as Chlorpromazine and [^{125}I]TID, show clear changes in labeling patterns in the various states, including closed, open, and desensitized [60–65]. A large number of ligand-binding studies on various channel blockers, including local anesthetics and detergents, show substantial changes in affinity upon desensitization [31, 66, 67]. This clearly demonstrates a change in affinity for channel blockers between the closed state and the desensitized state. Conversely, binding of these ligands also cause changes in the affinity of agonists: desensitizing ligands, such as ethidium, effectively increase the equilibrium affinity of agonists; channel blockers such as tetracaine, which preferentially bind the closed state, lower the effective acetylcholine affinity [30, 33].

Binding kinetics may provide further insight into the structure of the desensitized state. In one example, the binding affinity of the fluorescent channel blocker ethidium is increased by agonists, and decreased by α-bungarotoxin, which stabilizes the closed state [67, 68]. It binds rapidly to the open channel upon brief exposure to agonist, dramatically increasing the effective association rate [69]. However, once desensitized the dissociation rate for ethidium is slowed dramatically [67, 70]. This indicates that egress of the ligand from the channel is structurally impeded. The ethidium site has also been identified through photo-affinity labeling and binds at the mid- level (near 12′) of the M2-alpha helix [12, 71]. Notably, the current structures of the nAChR appear relatively open at the level of the binding site. So the kinetic data argues for a novel, as yet uncharacterized conformation with the channel impeded above the level of the ethidium site. Data from photoaffinity labeling with chlorpromazine in well-defined functional states further demonstrate structural changes in this region between the closed and desensitized states [63].

3.5 The Intermediate State of Desensitization

Desensitization has been characterized typically as "fast" and "slow" or alternatively "intermediate" and "full", reflecting the distinction between functional desensitization and the somewhat slower acquisition of slowly-reversible, high affinity agonist binding [28, 35]. Initial electrophysiological characterization of desensitization shows it taking place with multiple time constants [55, 72]. A detailed electrophysiological study of the kinetic constants of desensitization on the mouse muscle receptor showed up to five distinct time constants associated with desensitization [73].

Time-resolved photoaffinity labeling by a small hydrophobic 3-(trifluoromethyl)-3-(m-iodophenyl)diazirine (TID) further elucidated distinct structural changes at the level of the middle of M2 upon fast desensitization and near the cytoplasmic end upon slow desensitization (see Fig. 2.9). These indicated changes associated with fast and slow desensitization that were distinct within the channel domain and clearly distinct from the labeling patterns of the resting (closed) channel [74, 75]. Unlike intermediate closed states, the existence of intermediate desensitized states is clear. Early data based on the binding of fluorescent analogs of acetylcholine showed several exponential phases of binding [76]. Though these seemed to correlate with intermediate affinities of acetylcholine binding prior to full long-term desensitization the changes do not actually necessitate a change in agonist affinity, but likely reflect changes in the transmembrane domain alone. Both of these assays establish intermediate states before acquiring the slowly reversible high affinity for acetylcholine that is the hallmark of the fully desensitized state. A Monod-Wyman-Changeux (MWC) [77]-type model for these states is shown in Fig. 2.9.

3.6 The Role of Multiple Binding Sites

The heteropentameric structure of the embryonic muscle-type receptor ($\alpha_2\beta\gamma\delta$) results in highly distinct affinities for agonist at the two sites, which subsequently affects the dose-response curve of these proteins. The affinities for the agonists differ ~ 100 fold in the resting state, but only ~ threefold in the desensitized state [13]. For the *Torpedo* nAChR, the α-δ site has higher affinity for agonist in both states. In the resting state the α–δ site affinity is about 1 μM, as compared with ~ 100 μM for the α–γ site [13]. Since channel opening is dominated by biliganded opening events the dose response curve follows binding to the α–γ site, to a first approximation, as the α–δ site will be occupied at much lower concentrations. Because the affinities of the sites are closer in the desensitized state, the α–γ site undergoes a larger change in affinity upon conformational changes to the open and desensitized states. Therefore, most of the energy driving the conformational change is provided through the α–γ site [13, 78]. For the mouse muscle adult form of the nAChR, $\alpha_2\beta\varepsilon\delta$, the two affinities are nearly equal and the two sites contribute to the dose-response curve and the energetics more equally.

An interesting question arises for the homomeric receptors, such as α7, as to how many of the potentially five equivalent sites are required for activation. Work by Rayes et al. [79], in α7-$5HT_{3A}$ chimeric receptors suggests that two nonconsecutive sites are sufficient for effective activation (i.e. sites with an intervening subunit). Three sites are most efficacious, and this provides the rationale for the efficacy of allosteric activators such as benzodiazepines in $GABA_A$ receptors *via* binding to a third, nonconsensus site, that does not normally bind agonist [80]. Interestingly, this work found the same locus to be most efficacious and this site is equivalent to the δ–β interface on the muscle-type receptor (Fig. 2.1). It also suggests which interface may be a good target for allosteric therapeutic ligands.

4 Summary, Conclusions, and Unresolved Issues

The prevailing questions for understanding gating of the nAChRs are now focused on a detailed correlation between structure and function. Atomic resolution structures have provided a clear picture of gating at the level of the ligand-binding sites. This information is being brought to bear in drug development, particularly for treatment of disorders of the central nervous system [1]. In contrast, there are major outstanding questions as to whether the bacterial protein structures or those from the *Torpedo* nAChR better represent the gating transition that need to be resolved.

Although not unique to these channels nor to channels in general, it should be noted that proper ligand-gated channel function is highly dependent on the detailed kinetics of the mechanism. As one example, one instance of congenital myasthenia gravis turned out to be a mutation that had a modest two-fold effect on the channel gating efficiency [81]. In many proteins and enzyme such an effect on would have little to no impact on its functionality. In this respect, ion channels present an excellent model system for understanding conformational regulation and kinetic control of proteins at a detailed level. Functional studies, through the use of single-channel current measurements, presents great opportunities for detailed kinetic studies. Combined with single amino acid substitution they constitute a rich methodology for probing the roles of individual residues in function. Ultimately, the detailed structural knowledge must be correlated closely with these kinetic models to provide an overall understanding of these functions.

In summary, understanding and correlating the structure and function of ligand-gated ion channels presents a unique opportunity to probe deeply our understanding of the kinetic control of proteins in general. In addition, such detailed knowledge will provide additional information for the development of specific therapeutics targeting ligand-gated ion channels. The worth of allosteric ligands [1] has been proven in the GABA receptors, and therefore the nicotinic receptors provide a rich new area for similar types of intervention to help treat a number of disorders, the most prominent of which may be cigarette addiction.

Acknowledgements All molecular modeling figures were created using VMD [82].

References

1. Taly A, Corringer PJ, Guedin D, Lestage P, Changeux JP. Nicotinic receptors: allosteric transitions and therapeutic targets in the nervous system. Nat Rev Drug Discov. 2009;8(9):733–50.
2. Dacosta CJ, Baenziger JE. Gating of pentameric ligand-gated ion channels: structural insights and ambiguities. Structure. 2013;21(8):1271–83.
3. Smit AB, Syed NI, Schaap D, van Minnen J, Klumperman J, Kits KS, Lodder H, van der Schors RC, van Elk R, Sorgedrager B, Brejc K, Sixma TK, Geraerts WP. A glia-derived acetylcholine-binding protein that modulates synaptic transmission. Nature. 2001;411(6835): 261–8.

4. Brejc K, van Dijk WJ, Klaassen RV, Schuurmans M, van Der Oost J, Smit AB, Sixma TK. Crystal structure of an ACh-binding protein reveals the ligand-binding domain of nicotinic receptors. Nature. 2001;411(6835):269–76.
5. Hansen SB, Sulzenbacher G, Huxford T, Marchot P, Taylor P, Bourne Y. Structures of Aplysia AChBP complexes with nicotinic agonists and antagonists reveal distinctive binding interfaces and conformations. EMBO J. 2005;24(20):3635–46.
6. Li SX, Huang S, Bren N, Noridomi K, Dellisanti CD, Sine SM, Chen L. Ligand-binding domain of an alpha7-nicotinic receptor chimera and its complex with agonist. Nat Neurosci. 2011;14(10):1253–9.
7. Unwin N, Fujiyoshi Y. Gating movement of acetylcholine receptor caught by plunge-freezing. J Mol Biol. 2012;422(5):617–34.
8. Hilf RJ, Dutzler R. X-ray structure of a prokaryotic pentameric ligand-gated ion channel. Nature. 2008;452(7185):375–9.
9. Hilf RJ, Dutzler R. Structure of a potentially open state of a proton-activated pentameric ligand-gated ion channel. Nature. 2009;457(7225):115–8.
10. Hibbs RE, Gouaux E. Principles of activation and permeation in an anion-selective Cys-loop receptor. Nature. 2011;474(7349):54–60.
11. Neubig RR, Cohen JB. Equilibrium binding of [3H]tubocurarine and [3H]acetylcholine by Torpedo postsynaptic membranes: stoichiometry and ligand interactions. Biochemistry. 1979; 18(24):5464–75.
12. Pedersen SE. Site-selective photoaffinity labeling of the Torpedo californica nicotinic acetylcholine receptor by azide derivatives of ethidium bromide. Mol Pharmacol. 1995;47(1):1–9.
13. Andreeva IE, Nirthanan S, Cohen JB, Pedersen SE. Site specificity of agonist-induced opening and desensitization of the torpedo californica nicotinic acetylcholine receptor. Biochemistry. 2006;45(1):195–204.
14. Prince RJ, Sine SM. Molecular dissection of subunit interfaces in the acetylcholine receptor. Identification of residues that determine agonist selectivity. J Biol Chem. 1996;271(42):25770–7.
15. Pennington RA, Gao F, Sine SM, Prince RJ. Structural basis for epibatidine selectivity at desensitized nicotinic receptors. Mol Pharmacol. 2005;67(1):123–31.
16. Chiara DC, Xie Y, Cohen JB. Structure of the agonist-binding sites of the Torpedo nicotinic acetylcholine receptor: affinity-labeling and mutational analyses identify gamma Tyr-111/delta Arg-113 as antagonist affinity determinants. Biochemistry. 1999;38(20):6689–98.
17. Chiara DC, Cohen JB. Identification of amino acids contributing to high and Low affinity D-tubocurarine sites in the torpedo nicotinic acetylcholine receptor. J Biol Chem. 1997;272(52):32940–50.
18. Sine SM, Kreienkamp HJ, Bren N, Maeda R, Taylor P. Molecular dissection of subunit interfaces in the acetylcholine receptor: identification of determinants of alpha-conotoxin M1 selectivity. Neuron. 1995;15(1):205–11.
19. Bren N, Sine SM. Identification of residues in the adult nicotinic acetylcholine receptor that confer selectivity for curariform antagonists. J Biol Chem. 1997;272(49):30793–8.
20. Unwin N. Refined structure of the nicotinic acetylcholine receptor at 4A resolution. J Mol Biol. 2005;346(4):967–89.
21. Bocquet N, Prado de Carvalho L, Cartaud J, Neyton J, Le Poupon C, Taly A, Grutter T, Changeux JP, Corringer PJ. A Prokaryotic proton-gated ion channel from the nicotinic acetylcholine receptor family. Nature. 2007;445(7123):116–9.
22. Bocquet N, Nury H, Baaden M, Le Poupon C, Changeux JP, Delarue M, Corringer PJ. X-ray structure of a pentameric ligand-gated ion channel in an apparently open conformation. Nature. 2009;457(7225):111–4.
23. Sauguet L, Poitevin F, Murail S, Van Renterghem C, Moraga-Cid G, Malherbe L, Thompson AW, Koehl P, Corringer PJ, Baaden M, Delarue M. Structural basis for ion permeation mechanism in pentameric ligand-gated ion channels. EMBO J. 2013;32(5):728–41.
24. Prevost MS, Sauguet L, Nury H, Van Renterghem C, Huon C, Poitevin F, Baaden M, Delarue M, Corringer PJ. A locally closed conformation of a bacterial pentameric proton-gated ion channel. Nat Struct Mol Biol. 2012;19(6):642–9.

25. Grosman C, Salamone FN, Sine SM, Auerbach A. The extracellular linker of muscle acetylcholine receptor channels is a gating control element. J Gen Physiol. 2000;116(3):327–40.
26. Kash TL, Jenkins A, Kelley JC, Trudell JR, Harrison NL. Coupling of agonist binding to channel gating in the GABA(A) receptor. Nature. 2003;421(6920):272–5.
27. Katz B, Thesleff S. A study of the 'Desensitization' produced by acetylcholine at the motor endplate. J Physiol. 1957;138:63–80.
28. Neubig RR, Boyd ND, Cohen JB. Conformations of Torpedo acetylcholine receptor associated with ion transport and desensitization. Biochemistry. 1982;21(14):3460–7.
29. Boyd ND, Cohen JB. Desensitization of membrane-bound Torpedo acetylcholine receptor by amine noncompetitive antagonists and aliphatic alcohols: studies of [3H]acetylcholine binding and 22Na+ ion fluxes. Biochemistry. 1984;23(18):4023–33.
30. Cohen JB, Correll LA, Dreyer EB, Kuisk IR, Medynski DC, Strnad NP. Interactions of local anesthetics with torpedo nicotinic acetylcholine receptors. In: Roth SH, Miller KW, editors. Molecular and cellular mechanisms of anesthetics. New York: Plenum Publishing Corporation; 1986. p. 111–24.
31. Oswald RE. Binding of phencyclidine to the detergent solubilized acetylcholine receptor from Torpedo marmorata. Life Sci. 1983;32(10):1143–9.
32. Oswald RE, Heidmann T, Changeux JP. Multiple affinity states for noncompetitive blockers revealed by [3H]phencyclidine binding to acetylcholine receptor rich membrane fragments from Torpedo marmorata. Biochemistry. 1983;22(13):3128–36.
33. Middleton RE, Strnad NP, Cohen JB. Photoaffinity labeling the torpedo nicotinic acetylcholine receptor with [(3)H]tetracaine, a nondesensitizing noncompetitive antagonist. Mol Pharmacol. 1999;56(2):290–9.
34. Herz JM, Atherton SJ. Steric factors limit access to the noncompetitive inhibitor site of the nicotinic acetylcholine receptor. Fluorescence studies. Biophys J. 1992;62(1):74–6.
35. Boyd ND, Cohen JB. Kinetics of binding of [3H]acetylcholine to Torpedo postsynaptic membranes: association and dissociation rate constants by rapid mixing and ultrafiltration. Biochemistry. 1980;19(23):5353–8.
36. Boyd ND, Cohen JB. Kinetics of binding of [3H]acetylcholine and [3H]carbamoylcholine to Torpedo postsynaptic membranes: slow conformational transitions of the cholinergic receptor. Biochemistry. 1980;19(23):5344–53.
37. Grosman C, Zhou M, Auerbach A. Mapping the conformational wave of acetylcholine receptor channel gating. Nature. 2000;403(6771):773–6.
38. Auerbach A. The gating isomerization of neuromuscular acetylcholine receptors. J Physiol. 2010;588(Pt 4):573–86.
39. Cymes GD, Grosman C, Auerbach A. Structure of the transition state of gating in the acetylcholine receptor channel pore: a phi-value analysis. Biochemistry. 2002;41(17):5548–55.
40. Koshland Jr DE, Nemethy G, Filmer D. Comparison of experimental binding data and theoretical models in proteins containing subunits. Biochemistry. 1966;5(1):365–85.
41. Xiu X, Puskar NL, Shanata JA, Lester HA, Dougherty A. Nicotine binding to brain receptors requires a strong cation-pi interaction. Nature. 2009;458(7237):534–7.
42. Beene DL, Brandt GS, Zhong W, Zacharias NM, Lester HA, Dougherty DA. Cation-π interactions in ligand recognition by serotonergic (5-HT_{3A}) and nicotinic acetylcholine receptors: the anomalous binding properties of nicotine. Biochemistry. 2002;41(32):10262–9.
43. Auerbach A. The energy and work of a ligand-gated ion channel. J Mol Biol. 2013;425(9):1461–75.
44. Labarca C, Nowak MW, Zhang H, Tang L, Deshpande P, Lester HA. Channel gating governed symmetrically by conserved leucine residues in the M2 domain of nicotinic receptors. Nature. 1995;376(6540):514–6.
45. Unwin N. Acetylcholine receptor channel imaged in the open state. Nature. 1995;373(6509):37–43.
46. Unwin N. Nicotinic acetylcholine receptor at 9 A resolution. J Mol Biol. 1993;229(4):1101–24.
47. Miyazawa A, Fujiyoshi Y, Unwin N. Structure and gating mechanism of the acetylcholine receptor pore. Nature. 2003;423(6943):949–55.

48. Pascual JM, Karlin A. State-dependent accessibility and electrostatic potential in the channel of the acetylcholine receptor. Inferences from rates of reaction of thiosulfonates with substituted cysteines in the M2 segment of the alpha subunit. J Gen Physiol. 1998;111(6):717–39.
49. Beckstein O, Sansom MS. A hydrophobic gate in an ion channel: the closed state of the nicotinic acetylcholine receptor. Phys Biol. 2006;3(2):147–59.
50. Cymes GD, Ni Y, Grosman C. Probing Ion-channel pores One proton at a time. Nature. 2005;438(7070):975–80.
51. Burzomato V, Beato M, Groot-Kormelink PJ, Colquhoun D, Sivilotti LG. Single-channel behavior of heteromeric alpha1beta glycine receptors: an attempt to detect a conformational change before the channel opens. J Neurosci. 2004;24(48):10924–40.
52. Lape R, Colquhoun D, Sivilotti LG. On the nature of partial agonism in the nicotinic receptor superfamily. Nature. 2008;454(7205):722–7.
53. Lape R, Krashia P, Colquhoun D, Sivilotti LG. Agonist and blocking actions of choline and tetramethylammonium on human muscle acetylcholine receptors. J Physiol. 2009;587 (Pt 21):5045–72.
54. Purohit P, Auerbach A. Loop C and the mechanism of acetylcholine receptor-channel gating. J Gen Physiol. 2013;141(4):467–78.
55. Steinbach JH, Sine SM. Function of nicotinic acetylcholine receptors. Soc Gen Physiol Ser. 1987;41:19–42.
56. Papke RL. Enhanced inhibition of a mutant neuronal nicotinic acetylcholine receptor by agonists: protection of function by (E)-N-methyl-4-(3- pyridinyl)-3-butene-1-amine (TC-2403). J Pharmacol Exp Ther. 2002;301(2):765–73.
57. Papke RL, Thinschmidt JS. The correction of alpha7 nicotinic acetylcholine receptor concentration-response relationships in Xenopus oocytes. Neurosci Lett. 1998;256(3):163–6.
58. Jackson MB. Perfection of a synaptic receptor: kinetics and energetics of the acetylcholine receptor. Proc Natl Acad Sci U S A. 1989;86(7):2199–203.
59. Song XZ, Andreeva IE, Pedersen SE. Site-selective agonist binding to the nicotinic acetylcholine receptor from Torpedo californica. Biochemistry. 2003;42(14):4197–207.
60. White BH, Cohen JB. Agonist-induced changes in the structure of the acetylcholine receptor M2 regions revealed by photoincorporation of an uncharged nicotinic noncompetitive antagonist. J Biol Chem. 1992;267(22):15770–83.
61. White BH, Howard S, Cohen SG, Cohen JB. The hydrophobic photoreagent 3-(trifluoromethyl)-3-m-([125I] iodophenyl) diazirine is a novel noncompetitive antagonist of the nicotinic acetylcholine receptor. J Biol Chem. 1991;266(32):21595–607.
62. White BH, Cohen JB. Photolabeling of membrane-bound Torpedo nicotinic acetylcholine receptor with the hydrophobic probe 3-trifluoromethyl-3-(m-[125I]iodophenyl)diazirine. Biochemistry. 1988;27(24):8741–51.
63. Chiara DC, Hamouda AK, Ziebell MR, Mejia LA, Garcia 3rd G, Cohen JB. [(3)H]chlorpromazine photolabeling of the torpedo nicotinic acetylcholine receptor identifies two state-dependent binding sites in the ion channel. Biochemistry. 2009;48(42):10066–77.
64. Heidmann T, Changeux JP. Time-resolved photolabeling by the noncompetitive blocker chlorpromazine of the acetylcholine receptor in its transiently open and closed ion channel conformations. Proc Natl Acad Sci U S A. 1984;81(6):1897–901.
65. Oswald R, Changeux JP. Ultraviolet light-induced labeling by noncompetitive blockers of the acetylcholine receptor from Torpedo marmorata. Proc Natl Acad Sci U S A. 1981;78(6):3925–9.
66. Heidmann T, Oswald RE, Changeux JP. Multiple sites of action for noncompetitive blockers on acetylcholine receptor rich membrane fragments from torpedo marmorata. Biochemistry. 1983;22(13):3112–27.
67. Herz JM, Kolb SJ, Erlinger T, Schmid E. Channel permeant cations compete selectively with noncompetitive inhibitors of the nicotinic acetylcholine receptor. J Biol Chem. 1991; 266(25):16691–8.
68. Herz JM, Johnson DA, Taylor P. Interaction of noncompetitive inhibitors with the acetylcholine receptor. The site specificity and spectroscopic properties of ethidium binding. J Biol Chem. 1987;262(15):7238–47.

69. Rankin SE, Addona GH, Kloczewiak MA, Bugge B, Miller KW. The cholesterol dependence of activation and fast desensitization of the nicotinic acetylcholine receptor. Biophys J. 1997;73(5):2446–55.
70. Lurtz MM, Hareland ML, Pedersen SE. Quinacrine and ethidium bromide bind the same locus on the nicotinic acetylcholine receptor from Torpedo californica. Biochemistry. 1997;36(8): 2068–75.
71. Pratt MB, Pedersen SE, Cohen JB. Identification of the sites of incorporation of [3H]ethidium diazide within the Torpedo nicotinic acetylcholine receptor ion channel. Biochemistry. 2000;39(37):11452–62.
72. Sine SM, Steinbach JH. Activation of acetylcholine receptors on clonal mammalian BC3H-1 cells by high concentrations of agonist. J Physiol. 1987;385:325–59.
73. Elenes S, Auerbach A. Desensitization of diliganded mouse muscle nicotinic acetylcholine receptor channels. J Physiol. 2002;541(Pt 2):367–83.
74. Yamodo IH, Chiara DC, Cohen JB, Miller KW. Conformational changes in the nicotinic acetylcholine receptor during gating and desensitization. Biochemistry. 2010;49(1):156–65.
75. Arevalo E, Chiara DC, Forman SA, Cohen JB, Miller KW. Gating-enhanced accessibility of hydrophobic sites within the transmembrane region of the nicotinic acetylcholine receptor's {delta}-subunit. A time-resolved photolabeling study. J Biol Chem. 2005;280(14):13631–40.
76. Heidmann T, Bernhardt J, Neumann E, Changeux JP. Rapid kinetics of agonist binding and permeability response analyzed in parallel on acetylcholine receptor rich membranes from Torpedo marmorata. Biochemistry. 1983;22(23):5452–9.
77. Monod J, Wyman J, Changeux J-P. On the nature of allosteric transitions: a plausible model. J Mol Biol. 1965;12(1):88–118.
78. Purohit P, Bruhova I, Auerbach A. Sources of energy for gating by neurotransmitters in acetylcholine receptor channels. Proc Natl Acad Sci U S A. 2012;109(24):9384–9.
79. Rayes D, De Rosa MJ, Sine SM, Bouzat C. Number and locations of agonist binding sites required to activate homomeric Cys-loop receptors. J Neurosci. 2009;29(18):6022–32.
80. Campo-Soria C, Chang Y, Weiss DS. Mechanism of action of benzodiazepines on GABAA receptors. Br J Pharmacol. 2006;148(7):984–90.
81. Shen XM, Ohno K, Tsujino A, Brengman JM, Gingold M, Sine SM, Engel AG. Mutation causing severe myasthenia reveals functional asymmetry of AChR signature cystine loops in agonist binding and gating. J Clin Invest. 2003;111(4):497–505.
82. Humphrey W, Dalke A, Schulten K. VMD: visual molecular dynamics. J Mol Graph. 1996;14(1):33–8,27–8.

Chapter 3
Molecular Underpinnings of Neuronal Nicotinic Acetylcholine Receptor Expression

Michael D. Scofield and Paul D. Gardner

Abstract Neuronal nicotinic acetylcholine receptors (nAChRs) are critical signaling molecules in a broad variety of fundamental biological processes. In order for cholinergic signaling to function normally, the receptors must be expressed in the appropriate cells at the appropriate times. Expression of the receptors is regulated at many levels from transcription of the receptor subunit genes to posttranslational modifications of individual subunits. Regulating nAChR expression is further complicated because of the large number of genes encoding nAChR subunits, most of which, but not all, are located on distinct chromosomes. Here, we describe molecular events that underlie expression of nAChR subunit genes. We begin with a survey of the transcriptional mechanisms involved in nAChR subunit gene expression including a review of CHRNA5/A3/B4 cluster expression. An update on two emerging fields of investigation, microRNA and epigenetic regulation of nAChR expression, is provided followed by an overview of mechanisms involved in the nicotine-mediated upregulation of nAChR expression. Regulation of nAChR subunit expression is of fundamental importance as it underlies subunit availability, which impacts individual receptor subtype composition and thus, the biophysical properties of the nAChRs.

Keywords Nicotinic receptors • Transcription • miRNA • Receptor upregulation

1 Introduction

As described throughout this tome, nicotinic acetylcholine receptors (nAChRs) are key components in a wide variety of signal transduction cascades that are critical for normal behaviors ranging from muscle contraction to cognitive enhancement [1, 2].

M.D. Scofield, Ph.D.
Department of Neurosciences, Medical University of South Carolina, Charleston, SC, USA

P.D. Gardner, Ph.D. (✉)
Department of Psychiatry, Brudnick Neuropsychiatric Research Institute, University of Massachusetts Medical School, 303 Belmont Street, Worcester, MA 01604, USA
e-mail: paul.gardner@umassmed.edu

R.A.J. Lester (ed.), *Nicotinic Receptors*, The Receptors 26,
DOI 10.1007/978-1-4939-1167-7_3

In order to fulfill their roles in this vast array of biological functions, nAChRs must be expressed in the right places at the right times. If their temporal and/or spatial expression is compromised, cholinergic signaling will be disrupted leading to a plethora of health consequences [1, 3–9]. In this chapter, we focus upon the molecular events underlying expression of the genes encoding neuronal nAChR subunits with an emphasis on their expression in neuronal populations. The reader is directed to other references for reviews of regulatory mechanisms involved in nonneuronal nAChR expression [1, 2].

Neuronal nAChRs are evolutionarily conserved pentameric cation-selective channels assembled from a family of subunits, α2–α10 and β2–β4, as homomeric (α subunits only) or heteromeric complexes (α and β subunits) [1, 10]. Structurally, each subunit contains an approximately 200-residue extracellular amino-terminus, four membrane-spanning domains (referred to as M1–M4), an intracellular loop (100–200 residues depending on the subunit) and a relatively short extracellular carboxyl-terminus [11]. The ACh-binding domain is located on the amino-terminus with the interface between adjacent subunits forming the agonist-binding site [12]. The conducting pore of the ion channel is formed from each of the five M2 membrane-spanning domains with regions of the M1–M2 intercellular loop contributing to cation permeability and agonist-binding affinities [11, 13]. A critical point is that the unique biophysical and pharmacological properties of each nAChR subtype are determined by its subunit composition. Given the large number of subunits present in the nAChR family, there is the potential for a vast array of functionally distinct nAChR subtypes. The functional diversity exhibited by the neuronal nAChR family is due in large part to the temporally- and spatially regulated expression of the genes encoding the various subunits leading to the assembly of discrete subunits into mature nAChR subtypes.

Regulation of nAChR expression occurs at several levels beginning with gene transcription followed by a number of posttranscriptional processes including microRNA regulation, epigenetic mechanisms, posttranslational modifications, receptor assembly, and subsequent insertion in the plasma membrane. Here we focus upon transcriptional, epigenetic, and miRNA regulation of nAChR expression. For an excellent review of posttranslational control of nAChR expression, the reader is referred to Albuquerque et al. [1]. We finish with a current review of advances in our understanding of the molecular mechanisms underlying the phenomenon of nicotine-mediated nAChR upregulation.

2 Transcriptional Regulation of nAChR Subunit Genes

Regulation of nAChR subunit gene expression at the transcriptional level is required for the fidelity of cholinergic signaling and as a consequence, the complex physiological processes such signaling underlies. The cell type-specific enrichment or suppression of nAChR subunit gene expression is orchestrated through the interaction of noncoding regions of DNA with positive and/or negative transcriptional regulatory factors. The transcription factors recognize specific sequences (recognition elements) of DNA and bind to these regions thereby regulating transcription of nAChR subunit mRNA by impacting the efficiency of the RNA polymerase II complex either in a positive or negative manner. Importantly, the action of the transcription factors can by temporally and spatially restricted. Through this mechanism, regulatory information present in both proximal and distal regions of DNA, relative to the protein-coding sequences, directs expression of the required nAChR subunits that are assembled into mature receptor subtypes located in the specific brain regions and cell types that participate in cholinergic system function. Disruption of these mechanisms has been shown to contribute to addiction [14, 15], epilepsy [9, 16], schizophrenia [17] and risk of lung cancer [18].

2.1 *CHRNA2*

The nAChR α2 subunit gene (denoted *C*holinergic *R*eceptor for *N*icotine *A*lpha 2 or CHRNA2) displays a very restricted pattern of expression and is observed primarily in the retina and interpeduncular region in rodents, although α2 mRNA has been detected in other brain regions albeit at substantially lower levels [19–24]. Given its relatively low levels of expression it is likely subject to a large amount of transcriptional inhibition, however, limited information regarding the transcriptional regulation of CHRNA2 expression has been elucidated. It has been shown that transcription from the α2 promoter is activated by the transcription factor Brn-3b but not by other members of the family, namely Brn-3a or Brn-3c [25]. While it is possible that small RNAs may regulate CHRNA2 expression, such regulation is typically post-transcriptional (see below).

2.2 *CHRNA3*

In the nervous system, the α3 subunit is expressed at high levels in the periphery with a more restricted expression pattern centrally [26–30]. In the PNS, CHRNA3 expression is observed in trigeminal sensory neurons, the dorsal root ganglia, superior cervical ganglia, adrenal medulla, as well as in the spenopalatine and otic ganglia [28, 29, 31, 32]. In the CNS, the α3 subunit is expressed in the brainstem, cerebellum, spinal cord, substantia nigra, medial habenula, pineal gland, hippocampus, cortex, thalamus, ventral tegmental area, and interpeduncular nucleus [23, 29, 33–36]. α3-containing nAChRs are also expressed in nonneuronal cells such as lymphocytes, cells located in the gastrointestinal tract, vascular endothelial cells, polymorphonuclear cells, bronchial epithelial cells, and O2A progenitors [37–43]. Interestingly, the α3 subunit is also expressed in lung tissue and its expression increases as a result of the pathophysiology of lung cancer [44].

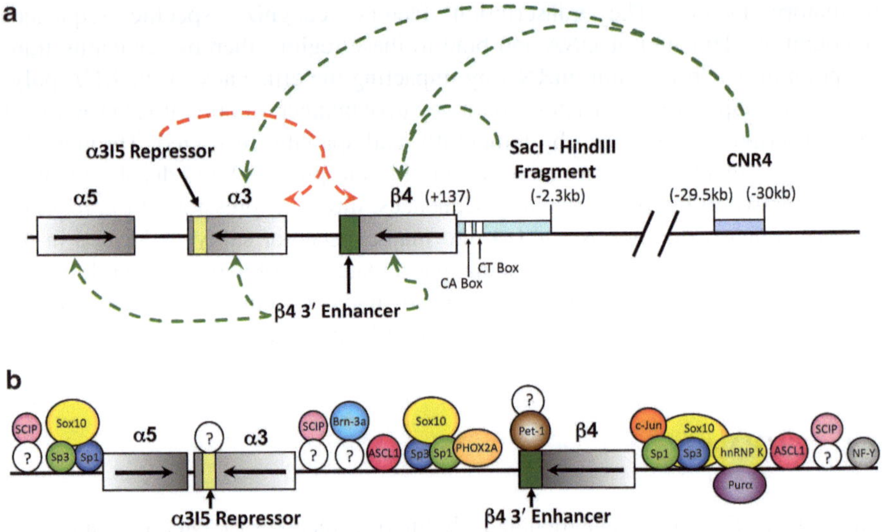

Fig. 3.1 Transcriptional regulation of the CHRNA5/A3/B4 gene cluster. (**a**) The CHRNA5/A3/B4 cluster is subject to positive and negative transcriptional control. Coding regions of the three subunit genes are shown as *grey boxes* with *arrows* indicating the directions of transcription. Four known regulatory elements/regions are presented: CNR4 (*dark blue*), the SacI/HindIII fragment (*light blue*), β43′ enhancer (*green*), and the α3I5 repressor (*yellow*). The 5′ and 3′ termini of CNR4 and the SacI/HindIII fragment are labeled relative to the CHRNB4 transcriptional start site. *Green arrows* indicate positive regulation whereas *red arrows* denote negative regulation. (**b**) Transcription factor–DNA interactions at the CHRNA5/A3/B4 cluster. The *solid black line* between the *grey boxes* (coding regions of the clustered genes) represents noncoding DNA and the *colored circles* represent transcription factors. Proteins thought to bind to DNA through an intermediate are depicted contacting an additional factor that directly binds to DNA. Transcription factors shown to physically interact are shown contacting each other. Purα is depicted below the solid line as it binds to the negative strand of DNA. Multiple Sp1-binding sites have been identified for each of the nAChR subunit genes, but for clarity, are shown as a single *green circle* at each promoter region. *Circles* labeled with a "?" represent transcription factors whose identities are unknown

In vitro experiments have shown that the paired-like homeodomain transcription factor, PHOX2A, regulates transcription from the α3 promoter [45]. PHOX2A does not appear to bind directly to DNA, however, as the DNA-binding domain does not need to be completely intact for PHOX2A to regulate expression of CHRNA3 [45]. Co-immunoprecipitation experiments demonstrate a physical interaction between Sp1 and PHOX2A, suggesting that PHOX2A is tethered to the α3 promoter through its interaction with Sp1, similar to the interactions of Sp1 with homeodomain transcription factors observed in other systems (Fig. 3.1b) [46].

As described in more detail below, the POU domain factor SCIP/Tst-1/Oct-6 positively regulates transcription from the α3 promoter in a cell-type-specific manner [47]. As is the case with PHOX2A, the POU domain factor SCIP/Tst-1/Oct-6 does not require DNA binding for transactivation of the α3 promoter [47]. However, transactivation of the α3 promoter by SCIP/Tst-1/Oct-6 does not depend on the presence of an Sp1 motif in the promoter region and instead likely mediates its effect via protein–protein interactions with the basal transcription machinery [48]. The transcription factor Brn-3a also transactivates the α3 promoter, while the other members of the Brn-3 family, Brn-3b and 3c, modestly repress α3 promoter activity [25]. The positive regulation by Brn-3a is also thought to be a result of protein–protein interaction as the α3 promoter lacks a classical octamer-related binding site for Brn-3 factors [25].

The Deneris group has identified a transcriptional enhancer upstream of the α3 promoter in a region that overlaps with a 3′-untranslated exon of the β4 gene [49]. This transcriptional enhancer contains two identical 37-base pair repeats separated by a 6-base pair spacer. The β43′ enhancer acts as a cell-type-specific enhancer and is capable of enhancing transcription from the α3 promoter in neuronal-like cells, as well as in cultured SCG neurons [50]. Functional analysis of the enhancer revealed several E-twenty six (ETS) factor-binding sites, that when mutated severely decrease α3 promoter activity. Moreover, the ETS-domain binding factor, Pet-1, has been shown to activate reporter gene transcription in a manner that is both cell type- and β43′ enhancer-dependent [51]. Taken together, these experiments suggest that Pet-1 interacts directly with the α3 promoter to activate transcription, likely in concert with additional cell-type-specific cofactors. In vivo experiments using transgenic mice demonstrated that a larger DNA fragment between coding regions of the α3 and β4 genes, containing both the β43′ enhancer and the α3 promoter, is capable of directing expression of a reporter gene to several areas of endogenous CHRNA3 expression in the brain [30, 52]. Surprisingly however, this DNA fragment was not able to direct reporter gene expression anywhere in the peripheral nervous system, in which CHRNA3 is highly expressed (see above). These data indicate that this region either contained additional negative regulatory elements or did not contain the sufficient regulatory information needed to provide peripheral expression of the CHRNA3.

In accordance with the possibility of additional negative regulatory elements, Fuentes Medel and Gardner located an intronic repressor element in the fifth intron of α3 gene [53]. The sequence of this α3 intron 5 repressor (α3I5) is evolutionarily conserved and is capable of repressor activity in vitro independent of its orientation. Given that the repression of promoter activity was observed to be more potent in nonneuronal cell lines than in neuronal cell lines, α3I5 is likely involved in maintaining cell type-specific expression of CHRNA3 [53]. The protein–DNA interactions that mediate this effect remain to be elucidated.

While regulation of CHRNA3 in neuronal cells has been extensively studied, the mechanisms regulating CHRNA3 expression in nonneuronal cells remain largely obscure. The Gardner laboratory was the first to report that a transcription factor, the achaete–scute complex homolog-1 (ASCL1), regulates the expression of CHRNA3 and CHRNB4 and to a lesser extent CHRNA5 in lung cancer cells (Fig. 3.1b) [54].

2.3 *CHRNA4*

The α4 subunit assembles with the β2 subunit to form the major high-affinity nAChR subtype expressed in the CNS and also assembles with several other nAChR subunits in additional unique receptor subtypes [1]. CHRNA4 expression is highest in CNS regions including the olfactory bulb, cortex, striatum, hippocampus, amygdala, substantia nigra, ventral tegmental area, thalamus hypothalamus, interpeduncular nucleus, medial habenula, pineal gland, raphe nuclei, and cerebellum [29, 30, 55]. Like most other nAChR subunit gene promoters, the α4 promoter region is GC-rich and lacks a TATA box suggesting that expression of CHRNA4 is positively regulated by factors such as Sp1/3 that recognize GC-rich targets [56]. Using transgenic mice, it was shown that, similar to the other nAChR subunit gene promoters discussed above, important regulatory elements are located both upstream from the transcription start site as well as in intronic regions. These combined regulatory elements are required for cell type-specific expression of CHRNA4 [56]. Despite the fact that the α4 subunit is widely expressed in the CNS and participates in a variety of nAChR subtypes, relatively little is known about the protein–DNA interactions that act to regulate expression from the α4 promoter highlighting the need for further research in this area.

A recent report describes an upstream open reading frame regulatory element that can act specifically to decrease expression of the shorter α4 isoform; however, this regulatory element appears to regulate CHRNA4 expression at the level of translation and not transcription [57].

2.4 *CHRNA5*

Similar to the α3 subunit, the α5 subunit is highly expressed in the PNS but is also expressed in a number of specific locations in the CNS where it assembles with several other nAChR subunits to form distinct nicotinic receptor subtypes [26, 27, 58–60]. Centrally, CHRNA5 expression is highest in the thalamus and cerebellum with lower levels in the cortex, hippocampus, habenula, interpeduncular nucleus, brainstem, and spinal cord [58, 61]. In the cortex, α5-containing nAChRs play a key role in the neurocircuitry underlying attention [62] while in the habenula, they play a pivotal role in nicotine-mediated behaviors [63, 64] (see Chap. 18). α5-containing nAChRs are also critical players in regulating dopamine transmission in the striatum [65]. Peripherally, α5 is expressed in the retina as well as in the majority of the autonomic ganglia [32, 39, 66]. Nonneuronal expression of CHRNA5 has been detected in the gastrointestinal tract, as well as in the thymus and testis [39, 40]. In addition, the α5 subunit is expressed in many of the same cell types as the α3 and β4 subunits including oral epithelium, vascular endothelial cells, bronchial epithelium, O2A progenitors, and a variety of immune cells [37, 41–43].

The α5 promoter region has been described in both the bovine and human genomic contexts [39, 67, 68]. The regulatory mechanisms that govern CHRNA5 expression remain to be completely elucidated but again, similar to other nAChR promoters, the α5 promoter is GC-rich and TATA-less [39, 67]. Sp1 interactions with GC boxes appears to be critical for CHRNA5 expression [67]. SCIP/Tst-1/ Oct-6 does not appear to regulate CHRNA5 expression though it regulates CHRNA3 and CHRNB4 (see above). Moreover, in lung cells, ASCL1 appears to regulate CHRNA3 and CHRNB4 but not CHRNA5 expression [54].

2.5 CHRNA6

In the CNS, expression of CHRNA6 is relatively restricted compared to CHRNA4 and CHRNA7 [69, 70]. Expression of CHRNA6 is observed in the retina, striatum, locus coeruleus, and substantia nigra, ventral tegmental area [19, 69, 70]. α6-containing nAChRs comprise 25–30 % of presynaptic nAChRs expressed on striatal dopaminergic terminals in rodents and 70 % in monkeys making receptors including this subunit potentially important for nicotine addiction [71]. The promoter region for the α6 gene, like all other neuronal nicotinic receptor subunit gene promoters, does not contain a TATA box [72]. However, the α6 promoter is not GC-rich and thus, is not subject to the same degree of regulation by transcription factors that recognize GC-rich sequences as the other nAChR gene promoters [72]. In-silico analysis revealed the presence of potential binding sites for AP-1, STATx, NF-kB, Oct-1, and Pax-2, while directed DNA mutational analysis demonstrated that promoter activity was increased when putative sites for Oct-1 and Pax-2 were removed, suggesting that these factors act to negatively regulate transcription of the CHRNA6 gene [72]. In addition to the predicted binding sites discussed above, the α6 promoter region also contains tandem *Alu* repeats shown to function as transcriptional repressor domains [72].

Interestingly, using a mouse model of the glaucomatose retina, it was shown that the pathophysiology of glaucoma causes a selective loss of CHRNA6 expression in retinal ganglion cells [73]. The mechanism for this downregulation has not yet been elucidated.

2.6 CHRNA7

The α7 subunit self-assembles to form the major homomeric receptor subtype in the nervous system. CHRNA7 is expressed throughout the CNS including the olfactory bulb, cortex, hippocampus, medial habenula, amygdala, hypothalamus, substantia nigra, ventral tegmental area, pineal gland interpeduncular nucleus, and cerebellum [69]. CHRNA7 is also expressed in non-neuronal cells including lymphoid tissue, lymphocytes, epidermal keratinocytes, bronchial epithelial cells, and vascular tissue [27]. For more information on this topic see Chap. 13.

Like most other nAChR subunit genes the α7 promoter contains an abundance of guanine and cytosine nucleotides yet lacks a TATA box [74, 75]. As expected, the α7 promoter contains multiple overlapping binding sites for transcription factors that recognize GC-rich targets such as Sp1/3, AP-2, and GCF [76] as well as binding sites for Egr-1 and CREB [77, 78]. Transcription of the α7 gene is positively regulated by transcription factors Sp1, Ap-2, Egr-1, GCF, TTF-1, and CREB [74, 75, 78, 79].

Expression of CHRNA7 has been shown to be upregulated by membrane depolarization [80]. When cultured rat superior cervical ganglion neurons were treated with KCl for 24-48 h, increased levels of homomeric α7 receptors were observed. KCl-mediated depolarization and subsequent calcium influx through L-type calcium channels resulted in increased α7 expression; this effect was CaM kinase-dependent [80]. Transcription of the α7 gene is also upregulated by BDNF activation of trkB receptors in ciliary ganglion cells (used as a model for parasympathetic neurons). This effect most likely occurs through activation of CREB via the BDNF-trkB pathway [81]. Akin to activity-dependent regulation, CHRNA7 expression is also upregulated by synapse formation. Studies have shown that expression of CHRNA7 can be enhanced by mimicking presynaptic input in vitro using a cysteine-rich isoform of neuregulin1 CRD-NRG [82]. Interestingly, expression of CHRNA7 is also upregulated by light in the rodent developing primary visual cortex, yet the mechanism for this upregulation has not been completely elucidated [83]. In addition, studies suggest that the pathophysiology of schizophrenia causes a decrease in expression of CHRNA7 [84] (see Chap. 20).

2.7 *CHRNA8*

The α8 subunit is expressed in avian tissue and forms receptors that have biophysical properties similar to the α7 homomeric receptor [85]. These receptors can be either homomeric or heteromeric receptors forming with the α7 subunit [27]. The transcriptional mechanisms underlying expression of this avian-specific nAChR subunit gene remain to be elucidated.

2.8 *CHRNA9 and CHRNA10*

The nAChR α9 and α10 subunits are co-expressed in a select few regions and their restricted pattern of expression suggests that they play very specific and unique roles in the regions in which they are present [86]. Studies have shown that α9- and α10-containing nAChRs are critical components of the efferent auditory system and act to tune the cochlea to improve signal detection [87, 88]. Accordingly, these receptor subunits are capable of co-assembly into an alpha-only α9α10 nAChR subtype [89]. Both α9 and α10 mRNA can be detected in the inner ear and tonsils as well as in dorsal root ganglion neurons, immortalized B-cells, cultured T-cells, and peripheral blood lymphocytes [86, 87, 90]. Analysis of the α9 and α10 promoter regions revealed putative binding sites for several transcription factors including AP-4, NFAT, MZF-1, BARBIE, STAF, and GATA-1 [91]. Criado and colleagues demonstrated that CHRNA9 expression is regulated by both positive and negative transcriptional regulatory elements [92]. The positive element contains Sox-binding elements and indeed interacts and is transactivated by members of the Sox family [92].

2.9 CHRNB2

As discussed above, the β2 subunit is part of the major high affinity nAChR α4β2 subtype and also assembles with several other subunits to form a variety of receptor subtypes [1]. Studies from knockout mice revealed that the β2 subunit is involved in self-administration of nicotine and is required for nicotine-mediated dopamine release from mesencephalic dopaminergic neurons [93]. The β2 subunit also appears to play a role in alcohol-mediated behaviors [94] and is associated with nocturnal frontal lobe epilepsy [95]. The β2 subunit is expressed in a variety of CNS locations including the olfactory bulb, cortex, striatum, hippocampus, amygdala, substantia nigra, ventral tegmental area, thalamus hypothalamus, interpeduncular nucleus, medial habenula, locus coeruleus, raphe nuclei, and cerebellum [29, 55].

The β2 promoter, like most other nAChR promoters is GC-rich, and contains an E-box as well as predicted binding sites for Sp1, CREB, and GATA-3 [96]. An approximately 1.2-kb genomic DNA fragment containing these elements is sufficient for driving reporter gene expression in a neuron-specific manner in transgenic mice [96]. Given the high levels of GC content and the predicted binding site for Sp1, it is likely that expression of CHRNB2 is positively regulated by Sp factors [97, 98]. In addition, the β2 promoter also contains a neuron-restrictive silencing element (NRSE), a regulatory element that can enhance or suppress expression in a neuronal cell type depending on where it is located in relation to the transcription start site. Interestingly, in nonneuronal cells the NRSE suppresses expression independent of its location in the promoter region [98]. These data demonstrate that both positive and negative regulatory forces must act in concert to direct cell type-specific expression of CHRNB2. More recently, single nucleotide polymorphisms in the CHRNB2 promoter region have been shown to be significantly associated with the subjective response to nicotine and may also be critical for CHRNB2 expression [99]. Whether these polymorphisms alter protein–DNA interactions remains to be determined.

2.10 CHRNB3

Expression of the nAChR β3 subunit is observed in relatively few areas in the CNS compared to other nAChR subunits. CHRNB3 is expressed in the striatum, medial habenula, interpeduncular nucleus, locus coeruleus, substantia nigra, and VTA [55]. Interestingly, CHRNB3 is also expressed in the auditory cortex and following injury (cochlea ablation), expression of CHRNB3 is increased [100]. To date, little is known regarding the β3 promoter and the mechanisms that govern its expression. One intriguing study, however, showed that a single nucleotide polymorphism in the CHRNB3 gene that is associated with nicotine dependence lies in close proximity to a potential binding-site for the transcription factor AP1 [101], but further work is required to determine whether the polymorphism alters AP1 function at the CHRNB3 promoter.

2.11 *CHRNB4*

As is the case with CHRNA3 and CHRNA5, CHRNB4 is widely expressed in the PNS with more limited expression centrally [27]. CHRNB4 is expressed at high levels in trigeminal sensory neurons as well as the superior cervical, dorsal root, spenopalatine, and otic ganglia [28, 29, 31, 32, 102, 103]. Expression of CHRNB4 is also observed in the adrenal medulla with lower levels of expression in the retina [66, 104]. In the CNS, CHRNB4 expression is particularly high in the olfactory bulb, pineal gland, medial habenula, and interpeduncular nucleus with lower expression in other thalamic nuclei, the cortex, piriform cortex, hippocampus, cerebellum, midbrain, and spinal cord [61]. In nonneuronal cells, CHRNB4 expression coincides with the expression of CHRNA3 and CHRNA5 in many cell types. For example, all three nAChR subunit genes are expressed in the intestine as well as in vascular endothelial cells, oral keratinocytes, polymorphonuclear cells, bronchial epithelium, and O2A progenitors [37–43]. Finally, CHRNB4 is co-expressed with CHRNA3 and CHRNA5 in lung and is also upregulated in lung cancer [44].

In addition to the Sp factors, Sox10, and SCIP/Tst-1/Oct-6 (see discussion of the nAChR gene cluster below), the β4 promoter is positively regulated by c-Jun [105]. Trans-activation by all of these factors is abolished when the Sp-binding site on the β4 promoter (referred to as a CA box) is mutated. Conversely, synergistic activation of the β4 promoter is observed when Sp1 is supplied in concert with Sox10, Sp3, or c-Jun [105, 106]. Co-immunoprecipitation experiments demonstrated that all of these factors physically interact [107]. Chromatin immunoprecipitation experiments confirmed that these interactions occur in the context of native chromatin (Fig. 3.1b) [108]. These findings suggest the existence of a positively-acting multi-subunit transcriptional regulatory complex that assembles on the β4 promoter. This result is consistent with the hypothesis that Sp1 is critical for transcription from the β4 promoter and likely nucleates the regulatory complex that drives expression of CHRNB4.

Two additional transcription factors have been shown to interact with the β4 promoter, Purα, and heterogeneous nuclear ribonucleoprotein K (hnRNPK) [109, 110]. These proteins interact with another motif, the CT box, located directly upstream of the CA box. hnRNP K is capable of repressing Sp factor-mediated trans-activation of the β4 promoter [110] and also physically interacts with Sox10 [107]. Similar to hnRNP K, Purα physically interacts with Sox10 [107]. Moreover, Purα and hnRNP

K themselves physically interact [107]. These proteins may participate in the multi-subunit complex described above to modulate expression of CHRNB4 in the appropriate cellular context. In vitro binding experiments demonstrated that each factor binds preferentially to the opposing single strand elements of the CT box, suggesting that some local DNA helix unwinding may occur (Fig. 3.1b) [111]. Interestingly, Purα and hnRNP K have been shown to function together to negatively impact transcription of genes in other systems and the same may be occurring at the β4 promoter [112].

Further studies indicated that a 2.3-kb fragment of the β4 promoter, containing the CT and CA boxes, is capable of directing reporter gene expression to an array of brain regions that endogenously express the β4 gene [113, 114]. Rather surprisingly, site-directed mutagenesis of the CA box essentially eliminated reporter gene activity in transgenic animals [114]. These experiments demonstrate that the 2.3-kb region of the β4 promoter plays a critical role in mediating β4 gene expression in vivo and that the CA box is crucial for CHRNB4 expression.

2.12 The CHRNA5/A3/B4 Gene Cluster

Three of the 11 mammalian neuronal nAChR subunit genes, CHRNA5/A3/B4, are located in a tight genomic cluster, likely arising from an ancient gene duplication event (Fig. 3.1) [10]. The clustering of these three nAChR subunit genes is conserved throughout the Kingdom Metazoa. The conservation at this locus coupled with the observation that the three receptor genes are co-expressed in a variety of cell types and tissues suggests that their expression is coordinately regulated (Fig. 3.1a) [61, 69]. Sequence analyses and functional characterization of the promoter regions of each of the three clustered subunit genes revealed that, like most nAChR subunit gene promoters, the α5, α3, and β4 gene promoters are GC-rich and lack classical CAAT and TATA boxes [115, 116]. Several transcription factors act to positively influence expression of each of the clustered genes including, Sp1, Sp3, Sox10, and SCIP/Tst-1/Oct-6 (Fig. 3.1b) [61]. Chromatin immunoprecipitation experiments demonstrated Sp1 binding activity in the context of native chromatin for all three promoters (Fig. 3.1b) [45, 114]. This is consistent with the idea that Sp1 is involved in tethering the basal transcription machinery to all GC-rich TATA-less nAChR subunit gene promoters, as has been described for other TATA-less promoters [117]. In addition to the Sp factors, the CHRNA3/A5/B4 promoter regions can directly interact with and be transactivated by the more spatially restricted regulatory factors Sox10 and SCIP/Tst-1/Oct-6 (Fig. 3.1b) [47, 48, 118]. Consistent with the overlapping mechanisms of regulation discussed above, the mRNA levels of the CHRNA3/A5/B4 genes are coordinately upregulated during neural development [119] and coordinately downregulated following denervation [120]. Furthermore, Deneris and colleagues have demonstrated that two transcriptional regulatory elements, the β43′ enhancer and the conserved noncoding region 4 (CNR4), play key roles in directing expression of the clustered nAChR genes in a tissue-specific manner. Specifically, the β43′ enhancer is involved in directing expression in the adrenal gland whereas the CNR4 element is critical for expression in the pineal gland and superior cervical ganglion [121]. In addition, it is likely that CNR4 plays an important role in directing nAChR gene expression in other regions of the brain [121].

In addition to the significant amount of coordinate regulation, it is important to note that the expression patterns of the clustered subunit genes do not completely overlap, indicating that in addition to coordinate regulation, unique regulatory mechanisms act to control expression of the individual subunit genes, likely involving the various transcription factors and regulatory sequences described earlier. Interestingly, transcription of CHRNA5 occurs in the opposite direction of CHRNA3 and CHRNB4 (see Fig. 3.1), again suggesting that in addition to shared transcriptional regulatory mechanisms, unique mechanisms are in place to govern CHRNA5 expression.

3 miRNA Regulation of nAChR Subunit Gene Expression

MicroRNAS (miRNAs) are 21–24 nucleotide long molecules predicted to regulate the majority of all mammalian protein coding genes [122]. These regulatory molecules are particularly abundant in the brain and play important roles in several aspects of nervous system development [123–125]. miRNAs typically function by binding to a miRNA-recognition element (MRE) in the 3′-untranslated region (UTR) of a target mRNA and then guiding the target mRNA to an RNA-induced silencing complex, which causes either inhibition of translation or mRNA degradation [126]. A number of recent reports have demonstrated that nicotine and cigarette smoke alter the expression of miRNAs. For example, Balaraman et al. demonstrated that nicotine-treatment, at concentrations attained by cigarette smokers, of cerebral cortical-derived neurosphere cultures induced a dose-related increase in several miRNAs, an effect that is blocked by the nonselective nAChR antagonist, mecamylamine [127]. Interestingly, given the comorbidity of alcohol abuse and nicotine dependence, three of the miRNAs regulated by nicotine are also ethanol-sensitive [127]. In another study, miRNA analysis of airway epithelium of smokers compared to nonsmokers demonstrated significant differences in miRNA expression [128] with similar findings observed in experiments performed by exposing rodent lung tissue to cigarette smoke [129]. Moreover, nicotine alone alters miRNA expression levels in canine atrial fibroblasts [130] as well as in the rat pheochromocytoma cell line, PC12 [131]. In vivo, it was shown that exposure to nicotine differentially regulates expression of miRNAs in the adult mouse brain [132].

Very little is known regarding the role of miRNAs in regulating nAChR expression. To date, one report has been published directly linking miRNAs to nAChR expression. A conserved muscle-specific miRNA, miR-1, was shown to regulate translation of two nAChR subunits, UNC-29 and UNC-63, in the body muscles of

C. elegans, ultimately leading to changes in the functional properties of the mature nAChR receptors [133]. While there are no reports of miRNA regulation of mammalian nAChR expression, in unpublished work from our laboratory, analysis of the 3′-UTR of each of the mammalian nAChR subunit genes revealed a plethora of potential miRNA-binding sites. A reporter gene approach coupled with site-directed mutagenesis of the MREs identified 14 miRNAs as regulators of nAChR expression. Moreover, the majority of these miRNAs are expressed in the brain and are downregulated by chronic nicotine (unpublished work from the Gardner laboratory). These findings may be relevant to the paradoxical upregulation of nAChRs by chronic nicotine (see below) as they suggest a novel mechanism for this phenomenon, as a decrease in miRNA expression would relieve their negative regulatory effects on nAChR expression.

4 Epigenetic Regulation of nAChR Subunit Gene Promoters

Despite the conventional view of evolution that requires natural selection and the refining of mutations in DNA sequences over many generations, it is now known that one's environment (including exposure to toxins and even drugs of abuse) can imprint information on DNA through the modification of chromatin structure [134, 135]. The term "epigenetic," originally coined by embryologist and geneticist Conrad Waddington [136], is used to describe regulatory forces acting on the genome without the alteration of DNA sequence. A commonly used analogy likens epigenetic regulation to the software controlling our genetic hardware [137]. In the case of nAChRs, several groups have discovered alterations in methylation of DNA regions controlling expression of nAChR subunit genes, leading to a suppression of gene expression [138].

Methylation can serve as a method of restricting expression of neuronal genes in nonneuronal cells in order to achieve cell-type specificity. The initial description of the chromatin state of the DNA comprising the CHRNA4 promoter region showed little methylation in preparations from brain tissue where CHRNA4 expression is abundant versus more substantial methylation in liver and muscle tissue where CHRNA4 expression is much lower [56]. Leonard and colleagues have also shown that cell-type specificity of CHRNA7 expression is also controlled in part by methylation of a proximal promoter region, with methylation levels correlating with CHRNA7 expression in several cell lines [139]. In another study, signaling through the α4β2 nAChR subtype was shown to underlie epigenetic-induced changes in glutamic acid decarboxylase expression, most likely as a consequence of decreased expression of DNA methyltransferase 1 [140].

In addition to restricting expression in nonneuronal cell types, it has been shown that differential methylation of nAChR subunit gene promoters can occur as a result of pathological conditions. To wit, studies performed on human lung cancer tissue show that the CHRNA3 promoter region is hypermethylated [141, 142]; whereas the CHRNB4 promoter region is hypomethylated [142]. These findings are particularly interesting given recent work linking this variability in these genomic regions to cancer susceptibility [143]. In addition, it has been shown that CHRNA7 expression is downregulated in the cortex of human postmortem brain tissue samples from both autism and Rett syndrome patients [144]. This effect was mediated by a loss of MeCP2, a protein that recognizes methylated DNA and can act to enhance or suppress gene expression [145]. Similar to miRNA regulation of nAChR expression, then, our understanding of the role of epigenetics in regulating nAChR expression is in its infancy. However, as the details of the molecular basis of nicotine dependence continue to be elucidated, undoubtedly the roles of miRNAs and epigenetics will emerge.

5 Nicotine and nAChR Expression

A substantial amount of work from numerous laboratories has shown that chronic exposure to nicotine differentially regulates the number, localization, stoichiometry, and subunit composition of nAChRs [19]. Investigations into the effects of chronic nicotine demonstrated an increase in high-affinity nicotine binding sites in studies performed using isotope-tagged nAChR ligands [146, 147]. In addition, autoradiographs of postmortem tissue also show increased ligand-binding sites in the brains of smokers compared to nonsmokers [44]. Furthermore, brains of smokers display increased nAChR binding sites when viewed using functional magnetic resonance imaging. As expected, the high-affinity α4β2 nAChR subtype exhibits the highest level of upregulation, with a four-fold increase upon chronic treatment [44, 148, 149]. Other subtypes such as α3β2, α3β4, and α7 nAChRs are also upregulated, while some, such as α6β3-containing nAChRs, may even be downregulated [44]. Interestingly, despite the upregulation of receptors at the surface, levels of nAChR subunit mRNA remain unchanged with upregulation occurring independent of protein synthesis [149–151]. These results indicate that posttranslational mechanisms of regulation mediate nicotine's effect on receptor expression levels.

Several plausible theories have emerged regarding the mechanism for nicotine-mediated upregulation of nAChR expression including increased receptor transport to the plasma membrane, decreased receptor internalization, increased resistance to lysosomal degradation, increased receptor assembly [19, 152] and as discussed above, miRNA regulation of nAChR mRNA levels. Evidence exists for all of these [147, 153–164], supporting the idea that it is not likely that any single mechanism exclusively accounts for all the observed changes in nAChR subtype expression during or after chronic nicotine exposure but it is more likely a consequence of multiple pathways functioning at distinct rates [165].

6 Summary

In summary, there are a wide variety of modalities contributing to the regulation of nAChR subunit gene expression ranging from conventional transcriptional regulation to more recently discovered epigenetic- and miRNA-mediated regulatory mechanisms. In addition, nicotine exposure itself profoundly impacts the expression and subcellular localization of nAChRs through a mechanism that appears to be distinct to nAChRs. As discussed above, the genetic regulation of nAChR subunit gene expression is a critical aspect of cholinergic system function as it acts to regulate subunit availability and as a result, the final subunit composition and hence, biophysical properties, of every nicotinic receptor subtype in each cell type in which they are expressed. Future advances in our understanding of how the expression of these genes is so precisely controlled will undoubtedly contribute to a better understanding of the biological processes that nAChRs participate in and is likely to provide insights to the rational design of therapeutic interventions for diseases in which nicotinic cholinergic signaling is compromised.

References

1. Albuquerque EX, et al. Mammalian nicotinic acetylcholine receptors: from structure to function. Physiol Rev. 2009;89(1):73–120.
2. Woolf NJ, Butcher LL. Cholinergic systems mediate action from movement to higher consciousness. Behav Brain Res. 2011;221(2):488–98.
3. Simon-Keller K, et al. Targeting the fetal acetylcholine receptor in rhabdomyosarcoma. Expert Opin Ther Targets. 2013;17(2):127–38.
4. de Almeida JP, Saldanha C. Nonneuronal cholinergic system in human erythrocytes: biological role and clinical relevance. J Membr Biol. 2010;234(3):227–34.
5. Wallace TL, Bertrand D. Alpha7 neuronal nicotinic receptors as a drug target in schizophrenia. Expert Opin Ther Targets. 2013;17(2):139–55.
6. Hurst R, Rollema H, Bertrand D. Nicotinic acetylcholine receptors: from basic science to therapeutics. Pharmacol Ther. 2013;137(1):22–54.
7. Buckingham SD, et al. Nicotinic acetylcholine receptor signalling: roles in Alzheimer's disease and amyloid neuroprotection. Pharmacol Rev. 2009;61(1):39–61.
8. Improgo MR, Tapper AR, Gardner PD. Nicotinic acetylcholine receptor-mediated mechanisms in lung cancer. Biochem Pharmacol. 2011;82(8):1015–21.
9. Steinlein OK. Animal models for autosomal dominant frontal lobe epilepsy: on the origin of seizures. Expert Rev Neurother. 2010;10(12):1859–67.
10. Le Novere N, Corringer PJ, Changeux JP. The diversity of subunit composition in nAChRs: evolutionary origins, physiologic and pharmacologic consequences. J Neurobiol. 2002;53(4): 447–56.
11. Corringer PJ, Le Novere N, Changeux JP. Nicotinic receptors at the amino acid level. Annu Rev Pharmacol Toxicol. 2000;40:431–58.
12. Eisele JL, et al. Chimaeric nicotinic-serotonergic receptor combines distinct ligand binding and channel specificities. Nature. 1993;366(6454):479–83.
13. McGehee DS, Role LW. Physiological diversity of nicotinic acetylcholine receptors expressed by vertebrate neurons. Annu Rev Physiol. 1995;57:521–46.

14. Wang JC, et al. Genetic variation in the CHRNA5 gene affects mRNA levels and is associated with risk for alcohol dependence. Mol Psychiatry. 2009;14(5):501–10.
15. Saccone NL, et al. Multiple distinct risk loci for nicotine dependence identified by dense coverage of the complete family of nicotinic receptor subunit (CHRN) genes. Am J Med Genet B Neuropsychiatr Genet. 2009;150B(4):453–66.
16. Rozycka A, et al. A transcript coding for a partially duplicated form of alpha7 nicotinic acetylcholine receptor is absent from the CD4+ T-lymphocytes of patients with autosomal dominant nocturnal frontal lobe epilepsy (ADNFLE). Folia Neuropathol. 2013;51(1):65–75.
17. Perl O, et al. The alpha7 nicotinic acetylcholine receptor in schizophrenia: decreased mRNA levels in peripheral blood lymphocytes. FASEB J. 2003;17(13):1948–50.
18. Falvella FS, et al. Transcription deregulation at the 15q25 locus in association with lung adenocarcinoma risk. Clin Cancer Res. 2009;15(5):1837–42.
19. Gaimarri A, et al. Regulation of neuronal nicotinic receptor traffic and expression. Brain Res Rev. 2007;55(1):134–43.
20. Son JH, Winzer-Serhan UH. Expression of neuronal nicotinic acetylcholine receptor subunit mRNAs in rat hippocampal GABAergic interneurons. J Comp Neurol. 2008;511(2):286–99.
21. Son JH, Winzer-Serhan UH. Postnatal expression of alpha2 nicotinic acetylcholine receptor subunit mRNA in developing cortex and hippocampus. J Chem Neuroanat. 2006;32(2–4): 179–90.
22. Ishii K, Wong JK, Sumikawa K. Comparison of alpha2 nicotinic acetylcholine receptor subunit mRNA expression in the central nervous system of rats and mice. J Comp Neurol. 2005;493(2):241–60.
23. Azam L, et al. Expression of neuronal nicotinic acetylcholine receptor subunit mRNAs within midbrain dopamine neurons. J Comp Neurol. 2002;444(3):260–74.
24. Keiger CJ, Walker JC. Individual variation in the expression profiles of nicotinic receptors in the olfactory bulb and trigeminal ganglion and identification of alpha2, alpha6, alpha9, and beta3 transcripts. Biochem Pharmacol. 2000;59(3):233–40.
25. Milton NG, et al. Differential regulation of neuronal nicotinic acetylcholine receptor subunit gene promoters by Brn-3 POU family transcription factors. Biochem J. 1996;317(Pt 2): 419–23.
26. Greenbaum L, Lerer B. Differential contribution of genetic variation in multiple brain nicotinic cholinergic receptors to nicotine dependence: recent progress and emerging open questions. Mol Psychiatry. 2009;14:912–45.
27. Gotti C, Clementi F. Neuronal nicotinic receptors: from structure to pathology. Prog Neurobiol. 2004;74(6):363–96.
28. Rust G, et al. Expression of neuronal nicotinic acetylcholine receptor subunit genes in the rat autonomic nervous system. Eur J Neurosci. 1994;6(3):478–85.
29. Zoli M, et al. Developmental regulation of nicotinic ACh receptor subunit mRNAs in the rat central and peripheral nervous systems. J Neurosci. 1995;15(3 Pt 1):1912–39.
30. Wada E, et al. Distribution of alpha 2, alpha 3, alpha 4, and beta 2 neuronal nicotinic receptor subunit mRNAs in the central nervous system: a hybridization histochemical study in the rat. J Comp Neurol. 1989;284(2):314–35.
31. Flores CM, et al. Neuronal nicotinic rceptor expression in sensory neurons of the rat trigeminal ganglion: demonstration of α3/β4, a novel subtype in the mammalian nervous system. J Neurosci. 1996;16:7892–901.
32. Liu L, et al. Neuronal nicotinic acetylcholine receptors in rat trigeminal ganglia. Brain Res. 1998;809(2):238–45.
33. Keiger CJ, et al. Developmental expression of nicotinic receptors in the chick and human spinal cord. J Comp Neurol. 2003;455(1):86–99.
34. Hellstrom-Lindahl E, et al. Regional distribution of nicotinic receptors during prenatal development of human brain and spinal cord. Brain Res Dev Brain Res. 1998;108(1–2):147–60.
35. Gahring LC, et al. Mouse strain-specific nicotinic acetylcholine receptor expression by inhibitory interneurons and astrocytes in the dorsal hippocampus. J Comp Neurol. 2004;468(3): 334–46.

36. Winzer-Serhan UH, Leslie FM. Codistribution of nicotinic acetylcholine receptor subunit alpha3 and beta4 mRNAs during rat brain development. J Comp Neurol. 1997;386(4): 540–54.
37. Rogers SW, et al. Neuronal nicotinic acetylcholine receptor expression by O2A/oligodendrocyte progenitor cells. Glia. 2001;33(4):306–13.
38. Benhammou K, et al. [(3)H]Nicotine binding in peripheral blood cells of smokers is correlated with the number of cigarettes smoked per day. Neuropharmacology. 2000;39(13): 2818–29.
39. Flora A, et al. Neuronal and extraneuronal expression and regulation of the human alpha5 nicotinic receptor subunit gene. J Neurochem. 2000;75(1):18–27.
40. Glushakov AV, et al. Distribution of neuronal nicotinic acetylcholine receptors containing different alpha-subunits in the submucosal plexus of the guinea-pig. Auton Neurosci. 2004; 110(1):19–26.
41. Macklin KD, et al. Human vascular endothelial cells express functional nicotinic acetylcholine receptors. J Pharmacol Exp Ther. 1998;287(1):435–9.
42. Wang Y, et al. Human bronchial epithelial and endothelial cells express alpha7 nicotinic acetylcholine receptors. Mol Pharmacol. 2001;60(6):1201–9.
43. Maus AD, et al. Human and rodent bronchial epithelial cells express functional nicotinic acetylcholine receptors. Mol Pharmacol. 1998;54(5):779–88.
44. Improgo MR, et al. From smoking to lung cancer: the CHRNA5/A3/B4 connection. Oncogene. 2010;29(35):4874–84.
45. Benfante R, et al. Transcription factor PHOX2A regulates the human α3 nicotinic receptor subunit gene promoter. J Biol Chem. 2007;282(18):13290–302.
46. Shimakura J, et al. The transcription factor Cdx2 regulates the intestine-specific expression of human peptide transporter 1 through functional interaction with Sp1. Biochem Pharmacol. 2006;71(11):1581–8.
47. Yang X, et al. Characterization of an acetylcholine receptor alpha 3 gene promoter and its activation by the POU domain factor SCIP/Tst-1. J Biol Chem. 1994;269(14):10252–64.
48. Fyodorov D, Deneris E. The POU domain of SCIP/Tst-1/Oct-6 is sufficient for activation of an acetylcholine receptor promoter. Mol Cell Biol. 1996;16(9):5004–14.
49. McDonough J, Deneris E. Beta43′: an enhancer displaying neural-restricted activity is located in the 3′-untranslated exon of the rat nicotinic acetylcholine receptor beta4 gene. J Neurosci. 1997;17(7):2273–83.
50. McDonough J, et al. Regulation of transcription in the neuronal nicotinic receptor subunit gene cluster by a neuron-selective enhancer and ETS domain factors. J Biol Chem. 2000; 275(37):28962–70.
51. Fyodorov D, Nelson T, Deneris E. Pet-1, a novel ETS domain factor that can activate neuronal nAchR gene transcription. J Neurobiol. 1998;34(2):151–63.
52. Yang X, et al. Elements between the protein-coding regions of the adjacent beta 4 and alpha 3 acetylcholine receptor genes direct neuron-specific expression in the central nervous system. J Neurobiol. 1997;32(3):311–24.
53. Fuentes Medel YF, Gardner PD. Transcriptional repression by a conserved intronic sequence in the nicotinic receptor alpha3 subunit gene. J Biol Chem. 2007;282(26):19062–70.
54. Improgo MR, et al. ASCL1 regulates the expression of the CHRNA5/A3/B4 lung cancer susceptibility locus. Mol Cancer Res. 2010;8(2):194–203.
55. Changeux J, Edelstein SJ. Allosteric mechanisms in normal and pathological nicotinic acetylcholine receptors. Curr Opin Neurobiol. 2001;11(3):369–77.
56. Watanabe H, Zoli M, Changeux JP. Promoter analysis of the neuronal nicotinic acetylcholine receptor alpha4 gene: methylation and expression of the transgene. Eur J Neurosci. 1998; 10(7):2244–53.
57. Eggert M, et al. Nicotinic acetylcholine receptor subunits alpha4 and alpha5 associated with smoking behaviour and lung cancer are regulated by upstream open reading frames. PLoS One. 2013;8(7):e66157.

58. Wada E, et al. The distribution of mRNA encoded by a new member of the neuronal nicotinic acetylcholine receptor gene family (alpha 5) in the rat central nervous system. Brain Res. 1990;526(1):45–53.
59. Boulter J, et al. Alpha 3, alpha 5, and beta 4: three members of the rat neuronal nicotinic acetylcholine receptor-related gene family form a gene cluster. J Biol Chem. 1990;265(8): 4472–82.
60. Conroy WG, Vernallis AB, Berg DK. The alpha 5 gene product assembles with multiple acetylcholine receptor subunits to form distinctive receptor subtypes in brain. Neuron. 1992;9(4):679–91.
61. Improgo MR, et al. The nicotinic acetylcholine receptor CHRNA5/A3/B4 gene cluster: Dual role in nicotine addiction and lung cancer. Prog Neurobiol. 2010;92:212–26.
62. Bailey CD, et al. The nicotinic acetylcholine receptor alpha5 subunit plays a key role in attention circuitry and accuracy. J Neurosci. 2010;30(27):9241–52.
63. Fowler CD, et al. Habenular α5 nicotinic receptor subunit signalling controls nicotine intake. Nature. 2011;471(7340):597–601.
64. Frahm S, et al. Aversion to nicotine is regulated by the balanced activity of β4 and α5 nicotinic receptor subunits in the medial habenula. Neuron. 2011;70(3):522–35.
65. Exley R, et al. Striatal α5 nicotinic receptor subunit regulates dopamine transmission in dorsal striatum. J Neurosci. 2012;32(7):2352–6.
66. Moretti M, et al. Nicotinic acetylcholine receptor subtypes expression during rat retina development and their regulation by visual experience. Mol Pharmacol. 2004;66(1):85–96.
67. Campos-Caro A, et al. Multiple functional Sp1 domains in the minimal promoter region of the neuronal nicotinic receptor alpha5 subunit gene. J Biol Chem. 1999;274(8):4693–701.
68. Flora A, et al. Transcriptional regulation of the human alpha5 nicotinic receptor subunit gene in neuronal and non-neuronal tissues. Eur J Pharmacol. 2000;393(1–3):85–95.
69. Gotti C, et al. Brain neuronal nicotinic receptors as new targets for drug discovery. Curr Pharm Des. 2006;12(4):407–28.
70. Champtiaux N, et al. Distribution and pharmacology of alpha 6-containing nicotinic acetylcholine receptors analyzed with mutant mice. J Neurosci. 2002;22(4):1208–17.
71. Salminen O, et al. Pharmacology of alpha-conotoxin MII-sensitive subtypes of nicotinic acetylcholine receptors isolated by breeding of null mutant mice. Mol Pharmacol. 2007;71(6): 1563–71.
72. Ebihara M, et al. Genomic organization and promoter analysis of the human nicotinic acetylcholine receptor alpha6 subunit (CHNRA6) gene: Alu and other elements direct transcriptional repression. Gene. 2002;298(1):101–8.
73. Munguba GC, et al. Effects of glaucoma on Chrna6 expression in the retina. Curr Eye Res. 2013;38(1):150–7.
74. Nagavarapu U, Danthi S, Boyd RT. Characterization of a rat neuronal nicotinic acetylcholine receptor alpha7 promoter. J Biol Chem. 2001;276(20):16749–57.
75. Carrasco-Serrano C, et al. GC- and E-box motifs as regulatory elements in the proximal promoter region of the neuronal nicotinic receptor alpha7 subunit gene. J Biol Chem. 1998; 273(32):20021–8.
76. Matter JM, et al. On the transcriptional regulation of neuronal nAChR genes. J Physiol Paris. 1998;92(3–4):245–8.
77. Gault J, et al. Genomic organization and partial duplication of the human alpha7 neuronal nicotinic acetylcholine receptor gene (CHRNA7). Genomics. 1998;52(2):173–85.
78. Criado M, et al. Differential expression of alpha-bungarotoxin-sensitive neuronal nicotinic receptors in adrenergic chromaffin cells: a role for transcription factor Egr-1. J Neurosci. 1997;17(17):6554–64.
79. Reynolds PR, Hoidal JR. Temporal-spatial expression and transcriptional regulation of alpha7 nicotinic acetylcholine receptor by thyroid transcription factor-1 and early growth response factor-1 during murine lung development. J Biol Chem. 2005;280(37):32548–54.
80. De Koninck P, Cooper E. Differential regulation of neuronal nicotinic ACh receptor subunit genes in cultured neonatal rat sympathetic neurons: specific induction of alpha 7 by

membrane depolarization through a Ca2+/calmodulin-dependent kinase pathway. J Neurosci. 1995;15(12):7966–78.
81. Zhou X, et al. Brain-derived neurotrophic factor and trkB signaling in parasympathetic neurons: relevance to regulating alpha7-containing nicotinic receptors and synaptic function. J Neurosci. 2004;24(18):4340–50.
82. Yang X, et al. A cysteine-rich isoform of neuregulin controls the level of expression of neuronal nicotinic receptor channels during synaptogenesis. Neuron. 1998;20(2):255–70.
83. Aztiria E, Gotti C, Domenici L. Alpha7 but not alpha4 AChR subunit expression is regulated by light in developing primary visual cortex. J Comp Neurol. 2004;480(4):378–91.
84. Mexal S, et al. Differential regulation of alpha7 nicotinic receptor gene (CHRNA7) expression in schizophrenic smokers. J Mol Neurosci. 2010;40(1–2):185–95.
85. Schoepfer R, et al. Brain alpha-bungarotoxin binding protein cDNAs and MAbs reveal subtypes of this branch of the ligand-gated ion channel gene superfamily. Neuron. 1990;5(1): 35–48.
86. Simmons DD, Morley BJ. Spatial and temporal expression patterns of nicotinic acetylcholine α9 and α10 subunits in the embryonic and early postnatal inner ear. Neuroscience. 2011;194: 326–36.
87. Lustig LR. Nicotinic acetylcholine receptor structure and function in the efferent auditory system. Anat Rec A Discov Mol Cell Evol Biol. 2006;288(4):424–34.
88. Vetter DE, et al. The alpha10 nicotinic acetylcholine receptor subunit is required for normal synaptic function and integrity of the olivocochlear system. Proc Natl Acad Sci U S A. 2007;104(51):20594–9.
89. Plazas PV, et al. Stoichiometry of the alpha9alpha10 nicotinic cholinergic receptor. J Neurosci. 2005;25(47):10905–12.
90. Lips KS, Pfeil U, Kummer W. Coexpression of alpha 9 and alpha 10 nicotinic acetylcholine receptors in rat dorsal root ganglion neurons. Neuroscience. 2002;115(1):1–5.
91. Lustig LR, et al. Molecular cloning and mapping of the human nicotinic acetylcholine receptor alpha10 (CHRNA10). Genomics. 2001;73(3):272–83.
92. Valor LM, et al. Transcriptional regulation by activation and repression elements located at the 5'-noncoding region of the human alpha9 nicotinic receptor subunit gene. J Biol Chem. 2003;278(39):37249–55.
93. Picciotto MR, et al. Acetylcholine receptors containing the beta2 subunit are involved in the reinforcing properties of nicotine. Nature. 1998;391(6663):173–7.
94. Dawson A, Miles MF, Damaj MI. The β2 nicotinic acetylcholine receptor subunit differentially influences ethanol behavioral effects in the mouse. Alcohol. 2013;47(2):85–94.
95. De Fusco M, et al. The nicotinic receptor beta 2 subunit is mutant in nocturnal frontal lobe epilepsy. Nat Genet. 2000;26(3):275–6.
96. Bessis A, et al. Promoter elements conferring neuron-specific expression of the β2-subunit of the neuronal nicotinic acetylcholine receptor studied in vitro and in transgenic mice. Neuroscience. 1995;69(3):807–19.
97. Lueders KK, et al. Genomic organization and mapping of the human and mouse neuronal beta2-nicotinic acetylcholine receptor genes. Mamm Genome. 1999;10(9):900–5.
98. Bessis A, et al. The neuron-restrictive silencer element: a dual enhancer/silencer crucial for patterned expression of a nicotinic receptor gene in the brain. Proc Natl Acad Sci U S A. 1997;94:5906–11.
99. Hoft NR, et al. CHRNB2 promoter region: association with subjective effects to nicotine and gene expression differences. Genes Brain Behav. 2011;10(2):176–85.
100. Oh SH, Kim CS, Song JJ. Gene expression and plasticity in the rat auditory cortex after bilateral cochlear ablation. Acta Otolaryngol. 2007;127(4):341–50.
101. Hoft NR, et al. Genetic association of the CHRNA6 and CHRNB3 genes with tobacco dependence in a nationally representative sample. Neuropsychopharmacology. 2009;34(3):698–706.
102. Mandelzys A, et al. The developmental increase in ACh current densities on rat sympathetic neurons correlates with changes in nicotinic ACh receptor alpha-subunit gene expression and occurs independent of innervation. J Neurosci. 1994;14(4):2357–64.

103. Dineley-Miller K, Patrick J. Gene transcripts for the nicotinic acetylcholine receptor subunit, beta4, are distributed in multiple areas of the rat central nervous system. Brain Res Mol Brain Res. 1992;16(3–4):339–44.
104. Di Angelantonio S, et al. Molecular biology and electrophysiology of neuronal nicotinic receptors of rat chromaffin cells. Eur J Neurosci. 2003;17(11):2313–22.
105. Melnikova IN, Gardner PD. The signal transduction pathway underlying ion channel gene regulation by SP1-C-Jun interactions. J Biol Chem. 2001;276(22):19040–5.
106. Melnikova IN, et al. Synergistic transcriptional activation by Sox10 and Sp1 family members. Neuropharmacology. 2000;39(13):2615–23.
107. Melnikova IN, Yang Y, Gardner PD. Interactions between regulatory proteins that bind to the nicotinic receptor beta4 subunit gene promoter. Eur J Pharmacol. 2000;393(1–3):75–83.
108. Scofield MD, et al. Transcription factor assembly on the nicotinic receptor beta4 subunit gene promoter. Neuroreport. 2008;19(6):687–90.
109. Du Q, Tomkinson AE, Gardner PD. Transcriptional regulation of neuronal nicotinic acetylcholine receptor genes. A possible role for the DNA-binding protein Puralpha. J Biol Chem. 1997;272(23):14990–5.
110. Du Q, Melnikova IN, Gardner PD. Differential effects of heterogeneous nuclear ribonucleoprotein K on Sp1- and Sp3-mediated transcriptional activation of a neuronal nicotinic acetylcholine receptor promoter. J Biol Chem. 1998;273(31):19877–83.
111. Krecic AM, Swanson MS. hnRNP complexes: composition, structure, and function. Curr Opin Cell Biol. 1999;11:363–71.
112. Da Silva N, Bharti A, Shelley CS. *hnRNP-K and Purα act together to repress the transcriptional activity of the CD43 gene promoter*. Blood. 2002;100(10):3536–44.
113. Bruschweiler-Li L, et al. Temporally- and spatially-regulated transcriptional activity of the nicotinic acetylcholine receptor beta4 subunit gene promoter. Neuroscience. 2010;166(3): 864–77.
114. Scofield MD, Tapper AR, Gardner PD. A transcriptional regulatory element critical for CHRNB4 promoter activity in vivo. Neuroscience. 2010;170(4):1056–64.
115. Fornasari D, et al. Transcriptional regulation of neuronal nicotinic receptor subunit genes. In: Arneric SP, Brioni JD, editors. Neuronal nicotinic receptors: pharmacology and therapeutic opportunities. Wilmington, DE: Wiley-Liss Inc.; 1998.
116. Fornasari D, et al. Structural and functional characterization of the human alpha3 nicotinic subunit gene promoter. Mol Pharmacol. 1997;51(2):250–61.
117. Pugh BF, Tjian R. Transcription from a TATA-less promoter requires a multisubunit TFIID complex. Genes Dev. 1991;5:1935–9145.
118. Liu Q, et al. Cell type-specific activation of neuronal nicotinic acetylcholine receptor subunit genes by Sox10. J Neurosci. 1999;19(22):9747–55.
119. Corriveau RA, Berg DK. Coexpression of multiple acetylcholine receptor genes in neurons: quantitation of transcripts during development. J Neurosci. 1993;13:2662–71.
120. Zhou Y, Deneris E, Zigmond RE. Differential regulation of levels of nicotinic receptor subunit transcripts in adult sympathetic neurons after axotomy. J Neurobiol. 1998;34(2):164–78.
121. Xu X, Scott MM, Deneris ES. Shared long-range regulatory elements coordinate expression of a gene cluster encoding nicotinic receptor heteromeric subtypes. Mol Cell Biol. 2006;26(15):5636–49.
122. Friedman RC, et al. Most mammalian mRNAs are conserved targets of microRNAs. Genome Res. 2009;19(1):92–105.
123. Cao X, et al. Noncoding RNAs in the mammalian central nervous system. Annu Rev Neurosci. 2006;29:77–103.
124. Bicker S, Schratt G. microRNAs: tiny regulators of synapse function in development and disease. J Cell Mol Med. 2008;12(5A):1466–76.
125. Fiore R, Siegel G, Schratt G. MicroRNA function in neuronal development, plasticity and disease. Biochim Biophys Acta. 2008;1779(8):471–8.
126. Ameres SL, Zamore PD. Diversifying microRNA sequence and function. Nat Rev Mol Cell Biol. 2013;14(8):475–88.

127. Balaraman S, Winzer-Serhan UH, Miranda RC. Opposing actions of ethanol and nicotine on microRNAs are mediated by nicotinic acetylcholine receptors in fetal cerebral cortical-derived neural progenitor cells. Alcohol Clin Exp Res. 2012;36(10):1669–77.
128. Schembri F, et al. MicroRNAs as modulators of smoking-induced gene expression changes in human airway epithelium. Proc Natl Acad Sci U S A. 2009;107(7):2319–24.
129. Izzotti A, et al. Downregulation of microRNA expression in the lungs of rats exposed to cigarette smoke. FASEB J. 2009;23(3):806–12.
130. Shan H, et al. Downregulation of miR-133 and miR-590 contributes to nicotine-induced atrial remodelling in canines. Cardiovasc Res. 2009;83(3):465–72.
131. Huang W, Li MD. Nicotine modulates expression of miR-140*, which targets the 3′-untranslated region of dynamin 1 gene (Dnm1). Int J Neuropsychopharmacol. 2009;12(4):537–46.
132. Lippi G, et al. Targeting of the Arpc3 actin nucleation factor by miR-29a/b regulates dendritic spine morphology. J Cell Biol. 2011;194(6):889–904.
133. Simon D, et al. The microRNA miR-1 regulates a MEF-2-dependent retrograde signal at neuromuscular junctions. Cell. 2008;2133(5):903–15.
134. Jablonka E. Epigenetic variations in heredity and evolution. Clin Pharmacol Ther. 2012;92(6):683–8.
135. Rando OJ, Verstrepen KJ. Timescales of genetic and epigenetic inheritance. Cell. 2007; 128(4):655–68.
136. Van Speybroeck L. From epigenesis to epigenetics: the case of C. H. Waddington. Ann N Y Acad Sci. 2002;981:61–81.
137. Brower V. Epigenetics: unravelling the cancer code. Nature. 2011;471(7339):S12–3.
138. Stower H. Epigenetics: dynamic DNA methylation. Nat Rev Genet. 2012;13(2):75.
139. Canastar A, et al. Promoter methylation and tissue-specific transcription of the alpha7 nicotinic receptor gene, CHRNA7. J Mol Neurosci. 2012;47(2):389–400.
140. Maloku E, et al. Selective α4β2 nicotinic acetylcholine receptor agonists target epigenetic mechanisms in cortical GABAergic neurons. Neuropsychopharmacology. 2011;36(7): 1366–74.
141. Paliwal A, et al. Aberrant DNA methylation links cancer susceptibility locus 15q25.1 to apoptotic regulation and lung cancer. Cancer Res. 2010;70(7):2779–88.
142. Scherf DB, et al. Epigenetic screen identifies genotype-specific promoter DNA methylation and oncogenic potential of CHRNB4. Oncogene. 2012;32:3329–38.
143. Caporaso N, et al. Genome-wide and candidate gene association study of cigarette smoking behaviors. PLoS One. 2009;4(2):e4653.
144. Yasui DH, et al. 15q11.2-13.3 chromatin analysis reveals epigenetic regulation of CHRNA7 with deficiencies in Rett and autism brain. Hum Mol Genet. 2011;20(22):4311–23.
145. Chahrour M, et al. MeCP2, a key contributor to neurological disease, activates and represses transcription. Science. 2008;320(5880):1224–9.
146. Bencherif M, et al. Mechanisms of up-regulation of neuronal nicotinic acetylcholine receptors in clonal cell lines and primary cultures of fetal rat brain. J Pharmacol Exp Ther. 1995;275(2):987–94.
147. Govind AP, Vezina P, Green WN. Nicotine-induced upregulation of nicotinic receptors: underlying mechanisms and relevance to nicotine addiction. Biochem Pharmacol. 2009; 78(7):756–65.
148. Fenster CP, et al. Upregulation of surface alpha4beta2 nicotinic receptors is initiated by receptor desensitization after chronic exposure to nicotine. J Neurosci. 1999;19(12): 4804–14.
149. Buisson B, Bertrand D. Chronic exposure to nicotine upregulates the human (alpha)4((beta)2 nicotinic acetylcholine receptor function. J Neurosci. 2001;21(6):1819–29.
150. Marks MJ, et al. Nicotine binding and nicotinic receptor subunit RNA after chronic nicotine treatment. J Neurosci. 1992;12(7):2765–84.
151. Huang LZ, Winzer-Serhan UH. Chronic neonatal nicotine upregulates heteromeric nicotinic acetylcholine receptor binding without change in subunit mRNA expression. Brain Res. 2006;1113(1):94–109.

152. Srinivasan R, et al. Nicotine up-regulates alpha4beta2 nicotinic receptors and ER exit sites via stoichiometry-dependent chaperoning. J Gen Physiol. 2010;137(1):59–79.
153. Peng X, et al. Nicotine-induced increase in neuronal nicotinic receptors results from a decrease in the rate of receptor turnover. Mol Pharmacol. 1994;46:523–30.
154. Darsow T, et al. Exocytic trafficking is required for nicotine-induced up-regulation of alpha 4 beta 2 nicotinic acetylcholine receptors. J Biol Chem. 2005;280(18):18311–20.
155. Nashmi R, et al. Assembly of α4β2 nicotinic acetylcholine receptors assessed with functional fluorescently labeled subunits: effects of localization, trafficking, and nicotine-induced upregulation in clonal mammalian cells and in cultured midbrain neurons. J Neurosci. 2003;23:11554–67.
156. Sallette J, et al. Nicotine upregulates its own receptors through enhanced intracellular maturation. Neuron. 2005;46(4):595–607.
157. Ficklin MB, Zhao S, Feng G. Ubiquilin-1 regulates nicotine-induced up-regulation of neuronal nicotinic acetylcholine receptors. J Biol Chem. 2005;280:34088–95.
158. Rezvani K, et al. Nicotine regulates multiple synaptic proteins by inhibiting proteosomal activity. J Neurosci. 2007;27:10508–19.
159. Nelson ME, Kuryatov A, Choi CH, Zhou Y, Lindstrom J. Alternate stoichiometries of alpha-4beta2 nicotinic acetylcholine receptors. J Mol Pharmacol. 2003;63(2):332–41.
160. Moroni M, et al. alpha4beta2 nicotinc receptors with high and low acetylcholine sensitivity: pharmacology, stoichiometry, and sensitivity to long-term exposure to nicotine. Mol Pharmacol. 2006;70:755–68.
161. Vallejo Y, et al. Chronic nicotine exposure upregulates nicotinic receptors by a novel mechanism. J Neurosci. 2005;25(23):5563–72.
162. Walsh H, et al. Up-regulation of nicotinic receptors by nicotine varies with receptor subtype. J Biol Chem. 2008;283(10):6022–32.
163. Columbo SF, et al. Biogenesis, trafficking and up-regulation of nicotinic ACh recpetors. Biochem Pharmacol. 2013;86:1063–73.
164. Kuryatov A, et al. Nicotine acts as a pharmacological chaperone to up-regulate human alpha-4beta2 acetylcholine receptors. Mol Pharmacol. 2005;68(6):1839–51.
165. Govind AP, Walsh H, Green WN. Nicotine-induced upregulation of native neuronal nicotinic receptors is caused by multiple mechanisms. J Neurosci. 2012;32(6):2227–38.

Chapter 4
Presynaptic Nicotinic Acetylcholine Receptors: Subtypes and Functions

Michael J. Marks, Sharon R. Grady, Tristan D. McClure-Begley, Heidi C. O'Neill, and Cristian A. Zambrano

Abstract Nicotinic acetylcholine receptors (nAChR) are widely distributed throughout the central nervous system. Much of the function of these central nAChRs appears to be modulatory mediating the release of several neurotransmitters and perhaps neuropeptides. Synaptosomal preparations have been widely used to investigate nAChR-mediated neurotransmitter release using pharmacological, immunochemical, and genetic approaches. This chapter summarizes results for nAChR-mediated $^{86}Rb^{+}$ efflux as well as nAChR-mediated dopamine, GABA, glutamate, norepinephrine, and ACh release. Studies with mice expressing mutated nAChR subunits (both null and gain-of function mutations) demonstrate that diverse nAChR subtypes contribute to presynaptic receptor activity.

Keywords Synaptosomes • Null mutant mice • Rubidium efflux • Dopamine • Gamma-aminobutyric acid • Glutamate • Neurotransmitter release • Nicotinic acetylcholine receptors • Presynaptic receptors

1 Introduction

Not long after the detection and localization of nicotinic acetylcholine receptors (nAChRs) in the CNS was measured by binding of radiolabeled agonists/antagonists [1], it was recognized that some nAChRs in CNS were localized to presynaptic regions [2]. Measuring activity of presynaptic nAChRs required different techniques

M.J. Marks, B.S., M.S., Ph.D. (✉)
Department of Psychology and Neuroscience, Institute for Behavioral Genetics,
University of Colorado, 1480 30th St., Boulder, CO 80303, USA
e-mail: marksm@colorado.edu

S.R. Grady, Ph.D. • H.C. O'Neill, Ph.D. • T.D. McClure-Begley, Ph.D.
C.A. Zambrano, Ph.D.
Institute for Behavioral Genetics, University of Colorado,
1480 30th St., 447 UCB, Boulder, CO 80303, USA
e-mail: sharon.grady@colorado.edu; heidi.oneill@colorado.edu;
mcclureb@colorado.edu; zambranc@colorado.edu

R.A.J. Lester (ed.), *Nicotinic Receptors*, The Receptors 26,
DOI 10.1007/978-1-4939-1167-7_4

than those used for measuring postsynaptic receptors. The isolation of nerve terminals (synaptosomes) was achieved in the 1960s [3, 4]. In the 1970s–1980s, the synaptosome as a system for measurement of neurotransmitter release was established [5–8]. Many investigators have used this approach as well as similar methods with brain slices or prisms to detect activity of nAChRs on terminals promoting release of dopamine (DA), GABA, glutamate, norepinephrine (NE), and acetylcholine (ACh) as well as other neurotransmitters, peptides, and hormones. The use of synaptosomal preparations with efflux of Rb^+ is another more general technique for measurement of presynaptic nAChR activity. This chapter presents methods of detection and functions of presynaptic nAChRs in a selection of more well-characterized systems.

2 Presynaptic nAChR-Mediated $^{86}Rb^+$ Efflux

Neuronal nicotinic receptors (nAChR) are ligand-gated cation channels and as such are amenable to electrophysiological measurements of their function. However, owing to the presynaptic location of many of these receptors, standard electrophysiological approaches are not easily applicable. Since the nAChR channel is relatively nonselective, both Na^+ and K^+ and, to some extent, Ca^{2+}, transit the open channel [9]. Thus, measurements of either Na^+ inflow or K^+ outflow using labeled tracers have been used to measure nAChR function [10–12]. Because efflux of K^+ (measured using $^{86}Rb^+$ to label the pool) affords greater sensitivity than Na^+ influx, the $^{86}Rb^+$ efflux assay has been more widely applied [13]. An advantage of this method over neurotransmitter release is that it is a more direct measure of receptor activity and does not depend on downstream function of calcium channels or synaptic vesicle fusion processes. However, since nAChR and NaK-ATPase are located on organelles in addition to synaptosomes it is possible that nAChR-mediated $^{86}Rb^+$ efflux is not exclusively synaptosomal.

2.1 *Initial Characterization of nAChR-Mediated $^{86}Rb^+$ Efflux*

The suitability of $^{86}Rb^+$ efflux assay to measure nAChR function was initially established by Lukas and colleagues using cell lines that expressed either native nAChR or had been transfected to express nAChR heterologously [13]. $^{86}Rb^+$ efflux was adapted to measure presynaptic nAChR function by superfusion of synaptosomal preparations. $^{86}Rb^+$ is transported into synaptosomes via Na^+K^+ATPase and will flow out through any open channel that allows K^+ efflux. Synaptosomal preparations can be somewhat purified or very crude. In either case, it appears that mostly presynaptic activity is measured. For example, it is known that habenula has a mixture of pre- and postsynaptic nAChRs and the postsynaptic receptors are mostly of the $\alpha 3\beta 4$*-nAChR subtype as measured by patch clamp electrophysiology [14];

however, primarily β2*-nAChR activity is seen with the synaptosomal $^{86}Rb^{+}$ assay [15].

The initial characterization of synaptosomal $^{86}Rb^{+}$ efflux established that nicotinic agonists differed in both potency and efficacy and nicotinic antagonists differed in potency [16]. The magnitude of the $^{86}Rb^{+}$ efflux varied among brain regions and correlated closely with the density of binding sites measured with [^{3}H]-nicotine, which is known to measure binding to α4β2*-nAChR sites [17, 18]. Nicotine-stimulated $^{86}Rb^{+}$ efflux is partially inhibited by the Na^{+} channel blockers tetrodotoxin and saxitoxin, indicating that some of the efflux is mediated by secondary activation of voltage-gated Na^{+} channels [19]. In contrast, specific K^{+} channel blockers affected only basal release. nAChR-mediated $^{86}Rb^{+}$ efflux is routinely measured in the presence of tetrodotoxin and cesium to minimize the contribution of Na^{+} and K^{+} channels to the response.

2.2 Desensitization of nAChR-Mediated $^{86}Rb^{+}$ Efflux

Synaptosomal $^{86}Rb^{+}$ efflux desensitizes following either prolonged stimulation with nicotine or by exposure to relatively low concentrations of nicotine [20]. All nicotinic agonists, while varying significantly in affinity and efficacy, elicit desensitization [20]. As assayed using thalamic synaptosomes, the concentration required to achieve desensitization by relatively low agonist concentrations correlated closely ($r=0.99$) to the affinity of the agonists as inhibitors of [^{3}H]-nicotine binding. This correlation to [^{3}H]-nicotine binding, a measure of the α4β2*-nAChR subtype, indicates that much of the $^{86}Rb^{+}$ efflux function is likely mediated by this subtype.

2.3 Using Mutant Mice to Identify nAChR Subtypes Mediating $^{86}Rb^{+}$ Efflux

The availability of null mutant mice has made the identification of the nAChR subtypes mediating agonist-stimulated $^{86}Rb^{+}$ efflux more precise. Deletion of the β2 nAChR subunit dramatically reduced activity measured from whole brain synaptosomal preparations [21]. The effect of deletion of the β2 subunit was also measured from synaptosomes prepared from 12 brain regions [22]. Activity was substantially reduced in the null mutants; however, significant activity persisted in several brain regions. Similar results were obtained for experiments examining the effects of the deletion of the α4 nAChR subunit [23]; significant reductions in activity were seen in each brain region of α4 null mutants tested with residual activity observed in several brain regions. These results demonstrate the presence of some responses that are not mediated by α4β2*-nAChR. These studies with null mutants are consistent with previous pharmacological analyses and confirm the dominant role of α4β2*-nAChR in nicotinic receptor-mediated $^{86}Rb^{+}$ efflux.

By continually monitoring nAChR-mediated $^{86}Rb^+$ efflux using on-line detection of radioactivity, a biphasic concentration effect curve for ACh-stimulated $^{86}Rb^+$ efflux was observed [21]. The higher sensitivity activity (HS), was also found to be more sensitive to inhibition by dihydro-β-erythroidine (DHβE) than the lower sensitivity activity (LS). This differential sensitivity to DHβE provided a second method for resolving these biphasic curves. Biphasic ACh concentration–effect curves for responses mediated by α4β2*-nAChR have been reported for heterologously expressed receptors and have been ascribed to the activity of receptors assembled with alternate stoichiometries [24–26] suggesting that the HS and LS activities in synaptosomal preparations are also mediated by α4β2*-nAChR with alternate stoichiometries. This possibility was examined using heterozygotic mice (+/–) to alter the relative expression of α4 and β2 mRNA thereby altering expression of α4 and β2 subunit proteins, respectively [27]. Increased relative expression of the β2 subunit and HS activity was observed for α4(+/–) mice, while increased relative expression of the α4 subunit and LS activity was observed for the β2(+/–) mice. Results of this study are consistent with the postulate that HS and LS components of $^{86}Rb^+$ efflux are mediated by α4β2-nACh with $(\alpha4)_2(\beta2)_3$ and $(\alpha4)_3(\beta2)_2$, respectively.

The effects of various nAChR subunit null mutations were assessed for both HS and LS activity (Fig. 4.1). These data confirm the absolute requirement for both β2 and α4 subunits for HS and LS components. The auxiliary nAChR subunit, α5, can co-assemble with α4 and β2 [15, 28]. Measurement of the effects of deletion of the α5 subunit confirm that α4β2α5-nAChRs mediate a significant fraction of ACh-stimulated $^{86}Rb^+$ efflux in thalamus (Fig. 4.1) and striatum, but not in cortex, hippocampus, or midbrain [29]. Deletion of the α5 subunit selectively reduced HS activity, with little effect on LS activity, indicating that α4β2α5 is an additional HS subtype, an observation consistent with results using heterologously expressed nAChR [30, 31]. A greater effect was observed for ACh-stimulated $^{86}Rb^+$ efflux than for total receptor expression measured with [^{125}I]-epibatidine binding, suggesting that α4β2α5-nAChRs may be more highly functional than other α4β2-nAChRs. Studies with transgenic mice indicate that the cortico-thalamic tract is the source of most of the α4β2α5-nAChR mediating $^{86}Rb^+$ efflux in thalamus [32]. The α5 subunit also is included in the nAChR populations that mediate $^{86}Rb^+$ efflux in habenula and interpeduncular nucleus (IPN) [33]. The presynaptic activity in habenula is primarily dependent on β2*-nAChR, while activity in IPN is mediated by a mixed population of β2*- and β4*-nAChR [15]. These data indicate that the presynaptic activity in habenula is largely mediated by α4α5β2-nAChR while in IPN, presynaptic activity could be mediated by populations of β2*- and β4*-nAChR, with either or both populations including the α5 subunit.

While most of the ACh-stimulated $^{86}Rb^+$ efflux in mouse brain synaptosomes is mediated by α4β2*-nAChR, given the general nature of the ion flux measured with $^{86}Rb^+$ efflux it is reasonable to expect that other nAChR subtypes elicit measurable $^{86}Rb^+$ efflux in select brain regions. Significant nAChR-mediated $^{86}Rb^+$ efflux was detected in the inferior colliculus and interpeduncular nucleus of β2 knockout mice.

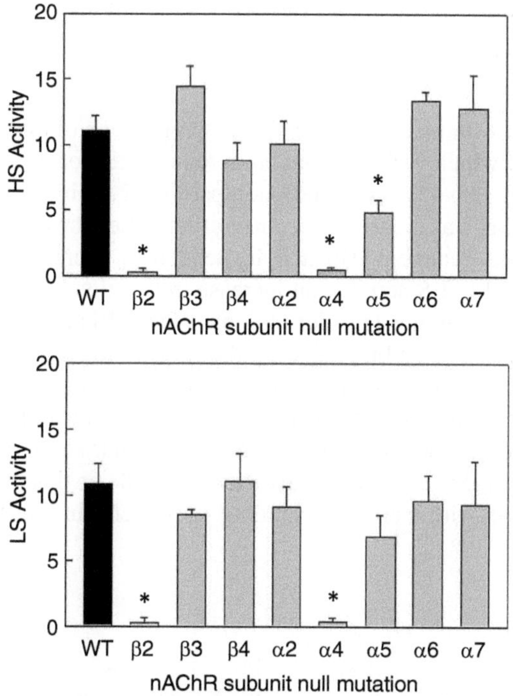

Fig. 4.1 Effect of null mutations on $^{86}Rb^+$ efflux from thalamic synaptosomes. Both HS (*upper panel*) and LS (*lower panel*) activity were assessed by $^{86}Rb^+$ efflux evoked by acetylcholine. Results were analyzed by one-way ANOVA followed by Dunnett's post hoc test (*, $P<0.01$)

These regions contain a significant population of [^{125}I]-epibatidine binding sites that are not blocked by A85380, a selective agonist for β2*-nAChRs [34, 35]. Pharmacological properties of nAChR-mediated $^{86}Rb^+$ efflux and [^{125}I]-epibatidine binding were subsequently measured in β2(−/−) mice in these two brain regions [36]. Full agonist activity of cytisine and inhibition by α-conotoxin AuIB indicated that the residual responses corresponded closely to the properties anticipated for α3β4*-nAChR.

The contribution of additional subtypes to nAChR-mediated $^{86}Rb^+$ efflux in wild-type mice is quite small. However, by inserting subunits with gain-of-function mutations, it has been possible to observe responses that are too low for reliable detection in wild-type mice. For example, insertion of α6 nAChR subunits carrying the L9'S channel mutation, allowed measurement of nicotine-stimulated $^{86}Rb^+$ efflux in superior colliculus mediated by α6β2*-nAChR, as determined by sensitivity to inhibition by α-conotoxin MII [37]. This result indicates that these receptors are likely present presynaptically in this brain region, but that their contribution to overall nAChR-mediated responses in the wild-type is too small to measure reliably.

2.4 nAChR-Mediated $^{86}Rb^{+}$ Efflux Following Chronic Nicotine Treatment

Chronic treatment with nicotine results in an upregulation of α4β2-nAChR binding sites, the extent of which varies among brain regions, and occurs with no change in amount of mRNA [38–43]. In a study comparing 12 brain regions in mice [44], function of receptors was measured by $^{86}Rb^{+}$ efflux, after chronic nicotine treatment compared to saline treatment. HS activity tended to decrease with a significant effect in thalamus. For LS activity, an increase in function was noted for olfactory bulbs, with no change in the 11 other regions. When HS activity was compared to the number of α4β2-nAChR binding sites, the slope of the regression line for all 12 regions assayed, decreased with increasing nicotine treatment dose. This study indicates that chronic treatment with nicotine changes overall functionality of α4β2-nAChRs in a dose-dependent manner in mice. A similar study of chronic nicotine treatment with rats [45] gave somewhat different results, in that $^{86}Rb^{+}$ efflux activity increased in parallel with increase in binding sites over four brain regions.

2.5 Summary

In summary, by using nicotinic agonist-stimulated $^{86}Rb^{+}$ efflux as a general measure of nAChR activity it has been possible to identify activity of several different subtypes. Under the conditions of these experiments, most of the response is mediated by α4β2*-nAChR, which corresponds to three major subtypes: $(\alpha4\beta2)_2\beta2$ (HS), $(\alpha4\beta2)_2\alpha5$ (HS), and $(\alpha4\beta2)_2\alpha4$(LS). In addition, nAChR with properties corresponding to α3β4*-nAChR has been identified in select brain regions. The activity of additional nAChR subtypes has been difficult to determine since the $^{86}Rb^{+}$ efflux assay either is not well suited for their measurement or they represent only a small fraction of the total response that is insufficient for adequate detection and characterization. The use of mutant subunits, such as the α6 L9'S, provides evidence for the activity of minor subunits by enhancing the amount of $^{86}Rb^{+}$ efflux mediated by these minor subtypes. Alternative methods such as neurotransmitter release are probably more appropriate for the study of the role of such subtypes in presynaptic nAChR function.

3 Presynaptic nAChR-Mediated Neurotransmitter Release

3.1 Dopamine

The importance of DA release to mechanisms of action of drugs of abuse was recognized several decades ago. The technique of in vivo microdialysis with probes placed in the DA terminal fields has shown that extracellular DA increases upon injection of many drugs of abuse [46]. The increase in DA following self-administered nicotine in rats was shown to be dependent on nAChRs in the VTA, rather than in the terminal fields [47]. Furthermore, a learned cue for a subsequent reward will also elicit dopamine release in trained monkeys, and DA decreases when the anticipated reward is withheld [48, 49]. In addition, aversive stimuli can also promote DA release [50]. With long-term nicotine treatment and withdrawal, nAChRs at both somatic sites in the VTA and axonal sites in NAc appear to play a role in determining levels of extracellular DA with possibly greater effects on basal levels [51].

Dopaminergic neurons of the substantia nigra (SN) and ventral tegmental area (VTA) project to the striatum (ST) and nucleus accumbens (NAc) and these axons are the only source of dopamine release in these regions allowing a clear division of presynaptic modulatory activity. In 1989, Mifsud et al. [52], using microdialysis techniques, showed that locally applied nicotine in the NAc increased DA in vivo leading to their early suggestion that nicotine stimulation of presynaptic receptors released DA and might be involved in addiction to tobacco. Early experiments also established that nicotinic agonists stimulate [^{3}H]-DA release from striatal preparations [53–57]. [^{3}H]-DA can be released from synaptosomes or slices by calcium-dependent mechanisms involving voltage-sensitive calcium channels (VSCC) or by calcium independent reversal of the dopamine transporter [2, 5, 58, 59] (see Fig. 4.2).

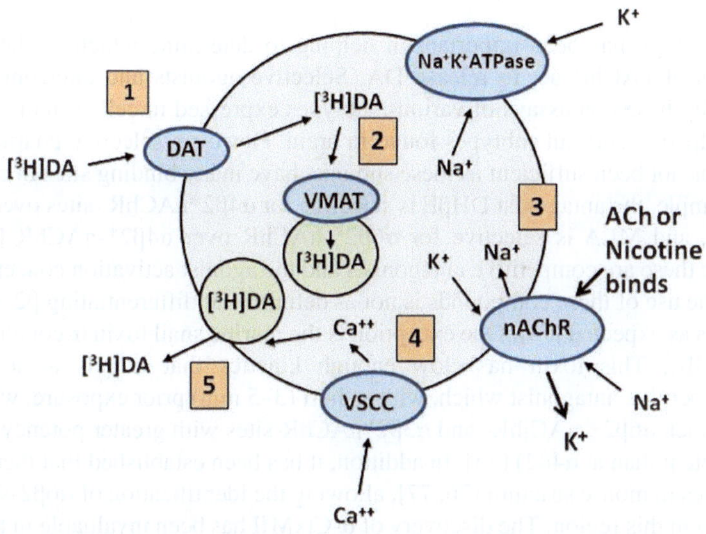

Fig. 4.2 Synaptosomal dopamine release assay. Synaptosomes are nerve terminals that retain mitochondria and can utilize glucose and maintain normal membrane potential via action of the sodium–potassium ATPase (Na^+K^+ATPase) in vitro for several hours. In this diagram, membrane proteins and represented by blue ovals and synaptic vesicles by green circles. The steps of the [^{3}H] dopamine (DA) release assay are illustrated as follows: (*1*) [^{3}H]DA enters via the dopamine transporter (DAT); (*2*) [^{3}H]DA enters synaptic vesicles via the vesicle monoamine transporter (VMAT); (*3*) agonist binds to nAChR and allows Na^+ entry causing local membrane potential change; (*4*) voltage-sensitive calcium channels (VSCC) open allowing Ca^{2+} influx; (*5*) increased Ca^{2+} supports vesicle fusion and release of [^{3}H]DA

For most nAChR agonists, DA release is dependent on external calcium and blocked by cadmium indicating that local ion flux through the activated receptors supports activation of VSCC promoting calcium influx that subsequently results in [^{3}H]-DA release [60, 61]. Using omega toxins selective for various subtypes of VSCC [62, 63], it has been shown that N- and P-type VSCC support nAChR-mediated [^{3}H]-DA release [60, 61, 64] and that the release evoked by α6β2*-nAChR (where * indicates possible presence of other subunits, [65]) appears to be coupled to P-type VSCC [61]. Fast scan cyclic voltammetry studies indicate that DA release elicited by a tonic or "discrete" stimulus is coupled to N- and P-type calcium channels to a greater extent than burst evoked release [66].

Localization of β2*-nAChRs at tyrosine hydroxylase positive (TH) terminals in the rat dorsal striatum by electron microscopy has shown that β2*-nAChRs are located at most (minimum 86 %) of these terminals [67]. About 15 % of this β2*-nAChR is associated with the plasma membrane; however, these receptors are rarely found at synapses, indicating volume transmission may be the mechanism of the modulatory activity of these presynaptic receptors [68].

3.2 *nAChR Subtypes in Striatum and Nucleus Accumbens*

Pharmacology has been important in helping to determine which of the various subtypes of nAChR act to release DA. Selective agonists and antagonists, with selectivity based on assays of various subtypes expressed in cell systems [69–71], have helped to sort out subtypes found in brain. However, selective pharmacology alone has not been sufficient as these subunits have many binding sites in common. For example, the antagonist DHβE is selective for α4β2*nAChR -sites over α6β2*-nAChR, and MLA is selective for α6β2*-nAChR over α4β2*-nAChR [72, 73]. Because these are competitive antagonists and the agonist activation concentrations differ, the use of these compounds is not as definitive at differentiating β2*-nAChR subtypes as expected [74]. One exception is the marine snail toxin α-conotoxin MII (α-CtxMII). This toxin has slow enough kinetics that it acts as a pseudo-noncompetitive antagonist which, with a short (3–5 min) prior exposure, will selectively block α6β2*-nACjhR- and α3β2*nAChR-sites with greater potency (1,000× more potent than at α4β2) [75]. In addition, it has been established that there are no α3β2 sites in mouse striatum [76, 77], allowing the identification of α6β2-sites with certainty in this region. The discovery of α-CtxMII has been invaluable in the study of DA release as the α6β2*-nAChR is one of two major classes of receptor found on these terminals [73, 78–80].

The availability of nAChR subunit null mutant mice has allowed characterization of functional nAChR subtypes at DA terminals [73, 81–83]. [^{3}H]-DA release can be measured from synaptosomal preparations of ST from various subunit knockout mice in combination with selective antagonists. Using these techniques, it has been determined that the population of receptors blocked by α-CtxMII (α-CtxMII-sensitive) in ST of mice is α6β2*-nAChR that include $(\alpha6\beta2)_2\beta3$-nAChR (where pairs in parentheses are binding sites and extra subunit is in the accessory position) and (α6β2)(α4β2)β3-nAChR subtypes [78, 82, 84, 85]. The latter is considered an α6β2* subtype despite having one α4β2 binding site, as this receptor is blocked from functioning by α-CtxMII. Small populations of $(\alpha6\beta2)_2\beta2$-nAChR and (α6β2)(α4β2)β2-nAChR may also be present; however, this determination is based on experiments with null mutant mice that could make small amounts of these subtypes when the β3 or both the β3 and α4 subunits are absent [85–87]. Receptor populations not blocked by α-CtxMII (α-CtxMII-resistant) include the $(\alpha4\beta2)_2\beta2$-nAChR, $(\alpha4\beta2)_2\alpha4$-nAChR, and $(\alpha4\beta2)_2\alpha5$-nAChR subtypes. Both the $(\alpha4\beta2)_2\beta2$-nAChR and the $(\alpha4\beta2)_2\alpha5$-nAChR subtypes are high-sensitivity (HS) forms, while the $(\alpha4\beta2)_2\alpha4$ form is low sensitivity (LS) meaning that higher concentrations of agonists are required for function [21, 88]. The combined techniques of selective pharmacology and subunit null mutations have shown that acetylcholine and nicotine are especially potent at activating the (α6β2)(α4β2)β3-nAChR subtype [86]. This subtype is likely of significance in the human smoking population where it may be activated by very low levels of nicotine.

While the availability of null mutations is mostly restricted to mice, other methods have been used to show that similar populations of nAChR subtypes exist in rats, monkeys, and humans. Taking advantage of selective lesioning of dopaminergic neurons by 6-hydroxydopamine, the loss of nAChR subtypes in rats was measured by immunoprecipitation methods to determine those subtypes associated with dopaminergic axons in striatum [72]. Data show significant losses of α4, α5, α6, β2, and β3 subunits, and gave estimates of $(\alpha4\beta2)_2\beta2$-nAChR (30 %), $(\alpha4\beta2)_2\alpha5$-nAChR (30 %), $(\alpha6\beta2)_2\beta3$-nAChR (25 %) and (α6β2)(α4β2)β3-nAChR (15 %) with the subtypes containing α5 or α6 subunits restricted to dopaminergic axons in striatum [72]. In a subsequent study, percentages of these subtypes were shown to vary somewhat with higher (α6β2)(α4β2)β3-nAChR and lower $(\alpha4\beta2)_2\alpha5$-nAChR in dorsal than in the ventral striatum of rats [89].

The selective neurotoxin MPTP has been used in monkey studies to look at loss of nAChR subtypes upon destruction of dopaminergic neurons. After MPTP treatment, significant losses of β2, β3, α3, α4, and α6 subunits in ST were found as measured by immunoprecipitation with selective antibodies [90]. The α5 subunit was not investigated in this study. In human brain, dopaminergic neurons are destroyed in Parkinson's Disease. Study of nAChR subunits by selective immunoprecipitation has shown loss of striatal α4, α6, β2, and β3 subunits in ST from postmortem human brains of PD patients [91]. The α5 subunit was detected at too low a level to see significant decreases.

4 Physiological Role of Presynaptic nAChR

The role of the presynaptic nAChRs on dopaminergic terminals in modulating the release of DA has been studied using fast scan cyclic voltammetry. This technique can be used both in brain slices and in vivo after implantation of carbon-fiber electrodes to detect DA [92–94]. The amount of DA released can be varied by strength and number of electrical pulses applied to a slice or to specific brain region by implanted microelectrodes. These measurements assess amount of DA available at the electrode which is influenced by activity of the DA transporter as well as presynaptic DA autoreceptors and various heteroreceptors including the nAChRs. DA has been shown to act by a volume transmission mode; that is to “spill over” from the synapse, bathing the area in DA and activating modulatory DA receptors [95, 96]. The source of ACh in the ST is the giant cholinergic interneurons that are tonically active [95, 96] and affect modulatory muscarinic and nicotinic receptors via volume transmission [95, 97].

The role of the presynaptic nAChRs on dopaminergic axons appears to be modulatory. They act to control the probability of DA release which differs for single-pulse (tonic firing) or bursts of activity (phasic firing) of the dopaminergic neurons [98–100]. When the nAChR are activated by tonic ACh release from the cholinergic interneurons, the amount of dopamine release is similar for tonic or phasic firing. However, when nAChRs are silenced either by pauses of the cholinergic interneurons, by nAChR antagonists, or by desensitization of the nAChRs in the presence of nicotine, the response to tonic stimulation decreases and to burst firing increases [101, 102]. This effect is elicited through α4β2*-nAChR in the dorsal ST, but via α6β2*-nAChR in the NAc [101] Studies using nAChR subunit null mutations have determined the important subtype for this regulation in the dorsal ST for wild type mice to be the $(\alpha4\beta2)_2\alpha5$, although with null mutations other subtypes, such as $(\alpha6\beta2)_2\beta3$-nAChR, can perform this function [103]. In contrast, in the NAc, the important functional subtype for modulation of response to tonic vs burst firing is (α6β2)(α4β2)β3, and no other subtype appears to be able to substitute [103]. This role of nAChR in modulating DA release could explain the importance of both α4, α6, and β2 subunits seen in models of nicotine reward [104–106]. The α4 subunit is necessary in the VTA for initiation of burst firing and both α4 and α6 together have modulatory effects in the nucleus accumbens [107].

In order to study the circuit-based role of cholinergic interneurons on control over dopamine release via ACh activation of the nAChRs on the dopaminergic axons in the ST, optogenetic techniques have been used [108]. The light-activated cation channel, channel rhodopsin2, can be selectively expressed in cholinergic neurons by using mice expressing cre-recombinase driven by the choline acetyltransferase (ChAT) promoter. Using these techniques, it has been shown that release of endogeneous ACh is sufficient to promote dopamine release via nAChR activity without activation of the dopamine neurons [109, 110]. The cholinergic interneurons are, in turn, under the control of thalamo-striatal glutamate neurons [109, 111]. Therefore, presynaptic nAChRs on dopamine terminals appear to have two different modulatory influences on dopamine release, one via activity of the dopamine neurons and the other via thalamo-striatal glutamate activity through cholinergic interneurons independent of actual dopamine neuron activity.

5 Gain-of-Function Mutations

In addition to the useful nAChR subunit null mutations in mice, there have been a number of nAChR channel mutations that confer a "gain-of-function". Some of these are models for human mutations that can cause autosomal dominant frontal lobe epilepsy (ADNFLE), while others have been useful as probes for nAChR function in mouse models [112, 113]. Two mutations that have been important for study of presynaptic nAChRs on dopamine terminals are the α4-L9'A knock-in [114] and the α6-L9'S BAC transgenic mice [37]. Assaying tissues from these mice as well as some of the subunit null mutations for agonist-stimulated synaptosomal dopamine release has shown that there is some amount of functional adaptation that occurs among nAChR subtypes on dopaminergic terminals. For subunit null mutations there appears to be no compensation when the subunits of the binding sites are removed; for example, removing α4 does not lead to increased α6 activity. However, removal of the accessory subunits α5 or β3 does increase activity of the remaining subtypes [73]. Compensatory changes in function at dopamine terminals are also found with some of the "gain-of-function" mutations. The α4-L9'A mouse shows a down-regulation of the function of α4β2*-nAChR and an increase in the function of the α6β2*-nAChR [74]. A shift in the same direction is seen in the α6-L9'S BAC transgenic mouse that has increased function of the α6β2*-nAChR and decreases in the α4β2*-nAChR [37]. This effect of balancing presynaptic dopamine release activity among subtypes may be unique to the populations of receptors on dopaminergic terminals which express a number of different subtypes.

6 Effect of Chronic Nicotine Treatment

Chronic treatment with nicotine in animal models has shown that an upregulation of the α4β2*-nAChRs, the extent of which varies among brain regions, occurs with no change in amounts of mRNA [38–43]. Dopaminergic neurons contain several populations of nAChR that have α4β2-nAChR binding sites. The α4α5β2-nAChR seems resistant to up-regulation in vivo [115, 116]. The α6β2*-nAChR, measured with [^{125}I]-α-CtxMII, decreases in striatum/nucleus accumbens with chronic nicotine treatment [89, 117–119]. Measurements with mice having YFP-tagged α4 subunits have shown that up-regulation of α4β2*-nAChR in ventral tegmental area (VTA) and substantia nigra pars compacta (SNC)/substantia nigra pars reticulata (SNR) are confined to the GABAergic neurons with no change in α4β2-nAChR binding sites seen in dopaminergic neurons [120]. If the total population is unchanged and the (α6β2)(α4β2)β3 population decreases, it is possible that the α4(non-α6)β2-nAChR population does upregulate. Owing to differences in receptor function and cellular distribution, nAChR-mediated dopaminergic function does not necessarily correspond directly to receptor number. Some nAChR expression measured by binding and immunological assays is to intracellular sites. Furthermore, α4β2*-nAChR are also expressed on non-dopaminergic neurons. Measurements of function at dopaminergic terminals after a few weeks of chronic treatment has generally shown a lack of effect overall [121] or a decrease [122, 123]. When components of dopamine release were assessed by measuring α-CtxMII resistant and -sensitive portions, generally no change was found for release mediated by α4β2*-nAChR, while there was a decrease in release mediated via α6β2*-nAChRs roughly paralleling the changes in α-CtxMII binding sites [117, 118]. This decrease may be more prominent for the (α4β2)(α6β2)β3-nAChR than the $(\alpha6\beta2)_2\beta3$-nAChR [124, 125].

There is recent evidence that changes in presynaptic function after chronic treatment may differ depending on how nicotine is given, on length of exposure, on cycles of withdrawal and on method used to assess function. [51, 126–128]. Greater upregulation of nAChR in the ST as well as larger effects on reward behavior was seen after several cycles of chronic nicotine treatment and withdrawal in mice [128]. In mice withdrawn from chronic nicotine administered in drinking water for 4 or 12 weeks, basal DA levels were decreased in the NAc. Following chronic nicotine, there was a greater inhibition of tonic dopamine release, which enhanced the contrast between tonic and phasic stimulation [51]. In rats treated with nicotine via minipump for 10 weeks, a loss of α6β2*-mediated synaptosomal DA release from ST was seen [129], while in monkeys given nicotine in drinking solution and food for 3–6 months, voltammetrically measured modulation of tonic vs phasic dopamine release via α6β2*-nAChRs was lost [127]. In the future, attempts such as these to model human smoking, quitting, and relapse may provide useful data on aspects of nicotine addiction and smoking cessation [130].

7 Variable Desensitization

It is well established that nAChRs are allosteric proteins that can desensitize when exposed to agonists for extended periods of time [131, 132]. Nicotine remains in the blood and brains of smokers for extended periods of time at concentrations of up to about 300 nM [133]. This level of nicotine is sufficient to produce significant desensitization of α4β2*-nAChRs [134]. There are indications that α4β2*-nAChRs that contain the α5 subunit may be more resistant to desensitization [30, 135–137]. In addition, the α6β2*-nAChRs may resist desensitization and retain more activity in the presence of smoking levels of nicotine [136, 138]. Variable desensitization induced by subtype differences, associated proteins, or posttranslational modifications of nAChRs may affect how each subtype is influenced by smoking levels of nicotine [132].

8 Dopamine Release in Other Brain Regions

Dopaminergic neurons project to regions other than the ST and NAc. The A9 neurons of the SN project in large part to the dorsal ST and the A10 neurons of the VTA project mainly to the NAc, olfactory tubercle (OT), and prefrontal cortex (PFC)

[139, 140]. As methods have improved, it has been shown that these designations are oversimplified and that there is some overlap in projection fields [140]. [^{3}H]-DA release from these four regions has been compared using mouse synaptosomes preparations [141]. In the OT, the nAChR-mediated [^{3}H]-DA release signal is strong, and similar to ST and NAc [37, 86, 141]. In the PFC, DA terminals are sparser and, therefore, harder to detect. The transporters for norepinephrine and serotonin can compete for uptake of [^{3}H]-DA, making use of various selective blockers necessary to selectively evaluate responses at dopaminergic terminals. Assayed in this way, nAChR-mediated [^{3}H]-DA release appears similar in all four regions. Only activation of β2*-nAChR subtypes was found to promote [^{3}H]-DA release with some blockade by α-CtxMII seen in all regions [141]. When measured from rats using PFC slices, no block by α-CtxMII was seen [142, 143] indicating that rats and mice may differ in this regard.

9 GABA

Over 30 years ago, studies in Aplysia demonstrated that application of snake venom toxins reduced ACh-evoked chloride currents in neurons, suggesting that AChRs were capable of enhancing inhibitory neurotransmission [144]. Additional electrophysiological studies in rodent and avian brain tissues indicated that nAChR agonists were capable of increasing GABAergic neurotransmission via activation of several nAChR subtypes [145–147] and that at least a portion of the enhanced GABAergic inhibitory postsynaptic currents (IPSCs) recorded were due to presynaptic nAChR evoked GABA release. As for nAChRs mediating presynaptic DA release, substantial heterogeneity in receptor subtypes has been demonstrated for nAChRs that evoke the release of GABA. Unlike DAergic nerve terminals, which exhibit anatomically defined expression in the mammalian CNS, the majority of GABA neurons that express nAChRs in the brain regions examined tend to be local inhibitory interneurons [148–150]. Dissection of the contribution of purely presynaptic nAChR-mediated GABA release on complex neurocircuitry from somatodendritic nAChR activation is challenging.

Assessment of GABA release from synaptosomes established the ability of nAChRs to act directly on presynaptic GABAergic nerve terminals [151, 152], although the nAChR subtypes present on GABAergic presynaptic terminals appear to vary by species. Recordings from chick dorsolateral geniculate nucleus (DGN) using subtype selective antagonists to block the effects of applied ACh indicate that only the α4β2 nAChR subtypes contribute to GABA release in that region [146]. The advent of nAChR subunit null mutant mice facilitated the detection of discreet nAChR receptor subtype contributions to synaptosomal [^{3}H]-GABA release. Measuring [^{3}H]-GABA release from multiple brain regions, in multiple lines of nAChR subunit knockout mice indicates that the β2 subunit is absolutely required for nAChR agonist-evoked release of GABA [151, 153, 154], with no evidence of α7 nAChR-mediated presynaptic GABA release. In mouse brain, nAChRs mediating presynaptic GABA release are largely the α4β2*-nAChR subtype [151, 154], with limited incorporation of the α5 accessory subunit in the cortex, hippocampus, and striatum [153], and extremely limited expression of an α3α4β2 nAChR in the superficial layers of the superior colliculus [155]. Evidence for α6β2* nAChRs expressed on GABAergic terminals in the rodent midbrain has also been reported [156]. In all brain regions and species examined, the release of GABA following activation of presynaptic nAChRs is Ca^{2+}-dependent, although the source of Ca^{2+} may vary according to species and the nAChR subtype being activated. In mouse brain, nAChR-evoked GABA release is entirely dependent on extracellular Ca^{2+}, and requires active recruitment of voltage-sensitive Ca^{2+} channels (VSCC) following nAChR activation [151]. In rat brain, by contrast, evoked release of GABA has been demonstrated following activation of presynaptic α7 nAChRs [157] through a functionally distinct mechanism. In rat brain, the release of GABA by α4β2 nAChR activation also requires VSCC contributions and extracellular Ca^{2+}, but α7-evoked GABA release appears to liberate Ca^{2+} from intracellular stores [157]. Additionally, presynaptic GABA release from rat cortical GABAergic interneurons has been found to involve activation of α3β4, T-type VSCCs, and the release of Ca^{2+} from intracellular stores [150].

10 Glutamate

Several components of glutamate neurotransmission are fundamental mediators of the rewarding actions of nicotine [158]. Nicotinic-glutamate interactions are believed to be critically important in the long-term effects of drugs of abuse as well as cognitive functioning. Although glutamatergic neurons are found throughout the brain, prominent populations are located in the hippocampus, outer layers of the cortex, ventral tegmental area, and the dorsal raphe nucleus [159–161]. In frontal cortex, fluorescently labeled α-bungarotoxin (Bgt) colocalizes with vesicular glutamate transporters in glutamate terminals [162]. In the hippocampus, α7 nAChRs have been detected in synaptic and perisynaptic locations of presynaptic terminals [163]. In the ventral tegmental area, α7 nAChRs are also localized to perisynaptic regions on glutamatergic terminals [164]. Owing to the ubiquitous nature of glutamate as a central player in nitrogen metabolism in the CNS, as well as being a neurotransmitter, direct detection of presynaptic release is difficult. The α7 nAChR rapidly desensitizes adding to detection difficulties in measuring direct release of glutamate using synaptosomal preparations [9]. Despite these problems, synaptosomal release has been detected using [^{3}H]-D-aspartate, a nonmetabolized analog of glutamate. While many forms of the vesicular glutamate transporter (VGLUT1,2,3) do not transport aspartate [165], one form (VEAT) found on a portion of hippocampal synaptic vesicles does transport aspartate [166]. The α7 nAChR mediates amino acid transmitter release when measuring [^{3}H]-D-aspartate release [162, 167–169].

In addition, electrophysiological studies have established a role for presynaptic nAChRs in the modulation of glutamate release [170, 171]. This approach has identified α-bungarotoxin (Bgt) and MLA-sensitive modulation of glutamate transmission in the rat hippocampus as well as olfactory bulb, suggesting a role for the α7 nAChR

[159, 171]. The role of α7-nAChR was confirmed using coculture of chick interpeduncular nucleus and medial habenula with antisense ablation of the α7 subunit [170]. α-Bgt blocked excitatory postsynaptic current frequency; following antisense treatment nicotine still modulated glutamate transmission, implying other subtype compositions were also relevant [170]. Presynaptic nicotinic modulation of glutamate release from cortico-striatal afferents in the rat striatum also appears to be mediated by α7 nAChRs; release in this system was blocked by α7-selective nicotinic antagonists α-Bgt, α-conotoxin IMI, MLA [172]. In the VTA, nicotine enhances glutamate release via activation of presynaptic α7-nAChRs [173]; the increased glutamate release contributes to the activation of the VTA DA neurons in reward pathways. While still evident in adults, the participation of the α7-mediated increase in glutamate release to DA neuron activation appears to be more dominant in young rats [160].

In contrast to the role for α7 nAChRs in many brain regions, glutamate release in the dorsal raphe nucleus is potentiated via activation of presynaptic α4β2*-nAChR [161]. In the prefrontal cortex activation of either presynaptic α7- or β2*-nAChRs leads to excitatory amino acid release [174]. Here, the two nAChR subtypes operate via distinct mechanisms. In response to α7-nAChR activation, Ca^{2+}-induced Ca^{2+} release is coupled to presynaptic ERK2 activation and synapsin-1 phosphorylation, while β2*-nAChR (likely α4β2*-nAChR) receptors facilitate [^{3}H]-D-aspartate release through a Ca^{2+} dependent mechanism that recruits VSCC [174]. These mechanisms suggest multiple routes for cholinergic modulation of glutamate release in the prefrontal cortex.

Acute nicotine treatment increases both the in vivo and in vitro release of glutamate in different brain regions [159, 173, 175–177]. Hippocampal synaptosomes prepared from rats given chronic nicotine treatment demonstrate increased glutamate overflow using KCl or 4-aminopyridine to depolarize neurons. However, the mechanism appeared to be transporter mediated rather than via an increase in synaptic vesicle release [178].

11 Norepinephrine

Noradrenergic neurons located in the locus coeruleus (LC) have diffuse and widespread ascending projections to the thalamus, hippocampus, and cerebral cortex as well as descending projections to the brain stem, cerebellum, and spinal cord [179, 180]. The LC in rodents expresses multiple nAChR subunits including α3, α4, α6, α7, β2, β3, and β4 [40, 181–184]. Norepinephrine (NE) release is elicited by local administration of nicotine to the hippocampus by in vivo microdialysis; this effect was ameliorated with subsequent locally administered mecamylamine [185] suggesting a local role for nAChRs. Direct presynaptic nAChR-mediated release of [^{3}H]-NE from slices or synaptosomal preparations of rat hippocampus differs pharmacologically from [^{3}H]-DA release from striatum. By the use of selective agonists and antagonists [186, 187] including the selective conotoxin, α-CtxAuIB, the nAChR subtype mediating this NE release was established as α3β4*-nAChR [188]. Release of [^{3}H]-NE depends upon external Ca^{2+}, partially via activation of VSCC and partially via direct Ca^{2+} flux through the nAChRs [61, 189]. Mice and rats differ somewhat in the nAChR subtypes mediating hippocampal presynaptic [^{3}H]-NE release as well as in developmental profile. Using subunit null mutant mice with and evaluating inhibition by selective α-conotoxins, a role for α6 subunits was detected in mice with data indicating complex subtypes α6α4β2β3β4-nAChR, α6β2β3β4-nAChR, α6β2β3-nAChR, and α6α4β2β3-nAChR all contributing to the hippocampal NE terminal nAChRs [190]. In addition, the presynaptic modulation of NE release by nAChRs in mouse hippocampus appears to be present in postnatal mice (2–3 weeks old) but absent in adults, while this activity is retained in rats [190], establishing a developmentally controlled difference between these rodent species..

Nicotine-evoked norepinephrine release is detectable in several brain regions in addition to hippocampus. Depolarization of dorsal raphe nucleus (DRN) using dimethylphenylpiperazinuim (DMPP) resulted in NE release in rats that was blocked by mecamylamine (50 μM), DHβE (500 μM) or methyllycaconitine (MLA) at 100 nM [191]. The authors suggest a role for α7-nAChR; however, this was reported before it was known that MLA at that concentration can also block α6β2*-nAChRs [192]. NE release from cortex [193–195] may be mediated by β2*-nAChRs, implying that different nAChR subtypes mediate NE release from cortex and DRN than from hippocampus. Other regions where nAChR-mediated NE release has been reported include cerebellum [193, 196] and thalamus [193].

12 Acetylcholine

Presynaptic autoregulation of neurotransmitter release is usually mediated by inhibitory rather than excitatory receptors. In this regard, most autoreceptors on cholinergic neurons and axons are muscarinic acetylcholine receptors of the M2 and M4 subtypes [194, 197–201]. However, cholinergic projection neurons and interneurons in many brain regions do express significant levels of mRNA for α7 and β2 nAChR subunits, and most express lower levels of α4 as well as minor levels of α2, α3, and β4 [202]. The habenula, source of the projection to the interpeduncular nucleus (IPN) was not explored in this study. The habenula projection to the IPN is the region where the strongest evidence for nicotinic presynaptic autoreceptors is found [203–205]. In the IPN, the magnitude of autoreceptor modulation appears to be unique. [^{3}H]-ACh release from IPN synaptosomal preparations is mediated by activation of presynaptic β4*-nAChRs [205]; this β4*-nAChR population is composed of α3β4- and α3β3β4-nAChR subtypes [15]. A comparison of 14 brain regions indicates that while all regions have substantial numbers of cholinergic terminals as measured by [^{3}H]-ACh release evoked by 50 mM K^+, only the IPN shows nicotine-stimulated (100 μM) release approaching the magnitude of K^+-evoked release (Table 4.1). There have been a few reports of presynaptic nAChR-mediated ACh release in other brain regions [206–208]. Minor amounts of nAChR-autoreceptor modulation (as in data of Table 4.1) could be significant if these nAChRs are restricted to particular terminals.

Table 4.1 Synaptosomal [^{3}H]-ACh release from various brain regions

Region	N[a]	K+response[b]	Nicotine response[c]
Interpeduncular nucleus	33	19.71 ± 1.44	17.37 ± 1.17
Habenula	10	13.44 ± 1.99	0.66 ± 0.20
Striatum	33	10.54 ± 1.11	0.09 ± 0.25
Olfactory tubercle	11	19.02 ± 1.72	0.50 ± 0.09
Hypothalamus	11	17.92 ± 2.36	0.61 ± 0.09
Hippocampus	11	11.56 ± 0.73	0.57 ± 0.13
Cerebral cortex	11	9.48 ± 1.14	0.35 ± 0.14
Superior colliculus	11	15.65 ± 1.38	1.21 ± 0.16
Inferior colliculus	11	15.68 ± 1.76	1.10 ± 0.15
Thalamus	11	12.47 ± 1.48	1.04 ± 0.36
Olfactory bulb	11	6.65 ± 0.75	0.40 ± 0.11
Midbrain	10	15.64 ± 0.92	1.12 ± 0.21
Cerebellum	11	12.97 ± 2.83	0.69 ± 0.38
Hindbrain	11	20.00 ± 3.38	0.62 ± 0.09

[a]Number of experiments
[b]Release evoked by 50 mM K^+ as cpm normalized to basal cpm released
[c]Release evoked by 100 μM nicotine as cpm normalized to basal cpm released

The α3β4*-nAChRs are known to be quite resistant to desensitization by nicotine [209, 210]. Possibly, this property of α3β4*-nAChR autoreceptors on the MHb to IPN cholinergic projection is responsible for the very selective degeneration of cholinergic axons projecting through the fasciculus retroflexus to the IPN in rats given chronic high doses of nicotine [211, 212] The α3β4*-nAChRs present on these cholinergic terminals may over-excite these neurons resulting in their destruction.

13 Summary and Future Directions

The role of diverse presynaptic nAChR in mediating the release of several neurotransmitters in the brain has been characterized using pharmacological, immunochemical, and genetic approaches. The diagram in Fig. 4.3 provides an overview of five neurotransmitters for which the role presynaptic nAChR has been well established. In addition, several lines of evidence indicate that nAChR, either directly or indirectly mediate the release of many biologically active compounds including other neurotransmitters, steroids, and neuropeptides that may be important modulators of the responses to acute and chronic nicotine exposure (for example: [213–217]). The understanding of the specific roles of presynaptic nAChR in the regulation of hormone and peptide release is currently incomplete. The continued investigation of the mechanisms by which nAChR modulate these diverse biological responses will further elucidate the role of this complex receptor family in both normal physiology as well as in nicotine dependence. Please see Chap. 7 for further discussion on the circuit implications of presynaptic nicotinic receptors.

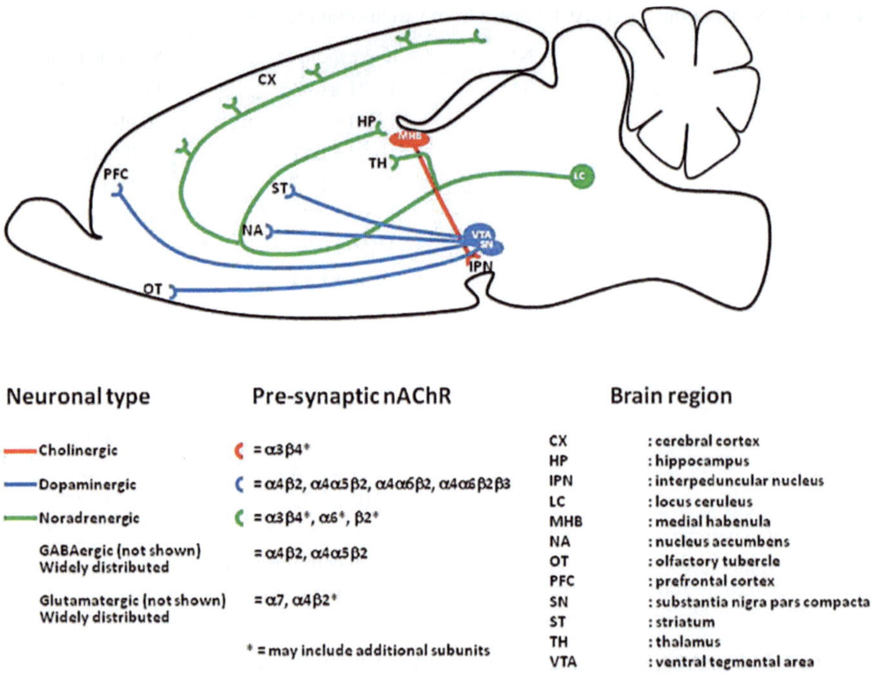

Fig. 4.3 Pathways and presynaptic nAChR subtypes. This diagram illustrates the pathways and presynaptic nAChR subtypes associated with particular neurotransmitters discussed in this chapter

Acknowledgements Supported by the following grants from the National Institute on Drug Abuse, National Institutes of Health: R01 DA003194, R01 DA012242, U19 DA019375, and P30 DA015663.

References

1. Collins AC, Salminen O, Marks MJ, Whiteaker P, Grady SR. The road to discovery of neuronal nicotinic cholinergic receptor subtypes. Handb Exp Pharmacol. 2009;192:85–112. PubMed PMID: 19184647, Epub 2009/02/03. eng.
2. Wonnacott S. Presynaptic nicotinic ACh receptors. Trends Neurosci. 1997;20(2):92–8. PubMed PMID: 9023878, Epub 1997/02/01. eng.
3. Gray EG, Whittaker VP. The isolation of nerve endings from brain: an electron-microscopic study of cell fragments derived by homogenization and centrifugation. J Anat. 1962;96:79–88. PubMed PMID: 13901297, Pubmed Central PMCID: PMC1244174, Epub 1962/01/01. eng.
4. Whittaker VP. Pharmacological studies with isolated cell components. Biochem Pharmacol. 1962;9:61–9. PubMed PMID: 14000417, Epub 1962/08/01. eng.
5. Raiteri M, Cerrito F, Cervoni AM, Levi G. Dopamine can be released by two mechanisms differentially affected by the dopamine transport inhibitor nomifensine. J Pharmacol Exp Ther. 1979;208(2):195–202. PubMed PMID: 216794, Epub 1979/02/01. eng.

6. Drapeau P, Blaustein MP. Initial release of [3H]dopamine from rat striatal synaptosomes: correlation with calcium entry. J Neurosci. 1983;3(4):703–13. PubMed PMID: 6834103, Epub 1983/04/01. eng.
7. Sanchez-Prieto J, Budd DC, Herrero I, Vazquez E, Nicholls DG. Presynaptic receptors and the control of glutamate exocytosis. Trends Neurosci. 1996;19(6):235–9. PubMed PMID: 8761959, Epub 1996/06/01. eng.
8. Kristensen EW, Bigelow JC, Wightman RM. Time resolved dopamine overflow from synaptosomes and chopped striatal tissue with rapid superfusion. Brain Res. 1988;461(1):44–52. PubMed PMID: 3224276, Epub 1988/09/27. eng.
9. Albuquerque EX, Pereira EF, Alkondon M, Rogers SW. Mammalian nicotinic acetylcholine receptors: from structure to function. Physiol Rev. 2009;89(1):73–120. PubMed PMID: 19126755, Pubmed Central PMCID: PMC2713585, Epub 2009/01/08. eng.
10. Wada A, Takara H, Izumi F, Kobayashi H, Yanagihara N. Influx of 22Na through acetylcholine receptor-associated Na channels: relationship between 22Na influx, 45Ca influx and secretion of catecholamines in cultured bovine adrenal medulla cells. Neuroscience. 1985;15(1):283–92. PubMed PMID: 2409474, Epub 1985/05/01. eng.
11. Stallcup WB. Sodium and calcium fluxes in a clonal nerve cell line. J Physiol. 1979;286:525–40. PubMed PMID: 571466, Pubmed Central PMCID: PMC1281587, Epub 1979/01/01. eng.
12. Robinson D, McGee Jr R. Agonist-induced regulation of the neuronal nicotinic acetylcholine receptor of PC12 cells. Mol Pharmacol. 1985;27(4):409–17. PubMed PMID: 3982388, Epub 1985/04/01. eng.
13. Lukas RJ, Cullen MJ. An isotopic rubidium ion efflux assay for the functional characterization of nicotinic acetylcholine receptors on clonal cell lines. Anal Biochem. 1988;175(1):212–8. PubMed PMID: 2469352, Epub 1988/11/15. eng.
14. Quick MW, Ceballos RM, Kasten M, McIntosh JM, Lester RA. Alpha3beta4 subunit-containing nicotinic receptors dominate function in rat medial habenula neurons. Neuropharmacology. 1999;38(6):769–83. PubMed PMID: 10465681, Epub 1999/08/28. eng.
15. Grady SR, Moretti M, Zoli M, Marks MJ, Zanardi A, Pucci L, et al. Rodent habenulo-interpeduncular pathway expresses a large variety of uncommon nAChR subtypes, but only the alpha3beta4* and alpha3beta3beta4* subtypes mediate acetylcholine release. J Neurosci. 2009;29(7):2272–82. PubMed PMID: 19228980, Pubmed Central PMCID: 2680386.
16. Marks MJ, Farnham DA, Grady SR, Collins AC. Nicotinic receptor function determined by stimulation of rubidium efflux from mouse brain synaptosomes. J Pharmacol Exp Ther. 1993;264(2):542–52. PubMed PMID: 8437106, Epub 1993/02/01. eng.
17. Zoli M, Lena C, Picciotto MR, Changeux JP. Identification of four classes of brain nicotinic receptors using beta2 mutant mice. J Neurosci. 1998;18(12):4461–72. PubMed PMID: 9614223, Epub 1998/06/10. eng.
18. Marubio LM, del Mar Arroyo-Jimenez M, Cordero-Erausquin M, Lena C, Le Novere N, de Kerchove d'Exaerde A, et al. Reduced antinociception in mice lacking neuronal nicotinic receptor subunits. Nature. 1999;398(6730):805–10. PubMed PMID: 10235262, Epub 1999/05/11. eng.
19. Marks MJ, Bullock AE, Collins AC. Sodium channel blockers partially inhibit nicotine-stimulated 86Rb+efflux from mouse brain synaptosomes. J Pharmacol Exp Ther. 1995;274(2):833–41. PubMed PMID: 7636747, Epub 1995/08/01. eng.
20. Marks MJ, Robinson SF, Collins AC. Nicotinic agonists differ in activation and desensitization of 86Rb+efflux from mouse thalamic synaptosomes. J Pharmacol Exp Ther. 1996;277(3):1383–96. PubMed PMID: 8667201, Epub 1996/06/01. eng.
21. Marks MJ, Whiteaker P, Calcaterra J, Stitzel JA, Bullock AE, Grady SR, et al. Two pharmacologically distinct components of nicotinic receptor-mediated rubidium efflux in mouse brain require the beta2 subunit. J Pharmacol Exp Ther. 1999;289(2):1090–103. PubMed PMID: 10215692, Epub 1999/04/24. eng.
22. Marks MJ, Stitzel JA, Grady SR, Picciotto MR, Changeux JP, Collins AC. Nicotinic-agonist stimulated (86)Rb(+) efflux and [(3)H]epibatidine binding of mice differing in beta2 genotype. Neuropharmacology. 2000;39(13):2632–45. PubMed PMID: 11044733, Epub 2000/10/25. eng.

23. Marks MJ, Meinerz NM, Drago J, Collins AC. Gene targeting demonstrates that alpha4 nicotinic acetylcholine receptor subunits contribute to expression of diverse [3H]epibatidine binding sites and components of biphasic 86Rb+efflux with high and low sensitivity to stimulation by acetylcholine. Neuropharmacology. 2007;53(3):390–405. PubMed PMID: 17631923, Pubmed Central PMCID: 2577786.
24. Zwart R, Vijverberg HP. Four pharmacologically distinct subtypes of alpha4beta2 nicotinic acetylcholine receptor expressed in Xenopus laevis oocytes. Mol Pharmacol. 1998;54(6): 1124–31. PubMed PMID: 9855643, Epub 1998/12/18. eng.
25. Nelson ME, Kuryatov A, Choi CH, Zhou Y, Lindstrom J. Alternate stoichiometries of alpha4beta2 nicotinic acetylcholine receptors. Mol Pharmacol. 2003;63(2):332–41. PubMed PMID: 12527804, Epub 2003/01/16. eng.
26. Moroni M, Bermudez I. Stoichiometry and pharmacology of two human alpha4beta2 nicotinic receptor types. J Mol Neurosci. 2006;30(1–2):95–6. PubMed PMID: 17192644, Epub 2006/12/29. eng.
27. Gotti C, Moretti M, Meinerz NM, Clementi F, Gaimarri A, Collins AC, et al. Partial deletion of the nicotinic cholinergic receptor alpha 4 or beta 2 subunit genes changes the acetylcholine sensitivity of receptor-mediated 86Rb+efflux in cortex and thalamus and alters relative expression of alpha 4 and beta 2 subunits. Mol Pharmacol. 2008;73(6):1796–807. PubMed PMID: 18337473.
28. Vernallis AB, Conroy WG, Berg DK. Neurons assemble acetylcholine receptors with as many as three kinds of subunits while maintaining subunit segregation among receptor subtypes. Neuron. 1993;10(3):451–64. PubMed PMID: 8461135, Epub 1993/03/01. eng.
29. Brown RW, Collins AC, Lindstrom JM, Whiteaker P. Nicotinic alpha5 subunit deletion locally reduces high-affinity agonist activation without altering nicotinic receptor numbers. J Neurochem. 2007;103(1):204–15. PubMed PMID: 17573823.
30. Kuryatov A, Berrettini W, Lindstrom J. Acetylcholine receptor (AChR) alpha5 subunit variant associated with risk for nicotine dependence and lung cancer reduces (alpha4beta2)(2) alpha5 AChR function. Mol Pharmacol. 2011;79(1):119–25. PubMed PMID: 20881005, Pubmed Central PMCID: 3014277.
31. Kuryatov A, Onksen J, Lindstrom J. Roles of accessory subunits in alpha4beta2(*) nicotinic receptors. Mol Pharmacol. 2008;74(1):132–43. PubMed PMID: 18381563, Epub 2008/04/03. eng.
32. Heath CJ, King SL, Gotti C, Marks MJ, Picciotto MR. Cortico-thalamic connectivity is vulnerable to nicotine exposure during early postnatal development through alpha4/beta2/alpha5 nicotinic acetylcholine receptors. Neuropsychopharmacology. 2010;35(12):2324–38. PubMed PMID: 20736992, Pubmed Central PMCID: 2955839.
33. Fowler CD, Lu Q, Johnson PM, Marks MJ, Kenny PJ. Habenular alpha5 nicotinic receptor subunit signalling controls nicotine intake. Nature. 2011;471(7340):597–601. PubMed PMID: 21278726, Pubmed Central PMCID: 3079537.
34. Whiteaker P, Jimenez M, McIntosh JM, Collins AC, Marks MJ. Identification of a novel nicotinic binding site in mouse brain using [(125)I]-epibatidine. Br J Pharmacol. 2000;131(4):729–39. PubMed PMID: 11030722, Pubmed Central PMCID: PMC1572375, Epub 2000/10/13. eng.
35. Mukhin AG, Gundisch D, Horti AG, Koren AO, Tamagnan G, Kimes AS, et al. 5-Iodo-A-85380, an alpha4beta2 subtype-selective ligand for nicotinic acetylcholine receptors. Mol Pharmacol. 2000;57(3):642–9. PubMed PMID: 10692507, Epub 2000/02/29. eng.
36. Marks MJ, Whiteaker P, Grady SR, Picciotto MR, McIntosh JM, Collins AC. Characterization of [(125) I]epibatidine binding and nicotinic agonist-mediated (86) Rb(+) efflux in interpeduncular nucleus and inferior colliculus of beta2 null mutant mice. J Neurochem. 2002;81(5):1102–15. PubMed PMID: 12065623, Epub 2002/06/18. eng.
37. Drenan RM, Grady SR, Whiteaker P, McClure-Begley T, McKinney S, Miwa JM, et al. In vivo activation of midbrain dopamine neurons via sensitized, high-affinity alpha 6 nicotinic acetylcholine receptors. Neuron. 2008;60(1):123–36. PubMed PMID: 18940593, Pubmed Central PMCID: 2632732.

38. Marks MJ, Burch JB, Collins AC. Effects of chronic nicotine infusion on tolerance development and nicotinic receptors. J Pharmacol Exp Ther. 1983;226(3):817–25. PubMed PMID: 6887012, Epub 1983/09/01. eng.
39. Schwartz RD, Kellar KJ. Nicotinic cholinergic receptor binding sites in the brain: regulation in vivo. Science. 1983;220(4593):214–6. PubMed PMID: 6828889, Epub 1983/04/08. eng.
40. Marks MJ, Pauly JR, Gross SD, Deneris ES, Hermans-Borgmeyer I, Heinemann SF, et al. Nicotine binding and nicotinic receptor subunit RNA after chronic nicotine treatment. J Neurosci. 1992;12(7):2765–84. PubMed PMID: 1613557, Epub 1992/07/01. eng.
41. McCallum SE, Collins AC, Paylor R, Marks MJ. Deletion of the beta 2 nicotinic acetylcholine receptor subunit alters development of tolerance to nicotine and eliminates receptor upregulation. Psychopharmacology (Berl). 2006;184(3–4):314–27. PubMed PMID: 16001112.
42. Marks MJ, McClure-Begley TD, Whiteaker P, Salminen O, Brown RW, Cooper J, et al. Increased nicotinic acetylcholine receptor protein underlies chronic nicotine-induced upregulation of nicotinic agonist binding sites in mouse brain. J Pharmacol Exp Ther. 2011;337(1):187–200. PubMed PMID: 21228066, Pubmed Central PMCID: 3063733.
43. Govind AP, Vezina P, Green WN. Nicotine-induced upregulation of nicotinic receptors: underlying mechanisms and relevance to nicotine addiction. Biochem Pharmacol. 2009;78(7):756–65. PubMed PMID: 19540212, Pubmed Central PMCID: PMC2728164, Epub 2009/06/23. eng.
44. Marks MJ, Rowell PP, Cao JZ, Grady SR, McCallum SE, Collins AC. Subsets of acetylcholine-stimulated 86Rb+efflux and [125I]-epibatidine binding sites in C57BL/6 mouse brain are differentially affected by chronic nicotine treatment. Neuropharmacology. 2004;46(8):1141–57. PubMed PMID: 15111021.
45. Nguyen HN, Rasmussen BA, Perry DC. Binding and functional activity of nicotinic cholinergic receptors in selected rat brain regions are increased following long-term but not short-term nicotine treatment. J Neurochem. 2004;90(1):40–9. PubMed PMID: 15198665, Epub 2004/06/17. eng.
46. Di Chiara G, Imperato A. Drugs abused by humans preferentially increase synaptic dopamine concentrations in the mesolimbic system of freely moving rats. Proc Natl Acad Sci U S A. 1988;85(14):5274–8. PubMed PMID: 2899326, Pubmed Central PMCID: PMC281732, Epub 1988/07/01. eng.
47. Corrigall WA, Coen KM, Adamson KL. Self-administered nicotine activates the mesolimbic dopamine system through the ventral tegmental area. Brain Res. 1994;653(1–2):278–84. PubMed PMID: 7982062, Epub 1994/08/08. eng.
48. Schultz W, Dayan P, Montague PR. A neural substrate of prediction and reward. Science. 1997;275(5306):1593–9. PubMed PMID: 9054347. Epub 1997/03/14. eng.
49. Dong Y, Zhang T, Li W, Doyon WM, Dani JA. Route of nicotine administration influences in vivo dopamine neuron activity: habituation, needle injection, and cannula infusion. J Mol Neurosci. 2010;40(1–2):164–71. PubMed PMID: 19714495, Pubmed Central PMCID: 2891747.
50. Horvitz JC. Mesolimbocortical and nigrostriatal dopamine responses to salient non-reward events. Neuroscience. 2000;96(4):651–6. PubMed PMID: 10727783, Epub 2000/03/23. eng.
51. Zhang L, Dong Y, Doyon WM, Dani JA. Withdrawal from chronic nicotine exposure alters dopamine signaling dynamics in the nucleus accumbens. Biol Psychiatry. 2012;71(3):184–91. PubMed PMID: 21872847, Pubmed Central PMCID: 3227792.
52. Mifsud JC, Hernandez L, Hoebel BG. Nicotine infused into the nucleus accumbens increases synaptic dopamine as measured by in vivo microdialysis. Brain Res. 1989;478(2):365–7. PubMed PMID: 2924134, Epub 1989/01/30. eng.
53. Rowell PP, Carr LA, Garner AC. Stimulation of [3H]dopamine release by nicotine in rat nucleus accumbens. J Neurochem. 1987;49(5):1449–54. PubMed PMID: 2889800, Epub 1987/11/01. eng.
54. Rapier C, Lunt GG, Wonnacott S. Stereoselective nicotine-induced release of dopamine from striatal synaptosomes: concentration dependence and repetitive stimulation. J Neurochem. 1988;50(4):1123–30. PubMed PMID: 3346670, Epub 1988/04/01. eng.

55. Rapier C, Lunt GG, Wonnacott S. Nicotinic modulation of [3H]dopamine release from striatal synaptosomes: pharmacological characterisation. J Neurochem. 1990;54(3):937–45. PubMed PMID: 2303820, Epub 1990/03/01. eng.
56. Dwoskin LP, Buxton ST, Jewell AL, Crooks PA. S(−)-nornicotine increases dopamine release in a calcium-dependent manner from superfused rat striatal slices. J Neurochem. 1993;60(6):2167–74. PubMed PMID: 8492124, Epub 1993/06/01. eng.
57. Grady SR, Marks MJ, Wonnacott S, Collins AC. Characterization of nicotinic receptor-mediated [3H]dopamine release from synaptosomes prepared from mouse striatum. J Neurochem. 1992;59:848–56.
58. Teng L, Crooks PA, Sonsalla PK, Dwoskin LP. Lobeline and nicotine evoke [3H]overflow from rat striatal slices preloaded with [3H]dopamine: differential inhibition of synaptosomal and vesicular [3H]dopamine uptake. J Pharmacol Exp Ther. 1997;280(3):1432–44. PubMed PMID: 9067333, Epub 1997/03/01. eng.
59. Carvalho CM, Ferreira IL, Duarte CB, Malva JO, Tretter L, Adam-Vizi V, et al. Relation of [Ca2+]i to dopamine release in striatal synaptosomes: role of Ca2+ channels. Brain Res. 1995;669(2):234–44. PubMed PMID: 7712179, Epub 1995/01/16. eng.
60. Soliakov L, Wonnacott S. Voltage-sensitive Ca2+ channels involved in nicotinic receptor-mediated [3H]dopamine release from rat striatal synaptosomes. J Neurochem. 1996;67(1):163–70. PubMed PMID: 8666987, Epub 1996/07/01. eng.
61. Kulak JM, McIntosh JM, Yoshikami D, Olivera BM. Nicotine-evoked transmitter release from synaptosomes: functional association of specific presynaptic acetylcholine receptors and voltage-gated calcium channels. J Neurochem. 2001;77(6):1581–9. PubMed PMID: 11413241, Epub 2001/06/20. eng.
62. Miller RJ. Rocking and rolling with Ca2+ channels. Trends Neurosci. 2001;24(8):445–9. PubMed PMID: 11476883, Epub 2001/07/31. eng.
63. Lewis RJ, Dutertre S, Vetter I, Christie MJ. Conus venom peptide pharmacology. Pharmacol Rev. 2012;64(2):259–98. PubMed PMID: 22407615, Epub 2012/03/13. eng.
64. Prince RJ, Fernandes KG, Gregory JC, Martyn ID, Lippiello PM. Modulation of nicotine-evoked [3H]dopamine release from rat striatal synaptosomes by voltage-sensitive calcium channel ligands. Biochem Pharmacol. 1996;52(4):613–8. PubMed PMID: 8759034, Epub 1996/08/23. eng.
65. Lukas RJ, Changeux JP, Le Novere N, Albuquerque EX, Balfour DJ, Berg DK, et al. International Union of Pharmacology. XX Current status of the nomenclature for nicotinic acetylcholine receptors and their subunits. Pharmacol Rev. 1999;51(2):397–401. PubMed PMID: 10353988, Epub 1999/06/03. eng.
66. Phillips PE, Stamford JA. Differential recruitment of N-, P- and Q-type voltage-operated calcium channels in striatal dopamine release evoked by 'regular' and 'burst' firing. Brain Res. 2000;884(1–2):139–46. PubMed PMID: 11082495, Epub 2000/11/18. eng.
67. Jones IW, Bolam JP, Wonnacott S. Presynaptic localisation of the nicotinic acetylcholine receptor beta2 subunit immunoreactivity in rat nigrostriatal dopaminergic neurones. J Comp Neurol. 2001;439(2):235–47. PubMed PMID: 11596051, Epub 2001/10/12. eng.
68. Lester RA. Activation and desensitization of heteromeric neuronal nicotinic receptors: implications for non-synaptic transmission. Bioorg Med Chem Lett. 2004;14(8):1897–900. PubMed PMID: 15050622, Epub 2004/03/31. eng.
69. Luetje CW, Patrick J. Both alpha- and beta-subunits contribute to the agonist sensitivity of neuronal nicotinic acetylcholine receptors. J Neurosci. 1991;11(3):837–45. PubMed PMID: 1705971, Epub 1991/03/01. eng.
70. Chavez-Noriega LE, Crona JH, Washburn MS, Urrutia A, Elliott KJ, Johnson EC. Pharmacological characterization of recombinant human neuronal nicotinic acetylcholine receptors h alpha 2 beta 2, h alpha 2 beta 4, h alpha 3 beta 2, h alpha 3 beta 4, h alpha 4 beta 2, h alpha 4 beta 4 and h alpha 7 expressed in Xenopus oocytes. J Pharmacol Exp Ther. 1997;280(1):346–56. PubMed PMID: 8996215, Epub 1997/01/01. eng.
71. Papke RL, Wecker L, Stitzel JA. Activation and inhibition of mouse muscle and neuronal nicotinic acetylcholine receptors expressed in Xenopus oocytes. J Pharmacol Exp Ther.

2010;333(2):501–18. PubMed PMID: 20100906, Pubmed Central PMCID: PMC2872959, Epub 2010/01/27. eng.
72. Zoli M, Moretti M, Zanardi A, McIntosh JM, Clementi F, Gotti C. Identification of the nicotinic receptor subtypes expressed on dopaminergic terminals in the rat striatum. J Neurosci. 2002;22(20):8785–9. PubMed PMID: 12388584, Epub 2002/10/22. eng.
73. Grady SR, Salminen O, Laverty DC, Whiteaker P, McIntosh JM, Collins AC, et al. The subtypes of nicotinic acetylcholine receptors on dopaminergic terminals of mouse striatum. Biochem Pharmacol. 2007;74(8):1235–46. PubMed PMID: 17825262, Pubmed Central PMCID: 2735219.
74. Soll LG, Grady SR, Salminen O, Marks MJ, Tapper AR. A role for alpha4(non-alpha6)* nicotinic acetylcholine receptors in motor behavior. Neuropharmacology. 2013;73C:19–30. PubMed PMID: 23688922, Epub 2013/05/22. Eng.
75. Cartier GE, Yoshikami D, Gray WR, Luo S, Olivera BM, McIntosh JM. A new alpha-conotoxin which targets alpha3beta2 nicotinic acetylcholine receptors. J Biol Chem. 1996;271(13):7522–8. PubMed PMID: 8631783, Epub 1996/03/29. eng.
76. Champtiaux N, Han ZY, Bessis A, Rossi FM, Zoli M, Marubio L, et al. Distribution and pharmacology of alpha 6-containing nicotinic acetylcholine receptors analyzed with mutant mice. J Neurosci. 2002;22(4):1208–17. PubMed PMID: 11850448, Epub 2002/02/19. eng.
77. Whiteaker P, Peterson CG, Xu W, McIntosh JM, Paylor R, Beaudet al, et al. Involvement of the alpha3 subunit in central nicotinic binding populations. J Neurosci. 2002;22(7):2522–9. PubMed PMID: 11923417, Epub 2002/03/30. eng.
78. Salminen O, Murphy KL, McIntosh JM, Drago J, Marks MJ, Collins AC, et al. Subunit composition and pharmacology of two classes of striatal presynaptic nicotinic acetylcholine receptors mediating dopamine release in mice. Mol Pharmacol. 2004;65(6):1526–35. PubMed PMID: 15155845, Epub 2004/05/25. eng.
79. Kulak JM, Nguyen TA, Olivera BM, McIntosh JM. Alpha-conotoxin MII blocks nicotine-stimulated dopamine release in rat striatal synaptosomes. J Neurosci. 1997;17(14):5263–70. PubMed PMID: 9204910, Epub 1997/07/15. eng.
80. Kaiser SA, Soliakov L, Harvey SC, Luetje CW, Wonnacott S. Differential inhibition by alpha-conotoxin-MII of the nicotinic stimulation of [3H]dopamine release from rat striatal synaptosomes and slices. J Neurochem. 1998;70(3):1069–76. PubMed PMID: 9489727, Epub 1998/03/07. eng.
81. Fowler CD, Kenny PJ. Utility of genetically modified mice for understanding the neurobiology of substance use disorders. Hum Genet. 2012;131(6):941–57. PubMed PMID: 22190154, Epub 2011/12/23. eng.
82. Champtiaux N, Gotti C, Cordero-Erausquin M, David DJ, Przybylski C, Lena C, et al. Subunit composition of functional nicotinic receptors in dopaminergic neurons investigated with knock-out mice. J Neurosci. 2003;23(21):7820–9. PubMed PMID: 12944511, Epub 2003/08/29. eng.
83. Whiteaker P, Wilking JA, Brown RW, Brennan RJ, Collins AC, Lindstrom JM, et al. Pharmacological and immunochemical characterization of alpha2* nicotinic acetylcholine receptors (nAChRs) in mouse brain. Acta Pharmacol Sin. 2009;30(6):795–804. PubMed PMID: 19498420, Epub 2009/06/06. eng.
84. Klink R, de Kerchove d'Exaerde A, Zoli M, Changeux JP. Molecular and physiological diversity of nicotinic acetylcholine receptors in the midbrain dopaminergic nuclei. J Neurosci. 2001;21(5):1452–63.
85. Cui C, Booker TK, Allen RS, Grady SR, Whiteaker P, Marks MJ, et al. The beta3 nicotinic receptor: a component of alpha-conotoxin MM-binding nicotinic acetylcholine receptors that modulate dopamine release and related behaviors. J Neurosci. 2003;23:11045–53.
86. Salminen O, Drapeau JA, McIntosh JM, Collins AC, Marks MJ, Grady SR. Pharmacology of alpha-conotoxin MII-sensitive subtypes of nicotinic acetylcholine receptors isolated by breeding of null mutant mice. Mol Pharmacol. 2007;71(6):1563–71. PubMed PMID: 17341654.
87. Gotti C, Moretti M, Clementi F, Riganti L, McIntosh JM, Collins AC, et al. Expression of nigrostriatal alpha 6-containing nicotinic acetylcholine receptors is selectively reduced, but

not eliminated, by beta 3 subunit gene deletion. Mol Pharmacol. 2005;67(6):2007–15. PubMed PMID: 15749993, Epub 2005/03/08. eng.

88. Grady SR, Salminen O, McIntosh JM, Marks MJ, Collins AC. Mouse striatal dopamine nerve terminals express alpha4alpha5beta2 and two stoichiometric forms of alpha4beta2*-nicotinic acetylcholine receptors. J Mol Neurosci. 2010;40(1–2):91–5. PubMed PMID: 19693710.
89. Gotti C, Guiducci S, Tedesco V, Corbioli S, Zanetti L, Moretti M, et al. Nicotinic acetylcholine receptors in the mesolimbic pathway: primary role of ventral tegmental area alpha-6beta2* receptors in mediating systemic nicotine effects on dopamine release, locomotion, and reinforcement. J Neurosci. 2010;30(15):5311–25. PubMed PMID: 20392953, Epub 2010/04/16. eng.
90. Quik M, Vailati S, Bordia T, Kulak JM, Fan H, McIntosh JM, et al. Subunit composition of nicotinic receptors in monkey striatum: effect of treatments with 1-methyl-4-phenyl-1,2,3,6-tetrahydropyridine or L-DOPA. Mol Pharmacol. 2005;67(1):32–41. PubMed PMID: 15470079, Epub 2004/10/08. eng.
91. Gotti C, Moretti M, Bohr I, Ziabreva I, Vailati S, Longhi R, et al. Selective nicotinic acetylcholine receptor subunit deficits identified in Alzheimer's disease, Parkinson's disease and dementia with Lewy bodies by immunoprecipitation. Neurobiol Dis. 2006;23(2):481–9. PubMed PMID: 16759874, Epub 2006/06/09. eng.
92. May LJ, Wightman RM. Heterogeneity of stimulated dopamine overflow within rat striatum as observed with in vivo voltammetry. Brain Res. 1989;487(2):311–20. PubMed PMID: 2786444, Epub 1989/05/22. eng.
93. Garris PA, Christensen JR, Rebec GV, Wightman RM. Real-time measurement of electrically evoked extracellular dopamine in the striatum of freely moving rats. J Neurochem. 1997;68(1):152–61. PubMed PMID: 8978721, Epub 1997/01/01. eng.
94. Schmitz Y, Benoit-Marand M, Gonon F, Sulzer D. Presynaptic regulation of dopaminergic neurotransmission. J Neurochem. 2003;87(2):273–89. PubMed PMID: 14511105, Epub 2003/09/27. eng.
95. Zhou FM, Wilson CJ, Dani JA. Cholinergic interneuron characteristics and nicotinic properties in the striatum. J Neurobiol. 2002;53(4):590–605. PubMed PMID: 12436423, Epub 2002/11/19. eng.
96. Rice ME, Cragg SJ. Dopamine spillover after quantal release: rethinking dopamine transmission in the nigrostriatal pathway. Brain Res Rev. 2008;58(2):303–13. PubMed PMID: 18433875, Pubmed Central PMCID: PMC2879278, Epub 2008/04/25. eng.
97. Crespo JA, Sturm K, Saria A, Zernig G. Activation of muscarinic and nicotinic acetylcholine receptors in the nucleus accumbens core is necessary for the acquisition of drug reinforcement. J Neurosci. 2006;26(22):6004–10. PubMed PMID: 16738243, Epub 2006/06/02. eng.
98. Zhang H, Sulzer D. Frequency-dependent modulation of dopamine release by nicotine. Nat Neurosci. 2004;7(6):581–2. PubMed PMID: 15146187, Epub 2004/05/18. eng.
99. Zhang T, Zhang L, Liang Y, Siapas AG, Zhou FM, Dani JA. Dopamine signaling differences in the nucleus accumbens and dorsal striatum exploited by nicotine. J Neurosci. 2009;29(13):4035–43. PubMed PMID: 19339599, Pubmed Central PMCID: PMC2743099, Epub 2009/04/03. eng.
100. Rice ME, Cragg SJ. Nicotine amplifies reward-related dopamine signals in striatum. Nat Neurosci. 2004;7(6):583–4. PubMed PMID: 15146188, Epub 2004/05/18. eng.
101. Exley R, Clements MA, Hartung H, McIntosh JM, Cragg SJ. Alpha6-containing nicotinic acetylcholine receptors dominate the nicotine control of dopamine neurotransmission in nucleus accumbens. Neuropsychopharmacology. 2008;33(9):2158–66. PubMed PMID: 18033235, Epub 2007/11/23. eng.
102. Threlfell S, Clements MA, Khodai T, Pienaar IS, Exley R, Wess J, et al. Striatal muscarinic receptors promote activity dependence of dopamine transmission via distinct receptor subtypes on cholinergic interneurons in ventral versus dorsal striatum. J Neurosci. 2010;30(9):3398–408. PubMed PMID: 20203199, Pubmed Central PMCID: PMC2866006, Epub 2010/03/06. eng.

103. Exley R, McIntosh JM, Marks MJ, Maskos U, Cragg SJ. Striatal alpha5 nicotinic receptor subunit regulates dopamine transmission in dorsal striatum. J Neurosci. 2012;32(7):2352–6. PubMed PMID: 22396410.
104. Picciotto MR, Zoli M, Rimondini R, Lena C, Marubio LM, Pich EM, et al. Acetylcholine receptors containing the beta2 subunit are involved in the reinforcing properties of nicotine. Nature. 1998;391(6663):173–7. PubMed PMID: 9428762, Epub 1998/01/15. eng.
105. Pons S, Fattore L, Cossu G, Tolu S, Porcu E, McIntosh JM, et al. Crucial role of alpha4 and alpha6 nicotinic acetylcholine receptor subunits from ventral tegmental area in systemic nicotine self-administration. J Neurosci. 2008;28(47):12318–27. PubMed PMID: 19020025, Pubmed Central PMCID: PMC2819191, Epub 2008/11/21. eng.
106. McGranahan TM, Patzlaff NE, Grady SR, Heinemann SF, Booker TK. Alpha4beta2 nicotinic acetylcholine receptors on dopaminergic neurons mediate nicotine reward and anxiety relief. J Neurosci. 2011;31(30):10891–902. PubMed PMID: 21795541.
107. Exley R, Maubourguet N, David V, Eddine R, Evrard A, Pons S, et al. Distinct contributions of nicotinic acetylcholine receptor subunit alpha4 and subunit alpha6 to the reinforcing effects of nicotine. Proc Natl Acad Sci U S A. 2011;108(18):7577–82. PubMed PMID: 21502501, Pubmed Central PMCID: PMC3088627, Epub 2011/04/20. eng.
108. Yizhar O, Fenno LE, Davidson TJ, Mogri M, Deisseroth K. Optogenetics in neural systems. Neuron. 2011;71(1):9–34. PubMed PMID: 21745635, Epub 2011/07/13. eng.
109. Threlfell S, Lalic T, Platt NJ, Jennings KA, Deisseroth K, Cragg SJ. Striatal dopamine release is triggered by synchronized activity in cholinergic interneurons. Neuron. 2012;75(1):58–64. PubMed PMID: 22794260, Epub 2012/07/17. eng.
110. Cachope R, Mateo Y, Mathur BN, Irving J, Wang HL, Morales M, et al. Selective activation of cholinergic interneurons enhances accumbal phasic dopamine release: setting the tone for reward processing. Cell Rep. 2012;2(1):33–41. PubMed PMID: 22840394, Pubmed Central PMCID: PMC3408582, Epub 2012/07/31. eng.
111. Ding JB, Guzman JN, Peterson JD, Goldberg JA, Surmeier DJ. Thalamic gating of corticostriatal signaling by cholinergic interneurons. Neuron. 2010;67(2):294–307. PubMed PMID: 20670836, Epub 2010/07/31. eng.
112. Drenan RM, Lester HA. Insights into the neurobiology of the nicotinic cholinergic system and nicotine addiction from mice expressing nicotinic receptors harboring gain-of-function mutations. Pharmacol Rev. 2012;64(4):869–79. PubMed PMID: 22885704, Pubmed Central PMCID: PMC3462994, Epub 2012/08/14. eng.
113. Steinlein OK, Bertrand D. Nicotinic receptor channelopathies and epilepsy. Pflugers Arch. 2010;460(2):495–503. PubMed PMID: 20016990, Epub 2009/12/18. eng.
114. Fonck C, Cohen BN, Nashmi R, Whiteaker P, Wagenaar DA, Rodrigues-Pinguet N, et al. Novel seizure phenotype and sleep disruptions in knock-in mice with hypersensitive alpha 4* nicotinic receptors. J Neurosci. 2005;25(49):11396–411. PubMed PMID: 16339034, Epub 2005/12/13. eng.
115. Mao D, Perry DC, Yasuda RP, Wolfe BB, Kellar KJ. The alpha4beta2alpha5 nicotinic cholinergic receptor in rat brain is resistant to up-regulation by nicotine in vivo. J Neurochem. 2008;104(2):446–56. PubMed PMID: 17961152.
116. Moretti M, Mugnaini M, Tessari M, Zoli M, Gaimarri A, Manfredi I, et al. A comparative study of the effects of the intravenous self-administration or subcutaneous minipump infusion of nicotine on the expression of brain neuronal nicotinic receptor subtypes. Mol Pharmacol. 2010;78(2):287–96. PubMed PMID: 20439469, Epub 2010/05/05. eng.
117. Lai A, Parameswaran N, Khwaja M, Whiteaker P, Lindstrom JM, Fan H, et al. Long-term nicotine treatment decreases striatal alpha 6* nicotinic acetylcholine receptor sites and function in mice. Mol Pharmacol. 2005;67(5):1639–47. PubMed PMID: 15681595, Epub 2005/02/01. eng.
118. Perry DC, Mao D, Gold AB, McIntosh JM, Pezzullo JC, Kellar KJ. Chronic nicotine differentially regulates alpha6- and beta3-containing nicotinic cholinergic receptors in rat brain. J Pharmacol Exp Ther. 2007;322(1):306–15. PubMed PMID: 17446303, Epub 2007/04/21. eng.

119. Mugnaini M, Garzotti M, Sartori I, Pilla M, Repeto P, Heidbreder CA, et al. Selective down-regulation of [(125)I]Y0-alpha-conotoxin MII binding in rat mesostriatal dopamine pathway following continuous infusion of nicotine. Neuroscience. 2006;137(2):565–72. PubMed PMID: 16289885, Epub 2005/11/18. eng.
120. Nashmi R, Xiao C, Deshpande P, McKinney S, Grady SR, Whiteaker P, et al. Chronic nicotine cell specifically upregulates functional alpha 4* nicotinic receptors: basis for both tolerance in midbrain and enhanced long-term potentiation in perforant path. J Neurosci. 2007;27(31):8202–18. PubMed PMID: 17670967, Epub 2007/08/03. eng.
121. Grilli M, Parodi M, Raiteri M, Marchi M. Chronic nicotine differentially affects the function of nicotinic receptor subtypes regulating neurotransmitter release. J Neurochem. 2005;93(5):1353–60. PubMed PMID: 15934954, Epub 2005/06/07. eng.
122. Marks MJ, Grady SR, Collins AC. Downregulation of nicotinic receptor function after chronic nicotine infusion. J Pharmacol Exp Ther. 1993;266(3):1268–76. PubMed PMID: 8371136, Epub 1993/09/01. eng.
123. Jacobs I, Anderson DJ, Surowy CS, Puttfarcken PS. Differential regulation of nicotinic receptor-mediated neurotransmitter release following chronic (–)-nicotine administration. Neuropharmacology. 2002;43(5):847–56. PubMed PMID: 12384170, Epub 2002/10/18. eng.
124. Perez XA, Bordia T, McIntosh JM, Grady SR, Quik M. Long-term nicotine treatment differentially regulates striatal alpha6alpha4beta2* and alpha6(nonalpha4)beta2* nAChR expression and function. Mol Pharmacol. 2008;74(3):844–53. PubMed PMID: 18583454, Pubmed Central PMCID: 2847502.
125. Quik M, Perez XA, Grady SR. Role of alpha6 nicotinic receptors in CNS dopaminergic function: relevance to addiction and neurological disorders. Biochem Pharmacol. 2011;82(8):873–82. PubMed PMID: 21684266, Pubmed Central PMCID: 3264546.
126. Baker LK, Mao D, Chi H, Govind AP, Vallejo YF, Iacoviello M, et al. Intermittent nicotine exposure upregulates nAChRs in VTA dopamine neurons and sensitises locomotor responding to the drug. Eur J Neurosci. 2013;37(6):1004–11. PubMed PMID: 23331514, Pubmed Central PMCID: PMC3604051, Epub 2013/01/22. eng.
127. Perez XA, Ly J, McIntosh JM, Quik M. Long-term nicotine exposure depresses dopamine release in nonhuman primate nucleus accumbens. J Pharmacol Exp Ther. 2012;342(2):335–44. PubMed PMID: 22562772, Pubmed Central PMCID: PMC3400796, Epub 2012/05/09. eng.
128. Hilario MR, Turner JR, Blendy JA. Reward sensitization: effects of repeated nicotine exposure and withdrawal in mice. Neuropsychopharmacology. 2012;37(12):2661–70. PubMed PMID: 22828747, Pubmed Central PMCID: PMC3473332, Epub 2012/07/26. eng.
129. Bordia T, McIntosh JM, Quik M. The nicotine-mediated decline in l-dopa-induced dyskinesias is associated with a decrease in striatal dopamine release. J Neurochem. 2013;1. PubMed PMID: 23373725, Epub 2013/02/05. Eng.
130. Sulzer D. How addictive drugs disrupt presynaptic dopamine neurotransmission. Neuron. 2011;69(4):628–49. PubMed PMID: 21338876, Pubmed Central PMCID: PMC3065181, Epub 2011/02/23. eng.
131. Changeux JP. The nicotinic acetylcholine receptor: the founding father of the pentameric ligand-gated ion channel superfamily. J Biol Chem. 2012;287(48):40207–15. PubMed PMID: 23038257, Pubmed Central PMCID: PMC3504736, Epub 2012/10/06. eng.
132. Araud T, Wonnacott S, Bertrand D. Associated proteins: The universal toolbox controlling ligand gated ion channel function. Biochem Pharmacol. 2010;80(2):160–9. PubMed PMID: 20346921, Epub 2010/03/30. eng.
133. Matta SG, Balfour DJ, Benowitz NL, Boyd RT, Buccafusco JJ, Caggiula AR, et al. Guidelines on nicotine dose selection for in vivo research. Psychopharmacology (Berl). 2007;190(3):269–319. PubMed PMID: 16896961.
134. Pidoplichko VI, DeBiasi M, Williams JT, Dani JA. Nicotine activates and desensitizes midbrain dopamine neurons. Nature. 1997;390(6658):401–4. PubMed PMID: 9389479, Epub 1997/12/06. eng.
135. Bailey CD, De Biasi M, Fletcher PJ, Lambe EK. The nicotinic acetylcholine receptor alpha5 subunit plays a key role in attention circuitry and accuracy. J Neurosci. 2010;30(27):9241–52. PubMed PMID: 20610759, Pubmed Central PMCID: 3004929.

136. Grady SR, Wageman CR, Patzlaff NE, Marks MJ. Low concentrations of nicotine differentially desensitize nicotinic acetylcholine receptors that include alpha5 or alpha6 subunits and that mediate synaptosomal neurotransmitter release. Neuropharmacology. 2012;62(5–6):1935–43. PubMed PMID: 22239849, Pubmed Central PMCID: 3278500.
137. Poorthuis RB, Bloem B, Verhoog MB, Mansvelder HD. Layer-specific interference with cholinergic signaling in the prefrontal cortex by smoking concentrations of nicotine. J Neurosci. 2013;33(11):4843–53.
138. Liu L, Zhao-Shea R, McIntosh JM, Gardner PD, Tapper AR. Nicotine persistently activates ventral tegmental area dopaminergic neurons via nicotinic acetylcholine receptors containing alpha4 and alpha6 subunits. Mol Pharmacol. 2012;81(4):541–8. PubMed PMID: 22222765, Pubmed Central PMCID: PMC3310415, Epub 2012/01/10. eng.
139. Voorn P, Vanderschuren LJ, Groenewegen HJ, Robbins TW, Pennartz CM. Putting a spin on the dorsal-ventral divide of the striatum. Trends Neurosci. 2004;27(8):468–74. PubMed PMID: 15271494, Epub 2004/07/24. eng.
140. Bjorklund A, Dunnett SB. Dopamine neuron systems in the brain: an update. Trends Neurosci. 2007;30(5):194–202. PubMed PMID: 17408759, Epub 2007/04/06. eng.
141. Grady SR, Murphy KA, Cao J, Marks MJ, McIntosh JM, Collins AC. Characterization of nicotinic agonist-induced [3H]dopamine release from synaptosomes prepared from four mouse brain regions. J Pharmacol Exp Ther. 2002;301:651–60.
142. Livingstone PD, Srinivasan J, Kew JN, Dawson LA, Gotti C, Moretti M, et al. Alpha7 and non-alpha7 nicotinic acetylcholine receptors modulate dopamine release in vitro and in vivo in the rat prefrontal cortex. Eur J Neurosci. 2009;29(3):539–50. PubMed PMID: 19187266. Epub 2009/02/04. eng.
143. Cao YJ, Surowy CS, Puttfarcken PS. Different nicotinic acetylcholine receptor subtypes mediating striatal and prefrontal cortical [3H]dopamine release. Neuropharmacology. 2005;48(1):72–9. PubMed PMID: 15617729, Epub 2004/12/25. eng.
144. Ono JK, Salvaterra PM. Snake alpha-toxin effects on cholinergic and noncholinergic responses of Aplysia californica neurons. J Neurosci. 1981;1(3):259–70. PubMed PMID: 6114995, Epub 1981/03/01. eng.
145. Frazier CJ, Rollins YD, Breese CR, Leonard S, Freedman R, Dunwiddie TV. Acetylcholine activates an alpha-bungarotoxin-sensitive nicotinic current in rat hippocampal interneurons, but not pyramidal cells. J Neurosci. 1998;18(4):1187–95. PubMed PMID: 9454829, Epub 1998/03/14. eng.
146. Guo JZ, Tredway TL, Chiappinelli VA. Glutamate and GABA release are enhanced by different subtypes of presynaptic nicotinic receptors in the lateral geniculate nucleus. J Neurosci. 1998;18(6):1963–9. PubMed PMID: 9482782, Epub 1998/04/18. eng.
147. Liu YB, Guo JZ, Chiappinelli VA. Nicotinic receptor-mediated biphasic effect on neuronal excitability in chick lateral spiriform neurons. Neuroscience. 2007;148(4):1004–14.
148. Luo R, Janssen MJ, Partridge JG, Vicini S. Direct and GABA-mediated indirect effects of nicotinic ACh receptor agonists on striatal neurones. J Physiol. 2013;591(Pt 1): 203–17. PubMed PMID: 23045343, Pubmed Central PMCID: PMC3630781, Epub 2012/10/10. eng.
149. Sooksawate T, Yanagawa Y, Isa T. Cholinergic responses in GABAergic and non-GABAergic neurons in the intermediate gray layer of mouse superior colliculus. Eur J Neurosci. 2012;36(4):2440–51. PubMed PMID: 22712760, Epub 2012/06/21. eng.
150. Tang AH, Karson MA, Nagode DA, McIntosh JM, Uebele VN, Renger JJ, et al. Nerve terminal nicotinic acetylcholine receptors initiate quantal GABA release from perisomatic interneurons by activating axonal T-type (Cav3) Ca(2)(+) channels and Ca(2)(+) release from stores. J Neurosci. 2011;31(38):13546–61. PubMed PMID: 21940446, Pubmed Central PMCID: PMC3353409, Epub 2011/09/24. eng.
151. Lu Y, Grady S, Marks MJ, Picciotto M, Changeux JP, Collins AC. Pharmacological characterization of nicotinic receptor-stimulated GABA release from mouse brain synaptosomes. J Pharmacol Exp Ther. 1998;287(2):648–57. PubMed PMID: 9808692, Epub 1998/11/10. eng.
152. Grilli M, Zappettini S, Raiteri L, Marchi M. Nicotinic and muscarinic cholinergic receptors coexist on GABAergic nerve endings in the mouse striatum and interact in modulating

GABA release. Neuropharmacology. 2009;56(3):610–4. PubMed PMID: 19027759, Epub 2008/11/26. eng.
153. McClure-Begley TD, King NM, Collins AC, Stitzel JA, Wehner JM, Butt CM. Acetylcholine-stimulated [3H]GABA release from mouse brain synaptosomes is modulated by alpha4beta2 and alpha4alpha5beta2 nicotinic receptor subtypes. Mol Pharmacol. 2009;75(4):918–26. PubMed PMID: 19139153, Pubmed Central PMCID: 2684932.
154. Whiteaker P, Marks MJ, Grady SR, Lu Y, Picciotto MR, Changeux JP, et al. Pharmacological and null mutation approaches reveal nicotinic receptor diversity. Eur J Pharmacol. 2000;393(1–3):123–35. PubMed PMID: 10771005, Epub 2000/04/20. eng.
155. McClure-Begley TD, Wageman CR, Grady SR, Marks MJ, McIntosh JM, Collins AC, et al. A novel alpha-conotoxin MII-sensitive nicotinic acetylcholine receptor modulates [(3) H]-GABA release in the superficial layers of the mouse superior colliculus. J Neurochem. 2012;122(1):48–57. PubMed PMID: 22506481.
156. Yang K, Buhlman L, Khan GM, Nichols RA, Jin G, McIntosh JM, et al. Functional nicotinic acetylcholine receptors containing alpha6 subunits are on GABAergic neuronal boutons adherent to ventral tegmental area dopamine neurons. J Neurosci. 2011;31(7):2537–48. PubMed PMID: 21325521, Pubmed Central PMCID: PMC3081713, Epub 2011/02/18. eng.
157. Zappettini S, Grilli M, Lagomarsino F, Cavallero A, Fedele E, Marchi M. Presynaptic nicotinic alpha7 and non-alpha7 receptors stimulate endogenous GABA release from rat hippocampal synaptosomes through two mechanisms of action. PLoS One. 2011;6(2):e16911. PubMed PMID: 21346795, Pubmed Central PMCID: PMC3034729, Epub 2011/02/25. eng.
158. Review MA. Neurobiology of nicotine dependence. Philos Trans R Soc Lond B Biol Sci. 2008;363(1507):3159–68. PubMed PMID: 18640919, Pubmed Central PMCID: PMC2607327, Epub 2008/07/22. eng.
159. Gray R, Rajan AS, Radcliffe KA, Yakehiro M, Dani JA. Hippocampal synaptic transmission enhanced by low concentrations of nicotine. Nature. 1996;383(6602):713–6. PubMed PMID: 8878480, Epub 1996/10/24. eng.
160. Mao D, Gallagher K, McGehee DS. Nicotine potentiation of excitatory inputs to ventral tegmental area dopamine neurons. J Neurosci. 2011;31(18):6710–20. PubMed PMID: 21543600, Pubmed Central PMCID: PMC3118498, Epub 2011/05/06. eng.
161. Garduno J, Galindo-Charles L, Jimenez-Rodriguez J, Galarraga E, Tapia D, Mihailescu S, et al. Presynaptic alpha4beta2 nicotinic acetylcholine receptors increase glutamate release and serotonin neuron excitability in the dorsal raphe nucleus. J Neurosci. 2012;32(43):15148–57. PubMed PMID: 23100436, Epub 2012/10/27. eng.
162. Rousseau SJ, Jones IW, Pullar IA, Wonnacott S. Presynaptic alpha7 and non-alpha7 nicotinic acetylcholine receptors modulate [3H]d-aspartate release from rat frontal cortex in vitro. Neuropharmacology. 2005;49(1):59–72. PubMed PMID: 15992581, Epub 2005/07/05. eng.
163. Fabian-Fine R, Skehel P, Errington ML, Davies HA, Sher E, Stewart MG, et al. Ultrastructural distribution of the alpha7 nicotinic acetylcholine receptor subunit in rat hippocampus. J Neurosci. 2001;21(20):7993–8003. PubMed PMID: 11588172, Epub 2001/10/06. eng.
164. Jones IW, Wonnacott S. Precise localization of alpha7 nicotinic acetylcholine receptors on glutamatergic axon terminals in the rat ventral tegmental area. J Neurosci. 2004;24(50):11244–52. PubMed PMID: 15601930, Epub 2004/12/17. eng.
165. Omote H, Miyaji T, Juge N, Moriyama Y. Vesicular neurotransmitter transporter: bioenergetics and regulation of glutamate transport. Biochemistry. 2011;50(25):5558–65. PubMed PMID: 21612282.
166. Miyaji T, Echigo N, Hiasa M, Senoh S, Omote H, Moriyama Y. Identification of a vesicular aspartate transporter. Proc Natl Acad Sci U S A. 2008;105(33):11720–4. PubMed PMID: 18695252, Pubmed Central PMCID: PMC2575331, Epub 2008/08/13. eng.
167. Fleck MW, Barrionuevo G, Palmer AM. Synaptosomal and vesicular accumulation of L-glutamate, L-aspartate and D-aspartate. Neurochem Int. 2001;39(3):217–25. PubMed PMID: 11434980, Epub 2001/07/04. eng.
168. Wonnacott S, Barik J, Dickinson J, Jones IW. Nicotinic receptors modulate transmitter cross talk in the CNS: nicotinic modulation of transmitters. J Mol Neurosci. 2006;30(1–2):137–40. PubMed PMID: 17192660, Epub 2006/12/29. eng.

169. Marchi M, Risso F, Viola C, Cavazzani P, Raiteri M. Direct evidence that release-stimulating alpha7* nicotinic cholinergic receptors are localized on human and rat brain glutamatergic axon terminals. J Neurochem. 2002;80(6):1071–8. PubMed PMID: 11953457, Epub 2002/04/16. eng.
170. McGehee DS, Heath MJ, Gelber S, Devay P, Role LW. Nicotine enhancement of fast excitatory synaptic transmission in CNS by presynaptic receptors. Science. 1995;269(5231):1692–6. PubMed PMID: 7569895. Epub 1995/09/22. eng.
171. Alkondon M, Rocha ES, Maelicke A, Albuquerque EX. Diversity of nicotinic acetylcholine receptors in rat brain. V alpha-Bungarotoxin-sensitive nicotinic receptors in olfactory bulb neurons and presynaptic modulation of glutamate release. J Pharmacol Exp Ther. 1996;278(3):1460–71. PubMed PMID: 8819534, Epub 1996/09/01. eng.
172. Kaiser S, Wonnacott S. Alpha-bungarotoxin-sensitive nicotinic receptors indirectly modulate [(3)H]dopamine release in rat striatal slices via glutamate release. Mol Pharmacol. 2000;58(2):312–8. PubMed PMID: 10908298. Epub 2000/07/25. eng.
173. Mansvelder HD, McGehee DS. Long-term potentiation of excitatory inputs to brain reward areas by nicotine. Neuron. 2000;27(2):349–57. PubMed PMID: 10985354, Epub 2000/09/14. eng.
174. Dickinson JA, Kew JN, Wonnacott S. Presynaptic alpha 7- and beta 2-containing nicotinic acetylcholine receptors modulate excitatory amino acid release from rat prefrontal cortex nerve terminals via distinct cellular mechanisms. Mol Pharmacol. 2008;74(2):348–59. PubMed PMID: 18445710, Epub 2008/05/01. eng.
175. Gioanni Y, Rougeot C, Clarke PB, Lepouse C, Thierry AM, Vidal C. Nicotinic receptors in the rat prefrontal cortex: increase in glutamate release and facilitation of mediodorsal thalamo-cortical transmission. Eur J Neurosci. 1999;11(1):18–30. PubMed PMID: 9987008, Epub 1999/02/13. eng.
176. Lopez E, Arce C, Vicente S, Oset-Gasque MJ, Gonzalez MP. Nicotinic receptors mediate the release of amino acid neurotransmitters in cultured cortical neurons. Cereb Cortex. 2001;11(2):158–63. PubMed PMID: 11208670. Epub 2001/02/24. eng.
177. Mansvelder HD, Keath JR, McGehee DS. Synaptic mechanisms underlie nicotine-induced excitability of brain reward areas. Neuron. 2002;33(6):905–19. PubMed PMID: 11906697, Epub 2002/03/22. eng.
178. Marchi M, Zappettini S, Olivero G, Pittaluga A, Grilli M. Chronic nicotine exposure selectively activates a carrier-mediated release of endogenous glutamate and aspartate from rat hippocampal synaptosomes. Neurochem Int. 2012;60(6):622–30. PubMed PMID: 22417725, Epub 2012/03/16. eng.
179. Szabadi E. Functional neuroanatomy of the central noradrenergic system. J Psychopharmacol. 2013;12. PubMed PMID: 23761387.
180. Eschenko O, Evrard HC, Neves RM, Beyerlein M, Murayama Y, Logothetis NK. Tracing of noradrenergic projections using manganese-enhanced MRI. Neuroimage. 2012;59(4):3252–65. PubMed PMID: 22119646.
181. Le Novere N, Zoli M, Changeux JP. Neuronal nicotinic receptor alpha 6 subunit mRNA is selectively concentrated in catecholaminergic nuclei of the rat brain. Eur J Neurosci. 1996;8(11):2428–39. PubMed PMID: 8950106, Epub 1996/11/01. eng.
182. Lena C, de D'Exaerde Kerchove A, Cordero-Erausquin M, Le Novere N, del Mar Arroyo-Jimenez M, Changeux JP. Diversity and distribution of nicotinic acetylcholine receptors in the locus ceruleus neurons. Proc Natl Acad Sci U S A. 1999;96(21):12126–31. PubMed PMID: 10518587, Pubmed Central PMCID: PMC18423, Epub 1999/10/16. eng.
183. Winzer-Serhan UH, Leslie FM. Codistribution of nicotinic acetylcholine receptor subunit alpha3 and beta4 mRNAs during rat brain development. J Comp Neurol. 1997;386(4):540–54. PubMed PMID: 9378850. Epub 1997/10/23 22:34. eng.
184. Vincler MA, Eisenach JC. Immunocytochemical localization of the alpha3, alpha4, alpha5, alpha7, beta2, beta3 and beta4 nicotinic acetylcholine receptor subunits in the locus coeruleus of the rat. Brain Res. 2003;974(1–2):25–36. PubMed PMID: 12742621, Epub 2003/05/14. eng.
185. Mitchell SN. Role of the locus coeruleus in the noradrenergic response to a systemic administration of nicotine. Neuropharmacology. 1993;32(10):937–49. PubMed PMID: 8295716, Epub 1993/10/01. eng.

186. Sacaan AI, Dunlop JL, Lloyd GK. Pharmacological characterization of neuronal acetylcholine gated ion channel receptor-mediated hippocampal norepinephrine and striatal dopamine release from rat brain slices. J Pharmacol Exp Ther. 1995;274(1):224–30. PubMed PMID: 7542334, Epub 1995/07/01. eng.
187. Clarke PB, Reuben M. Release of [3H]-noradrenaline from rat hippocampal synaptosomes by nicotine: mediation by different nicotinic receptor subtypes from striatal [3H]-dopamine release. Br J Pharmacol. 1996;117(4):595–606. PubMed PMID: 8646402, Pubmed Central PMCID: PMC1909326, Epub 1996/02/01. eng.
188. Luo S, Kulak JM, Cartier GE, Jacobsen RB, Yoshikami D, Olivera BM, et al. Alpha-conotoxin AuIB selectively blocks alpha3 beta4 nicotinic acetylcholine receptors and nicotine-evoked norepinephrine release. J Neurosci. 1998;18(21):8571–9. PubMed PMID: 9786965. Epub 1998/10/24. eng.
189. Sershen H, Balla A, Lajtha A, Vizi ES. Characterization of nicotinic receptors involved in the release of noradrenaline from the hippocampus. Neuroscience. 1997;77(1):121–30. PubMed PMID: 9044380, Epub 1997/03/01. eng.
190. Azam L, McIntosh JM. Characterization of nicotinic acetylcholine receptors that modulate nicotine-evoked [3H]norepinephrine release from mouse hippocampal synaptosomes. Mol Pharmacol. 2006;70(3):967–76. PubMed PMID: 16735605.
191. Li X, Rainnie DG, McCarley RW, Greene RW. Presynaptic nicotinic receptors facilitate monoaminergic transmission. J Neurosci. 1998;18(5):1904–12. PubMed PMID: 9465015, Epub 1998/03/07. eng.
192. Mogg AJ, Whiteaker P, McIntosh JM, Marks M, Collins AC, Wonnacott S. Methyllycaconitine is a potent antagonist of alpha-conotoxin-MII-sensitive presynaptic nicotinic acetylcholine receptors in rat striatum. J Pharmacol Exp Ther. 2002;302(1):197–204. PubMed PMID: 12065717, Epub 2002/06/18. eng.
193. Anderson DJ, Puttfarcken PS, Jacobs I, Faltynek C. Assessment of nicotinic acetylcholine receptor-mediated release of [(3)H]-norepinephrine from rat brain slices using a new 96-well format assay. Neuropharmacology. 2000;39(13):2663–72. PubMed PMID: 11044736, Epub 2000/10/25. eng.
194. Amtage F, Neughebauer B, McIntosh JM, Freiman T, Zentner J, Feuerstein TJ, et al. Characterization of nicotinic receptors inducing noradrenaline release and absence of nicotinic autoreceptors in human neocortex. Brain Res Bull. 2004;62(5):413–23. PubMed PMID: 15168907, Epub 2004/06/01. eng.
195. Kennett A, Heal DJ, Wonnacott S. Pharmacological differences between rat frontal cortex and hippocampus in the nicotinic modulation of noradrenaline release implicate distinct receptor subtypes. Nicotine Tob Res. 2012;14(11):1339–45. PubMed PMID: 22614547, Epub 2012/05/23. eng.
196. O'Leary KT, Leslie FM. Developmental regulation of nicotinic acetylcholine receptor-mediated [3H]norepinephrine release from rat cerebellum. J Neurochem. 2003;84(5):952–9. PubMed PMID: 12603820, Epub 2003/02/27. eng.
197. Wess J, Eglen RM, Gautam D. Muscarinic acetylcholine receptors: mutant mice provide new insights for drug development. Nat Rev Drug Discov. 2007;6(9):721–33. PubMed PMID: 17762886.
198. Picciotto MR, Higley MJ, Mineur YS. Acetylcholine as a neuromodulator: cholinergic signaling shapes nervous system function and behavior. Neuron. 2012;76(1):116–29. PubMed PMID: 23040810, Pubmed Central PMCID: 3466476.
199. Yan Z, Surmeier DJ. Muscarinic (m2/m4) receptors reduce N- and P-type Ca2+ currents in rat neostriatal cholinergic interneurons through a fast, membrane-delimited, G-protein pathway. J Neurosci. 1996;16(8):2592–604. PubMed PMID: 8786435, Epub 1996/04/15. eng.
200. Zhang W, Basile AS, Gomeza J, Volpicelli LA, Levey AI, Wess J. Characterization of central inhibitory muscarinic autoreceptors by the use of muscarinic acetylcholine receptor knock-out mice. J Neurosci. 2002;22(5):1709–17. PubMed PMID: 11880500, Epub 2002/03/07. eng.
201. Marchi M, Paudice P, Raiteri M. Autoregulation of acetylcholine release in isolated hippocampal nerve endings. Eur J Pharmacol. 1981;73(1):75–9. PubMed PMID: 7318890, Epub 1981/07/17. eng.

202. Azam L, Winzer-Serhan U, Leslie FM. Co-expression of alpha7 and beta2 nicotinic acetylcholine receptor subunit mRNAs within rat brain cholinergic neurons. Neuroscience. 2003;119(4):965–77. PubMed PMID: 12831856, Epub 2003/07/02. eng.
203. Brown DA, Docherty RJ, Halliwell JV. The action of cholinomimetic substances on impulse conduction in the habenulointerpeduncular pathway of the rat in vitro. J Physiol. 1984;353:101–9. PubMed PMID: 6481621, Pubmed Central PMCID: PMC1193296, Epub 1984/08/01. eng.
204. Mulle C, Vidal C, Benoit P, Changeux JP. Existence of different subtypes of nicotinic acetylcholine receptors in the rat habenulo-interpeduncular system. J Neurosci. 1991;11(8):2588–97. PubMed PMID: 1869929, Epub 1991/08/01. eng.
205. Grady SR, Meinerz NM, Cao J, Reynolds AM, Picciotto MR, Changeux J-P, et al. Nicotinic agonists stimulate acetylcholine release from mouse interpeduncular nucleus: a function mediated by a different nAChR than dopamine release from striatum. J Neurochem. 2001; 76:258–68.
206. Wilkie GI, Hutson PH, Stephens MW, Whiting P, Wonnacott S. Hippocampal nicotinic autoreceptors modulate acetylcholine release. Biochem Soc Trans. 1993;21(2):429–31. PubMed PMID: 8359505, Epub 1993/05/01. eng.
207. Araujo DM, Lapchak PA, Collier B, Quirion R. Characterization of N-[3H]methylcarbamylcholine binding sites and effect of N-methylcarbamylcholine on acetylcholine release in rat brain. J Neurochem. 1988;51(1):292–9. PubMed PMID: 3379410, Epub 1988/07/01. eng.
208. Lapchak PA, Araujo DM, Quirion R, Collier B. Effect of chronic nicotine treatment on nicotinic autoreceptor function and N-[3H]methylcarbamylcholine binding sites in the rat brain. J Neurochem. 1989;52(2):483–91. PubMed PMID: 2911027, Epub 1989/02/01. eng.
209. Fenster CP, Rains MF, Noerager B, Quick MW, Lester RA. Influence of subunit composition on desensitization of neuronal acetylcholine receptors at low concentrations of nicotine. J Neurosci. 1997;17(15):5747–59. PubMed PMID: 9221773, Epub 1997/08/01. eng.
210. Giniatullin R, Nistri A, Yakel JL. Desensitization of nicotinic ACh receptors: shaping cholinergic signaling. Trends Neurosci. 2005;28(7):371–8. PubMed PMID: 15979501, Epub 2005/06/28. eng.
211. Carlson J, Armstrong B, Switzer III RC, Ellison G. Selective neurotoxic effects of nicotine on axons in fasciculus retroflexus further support evidence that this a weak link in brain across multiple drugs of abuse. Neuropharmacology. 2000;39(13):2792–8. PubMed PMID: 11044749, Epub 2000/10/25. eng.
212. Ellison G. Neural degeneration following chronic stimulant abuse reveals a weak link in brain, fasciculus retroflexus, implying the loss of forebrain control circuitry. Eur Neuropsychopharmacol. 2002;12(4):287–97. PubMed PMID: 12126867, Epub 2002/07/20. eng.
213. Britt JP, McGehee DS. Presynaptic opioid and nicotinic receptor modulation of dopamine overflow in the nucleus accumbens. J Neurosci. 2008;28(7):1672–81. PubMed PMID: 18272687, Epub 2008/02/15. eng.
214. Marchi M, Grilli M. Presynaptic nicotinic receptors modulating neurotransmitter release in the central nervous system: functional interactions with other coexisting receptors. Prog Neurobiol. 2010;92(2):105–11.
215. Corrigall WA. Hypocretin mechanisms in nicotine addiction: evidence and speculation. Psychopharmacology (Berl). 2009;206(1):23–37. PubMed PMID: 19529922.
216. Bruijnzeel AW. Tobacco addiction and the dysregulation of brain stress systems. Neurosci Biobehav Rev. 2012;36(5):1418–41. PubMed PMID: 22405889, Pubmed Central PMCID: 3340450.
217. Tweed JO, Hsia SH, Lutfy K, Friedman TC. The endocrine effects of nicotine and cigarette smoke. Trends Endocrinol Metab. 2012;23(7):334–42. PubMed PMID: 22561025, Pubmed Central PMCID: 3389568.

Chapter 5
Functional Distribution and Regulation of Neuronal Nicotinic ACh Receptors in the Mammalian Brain

Jerrel L. Yakel

Abstract The neurotransmitter acetylcholine (ACh) can regulate neuronal excitability throughout the nervous system by acting on the cys-loop cation-conducting ligand-gated nicotinic ACh receptor channels (nAChRs). These receptors are widely distributed throughout the nervous system, being expressed on neurons and nonneuronal cells where they participate in a variety of physiological responses. In the mammalian brain, nine different subunits have been discovered thus far, which assemble into pentameric complexes with much diversity. The neuronal subtypes of these receptors, primarily composed of the α7 and non-α7 subtypes (e.g. α4β2 and α3β4), are involved in a variety of neurobehavioral processes such as anxiety, the central processing of pain, food intake, nicotine-seeking behavior, and cognitive functions. Neuronal nAChR dysfunction is involved in the pathophysiology of many neurological disorders and diseases including (but not limited to) Alzheimer's and Parkinson's diseases, schizophrenia, and epilepsy. Here I will briefly discuss the functional makeup and expression of nAChRs in the mammalian brain, and the role that they play in these various circuits, in normal function, and in disease.

Keywords Acetylcholine • Synaptic plasticity • Neurotransmitter • Allosteric modulation • Neurological disease

1 Introduction

The nicotinic acetylcholine receptors (nAChRs) are in the superfamily of cys-loop receptors, which also includes the serotonin 5-HT_3, $GABA_A$ and $GABA_C$, and glycine receptors [1, 2]. The nAChRs are widely expressed in the nervous system where they participate in a variety of physiological functions, including regulating excitability

J.L. Yakel, Ph.D. (✉)
Laboratory of Neurobiology, National Institute of Environmental Health Sciences, National Institutes of Health, Department of Health and Human Services, 111 T.W. Alexander Drive, P.O Box 12233, Research Triangle Park, NC 27709, USA
e-mail: yakel@niehs.nih.gov

R.A.J. Lester (ed.), *Nicotinic Receptors*, The Receptors 26,
DOI 10.1007/978-1-4939-1167-7_5

and neurotransmitter release. In the mammalian brain, all least nine different nAChR subunits are known to exist (α2-7 and β2-4), which combine into functional pentameric complexes of either homo- or heteromeric receptors [1, 3–11]. While there is much subtype diversity due to the assembly of various combinations of subunits, the primary subtypes that have been found to be functionally expressed are composed of the α7 subunit (which is usually thought to be homomeric but can combine with other subunits under certain conditions; see below), and the α4β2 and α3β4 subtypes of non-α7 receptors [11–16]. In addition the α2, α5, α6, and β3 subunits play various roles in various brain regions (discussed below). A major endeavor in the nAChR field is to determine the precise molecular makeup of these receptors since this is the critical determinant of pharmacological and physiological properties of these channels, and how they regulate brain circuit excitability.

While nAChRs in the brain were initially thought to serve primarily presynaptic roles [11, 13, 17, 18], clear evidence has existed for over 15 years [12] for the somatodendritic (i.e. postsynaptic and/or extrasynaptic) localization of these receptors, and that they mediate nAChR-mediated postsynaptic responses (both α7 and non-α7; [19]). While α7 nAChR-mediated fast synaptic responses have previously been observed in hippocampal interneurons [20, 21], using optogenetics to stimulate cholinergic input directly into the hippocampus has only provided evidence to date for slower α4β2 nAChR-mediated postsynaptic responses in recordings from either interneurons [22] or pyramidal cells [19]. Various subtypes of nAChRs are thought to mediate synaptic transmission in other areas of the brain as well, including (but not limited to) the visual cortex, cortical interneurons, supraoptic nucleus, and thalamic nuclei [23–26]. Therefore the nAChRs are located both presynaptically, where they can mediate the release of other neurotransmitters, and postsynaptically, where they can directly mediate synaptic transmission. In addition as discussed below, nAChRs are expressed in a variety of nonneuronal cells in the brain (e.g. astrocytes and microglia) and elsewhere [27].

2 Ligand Binding and Gating

The nAChRs are pentameric assemblies of five subunits, with each subunit arranged around the central cation-conducting pore. Each subunit has a long extracellular N-terminal domain, four transmembrane segments (with the second segment of each subunit lining the pore of the channel), an intracellular loop between the third and fourth segments, and a short C-terminal end. Ligands bind at the interface between two subunits in a hydrophobic pocket, and at least two ligands are needed to bind to and open the channel. While these channels pass monovalent cations, they have a differential permeability to calcium [28]. With the 4 Å resolution of the *Torpedo* nAChR, and the crystal structure of the related ACh binding proteins (originally isolated from molluscs), much has been learned about the structure of the ligand binding domain and the channel pore, as well as major structural rearrangements that may confer channel opening. Many recent reviews have discussed these details, which I will not go into here [2, 9, 29]. To understand how these receptors function, it is important to know the structure and transition of the receptor in its various states, including the closed (in the absence of agonist), the open (in the presence of agonist) and the desensitized states (high-affinity ligand-bound but nonconducting state of the channel). For additional information please refer to Chap. 2.

3 Regional Expression and Functional Assembly of Various Subtypes in the Brain

The major subtypes of nAChRs expressed in the mammalian brain consist of the α7, α4β2, and α3β4 subtypes. However, there is much diversity in the functional expression of the various subtypes of receptors due to co-assembly with other subunits that are expressed differentially throughout the brain, and different receptor stoichiometries. For example, native α4β2 and α3β4 receptors both have different stoichiometries, either with two α/three β subunits, or three α/two β subunits [11]. For the α4β2 receptors, the configuration with the two α4 subunits has a higher affinity for agonists than the three α4 subunit configuration [30, 31], and different calcium permeability [32]. For the α3β4 receptor, the two stoichiometries have similar affinity, but differ in zinc sensitivity and single channel conductance [33, 34]. One of the major challenges in the field is to identify the molecular makeup of functional nAChRs; the potential combinations are high, and the pharmacological tools are lacking at this point to unambiguously identify the subunit makeup of particular nAChR subtypes.

The different nAChR subunits have different patterns of functional expression throughout the brain. Although the α2 subunit is sparsely expressed in the brain, it can be found in many regions, including (but not limited to) the interpeduncular nucleus (IPN), amygdala, hippocampal interneurons, and cortex [8, 35, 36]. Functional α2-containing receptors appear to be expressed on GABAergic interneurons (particularly in the stratum oriens layer) in the hippocampus [15, 37, 38] and IPN [39], and have been found to be involved in synaptic plasticity in the hippocampus [40, 41].

The α3β4 nAChR is primarily known as a ganglionic receptor in the peripheral nervous system, but it is also expressed in the brain in a variety of areas, including (but not limited to) the IPN and median habenula [11, 16]. Recent data has suggested that functional nAChRs in the ventrolateral median habenula also contain the α4 subunit [42]. Interestingly, genes encoding the α3, α5, and β4 nAChR subunits form a cluster on chromosome 15q24-25.1, allowing for highly coordinated regulation [10, 11, 29]. The α4β2 receptor subtype was initially found (through immunoprecipitation in a variety of species) to be the major subtype of nAChR in the brain, where it comprises 90 % of the high nicotine affinity binding sites [11, 43]. As will be discussed below, many other nAChR subunits appear to combine with α3β4 and α4β2 receptors.

Some nAChR subunits, including the α5 and β3 subunits, have been referred to as orphan or accessory subunits since they do not appear to participate in forming the binding site [11, 44, 45]. Instead, these subunits are thought to form functional triplet nAChRs when co-assembled with another α and β subunit [43]. For example, the α5 subunit can co-assemble with either the α3β4 or α4β2 nAChRs and alter functional properties, including single-channel conductance and calcium permeability [15, 28, 43, 46–49]. More recently, a prominent role for the α5 subunit has been found relating to smoking behavior; a particular mutation in this subunit in humans is linked with (in animal models) increased risk of nicotine dependence and smoking-related diseases [11, 50]. The β3 subunit is also thought to have a dominant-negative role in the expression of several other nAChR subtypes [43, 51].

The α6 nAChR subunit, which is expressed mainly in the mesostriatal dopamine neurons [11, 18, 29, 46, 52], was initially thought to be an orphan subunit due to the difficulties in expressing α6-containing receptors in heterologous expression systems [29, 43, 53]. The α6 subunit combines with many other subunits to form a wide array of diverse nAChR subtypes [18], which (combined with the lack of specific pharmacological agents) makes it extremely difficult to pharmacologically determine the molecular makeup of these receptors. Nevertheless, α6β2 receptors (along with α4β2) have been shown to play an important role in dopamine neurons and in the addictive and rewarding aspect of nicotine [4, 54–58]. In the basal ganglia, including the ventral tegmental area (VTA) and substantia nigra, the α6 and possibly the β3 nAChR subunits co-assemble in α4β2 nAChR complexes to generate high affinity receptors [10].

The α7 nAChR subunit is a particularly intriguing subunit in the mammalian brain (see Chap. 13). While the α7 receptor is thought to be functionally expressed mostly as homomeric receptors, it has been shown to be capable of co-assembling with other subunits. It was observed that the properties of α7-containing receptors were not identical to those observed for expressed homomeric α7 receptors in terms of pharmacology, gating properties, and single-channel conductance [47, 59–62]. We initially found a link between the α7 and β2 subunits, and later that they co-assembled in heterologous expression systems [14, 15, 63]. Recently it was discovered that basal forebrain cholinergic neurons express functional α7β2 receptors with an enhanced sensitivity to the Aβ peptide associated with Alzheimer's disease [64].

The α7 receptors are also expressed on both neurons and nonneuronal cells in the brain [27], including astrocytes, microglia, oligodendrocyte precursor cells, endothelial cells, and NG2 cells [65–71]. The expression in these nonneuronal cells suggests a possible role of α7 receptor function in neuroprotection [72, 73] and inflammation. Immune cell expression of α7 receptors has been shown to modulate inflammatory responses by regulating the production of inflammatory cytokines and chemokines [74, 75]. Lastly, as discussed below, this nAChR subtype is one of the most calcium permeable of all the nAChR subunits expressed in the mammalian brain [73, 76], which has ramifications for synaptic excitability and plasticity, and intracellular signaling cascades.

As mentioned above, nAChRs (including α7 receptors) are not only located at the synapse (both pre- and postsynaptically), but also extrasynaptically where they participate in nonsynaptic communication [10, 77]. Initially it was observed that while most cholinergic boutons do not make synaptic contacts, they are able to release ACh [78–81]. Since extrasynaptic nAChRs may only sense slowly increasing ACh levels (and well below those at the synapse due to diffusion and breakdown by acetylcholinesterase), they may be mainly modulating neuronal excitability and/or intracellular processes [9, 73]. The precise role that extrasynaptic nAChRs are playing in brain circuit excitability remains to be determined.

4 Regulation by Signal Transduction Cascades and Lipid Rafts

There is much evidence indicating that not only do signal transduction cascades regulate the function of nAChRs, but that the activation of these receptors can also activate signal transduction cascades. Calcium ions are one of the most prevalent and versatile signal transducers [73], with the ability to activate and regulate many calcium-dependent intracellular mechanisms. In addition, the α7 receptor has one of the highest calcium permeabilities among the ligand-gated ion channel family [66, 76, 82]. There are three basic ways in which activation of nAChRs increase cytoplasmic calcium levels; (1) direct calcium influx through the channel itself [83, 84], (2) indirect calcium influx through voltage-dependent calcium channels (VDCCs) due to the nAChR-mediated depolarization [83, 84], and (3) calcium-induced calcium release (CICR) from the endoplasmic reticulum (ER) through the ryanodine [66, 76, 85] and inositol (1,4,5)-triphosphate receptors (IP_3Rs) [66, 76, 86, 87].

The relative permeability ratio of calcium to sodium ions (P_{Ca}/P_{Na}) was estimated using the Goldmann-Hodgkin-Katz constant field equation, and found to be ~2 for non-α7 nAChRs, and >10 for α7 nAChRs [28, 88–91]. When calcium permeability was measured instead using fluorescent calcium indicators to determine the fractional calcium current (Pf, the percentage of the total current flowing through the channel that is carried by calcium ions; [28, 92–95]), the non-α7 nAChRs had a Pf value of 2–5 %, whereas the α7 nAChRs had a Pf value of 6–12 % [28, 95]. Intriguingly, incorporation of the α5 subunit into the α3 subunit-containing human nAChRs significantly increased the calcium permeability [48], whereas the different stoichiometries of the α4β2 receptors have different calcium permeabilities [32].

In addition to the influx of calcium through the nAChRs or indirectly by the activation of the VDCCs, the release of calcium from intracellular stores (via CICR) can play an important role in the calcium responses. For example the influx of calcium through the α7 nAChRs can activate CICR from ryanodine- or IP_3 receptor (IP_3R)-dependent stores [66, 76, 85, 87]. In neurons of the substantia nigra pars compacta, depletion of internal calcium stores inhibits the increase in cytoplasmic calcium levels induced by nicotine and the α7 nAChR-selective agonist choline [85]. Blockade of ryanodine receptors in neuroblastoma cells also significantly reduces the increase in cytoplasmic calcium induced by activation of β2- and α7 subunit-containing nAChRs [76], while IP_3R-selective antagonists reduced nAChR-induced calcium responses in neurons [66, 76, 87]. Functional coupling between α7 nAChRs and ryanodine receptors has also been observed in cultured hippocampal astrocytes, where α7 nAChR-mediated calcium signals arise primarily from CICR [66].

The ability to activate different sources of calcium, either from extracellular sources or intracellular stores, confers a further spatial and temporal dimension to the calcium signals evoked by nAChR activation. By converting acute nAChR stimulation into sustained cellular events, calcium signals are a critical link between nAChRs and downstream signaling cascades. For nAChRs located on presynaptic terminals, the increase in intraterminal calcium levels (either through direct influx or via CICR) will induce neurotransmitter release directly (e.g. glutamate release in the hippocampus) [96, 97]. In hippocampal synaptosomes, the activation of α3β4 nAChRs induces the release of noradrenaline without the involvement of VDCCs [98], whereas in striatal dopamine synaptosomes with β2 subunit-containing nAChRs [99], the nAChR-induced release of dopamine was mediated by VDCCs [98, 100]. Calcium can also indirectly regulate neurotransmitter release via calcium-dependent signaling cascades such as protein kinase C (PKC), which has been proposed to modulate striatal dopamine release by nAChR activation [101].

Somatodendritic (i.e. postsynaptic and extrasynaptic) nAChRs are also regulated by calcium and calcium-dependent signal transduction cascades. Desensitization is important in cholinergic signaling and synaptic efficacy, and in some nAChR-related diseases (e.g. some forms of epilepsy and congenital myasthenic syndrome [102]). In addition, differences in nAChR desensitization kinetics are thought to play a role in nicotine addiction; low doses of nicotine desensitize non-α7 nAChRs on dopaminergic and GABAergic neurons, while activating α7 receptors and enhancing glutamate-mediated excitatory inputs to the dopaminergic neurons and facilitating dopamine release onto neurons in the NAc [103]. For α7-containing nAChRs on rat hippocampal interneurons [104], the recovery from desensitization is delayed by elevated cytoplasmic calcium levels, probably through a mechanism involving either PKC or calcineurin [102, 105]. In addition, α7 nAChRs on chick ciliary ganglion neurons can undergo substantial activity-dependent inactivation through a mechanism involving calcium, CaMKII, and calcineurin [106, 107]. Recently it was found that for cultured hippocampal interneurons, the α7 receptor-mediated intracellular calcium dynamics involves CaMKII, PSD-95, and the calcium-ATPase pump isoform 2 (PMCA2) [108]. Lastly, the potentiation that we observed due to the close repetitive activation of α7 receptors on interneurons in hippocampal slices was regulated by calcineurin, PKC, and CaMKII [109].

The non-α7 nAChRs are also involved in calcium-dependent signal transduction cascades. For example for the α4β2 nAChRs expressed in *Xenopus* oocytes, the increase in cytoplasmic calcium levels and activation of PKC promotes recovery from desensitization [110], in contrast to the effect on the α7 nAChRs. In addition, activation of β2 subunit-containing receptors induces dendritic spine formation in the hippocampus through a calcium-dependent pathway requiring CaMKII [111].

Lipid rafts are areas of the cell membrane enriched in cholesterol and sphingolipids (which may serve as structures relevant for receptor regulation), and α7 nAChRs have been reported to be localized into lipid rafts [112–114].

Previously we showed that the disruption of lipid rafts in rat hippocampal neurons, through cholesterol-scavenging drugs and the enzymatic breakdown of sphingomyelin, alters the desensitization kinetics of α7 nAChRs, supporting the idea that lipid raft integrity is critical for proper receptor function [115]. We also showed that the disruption of lipid rafts also affects desensitization of the α3β2 subunit-containing nAChRs, suggesting that lipid rafts may play an important role in the modulation of nAChRs in general. Presently it is unclear whether these modulatory effects are mediated by changes in channel localization or by a loss of lipid raft-mediated interactions.

5 Regulation of Gene Transcription

It has been known for some time that nAChRs are involved in the regulation of gene transcription [116]. More recently, activation of α7 receptors was found to regulate tyrosine hydroxylase and dopamine beta-hydroxylase gene expression in PC12 cells [87], and the maturation and integration of adult-born neurons in the hippocampus [117, 118]. In neuroblastoma cells, exposure to nicotine influences the expression of many genes and proteins associated with RNA binding and the plasma membrane [119]. Lastly in the chick ciliary ganglion, the nAChR-mediated control of transcription relies on calcium influx and CICR to activate first CaMKII/IV, then ERK/MAPK, and finally the transcription factor CREB (the cAMP response element-binding protein), which can alter gene expression [120]. The nAChRs mediate the calcium-dependent activation of ERK/MAPK and CREB in multiple systems, including the hippocampus, a key area for memory processing [117, 121, 122]. Activation of the hippocampal ERK/MAPK pathway is required for the formation of contextual and spatial memories in mammals [123].

6 Involvement of nAChRs in Brain Circuits; Excitability, Synaptic Plasticity, and Oscillations

The hippocampus is an important area in the brain for learning and memory [124–127]. The hippocampus receives the majority (up to 90 %) of its cholinergic inputs from the medial septum via the fimbria/fornix, which enters the hippocampus through the stratum oriens [128]. Dysfunction in the hippocampal cholinergic system has been linked to cognitive deficits and a variety of neurological disorders and diseases, including Alzheimer's disease, nicotine addiction, and schizophrenia [4, 129]. Multiple forms of synaptic plasticity have previously been shown to be regulated by activation of both the nAChRs and G protein-coupled muscarinic AChRs [130–134]. Activation of the α7 nAChRs with exogenous ligands in the CA1 and dentate gyrus regions of the hippocampus enhances synaptic plasticity [135–138], an effect that not only depends on the location of these receptors, but on the timing of their activation as well. For example, the activation of the α7 nAChRs on hippocampal interneurons can block concurrent short-term and long-term potentiation (i.e. STP and LTP) in pyramidal cells, whereas presynaptic nAChRs can enhance the release of glutamate and increase the probability of inducing LTP [131]. The timing of exogenously applied ACh is also important in modulating high frequency stimulation (HFS)-induced hippocampal synaptic plasticity [131, 139]. The regulation of nAChRs in the hippocampus has also been linked to long-term depression (LTD) [130, 139].

We investigated how the activation of the endogenous cholinergic inputs from the septum to the hippocampus [19], either electrically or through an optogenetic approach, can regulate hippocampal synaptic plasticity. We found that activation of the cholinergic input to the hippocampus can induce different forms of hippocampal synaptic plasticity with a timing precision in the millisecond range. When the cholinergic input to the CA1 hippocampal region was activated 100 ms prior to activation of the Schaffer collateral (SC) pathway, this induced an α7 nAChR-dependent LTP. When the cholinergic input was activated only 10 msec prior to the SC pathway, this induced an α7 nAChR-dependent short-term depression (STD). If however the cholinergic input was activated 10 ms *after* the SC pathway, this induced a mAChR-dependent LTP. Therefore altering the timing of activation of the septal cholinergic input to the hippocampus induced three different forms of plasticity that depended solely on the timing of the input relative to the stimulation of the SC pathway.

Next we combined the use of genetically encoded calcium indicators (GECIs) to directly monitor neuronal activities (by measuring changes in cytoplasmic calcium levels) at either the synapse or network level, with a septo-hippocampal co-culture system [140–143]. In this way we could monitor the pre- and postsynaptic activities of hippocampal SC to CA1 synapses during the α7 nAChR-dependent LTP and STD protocols [144]. During the LTP, we observed a prolonged enhancement of the SC-induced calcium responses both post- and presynaptically, while during the STD we observed a short-term depression of the calcium responses both pre- and postsynaptically. Next we found that the presence of the α7 nAChRs to both pre- and postsynaptic sites appeared to be required to induce both LTP and STD. Dual-color calcium imaging revealed a differential time course and pattern of post- versus presynaptic modulation during both LTP and STD, suggesting the existence of independent postsynaptic modulatory mechanisms. Therefore α7 nAChRs appear to be able to coordinate pre- and postsynaptic activities to induce glutamatergic synaptic plasticity, and thus provide a novel mechanism underlying physiological neuronal communication that could lead to timing-dependent synaptic plasticity [144].

As noted above, the non-α7 nAChRs in the hippocampus have also been linked to cognitive function. To gain more insights into which regions in the hippocampal complex are responsible for the initiation and spreading of information involving cholinergic receptors during smoking, we utilized voltage-sensitive dye imaging (VSDI) techniques in combination with electrophysiological recordings to investigate spatial-temporal aspects of cholinergic responses in the hippocampus [145]. The bath application of nicotine depolarized neurons in the deep EC cortical layers (layer VI) via activation of the α4β2 nAChRs. We found that subicular neurons also contained functional non-α7 nAChRs that were activated by the bath-applied nicotine. Interestingly both of these nAChR-expressing ECVI and Sb groups of neurons were primarily glutamatergic, and nicotine enhanced the glutamatergic synaptic plasticity in the ECVI neurons, suggesting that this nicotine-induced plasticity could help in understanding the pro-cognitive effects of nicotine.

The cholinergic and GABAergic inputs from the septum are known to initiate and sustain network oscillations (e.g. hippocampal theta rhythm) in vivo and in vitro [133, 146–151]. Additionally, inputs to the hippocampus from the entorhinal cortex (EC) are thought to regulate hippocampal theta rhythm [149, 151]. Presently it is unclear precisely how the activation of both mAChRs and nAChRs, working in concert, can modulate the oscillatory properties of neurons within the hippocampus. Understanding how cholinergic receptor signaling regulates hippocampal network activity is critical since dysregulation of normal oscillations may induce seizures [152–154], and cognitive deficits linked with Alzheimer's disease [155].

As noted above, the function of nAChRs in the mesostriatal dopamine neurons is thought to play an important role in the addictive and rewarding aspect of nicotine [4, 54–58], effects that are thought to involve the modulation of synaptic plasticity and changes in gene expression [156, 157]. Similar to other drugs of abuse, nicotine increases the release of dopamine from the mesolimbic projections to the NAc [9, 83, 158]. While somatodendritic nAChRs on VTA dopaminergic neurons excite them directly, which results in transient responses that are terminated by desensitization of the nAChRs [159], the stimulation and subsequent desensitization of GABAergic neurons in the VTA also contributes to an excitation through removal of the inhibitory influence of GABA [160]. Furthermore in rat brain slices with both the VTA and NAc, activation of presynaptic α7 nAChRs induces LTP of the excitatory input to the VTA if nicotine application is paired with postsynaptic stimulation [161].

7 Involvement in Neuroprotection

Nicotine and other nAChR agonists, including those acting at the α7 nAChR, are neuroprotective in various models of neuronal death [73, 162, 163]. The nAChR-mediated neuroprotection against excitotoxicity is calcium-dependent, and does not involve blockade of glutamate receptors [164–167]. In hippocampal slices, the nicotine-mediated protection against acute NMDA excitotoxicity is mediated by the activation of phosphatidylinositol 3-kinase (PI3K) and the ERK/MAPK pathway, and may involve calbindin-D28K [166, 167]. In cortical cultures, the nicotine-induced calcium-dependent activation of the phosphatase calcineurin is proposed to mediate the protection afforded by nicotine against glutamate excitotoxicity [168]. In addition, the nicotine-mediated neuroprotection against the Aβ peptide is thought to be acting through the α7 nAChR [72, 163, 169, 170].

8 Therapeutics and Disease- Competitive Ligands and Allosteric Modulators

Understanding the basic pharmacology and physiology of nAChRs is critical information that will aid in the development of therapeutics to treat and/or mitigate the symptoms of nAChR dysfunction. For example, cholinergic dysfunction (particularly in the hippocampus) produces an array of disorders in learning and memory, and have been linked with a variety of neurological disorders and diseases, including Alzheimer's disease, schizophrenia, epilepsy, and addiction [9, 171]. Agonists for nAChRs have been under investigation for some time, and behavioral studies with nAChR agonists have primarily focused on animal models of cognition, depression, and neuropathic pain [4, 172]. However to date, the only approved drug is the partial agonist of the α4β2 nicotinic acetylcholine receptor (nAChR), varenicline, for smoking cessation treatment [29, 173]. Varenicline is a high-affinity partial agonist of the α4β2 receptor that activates (with lower efficacy than either ACh or nicotine) and then inactivates the receptor due to desensitization. In addition, it will compete with nicotine and block the rewarding aspect of nicotine [174, 175]. Since understanding the structural basis of the interaction between varenicline (and other similar molecules) and the nAChR has huge potential in the design of therapeutics for smoking cessation and other disorders and diseases, high-resolution structures are needed. Therefore we determined the X-ray crystal structure of varenicline with the acetylcholine binding protein (AChBP; a soluble protein with a similar structure to the ligand binding domain of nAChRs) from the annelid *Capitella teleta* [176]. While this structure pinpoints contact residues that potentially mediate their molecular actions with α4β2 nAChRs, it is important to understand that AChBPs are not ligand-gated channels and do not have a channel pore. Therefore caution must be used when interpreting these structures. To address this, we made mutations in the α4β2 nAChRs (which were expressed in heterologous expression systems) and confirmed crucial interactions of varenicline with residues on the complementary face of the binding site in α4β2 nAChRs; in particular we found that loops D and E were determinants of desensitization and channel opening with limited efficacy by the partial agonist varenicline [176]. Several partial agonists and antagonists of nAChRs have shown promise not only in nicotine addiction and alcoholism [177], but a variety of conditions, including (but not limited to) depression, cognition, schizophrenia, Parkinson's disease, and neuropathic pain [29].

The nAChRs are allosteric proteins with multiple, interconvertible conformations [178–183]. While competitive ligands (both agonists and antagonists) bind to the orthosteric site, there are many ligands which are known to bind to the receptor at sites distinct from the orthosteric site; these are referred to as allosteric sites [183–191], and the ligands that bind to the nAChRs at these sites are referred to as allosteric modulators [185, 186, 192]. Another promising avenue for drug design is allosteric modulators [172]; positive allosteric modulators (PAMs) are compounds that increase the receptor response induced by an agonist [185, 186], while negative allosteric modulators (NAMs) are compounds which decrease the receptor response [185, 186, 192, 193]. In addition, PAMs do not have any agonist activity on their own, but they change the ability of the orthosteric ligand to affect channel opening [185, 186, 192].

Since allosteric modulators bind to different sites and affect channel function differently than orthosteric ligands, they provide an alternative approach to manipulate nAChR function [194–197]. For example, PAMs can increase the effectiveness of endogenous ACh, and strengthen the cholinergic tone without directly activating the receptors [186, 192]. Partial and full agonists cause desensitization of nAChRs and an upregulation of receptor expression levels [198–201], both of which may be problematic for therapeutic intervention; PAMs do not activate the nAChRs directly, therefore they do not induce either desensitization or upregulation. Therefore the nAChR-driven cholinergic synapse will remain under the control of the released endogenous ACh in the presence of PAMs.

While conventional ligands for the nAChRs have been studied for their behavioral effects in animals, allosteric modulators for these receptors have only recently gained attention, and research on their behavioral effects is growing rapidly [172]. Behavioral studies with allosteric modulators of the α7 receptors have focused on PAMs since there is evidence that increasing cholinergic output through these receptors enhances cognition and reduces pain [185, 202]. The α7 receptor PAMs are divided into two categories; type I PAMs increase the receptor response evoked by agonists and maintains the response kinetics (e.g. desensitization), and type II PAMs which enhance the receptor response and dramatically reduces desensitization [185]. The type II α7 PAM, PNU-120596, was the first to be tested on rodents, where it was found to improve the auditory gating deficit caused by amphetamine in a model of schizophrenia [203], and ameliorate the MK-801-induced auditory gating deficits in the pre-pulse inhibition test [204]. Another type II α7 PAM for α7 receptors, JNJ-1930942, improved sensory gating in DBA/2 mice [205].

There have been many behavioral studies indicating that type II α7 receptor PAMs enhance cognitive function; in a social discrimination test [206], an attentional set-shifting test with phencyclidine (PCP)-treated female rats [207], and in the radial arm maze test (to measure spatial-learning memory) and the novel object recognition test (to measure episodic memory) [208]. Type I α7 receptors PAMs have also enhanced cognitive function in the (−)-scopolamine-induced deficit in acquisition of a water-maze learning task [209], in a social recognition test similar to nicotine [209], in the MK-801-induced impairment in pre-pulse inhibition [204], and in improving pre-attention, working memory, short-term recognition memory, and in inducing memory consolidation [210, 211].

Type I and II α7 receptor PAMs are known to have effects on both acute and chronic pain models. For example, PNU-120596 reduced mechanical hyperalgesia and attenuated carrageenan-induced increases in levels of TNF-α and IL-6 [212], and decreased formalin-induced pain [213]. In addition, both the type I α7 receptor PAM NS-1738 and type II PAM PNU-120596 reduced the carrageenan-induced inflammatory pain and chronic constriction injury models for neuropathic pain, and heat-induced hyperalgesia [214].

For the α4β2 nAChRs, many ligands are known to be allosteric modulators, including (but not limited to) 17-β-estradiol [184], NS-9283 [215–217], desformylflustrabromine [187, 188], and LY-2087101(a (2-amino-5-keto) thiazole compound) [218] which act as PAMs, and KAB-18 [190], UCI-30002 [219] and progesterone [220] which act as NAMs. Behavioral studies with α4β2 receptor PAMs have mainly focused on their ability to reduce pain and improve cognitive function. Studies using NS-9283 to test its effect on pain have been done in combination with agonists (full or partial); the acute administration of NS-9283 alone in vivo did not affect mechanical allodynia in the spinal nerve ligation test, however co-administration of NS-9283 with ABT-594, an agonist for nAChRs, increased the anti-allodynic effects of ABT-594 [216]. In addition, NS-9283 has been studied for its effects on cognitive performance in a variety of behavioral tests in rodents. In oxidopamine-lesioned rats (a model of Parkinson's disease), NS-9283 alone did not induce rotational behavior, however in combination with the partial agonist NS-3956, NS-9283 was able to block rotational behaviors in the rats [218]. NS-9283 was found to reverse the PCP-induced impairment in rats, and improved social recognition memory [221].

PAMs for other nAChR subtypes have been identified; the anthelmintic compounds levamisole [222] and morantel [223] are PAMs for human α3β2 and α3β4 nAChR subtypes, however no behavioral studies with either of these compounds have been reported thus far. Hypothalamic α3β4 receptors have been identified as the nAChR subtype that is involved in the regulation of food intake in mice [224], therefore nAChR PAMs for subtypes other than the α4β2 and α7 receptors may have important clinical applications.

9 Conclusion

Various subtypes of neuronal nAChRs are expressed throughout the nervous system where they participate in a variety of physiological responses and synaptic excitability and plasticity. Furthermore, dysfunction in the cholinergic system has been linked to a wide variety of neurological disorders and diseases (see Chaps. 19 and 20). Despite the complexity of the nAChR system, it is a very exciting time in the field, in part because of the advances in understanding receptor subtypes and their functional and pharmacological properties, but also the advent of new tools (e.g. animal models and optogenetics) that is allowing unprecedented investigations that we could only dream about 10 years ago. There is no doubt that new discoveries, and therapeutics to treat diseases and disorders, are just around the corner (see Chap. 21).

References

1. Yakel JL. Gating of nicotinic ACh receptors: latest insights into ligand binding and function. J Physiol. 2010;588(Pt 4):597–602.
2. Gay EA, Yakel JL. Gating of nicotinic ACh receptors; new insights into structural transitions triggered by agonist binding that induce channel opening. J Physiol. 2007;584(Pt 3):727–33.

3. Yakel JL. Cholinergic receptors: functional role of nicotinic ACh receptors in brain circuits and disease. Pflugers Arch. 2013;465(4):441–50.
4. Hurst R, Rollema H, Bertrand D. Nicotinic acetylcholine receptors: From basic science to therapeutics. Pharmacol Ther. 2012;137:22–54.
5. Alkondon M, Albuquerque EX. Diversity of nicotinic acetylcholine receptors in rat hippocampal neurons. I. Pharmacological and functional evidence for distinct structural subtypes. J Pharmacol Exp Ther. 1993;265(3):1455–73.
6. Alkondon M, Albuquerque EX. The nicotinic acetylcholine receptor subtypes and their function in the hippocampus and cerebral cortex. Prog Brain Res. 2004;145:109–20.
7. Sargent PB. The diversity of neuronal nicotinic acetylcholine receptors. Annu Rev Neurosci. 1993;16:403–43.
8. Wada E, et al. Distribution of alpha 2, alpha 3, alpha 4, and beta 2 neuronal nicotinic receptor subunit mRNAs in the central nervous system: a hybridization histochemical study in the rat. J Comp Neurol. 1989;284(2):314–35.
9. Dani JA, Bertrand D. Nicotinic acetylcholine receptors and nicotinic cholinergic mechanisms of the central nervous system. Annu Rev Pharmacol Toxicol. 2007;47:699–729.
10. Albuquerque EX, et al. Mammalian nicotinic acetylcholine receptors: from structure to function. Physiol Rev. 2009;89(1):73–120.
11. Colombo SF, et al. Biogenesis, trafficking and up-regulation of nicotinic ACh receptors. Biochem Pharmacol. 2013;86:1063–73.
12. Jones S, Sudweeks S, Yakel JL. Nicotinic receptors in the brain: correlating physiology with function. Trends Neurosci. 1999;22(12):555–61.
13. Jones S, Yakel JL. Functional nicotinic ACh receptors on interneurones in the rat hippocampus. J Physiol. 1997;504(Pt 3):603–10.
14. Khiroug SS, et al. Rat nicotinic ACh receptor alpha7 and beta2 subunits co-assemble to form functional heteromeric nicotinic receptor channels. J Physiol. 2002;540(Pt 2):425–34.
15. Sudweeks SN, Yakel JL. Functional and molecular characterization of neuronal nicotinic ACh receptors in rat CA1 hippocampal neurons. J Physiol. 2000;527(Pt 3):515–28.
16. Quick MW, et al. Alpha3beta4 subunit-containing nicotinic receptors dominate function in rat medial habenula neurons. Neuropharmacology. 1999;38(6):769–83.
17. Wonnacott S. Presynaptic nicotinic ACh receptors. Trends Neurosci. 1997;20(2):92–8.
18. Gotti C, Zoli M, Clementi F. Brain nicotinic acetylcholine receptors: native subtypes and their relevance. Trends Pharmacol Sci. 2006;27(9):482–91.
19. Gu Z, Yakel JL. Timing-dependent septal cholinergic induction of dynamic hippocampal synaptic plasticity. Neuron. 2011;71(1):155–65.
20. Alkondon M, Pereira EF, Albuquerque EX. Alpha-bungarotoxin- and methyllycaconitine-sensitive nicotinic receptors mediate fast synaptic transmission in interneurons of rat hippocampal slices. Brain Res. 1998;810(1–2):257–63.
21. Frazier CJ, et al. Synaptic potentials mediated via alpha-bungarotoxin-sensitive nicotinic acetylcholine receptors in rat hippocampal interneurons. J Neurosci. 1998;18(20):8228–35.
22. Bell KA, et al. Nicotinic excitatory postsynaptic potentials in hippocampal CA1 interneurons are predominantly mediated by nicotinic receptors that contain alpha4 and beta2 subunits. Neuropharmacology. 2011;61(8):1379–88.
23. Roerig B, Nelson DA, Katz LC. Fast synaptic signaling by nicotinic acetylcholine and serotonin 5-HT3 receptors in developing visual cortex. J Neurosci. 1997;17(21):8353–62.
24. Sun YG, et al. Biphasic cholinergic synaptic transmission controls action potential activity in thalamic reticular nucleus neurons. J Neurosci. 2013;33(5):2048–59.
25. Hatton GI, Yang QZ. Synaptic potentials mediated by alpha 7 nicotinic acetylcholine receptors in supraoptic nucleus. J Neurosci. 2002;22(1):29–37.
26. Bennett C, et al. Mechanisms generating dual-component nicotinic EPSCs in cortical interneurons. J Neurosci. 2012;32(48):17287–96.
27. Wessler I, Kirkpatrick CJ. Acetylcholine beyond neurons: the non-neuronal cholinergic system in humans. Br J Pharmacol. 2008;154(8):1558–71.

28. Fucile S. Ca2+ permeability of nicotinic acetylcholine receptors. Cell Calcium. 2004;35(1): 1–8.
29. Hurst R, Rollema H, Bertrand D. Nicotinic acetylcholine receptors: from basic science to therapeutics. Pharmacol Ther. 2013;137(1):22–54.
30. Nelson ME, et al. Alternate stoichiometries of alpha4beta2 nicotinic acetylcholine receptors. Mol Pharmacol. 2003;63(2):332–41.
31. Moroni M, et al. alpha4beta2 nicotinic receptors with high and low acetylcholine sensitivity: pharmacology, stoichiometry, and sensitivity to long-term exposure to nicotine. Mol Pharmacol. 2006;70(2):755–68.
32. Tapia L, Kuryatov A, Lindstrom J. Ca2+ permeability of the (alpha4)3(beta2)2 stoichiometry greatly exceeds that of (alpha4)2(beta2)3 human acetylcholine receptors. Mol Pharmacol. 2007;71(3):769–76.
33. Krashia P, et al. Human alpha3beta4 neuronal nicotinic receptors show different stoichiometry if they are expressed in Xenopus oocytes or mammalian HEK293 cells. PLoS One. 2010;5(10):e13611.
34. Grishin AA, et al. Alpha-conotoxin AuIB isomers exhibit distinct inhibitory mechanisms and differential sensitivity to stoichiometry of alpha3beta4 nicotinic acetylcholine receptors. J Biol Chem. 2010;285(29):22254–63.
35. Wada K, et al. Functional expression of a new pharmacological subtype of brain nicotinic acetylcholine receptor. Science. 1988;240(4850):330–4.
36. Lotfipour S, et al. Targeted deletion of the mouse alpha2 nicotinic acetylcholine receptor subunit gene (Chrna2) potentiates nicotine-modulated behaviors. J Neurosci. 2013;33(18): 7728–41.
37. McQuiston AR, Madison DV. Nicotinic receptor activation excites distinct subtypes of interneurons in the rat hippocampus. J Neurosci. 1999;19(8):2887–96.
38. Nakauchi S, et al. Nicotine gates long-term potentiation in the hippocampal CA1 region via the activation of alpha2* nicotinic ACh receptors. Eur J Neurosci. 2007;25(9):2666–81.
39. Mulle C, et al. Existence of different subtypes of nicotinic acetylcholine receptors in the rat habenulo-interpeduncular system. J Neurosci. 1991;11(8):2588–97.
40. Jia Y, et al. Nicotine facilitates long-term potentiation induction in oriens-lacunosum moleculare cells via Ca2+ entry through non-alpha7 nicotinic acetylcholine receptors. Eur J Neurosci. 2010;31(3):463–76.
41. Jia Y, et al. Alpha2 nicotine receptors function as a molecular switch to continuously excite a subset of interneurons in rat hippocampal circuits. Eur J Neurosci. 2009;29(8):1588–603.
42. Fonck C, et al. Demonstration of functional alpha4-containing nicotinic receptors in the medial habenula. Neuropharmacology. 2009;56(1):247–53.
43. Millar NS, Harkness PC. Assembly and trafficking of nicotinic acetylcholine receptors (review). Mol Membr Biol. 2008;25(4):279–92.
44. Ramirez-Latorre J, et al. Functional contributions of alpha5 subunit to neuronal acetylcholine receptor channels. Nature. 1996;380(6572):347–51.
45. Groot-Kormelink PJ, et al. A reporter mutation approach shows incorporation of the "orphan" subunit beta3 into a functional nicotinic receptor. J Biol Chem. 1998;273(25):15317–20.
46. Gotti C, et al. Heterogeneity and complexity of native brain nicotinic receptors. Biochem Pharmacol. 2007;74(8):1102–11.
47. Shao Z, Yakel JL. Single channel properties of neuronal nicotinic ACh receptors in stratum radiatum interneurons of rat hippocampal slices. J Physiol. 2000;527(Pt 3):507–13.
48. Gerzanich V, et al. alpha 5 Subunit alters desensitization, pharmacology, Ca++ permeability and Ca++ modulation of human neuronal alpha 3 nicotinic receptors. J Pharmacol Exp Ther. 1998;286(1):311–20.
49. Groot-Kormelink PJ, Boorman JP, Sivilotti LG. Formation of functional alpha3beta4alpha5 human neuronal nicotinic receptors in Xenopus oocytes: a reporter mutation approach. Br J Pharmacol. 2001;134(4):789–96.
50. Dani JA, De Biasi M. Mesolimbic dopamine and habenulo-interpeduncular pathways in nicotine withdrawal. In: Kenny PJ, Christopher Pierce R, editors. Addiction, Cold spring

harbor perspectives in medicine. New York: Cold Spring Harbor Laboratory Press; 2013. p. 237–50. doi:10.1101/cshperspect.a012138.
51. Broadbent S, et al. Incorporation of the beta3 subunit has a dominant-negative effect on the function of recombinant central-type neuronal nicotinic receptors. Mol Pharmacol. 2006; 70(4):1350–7.
52. Exley R, et al. Alpha6-containing nicotinic acetylcholine receptors dominate the nicotine control of dopamine neurotransmission in nucleus accumbens. Neuropsychopharmacology. 2008;33(9):2158–66.
53. Gerzanich V, et al. "Orphan" alpha6 nicotinic AChR subunit can form a functional heteromeric acetylcholine receptor. Mol Pharmacol. 1997;51(2):320–7.
54. Changeux JP. Nicotine addiction and nicotinic receptors: lessons from genetically modified mice. Nat Rev Neurosci. 2010;11(6):389–401.
55. Klink R, et al. Molecular and physiological diversity of nicotinic acetylcholine receptors in the midbrain dopaminergic nuclei. J Neurosci. 2001;21(5):1452–63.
56. Champtiaux N, et al. Subunit composition of functional nicotinic receptors in dopaminergic neurons investigated with knock-out mice. J Neurosci. 2003;23(21):7820–9.
57. Grady SR, et al. The subtypes of nicotinic acetylcholine receptors on dopaminergic terminals of mouse striatum. Biochem Pharmacol. 2007;74(8):1235–46.
58. Drenan RM, et al. In vivo activation of midbrain dopamine neurons via sensitized, high-affinity alpha 6 nicotinic acetylcholine receptors. Neuron. 2008;60(1):123–36.
59. Anand R, Peng X, Lindstrom J. Homomeric and native alpha 7 acetylcholine receptors exhibit remarkably similar but non-identical pharmacological properties, suggesting that the native receptor is a heteromeric protein complex. FEBS Lett. 1993;327(2):241–6.
60. Yu CR, Role LW. Functional contribution of the alpha7 subunit to multiple subtypes of nicotinic receptors in embryonic chick sympathetic neurones. J Physiol. 1998;509(Pt 3):651–65.
61. Girod R, et al. Heteromeric complexes of alpha 5 and/or alpha 7 subunits. Effects of calcium and potential role in nicotine-induced presynaptic facilitation. Ann N Y Acad Sci. 1999; 868:578–90.
62. Palma E, et al. Nicotinic acetylcholine receptors assembled from the alpha7 and beta3 subunits. J Biol Chem. 1999;274(26):18335–40.
63. Murray TA, et al. alpha7beta2 nicotinic acetylcholine receptors assemble, function, and are activated primarily via their alpha7-alpha7 interfaces. Mol Pharmacol. 2012;81(2):175–88.
64. Liu Q, et al. A novel nicotinic acetylcholine receptor subtype in basal forebrain cholinergic neurons with high sensitivity to amyloid peptides. J Neurosci. 2009;29(4):918–29.
65. Shen JX, Yakel JL. Functional alpha7 nicotinic ACh receptors on astrocytes in rat hippocampal CA1 slices. J Mol Neurosci. 2012;48(1):14–21.
66. Sharma G, Vijayaraghavan S. Nicotinic cholinergic signaling in hippocampal astrocytes involves calcium-induced calcium release from intracellular stores. Proc Natl Acad Sci U S A. 2001;98(7):4148–53.
67. Shytle RD, et al. Cholinergic modulation of microglial activation by alpha 7 nicotinic receptors. J Neurochem. 2004;89(2):337–43.
68. Velez-Fort M, Audinat E, Angulo MC. Functional alpha 7-containing nicotinic receptors of NG2-expressing cells in the hippocampus. Glia. 2009;57(10):1104–14.
69. De Simone R, et al. Activation of alpha7 nicotinic acetylcholine receptor by nicotine selectively up-regulates cyclooxygenase-2 and prostaglandin E2 in rat microglial cultures. J Neuroinflammation. 2005;2(1):4.
70. Suzuki T, et al. Microglial alpha7 nicotinic acetylcholine receptors drive a phospholipase C/IP3 pathway and modulate the cell activation toward a neuroprotective role. J Neurosci Res. 2006;83(8):1461–70.
71. Hawkins BT, Egleton RD, Davis TP. Modulation of cerebral microvascular permeability by endothelial nicotinic acetylcholine receptors. Am J Physiol Heart Circ Physiol. 2005;289(1): H212–9.
72. Hernandez CM, Dineley KT. Alpha7 nicotinic acetylcholine receptors in Alzheimer's disease: neuroprotective, neurotrophic or both? Curr Drug Targets. 2012;13(5):613–22.

73. Shen JX, Yakel JL. Nicotinic acetylcholine receptor-mediated calcium signaling in the nervous system. Acta Pharmacol Sin. 2009;30(6):673–80.
74. Gahring LC, et al. Nicotinic receptor alpha7 expression identifies a novel hematopoietic progenitor lineage. PLoS One. 2013;8(3):e57481.
75. Olofsson PS, et al. Rethinking inflammation: neural circuits in the regulation of immunity. Immunol Rev. 2012;248(1):188–204.
76. Dajas-Bailador FA, Mogg AJ, Wonnacott S. Intracellular Ca2+ signals evoked by stimulation of nicotinic acetylcholine receptors in SH-SY5Y cells: contribution of voltage-operated Ca2+ channels and Ca2+ stores. J Neurochem. 2002;81(3):606–14.
77. Lendvai B, Vizi ES. Nonsynaptic chemical transmission through nicotinic acetylcholine receptors. Physiol Rev. 2008;88(2):333–49.
78. Descarries L, Gisiger V, Steriade M. Diffuse transmission by acetylcholine in the CNS. Prog Neurobiol. 1997;53(5):603–25.
79. Kasa P, et al. Synaptic and non-synaptic cholinergic innervation of the various types of neurons in the main olfactory bulb of adult rat: immunocytochemistry of choline acetyltransferase. Neuroscience. 1995;67(3):667–77.
80. Kiss JP, Vizi ES, Westerink BH. Effect of neostigmine on the hippocampal noradrenaline release: role of cholinergic receptors. Neuroreport. 1999;10(1):81–6.
81. Umbriaco D, et al. Relational features of acetylcholine, noradrenaline, serotonin and GABA axon terminals in the stratum radiatum of adult rat hippocampus (CA1). Hippocampus. 1995; 5(6):605–20.
82. Gilbert D, et al. Local and global calcium signals associated with the opening of neuronal alpha7 nicotinic acetylcholine receptors. Cell Calcium. 2009;45(2):198–207.
83. Dajas-Bailador F, Wonnacott S. Nicotinic acetylcholine receptors and the regulation of neuronal signalling. Trends Pharmacol Sci. 2004;25(6):317–24.
84. Rathouz MM, Berg DK. Synaptic-type acetylcholine receptors raise intracellular calcium levels in neurons by two mechanisms. J Neurosci. 1994;14(11 Pt 2):6935–45.
85. Tsuneki H, et al. Calcium mobilization elicited by two types of nicotinic acetylcholine receptors in mouse substantia nigra pars compacta. Eur J Neurosci. 2000;12(7):2475–85.
86. Rizzoli S, Sharma G, Vijayaraghavan S. Calcium rise in cultured neurons from medial septum elicits calcium waves in surrounding glial cells. Brain Res. 2002;957(2):287–97.
87. Gueorguiev VD, et al. Involvement of alpha7 nicotinic acetylcholine receptors in activation of tyrosine hydroxylase and dopamine beta-hydroxylase gene expression in PC12 cells. J Neurochem. 2000;75(5):1997–2005.
88. McKay BE, Placzek AN, Dani JA. Regulation of synaptic transmission and plasticity by neuronal nicotinic acetylcholine receptors. Biochem Pharmacol. 2007;74(8):1120–33.
89. Szabo SI, Zelles T, Lendvai B. Intracellular Ca2+ dynamics of hippocampal interneurons following nicotinic acetylcholine receptor activation. Neurochem Int. 2008;52(1–2):135–41.
90. Fayuk D, Yakel JL. Dendritic Ca2+ signalling due to activation of alpha 7-containing nicotinic acetylcholine receptors in rat hippocampal neurons. J Physiol. 2007;582(Pt 2): 597–611.
91. Fayuk D, Yakel JL. Ca2+ permeability of nicotinic acetylcholine receptors in rat hippocampal CA1 interneurones. J Physiol. 2005;566(Pt 3):759–68.
92. Grynkiewicz G, Poenie M, Tsien RY. A new generation of Ca2+ indicators with greatly improved fluorescence properties. J Biol Chem. 1985;260(6):3440–50.
93. Zhou Z, Neher E. Calcium permeability of nicotinic acetylcholine receptor channels in bovine adrenal chromaffin cells. Pflugers Arch. 1993;425(5–6):511–7.
94. Neher E. The use of fura-2 for estimating Ca buffers and Ca fluxes. Neuropharmacology. 1995;34(11):1423–42.
95. Fucile S, et al. Fractional Ca(2+) current through human neuronal alpha7 nicotinic acetylcholine receptors. Cell Calcium. 2003;34(2):205–9.
96. Gray R, et al. Hippocampal synaptic transmission enhanced by low concentrations of nicotine. Nature. 1996;383(6602):713–6.

97. Sharma G, Vijayaraghavan S. Modulation of presynaptic store calcium induces release of glutamate and postsynaptic firing. Neuron. 2003;38(6):929–39.
98. Kulak JM, et al. Nicotine-evoked transmitter release from synaptosomes: functional association of specific presynaptic acetylcholine receptors and voltage-gated calcium channels. J Neurochem. 2001;77(6):1581–9.
99. Zoli M, et al. Identification of the nicotinic receptor subtypes expressed on dopaminergic terminals in the rat striatum. J Neurosci. 2002;22(20):8785–9.
100. Soliakov L, Wonnacott S. Voltage-sensitive Ca2+ channels involved in nicotinic receptor-mediated [3H]dopamine release from rat striatal synaptosomes. J Neurochem. 1996;67(1):163–70.
101. Soliakov L, Wonnacott S. Involvement of protein kinase C in the presynaptic nicotinic modulation of [(3)H]-dopamine release from rat striatal synaptosomes. Br J Pharmacol. 2001;132(3):785–91.
102. Giniatullin R, Nistri A, Yakel JL. Desensitization of nicotinic ACh receptors: shaping cholinergic signaling. Trends Neurosci. 2005;28(7):371–8.
103. Mansvelder HD, McGehee DS. Cellular and synaptic mechanisms of nicotine addiction. J Neurobiol. 2002;53(4):606–17.
104. Khiroug L, et al. Functional mapping and Ca2+ regulation of nicotinic acetylcholine receptor channels in rat hippocampal CA1 neurons. J Neurosci. 2003;23(27):9024–31.
105. Quick MW, Lester RA. Desensitization of neuronal nicotinic receptors. J Neurobiol. 2002;53(4):457–78.
106. Berg DK, Conroy WG. Nicotinic alpha 7 receptors: synaptic options and downstream signaling in neurons. J Neurobiol. 2002;53(4):512–23.
107. Liu Q, Berg DK. Actin filaments and the opposing actions of CaM kinase II and calcineurin in regulating alpha7-containing nicotinic receptors on chick ciliary ganglion neurons. J Neurosci. 1999;19(23):10280–8.
108. Gomez-Varela D, et al. PMCA2 via PSD-95 controls calcium signaling by alpha7-containing nicotinic acetylcholine receptors on aspiny interneurons. J Neurosci. 2012;32(20):6894–905.
109. Klein RC, Yakel JL. Paired-pulse potentiation of alpha7-containing nAChRs in rat hippocampal CA1 stratum radiatum interneurones. J Physiol. 2005;568(Pt 3):881–9.
110. Fenster CP, et al. Desensitization of nicotinic receptors in the central nervous system. Ann N Y Acad Sci. 1999;868:620–3.
111. Lozada AF, et al. Induction of dendritic spines by beta2-containing nicotinic receptors. J Neurosci. 2012;32(24):8391–400.
112. Fernandes CC, Berg DK, Gomez-Varela D. Lateral mobility of nicotinic acetylcholine receptors on neurons is determined by receptor composition, local domain, and cell type. J Neurosci. 2010;30(26):8841–51.
113. Bruses JL, Chauvet N, Rutishauser U. Membrane lipid rafts are necessary for the maintenance of the (alpha)7 nicotinic acetylcholine receptor in somatic spines of ciliary neurons. J Neurosci. 2001;21(2):504–12.
114. Oshikawa J, et al. Nicotinic acetylcholine receptor alpha 7 regulates cAMP signal within lipid rafts. Am J Physiol Cell Physiol. 2003;285(3):C567–74.
115. Colon-Saez JO, Yakel JL. The alpha7 nicotinic acetylcholine receptor function in hippocampal neurons is regulated by the lipid composition of the plasma membrane. J Physiol. 2011;589(Pt 13):3163–74.
116. Greenberg ME, Ziff EB, Greene LA. Stimulation of neuronal acetylcholine receptors induces rapid gene transcription. Science. 1986;234(4772):80–3.
117. Hu M, et al. Nicotinic regulation of CREB activation in hippocampal neurons by glutamatergic and nonglutamatergic pathways. Mol Cell Neurosci. 2002;21(4):616–25.
118. Campbell NR, et al. Endogenous signaling through alpha7-containing nicotinic receptors promotes maturation and integration of adult-born neurons in the hippocampus. J Neurosci. 2010;30(26):8734–44.
119. Dunckley T, Lukas RJ. Nicotine modulates the expression of a diverse set of genes in the neuronal SH-SY5Y cell line. J Biol Chem. 2003;278(18):15633–40.

120. Chang KT, Berg DK. Voltage-gated channels block nicotinic regulation of CREB phosphorylation and gene expression in neurons. Neuron. 2001;32(5):855–65.
121. Nakayama H, et al. Nicotine-induced phosphorylation of extracellular signal-regulated protein kinase and CREB in PC12h cells. J Neurochem. 2001;79(3):489–98.
122. Dajas-Bailador FA, Soliakov L, Wonnacott S. Nicotine activates the extracellular signal-regulated kinase 1/2 via the alpha7 nicotinic acetylcholine receptor and protein kinase A, in SH-SY5Y cells and hippocampal neurones. J Neurochem. 2002;80(3):520–30.
123. Sweatt JD. The neuronal MAP kinase cascade: a biochemical signal integration system subserving synaptic plasticity and memory. J Neurochem. 2001;76(1):1–10.
124. Blozovski D. Deficits in passive avoidance learning in young rats following mecamylamine injections in the hippocampo-entorhinal area. Exp Brain Res. 1983;50(2–3):442–8.
125. Blozovski D. Mediation of passive avoidance learning by nicotinic hippocampo-entorhinal components in young rats. Dev Psychobiol. 1985;18(4):355–66.
126. Davis JA, Kenney JW, Gould TJ. Hippocampal alpha4beta2 nicotinic acetylcholine receptor involvement in the enhancing effect of acute nicotine on contextual fear conditioning. J Neurosci. 2007;27(40):10870–7.
127. Izquierdo I, et al. The evidence for hippocampal long-term potentiation as a basis of memory for simple tasks. An Acad Bras Cienc. 2008;80(1):115–27.
128. Dutar P, et al. The septohippocampal pathway: structure and function of a central cholinergic system. Physiol Rev. 1995;75(2):393–427.
129. Kenney JW, Gould TJ. Modulation of hippocampus-dependent learning and synaptic plasticity by nicotine. Mol Neurobiol. 2008;38(1):101–21.
130. Fujii S, Sumikawa K. Nicotine accelerates reversal of long-term potentiation and enhances long-term depression in the rat hippocampal CA1 region. Brain Res. 2001;894(2):340–6.
131. Ji D, Lape R, Dani JA. Timing and location of nicotinic activity enhances or depresses hippocampal synaptic plasticity. Neuron. 2001;31(1):131–41.
132. McGehee DS. Nicotinic receptors and hippocampal synaptic plasticity … it's all in the timing. Trends Neurosci. 2002;25(4):171–2.
133. Cobb SR, Davies CH. Cholinergic modulation of hippocampal cells and circuits. J Physiol. 2005;562(Pt 1):81–8.
134. Maylie J, Adelman JP. Cholinergic signaling through synaptic SK channels: it's a protein kinase but which one? Neuron. 2010;68(5):809–11.
135. Fujii S, et al. Acute and chronic nicotine exposure differentially facilitate the induction of LTP. Brain Res. 1999;846(1):137–43.
136. Mann EO, Greenfield SA. Novel modulatory mechanisms revealed by the sustained application of nicotine in the guinea-pig hippocampus in vitro. J Physiol. 2003;551(Pt 2):539–50.
137. Welsby P, Rowan M, Anwyl R. Nicotinic receptor-mediated enhancement of long-term potentiation involves activation of metabotropic glutamate receptors and ryanodine-sensitive calcium stores in the dentate gyrus. Eur J Neurosci. 2006;24(11):3109–18.
138. Welsby PJ, Rowan MJ, Anwyl R. Beta-amyloid blocks high frequency stimulation induced LTP but not nicotine enhanced LTP. Neuropharmacology. 2007;53(1):188–95.
139. Ge S, Dani JA. Nicotinic acetylcholine receptors at glutamate synapses facilitate long-term depression or potentiation. J Neurosci. 2005;25(26):6084–91.
140. Gahwiler BH, Brown DA. Functional innervation of cultured hippocampal neurones by cholinergic afferents from co-cultured septal explants. Nature. 1985;313(6003):577–9.
141. Gahwiler BH, Hefti F. Guidance of acetylcholinesterase-containing fibres by target tissue in co-cultured brain slices. Neuroscience. 1984;13(3):681–9.
142. Rimvall K, Keller F, Waser PG. Development of cholinergic projections in organotypic cultures of rat septum, hippocampus and cerebellum. Brain Res. 1985;351(2):267–78.
143. Fischer Y, Gahwiler BH, Thompson SM. Activation of intrinsic hippocampal theta oscillations by acetylcholine in rat septo-hippocampal cocultures. J Physiol. 1999;519(Pt 2):405–13.
144. Gu Z, Lamb PW, Yakel JL. Cholinergic coordination of presynaptic and postsynaptic activity induces timing-dependent hippocampal synaptic plasticity. J Neurosci. 2012;32(36):12337–48.

145. Tu B, et al. Characterization of a nicotine-sensitive neuronal population in rat entorhinal cortex. J Neurosci. 2009;29(33):10436–48.
146. Frotscher M, Leranth C. Cholinergic innervation of the rat hippocampus as revealed by choline acetyltransferase immunocytochemistry: a combined light and electron microscopic study. J Comp Neurol. 1985;239(2):237–46.
147. Lawrence JJ, et al. Muscarinic receptor activation tunes mouse stratum oriens interneurones to amplify spike reliability. J Physiol. 2006;571(Pt 3):555–62.
148. Lawrence JJ, et al. Cell type-specific dependence of muscarinic signalling in mouse hippocampal stratum oriens interneurones. J Physiol. 2006;570(Pt 3):595–610.
149. Buzsaki G. Theta oscillations in the hippocampus. Neuron. 2002;33(3):325–40.
150. Royer S, et al. Control of timing, rate and bursts of hippocampal place cells by dendritic and somatic inhibition. Nat Neurosci. 2012;15(5):769–75.
151. Barry C, Heys JG, Hasselmo ME. Possible role of acetylcholine in regulating spatial novelty effects on theta rhythm and grid cells. Front Neural Circuits. 2012;6:5.
152. Turski L, et al. Review: cholinergic mechanisms and epileptogenesis. The seizures induced by pilocarpine: a novel experimental model of intractable epilepsy. Synapse. 1989;3(2):154–71.
153. Bertrand S, et al. Properties of neuronal nicotinic acetylcholine receptor mutants from humans suffering from autosomal dominant nocturnal frontal lobe epilepsy. Br J Pharmacol. 1998;125(4):751–60.
154. Damaj MI, et al. Pharmacological characterization of nicotine-induced seizures in mice. J Pharmacol Exp Ther. 1999;291(3):1284–91.
155. Fodale V, et al. Alzheimer's disease and anaesthesia: implications for the central cholinergic system. Br J Anaesth. 2006;97(4):445–52.
156. Dani JA, Ji D, Zhou FM. Synaptic plasticity and nicotine addiction. Neuron. 2001;31(3):349–52.
157. Placzek AN, Dani JA. Synaptic plasticity within midbrain dopamine centers contributes to nicotine addiction. Nebr Symp Motiv. 2009;55:5–15.
158. Di Chiara G, Imperato A. Drugs abused by humans preferentially increase synaptic dopamine concentrations in the mesolimbic system of freely moving rats. Proc Natl Acad Sci U S A. 1988;85(14):5274–8.
159. Pidoplichko VI, et al. Nicotine activates and desensitizes midbrain dopamine neurons. Nature. 1997;390(6658):401–4.
160. Laviolette SR, van der Kooy D. The neurobiology of nicotine addiction: bridging the gap from molecules to behaviour. Nat Rev Neurosci. 2004;5(1):55–65.
161. Mansvelder HD, McGehee DS. Long-term potentiation of excitatory inputs to brain reward areas by nicotine. Neuron. 2000;27(2):349–57.
162. O'Neill MJ, et al. The role of neuronal nicotinic acetylcholine receptors in acute and chronic neurodegeneration. Curr Drug Targets CNS Neurol Disord. 2002;1(4):399–411.
163. Parri HR, Hernandez CM, Dineley KT. Research update: alpha7 nicotinic acetylcholine receptor mechanisms in Alzheimer's disease. Biochem Pharmacol. 2011;82(8):931–42.
164. Donnelly-Roberts DL, et al. In vitro neuroprotective properties of the novel cholinergic channel activator (ChCA), ABT-418. Brain Res. 1996;719(1–2):36–44.
165. Dajas-Bailador FA, Lima PA, Wonnacott S. The alpha7 nicotinic acetylcholine receptor subtype mediates nicotine protection against NMDA excitotoxicity in primary hippocampal cultures through a Ca(2+) dependent mechanism. Neuropharmacology. 2000;39(13):2799–807.
166. Ferchmin PA, et al. Nicotinic receptors differentially regulate N-methyl-D-aspartate damage in acute hippocampal slices. J Pharmacol Exp Ther. 2003;305(3):1071–8.
167. Prendergast MA, et al. Chronic nicotine exposure reduces N-methyl-D-aspartate receptor-mediated damage in the hippocampus without altering calcium accumulation or extrusion: evidence of calbindin-D28K overexpression. Neuroscience. 2001;102(1):75–85.
168. Stevens TR, et al. Neuroprotection by nicotine in mouse primary cortical cultures involves activation of calcineurin and L-type calcium channel inactivation. J Neurosci. 2003;23(31):10093–9.

169. Kihara T, et al. Nicotinic receptor stimulation protects neurons against beta-amyloid toxicity. Ann Neurol. 1997;42(2):159–63.
170. Svensson AL, Nordberg A. Beta-estradiol attenuate amyloid beta-peptide toxicity via nicotinic receptors. Neuroreport. 1999;10(17):3485–9.
171. Terry Jr AV, Buccafusco JJ. The cholinergic hypothesis of age and Alzheimer's disease-related cognitive deficits: recent challenges and their implications for novel drug development. J Pharmacol Exp Ther. 2003;306(3):821–7.
172. Pandya AA, Yakel JL. Effects of neuronal nicotinic acetylcholine receptor allosteric modulators in animal behavior studies. Biochem Pharmacol. 2013;86:1054–62.
173. Coe JW, et al. Varenicline: an alpha4beta2 nicotinic receptor partial agonist for smoking cessation. J Med Chem. 2005;48(10):3474–7.
174. Rollema H, et al. Pre-clinical properties of the alpha4beta2 nicotinic acetylcholine receptor partial agonists varenicline, cytisine and dianicline translate to clinical efficacy for nicotine dependence. Br J Pharmacol. 2010;160(2):334–45.
175. Foulds J. The neurobiological basis for partial agonist treatment of nicotine dependence: varenicline. Int J Clin Pract. 2006;60(5):571–6.
176. Billen B, et al. Molecular actions of smoking cessation drugs at alpha4beta2 nicotinic receptors defined in crystal structures of a homologous binding protein. Proc Natl Acad Sci U S A. 2012;109(23):9173–8.
177. Rahman S. Nicotinic receptors as therapeutic targets for drug addictive disorders. CNS Neurol Disord Drug Targets. 2013;12(5):633–40.
178. Chang YC, et al. Allosteric activation mechanism of the cys-loop receptors. Acta Pharmacol Sin. 2009;30(6):663–72.
179. Taly A, et al. Implications of the quaternary twist allosteric model for the physiology and pathology of nicotinic acetylcholine receptors. Proc Natl Acad Sci U S A. 2006;103(45): 16965–70.
180. Edelstein SJ, et al. A kinetic mechanism for nicotinic acetylcholine receptors based on multiple allosteric transitions. Biol Cybern. 1996;75(5):361–79.
181. Lena C, Changeux JP. Allosteric modulations of the nicotinic acetylcholine receptor. Trends Neurosci. 1993;16(5):181–6.
182. Grosman C, Auerbach A. The dissociation of acetylcholine from open nicotinic receptor channels. Proc Natl Acad Sci U S A. 2001;98(24):14102–7.
183. Hille B. Ion channels of excitable membranes. 3rd ed. Sunderland, MA: Sinauer; 2001. p. 814. xviii.
184. Paradiso K, Zhang J, Steinbach JH. The C terminus of the human nicotinic alpha4beta2 receptor forms a binding site required for potentiation by an estrogenic steroid. J Neurosci. 2001;21(17):6561–8.
185. Bertrand D, Gopalakrishnan M. Allosteric modulation of nicotinic acetylcholine receptors. Biochem Pharmacol. 2007;74(8):1155–63.
186. Pandya A, Yakel JL. Allosteric modulators of the alpha4beta2 subtype of neuronal nicotinic acetylcholine receptors. Biochem Pharmacol. 2011;82(8):952–8.
187. Kim JS, et al. Synthesis of desformylflustrabromine and its evaluation as an alpha4beta2 and alpha7 nACh receptor modulator. Bioorg Med Chem Lett. 2007;17(17):4855–60.
188. German N, et al. Deconstruction of the alpha4beta2 nicotinic acetylcholine receptor positive allosteric modulator desformylflustrabromine. J Med Chem. 2011;54(20):7259–67.
189. Collins T, Young GT, Millar NS. Competitive binding at a nicotinic receptor transmembrane site of two alpha7-selective positive allosteric modulators with differing effects on agonist-evoked desensitization. Neuropharmacology. 2011;61(8):1306–13.
190. Henderson BJ, et al. Negative allosteric modulators that target human alpha4beta2 neuronal nicotinic receptors. J Pharmacol Exp Ther. 2011;334(3):761–74.
191. Young GT, et al. Potentiation of alpha7 nicotinic acetylcholine receptors via an allosteric transmembrane site. Proc Natl Acad Sci U S A. 2008;105(38):14686–91.
192. Arias HR, et al. Novel positive allosteric modulators of the human alpha7 nicotinic acetylcholine receptor. Biochemistry. 2011;50(23):5263–78.

193. Pavlovicz RE, et al. Identification of a negative allosteric site on human alpha4beta2 and alpha3beta4 neuronal nicotinic acetylcholine receptors. PLoS One. 2011;6(9):e24949.
194. Taly A, et al. Nicotinic receptors: allosteric transitions and therapeutic targets in the nervous system. Nat Rev Drug Discov. 2009;8(9):733–50.
195. Changeux JP, Taly A. Nicotinic receptors, allosteric proteins and medicine. Trends Mol Med. 2008;14(3):93–102.
196. Maelicke A, Albuquerque EX. Allosteric modulation of nicotinic acetylcholine receptors as a treatment strategy for Alzheimer's disease. Eur J Pharmacol. 2000;393(1–3):165–70.
197. Romanelli MN, Gualtieri F. Cholinergic nicotinic receptors: competitive ligands, allosteric modulators, and their potential applications. Med Res Rev. 2003;23(4):393–426.
198. Whiteaker P, Sharples CG, Wonnacott S. Agonist-induced up-regulation of alpha4beta2 nicotinic acetylcholine receptors in M10 cells: pharmacological and spatial definition. Mol Pharmacol. 1998;53(5):950–62.
199. Buccafusco JJ, Beach JW, Terry Jr AV. Desensitization of nicotinic acetylcholine receptors as a strategy for drug development. J Pharmacol Exp Ther. 2009;328(2):364–70.
200. Lu Y, Marks MJ, Collins AC. Desensitization of nicotinic agonist-induced [3H]gamma-aminobutyric acid release from mouse brain synaptosomes is produced by subactivating concentrations of agonists. J Pharmacol Exp Ther. 1999;291(3):1127–34.
201. Sallette J, et al. Nicotine upregulates its own receptors through enhanced intracellular maturation. Neuron. 2005;46(4):595–607.
202. Miwa JM, Lester HA, Walz A. Optimizing cholinergic tone through lynx modulators of nicotinic receptors: implications for plasticity and nicotine addiction. Physiology (Bethesda). 2012;27(4):187–99.
203. Hurst RS, et al. A novel positive allosteric modulator of the alpha7 neuronal nicotinic acetylcholine receptor: in vitro and in vivo characterization. J Neurosci. 2005;25(17):4396–405.
204. Dunlop J, et al. Old and new pharmacology: positive allosteric modulation of the alpha7 nicotinic acetylcholine receptor by the 5-hydroxytryptamine(2B/C) receptor antagonist SB-206553 (3,5-dihydro-5-methyl-N-3-pyridinylbenzo[1,2-b:4,5-b′]di pyrrole-1(2H)-carboxamide). J Pharmacol Exp Ther. 2009;328(3):766–76.
205. Dinklo T, et al. Characterization of 2-[[4-fluoro-3-(trifluoromethyl)phenyl]amino]-4-(4-pyridinyl)-5-thiazolemethanol (JNJ-1930942), a novel positive allosteric modulator of the {alpha}7 nicotinic acetylcholine receptor. J Pharmacol Exp Ther. 2011;336(2):560–74.
206. Thomsen MS, El-Sayed M, Mikkelsen JD. Differential immediate and sustained memory enhancing effects of alpha7 nicotinic receptor agonists and allosteric modulators in rats. PLoS One. 2011;6(11):e27014.
207. McLean SL, et al. PNU-120596, a positive allosteric modulator of alpha7 nicotinic acetylcholine receptors, reverses a sub-chronic phencyclidine-induced cognitive deficit in the attentional set-shifting task in female rats. J Psychopharmacol. 2012;26(9):1265–70.
208. Pandya AA, Yakel JL. Activation of the alpha7 nicotinic ACh receptor induces anxiogenic effects in rats which is blocked by a 5-HT(1a) receptor antagonist. Neuropharmacology. 2013;70C:35–42.
209. Timmermann DB, et al. An allosteric modulator of the alpha7 nicotinic acetylcholine receptor possessing cognition-enhancing properties in vivo. J Pharmacol Exp Ther. 2007; 323(1):294–307.
210. Ng HJ, et al. Nootropic alpha7 nicotinic receptor allosteric modulator derived from GABAA receptor modulators. Proc Natl Acad Sci U S A. 2007;104(19):8059–64.
211. Johnstone TB, et al. Allosteric modulation of related ligand-gated ion channels synergistically induces long-term potentiation in the hippocampus and enhances cognition. J Pharmacol Exp Ther. 2010;336(3):908–15.
212. Munro G, et al. The alpha7 nicotinic ACh receptor agonist compound B and positive allosteric modulator PNU-120596 both alleviate inflammatory hyperalgesia and cytokine release in the rat. Br J Pharmacol. 2012;167(2):421–35.
213. Freitas K, et al. In vivo pharmacological interactions between a type ii positive allosteric modulator of alpha7 nicotinic acetylcholine receptors and nicotinic agonists in a murine tonic pain model. Br J Pharmacol. 2012;169:567–79.

214. Freitas K, et al. Effects of alpha 7 positive allosteric modulators in murine inflammatory and chronic neuropathic pain models. Neuropharmacology. 2013;65:156–64.
215. Zhu CZ, et al. Potentiation of analgesic efficacy but not side effects: co-administration of an alpha4beta2 neuronal nicotinic acetylcholine receptor agonist and its positive allosteric modulator in experimental models of pain in rats. Biochem Pharmacol. 2011;82(8):967–76.
216. Lee CH, et al. alpha4beta2 neuronal nicotinic receptor positive allosteric modulation: an approach for improving the therapeutic index of alpha4beta2 nAChR agonists in pain. Biochem Pharmacol. 2011;82(8):959–66.
217. Rode F, et al. Positive allosteric modulation of alpha4beta2 nAChR agonist induced behaviour. Brain Res. 2012;1458:67–75.
218. Broad LM, et al. Identification and pharmacological profile of a new class of selective nicotinic acetylcholine receptor potentiators. J Pharmacol Exp Ther. 2006;318(3):1108–17.
219. Yoshimura RF, et al. Negative allosteric modulation of nicotinic acetylcholine receptors blocks nicotine self-administration in rats. J Pharmacol Exp Ther. 2007;323(3):907–15.
220. Valera S, Ballivet M, Bertrand D. Progesterone modulates a neuronal nicotinic acetylcholine receptor. Proc Natl Acad Sci U S A. 1992;89(20):9949–53.
221. Timmermann DB, et al. Augmentation of cognitive function by NS9283, a stoichiometry-dependent positive allosteric modulator of alpha2- and alpha4-containing nicotinic acetylcholine receptors. Br J Pharmacol. 2012;167(1):164–82.
222. Levandoski MM, Piket B, Chang J. The anthelmintic levamisole is an allosteric modulator of human neuronal nicotinic acetylcholine receptors. Eur J Pharmacol. 2003;471(1):9–20.
223. Wu TY, et al. Morantel allosterically enhances channel gating of neuronal nicotinic acetylcholine alpha 3 beta 2 receptors. Mol Pharmacol. 2008;74(2):466–75.
224. Mineur YS, et al. Nicotine decreases food intake through activation of POMC neurons. Science. 2011;332(6035):1330–2.

Chapter 6
Nicotinic Signaling in Development

Catarina C. Fernandes, Adrian F. Lozada, and Darwin K. Berg

Abstract Nicotinic acetylcholine receptors (nAChRs) appear early in development, reaching their highest relative levels in early postnatal life. They are expressed on nearly every neuron in the central nervous system (CNS) and in many nonneuronal cell-types as well. Cholinergic neurons not only appear early on, but also project to many brain regions at this time. These events largely precede the bulk of glutamatergic synapse formation and the maturation of GABAergic transmission. As a result cholinergic nicotinic signaling is temporally and spatially positioned to have a substantial impact on maturation of the nervous system and the formation of neural nets. This chapter will review recent findings indicating that endogenous nicotinic input is required for normal maturation of the nervous system and that excessive or altered nicotinic signaling at early times can produce significant aberrations in the synaptic pathways that form. First we summarize the nAChR subtypes, their appearance and distribution during development, and discuss the positioning and abundance of cholinergic neurons and their projections to potential synaptic targets. Next we consider the kinds of nicotinic signaling found early in development, including spontaneous waves extending across large regions, and discuss the organizational impact this is likely to have. We then address the role that nicotinic signaling has in driving the conversion of GABAergic signaling from the excitatory mode found in early postnatal life to the inhibitory mode characteristic of the adult. Lastly we review recent results demonstrating that endogenous nicotinic signaling is required during early postnatal life to achieve normal numbers of glutamatergic synapses in the adult and shape the neural networks that form. Disruption of these events is likely to have long-lasting consequences, perhaps accounting for many of the behavioral deficits found in adults after early disruption or abuse of nicotinic cholinergic signaling.

Keywords Nicotinic • Nicotinic receptors • Synaptogenesis • Synaptic plasticity • Acetylcholine • Development • Cholinergic • Spontaneous activity • Waves • Excitatory GABA • Hippocampus • Critical period • Nicotine

---

C.C. Fernandes, Ph.D. • A.F. Lozada, Ph.D. • D.K. Berg, Ph.D. (✉)
Section of Neurobiology, Division of Biological Sciences, University of California, San Diego, 9500 Gilman Drive, La Jolla, San Diego, CA 92093, USA
e-mail: ccfernandes@ucsd.edu; aflozada@ucsde.edu; dberg@uscd.edu

R.A.J. Lester (ed.), *Nicotinic Receptors*, The Receptors 26,
DOI 10.1007/978-1-4939-1167-7_6

1 Appearance of Nicotinic Components

Nicotinic cholinergic signaling is mediated by nAChRs, which constitute a heterogeneous family of ligand-gated ion channels widely expressed in the brain (also see Chap. 5). The receptors are pentameric transmembrane proteins that function as relatively nonselective cation channels. Binding of acetylcholine (ACh) or other agonists allows a net ion flux that can depolarize the neuronal membrane ([24, 99]; see also Chap. 2) and cause direct or indirect rises in cytoplasmic calcium levels. Because calcium is one of the key regulators of neuronal development, nAChRs are in a position to influence the formation of neural networks. A variety of nAChR subtypes are expressed in the vertebrate nervous system, composed of various combinations of α- and β-type subunits. A total of nine different α subunit (α2–α10) and three different β subunit (β2–β4) proteins can be found in the CNS ([98]; see also Chap. 3). All neuronal nAChRs contain at least two α subunits but have a variable number of β subunits. Most abundant are homopentamers containing only α7 subunits (α7-nAChRs) and heteropentamers containing at least one β2 subunit along with α4 and other subunits (β2*-nAChRs; [3, 26, 36, 48, 119, 131]).

Many studies indicate that nAChRs appear early during CNS development (Fig. 6.1). Functional nAChRs can be detected by patch-clamp recording of agonist-evoked currents in fetal mouse cerebral cortex as early as embryonic day (E)10 [6]. Expression of α4 and β2 mRNAs have been detected as early as E11 in rat spinal cord and in more rostral CNS structures by E12; expression extends to the neocortex by E17–E19 [131]. Detection of nAChRs by high affinity binding of [^{3}H]nicotine reveals a caudal-to-rostral pattern of receptor appearance developmentally; binding is apparent in the neocortex by E20 [86]. Both transcript and binding sites for α7-nAChRs are also detectable in the developing rat brain early on, first appearing in cortical and thalamic regions at E13–E15 [18]. Binding of [^{125}I]α-bungarotoxin has been used to detect α7-nAChRs, and this marker reveals the receptors as early as E16 throughout the embryonic brain and spinal cord [116]. Both α3 and β4 mRNA transcripts are also widely distributed in rat brain between E19-E21, but expression levels decline somewhat as the brain matures [122]. The α5 subunit, appearing transiently in some regions during embryonic development and more stably elsewhere, usually combines with α4 and β2 subunits [120]; it is expressed by E18–E20 in embryonic cortex and hippocampus, as well as more broadly in catecholaminergic neurons [7, 90, 123]. In contrast, α6 and β3 subunit mRNAs are strongly expressed in adult catecholaminergic neurons, but are barely detectable during the prenatal period [7, 90].

The nAChR agonist ACh comes from two primary sources in the adult brain: projection neurons that innervate distal regions, and local interneurons that are interspersed among their neuronal targets [125]. Central cholinergic projection neurons are found in nuclei throughout the brain, notably the basal forebrain complex [77, 125, 128, 129], the medial septum [125], the medial habenula [103, 125], and the pedunculopontine and laterodorsal tegmental areas [59, 125].

Projection patterns of these cholinergic neurons are widespread and diffuse as revealed by visualizing their terminal branches [125].

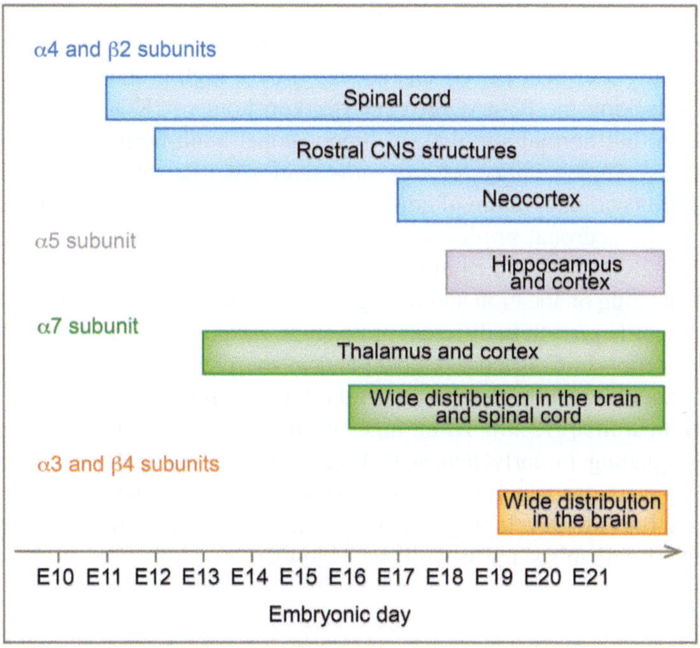

Fig. 6.1 Expression of nAChRs subunits in the nervous system during rodent embryonic development. Diagrammed here are the primary sites of major nAChR prenatal expression

Cholinergic neuronal populations can be detected by immunostaining for the enzyme choline acetyltransferase (ChAT), which synthesizes ACh and has been detected as early as neural plate stages in the presumptive crest [114]. This marker reveals a caudorostral gradient for appearance of cholinergic neurons during development [5, 95, 106]. ChAT-positive neurons first appear in the spinal cord (somatic and then visceral motoneurons at E12-E14) and then in more rostral structures such as the basal forebrain (E17–E18; [94, 108]). Cholinergic neurons are among the first to differentiate in the CNS regions studied, occurring even prior to any evidence of synaptogenesis [10, 94, 108, 109].

Nicotinic signaling influences neuronal survival. In the chick ciliary ganglion (CG), 50 % of the neurons are eliminated due to programmed cell death between E8–E14 [61]. All CG neurons receive functional nicotinic innervation by E7 [60]. Surprisingly, blockade of α7-nAChRs between E8–E14 prevents nearly all of the naturally-occurring cell death [19, 52, 76]. The ability of α7-nAChRs to mediate cell death in this case may reflect their ability to elevate intracellular calcium levels, perhaps exceeding some critical threshold for cell survival [14, 31, 107]. Indeed, excessive α7-nAChR function can produce abnormal and massive cell death, as in transgenic mice homozygous for a gain-of-function mutation in α7-nAChRs [92].

Nicotinic signaling can also regulate neuronal survival in a positive way. Cholinergic forebrain lesion decreases adult-born neuron survival in the hippocampus, while enhancing cholinergic activity increases survival of the neurons [27, 55]. Contrary to the effect on CG neurons, α7-nAChRs protect adult-born hippocampal neurons during a critical period characterized by a high probability of cell death [21]. Mice lacking the β2-nAChR gene (β2 knockouts, β2KOs) show decreased proliferation, but normal survival, of hippocampal adult-born neurons [46]. This indicates the complementary roles of individual nAChRs subtypes in determining the maturation and integration of adult-born neurons into existing hippocampal circuits. Notably, neuronal progenitors from the subventricular zone show reduced survival in β2KOs [74], indicating that a given nAChR subtype can have different actions depending on location and timing. Another example comes from α7-nAChR effects on neurite growth in different brain areas. Activation of α7-nAChRs induces neurite retraction in CG neurons [102] and in the pheochromocytoma line PC12 [88], but promotes neurite elongation in rat olfactory bulb neurons and hippocampal adult-born neurons [21, 28]. Taken together, these observations support a role for nicotinic signaling in early neuronal development. Once differentiated, neurons establish connections which allow the propagation of spontaneous waves of depolarization across the developing circuitry. Remarkably, nicotinic signaling plays a crucial role during this stage as well (see below).

2 Nicotinic Activity Patterns Early in Development

Spontaneous neuronal oscillations are propagating bursts of action potentials that occur early in development and constitute a hallmark of immature networks. They initially are restricted to a few pairs of neurons, but more neurons become engaged in each event as the network matures. These neuronal oscillations (or waves) appear conserved in evolution, having been described in a number of structures ([15, 56]; see [30] for references), and they can provide most of the activity during a restricted time window of development. Because the waves employ calcium signaling, they are thought to alter synaptic transmission and promote the formation and structural refinement of neuronal networks. This is likely to include changes in gene expression which activity and calcium influx are known to regulate [16, 23, 33, 44, 53].

Due to the breadth of this phenomenon, the mechanisms that underlie the occurrence of spontaneous waves of activity have been studied intensely. Patterned spontaneous bursting activity is remarkably similar across systems [15]. Each brain area comprises a unique circuit, but some aspects of the mechanisms used to generate spontaneous activity are similar in different systems. In general, all immature networks follow a specific developmental sequence initially characterized by intrinsic, synapse-independent voltage-gated calcium currents, followed by large calcium plateaus in small neuronal populations connected by gap junctions. Subsequently, primitive spontaneous transmitter-driven patterns appear, which can even occur before synapses are formed. In this section, we focus on the retina, hippocampus, and spinal cord, where nicotinic signaling has been shown to have important instructive or permissive roles in the generation of patterned neuronal oscillations.

2.1 Retina

Spontaneous neuronal oscillations have been best studied in the developing visual system, where waves of spontaneous activity originate in the retina [75, 124] and dictate firing patterns up to the primary visual cortex [1, 81]. Isolated preparations of retina exhibit propagating bursts of action potentials (termed "retinal waves") among neighboring retinal ganglion cells (RGCs). Spontaneous retinal waves propagate spatially, coordinating the firing of neighboring cells (Fig. 6.2a). As RGCs relay visual information to higher-order structures in the CNS, retinal waves are thought to have a key role in the activity-dependent refinement of topographic neural maps in the superior colliculus (SC), dorsal lateral geniculate nucleus (dLGN), and visual cortex, which exhibit functional connectivity before the onset of visual experience. More recently, Ackman and colleagues reported that these waves are preferentially initiated in the binocular retina and exhibit spatiotemporal correlations between the two hemispheres [1].

In rodents, spontaneous retinal waves are present for an extended period of development in vivo and exhibit a pattern of activity appropriate for communicating retinal organization to circuits throughout the visual system. As retinal circuits change with age, so does the wave generation mechanism. Stage I retinal waves occur independent of fast synaptic transmission; instead, these waves require gap junctions and adenosine for propagation. Stage II retinal waves are mediated by nAChRs, and stage III retinal waves are mediated by ionotropic glutamate receptors. The best understood wave-generating circuit is based on nicotinic cholinergic signaling, which is thought to segregate RGC axons from the two eyes and also to segregate axons arising from RGCs in the same eye and of the same subtype. Cholinergic retinal waves are present in rodents up through P10 [9]. They are initiated by a network of cholinergic amacrine cells called starburst amacrine cells (SAC), which display spontaneous depolarizations that eventually lead to a diffuse release of ACh onto neighboring RGCs and other SACs [34, 37, 130]. At this stage, SACs express β2*-nAChRs, which will promote the amplification of the signal and propagation of the wave. Transgenic mice have been useful for dissecting the role of cholinergic waves and nicotinic signaling in the establishment and refinement of the visual system. Mice lacking ChAT do not exhibit cholinergic- or other transmitter-mediated waves [115]. Instead, they display compensatory waves mediated by gap junctions, as an extension of an earlier, nonsynaptic wave-generating mechanism that has been observed in embryonic mice. This observation suggests a homeostatic regulation of spontaneous oscillations in the retina at this early stage and that cholinergic signaling may act to suppress neuronal coupling as chemical synaptic connectivity matures.

The β2KO mouse has been useful for assessing the role of nicotinic cholinergic signaling in generating patterned spontaneous waves. These mice show altered patterns of spontaneous retinal activity up through P8 RGCs spiking randomly and showing little correlation with neighboring RGCs [73]. The absence of cholinergic waves culminates in abnormal retinofugal projections in the dLGL and SC with loss of eye-specific segregation, but has no influence on the development of retinotopic

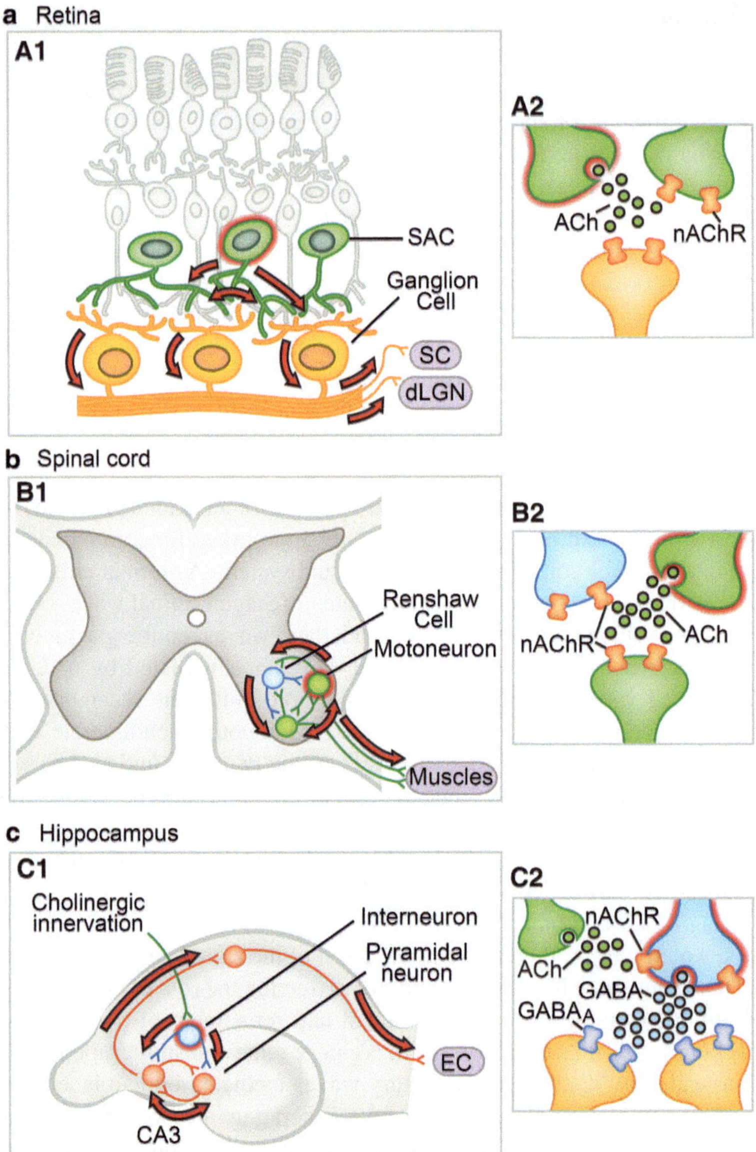

Fig. 6.2 Spontaneous network activity during development depends on nAChRs. Cholinergic neurons (*green*) can activate nAChRs on glutamatergic (*orange*) and GABAergic (*blue*) neurons. Patterned spontaneous waves of excitation are initiated by a depolarization in a particular cell type (outlined in *red*) for a given region, and are then amplified by recurrent connections (*red double-headed arrows*) and propagated to their targets (*red arrows*). (**a**) Retina. A diagram of the overall circuit is shown (*A1*) together with a blow-up of transmitter/receptor interactions (ACh and nAChRs) mediating the waves (*A2*). (**b**) Spinal Cord. Overall circuit (*B1*) and blow-up (*B2*) as in panel (**a**). (**c**) Hippocampus. Circuit (*C1*) and blow-up with GABA being key for propagation in this case (*C2*). Although ACh does not mediate patterned waves in the hippocampus, it does determine the temporal characteristics because interneurons express nAChRs that are activated by cholinergic neurons that project from the septum

maps in the monocular zone of the dLGN and SC [22, 73, 84, 126]. Patterned RGC waves return to normal at P8 in β2KO mice [9, 73] when glutamatergic waves appear as a compensatory mechanism. Columnar specificity, however, remains severely abnormal in adult β2KOs, revealing a critical period for columnar segregation of RGC projections that is mediated by β2*-nAChR-dependent spontaneous retinal activity during the first postnatal week [51]. These observations also demonstrate that homeostatically generated glutamatergic activity does not serve the same function as normal cholinergic activity.

2.2 Spinal Cord

Embryonic spinal cord both in mouse and chick displays patterned spontaneous activity that has many similarities to the retinal waves described above. Patterned waves of activity are generated in isolated lumbar cords even in the absence of descending and afferent input and consist of propagating depolarizing events that cause near synchronous activation of most motor neurons on both sides of the cord. This spontaneous activity seems to be required for accurate motoneuron pathfinding and for the formation of the central pattern generator, which produces oscillatory rhythms for locomotion into adulthood [44, 85].

Spontaneous rhythmic activity occurs as early as E3–E4 in embryonic chicks and E11–E12 in embryonic mice. Two phases of spinal cord development have been identified: phase I is dependent upon electrical transmission and cholinergic signaling (E12.5–E14.5 in mice), and phase II is defined by glutamatergic activity (from E15.5 onward in mice; [43, 79]). In contrast to the retina and hippocampus, no pacemaker-like neuron has been conclusively identified in the developing spinal cord. Motor neurons, however, are thought to be responsible for triggering the spontaneous waves of activity since they are the first to be active in each synchronized episode (Fig. 6.2b). During phase I, motor neurons (which are cholinergic) may form transient nAChR-mediated synapses onto other motor neurons and onto local GABAergic interneurons (Renshaw cells), which are depolarized by ACh and contribute to the propagation of a synchronized wave [45]. Blockade of nAChRs with antagonists does prevent spontaneous activity early in spinal cord development [43]. Interestingly, the immature network can overcome nAChR blockade; a compensatory glutamatergic-driven network appears a few minutes after the nAChR blockade. Consistent with this, Myers and colleagues showed that ChAT mutants also exhibit a different pattern of spontaneous activity during development [85]. In these mutants, motor activity is reduced in both phase I and II, and spontaneous network activity is prematurely mediated by glutamatergic signaling. Further, the duration of each cycle is elongated, and both right-left and flexor-extensor coordinations are abnormal. Blockade of nAChRs after the central pattern generator is formed, however, does not affect right-left or flexor-extensor coordination, suggesting that cholinergic signaling is required during a transient period of development.

2.3 *Hippocampus*

In the rodent hippocampus, spontaneous correlated activity appears during the first postnatal week. These GDPs [12], which are network-driven synaptic events generated by GABA and NMDA receptors, are usually present between P2 and P10 [30]. GDPs may be the in vitro counterpart of sharp waves, which occur in rat pups during periods of immobility periods such as during sleep and feeding [63], and occur at a similar frequency as GDPs in acute slices (0.3–0.1 Hz). This primitive form of network oscillation may be a primordial form of synchrony between neurons, which precedes more organized forms of activity in the hippocampus, such as the theta and gamma rhythms [20].

As noted above, transmitter-driven synchronized oscillations are first dependent on cholinergic transmission in the spinal cord and retina, while in the hippocampus they are initiated by GABA ([69]; Fig. 6.2c). Nevertheless, nicotinic signaling may still play an important role in modulating hippocampal GDPs, because the interneurons have nAChRs and receive cholinergic innervation from the medial septum–diagonal band complex of the basal forebrain [38]. Though not required for GDP initiation, nAChRs may help shape the spatiotemporal characteristics of hippocampal GDPs. This has yet to be tested, but strong evidence suggests that signaling via α7- and β2*-nAChRs exerts powerful regulatory actions on network-driven GDPs. Blockade of β2*-nAChRs with a selective antagonist decreases the frequency of GDPs in the hippocampus [70], while antagonizing α7-nAChRs can either increase or decrease the frequency of GDPs [70]. In addition, application of nicotine induces an increase in the frequency of GDPs that is dependent on both α7- and β2*-nAChRs [62, 70]. Maggi and colleagues propose that activation of α7- and β2*-nAChRs in interneurons synapsing directly onto pyramidal cells leads to an increase in the release of GABA, stepping up the frequency of GDPs; activation of α7-nAChRs on interneurons synapsing onto other interneurons could decrease GDP frequency. This model does not rule out the possibility that α7- and β2*-nAChRs expressed by pyramidal cells at this age could also increase the release of glutamate to promote the occurrence of patterned waves. GDPs disappear toward the end of the first postnatal week, when GABA becomes inhibitory (see below).

3 Nicotinic Control of Gabaergic Maturation

In immature networks, spontaneous patterned waves rely on the hyperexcitable nature of recurrently connected synaptic circuits. In several regions of the nervous system, including the spinal cord and hippocampus, hyperexcitability is caused by an excitatory action of GABA and glycine [12, 13, 104], which in the adult brain act as inhibitory neurotransmitters. Both GABA and glycine can activate chloride-permeable channels, allowing the flow of Cl^- down its electrochemical gradient. Immature and mature neurons show different $[Cl^-]_i$ due to differential expression of two Cl^- cotransporters, NKCC1 and KCC2, which enhance and lower intracellular $[Cl^-]_i$, respectively. In immature neurons, a higher expression of NKCC1 leads to the accumulation $[Cl^-]_i$, setting its electrochemical gradient to more positive values than the resting membrane potential (Fig. 6.3a). These neurons are depolarized by GABA, supporting episodes of bursting activity and underlying waves of depolarization. Depolarizing GABA can, however, also produce inhibition by shunting excitatory currents, offering an inhibitory mechanism early in development [57]. As the network matures, the expression of KCC2 is favored, resulting in the extrusion of Cl^- from the internal milieu and hyperpolarization of the Cl^- reversal potential (Fig. 6.3b).

Endogenous nicotinic cholinergic signaling appears to regulate the expression of chloride transporters during development and, as a result, helps determine when GABAergic signaling becomes inhibitory ([66]; Fig. 6.3c). Loss or blockade of nAChRs has been shown to delay the switch from GABA-mediated excitation to inhibition in CG, spinal cord, and hippocampus [21, 66]. In the developing retina, however, it is not clear whether GABA signaling is required for cholinergic retinal wave generation, though activation of $GABA_A$ receptors on RGCs is initially depolarizing [110, 121]. $GABA_A$ receptor antagonists block retinal waves in turtles, but not in ferrets or mice where they only modulate wave properties.

3.1 *Spinal Cord*

GABAergic signaling is depolarizing in the spinal cord of chick embryos by E6 and converts to hyperpolarizing at E10 [89]. In the mouse embryonic spinal cord, both NKCC1 and KCC2 are expressed and functional early in development (E11.5–E13.5) when $GABA_A$ and glycine receptor activation induces strong excitatory action [32]. At this stage, blockade of $GABA_A$ receptors markedly diminishes the frequency of spontaneous motor bursts. After E15.5, a switch occurs rendering GABA and glycine unable to provide excitation. Thus, the effects of GABA and glycine on motor activity switch from excitatory to inhibitory as development progresses. Blockade of α7- and β2*-nAChRs in ovo at E2 delays the maturation of a mature reversal potential for Cl^- in spinal cord neurons [66].

3.2 *Hippocampus*

In the hippocampus, the initial period of depolarizing GABAergic signaling is necessary both for early postnatal and adult-born neurons to develop properly and integrate into circuits [13, 39, 93, 104]. Interestingly, at birth there is a dramatic and

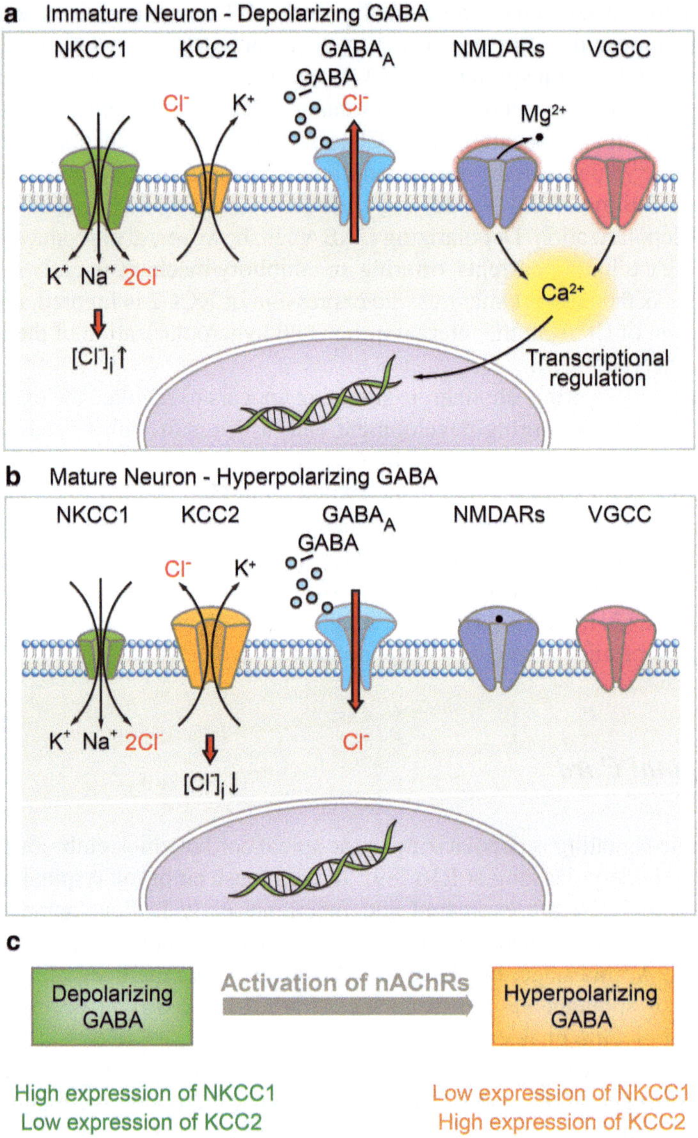

Fig. 6.3 Endogenous nicotinic cholinergic signaling regulates the timing for GABAergic conversion from depolarizing to hyperpolarizing. (**a**) Immature neurons have higher levels of NKCC1, which increases intracellular chloride $[Cl^-]_i$ and elevates the chloride reversal potential relative to the membrane potential. Activation of $GABA_A$ receptors then depolarizes the membrane, allowing glutamate activation of NMDA receptors (NMDARs) and/or opening of voltage-gated calcium channels (VGCC). (**b**) As the network matures, KCC2 levels increase, resulting in the chloride extrusion and creation of a more negative chloride reversal potential. (**c**) The switch in the chloride gradient is driven at least in part by α7-nAChR activity, which promotes elevated KCC2 levels along with decreased NKCC1 levels

temporary hyperpolarizing shift in the reversal potential of Cl^-, providing inhibitory GABAergic signaling to the immature network which is thought to minimize the severity of anoxic episodes that are frequent during delivery [117]. This initial GABA switch is transient, supporting the idea that a period of depolarizing GABA during early development is needed for proper maturation of neurons and neural networks [91]. For instance, depolarizing GABA is crucial for the generation of GDPs in the developing hippocampus. Pharmacological blockade of NKCC1 transporters causes the inhibition of GDPs [87, 111]. In freely moving rats, NKCC1 inhibitor bumetanide blocks sharp waves, supporting the view that sharp waves and GDPs are homologous [111]. By P7-P10, GABA becomes inhibitory due to a shift of the Cl^- reversal potential to hyperpolarized values. As in the spinal cord, this developmental switch appears to depend on nAChRs [66]. Mice lacking α7-nAChRs (α7 knockouts, α7KOs) show a delayed transition from depolarizing GABA to hyperpolarizing in the hippocampus. Cholinergic signaling through α7-nAChRs is also required for maturation of the Cl^- gradient of adult-born neurons in the dentate gyrus [21]. Adult-born neurons in α7KOs show a prolonged period of GABA being depolarizing, and GABAergic currents that have slow kinetics under these conditions [21] as found in immature neurons. For both embryonic and adult neurogenesis, nicotinic signaling dictates the timing for the transition from an immature network to a more mature stage, in which GABA plays a critical role in inhibition.

4 Nicotinic Promotion of Glutamatergic Synapse Formation

Studies involving nicotine exposure during development suggest roles that nAChRs may have in modulating circuit formation in critical brain regions. For example, nicotine application to hippocampal slices from P2 to P6 rats can increase GDPs by activating α7-nAChRs and modulating GABA release in the CA3 region [70]. Nicotine application can also affect the composition of synaptic receptors. In hippocampal slices from P1-P5 rats, nicotine application to immature Schaffer collateral-CA1 connections can convert presynaptically silent synapses to functional status [71]. Application of nicotine also selectively enhances the NMDA receptor-mediated component of excitatory postsynaptic potentials (EPSPs) in thalamocortical neurons from P8–P16 rats. This was not observed, however, at P19–P24, suggesting a limited period of vulnerability to nicotine [4]. The effect at P8–P16 depended on α7-nAChRs and involved enhanced glutamate release [4, 71, 78].

Though nicotinic signaling has long been known to promote synaptic plasticity and alter circuit function in the adult CNS, it was less clear whether it might also have a more fundamental role early in development, perhaps determining the number of synapses comprising circuits. Recently it has been shown that signaling through α7-nAChRs is necessary to achieve normal numbers of glutamatergic synapses and to establish the appropriate ratio of glutamatergic-to-GABAergic input capability for neurons [67]. This was demonstrated by analyzing synaptic contacts in α7KO mice and comparing the outcome with age-matched wild-type (WT) mice. The α7KOs had fewer glutamatergic synapses both in the hippocampal CA1 region and in the visual cortex at P12 than did WT mice. Moreover, the deficits were maintained into the adult, e.g. P60, demonstrating their persistence. Surprisingly, no deficits were seen in GABAergic synapses, and electrophysiological analysis confirmed an altered ratio of glutamatergic/GABAergic input to CA1 pyramidal neurons in α7KO vs WT mice. This raises the prospect of long-lasting important changes in network function and may well account for some of the behavioral deficits previously reported for α7KOs. These include attention deficits, impaired spatial discrimination, and diminished working/episodic memory [35, 49, 62, 65, 127].

A quite different kind of effect on glutamatergic synapses was discovered in β2KO mice. Initial reports indicated that the mice had altered numbers of dendritic spines, the primary location of glutamatergic synapses in the adult. At 4–5 months of age, β2KOs have significant reductions in spine density and in total numbers of spines in the prelimbic/infralimbic cortex and in the M1 field of motor cortex [8]. In contrast α7KOs have spine numbers comparable to WT littermates both in the visual cortex and in the hippocampal CA1 region [67], with an increase actually being reported for basal dendrites in the CA1 [83]. The spine deficits in β2KOs are apparent as early as P4-P12 and persist at least to P40 [68]. The requirement for β2*-nAChRs is cell-autonomous and appears to depend on intracellular calcium and activation of calcium, calmodulin-dependent protein kinase II [68]. Unexpectedly, β2KOs have normal numbers of glutamatergic synapses. The explanation is that more glutamatergic synapses are located on the dendritic shafts in β2KOs than found in WTs, thereby compensating for the deficit in spines [68]. Also noteworthy was the finding that acute activation of β2*-nAChRs by local stereotaxic injection of minute amounts of nicotine can quickly induce spines de novo, e.g. within an hour [68].

These results raise interesting questions about the possible roles of nicotinic signaling via β2*-nAChRs in determining the type and location of glutamatergic synapses formed on a neuron (Fig. 6.4). The synaptic rearrangements found in β2KOs could account in part for the numerous behavioral deficits reported for the mice. These include abnormal passive avoidance, impaired nicotine self-administration and drug discrimination, reduced nociceptive response to nicotine, decreased visual acuity, reduced locomotion in a familiar environment, and deficits in executive functions and social behavior reported in β2KOs [42, 72, 96, 97, 99, 105, 112].

5 Nicotinic Contribution to Neural Net Formation

Numerous studies indicate that early exposure to nicotine produces long-lasting changes in behavior [47]. This emerges both from experiments with rodents and from data collection on human populations. Unusual vulnerability appears to extend

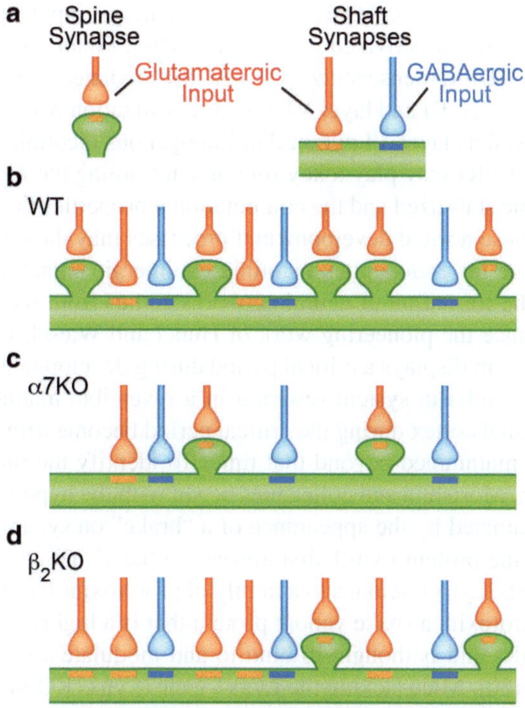

Fig. 6.4 Endogenous nicotinic cholinergic signaling helps determine the pattern of glutamatergic synapses and network formation during development. (**a**) Dendritic shaft and spine synapses. (**b**) Relative distributions of synaptic types in early postnatal WT. (**c**) Reduced numbers of glutamatergic synapses in α7KOs compared to WT. (**d**) Equivalent numbers of glutamatergic synapses in β2KOs and WTs, but the reduced number of dendritic spines in β2KOs apparently causes a greater number of glutamatergic synapses to locate on dendritic shafts. This difference persists into adulthood, indicating that the animal is unable to compensate for the deficits in spine number resulting from β2*-nAChR loss

to the adolescent brain as well when nicotine apparently can have major effects on the nervous system that is yet to achieve mature status [41]. This is likely due to the pervasive distribution of nAChRs in key brain regions associated with reward and cognitive in the adolescent brain both in humans [25, 54] and in animal models [2, 11, 17, 64, 113, 118]. In the prefrontal cortex (PFC), nAChRs are found across all layers [40, 101]. Signaling through nAChRs can alter pyramidal neuron activity by enhancing GABA and glutamatergic input or by activating the postsynaptic cell directly [100, 101]. In PFC layer V, activation of presynaptic β2*-nAChRs on glutamatergic inputs to pyramidal neurons from the thalamus strongly enhances activity of these cells [29, 40, 58, 101]; layer V pyramidal neurons also have post-synaptic α7-nAChRs [101]. In contrast, layer VI of the same brain area contains pyramidal neurons with β2*-nAChRs while glutamatergic inputs to those same neurons are only moderately activated by nAChRs [101]. Finally, smoking concentrations of nicotine (300 nM) densensitize β2*-nAChR-mediated current in interneurons located in PFC layer II–III and layer VI; less desensitization was detected in layer V interneurons as well as layer VI pyramidal. Endogenous nicotinic cholinergic input through these nAChRs may play a key role in determining the kinds of neural networks that become stabilized and the computational properties that result. Nicotinic perturbation of these normal developmental processes may then alter the outcome.

Direct evidence for endogenous nicotinic cholinergic signaling playing fundamental roles in the formation of neural networks comes from recent studies on the visual system. Since the pioneering work of Hubel and Wiesel, it has been known that the visual system displays a critical period during development in which unilateral eye closure results in system rewiring in a reversible manner [50]. Changes induced in the visual cortex during the critical period become irreversible, however, if eye closure is maintained beyond that time. To identify the molecular basis for this phenomenon, Morishita and colleagues began with the hypothesis that the critical period is terminated by the appearance of a "brake" on synaptic plasticity [82]. They found that the protein Lynx1 first appears in the visual cortex of mice as the critical period ends. Lynx1 belongs to a family of prototoxins having sequence similarity to α-bungarotoxin, a snake venom protein that is a highly efficient antagonist of muscle nAChRs and is thought to bind to and modulate a number of neuronal nAChRs as well [80]. Mice lacking the Lynx1 gene (Lynx1KO mice) retained plasticity in the visual system: multiday eye closure applied to adult Lynx1KO mice produced shifts in eye dominance away from the closed eye when tested after opening. Moreover, if eye closure was initiated during the critical period and extended beyond it, opening of the eye subsequently allowed recovery of normal vision within days, something that was never seen in WT mice.

These observations led Morishita and colleagues to propose that nicotinic signaling mediates plasticity in the developing visual system and that the appearance of Lynx1 terminates the plasticity by blocking nAChRs, thereby establishing the end of the critical period. In support of this, they demonstrated that either the relatively global nAChR blocker mecamylamine or the more specific blockers dihydro-β-erythroidine and methyllycaconitine (specific for α4β2- and α7-nAChRs, respectively) applied to Lynx1KO mice rendered them equivalent to WT mice with respect to critical period constraints. Conversely, blockade of acetylcholinesterase in WT mice to increase endogenous ACh levels extended the critical period, making them more like the Lynx1KO mice in this respect. The additional ACh was apparently able to compete at least partially the blocking effects of endogenous Lynx1. The results strongly suggest that endogenous nicotinic cholinergic signaling promotes plasticity and rewiring of circuits during development, and that Lynx1 acts to terminate this process subsequently, thereby securing the "mature" configuration. Which nAChR subtypes are responsible and whether this process extends widely across the CNS are questions for the future.

6 Summary and the Future

Nicotinic cholinergic signaling appears early in development and is widely distributed. It drives waves of excitatory activity across numerous CNS regions prior to and during the major period of de novo synapse formation. Activity through α7-nAChRs promotes glutamatergic synapse formation while activity through β2*-nAChRs increases dendritic spine number and the location of synapses at these sites. In the absence of α7-nAChRs, GABAergic development proceeds more slowly, retaining the early depolarizing/excitatory mode much longer in what probably reflects a delay in reaching some critical level of innervation required for the switch in chloride transporters that makes possible the GABAergic conversion. The relevance of nicotinic signaling for network plasticity is perhaps best demonstrated in the visual system where it appears to be responsible for the famed critical period; Lynx1 terminates the critical period apparently by blocking endogenous nicotinic cholinergic signaling. Important questions for the future include the molecular mechanisms by which nicotinic signaling achieves its effects on synaptic plasticity and how these mechanisms influence network capacity overall. Also important will be determining the breadth of nicotine-mediated plasticity across the CNS and whether some of these mechanisms remain operational in adult circuits. A final challenge will be the question of whether plasticity controlled by nicotinic signaling can be manipulated therapeutically to reverse or compensate neurological deficits.

References

1. Ackman JB, Burbridge TJ, Crair MC. Retinal waves coordinate patterned activity throughout the developing visual system. Nature. 2012;490:219–25.
2. Adriani W, Spijker S, Deroche-Gamonet V, Laviola G, Le Moal M, Smit AB, Piazza PV. Evidence for enhanced neurobehavioral vulnerability to nicotine during periadolescence in rats. J Neurosci. 2003;23:4712–6.
3. Anand R, Conroy WG, Schoepfer R, Whiting P, Lindstrom J. Neuronal nicotinic acetylcholine receptors expressed in Xenopus oocytes have a pentameric quaternary structure. J Biol Chem. 1991;266:11192–8.
4. Aramakis VB, Metherate R. Nicotine selectively enhances NMDA receptor-mediated synaptic transmission during postnatal development in sensory neocortex. J Neurosci. 1998;18:8485–95.
5. Armstrong DM, Bruce G, Hersh LB, Gage FH. Development of cholinergic neurons in the septal/diagonal band complex of the rat. Brain Res. 1987;433:249–56.
6. Atluri P, Fleck MW, Shen Q, Mah SJ, Stadfelt D, Barnes W, Goderie SK, Temple S, Schneider AS. Functional nicotinic acetylcholine receptor expression in stem and progenitor cells of the early embryonic mouse cerebral cortex. Dev Biol. 2001;240:143–56.
7. Azam L, Chen Y, Leslie FM. Developmental regulation of nicotinic acetylcholine receptors within midbrain dopamine neurons. Neuroscience. 2007;144:1347–60.
8. Ballesteros-Yáñez I, Benavides-Piccione R, Bourgeois JP, Changeux JP, DeFelipe J. Alterations of cortical pyramidal neurons in mice lacking high-affinity nicotinic receptors. Proc Natl Acad Sci U S A. 2010;107:11567–72.

9. Bansal A, Singer JH, Hwang BJ, Xu W, Beaudet A, Feller MB. Mice lacking specific nicotinic acetylcholine receptor subunits exhibit dramatically altered spontaneous activity patterns and reveal a limited role for retinal waves in forming ON and OFF circuits in the inner retina. J Neurosci. 2000;20:7672–81.
10. Barber RP, Phelps PE, Vaughn JE. Generation patterns of immunocytochemically identified cholinergic neurons at autonomic levels of the rat spinal cord. J Comp Neurol. 1991;311: 509–19.
11. Belluzzi JD, Lee AG, Oliff HS, Leslie FM. Age-dependent effects of nicotine on locomotor activity and conditioned place preference in rats. Psychopharmacology (Berl). 2004;174: 389–95.
12. Ben-Ari Y, Cherubini E, Corradetti R, Gaiarsa JL. Giant synaptic potentials in immature rat CA3 hippocampal neurones. J Physiol. 1989;416:303–25.
13. Ben-Ari Y. Excitatory actions of gaba during development: the nature of the nurture. Nat Rev Neurosci. 2002;3:728–39.
14. Bertrand D, Galzi JL, Devillers-Thiery A, Bertrand S, Changeux JP. Mutations at two distinct sites within the channel domain M2 alter calcium permeability of neuronal α7 nicotinic receptor. Proc Natl Acad Sci U S A. 1993;90:6971–5.
15. Blankenship AG, Feller MB. Mechanisms underlying spontaneous patterned activity in developing neural circuits. Nat Rev Neurosci. 2010;11:18–29.
16. Borodinsky LN, Root CM, Cronin JA, Sann SB, Gu X, Spitzer NC. Activity-dependent homeostatic specification of transmitter expression in embryonic neurons. Nature. 2004;429: 523–30.
17. Brielmaier JM, McDonald CG, Smith RF. Immediate and long-term behavioral effects of a single nicotine injection in adolescent and adult rats. Neurotoxicol Teratol. 2007;29:74–80.
18. Broide RS, O'Connor LT, Smith MA, Smith JA, Leslie FM. Developmental expression of alpha 7 neuronal nicotinic receptor messenger RNA in rat sensory cortex and thalamus. Neuroscience. 1995;67:83–94.
19. Bunker GL, Nishi R. Developmental cell death in vivo: rescue of neurons independently of changes at target tissues. J Comp Neurol. 2002;452:80–92.
20. Buzsáki G, Draguhn A. Neuronal oscillations in cortical networks. Science. 2004;304:1926–9.
21. Campbell NR, Fernandes CC, Halff AW, Berg DK. Endogenous signaling through alpha7-containing nicotinic receptors promotes maturation and integration of adult-born neurons in the hippocampus. J Neurosci. 2010;30:8734–44.
22. Chandrasekaran AR, Plas DT, Gonzalez E, Crair MC. Evidence for an instructive role of retinal activity in retinotopic map refinement in the superior colliculus of the mouse. J Neurosci. 2005;25:6929–38.
23. Chang KT, Berg DK. Voltage-gated channels block nicotinic regulation of CREB phosphorylation and gene expression in neurons. Neuron. 2001;32:855–65.
24. Changeux JP, Bertrand D, Corringer PJ, Dehaene S, Edelstein S, Léna C, Le Novère N, Marubio L, Picciotto M, Zoli M. Brain nicotinic receptors: structure and regulation, role in learning and reinforcement. Brain Res Brain Res Rev. 1998;26:198–216.
25. Colby SM, Tiffany ST, Shiffman S, Niaura RS. Are adolescent smokers dependent on nicotine? A review of the evidence. Drug Alcohol Depend. 2000;59 Suppl 1:S83–95.
26. Cooper E, Couturier S, Ballivet M. Pentameric structure and subunit stoichiometry of a neuronal nicotinic acetylcholine receptor. Nature. 1991;350:235–8.
27. Cooper-Kuhn CM, Winkler J, Kuhn HG. Decreased neurogenesis after cholinergic forebrain lesion in the adult rat. J Neurosci Res. 2004;77:155–65.
28. Coronas V, Durand M, Chabot JG, Jourdan F, Quirion R. Acetylcholine induces neuritic outgrowth in rat primary olfactory bulb cultures. Neuroscience. 2000;98:213–9.
29. Couey JJ, Meredith RM, Spijker S, Poorthuis RB, Smit AB, Brussaard AB, Mansvelder HD. Distributed network actions by nicotine increase the threshold for spike-timing-dependent plasticity in prefrontal cortex. Neuron. 2007;54:73–87.

30. Crépel V, Aronov D, Jorquera I, Represa A, Ben-Ari Y, Cossart R. A parturition-associated nonsynaptic coherent activity pattern in the developing hippocampus. Neuron. 2007;54: 105–20.
31. Dajas-Bailador F, Wonnacott S. Nicotinic acetylcholine receptors and the regulation of neuronal signalling. Trends Pharmacol Sci. 2004;25:317–24.
32. Delpy A, Allain AE, Meyrand P, Branchereau P. NKCC1 cotransporter inactivation underlies embryonic development of chloride-mediated inhibition in mouse spinal motoneuron. J Physiol. 2007;586:1059–75.
33. Dolmetsch RE, Lewis RS, Goodnow CC, Healy JI. Differential activation of transcription factors induced by Ca2+ response amplitude and duration. Nature. 1997;386:855–8.
34. Feller MB, Wellis DP, Stellwagen D, Werblin FS, Shatz CJ. Requirement for cholinergic synaptic transmission in the propagation of spontaneous retinal waves. Science. 1996;272: 1182–7.
35. Fernandes C, Hoyle E, Dempster E, Schalkwyk LC, Collier DA. Performance deficit of alpha7 nicotinic receptor knockout mice in a delayed matching-to-place task suggests a mild impairment of working/episodic-like memory. Genes Brain Behav. 2006;5:433–40.
36. Flores CM, Rogers SW, Pabreza LA, Wolfe BB, Kellar KJ. A subtype of nicotinic cholinergic receptor in rat brain is composed of alpha 4 and beta 2 subunits and is up-regulated by chronic nicotine treatment. Mol Pharmacol. 1992;41:31–7.
37. Ford KJ, Félix AL, Feller MB. Cellular mechanisms underlying spatiotemporal features of cholinergic retinal waves. J Neurosci. 2012;32:850–63.
38. Frotscher M, Léránth C. Cholinergic innervation of the rat hippocampus as revealed by choline acetyltransferase immunocytochemistry: a combined light and electron microscopic study. J Comp Neurol. 1985;239:237–46.
39. Ge S, Goh EL, Sailor KA, Kitabatake Y, Ming GL, Song H. GABA regulates synaptic integration of newly generated neurons in the adult brain. Nature. 2006;439:589–93.
40. Gioanni Y, Rougeot C, Clarke PB, Lepousé C, Thierry AM, Vidal C. Nicotinic receptors in the rat prefrontal cortex: increase in glutamate release and facilitation of mediodorsal thalamo-cortical transmission. Eur J Neurosci. 1999;11:18–30.
41. Goriounova NA, Mansvelder HD. Nicotine exposure during adolescence leads to short- and long-term changes in spike timing-dependent plasticity in rat prefrontal cortex. J Neurosci. 2012;32:10484–93.
42. Granon S, Faure P, Changeux JP. Executive and social behaviors under nicotinic receptor regulation. Proc Natl Acad Sci U S A. 2003;100:9596–601.
43. Hanson MG, Landmesser LT. Characterization of the circuits that generate spontaneous episodes of activity in the early embryonic mouse spinal cord. J Neurosci. 2003;23:587–600.
44. Hanson MG, Landmesser LT. Normal patterns of spontaneous activity are required for correct motor axon guidance and the expression of specific guidance molecules. Neuron. 2004;43:687–701.
45. Hanson MG, Milner LD, Landmesser LT. Spontaneous rhythmic activity in early chick spinal cord influences distinct motor axon pathfinding decisions. Brain Res Rev. 2008;57:77–85.
46. Harrist A, Beech RD, King SL, Zanardi A, Cleary MA, Caldarone BJ, Eisch A, Zoli M, Picciotto MR. Alteration of hippocampal cell proliferation in mice lacking the beta 2 subunit of the neuronal nicotinic acetylcholine receptor. Synapse. 2004;54:200–6.
47. Heath CH, Picciotto MR. Nicotine-induced plasticity during development: modulation of the cholinergic system and long-term consequences for circuits involved in attention and sensory processing. Neuropharmacology. 2009;56:254–62.
48. Hill Jr JA, Zoli M, Bourgeois JP, Changeux JP. Immunocytochemical localization of a neuronal nicotinic receptor: the beta 2-subunit. J Neurosci. 1993;13:1551–68.
49. Hoyle E, Genn RF, Fernandes C, Stolerman IP. Impaired performance of alpha7 nicotinic receptor knockout mice in the five-choice serial reaction time task. Psychopharmacology (Berl). 2006;189:211–23.
50. Hubel DH, Wiesel TN. The period of susceptibility to the physiologicl effects of unilateral eye closure in kittens. J Physiol. 1970;206:419–36.

51. Huberman AD, Manu M, Koch SM, Susman MW, Lutz AB, Ullian EM, Baccus SA, Barres BA. Architecture and activity-mediated refinement of axonal projections from a mosaic of genetically identified retinal ganglion cells. Neuron. 2008;59:425–38.
52. Hruska M, Nishi R. Cell-autonomous inhibition of alpha 7-containing nicotinic acetylcholine receptors prevents death of parasympathetic neurons during development. J Neurosci. 2007;27:11501–9.
53. Itoh K, Stevens B, Schachner M, Fields RD. Regulated expression of the neural cell adhesion molecule L1 by specific patterns of neural impulses. Science. 1995;270:1369–72.
54. Kandel DB, Chen K. Extent of smoking and nicotine dependence in the United States: 1991–1993. Nicotine Tob Res. 2000;2:263–74.
55. Kaneko N, Okano H, Sawamoto K. Role of the cholinergic system in regulating survival of newborn neurons in the adult mouse dentate gyrus and olfactory bulb. Genes Cells. 2006; 11:1145–59.
56. Katz LC, Shatz CJ. Synaptic activity and the construction of cortical circuits. Science. 1996;274:1133–8.
57. Khalilov I, Dzhala V, Ben-Ari Y, Khazipov R. Dual role of GABA in the neonatal rat hippocampus. Dev Neurosci. 1999;21:310–9.
58. Lambe EK, Picciotto MR, Aghajanian GK. Nicotine induces glutamate release from thalamocortical terminals in prefrontal cortex. Neuropsychopharmacology. 2003;28:216–25.
59. Lammel S, Lim BK, Ran C, Huang KW, Betley MJ, Tye KM, Deisseroth K, Malenka RC. Input-specific control of reward and aversion in the ventral tegmental area. Nature. 2012;491:212–7.
60. Landmesser L, Pilar G. The onset and development of transmission in the chick ciliary ganglion. J Physiol. 1972;222:691–713.
61. Landmesser L, Pilar G. Synaptic transmission and cell death during normal ganglionic development. J Physiol. 1974;241:737–49.
62. Le Magueresse C, Safiulina V, Changeux JP, Cherubini E. Nicotinic modulation of network and synaptic transmission in the immature hippocampus investigated with genetically modified mice. J Physiol. 2006;576:533–46.
63. Leinekugel X, Khazipov R, Cannon R, Hirase H, Ben-Ari Y, Buzsáki G. Correlated bursts of activity in the neonatal hippocampus in vivo. Science. 2002;296:2049–52.
64. Levin ED, Rezvani AH, Montoya D, Rose JE, Swartzwelder HS. Adolescent-onset nicotine self-administration modeled in female rats. Psychopharmacology (Berl). 2003;169:141–9.
65. Levin ED, Petro A, Rezvani AH, Pollard N, Christopher NC, Strauss M, Avery J, Nicholson J, Rose JE. Nicotinic alpha7- or beta2-containing receptor knockout: effects on radial-arm maze learning and long-term nicotine consumption in mice. Behav Brain Res. 2009;196: 207–13.
66. Liu Z, Neff RA, Berg DK. Sequential interplay of nicotinic and GABAergic signaling guides neuronal development. Science. 2006;314:1610–3.
67. Lozada AF, Wang X, Gounko NV, Massey KA, Duan J, Liu Z, Berg DK. Glutamatergic synapse formation is promoted by α7-containing nicotinic acetylcholine receptors. J Neurosci. 2012;32:7651–61.
68. Lozada AF, Wang X, Gounko NV, Massey KA, Duan J, Liu Z, Berg DK. Induction of dendritic spines by β2-containing nicotinic receptors. J Neurosci. 2012;32:8391–400.
69. Maccaferri G, McBain CJ. The hyperpolarization-activated current (Ih) and its contribution to pacemaker activity in rat CA1 hippocampal stratum oriens-alveus interneurones. J Physiol. 1996;497:119–30.
70. Maggi L, Sher E, Cherubini E. Regulation of GABA release by nicotinic acetylcholine receptors in the neonatal rat hippocampus. J Physiol. 2001;536:89–100.
71. Maggi L, Le Magueresse C, Changeux JP, Cherubini E. Nicotine activates immature "silent" connections in the developing hippocampus. Proc Natl Acad Sci U S A. 2003;100:2059–64.
72. Marubio LM, del Mar Arroyo-Jimenez M, Cordero-Erausquin M, Léna C, Le Novère N, de Kerchove d'Exaerde A, Huchet M, Damaj MI, Changeux JP. Reduced antinociception in mice lacking neuronal nicotinic receptor subunits. Nature. 1999;398:805–10.

73. McLaughlin T, Torborg CL, Feller MB, O'Leary DD. Retinotopic map refinement requires spontaneous retinal waves during a brief critical period of development. Neuron. 2003;40: 1147–60.
74. Mechawar N, Saghatelyan A, Grailhe R, Scoriels L, Gheusi G, Gabellec MM, Lledo PM, Changeux JP. Nicotinic receptors regulate the survival of newborn neurons in the adult olfactory bulb. Proc Natl Acad Sci U S A. 2004;101:9822–6.
75. Meister M, Wong RO, Baylor DA, Shatz CJ. Synchronous bursts of action potentials in ganglion cells of the developing mammalian retina. Science. 1991;252:939–43.
76. Meriney SD, Pilar G, Ogawa M, Nunez R. Differential neuronal survival in the avian ciliary ganglion after chronic acetylcholine receptor blockade. J Neurosci. 1987;7:3840–9.
77. Mesulam MM. Cholinergic pathways and the ascending reticular activating system of the human brain. Ann N Y Acad Sci. 1995;757:169–79.
78. Metherate R. Nicotinic acetylcholine receptors in sensory cortex. Learn Mem. 2004;11:50–9.
79. Milner LD, Landmesser LT. Cholinergic and GABAergic inputs drive patterned spontaneous motoneuron activity before target contact. J Neurosci. 1999;19:3007–22.
80. Miwa JM, Ibanez-Tallon I, Crabtree GW, Sánchez R, Sali A, Role LW, Heintz N. Lynx1, an endogenous toxin-like modulator of nicotinic acetylcholine receptors in the mammalian CNS. Neuron. 1999;23:105–14.
81. Mooney R, Penn AA, Gallego R, Shatz CJ. Thalamic relay of spontaneous retinal activity prior to vision. Neuron. 1996;17:863–74.
82. Morishita H, Miwa JM, Heintz N, Hensch TK. Lynx1, a cholinergic brake, limits plasticity in adult visual cortex. Science. 2010;330:1238–40.
83. Morley BJ, Mervis RF. Dendritic spine alterations in the hippocampus and parietal cortex of alpha7 nicotinic acetylcholine receptor knockout mice. Neuroscience. 2013;233:54–63.
84. Muir-Robinson G, Hwang BJ, Feller MB. Retinogeniculate axons undergo eye-specific segregation in the absence of eye-specific layers. J Neurosci. 2002;22:5259–64.
85. Myers CP, Lewcock JW, Hanson MG, Gosgnach S, Aimone JB, Gage FH, Lee KF, Landmesser LT, Pfaff SL. Cholinergic input is required during embryonic development to mediate proper assembly of spinal locomotor circuits. Neuron. 2005;46:37–49.
86. Naeff B, Schlumpf M, Lichtensteiger W. Pre- and postnatal development of high-affinity [3H]nicotine binding sites in rat brain regions: an autoradiographic study. Brain Res Dev Brain Res. 1992;68:163–74.
87. Nardou R, Ben-Ari Y, Khalilov I. Bumetanide, an NKCC1 antagonist, does not prevent formation of epileptogenic focus but blocks epileptic focus seizures in immature rat hippocampus. J Neurophysiol. 2009;101:2878–88.
88. Nordman JC, Kabbani N. An interaction between α7 nicotinic receptors and a G-protein pathway complex regulates neurite growth in neural cells. J Cell Sci. 2012;125:5502–13.
89. Obata K, Oide M, Tanaka H. Excitatory and inhibitory actions of GABA and glycine on embryonic chick spinal neurons in culture. Brain Res. 1978;144:179–84.
90. O'Leary KT, Loughlin SE, Chen Y, Leslie FM. Nicotinic acetylcholine receptor subunit mRNA expression in adult and developing rat medullary catecholamine neurons. J Comp Neurol. 2008;510:655–72.
91. Owens DF, Kriegstein AR. Is there more to GABA than synaptic inhibition? Nat Rev Neurosci. 2002;3:715–27.
92. Orr-Urtreger A, Broide RS, Kasten MR, Dang H, Dani JA, Beaudet AL, Patrick JW. Mice homozygous for the L250T mutation in the alpha7 nicotinic acetylcholine receptor show increased neuronal apoptosis and die within 1 day of birth. J Neurochem. 2000;74:2154–66.
93. Payne JA, Rivera C, Voipio J, Kaila K. Cation-chloride co-transporters in neuronal communication, development and trauma. Trends Neurosci. 2003;26:199–206.
94. Phelps PE, Brennan LA, Vaughn JE. Generation patterns of immunocytochemically identified cholinergic neurons in rat brainstem. Brain Res Dev Brain Res. 1990;56:63–74.
95. Phelps PE, Barber RP, Vaughn JE. Embryonic development of choline acetyltransferase in thoracic spinal motor neurons: somatic and autonomic neurons may be derived from a common cellular group. J Comp Neurol. 1991;307:77–86.

96. Picciotto MR, Zoli M, Léna C, Bessis A, Lallemand Y, Le Novère N, Vincent P, Pich EM, Brûlet P, Changeux JP. Abnormal avoidance learning in mice lacking functional high-affinity nicotine receptor in the brain. Nature. 1995;374:65–7.
97. Picciotto MR, Zoli M, Rimondini R, Léna C, Marubio LM, Pich EM, Fuxe K, Changeux JP. Acetylcholine receptors containing the beta2 subunit are involved in the reinforcing properties of nicotine. Nature. 1998;391:173–7.
98. Picciotto MR, Caldarone BJ, King SL, Zachariou V. Nicotinic receptors in the brain. Links between molecular biology and behavior. Neuropsychopharmacology. 2000;22:451–65.
99. Picciotto MR, Caldarone BJ, Brunzell DH, Zachariou V, Stevens TR, King SL. Neuronal nicotinic acetylcholine receptor subunit knockout mice: physiological and behavioral phenotypes and possible clinical implications. Pharmacol Ther. 2001;92:89–108.
100. Poorthuis RB, Goriounova NA, Couey JJ, Mansvelder HD. Nicotinic actions on neuronal networks for cognition: general principles and long-term consequences. Biochem Pharmacol. 2009;78:668–76.
101. Poorthuis RB, Bloem B, Verhoog MB, Mansvelder HD. Layer-specific interference with cholinergic signaling in the prefrontal cortex by smoking concentrations of nicotine. J Neurosci. 2013;33:4843–53.
102. Pugh PC, Berg DK. Neuronal acetylcholine receptors that bind alpha-bungarotoxin mediate neurite retraction in a calcium-dependent manner. J Neurosci. 1994;14:889–96.
103. Ren J, Qin C, Hu F, Tan J, Qiu L, Zhao S, Feng G, Luo M. Habenula “cholinergic” neurons co-release glutamate and acetylcholine and activate postsynaptic neurons via distinct transmission modes. Neuron. 2011;69:445–52.
104. Rivera C, Voipio J, Payne JA, Ruusuvuori E, Lahtinen H, Lamsa K, Pirvola U, Saarma M, Kaila K. The K+/Cl- co-transporter KCC2 renders GABA hyperpolarizing during neuronal maturation. Nature. 1999;397:251–5.
105. Rossi FM, Pizzorusso T, Porciatti V, Marubio LM, Maffei L, Changeux JP. Requirement of the nicotinic acetylcholine receptor beta 2 subunit for the anatomical and functional development of the visual system. Proc Natl Acad Sci U S A. 2001;98:6453–8.
106. Schambra UB, Sulik KK, Petrusz P, Lauder JM. Ontogeny of cholinergic neurons in the mouse forebrain. J Comp Neurol. 1989;288:101–22.
107. Séguéla P, Wadiche J, Dineley-Miller K, Dani JA, Patrick JW. Molecular cloning, functional properties, and distribution of rat brain alpha 7: a nicotinic cation channel highly permeable to calcium. J Neurosci. 1993;13:596–604.
108. Semba K, Fibiger HC. Time of origin of cholinergic neurons in the rat basal forebrain. J Comp Neurol. 1988;269:87–95.
109. Semba K, Vincent SR, Fibiger HC. Different times of origin of choline acetyltransferase- and somatostatin-immunoreactive neurons in the rat striatum. J Neurosci. 1988;8:3937–44.
110. Sernagor E, Young C, Eglen SJ. Developmental modulation of retinal wave dynamics: shedding light on the GABA saga. J Neurosci. 2003;23:7621–9.
111. Sipilä ST, Schuchmann S, Voipio J, Yamada J, Kaila K. The cation-chloride cotransporter NKCC1 promotes sharp waves in the neonatal rat hippocampus. J Physiol. 2006;573:765–73.
112. Shoaib M, Gommans J, Morley A, Stolerman IP, Grailhe R, Changeux JP. The role of nicotinic receptor beta-2 subunits in nicotine discrimination and conditioned taste aversion. Neuropharmacology. 2002;42:530–9.
113. Slotkin TA. Nicotine and the adolescent brain: insights from an animal model. Neurotoxicol Teratol. 2002;24:369–84.
114. Smith J, Fauquet M, Ziller C, Le Douarin NM. Acetylcholine synthesis by mesencephalic neural crest cells in the process of migration in vivo. Nature. 1979;282:853–5.
115. Stacy RC, Demas J, Burgess RW, Sanes JR, Wong RO. Disruption and recovery of patterned retinal activity in the absence of acetylcholine. J Neurosci. 2005;25:9347–57.
116. Tribollet E, Bertrand D, Marguerat A, Raggenbass M. Comparative distribution of nicotinic receptor subtypes during development, adulthood and aging: an autoradiographic study in the rat brain. Neuroscience. 2004;124:405–20.

117. Tyzio R, Cossart R, Khalilov I, Minlebaev M, Hübner CA, Represa A, Ben-Ari Y, Khazipov R. Maternal oxytocin triggers a transient inhibitory switch in GABA signaling in the fetal brain during delivery. Science. 2006;314:1788–92.
118. Vastola BJ, Douglas LA, Varlinskaya EI, Spear LP. Nicotine-induced conditioned place preference in adolescent and adult rats. Physiol Behav. 2002;77:107–14.
119. Wada E, Wada K, Boulter J, Deneris E, Heinemann S, Patrick J, Swanson LW. Distribution of alpha 2, alpha 3, alpha 4, and beta 2 neuronal nicotinic receptor subunit mRNAs in the central nervous system: a hybridization histochemical study in the rat. J Comp Neurol. 1989;284:314–35.
120. Wang F, Gerzanich V, Wells GB, Anand R, Peng X, Keyser K, Lindstrom J. Assembly of human neuronal nicotinic receptor alpha5 subunits with alpha3, beta2, and beta4 subunits. J Biol Chem. 1996;271:17656–65.
121. Wang CT, Blankenship AG, Anishchenko A, Elstrott J, Fikhman M, Nakanishi S, Feller MB. GABA(A) receptor-mediated signaling alters the structure of spontaneous activity in the developing retina. J Neurosci. 2007;27:9130–40.
122. Winzer-Serhan UH, Leslie FM. Codistribution of nicotinic acetylcholine receptor subunit alpha3 and beta4 mRNAs during rat brain development. J Comp Neurol. 1997;386:540–54.
123. Winzer-Serhan UH, Leslie FM. Expression of alpha5 nicotinic acetylcholine receptor subunit mRNA during hippocampal and cortical development. J Comp Neurol. 2005;481:19–30.
124. Wong RO. Retinal waves: stirring up a storm. Neuron. 1999;24:493–5.
125. Woolf NJ. Cholinergic systems in mammalian brain and spinal cord. Prog Neurobiol. 1991;37:475–524.
126. Xu H, Khakhalin AS, Nurmikko AV, Aizenman CD. Visual experience-dependent maturation of correlated neuronal activity patterns in a developing visual system. J Neurosci. 2011; 31:8025–36.
127. Young JW, Crawford N, Kelly JS, Kerr LE, Marston HM, Spratt C, Finlayson K, Sharkey J. Impaired attention is central to the cognitive deficits observed in alpha 7 deficient mice. Eur Neuropsychopharmacol. 2007;17:145–55.
128. Zaborszky L. The modular organization of brain systems. Basal forebrain: the last frontier. Prog Brain Res. 2002;136:359–72.
129. Zaborszky L, Hoemke L, Mohlberg H, Schleicher A, Amunts K, Zilles K. Stereotaxic probabilistic maps of the magnocellular cell groups in human basal forebrain. Neuroimage. 2008; 42:1127–41.
130. Zhou ZJ. Direct participation of starburst amacrine cells in spontaneous rhythmic activities in the developing mammalian retina. J Neurosci. 1998;18:4155–65.
131. Zoli M, Le Novère N, Hill Jr JA, Changeux JP. Developmental regulation of nicotinic ACh receptor subunit mRNAs in the rat central and peripheral nervous systems. J Neurosci. 1995;15:1912–39.

Chapter 7
Presynaptic Nicotinic Acetylcholine Receptors and the Modulation of Circuit Excitability

Chongbo Zhong, Gretchen Y. López-Hernández, David A. Talmage, and Lorna W. Role

Abstract Cholinergic modulation of circuit excitability by activation of nicotinic acetylcholine receptors (nAChRs) is involved in essential aspects of motivated behaviors, attention, and affect. In this review we focus entirely on the contribution of *pre*synaptic nAChRs to the regulation and dysregulation of synapses and circuits in the CNS. In particular we highlight recent insights into the regulation of presynaptic nAChR targeting and advances into discerning the mechanisms by which presynaptic nAChRs regulate neurotransmitter release and synaptic transmission in the brain.

Keywords Acetylcholine • Cholinergic system • Nicotine • Nicotinic acetylcholine receptors • Presynaptic modulation • Neurotransmitter release • Synaptic transmission

1 Introduction

The activation and inactivation (or desensitization) of nicotinic acetylcholine receptors (nAChRs) contribute to fundamental aspects of behaviors related to memory, motivation, and mood, and have been implicated in neurodegenerative and neuropsychiatric disorders including Alzheimer's disease, Parkinson's disease, schizophrenia, and addiction. Our major goal in this chapter is to provide an overview of the recent literature on the contributions of presynaptic nAChRs to shaping circuit excitability in the central nervous system (CNS). We believe that the evidence for presynaptic nAChR modulation of CNS circuits is compelling and, as such supports the idea that these unique modulatory sites constitute important, novel, therapeutic targets.

C. Zhong, M.D. Ph.D. (✉) • G.Y. López-Hernández, Ph.D.
D.A. Talmage, Ph.D. • L.W. Role, Ph.D.
Department of Neurobiology and Behavior, Center for Brain and Spinal Cord Research, Stony Brook University, 100 Nicolls Road, Stony Brook, NY 11794, USA
e-mail: chongbo.zhong@stonybrook.edu

R.A.J. Lester (ed.), *Nicotinic Receptors*, The Receptors 26,
DOI 10.1007/978-1-4939-1167-7_7

Progress over the last 10 years has provided important insights to the identity and subunit composition of presynaptic nAChRs (Part IA and 1B), to their role in modulation of transmitter release (Part II), as well as their precise location, targeting, and intracellular signaling mechanisms (Part III). Recent studies build on more than two decades of prior thought and experimentation that originally established the fundamental concept of presynaptic ionotropic receptors and introduced nAChRs as key players in the modulation of circuit excitability.

1.1 Early Work (Pre-Twenty-First Century)

Studies of presynaptic nAChRs have been subject to considerable controversy from the get go. Early work implicating the existence of presynaptic nAChRs at neuromuscular and ganglionic synapses emphasized their potential role as autoreceptors, with perhaps the first formal proposal put forth by Koelle in 1961 [1–3]. It was not so much the existence of presynaptic nAChRs, which culled support from many of the most preeminent synaptic and receptor biologists of the twentieth century, but rather the issue of a physiological role of presynaptic nAChRs that has elicited (and still elicits) such intense debate [2, 4–6].

Both autonomic ganglia and the neuromuscular junction constitute examples of possible sites of homosynaptic modulation (as illustrated in Fig. 7.1a). That is, these

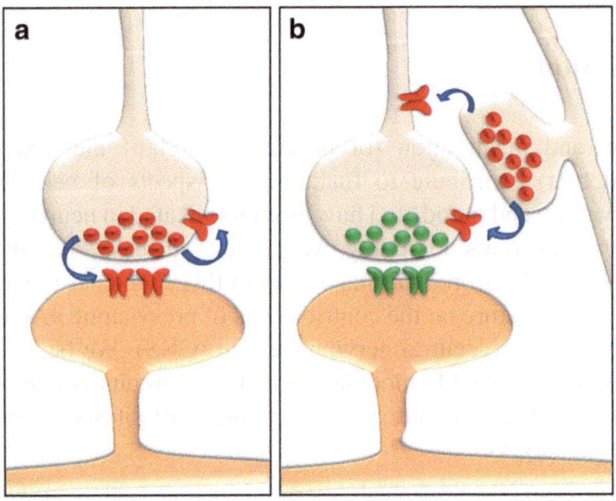

Fig. 7.1 Possible configurations for homosynaptic (**a**) vs. heterosynaptic (**b**) modulation via presynaptic nAChRs. (**a**) Homosynaptic modulation by ACh: At cholinergic synapses (such as those within sympathetic ganglia) released ACh could interact with both pre- and postsynaptic ACh receptors. ACh activation of the presynaptic nicotinic AChRs modulates probability of ACh release. (**b**) Heterosynaptic modulation by ACh: At non-cholinergic synapses (e.g. CNS glutamatergic or GABAergic synapses) the probability of release of the transmitter is modulated by presynaptic nAChRs. Activation of these presynaptic AChRs is due to ACh released from neighboring cholinergic (axo-axonic) synapses

are sites at which the released acetylcholine (ACh) was proposed to up and/or down regulate subsequent ACh release via interaction with autoreceptors. The same studies that established the presence of presynaptic nAChRs and noted the distinct pharmacology of the pre- from postsynaptic nAChRs, also presented convincing arguments against a physiological contribution of these nAChRs to normal transmission [6]. But then, more than a decade later, with increasing focus on CNS synapses, some of these same investigators provided the most compelling evidence in support of a physiological role of presynaptic nAChRs in the brain [7]. This and other early electrophysiological demonstrations of the modulation of glutamatergic synaptic transmission by presynaptic nicotinic AChRs in the brain [8–10], provided the essential compliment to parallel studies using synaptosomal preparations (see below) to establish the existence of presynaptic nAChRs. Numerous studies followed in the 1990s, documenting many CNS synapses and circuits where cholinergic, GABA-ergic, and aminergic transmission appeared to be fine-tuned by the activation and/or inactivation of *presynaptic* nAChRs [11].

These early studies in the CNS emphasized the idea that presynaptic nAChRs control the extent of release of *non*-cholinergic transmitters. That is, unlike the early discussion of presynaptic nAChRs in *homosynaptic* plasticity of ganglionic transmission, the CNS studies highlighted examples of hetero-synaptic plasticity, as illustrated in Fig. 7.1b. However, such studies also raised the important question of "what is the usual source of cholinergic ligand(s) that interact with these presynaptic receptors?" This is an active and very much ongoing debate that we discuss in more detail in Part 1B (see below).

Some of the most compelling early work showing that nicotinic receptors are localized to presynaptic sites began in the 1980s with studies of synaptosomal preparations [12, 13]. A particularly important focus of early synaptosome work was the detailed analysis of nicotinic modulation of dopamine release from striatal preparations [14–17]. These studies presaged the recent optogenetic demonstrations that presynaptic nAChRs are uniquely positioned to control dopamine release from ventral tegmental area (VTA) terminals in striatum, independent of action potential evoked from VTA (see Part II).

Despite the important findings by investigators using synaptosomal preparations, the existence and role of presynaptic nAChRs remained controversial because of the potential confounds of such approaches. Nevertheless, a careful search of the literature and an objective reflection on the contribution of synaptosome preparations to our current understanding of the physiological roles of presynaptic nAChRs reveals the pivotal role of these studies. Synaptosome studies since the 1970s established presynaptic nAChRs as direct modulators of serotonin [18], dopamine [19, 20], and ACh [21, 22] release. More recent work with synaptosomes of higher purity confirms and extends prior studies to show that presynaptic nAChRs fine tune the release of glutamate and GABA at numerous CNS synapses [23, 24].

1.2 *Fundamental Issues That Must Be Addressed to Establish Whether Presynaptic nAChRs Contribute to Neural Signaling*

There are three major issues that must be addressed to demonstrate a physiological vs. pharmacological role of presynaptic ionotropic receptors in general, and for presynaptic nAChRs, in particular [11]. The first hurdle is to demonstrate that nAChRs are actually located at presynaptic sites, along axons or at axon terminals per se. These data come in two forms: direct (typically post hoc), immunochemical assays of nAChR subunits or histological assays with toxin probes or genetically labeled nAChRs (Part IA) and indirect—though often quite compelling—evidence from functional studies of nAChR modulated transmitter release (Part II). Key pre-twenty-first century contributions to demonstrating the presynaptic localization of nAChRs evolved from work of Changeux, Clarke, Dani, and Wonnacott and their colleagues, as well as our laboratory [8–10, 17, 25].

The second critical issue to address vis *a* vis functional significance of presynaptic nAChRs is to identify the source(s) of available agonist or antagonist(s). We summarize recent evidence for a physiological and/or pharmacological role of presynaptic nAChRs that consider both the potential sources of endogenous (e.g. ACh, choline, lynx) and exogenous (e.g. nicotine, specific nerve gases, and numerous snail and snake toxins) nAChR ligands (Part IB). Finally, we consider evidence for the mechanisms that link activation or inhibition of these presynaptic nAChRs and the downstream effects on axonal excitability and/or the probability of release of the stored neurotransmitter(s).

2 Part IA: nAChRs Subtypes and Presynaptic Localization of nAChRs in the Brain

The diverse subunit composition of nAChRs in the brain has been extensively documented by assessing nAChR subunit mRNA expression at the single cell level and from assays of subunit protein expression [26–32]. The heteromeric $\alpha4\beta2^*$ and homomeric $\alpha7^*$ nAChRs are the two predominant nAChR subtypes in the brain whereas $\alpha3\beta4^*$ nAChRs are more common in the peripheral nervous system (where * signifies the inclusion of the noted subunit(s) in a pentameric assembly with or without other α and/ or β subunits) [33, 34].

The functions of nAChRs critically depend on both the subunit composition and subcellular location (Fig. 7.2). Knowledge of the precise subcellular localization of specific nAChRs subtypes in the brain is essential to understanding how nicotinic signaling affects circuits and behavior and, ultimately, is required for the design of targeted pharmacological tools [26, 31]. Presynaptic localization of nAChRs has been demonstrated with receptor ligand binding and immunocytochemical methods, most convincingly with immuno-gold labeling of nAChRs at the electron microscopic

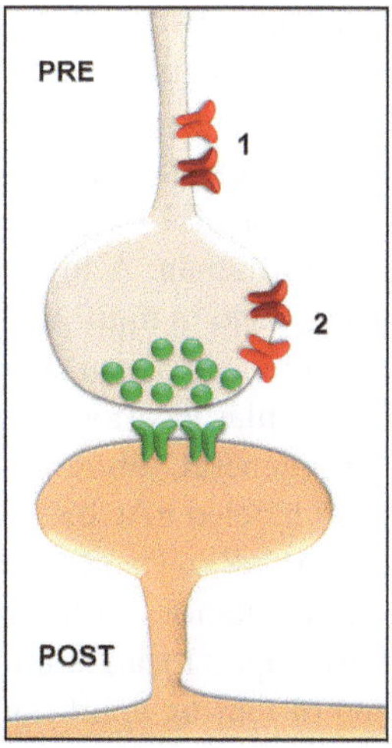

Fig. 7.2 Multiple subtypes of nAChRs are located within pre-terminal and terminal zones of non-cholinergic axons. Presynaptic nAChRs may be targeted to pre-terminal domains (*1*) or to synaptic boutons (*2*). Depending on the subtype(s) of nAChRs at each location (indicated by the varying shades of *red*) the effects of nAChR activation on transmitter release may be brief or prolonged and may involve distinct signaling pathways (See text for discussion)

(EM) level. At the EM level, α7* nAChRs, identified by either biotin- or gold-conjugated alpha-bungarotoxin (αBgTX), are found on axon terminals forming synapses onto dendrites from rat VTA, prefrontal cortex (PFC), and hippocampus [35–38]. Alpha4 and β2 subunits, identified by anti- α4 or β2 antibodies, are also located at presynapic sites in rat cerebellar cortex and striatum [39–41].

To more directly address the issue of nAChR localization, subunit-fluorescent protein chimeras have been constructed for the α3, α4, α6, α7, β2, β3, and β4 nAChR subunits (these subunit-fluorescent protein fusions produce functionally normal receptors as measured by whole cell electrophysiology recordings and the calcium-sensitive indicator fura-2). Expressing these chimeric, fluorescent proteins in cultured neurons and in mice [42] allows detection of nAChRs subunit locations as well as studies of receptor trafficking and targeting, with confocal microscopy and Förster Resonance Energy Transfer (FRET) and/or Total Internal Reflection Fluorescence (TIRF) techniques [42–51].

The presence of functional, presynaptic nAChRs on nerve terminals is supported by the ability of various subtype-selective nAChRs agonists to evoke neurotransmitter release from synaptosomal preparations, from acute brain slices and in in vivo preparations, neurochemical, neuroimaging, and/or electrophysiological assays of neurotransmitter release from synaptosomes and/or brain slices that provide important functional evidence for presynaptic nAChRs will be addressed in detail in Part II.

A major challenge to developing pharmaceuticals targeting presynaptic nAChRs is the in vivo demonstration, of presynaptic nAChRs in humans. Receptor subtype imaging by autoradiography and positron emission tomography (PET), Single Photon Emission Computed Tomography (SPECT) and functional magnetic resonance imaging (fMRI) with radioactively labeled ligands have provided a more thorough profile of the distribution of nAChRs in human brain in vivo [52, 53].

Currently, two PET ligands, including [^{11}C]nicotine and 2-[^{18}F]fluoro-3-(2 (S) azetidinylmethoxy)pyridine (2-[^{18}F]F-A-85380), and a SPECT ligand, 5-[^{123}I]iodo-3-(2 (S)-2-azetidinylmethoxy)pyridine (5-[^{123}I]I-A-85380), have been used for in vivo detection of α4β2* nAChR [54–57], whereas [^{11}C]CHIBA-1001 [58–60] is the major PET ligand currently available for in vivo study of α7* nAChRs distribution in human. A high density of nAChRs (including both α4β2* and α7*) have been found in multiple brain regions with these in vivo technologies, but the spatial resolution of these imaging techniques is still insufficient to identify the precise subcellular location (i.e. pre- vs. postsynaptic) [61–66].

In general, immunochemical assays with toxin probes, antibodies, and/or genetically labeled nAChRs subunits provide direct evidence for presynaptic localization of nAChRs. These studies convincingly corroborate biochemical and functional studies demonstrating presynaptic nAChRs but fall short of verifying the precise presynaptic localization of these receptors (i.e. axonal *vs*. terminal *vs*. pre-terminal) and of teasing apart the relative functional implications of these receptors in vivo, especially in humans.

3 Part IB: Endogenous and Exogenous Ligands of Presynaptic nAChRs

Appreciating the role that presynaptic nAChRs play in the brain requires an understanding of where, when, and in what manner these receptors are activated (or inactivated) by endogenous and exogenous ligands. The major endogenous nAChR agonist in the brain is ACh. In addition, there is increasing evidence in support of other endogenous nAChR ligands, such as choline and lynx, and perhaps Aβ peptides, that may also interact with presynaptic nAChRs. Exogenous ligands for nAChRs abound, we will focus our discussion of endogenous/exogenous ligand interactions at presynaptic nAChRs to the effects of nicotine at concentrations that are achieved in brain by smoking tobacco.

3.1 Endogenous Ligands

Although the mammalian brain contains relatively few cholinergic neurons, these neurons project widely to nearly all major neocortical areas, as well as to the hippocampus and amygdala. In addition there are cholinergic interneurons in some cortical regions and throughout striatum that modulate circuit activity to an extent that far exceeds predictions based on their modest numbers. The relative paucity of cholinergic neurons, the diffuse nature of CNS cholinergic projection and the lack of classical point-to-point cholinergic synapses raise challenges to deciphering the conditions under which ACh, and other endogenous nAChR ligands, activate presynaptic nAChRs.

3.1.1 Acetylcholine (ACh)

Measuring ACh levels in the CNS have traditionally relied on two methodologies: microdialysis, followed by remote online detection and local electrochemical/enzymatic detection. Much of what is known about ACh release has come from the past three decades of microdialysis studies; however, the temporal dynamics of ACh release is not well resolved using this tool. Early studies using microdialysis report extracellular ACh concentrations ranging from low nanomolar levels in the rat striatum [67, 68] to pico—or even femtomolar concentrations in the rat cortex [69, 70]. The reported low nanomolar ambient ACh levels were proposed to tonically activate or inhibit high affinity AChRs influencing circuit activity and behavior [67, 71].

Additional spatial and temporal resolution of ACh release has been obtained through the development of ACh- and choline-sensitive biosensors [72–79]. Enzyme-based microelectrode arrays (MEA) significantly improve assays of local ACh concentrations, yielding measurements with subsecond time resolution. These enzyme-based microelectrodes convert a non-electroactive agent into an electroactive substance that can be measure by amperometric detection [72]. Thus, MEAs provide accurate assessment of dynamic, nearly real time assays of stimulated ACh release during normal behaviors and precise assays of the effects of pharmacological manipulations on ACh release. MEA based measures of basal rat cortical ACh levels fall into the micromolar range—0.6 to 1 μM and increase ~tenfold following KCl evoked release [74] or during performance of cued reward tasks [78, 79].

The studies highlight the role of ACh in controlling cue detection and attentional performance and also provide evidence that ACh operates at different timescales ranging from subseconds to minutes [75]. Combining these techniques with targeted pharmacological tools revealed that enhancement of attentional performance can be achieved by activation of ACh release and selective stimulation of α4β2-containing nAChRs [80].

A major barrier to dissecting the role of presynaptic nAChRs to the modulation of CNS transmission under physiological conditions has been the inability to selectively stimulate cholinergic inputs. Electrical stimulation of "cholinergic" basal forebrain nuclei or projection pathways including cholinergic fibers is obviously confounded by the coactivation of non-cholinergic inputs. Nevertheless such approaches have been used successfully to implicate endogenous cholinergic activity in the facilitation of action potential evoked dopamine release in mouse striatum by activation of β2 containing nAChRs but not by α7 nAChRs [81], to increase glutamate release and enhance glutamatergic transmission by activation of presynaptic α4β2 but not α7 nAChRs in the dorsal raphe nucleus [82] and in lateral geniculate nucleus by activation of presynaptic α3*- and/or α6* nAChRs [83].

Combining selective optogenetic stimulation of ACh release with MEA measured local changes in ACh release, is beginning to reveal the contribution of *pre*-synaptic nAChRs to specific circuits and behaviors. For example, Cachope et al. used such combinatorial technology to demonstrate collaboration between ACh and glutamate in the modulation of dopamine release in the nucleus accumbens (nAcc) [84]. By optogenetically stimulating cholinergic interneurons in the nAcc, and measuring dopamine release by fast-scan cyclic voltammetry, the authors demonstrated that endogenous ACh directly enhances dopamine release, at least in part, via the activation of β2* presynaptic nAChRs [84].

To date the use of optogenetic approaches to selectively elicit endogenous ACh release has helped establish that nAChRs: (1) trigger GABA release from hippocampal inhibitory interneurons (recorded as bursts of mIPSCs in pyramidal neurons) by activating presynaptic non-α7 nAChRs (α3β4) [85]; (2) modulate hippocampal synaptic plasticity by activation of presynaptic α7 nAChRs [86, 87]; and (3) trigger striatal dopamine release by activation of axonal nAChRs [88].

3.1.2 Other Endogenous Ligands That Interact with Presynaptic nAChRs: Choline, Lynx

The overall level of cholinergic "tone" is not determined solely by the local concentrations of ACh. In fact the primary pathway for the hydrolytic cleavage of ACh generates another (albeit very low affinity) α7* nAChR agonist: choline. As low affinity an agonist as choline is, its effects cannot be discounted as it is stable and highly diffusible. In fact choline levels in brain cerebrospinal fluid (CSF) (at rest) are ~10 μM and could easily rise to the Kda of ~100 μM following ACh release and hydrolysis [89].

Modulation of circuit activity by presynaptic nAChRs may also be controlled through nAChR interaction with nicotinic receptor "endotoxins" such as lynx [90–96]. Lynx genes are evolutionary antecedents of α-neurotoxins (e.g. αBgTx), sharing structural similarities including a conserved three-looped motif [92, 94]. Several members of the lynx gene family are expressed in the mammalian brain where they are proposed to serve as a brake of nAChR-mediated plasticity [90–96].

Lynx1 is highly expressed in the hippocampus, cerebellum, and cortex [93]. Double-immunofluorescence and immunoprecipitation studies indicated that lynx1 colocalized and co-immunoprecipitated with α4β2 and α7 nAChRs [91]. When functionally analyzed in vitro, lynx1 exhibited predominantly inhibitory effects on nAChRs, reducing ACh sensitivity and enhancing desensitization [91, 92, 94] although increased amplitude of α7* nAChR-mediated macroscopic currents are also seen [93].

The role of lynx1 in the regulation of nAChR function and thus cholinergic tone was demonstrated in vivo through targeted gene deletion in mice [95]. Deletion of the lynx1 gene in mice increased agonist sensitivity to nicotine and reduced the paired pulse ratio, possibly reflecting an increase in the neurotransmitter release probability [95]. These results suggested that through its interaction with presynaptic nAChRs, lynx1 regulates synaptic efficacy; changes associated with enhanced associative learning and nicotine-mediated motor learning [95]. On the down side, lynx1 deletion resulted in vacuolating neurodegeneration in aged mice, predominantly of axonal tracts, associated with increased activity of nAChRs [95]. Taken together, these results suggest that lynx1 is an endogenous regulator of nAChRs, serving as a molecular brake of synaptic plasticity [90, 92, 94, 95]. Further evidence for the role of lynx1 as a cholinergic brake came from studies evaluating the effects of monocular deprivation on synaptic plasticity in adult visual cortex of lynx1 KO mice [96]. Morishita et al. demonstrated that lynx1 expression increases as the critical period for plasticity in the primary visual cortex closes; removal of this molecular brake was achieved by enhanced activation of nAChRs through treatment with the acetylcholinesterase (AChE) inhibitor physostigmine [96].

3.2 Interactions of Endogenous and Exogenous nAChR Ligands: Smoking, Nicotine, and nAChR Activity

Consideration of how the activation and/or inactivation of presynaptic nAChRs might contribute to behavior must include assessment of ACh–nicotine interactions (Fig. 7.3), as approximately 1 billion people worldwide self-administer nicotine by smoking cigarettes (WHO Fact sheet #339, July 2013). Although the adult levels of smoking have declined in the last decade, the Centers for Disease Control and Prevention estimates that ~10 % of pregnant women and 18 % of high school students smoke. In view of the considerable evidence that early age smokers are the most prone to long term addiction, it is particularly important to assess nicotine–ACh interactions throughout development [31].

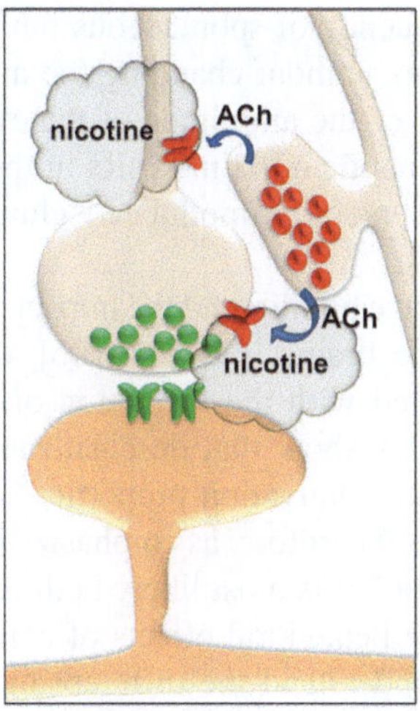

Fig. 7.3 Interaction of endogenous (e.g. ACh) and exogenous (e.g. nicotine) ligands at presynaptic nAChRs. Schematic of potential interactions of endogenous and exogenous ligands in the activation and/or desensitization of presynaptic nAChRs. Different nAChR subtypes (indicated by different *red* color) may normally be activated by basal levels of ACh release, but desensitized in the presence of smokers' concentrations of nicotine (See text Part 1B for discussion)

Nicotine levels in tobacco smokers reach 500 nM immediately after smoking a cigarette, and with a half-life of >2 h the average smoker maintains a relatively steady low concentration of nicotine through the day [97, 98]. Such prolonged exposure to nicotine has been shown to induce desensitization of nAChRs in brain areas such as the VTA. However, the extent of nicotine-induced desensitization of nAChRs is dependent not only on the rate of induction of desensitization, but also the rate of recovery [99].

The effects of nicotine on presynaptic nAChRs, like the effects on postsynaptic nAChRs depend on the nicotinic receptor composition (and hence pharmacology), their conductance, activation, and inactivation kinetics and the relative ionic permeability profile [92, 100]. What is often not well appreciated about the physiological role of presynaptic nAChRs per se is the profound difference in the impedance (or volume) of a presynaptic terminal vs. that of a soma or dendrite. This difference in impedance can greatly amplify the effect of activation of relatively few receptors if they are localized to a synaptic bouton, coupled to intracellular signaling cascades and/or directly to transmitter release (see Part III).

The variety of nAChR subtypes targeted to presynaptic domains is impressive. For example, studies by McGehee and colleagues demonstrated that GABAergic and glutamatergic synaptic transmission to the dopaminergic (DA) neurons of the VTA are modulated by different nAChRs with different desensitization properties [101]. GABAergic input to DA neurons in the VTA include $\alpha4\beta2^*$ nAChRs, whereas glutamatergic input to the VTA are distinguished by their expression of presynaptic $\alpha7^*$ receptors [101, 102]. Long term nicotine exposure appears to enhance dopamine release in the VTA through two distinct presynaptic mechanisms [103]. First, nicotine augments GABAergic inhibition to substantia nigra compacta (SNc) DA neurons by altering the frequency of spontaneous inhibitory postsynaptic currents (sIPSCs) in SNc DA neurons without changing the amplitude of sIPSCs, the frequency of miniature IPSCs, or the amplitude of the evoked IPSCs [103]. Second, through its actions on presynaptic $\alpha4\beta2$ nAChRs in the dopaminergic terminals of the dorsal striatum, chronic nicotine modulated glutamate release onto medium spiny neurons (MSNs) [103].

Although many studies have highlighted the impact of nAChR desensitization on the effects of nicotine in the brain [101, 104–106], certain behavioral effects of nicotine are indeed associated with the activation of nAChRs [107]. Although a generalization, several studies show that desensitization of nAChRs is linked to nicotine tolerance whereas the activation properties of nAChRs contribute to the rewarding effects of nicotine. Therefore, as emphasized recent work by Piciotto and colleagues, “it is not either/or,” it is most likely both activation and desensitization of nAChRs that underlie the behavioral effects of chronic nicotine and both must be considered in order to understand the different behavioral profiles in tobacco smokers [31, 107].

Desensitization induced by chronic exposure to nicotine has been hypothesized to trigger the up-regulation of the $\alpha4\beta2$ nAChRs [108–111]. Nicotine-induced up-regulation is an important molecular mechanism affecting nAChR-containing circuits by potentially magnifying the activation and desensitization effects of ACh and nicotine [92]. Nicotine-induced up-regulation might predominantly favor a higher sensitivity α4β2 nAChR stoichiometry [103, 112–114]. Furthermore, chronic nicotine up-regulation of α4β2 nAChRs was showed to display selectivity not only in terms of the nAChR subtype and cell type affected but also exhibited preference at the level of cellular compartment [82, 103, 115].

Overall, the discussed studies reinforce the idea that nicotine at the level experienced by chronic smokers is sufficient to both activate and inactivate presynaptic nAChRs. The net effect of nicotine, coupled with varying levels of ACh, will be even more complex and critically dependent on the receptor subtype/stoichiometry, receptor number, and location and, from recent work of Berg and colleagues, the degree of receptor clustering.

3.3 Partial Agonists as Therapeutic Interventions for Smoking Cessation and Their Effects on Presynaptic nAChRs: Varenicline

Smoking cessation therapy strategies have focused in recent years in the development of partial agonists of nAChRs. A partial agonist has dual actions, activating nAChRs when the concentration of ACh and/or nicotine are very low but decreasing the efficacy of ACh and nicotine when these "full" agonists are present at high concentration [116, 117]. At least in theory, the state dependence of partial agonists makes them particularly attractive as potential therapeutic agents for smoking cessation. If specific subtypes of presynaptic nAChRs are important mediators of nicotine dependence, then development of partial agonists that interact with these presynaptic nAChRs could provide the proverbial silver bullet for nicotine addiction. The development of Varenicline, a partial agonist of nAChRs with activity at presynaptic nAChRs highlights the advantages and complexities of this strategy. As delineated above, presynaptic α4β2* nAChRs in the nucleus accumbens contribute to the behavioral actions of nicotine in terms of addiction and dependence [100]. As such, considerable interest in the development of partial agonists as smoking cessation agents has focused on the targeting of α4β2* nAChRs, e.g. cytisine and varenicline [116, 118–120]. Varenicline (Chantix) is a cytisine-related compound and a weak partial agonist of α4β2 nAChRs [116, 118, 119]. In vivo studies showed that varenicline attenuated the central dopaminergic response to nicotine [118]. Despite the impressive efficacy of varenicline in smoking cessation and the fact that it is currently approved by regulatory agencies in the U.S. [121], it appears that varenicline also has significant activity at α7* nAChRs. In view of the numerous effects of activation of presynaptic α7* nAChRs in memory and mood, undesirable side effects may well limit its use [116, 117].

In conclusion, the combinations of optogenetic techniques and advances in the measurement of neurotransmitter release have highlighted the intricate framework of presynaptic nAChRs and the mechanistic complexities of their actions. The endogenous agonists of nAChRs are ACh, and for α7*nAChRs, choline as well, both of which contribute to the overall cholinergic tone. Cholinergic tone is further modulated by proteins such as the lynx1-like endotoxins and perhaps Aβ peptide. Indeed, genetic disruption of lynx1 provides insight into the consequence of altered cholinergic tone [95]. Additionally exogenous ligands, notably nicotine, have profound effects on cholinergic signaling via complex interaction with endogenous ligands and via nicotine-induced activation and desensitization of presynaptic nAChRs. The current strategies in the development of smoking cessation agents have also highlighted the central role of presynaptic nAChRs in regulating cholinergic tone and in mediating the actions of nicotine in the brain.

4 Part II: Activation of Presynaptic nAChRs Modulates Synaptic Transmission by Affecting Neurotransmitter Release

Presynaptic nAChRs, as auto- and/or hetero-receptors, modify the release of nearly every neurotransmitter examined [11, 122–128]. Modulation of the release of neurotransmitters by activation of presynaptic nAChRs is the most prevalent mechanism of nicotinic facilitation of synaptic transmission in the central nervous system. Glutamatergic and GABAergic transmission in hippocampal and cortical circuits and dopaminergic transmission in mesolimbic circuits (including SNc, VTA, and nAcc) are the best documented examples of nicotinic modulation of synaptic transmission. In this section, we highlight recent advances in demonstrating that presynaptic nAChRs provide a powerful and sensitive means to modulate neurotransmitter release.

4.1 Nicotine Enhances Neurotransmitter Release and Elicits Both Short and Long-Term Potentiation of Synaptic Transmission by Activating Presynaptic nAChRs

Dopamine release from striatal synaptosomes is facilitated by activating presynaptic nAChRs [129–131]. The pharmacology of nAChR modulation of dopamine release has been studied in considerable detail: release is inhibited by specific nAChR subtype antagonists such as α-conotoxins (for α3β2* and α6* nAChRs [132–138]), dihydro-β-erythroidine (for α4β2* nAChRs [16, 137, 139]), and methyllycaconitine but not by α-BgTx (for α7* nAChRs [140]). Studies of striatal synaptosomes prepared from mutant mice lacking individual nAChR subunit genes implicate combinations of α4, α5, α6, β2, and β3 subunits (e.g. α4β2, α6β2β3, α4α6β2β3, and α4α5β2) in mediating the nicotinic modulation of dopamine release. Of particular note is the equally compelling evidence that α7 and β4 subunits are not participating in the complex array of nAChR subtypes that regulate dopamine release in striatum [137, 141–144]. Additional studies that combined subtype-selective antagonists with nAChRs subunit-deficient mice, implicate α6β2* and α3β4* presynaptic nAChRs in the modulation of nicotine-evoked [^{3}H]-norepinephrine release from mouse hippocampal synaptosomes [145–147].

Presynaptic α7*nAChRs appear to play an indirect role in stimulating striatal dopamine release—activation of α7*nAChRs on glutamatergic axons stimulated glutamate release which subsequently acts via presynaptic glutamate receptors on dopamine terminals to stimulate dopamine release [148]. Functional presynaptic nAChRs, both α7 and non-α7 (including α4, α5, and β2) containing nAChRs, have been found on hippocampal and cortical (including neocortex and prefrontal cortex) synaptosomal preparations and have been shown to increase release of GABA, glutamate, and several other amino acid transmitters (including aspartate, glycine) [149–158].

Electrophysiological analyses of in vitro preparations (e.g. brain slices, cocultures) have provided considerable evidence that presynaptic nAChRs are important physiological modulators of synaptic transmission in the brain [26, 104, 159–161], and have implicated the activation and/or inactivation of presynaptic nAChR in nicotine-induced potentiation of glutamatergic [162–165], GABAergic [166–168], and dopaminergic transmission [102, 105, 129, 169–173] in the brain. Activation of presynaptic nAChRs by nicotine boosts short term potentiation (STP) to long term potentiation (LTP) and facilitates the induction of LTP in different brain regions (including hippocampus, prefrontal cortex, and amygdala) [87, 162, 174–179].

Microdialysis of transmitter "over flow" in freely moving animals provides additional in vivo evidence that activation of nAChRs modulates neurotransmitter release. Local application of nAChRs subtype specific agonists and antagonists to specific brain regions in vivo modulates: (1) dopamine release in nigrostriatal pathways (via α4, α6, and β3 subunits) [180–184], in the mesolimbic dopamine system, where neurons of the VTA project to the nucleus accumbens (α7, α3β2, α6, β3) [185], and in prefrontal cortex (α7 and β2) [186]; (2) noradrenaline (NA) release in frontal cortex (β2) and hippocampus (α3 and β4) [187]; (3) glutamate, aspartate, glycine, and GABA release in hippocampus (α7, α4β2) [155, 188]. By recording nicotinic modulation of long-term synaptic plasticity (i.e. LTP) in vivo, both α7 and α4β2 containing nAChRs have been shown to modify synaptic plasticity in the hippocampus, prefontal cortex, striatum, and cerebellum [189–192].

Optical live cell imaging techniques with fluorescent indicators provide a means to directly monitor presynaptic activity [193]. Depolarization induced synaptic vesicle release viewed by styryl amphipathic FM dyes (including FM1-43 and FM4-64) and/or synapto-pHluorin has been reported at glutamatergic and dopaminergic presynaptic terminals [194–197]. More recently developed cell-based fluorescent neurotransmitter reporters, such as CNiFERs for ACh [198] and iGluSnFer for glutamate [199, 200], allow direct monitoring of neurotransmitter release at the synaptic terminal. Using these assays, carbachol and nicotine were shown to increase vesicle release (with FM1-43 and FM 2-10 as indicator) from hippocampal neurons [201, 202].

The powerful combination of pharmacological and genetic manipulations with both more traditional synaptosomal and electrophysiological methods, and with newer optogenetic or live imaging approaches is yielding an increasingly detailed picture of which nAChR subunit combinations are functionally important at different synapses throughout the CNS.

5 Part III: Signaling Mechanisms Underlying the Effects of Presynaptic nAChRs

As reviewed above (Part II), there is ample evidence for presynaptic action of a wide combination of nAChRs. Although most (perhaps all) nAChR subunits have been implicated in presynaptic functions, they appear to do so in different circuits and via different mechanisms. Elucidating what processes control the presynaptic targeting of different nAChRs and the mechanism by which each nAChR type functions at presynaptic sites will help clarify the molecular logic of this complex modulatory system. In the following section we review the current understanding of these two issues.

5.1 Trafficking of nAChRs

In order to maintain their remarkable functional polarity, neurons must tightly regulate the targeting of proteins to specific subcellular compartments. How the targeting of nAChRs to presynaptic vs. somato-dendritic domains is differentially regulated in different neuronal types is largely unknown. Attempts by a number of groups to express functional nAChRs in various nonneuronal cell lines revealed that neuron specific chaperones are required for the efficient assembly and trafficking of nAChRs through the exocytic pathway [203, 204]. The identity of the full repertoire of chaperones acting in CNS neurons, and whether there are distinct assembly factors for axonally destined nAChRs is not known. Ric-3, a nAChR chaperone first identified in *C. elegans*, is an important regulator of α7*nAChR assembly and trafficking, both in nonneuronal cells and in neurons. Ric-3 interacts with α7* via sites in the M3-M4 cytosolic loop adjacent to an endoplasmic reticulum (ER) retention signal and a dendrite targeting signal. Thus the availability of Ric-3 during α7*nAChR synthesis and assembly might serve to "guide" these receptors through the ER to dendritic targets; disrupting these interactions might be a prerequisite for axonal / presynaptic targeting of these nAChRs [51, 205, 206]. The α4, α3 and β2 subunits contain intrinsic targeting domains within their M3-M4 cytoplasmic loops [51, 207], although the partners with which these signals interact are currently unknown. In the case of the α3 subunit, this domain targets α3* nAChRs to postsynaptic subdomains in ciliary ganglion neurons, possibly by mediating interactions with postsynaptic density (PSDs) and cytoskeletal components [207–209]. Within the α4β2 complex, the β2 axonal signal dominates, perhaps via interactions with neurexins [210], underscoring that the actual subunit combinations could also determine subcellular targeting as seen in α5 null mice [211].

In micro-explant cultures from the mouse ventral hippocampus, we readily detect functional presynaptic α7* and α4β2* nAChRs [165, 212]. By relocalizing surface pools of α7*nAChR relative to either sites of nicotine induced increases in intracellular Ca^{2+}, or nicotine induced vesicle fusion, we were able to demonstrate clusters of α7*nAChR spatially associated with presynaptic specializations. In ventral hippocampal explants the levels of functional α7*nAChRs (but not α4β2*nAChRs) is dependent on axonal signaling by the Type III isoforms of neuregulin1 (Nrg1). Mice that are heterozygous for a disruption of the Type III Nrg1 gene have altered levels of α7 expression in the ventral hippocampus [213], and lack functional surface α7*nAChRs along ventral hippocampal projections in vitro [165] and on cortical projections to the basal lateral amygdala in vivo [214]. These deficits can be rescued by stimulating Type III Nrg1 back-signaling, at least in part via mobilizing the trafficking of intracellular pools of α7*nAChRs to the axonal surface [165, 214, 215].

Berg and colleagues have also reported α7*nAChRs at presynaptic specialization in dispersed hippocampal neuronal cultures, and have used elegant single particle tracking approaches to demonstrate that the presynaptically localized pool is relatively immobile, most likely the result of trapping by interaction with CAST/ELKS (calpastatin/glutamine, leucine, lysine, and serine-rich protein) or other proteins in the active zone [216]. Induced clustering of these presynaptic α7*nAChRs results in a significant enhancement of transmitter release and an increase in the size of the readily releasable pool of vesicles. These results, along with our findings of clustered nAChRs at presynaptic specializations [165, 212, 215] raises questions about the importance of high density packing of presynaptic nAChRs and the possibility that there is a degree of cooperativity between receptors within these clusters that could enhance their ability to modulate synaptic transmission.

5.2 *Mechanisms by Which Presynaptic nAChRs Can Alter Neurotransmitter Release*

Most studies provide evidence that changes in intracellular calcium levels underlie cholinergic (endogenous and/or exogenous) activation of neurotransmitter release and facilitation of synaptic transmission [86, 165, 217]. In this section, we highlight recent advances in our understanding of the cellular mechanisms that underlie presynaptic nAChR function. Fusion of synaptic vesicles and release of neurotransmitter are calcium dependent events and local concentration of voltage dependent calcium channels at presynaptic active zones allows the translation of incoming action potential dependent voltage changes into neurotransmitter release. The magnitude of the local rise in intracellular calcium is the major contributor in determining the probability that an electrical signal will be converted into a chemical one at the synapse. Local, presynaptic events that alter intracellular calcium will alter the

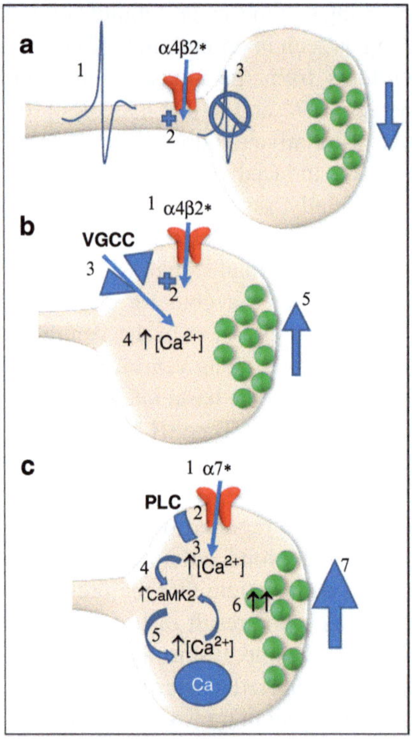

Fig. 7.4 Mechanisms by which activation of presynaptic nAChRs modulates transmitter release (**a**) Modulation of TTX-sensitive transmitter release by pre-terminal nAChRs. Location of α4β2* nAChRs within pre-terminal domains (*2*) may cause depolarization block of action potential invasion (*1* and *3*), thereby decreasing TTX-sensitive transmitter release. (**b**) Modulation of transmitter release by synaptic bouton targeted α4β2* nAChRs. Activation of α4β2* nAChRs on presynaptic boutons (*1*) elicits depolarization (*2*) and activation of local voltage-gated Ca^{2+} channels (*3*). The consequent increase in intrasynaptic Ca^{2+} (*4*) increases the probability of vesicular fusion and exocytosis, thereby increasing transmitter release (*5*). (**c**) Modulation of transmitter release by activation of synaptic bouton targeted α7* nAChRs. Activation of α7* nAChRs (*1*) on presynaptic boutons elicits Ca^{2+} entry through α7* nAChRs (*2*). This increase in Ca^{2+} is linked through as yet unidentified mechanisms to the activation of PLC (*3*) and of CaMKII (*4*). A positive feedback loop between CaMKII activation and additional release of Ca^{2+} from intracellular stores (*5*) supports a prolonged increase in the probability of vesicular fusion and exocytosis (*6*) and, as such, supports sustained enhancement of synaptic transmission (*7*)

probability of neurotransmitter release, and therefore targeting proteins such as the nAChRs, to presynaptic or perisynaptic sites provides a powerful means of presynaptic modulation of synaptic transmission.

Three examples are illustrated in Fig. 7.4. Activation of nAChRs (α4β2* in the example in Fig. 7.4a) localized to axonal membrane proximal to the synaptic bouton can result in local depolarization induced block of the incoming action potential thereby inhibiting action potential dependent neurotransmitter release. Experimentally this would affect tetrodotoxin (TTX) sensitive but not TTX resistant spontaneous events. In contrast, activation of presynaptic nAChRs could increase transmitter release via at least two mechanisms (note that these are not mutually exclusive possibilities). Agonist binding to nAChRs leading to local depolarization can activate voltage-gated Ca^{2+} channels (VGCC) increasing either spontaneous vesicle fusion (detected experimentally as an increase in TTX resistant, or miniature events) or increase the probability of release in response to an incoming action potential (altered paired pulse facilitation), and indeed at some synapses result in sustained facilitation [162, 214]. Nicotine-induced calcium transients have been observed in cultured neurons and in several cell types that express nAChRs [218–220]. Specific subtypes of nAChRs have been associated with defined Ca^{2+} signaling pathways: non-α7*nAChRs are mainly associated with Ca^{2+} signals mediated by activation of VGCC (Fig. 7.4b). Different subtypes of neuronal nAChR are differentially permeable to Ca^{2+}, the α7*nAChR is the most permeable to Ca^{2+} [221, 222]. As a result, activation of presynaptic α7*nAChR can increase Ca^{2+} levels in the bouton both by local depolarization events and by directly gating calcium. Indeed activation of α7*nAChRs can activate both VGCC and calcium induced calcium release (CICR) from internal stores (Fig. 7.4c) [122, 223, 224]. Studies from neuroblastoma cells and hippocampal astrocytes have shown that α7*nAChR-mediates Ca^{2+} signaling primarily from ryanodine receptor dependent CICR [225–227], whereas in ventral hippocampal axons, the inositol trisphosphate (IP_3) receptor-dependent Ca^{2+} stores rather are required for the α7*nAChR-mediated Ca^{2+} response [212].

In a series of recent studies we have demonstrated that both mechanisms of presynaptic facilitation can occur in the same system. Brief application of nicotine to ventral hippocampal inputs to medium spiny neurons results in a two component response: there is a transient increase in $[Ca^{2+}]_i$ and glutamate release that is mediated by non-α7*nAChRs (presumably α4β2*nAChRs) and a sustained (lasting ≥10–30 min) increase in $[Ca^{2+}]_i$ and glutamate release that is mediated by α7*nAChRs [165, 212]. Further analysis of the α7*nAChR-dependent effect revealed that transient activation of presynaptic α7*nAChR triggers a sustained CICR from IP_3 receptor regulated stores, and involved activation of presynaptic phospholipase C (PLC) and calcium calmodulin dependent kinase signaling [212].

It is well documented that nicotine activates several second messenger cascades, involving protein kinase A (PKA) [224], protein kinase C (PKC) [228], phosphatidylinositol 3-kinase (PI3K) [229], mitogen-activated protein kinase (MAPK) [230] and calcium/calmodulin-dependent protein kinase II (CaMKII) [231, 232]. Of note, activated CaMKII has been associated with presynaptic neurotransmitter vesicles at synapses [233] and with the modulation of neurotransmitter release and synaptic transmission [234]. In our recent studies α7*nAChR-mediated calcium signaling and the subsequent IP_3 receptor-mediated CICR appeared to be requisite steps in the nicotinic activation of CaMKII at presynaptic sites. In addition, α7*nAChR triggered IP_3 receptor-mediated CICR is both an *activator* of CaMKII and a key *substrate* of CaMKII that is necessary for prolonged enhancement of neurotransmitter release [212]. Direct interactions between α7*nAChRs and a G-protein signaling complex have been reported in PC12 cells [235], raising the possibility that ventral hippocampal axonal α7*nAChRs also functionally interact with a G protein coupled signaling complex.

In conclusion, activation of presynaptic nAChRs by nicotine induces calcium influx into presynaptic axons. Activation of non-α7*nAChR (low Ca^{2+}-permeability) elicits small and short-term Ca^{2+} signals along presynaptic axons and thus induces a short-term facilitation of neurotransmitter release and synaptic transmission [165]. By converting acute α7*nAChR activation into sustained cellular signaling, the increase in Ca^{2+} signals along presynaptic axons appears to be a crucial link between nAChRs and the downstream processes that participate in nicotine induced neurotransmitter release and potentiation of synaptic transmission [165, 212].

6 Summary of Recent Advances and Remaining Challenges

Based on a vast number of studies over the last four decades, using an increasingly diverse array of technical approaches, most nicotinic receptor biologists embrace the physiological role of presynaptic nAChRs in regulating synaptic gain. Modulation of circuit excitability by activation and/or inactivation of presynaptic nAChRs appears to be a major mechanism by which ACh exerts its influence on CNS circuits, although it is certainly not the only means of read out of ACh signaling via ligand-gated AChRs. Analysis of presynaptic nAChR subunit composition, pharmacology, and physiological effects has gained considerable precision over the last decade, as have our techniques for assessing the levels of ACh released and our methods for selective stimulation of cholinergic inputs. Major mysteries still remain in the domain of nAChR ligand interactions—such as those that inevitably govern the responses to ACh in a smoker compared with individuals who are and have always been nicotine free. The greatest areas of exploration yet to be tackled are at both ends of the spectrum of scientific inquiry: little is known of the subcellular mechanisms by which presynaptic nAChRs elicit sustained changes in synaptic plasticity or of the precise contribution of presynaptic (as opposed to postsynaptic) nAChRs in the modulation of complex behaviors. As we continue to develop higher spatial and temporal resolution methods for measuring ligand–nAChR interactions, for dissecting downstream signaling cascades and for visualizing nAChR activity in the living animal, we will get closer to intelligent design of nAChR drugs that can correct deficits in cholinergic signaling in disorders as mechanistically diverse Alzheimer's dementia and nicotine addiction.

References

1. Koelle GB. A proposed dual neurohumoral role of acetylcholine: its functions at the pre- and post-synaptic sites. Nature. 1961;190:208–11. Epub 1961/04/15.
2. Koelle GB. Early evidence of presynaptic receptors. Ann N Y Acad Sci. 1990;604:488–91. Epub 1990/01/01.

3. Langer SZ. 25 years since the discovery of presynaptic receptors: present knowledge and future perspectives. Trends Pharmacol Sci. 1997;18(3):95–9. Epub 1997/03/01.
4. Collier B, Katz HS. The synthesis, turnover and release of surplus acetylcholine in a sympathetic ganglion. J Physiol. 1971;214(3):537–52. Epub 1971/05/01.
5. Kalsner S. Heteroreceptors, autoreceptors, and other terminal sites. Ann N Y Acad Sci. 1990;604:1–6. Epub 1990/01/01.
6. Brown DA, Jones KB, Halliwell JV, Quilliam JP. Evidence against a presynaptic action of acetylcholine during ganglionic transmission. Nature. 1970;226(5249):958–9. Epub 1970/06/06.
7. Brown DA, Docherty RJ, Halliwell JV. The action of cholinomimetic substances on impulse conduction in the habenulointerpeduncular pathway of the rat in vitro. J Physiol. 1984;353:101–9. Epub 1984/08/01.
8. Gray R, Rajan AS, Radcliffe KA, Yakehiro M, Dani JA. Hippocampal synaptic transmission enhanced by low concentrations of nicotine. Nature. 1996;383(6602):713–6. Epub 1996/10/24.
9. McGehee DS, Heath MJ, Gelber S, Devay P, Role LW. Nicotine enhancement of fast excitatory synaptic transmission in CNS by presynaptic receptors. Science. 1995;269(5231):1692–6. Epub 1995/09/22.
10. Vidal C, Changeux JP. Nicotinic and muscarinic modulations of excitatory synaptic transmission in the rat prefrontal cortex in vitro. Neuroscience. 1993;56(1):23–32. Epub 1993/09/01.
11. MacDermott AB, Role LW, Siegelbaum SA. Presynaptic ionotropic receptors and the control of transmitter release. Annu Rev Neurosci. 1999;22:443–85. Epub 1999/04/15.
12. Clarke PB, Hamill GS, Nadi NS, Jacobowitz DM, Pert A. 3H-nicotine- and 125I-alpha-bungarotoxin-labeled nicotinic receptors in the interpeduncular nucleus of rats. II. Effects of habenular deafferentation. J Comp Neurol. 1986;251(3):407–13. Epub 1986/09/15.
13. Wonnacott S, Irons J, Rapier C, Thorne B, Lunt GG. Presynaptic modulation of transmitter release by nicotinic receptors. Prog Brain Res. 1989;79:157–63. Epub 1989/01/01.
14. Kaiser SA, Soliakov L, Harvey SC, Luetje CW, Wonnacott S. Differential inhibition by alpha-conotoxin-MII of the nicotinic stimulation of [3H]dopamine release from rat striatal synaptosomes and slices. J Neurochem. 1998;70(3):1069–76. Epub 1998/03/07.
15. Rapier C, Lunt GG, Wonnacott S. Nicotinic modulation of [3H]dopamine release from striatal synaptosomes: pharmacological characterisation. J Neurochem. 1990;54(3):937–45. Epub 1990/03/01.
16. Rowell PP. Nanomolar concentrations of nicotine increase the release of [3H]dopamine from rat striatal synaptosomes. Neurosci Lett. 1995;189(3):171–5. Epub 1995/04/21.
17. Wonnacott S, Kaiser S, Mogg A, Soliakov L, Jones IW. Presynaptic nicotinic receptors modulating dopamine release in the rat striatum. Eur J Pharmacol. 2000;393(1–3):51–8. Epub 2000/04/20.
18. Reuben M, Clarke PB. Nicotine-evoked [3H]5-hydroxytryptamine release from rat striatal synaptosomes. Neuropharmacology. 2000;39(2):290–9. Epub 2000/02/12.
19. Grady S, Marks MJ, Wonnacott S, Collins AC. Characterization of nicotinic receptor-mediated [3H]dopamine release from synaptosomes prepared from mouse striatum. J Neurochem. 1992;59(3):848–56. Epub 1992/09/01.
20. Takano Y, Sakurai Y, Kohjimoto Y, Honda K, Kamiya HO. Presynaptic modulation of the release of dopamine from striatal synaptosomes: differences in the effects of high K+ stimulation, methamphetamine and nicotinic drugs. Brain Res. 1983;279(1–2):330–4. Epub 1983/11/21.
21. Marchi M, Raiteri M. Nicotinic autoreceptors mediating enhancement of acetylcholine release become operative in conditions of "impaired" cholinergic presynaptic function. J Neurochem. 1996;67(5):1974–81. Epub 1996/11/01.
22. Meyer EM, Arendash GW, Judkins JH, Ying L, Wade C, Kem WR. Effects of nucleus basalis lesions on the muscarinic and nicotinic modulation of [3H]acetylcholine release in the rat cerebral cortex. J Neurochem. 1987;49(6):1758–62. Epub 1987/12/01.
23. McGehee DS, Role LW. Presynaptic ionotropic receptors. Curr Opin Neurobiol. 1996;6(3):342–9. Epub 1996/06/01.
24. Wonnacott S. Presynaptic nicotinic ACh receptors. Trends Neurosci. 1997;20(2):92–8. Epub 1997/02/01.

25. Clarke PB, Reuben M. Release of [3H]-noradrenaline from rat hippocampal synaptosomes by nicotine: mediation by different nicotinic receptor subtypes from striatal [3H]-dopamine release. Br J Pharmacol. 1996;117(4):595–606. Epub 1996/02/01.
26. Albuquerque EX, Pereira EF, Alkondon M, Rogers SW. Mammalian nicotinic acetylcholine receptors: from structure to function. Physiol Rev. 2009;89(1):73–120. Epub 2009/01/08.
27. Gaimarri A, Moretti M, Riganti L, Zanardi A, Clementi F, Gotti C. Regulation of neuronal nicotinic receptor traffic and expression. Brain Res Rev. 2007;55(1):134–43. Epub 2007/03/27.
28. Gotti C, Clementi F, Fornari A, Gaimarri A, Guiducci S, Manfredi I, et al. Structural and functional diversity of native brain neuronal nicotinic receptors. Biochem Pharmacol. 2009;78(7):703–11. Epub 2009/06/02.
29. Gotti C, Riganti L, Vailati S, Clementi F. Brain neuronal nicotinic receptors as new targets for drug discovery. Curr Pharm Des. 2006;12(4):407–28. Epub 2006/02/14.
30. Hogg RC, Raggenbass M, Bertrand D. Nicotinic acetylcholine receptors: from structure to brain function. Rev Physiol Biochem Pharmacol. 2003;147:1–46. Epub 2003/06/05.
31. Picciotto MR, Higley MJ, Mineur YS. Acetylcholine as a neuromodulator: cholinergic signaling shapes nervous system function and behavior. Neuron. 2012;76(1):116–29. Epub 2012/10/09.
32. Tammimaki A, Horton WJ, Stitzel JA. Recent advances in gene manipulation and nicotinic acetylcholine receptor biology. Biochem Pharmacol. 2011;82(8):808–19. Epub 2011/06/28.
33. Hellstrom-Lindahl E, Mousavi M, Zhang X, Ravid R, Nordberg A. Regional distribution of nicotinic receptor subunit mRNAs in human brain: comparison between Alzheimer and normal brain. Brain Res Mol Brain Res. 1999;66(1–2):94–103. Epub 1999/03/30.
34. Yeh JJ, Yasuda RP, Davila-Garcia MI, Xiao Y, Ebert S, Gupta T, et al. Neuronal nicotinic acetylcholine receptor alpha3 subunit protein in rat brain and sympathetic ganglion measured using a subunit-specific antibody: regional and ontogenic expression. J Neurochem. 2001;77(1):336–46. Epub 2001/03/30.
35. Duffy AM, Fitzgerald ML, Chan J, Robinson DC, Milner TA, Mackie K, et al. Acetylcholine alpha7 nicotinic and dopamine D2 receptors are targeted to many of the same postsynaptic dendrites and astrocytes in the rodent prefrontal cortex. Synapse. 2011;65(12):1350–67. Epub 2011/08/23.
36. Fabian-Fine R, Skehel P, Errington ML, Davies HA, Sher E, Stewart MG, et al. Ultrastructural distribution of the alpha7 nicotinic acetylcholine receptor subunit in rat hippocampus. J Neurosci. 2001;21(20):7993–8003. Epub 2001/10/06.
37. Jones IW, Barik J, O'Neill MJ, Wonnacott S. Alpha bungarotoxin-1.4 nm gold: a novel conjugate for visualising the precise subcellular distribution of alpha 7* nicotinic acetylcholine receptors. J Neurosci Methods. 2004;134(1):65–74.
38. Jones IW, Wonnacott S. Precise localization of alpha7 nicotinic acetylcholine receptors on glutamatergic axon terminals in the rat ventral tegmental area. J Neurosci. 2004; 24(50):11244–52. Epub 2004/12/17.
39. Commons KG. Alpha4 containing nicotinic receptors are positioned to mediate postsynaptic effects on 5-HT neurons in the rat dorsal raphe nucleus. Neuroscience. 2008;153(3):851–9. Epub 2008/04/12.
40. Jones IW, Bolam JP, Wonnacott S. Presynaptic localisation of the nicotinic acetylcholine receptor beta2 subunit immunoreactivity in rat nigrostriatal dopaminergic neurones. J Comp Neurol. 2001;439(2):235–47. Epub 2001/10/12.
41. Nakayama H, Shioda S, Nakajo S, Ueno S, Nakashima T, Nakai Y. Immunocytochemical localization of nicotinic acetylcholine receptor in the rat cerebellar cortex. Neurosci Res. 1997;29(3):233–9. Epub 1998/01/22.
42. Nashmi R, Dickinson ME, McKinney S, Jareb M, Labarca C, Fraser SE, et al. Assembly of alpha4beta2 nicotinic acetylcholine receptors assessed with functional fluorescently labeled subunits: effects of localization, trafficking, and nicotine-induced upregulation in clonal mammalian cells and in cultured midbrain neurons. J Neurosci. 2003;23(37):11554–67. Epub 2003/12/20.

43. Bruses JL. Cell surface localization of alpha3beta4 nicotinic acetylcholine receptors is regulated by N-cadherin homotypic binding and actomyosin contractility. PLoS One. 2013;8(4):e62435. Epub 2013/04/30.
44. Chatterjee S, Santos N, Holgate J, Haass-Koffler CL, Hopf FW, Kharazia V, et al. The alpha5 subunit regulates the expression and function of alpha4*-containing neuronal nicotinic acetylcholine receptors in the ventral-tegmental area. PLoS One. 2013;8(7):e68300. Epub 2013/07/23.
45. Colombo SF, Mazzo F, Pistillo F, Gotti C. Biogenesis, trafficking and up-regulation of nicotinic ACh receptors. Biochem Pharmacol. 2013;86(8):1063–73. Epub 2013/07/09.
46. Drenan RM, Nashmi R, Imoukhuede P, Just H, McKinney S, Lester HA. Subcellular trafficking, pentameric assembly, and subunit stoichiometry of neuronal nicotinic acetylcholine receptors containing fluorescently labeled alpha6 and beta3 subunits. Mol Pharmacol. 2008;73(1):27–41. Epub 2007/10/13.
47. Grailhe R, de Carvalho LP, Paas Y, Le Poupon C, Soudant M, Bregestovski P, et al. Distinct subcellular targeting of fluorescent nicotinic alpha 3 beta 4 and serotoninergic 5-HT3A receptors in hippocampal neurons. Eur J Neurosci. 2004;19(4):855–62. Epub 2004/03/11.
48. Murray TA, Liu Q, Whiteaker P, Wu J, Lukas RJ. Nicotinic acetylcholine receptor alpha7 subunits with a C2 cytoplasmic loop yellow fluorescent protein insertion form functional receptors. Acta Pharmacol Sin. 2009;30(6):828–41. Epub 2009/06/06.
49. Richards CI, Srinivasan R, Xiao C, Mackey ED, Miwa JM, Lester HA. Trafficking of alpha4* nicotinic receptors revealed by superecliptic phluorin: effects of a beta4 amyotrophic lateral sclerosis-associated mutation and chronic exposure to nicotine. J Biol Chem. 2011; 286(36):31241–9. Epub 2011/07/20.
50. St John PA. Cellular trafficking of nicotinic acetylcholine receptors. Acta Pharmacol Sin. 2009;30(6):656–62. Epub 2009/06/06.
51. Xu J, Zhu Y, Heinemann SF. Identification of sequence motifs that target neuronal nicotinic receptors to dendrites and axons. J Neurosci. 2006;26(38):9780–93. Epub 2006/09/22.
52. Graham AJ, Martin-Ruiz CM, Teaktong T, Ray MA, Court JA. Human brain nicotinic receptors, their distribution and participation in neuropsychiatric disorders. Curr Drug Targets CNS Neurol Disord. 2002;1(4):387–97. Epub 2003/05/29.
53. Paterson D, Nordberg A. Neuronal nicotinic receptors in the human brain. Prog Neurobiol. 2000;61(1):75–111. Epub 2000/04/12.
54. Colloby SJ, Firbank MJ, Pakrasi S, Perry EK, Pimlott SL, Wyper DJ, et al. Alterations in nicotinic alpha4beta2 receptor binding in vascular dementia using (1)(2)(3)I-5IA-85380 SPECT: comparison with regional cerebral blood flow. Neurobiol Aging. 2011;32(2): 293–301. Epub 2009/03/10.
55. Ellis JR, Nathan PJ, Villemagne VL, Mulligan RS, Ellis KA, Tochon-Danguy HJ, et al. The relationship between nicotinic receptors and cognitive functioning in healthy aging: An in vivo positron emission tomography (PET) study with 2-[(18)F]fluoro-A-85380. Synapse. 2009;63(9):752–63. Epub 2009/06/02.
56. Gallezot JD, Bottlaender M, Gregoire MC, Roumenov D, Deverre JR, Coulon C, et al. In vivo imaging of human cerebral nicotinic acetylcholine receptors with 2-18F-fluoro-A-85380 and PET. J Nucl Med. 2005;46(2):240–7. Epub 2005/02/08.
57. Kimes AS, Horti AG, London ED, Chefer SI, Contoreggi C, Ernst M, et al. 2-[18F]F-A-85380: PET imaging of brain nicotinic acetylcholine receptors and whole body distribution in humans. FASEB J. 2003;17(10):1331–3. Epub 2003/05/22.
58. Hashimoto K, Nishiyama S, Ohba H, Matsuo M, Kobashi T, Takahagi M, et al. CHIBA-1001 as a novel PET ligand for alpha7 nicotinic receptors in the brain: a PET study in conscious monkeys. PLoS One. 2008;3(9):e3231. Epub 2008/09/19.
59. Toyohara J, Sakata M, Wu J, Ishikawa M, Oda K, Ishii K, et al. Preclinical and the first clinical studies on [11C]CHIBA-1001 for mapping alpha7 nicotinic receptors by positron emission tomography. Ann Nucl Med. 2009;23(3):301–9. Epub 2009/04/02.
60. Yin L, Zhao Q, Li L, Zhang SL, Chen XQ, Ma C, et al. An experimental study on (131) I-CHIBA-1001: a radioligand for alpha7 nicotinic acetylcholine receptors. PLoS One. 2013;8(7):e70188. Epub 2013/08/13.

61. Brasic JR, Zhou Y, Musachio JL, Hilton J, Fan H, Crabb A, et al. Single photon emission computed tomography experience with (S)-5-[(123)I]iodo-3-(2-azetidinylmethoxy)pyridine in the living human brain of smokers and nonsmokers. Synapse. 2009;63(4):339–58. Epub 2009/01/14.
62. Kimes AS, Chefer SI, Matochik JA, Contoreggi CS, Vaupel DB, Stein EA, et al. Quantification of nicotinic acetylcholine receptors in the human brain with PET: bolus plus infusion administration of 2-[18F]F-A85380. Neuroimage. 2008;39(2):717–27. Epub 2007/10/27.
63. Picard F, Bruel D, Servent D, Saba W, Fruchart-Gaillard C, Schollhorn-Peyronneau MA, et al. Alteration of the in vivo nicotinic receptor density in ADNFLE patients: a PET study. Brain. 2006;129(Pt 8):2047–60. Epub 2006/07/04.
64. Picard F, Sadaghiani S, Leroy C, Courvoisier DS, Maroy R, Bottlaender M. High density of nicotinic receptors in the cingulo-insular network. Neuroimage. 2013;79:42–51. Epub 2013/05/02.
65. Mukhin AG, Kimes AS, Chefer SI, Matochik JA, Contoreggi CS, Horti AG, et al. Greater nicotinic acetylcholine receptor density in smokers than in nonsmokers: a PET study with 2-18F-FA-85380. J Nucl Med. 2008;49(10):1628–35. Epub 2008/09/17.
66. Oishi N, Hashikawa K, Yoshida H, Ishizu K, Ueda M, Kawashima H, et al. Quantification of nicotinic acetylcholine receptors in Parkinson's disease with (123)I-5IA SPECT. J Neurol Sci. 2007;256(1–2):52–60. Epub 2007/03/21.
67. Descarries L, Gisiger V, Steriade M. Diffuse transmission by acetylcholine in the CNS. Prog Neurobiol. 1997;53(5):603–25. Epub 1998/01/09.
68. Vinson PN, Justice Jr JB. Effect of neostigmine on concentration and extraction fraction of acetylcholine using quantitative microdialysis. J Neurosci Methods. 1997;73(1):61–7. Epub 1997/04/25.
69. Quirion R, Richard J, Wilson A. Muscarinic and nicotinic modulation of cortical acetylcholine release monitored by in vivo microdialysis in freely moving adult rats. Synapse. 1994;17(2):92–100. Epub 1994/06/01.
70. Testylier G, Dykes RW. Acetylcholine release from frontal cortex in the waking rat measured by microdialysis without acetylcholinesterase inhibitors: effects of diisopropylfluorophosphate. Brain Res. 1996;740(1–2):307–15. Epub 1996/11/18.
71. Vizi ES, Fekete A, Karoly R, Mike A. Non-synaptic receptors and transporters involved in brain functions and targets of drug treatment. Br J Pharmacol. 2010;160(4):785–809. Epub 2010/02/09.
72. Bruno JP, Gash C, Martin B, Zmarowski A, Pomerleau F, Burmeister J, et al. Second-by-second measurement of acetylcholine release in prefrontal cortex. Eur J Neurosci. 2006;24(10):2749–57. Epub 2006/12/13.
73. Giuliano C, Parikh V, Ward JR, Chiamulera C, Sarter M. Increases in cholinergic neurotransmission measured by using choline-sensitive microelectrodes: enhanced detection by hydrolysis of acetylcholine on recording sites? Neurochem Int. 2008;52(7):1343–50. Epub 2008/03/19.
74. Mattinson CE, Burmeister JJ, Quintero JE, Pomerleau F, Huettl P, Gerhardt GA. Tonic and phasic release of glutamate and acetylcholine neurotransmission in sub-regions of the rat prefrontal cortex using enzyme-based microelectrode arrays. J Neurosci Methods. 2011;202(2):199–208. Epub 2011/09/08.
75. Parikh V, Kozak R, Martinez V, Sarter M. Prefrontal acetylcholine release controls cue detection on multiple timescales. Neuron. 2007;56(1):141–54. Epub 2007/10/09.
76. Parikh V, Pomerleau F, Huettl P, Gerhardt GA, Sarter M, Bruno JP. Rapid assessment of in vivo cholinergic transmission by amperometric detection of changes in extracellular choline levels. Eur J Neurosci. 2004;20(6):1545–54. Epub 2004/09/10.
77. Fadel JR. Regulation of cortical acetylcholine release: insights from in vivo microdialysis studies. Behav Brain Res. 2011;221(2):527–36. Epub 2010/02/23.
78. Sarter M, Parikh V, Howe WM. nAChR agonist-induced cognition enhancement: integration of cognitive and neuronal mechanisms. Biochem Pharmacol. 2009;78(7):658–67. Epub 2009/05/02.

79. Sarter M, Parikh V, Howe WM. Phasic acetylcholine release and the volume transmission hypothesis: time to move on. Nat Rev Neurosci. 2009;10(5):383–90. Epub 2009/04/21.
80. Howe WM, Ji J, Parikh V, Williams S, Mocaer E, Trocme-Thibierge C, et al. Enhancement of attentional performance by selective stimulation of alpha4beta2(*) nAChRs: underlying cholinergic mechanisms. Neuropsychopharmacology. 2010;35(6):1391–401. Epub 2010/02/12.
81. Zhou FM, Liang Y, Dani JA. Endogenous nicotinic cholinergic activity regulates dopamine release in the striatum. Nat Neurosci. 2001;4(12):1224–9. Epub 2001/11/20.
82. Garduno J, Galindo-Charles L, Jimenez-Rodriguez J, Galarraga E, Tapia D, Mihailescu S, et al. Presynaptic alpha4beta2 nicotinic acetylcholine receptors increase glutamate release and serotonin neuron excitability in the dorsal raphe nucleus. J Neurosci. 2012;32(43):-15148–57. Epub 2012/10/27.
83. Guo JZ, Liu Y, Sorenson EM, Chiappinelli VA. Synaptically released and exogenous ACh activates different nicotinic receptors to enhance evoked glutamatergic transmission in the lateral geniculate nucleus. J Neurophysiol. 2005;94(4):2549–60. Epub 2005/06/24.
84. Cachope R, Mateo Y, Mathur BN, Irving J, Wang HL, Morales M, et al. Selective activation of cholinergic interneurons enhances accumbal phasic dopamine release: setting the tone for reward processing. Cell Rep. 2012;2(1):33–41. Epub 2012/07/31.
85. Tang AH, Karson MA, Nagode DA, McIntosh JM, Uebele VN, Renger JJ, et al. Nerve terminal nicotinic acetylcholine receptors initiate quantal GABA release from perisomatic interneurons by activating axonal T-type (Cav3) Ca(2)(+) channels and Ca(2)(+) release from stores. J Neurosci. 2011;31(38):13546–61. Epub 2011/09/24.
86. Gu Z, Yakel JL. Timing-dependent septal cholinergic induction of dynamic hippocampal synaptic plasticity. Neuron. 2011;71(1):155–65. Epub 2011/07/13.
87. Nakauchi S, Sumikawa K. Endogenously released ACh and exogenous nicotine differentially facilitate long-term potentiation induction in the hippocampal CA1 region of mice. Eur J Neurosci. 2012;35(9):1381–95. Epub 2012/04/03.
88. Threlfell S, Lalic T, Platt NJ, Jennings KA, Deisseroth K, Cragg SJ. Striatal dopamine release is triggered by synchronized activity in cholinergic interneurons. Neuron. 2012;75(1):58–64. Epub 2012/07/17.
89. Alkondon M, Pereira EF, Cortes WS, Maelicke A, Albuquerque EX. Choline is a selective agonist of alpha7 nicotinic acetylcholine receptors in the rat brain neurons. Eur J Neurosci. 1997;9(12):2734–42. Epub 1998/03/28.
90. Higley MJ, Strittmatter SM. Neuroscience. Lynx for braking plasticity. Science. 2010;330(6008):1189–90. Epub 2010/11/27.
91. Ibanez-Tallon I, Miwa JM, Wang HL, Adams NC, Crabtree GW, Sine SM, et al. Novel modulation of neuronal nicotinic acetylcholine receptors by association with the endogenous prototoxin lynx1. Neuron. 2002;33(6):893–903. Epub 2002/03/22.
92. Miwa JM, Freedman R, Lester HA. Neural systems governed by nicotinic acetylcholine receptors: emerging hypotheses. Neuron. 2011;70(1):20–33. Epub 2011/04/13.
93. Miwa JM, Ibanez-Tallon I, Crabtree GW, Sanchez R, Sali A, Role LW, et al. lynx1, an endogenous toxin-like modulator of nicotinic acetylcholine receptors in the mammalian CNS. Neuron. 1999;23(1):105–14. Epub 1999/07/13.
94. Miwa JM, Lester HA, Walz A. Optimizing cholinergic tone through lynx modulators of nicotinic receptors: implications for plasticity and nicotine addiction. Physiology (Bethesda). 2012;27(4):187–99. Epub 2012/08/10.
95. Miwa JM, Stevens TR, King SL, Caldarone BJ, Ibanez-Tallon I, Xiao C, et al. The prototoxin lynx1 acts on nicotinic acetylcholine receptors to balance neuronal activity and survival in vivo. Neuron. 2006;51(5):587–600. Epub 2006/09/05.
96. Morishita H, Miwa JM, Heintz N, Hensch TK. Lynx1, a cholinergic brake, limits plasticity in adult visual cortex. Science. 2010;330(6008):1238–40. Epub 2010/11/13.
97. Benowitz NL, Porchet H, Jacob 3rd P. Nicotine dependence and tolerance in man: pharmacokinetic and pharmacodynamic investigations. Prog Brain Res. 1989;79:279–87. Epub 1989/01/01.

98. Henningfield JE, Stapleton JM, Benowitz NL, Grayson RF, London ED. Higher levels of nicotine in arterial than in venous blood after cigarette smoking. Drug Alcohol Depend. 1993;33(1):23–9. Epub 1993/06/01.
99. Pidoplichko VI, DeBiasi M, Williams JT, Dani JA. Nicotine activates and desensitizes midbrain dopamine neurons. Nature. 1997;390(6658):401–4. Epub 1997/12/06.
100. Picciotto MR, Kenny PJ. Molecular mechanisms underlying behaviors related to nicotine addiction. Cold Spring Harb Perspect Med. 2013;3(1):a012112. Epub 2012/11/13.
101. Mansvelder HD, Keath JR, McGehee DS. Synaptic mechanisms underlie nicotine-induced excitability of brain reward areas. Neuron. 2002;33(6):905–19. Epub 2002/03/22.
102. Mansvelder HD, McGehee DS. Long-term potentiation of excitatory inputs to brain reward areas by nicotine. Neuron. 2000;27(2):349–57. Epub 2000/09/14.
103. Xiao C, Nashmi R, McKinney S, Cai H, McIntosh JM, Lester HA. Chronic nicotine selectively enhances alpha4beta2* nicotinic acetylcholine receptors in the nigrostriatal dopamine pathway. J Neurosci. 2009;29(40):12428–39. Epub 2009/10/09.
104. Dani JA, Bertrand D. Nicotinic acetylcholine receptors and nicotinic cholinergic mechanisms of the central nervous system. Annu Rev Pharmacol Toxicol. 2007;47:699–729. Epub 2006/10/03.
105. Pidoplichko VI, Noguchi J, Areola OO, Liang Y, Peterson J, Zhang T, et al. Nicotinic cholinergic synaptic mechanisms in the ventral tegmental area contribute to nicotine addiction. Learn Mem. 2004;11(1):60–9. Epub 2004/01/30.
106. Poorthuis RB, Mansvelder HD. Nicotinic acetylcholine receptors controlling attention: behavior, circuits and sensitivity to disruption by nicotine. Biochem Pharmacol. 2013; 86(8):1089–98. Epub 2013/07/17.
107. Picciotto MR, Addy NA, Mineur YS, Brunzell DH. It is not "either/or": activation and desensitization of nicotinic acetylcholine receptors both contribute to behaviors related to nicotine addiction and mood. Prog Neurobiol. 2008;84(4):329–42. Epub 2008/02/05.
108. Fenster CP, Whitworth TL, Sheffield EB, Quick MW, Lester RA. Upregulation of surface alpha4beta2 nicotinic receptors is initiated by receptor desensitization after chronic exposure to nicotine. J Neurosci. 1999;19(12):4804–14. Epub 1999/06/15.
109. Marks MJ, Burch JB, Collins AC. Effects of chronic nicotine infusion on tolerance development and nicotinic receptors. J Pharmacol Exp Ther. 1983;226(3):817–25. Epub 1983/09/01.
110. Quick MW, Lester RA. Desensitization of neuronal nicotinic receptors. J Neurobiol. 2002;53(4):457–78. Epub 2002/11/19.
111. Schwartz RD, Kellar KJ. In vivo regulation of [3H]acetylcholine recognition sites in brain by nicotinic cholinergic drugs. J Neurochem. 1985;45(2):427–33. Epub 1985/08/01.
112. Lopez-Hernandez GY, Sanchez-Padilla J, Ortiz-Acevedo A, Lizardi-Ortiz J, Salas-Vincenty J, Rojas LV, et al. Nicotine-induced up-regulation and desensitization of alpha4beta2 neuronal nicotinic receptors depend on subunit ratio. J Biol Chem. 2004;279(36):38007–15. Epub 2004/07/13.
113. Moroni M, Zwart R, Sher E, Cassels BK, Bermudez I. alpha4beta2 nicotinic receptors with high and low acetylcholine sensitivity: pharmacology, stoichiometry, and sensitivity to long-term exposure to nicotine. Mol Pharmacol. 2006;70(2):755–68. Epub 2006/05/25.
114. Nelson ME, Kuryatov A, Choi CH, Zhou Y, Lindstrom J. Alternate stoichiometries of alpha4beta2 nicotinic acetylcholine receptors. Mol Pharmacol. 2003;63(2):332–41. Epub 2003/01/16.
115. Nashmi R, Xiao C, Deshpande P, McKinney S, Grady SR, Whiteaker P, et al. Chronic nicotine cell specifically upregulates functional alpha 4* nicotinic receptors: basis for both tolerance in midbrain and enhanced long-term potentiation in perforant path. J Neurosci. 2007;27(31):8202–18. Epub 2007/08/03.
116. Papke RL, Trocme-Thibierge C, Guendisch D, Al Rubaiy SA, Bloom SA. Electrophysiological perspectives on the therapeutic use of nicotinic acetylcholine receptor partial agonists. J Pharmacol Exp Ther. 2011;337(2):367–79. Epub 2011/02/03.
117. Peng C, Stokes C, Mineur YS, Picciotto MR, Tian C, Eibl C, et al. Differential modulation of brain nicotinic acetylcholine receptor function by cytisine, varenicline, and two novel bispi-

dine compounds: emergent properties of a hybrid molecule. J Pharmacol Exp Ther. 2013; 347(2):424–37. Epub 2013/08/21.

118. Coe JW, Brooks PR, Vetelino MG, Wirtz MC, Arnold EP, Huang J, et al. Varenicline: an alpha4beta2 nicotinic receptor partial agonist for smoking cessation. J Med Chem. 2005;48(10):3474–7. Epub 2005/05/13.
119. Mihalak KB, Carroll FI, Luetje CW. Varenicline is a partial agonist at alpha4beta2 and a full agonist at alpha7 neuronal nicotinic receptors. Mol Pharmacol. 2006;70(3):801–5. Epub 2006/06/13.
120. Papke RL, Heinemann SF. Partial agonist properties of cytisine on neuronal nicotinic receptors containing the beta 2 subunit. Mol Pharmacol. 1994;45(1):142–9. Epub 1994/01/01.
121. Rollema H, Coe JW, Chambers LK, Hurst RS, Stahl SM, Williams KE. Rationale, pharmacology and clinical efficacy of partial agonists of alpha4beta2 nACh receptors for smoking cessation. Trends Pharmacol Sci. 2007;28(7):316–25. Epub 2007/06/19.
122. Dajas-Bailador F, Wonnacott S. Nicotinic acetylcholine receptors and the regulation of neuronal signalling. Trends Pharmacol Sci. 2004;25(6):317–24. Epub 2004/05/29.
123. De Filippi G, Baldwinson T, Sher E. Nicotinic receptor modulation of neurotransmitter release in the cerebellum. Prog Brain Res. 2005;148:307–20. Epub 2005/01/22.
124. Hurst R, Rollema H, Bertrand D. Nicotinic acetylcholine receptors: from basic science to therapeutics. Pharmacol Ther. 2013;137(1):22–54. Epub 2012/08/29.
125. Lawrence JJ. Cholinergic control of GABA release: emerging parallels between neocortex and hippocampus. Trends Neurosci. 2008;31(7):317–27. Epub 2008/06/17.
126. Marchi M, Grilli M. Presynaptic nicotinic receptors modulating neurotransmitter release in the central nervous system: functional interactions with other coexisting receptors. Prog Neurobiol. 2010;92(2):105–11. Epub 2010/06/19.
127. Sher E, Chen Y, Sharples TJ, Broad LM, Benedetti G, Zwart R, et al. Physiological roles of neuronal nicotinic receptor subtypes: new insights on the nicotinic modulation of neurotransmitter release, synaptic transmission and plasticity. Curr Top Med Chem. 2004;4(3):283–97. Epub 2004/02/03.
128. Wonnacott S, Barik J, Dickinson J, Jones IW. Nicotinic receptors modulate transmitter cross talk in the CNS: nicotinic modulation of transmitters. J Mol Neurosci. 2006;30(1–2):137–40. Epub 2006/12/29.
129. Exley R, Cragg SJ. Presynaptic nicotinic receptors: a dynamic and diverse cholinergic filter of striatal dopamine neurotransmission. Br J Pharmacol. 2008;153 Suppl 1:S283–97. Epub 2007/11/27.
130. Livingstone PD, Wonnacott S. Nicotinic acetylcholine receptors and the ascending dopamine pathways. Biochem Pharmacol. 2009;78(7):744–55. Epub 2009/06/16.
131. Wonnacott S. Gates and filters: unveiling the physiological roles of nicotinic acetylcholine receptors in dopaminergic transmission. Br J Pharmacol. 2008;153 Suppl 1:S2–4. Epub 2008/02/05.
132. Azam L, McIntosh JM. Effect of novel alpha-conotoxins on nicotine-stimulated [3H]dopamine release from rat striatal synaptosomes. J Pharmacol Exp Ther. 2005;312(1):231–7. Epub 2004/08/19.
133. Cui C, Booker TK, Allen RS, Grady SR, Whiteaker P, Marks MJ, et al. The beta3 nicotinic receptor subunit: a component of alpha-conotoxin MII-binding nicotinic acetylcholine receptors that modulate dopamine release and related behaviors. J Neurosci. 2003; 23(35):11045–53. Epub 2003/12/06.
134. Drenan RM, Grady SR, Steele AD, McKinney S, Patzlaff NE, McIntosh JM, et al. Cholinergic modulation of locomotion and striatal dopamine release is mediated by alpha6alpha4* nicotinic acetylcholine receptors. J Neurosci. 2010;30(29):9877–89. Epub 2010/07/28.
135. Kulak JM, Nguyen TA, Olivera BM, McIntosh JM. Alpha-conotoxin MII blocks nicotine-stimulated dopamine release in rat striatal synaptosomes. J Neurosci. 1997;17(14):5263–70. Epub 1997/07/15.

136. Salminen O, Drapeau JA, McIntosh JM, Collins AC, Marks MJ, Grady SR. Pharmacology of alpha-conotoxin MII-sensitive subtypes of nicotinic acetylcholine receptors isolated by breeding of null mutant mice. Mol Pharmacol. 2007;71(6):1563–71. Epub 2007/03/08.
137. Salminen O, Murphy KL, McIntosh JM, Drago J, Marks MJ, Collins AC, et al. Subunit composition and pharmacology of two classes of striatal presynaptic nicotinic acetylcholine receptors mediating dopamine release in mice. Mol Pharmacol. 2004;65(6):1526–35. Epub 2004/05/25.
138. Yang K, Buhlman L, Khan GM, Nichols RA, Jin G, McIntosh JM, et al. Functional nicotinic acetylcholine receptors containing alpha6 subunits are on GABAergic neuronal boutons adherent to ventral tegmental area dopamine neurons. J Neurosci. 2011;31(7):2537–48. Epub 2011/02/18.
139. Soll LG, Grady SR, Salminen O, Marks MJ, Tapper AR. A role for alpha4(non-alpha6)* nicotinic acetylcholine receptors in motor behavior. Neuropharmacology. 2013;73:19–30. Epub 2013/05/22.
140. Mogg AJ, Whiteaker P, McIntosh JM, Marks M, Collins AC, Wonnacott S. Methyllycaconitine is a potent antagonist of alpha-conotoxin-MII-sensitive presynaptic nicotinic acetylcholine receptors in rat striatum. J Pharmacol Exp Ther. 2002;302(1):197–204. Epub 2002/06/18.
141. Cohen BN, Mackey ED, Grady SR, McKinney S, Patzlaff NE, Wageman CR, et al. Nicotinic cholinergic mechanisms causing elevated dopamine release and abnormal locomotor behavior. Neuroscience. 2012;200:31–41. Epub 2011/11/15.
142. Grady SR, Meinerz NM, Cao J, Reynolds AM, Picciotto MR, Changeux JP, et al. Nicotinic agonists stimulate acetylcholine release from mouse interpeduncular nucleus: a function mediated by a different nAChR than dopamine release from striatum. J Neurochem. 2001;76(1):258–68. Epub 2001/01/09.
143. Grady SR, Murphy KL, Cao J, Marks MJ, McIntosh JM, Collins AC. Characterization of nicotinic agonist-induced [(3)H]dopamine release from synaptosomes prepared from four mouse brain regions. J Pharmacol Exp Ther. 2002;301(2):651–60. Epub 2002/04/19.
144. Parish CL, Nunan J, Finkelstein DI, McNamara FN, Wong JY, Waddington JL, et al. Mice lacking the alpha4 nicotinic receptor subunit fail to modulate dopaminergic neuronal arbors and possess impaired dopamine transporter function. Mol Pharmacol. 2005;68(5):1376–86. Epub 2005/08/04.
145. Azam L, McIntosh JM. Characterization of nicotinic acetylcholine receptors that modulate nicotine-evoked [3H]norepinephrine release from mouse hippocampal synaptosomes. Mol Pharmacol. 2006;70(3):967–76. Epub 2006/06/01.
146. Grilli M, Parodi M, Raiteri M, Marchi M. Chronic nicotine differentially affects the function of nicotinic receptor subtypes regulating neurotransmitter release. J Neurochem. 2005;93(5):1353–60. Epub 2005/06/07.
147. Scholze P, Orr-Urtreger A, Changeux JP, McIntosh JM, Huck S. Catecholamine outflow from mouse and rat brain slice preparations evoked by nicotinic acetylcholine receptor activation and electrical field stimulation. Br J Pharmacol. 2007;151(3):414–22. Epub 2007/04/03.
148. Kaiser S, Wonnacott S. Alpha-bungarotoxin-sensitive nicotinic receptors indirectly modulate [(3)H]dopamine release in rat striatal slices via glutamate release. Mol Pharmacol. 2000; 58(2):312–8. Epub 2000/07/25.
149. Dickinson JA, Kew JN, Wonnacott S. Presynaptic alpha 7- and beta 2-containing nicotinic acetylcholine receptors modulate excitatory amino acid release from rat prefrontal cortex nerve terminals via distinct cellular mechanisms. Mol Pharmacol. 2008;74(2):348–59. Epub 2008/05/01.
150. Livingstone PD, Dickinson JA, Srinivasan J, Kew JN, Wonnacott S. Glutamate-dopamine crosstalk in the rat prefrontal cortex is modulated by Alpha7 nicotinic receptors and potentiated by PNU-120596. J Mol Neurosci. 2010;40(1–2):172–6. Epub 2009/08/19.
151. Marchi M, Risso F, Viola C, Cavazzani P, Raiteri M. Direct evidence that release-stimulating alpha7* nicotinic cholinergic receptors are localized on human and rat brain glutamatergic axon terminals. J Neurochem. 2002;80(6):1071–8. Epub 2002/04/16.

152. Marchi M, Zappettini S, Olivero G, Pittaluga A, Grilli M. Chronic nicotine exposure selectively activates a carrier-mediated release of endogenous glutamate and aspartate from rat hippocampal synaptosomes. Neurochem Int. 2012;60(6):622–30. Epub 2012/03/16.
153. Patti L, Raiteri L, Grilli M, Zappettini S, Bonanno G, Marchi M. Evidence that alpha7 nicotinic receptor modulates glutamate release from mouse neocortical gliosomes. Neurochem Int. 2007;51(1):1–7. Epub 2007/04/28.
154. Rousseau SJ, Jones IW, Pullar IA, Wonnacott S. Presynaptic alpha7 and non-alpha7 nicotinic acetylcholine receptors modulate [3H]d-aspartate release from rat frontal cortex in vitro. Neuropharmacology. 2005;49(1):59–72. Epub 2005/07/05.
155. Zappettini S, Grilli M, Lagomarsino F, Cavallero A, Fedele E, Marchi M. Presynaptic nicotinic alpha7 and non-alpha7 receptors stimulate endogenous GABA release from rat hippocampal synaptosomes through two mechanisms of action. PLoS One. 2011;6(2):e16911. Epub 2011/02/25.
156. Zappettini S, Grilli M, Olivero G, Mura E, Preda S, Govoni S, et al. Beta amyloid differently modulate nicotinic and muscarinic receptor subtypes which stimulate in vitro and in vivo the release of glycine in the rat hippocampus. Front Pharmacol. 2012;3:146. Epub 2012/08/07.
157. Zappettini S, Grilli M, Salamone A, Fedele E, Marchi M. Pre-synaptic nicotinic receptors evoke endogenous glutamate and aspartate release from hippocampal synaptosomes by way of distinct coupling mechanisms. Br J Pharmacol. 2010;161(5):1161–71. Epub 2010/07/17.
158. Zappettini S, Mura E, Grilli M, Preda S, Salamone A, Olivero G, et al. Different presynaptic nicotinic receptor subtypes modulate in vivo and in vitro the release of glycine in the rat hippocampus. Neurochem Int. 2011;59(5):729–38. Epub 2011/07/19.
159. Mansvelder HD, Mertz M, Role LW. Nicotinic modulation of synaptic transmission and plasticity in cortico-limbic circuits. Semin Cell Dev Biol. 2009;20(4):432–40. Epub 2009/06/30.
160. McKay BE, Placzek AN, Dani JA. Regulation of synaptic transmission and plasticity by neuronal nicotinic acetylcholine receptors. Biochem Pharmacol. 2007;74(8):1120–33. Epub 2007/08/11.
161. Yakel JL. Nicotinic ACh receptors in the hippocampus: role in excitability and plasticity. Nicotine Tob Res. 2012;14(11):1249–57. Epub 2012/04/05.
162. Jiang L, Role LW. Facilitation of cortico-amygdala synapses by nicotine: activity-dependent modulation of glutamatergic transmission. J Neurophysiol. 2008;99(4):1988–99. Epub 2008/02/15.
163. Kalappa BI, Feng L, Kem WR, Gusev AG, Uteshev VV. Mechanisms of facilitation of synaptic glutamate release by nicotinic agonists in the nucleus of the solitary tract. Am J Physiol Cell Physiol. 2011;301(2):C347–61. Epub 2011/05/27.
164. Parikh V, Ji J, Decker MW, Sarter M. Prefrontal beta2 subunit-containing and alpha7 nicotinic acetylcholine receptors differentially control glutamatergic and cholinergic signaling. J Neurosci. 2010;30(9):3518–30. Epub 2010/03/06.
165. Zhong C, Du C, Hancock M, Mertz M, Talmage DA, Role LW. Presynaptic type III neuregulin 1 is required for sustained enhancement of hippocampal transmission by nicotine and for axonal targeting of alpha7 nicotinic acetylcholine receptors. J Neurosci. 2008;28(37):9111–6. Epub 2008/09/12.
166. Barazangi N, Role LW. Nicotine-induced enhancement of glutamatergic and GABAergic synaptic transmission in the mouse amygdala. J Neurophysiol. 2001;86(1):463–74. Epub 2001/06/30.
167. Radcliffe KA, Fisher JL, Gray R, Dani JA. Nicotinic modulation of glutamate and GABA synaptic transmission of hippocampal neurons. Ann N Y Acad Sci. 1999;868:591–610. Epub 1999/07/22.
168. Seddik R, Schlichter R, Trouslard J. Modulation of GABAergic synaptic transmission by terminal nicotinic acetylcholine receptors in the central autonomic nucleus of the neonatal rat spinal cord. Neuropharmacology. 2006;51(1):77–89. Epub 2006/05/09.
169. Exley R, Clements MA, Hartung H, McIntosh JM, Franklin M, Bermudez I, et al. Striatal dopamine transmission is reduced after chronic nicotine with a decrease in alpha6-nicotinic receptor control in nucleus accumbens. Eur J Neurosci. 2013;38:3036–43. Epub 2013/07/12.

170. Exley R, Maubourguet N, David V, Eddine R, Evrard A, Pons S, et al. Distinct contributions of nicotinic acetylcholine receptor subunit alpha4 and subunit alpha6 to the reinforcing effects of nicotine. Proc Natl Acad Sci U S A. 2011;108(18):7577–82. Epub 2011/04/20.
171. Exley R, McIntosh JM, Marks MJ, Maskos U, Cragg SJ. Striatal alpha5 nicotinic receptor subunit regulates dopamine transmission in dorsal striatum. J Neurosci. 2012;32(7):2352–6. Epub 2012/03/08.
172. Gao M, Jin Y, Yang K, Zhang D, Lukas RJ, Wu J. Mechanisms involved in systemic nicotine-induced glutamatergic synaptic plasticity on dopamine neurons in the ventral tegmental area. J Neurosci. 2010;30(41):13814–25. Epub 2010/10/15.
173. Schilstrom B, Ivanov VB, Wiker C, Svensson TH. Galantamine enhances dopaminergic neurotransmission in vivo via allosteric potentiation of nicotinic acetylcholine receptors. Neuropsychopharmacology. 2007;32(1):43–53. Epub 2006/04/28.
174. Griguoli M, Cellot G, Cherubini E. In hippocampal oriens interneurons anti-Hebbian long-term potentiation requires cholinergic signaling via alpha7 nicotinic acetylcholine receptors. J Neurosci. 2013;33(3):1044–9. Epub 2013/01/18.
175. Jia Y, Yamazaki Y, Nakauchi S, Ito K, Sumikawa K. Nicotine facilitates long-term potentiation induction in oriens-lacunosum moleculare cells via Ca2+ entry through non-alpha7 nicotinic acetylcholine receptors. Eur J Neurosci. 2010;31(3):463–76. Epub 2010/02/02.
176. Nakauchi S, Brennan RJ, Boulter J, Sumikawa K. Nicotine gates long-term potentiation in the hippocampal CA1 region via the activation of alpha2* nicotinic ACh receptors. Eur J Neurosci. 2007;25(9):2666–81. Epub 2007/05/01.
177. Prestori F, Bonardi C, Mapelli L, Lombardo P, Goselink R, De Stefano ME, et al. Gating of long-term potentiation by nicotinic acetylcholine receptors at the cerebellum input stage. PLoS One. 2013;8(5):e64828. Epub 2013/06/07.
178. Yamazaki Y, Fujii S, Jia Y, Sumikawa K. Nicotine withdrawal suppresses nicotinic modulation of long-term potentiation induction in the hippocampal CA1 region. Eur J Neurosci. 2006;24(10):2903–16. Epub 2006/12/13.
179. Le Magueresse C, Safiulina V, Changeux JP, Cherubini E. Nicotinic modulation of network and synaptic transmission in the immature hippocampus investigated with genetically modified mice. J Physiol. 2006;576(Pt 2):533–46. Epub 2006/08/12.
180. Perez XA, Bordia T, McIntosh JM, Quik M. α6ß2* and α4ß2* nicotinic receptors both regulate dopamine signaling with increased nigrostriatal damage: relevance to Parkinson's disease. Mol Pharmacol. 2010;78(5):971–80. doi:10.1124/mol.110.067561. Epub 2010/08/23.
181. Champtiaux N, Gotti C, Cordero-Erausquin M, David DJ, Przybylski C, Lena C, et al. Subunit composition of functional nicotinic receptors in dopaminergic neurons investigated with knock-out mice. J Neurosci. 2003;23(21):7820–9. Epub 2003/08/29.
182. Lecca D, Shim I, Costa E, Javaid JI. Striatal application of nicotine, but not of lobeline, attenuates dopamine release in freely moving rats. Neuropharmacology. 2000;39(1):88–98. Epub 2000/02/09.
183. Marubio LM, Gardier AM, Durier S, David D, Klink R, Arroyo-Jimenez MM, et al. Effects of nicotine in the dopaminergic system of mice lacking the alpha4 subunit of neuronal nicotinic acetylcholine receptors. Eur J Neurosci. 2003;17(7):1329–37. Epub 2003/04/26.
184. Visanji NP, Mitchell SN, O'Neill MJ, Duty S. Chronic pre-treatment with nicotine enhances nicotine-evoked striatal dopamine release and alpha6 and beta3 nicotinic acetylcholine receptor subunit mRNA in the substantia nigra pars compacta of the rat. Neuropharmacology. 2006;50(1):36–46. Epub 2005/09/13.
185. Zanetti L, Picciotto MR, Zoli M. Differential effects of nicotinic antagonists perfused into the nucleus accumbens or the ventral tegmental area on cocaine-induced dopamine release in the nucleus accumbens of mice. Psychopharmacology (Berl). 2007;190(2):189–99. Epub 2006/10/25.
186. Livingstone PD, Srinivasan J, Kew JN, Dawson LA, Gotti C, Moretti M, et al. alpha7 and non-alpha7 nicotinic acetylcholine receptors modulate dopamine release in vitro and in vivo in the rat prefrontal cortex. Eur J Neurosci. 2009;29(3):539–50. Epub 2009/02/04.

187. Kennett A, Heal DJ, Wonnacott S. Pharmacological differences between rat frontal cortex and hippocampus in the nicotinic modulation of noradrenaline release implicate distinct receptor subtypes. Nicotine Tob Res. 2012;14(11):1339–45. Epub 2012/05/23.
188. Mura E, Zappettini S, Preda S, Biundo F, Lanni C, Grilli M, et al. Dual effect of beta-amyloid on alpha7 and alpha4beta2 nicotinic receptors controlling the release of glutamate, aspartate and GABA in rat hippocampus. PLoS One. 2012;7(1):e29661. Epub 2012/01/19.
189. Bueno-Junior LS, Lopes-Aguiar C, Ruggiero RN, Romcy-Pereira RN, Leite JP. Muscarinic and nicotinic modulation of thalamo-prefrontal cortex synaptic plasticity [corrected] in vivo. PLoS One. 2012;7(10):e47484. Epub 2012/11/03.
190. Mao D, Gallagher K, McGehee DS. Nicotine potentiation of excitatory inputs to ventral tegmental area dopamine neurons. J Neurosci. 2011;31(18):6710–20. Epub 2011/05/06.
191. Matsuyama S, Matsumoto A, Enomoto T, Nishizaki T. Activation of nicotinic acetylcholine receptors induces long-term potentiation in vivo in the intact mouse dentate gyrus. Eur J Neurosci. 2000;12(10):3741–7. Epub 2000/10/13.
192. Wang Y, Sherwood JL, Lodge D. The alpha4beta2 nicotinic acetylcholine receptor agonist TC-2559 impairs long-term potentiation in the dentate gyrus in vivo. Neurosci Lett. 2006;406(3):183–8. Epub 2006/08/29.
193. Keighron JD, Ewing AG, Cans AS. Analytical tools to monitor exocytosis: a focus on new fluorescent probes and methods. Analyst. 2012;137(8):1755–63. Epub 2012/02/22.
194. Klingauf J, Kavalali ET, Tsien RW. Kinetics and regulation of fast endocytosis at hippocampal synapses. Nature. 1998;394(6693):581–5. Epub 1998/08/26.
195. Mani M, Ryan TA. Live imaging of synaptic vesicle release and retrieval in dopaminergic neurons. Front Neural Circuits. 2009;3:3. Epub 2009/06/13.
196. Moulder KL, Jiang X, Taylor AA, Shin W, Gillis KD, Mennerick S. Vesicle pool heterogeneity at hippocampal glutamate and GABA synapses. J Neurosci. 2007;27(37):9846–54. Epub 2007/09/15.
197. Welzel O, Henkel AW, Stroebel AM, Jung J, Tischbirek CH, Ebert K, et al. Systematic heterogeneity of fractional vesicle pool sizes and release rates of hippocampal synapses. Biophys J. 2011;100(3):593–601. Epub 2011/02/02.
198. Nguyen QT, Schroeder LF, Mank M, Muller A, Taylor P, Griesbeck O, et al. An in vivo biosensor for neurotransmitter release and in situ receptor activity. Nat Neurosci. 2010;13(1):127–32. Epub 2009/12/17.
199. Borghuis BG, Marvin JS, Looger LL, Demb JB. Two-photon imaging of nonlinear glutamate release dynamics at bipolar cell synapses in the mouse retina. J Neurosci. 2013;33(27):10972–85. Epub 2013/07/05.
200. Marvin JS, Borghuis BG, Tian L, Cichon J, Harnett MT, Akerboom J, et al. An optimized fluorescent probe for visualizing glutamate neurotransmission. Nat Methods. 2013;10(2):162–70. Epub 2013/01/15.
201. Bouron A, Reuter H. Muscarinic stimulation of synaptic activity by protein kinase C is inhibited by adenosine in cultured hippocampal neurons. Proc Natl Acad Sci U S A. 1997;94(22):12224–9. Epub 1997/10/29.
202. Breeze LJ, Vijayaraghavan S. Optical measurements of presynaptic nicotinic effects. Neuropharmacology. 2000;39(13):2706–14. Epub 2000/10/25.
203. Blumenthal EM, Conroy WG, Romano SJ, Kassner PD, Berg DK. Detection of functional nicotinic receptors blocked by alpha-bungarotoxin on PC12 cells and dependence of their expression on post-translational events. J Neurosci. 1997;17(16):6094–104. Epub 1997/08/15.
204. Wanamaker CP, Green WN. Endoplasmic reticulum chaperones stabilize nicotinic receptor subunits and regulate receptor assembly. J Biol Chem. 2007;282(43):31113–23. Epub 2007/08/31.
205. Alexander JK, Sagher D, Krivoshein AV, Criado M, Jefford G, Green WN. Ric-3 promotes alpha7 nicotinic receptor assembly and trafficking through the ER subcompartment of dendrites. J Neurosci. 2010;30(30):10112–26. Epub 2010/07/30.
206. Castillo M, Mulet J, Gutierrez LM, Ortiz JA, Castelan F, Gerber S, et al. Dual role of the RIC-3 protein in trafficking of serotonin and nicotinic acetylcholine receptors. J Biol Chem. 2005;280(29):27062–8. Epub 2005/06/02.

207. Williams BM, Temburni MK, Levey MS, Bertrand S, Bertrand D, Jacob MH. The long internal loop of the alpha 3 subunit targets nAChRs to subdomains within individual synapses on neurons in vivo. Nat Neurosci. 1998;1(7):557–62. Epub 1999/04/10.
208. Fernandes CC, Berg DK, Gomez-Varela D. Lateral mobility of nicotinic acetylcholine receptors on neurons is determined by receptor composition, local domain, and cell type. J Neurosci. 2010;30(26):8841–51. Epub 2010/07/02.
209. Temburni MK, Rosenberg MM, Pathak N, McConnell R, Jacob MH. Neuronal nicotinic synapse assembly requires the adenomatous polyposis coli tumor suppressor protein. J Neurosci. 2004;24(30):6776–84. Epub 2004/07/30.
210. Cheng SB, Amici SA, Ren XQ, McKay SB, Treuil MW, Lindstrom JM, et al. Presynaptic targeting of alpha4beta 2 nicotinic acetylcholine receptors is regulated by neurexin-1beta. J Biol Chem. 2009;284(35):23251–9. Epub 2009/07/02.
211. Fischer H, Orr-Urtreger A, Role LW, Huck S. Selective deletion of the alpha5 subunit differentially affects somatic-dendritic versus axonally targeted nicotinic ACh receptors in mouse. J Physiol. 2005;563(Pt 1):119–37. Epub 2004/12/22.
212. Zhong C, Talmage DA, Role LW. Nicotine elicits prolonged calcium signaling along ventral hippocampal axons. PLoS One. 2013;8(12):e82719.
213. Chen YJ, Johnson MA, Lieberman MD, Goodchild RE, Schobel S, Lewandowski N, et al. Type III neuregulin-1 is required for normal sensorimotor gating, memory-related behaviors, and corticostriatal circuit components. J Neurosci. 2008;28(27):6872–83. Epub 2008/07/04.
214. Jiang L, Emmetsberger J, Talmage DA, Role LW. Type III neuregulin 1 is required for multiple forms of excitatory synaptic plasticity of mouse cortico-amygdala circuits. J Neurosci. 2013;33(23):9655–66. Epub 2013/06/07.
215. Hancock ML, Canetta SE, Role LW, Talmage DA. Presynaptic type III neuregulin1-ErbB signaling targets {alpha}7 nicotinic acetylcholine receptors to axons. J Cell Biol. 2008; 181(3):511–21. Epub 2008/05/07.
216. Gomez-Varela D, Berg DK. Lateral mobility of presynaptic alpha7-containing nicotinic receptors and its relevance for glutamate release. J Neurosci. 2013;33(43):17062–71. Epub 2013/10/25.
217. Gu Z, Lamb PW, Yakel JL. Cholinergic coordination of presynaptic and postsynaptic activity induces timing-dependent hippocampal synaptic plasticity. J Neurosci. 2012;32(36):12337–48. Epub 2012/09/08.
218. del Barrio L, Egea J, Leon R, Romero A, Ruiz A, Montero M, et al. Calcium signalling mediated through alpha7 and non-alpha7 nAChR stimulation is differentially regulated in bovine chromaffin cells to induce catecholamine release. Br J Pharmacol. 2011;162(1):94–110. Epub 2010/09/16.
219. Gilbert D, Lecchi M, Arnaudeau S, Bertrand D, Demaurex N. Local and global calcium signals associated with the opening of neuronal alpha7 nicotinic acetylcholine receptors. Cell Calcium. 2009;45(2):198–207. Epub 2008/11/29.
220. Szabo SI, Zelles T, Vizi ES, Lendvai B. The effect of nicotine on spiking activity and Ca2+ dynamics of dendritic spines in rat CA1 pyramidal neurons. Hippocampus. 2008;18(4):376–85. Epub 2008/01/15.
221. Fayuk D, Yakel JL. Ca2+ permeability of nicotinic acetylcholine receptors in rat hippocampal CA1 interneurones. J Physiol. 2005;566(Pt 3):759–68. Epub 2005/06/04.
222. Fucile S. Ca2+ permeability of nicotinic acetylcholine receptors. Cell Calcium. 2004;35(1): 1–8. Epub 2003/12/13.
223. Dajas-Bailador FA, Mogg AJ, Wonnacott S. Intracellular Ca2+ signals evoked by stimulation of nicotinic acetylcholine receptors in SH-SY5Y cells: contribution of voltage-operated Ca2+ channels and Ca2+ stores. J Neurochem. 2002;81(3):606–14. Epub 2002/06/18.
224. Dajas-Bailador FA, Soliakov L, Wonnacott S. Nicotine activates the extracellular signal-regulated kinase 1/2 via the alpha7 nicotinic acetylcholine receptor and protein kinase A, in SH-SY5Y cells and hippocampal neurones. J Neurochem. 2002;80(3):520–30. Epub 2002/03/22.

225. Sharma G, Vijayaraghavan S. Nicotinic cholinergic signaling in hippocampal astrocytes involves calcium-induced calcium release from intracellular stores. Proc Natl Acad Sci U S A. 2001;98(7):4148–53. Epub 2001/03/22.
226. Sharma G, Vijayaraghavan S. Nicotinic receptors containing the alpha7 subunit: a model for rational drug design. Curr Med Chem. 2008;15(28):2921–32. Epub 2008/12/17.
227. Welsby P, Rowan M, Anwyl R. Nicotinic receptor-mediated enhancement of long-term potentiation involves activation of metabotropic glutamate receptors and ryanodine-sensitive calcium stores in the dentate gyrus. Eur J Neurosci. 2006;24(11):3109–18. Epub 2006/12/13.
228. Cox ME, Parsons SJ. Roles for protein kinase C and mitogen-activated protein kinase in nicotine-induced secretion from bovine adrenal chromaffin cells. J Neurochem. 1997; 69(3):1119–30. Epub 1997/09/01.
229. Kihara T, Shimohama S, Sawada H, Honda K, Nakamizo T, Shibasaki H, et al. alpha 7 nicotinic receptor transduces signals to phosphatidylinositol 3-kinase to block A beta-amyloid-induced neurotoxicity. J Biol Chem. 2001;276(17):13541–6. Epub 2001/03/30.
230. Dineley KT, Westerman M, Bui D, Bell K, Ashe KH, Sweatt JD. Beta-amyloid activates the mitogen-activated protein kinase cascade via hippocampal alpha7 nicotinic acetylcholine receptors: In vitro and in vivo mechanisms related to Alzheimer's disease. J Neurosci. 2001;21(12):4125–33. Epub 2001/06/19.
231. Chang KT, Berg DK. Voltage-gated channels block nicotinic regulation of CREB phosphorylation and gene expression in neurons. Neuron. 2001;32(5):855–65. Epub 2001/12/12.
232. Damaj MI. Nicotinic regulation of calcium/calmodulin-dependent protein kinase II activation in the spinal cord. J Pharmacol Exp Ther. 2007;320(1):244–9. Epub 2006/10/17.
233. Salin PA, Scanziani M, Malenka RC, Nicoll RA. Distinct short-term plasticity at two excitatory synapses in the hippocampus. Proc Natl Acad Sci U S A. 1996;93(23):13304–9. Epub 1996/11/12.
234. Wang BW, Liao WN, Chang CT, Wang SJ. Facilitation of glutamate release by nicotine involves the activation of a Ca2+/calmodulin signaling pathway in rat prefrontal cortex nerve terminals. Synapse. 2006;59(8):491–501. Epub 2006/03/28.
235. Nordman JC, Kabbani N. An interaction between alpha7 nicotinic receptors and a G-protein pathway complex regulates neurite growth in neural cells. J Cell Sci. 2012;125(Pt 22): 5502–13. Epub 2012/09/08.

Chapter 8
Autonomic Nervous System Transmission

Peter B. Sargent

Abstract Rapid synaptic transmission within mammalian autonomic ganglia is mediated by nicotinic receptors (nAChRs) containing both the α3 and β4 subunits. Some nAChRs additionally have β2 or α5 subunits, and in embryonic chicken ciliary ganglia some nAChRs have all four subunits. While autonomic neurons express multiple α3-containing nAChR types, including ones that are expressed presynaptically on terminals, it is not known directly which forms are targeted to which domains of the cell. α7 nAChRs are expressed in both mammalian and embryonic chicken ganglia, but only in the chicken are they involved in synaptic transmission. The autonomic nervous system remains a good model system in which to study nAChR diversity as well as the properties of nAChRs derived from the CHRNA5/A3/B4 cluster, which is linked to nicotine dependence.

Keywords α3 • β2 • β4 • α5 • α7 • Autonomic • Sympathetic • Parasympathetic • Enteric • Superior cervical ganglion • Ciliary ganglion • EPSC (excitatory postsynaptic current) • Channel open time • Channel burst time

1 Introduction

This chapter will focus on the role that nicotinic receptors (nAChRs) play in the transmission of information from the CNS to autonomic end organs (cardiac muscle, smooth muscle, or glands). Unlike the somatic motor system, the autonomic motor system has at least one neuron, an autonomic motor neuron (aka a postganglionic neuron or a ganglion cell), interposed between cholinergic preganglionic neurons in the CNS and the target cell. Ionotropic synaptic transmission onto these neurons is mediated by nAChRs, and metabotropic transmission, when present, is mediated by muscarinic receptors and neuropeptide receptors. Transmission onto autonomic target cells is mediated by a host of non-nicotinic receptors, including muscarinic, adrenergic, purinergic, and neuropeptide receptors.

P.B. Sargent, Ph.D. (✉)
Department of Cell and Tissue Biology, University of California at San Francisco,
521 Parnassus Avenue, San Francisco, CA 94143, USA
e-mail: peter.sargent@ucsf.edu

R.A.J. Lester (ed.), *Nicotinic Receptors*, The Receptors 26,
DOI 10.1007/978-1-4939-1167-7_8

NAChRs expressed by autonomic ganglion cells were among the first neuronal nAChRs to be studied, and they remain among the best characterized neuronal nAChRs; however, interest in their role in synaptic transmission has waned with the advent of methods that facilitate conducting functional studies in the CNS (patch clamp, brain slice) and with a growing interest in the role that nAChRs play in brain circuits, especially those that are "hijacked" by nicotine. Nonetheless, there is still much to learn from examining the role that nAChRs and autonomic transmission play in health and disease (e.g., [1–3]). Moreover, ganglionic nAChRs can still serve as a useful model for central nAChRs, especially inasmuch as variants in the CHRNA5/A3/B4 gene cluster are associated with increased risk for nicotine dependence and lung cancer (reviewed in [4]; see also Chaps. 3, 16 and 17) and encode a heteromeric nAChR commonly expressed by autonomic ganglion cells.

Despite the fact that the autonomic nervous system was first described more than a century ago (reviewed in [5]), transmission within the autonomic nervous system remains poorly characterized. What benefits derive from having interposed ganglion cells, other than the obvious transmitter switch it affords for sympathetic pathways? McLachlan [6] addressed this question by asking whether ganglion cells serve to integrate input from preganglionic neurons, to modulate it, or to distribute (disperse) it. At least for the sympathetic and parasympathetic divisions of the autonomic nervous system, ganglion cells are usually not spontaneously active; transmission must therefore rely on their ability to integrate synaptic inputs from preganglionic axons [7–10]. Paravertebral ganglion cells receive convergent innervation from preganglionic neurons; in mice and in rabbits an average of about 5 and 15 preganglionic axons innervate each ganglion cell, respectively [11]. However, typically only one or two of these inputs is "strong" and produces a suprathreshold EPSP when stimulated; the remaining inputs elicit smaller, subthreshold EPSPs [8]. Long-term recordings from neurons in anesthetized rats suggest that ganglion cells only fire as a result of activity of the strong preganglionic inputs [7, 8, 10], which raises questions about the functional significance of the weaker inputs (see [12]). While weaker inputs can drive ganglion cells to threshold [13], their ability to do so in vivo is likely limited by rates of preganglionic firing [8, 9, 14]. Thus, to date, we have little evidence that ganglion cells "integrate" inputs from preganglionic neurons in the spatial or temporal sense; rather, they "follow" activation of strong inputs.

Long-term potentiation of fast nicotinic EPSPs has been described in the rat SCG (e.g., [15]), but the role that this plays in ganglionic transmission in unknown. Another form of modulation that exists in sympathetic ganglia is that produced by activation of muscarinic receptors and of receptors to neuropeptides. Activation of m1 muscarinic receptors leads to inhibition of potassium M-currents, which would be expected to increase sympathetic neuron excitability. Targeted deletion of the m1 muscarinic receptor yields viable mice [16], but the consequences of loss of the m1 receptor on ganglionic transmission has not been yet explored. A variety of neuropeptides are expressed by preganglionic neurons and are released in a frequency-dependent manner onto their targets; these peptides can alter neuronal excitability [17], mimicking the consequences of m1 receptor activation.

Exogenously supplied neuropeptides can also alter the quantal parameters of nicotinic synaptic transmission [18] and even the behavior of the nAChR channels [19], but the role of nerve-released peptides has not been systematically explored. Generally, we do not yet have a clear understanding of how modulation serves to regulate transmission at ganglionic synapses.

One of the more striking features about autonomic motor pathways is that ganglion cells greatly outnumber preganglionic neurons [20]. Purves et al. [11] counted the number of sympathetic preganglionic neurons and ganglion cells in a number of small mammals (body mass ranging from ~20 to ~2,000 g) and found that the number of preganglionic neurons did not keep pace with body size or autonomic end organ size. They found that ganglion cell *number* was better correlated with body/target size, and that the divergence ratio (# of ganglion cells divided by # of preganglionic neurons) steadily increased with size, reaching 25:1 in rabbits (and 200:1 in humans [20]). However, even ganglion cell number does not keep pace with increasing body size; what does "track" body size is total ganglion cell volume [21]. When coupled with physiological data on convergence [11], these measurements reveal that each preganglionic neuron innervates an increasing number of target cells with increasing body size; the number may be as large as 4,000 in humans [6]. This suggests that one role of ganglia in at least the sympathetic division is to "distribute" and concomitantly amplify the motor pathway to enable a small population of preganglionic neurons to drive the target "load" effectively [6, 11].

2 Mammalian Superior Cervical Ganglion (SCG)

The discovery that sympathetic ganglion cells possess nicotinic receptors was made by Langley [22] and rooted in the very beginnings of the work that he, Elliott, Bernard, and others conducted on the basis of the responses of nervous tissue to "poisons" such as nicotine. More than a century later, we know that neonatal rat SCG contains mRNA corresponding to α3, α5, α7, β2, and β4 nAChR subunits [23, 24]. Newborn mice, but not rat, SCGs also have α4 mRNA [25], but α4 mRNA levels decline over the first week after birth and have not been reported consistently in older rodent species (but see [26]). (In cultured embryonic *chicken* sympathetic neurons, Listerud et al. [27] found expression of α4 in addition to α3, α5, α7, β2, and β4 genes.) The presence of five nAChR subunit mRNAs in the SCG (the "odd" αs and the "even" βs) potentially allows for the expression of several nAChR pentamers; including α3β4, which should show greater sensitivity to cytisine than acetylcholine (ACh), and α3β2, which should show the opposite pattern [28]. Covernton et al. [29] found that the relative agonist potency for responses elicited from acutely dissociated SCG neurons in young adult rats was similar, but not identical, to that of α3β4 nAChRs expressed in *Xenopus* oocytes; agonist potency was also distinct from that of heterologously expressed α3β2 nAChRs. These results might be explained if native nAChRs on SCG neurons express nAChRs containing more than two distinct subunits (e.g., both β2 and β4 in addition to α3 [23]), or if heterologous expression systems do not assemble native nAChR oligomers correctly and/or do not appropriately posttranslationally modify nAChRs subunits [30, 31]. Krashia et al. [32] found that α3 and β4 subunits can assemble in different ratios (as is known for α4 and β2 [33]); the presumptive $\alpha3_3\beta4_2$ form had a conductance of ~33 pS and short channel bursts (~15 ms) while the $\alpha3_2\beta4_3$ form had a lower conductance (26 pS) with longer channel bursts (100 ms or more). It has yet to be determined if native α3β4 nAChRs exist in multiple states, but this result offers an explanation for why native nAChRs in SCG neurons might indeed be α3β4 and yet not resemble α3β4 channels expressed heterologously [30].

Ciuraskiewicz et al. [34] studied single channel properties in SCG neurons from WT animals as well as from single (α5, β2), double (α5β2), or triple (α5β2α7) KO mice. In WT, α5β2 KO, and α5β2α7 KO animals, the principal conductance state of nAChR channels was ~33 pS; this is likely to arise from α3β4 nAChRs and is similar to the conductance states for SCG cell nAChRs reported by others [29, 30, 35] as well as for heterologously expressed α3β4 nAChRs [36, 37]. α3β4α5 nAChR channels had a similar principal conductance state but longer channel open times and burst lengths, while α3β4β2 channels had a lower principal conductance state (15 pS) and longer open times and burst lengths. The association of α3, β4, and possibly α5 subunits is of interest given the fact that these three genes are tightly clustered [38] and can be regulated by common factors, such as Sox10 [39]. Moreover, variants in the CHRNA5/A3/β4 gene complex are associated with increased risk for nicotine dependence (reviewed in [4]).

Both Mao et al. [40] and David et al. [41] used immunoprecipitation and blot methods to systematically catalogue high-affinity ^{3}H-epibatidine binding to SCG extracts in rat and mouse, respectively. They both found that slightly more than half of the binding sites are due to an α3β4 species (lacking α5 and β2); the remainder of the binding sites were distributed between α3β4α5 (accounting for a majority of the remaining sites) and α3β4β2. Neither study found evidence for an α3β4α5β2 species, which is found in the embryonic chicken ciliary ganglion (see below).

Mandelzys et al. [23] found no evidence for functional nicotinic currents with the pharmacology expected of α7 homo-oligomeric nAChRs; similar results have been reported by others: α7-mediated currents, if present, are small [41–44]. By contrast, Cuevas et al. [45] readily detected α7-mediated currents elicited from acutely dissociated rat neonatal SCG neurons and found evidence for two types of α7-mediated currents: a "classical" current (α7-1) that rapidly desensitizes and that is irreversibly blocked by α-BuTx and a second current (α7-2) that desensitizes slowly and that is reversibly blocked by α-BuTx. The reason for the difference in findings among laboratories is not known. Adult mice do not express detectable levels of α7-mediated nicotinic currents, even in the presence of the positive allosteric mediator PNU-120596 [32, 41]. Yu and Role [46] reported that embryonic *chick* sympathetic neurons maintained in culture express α-BuTx and MLA sensitive currents, indicative of α7 nAChRs, and the results of antisense nucleotide knockdown experiments suggested that α7 subunits form heteromeric nAChR pentamers with other α and/or β subunits. There is no evidence in any sympathetic ganglia that alpha7 nAChRs are involved in synaptic transmission; the role of cell surface receptors binding α-BuTx in these ganglia [47–49] is not known.

One approach to exploring which of the three nAChRs detected in rat SCG cells (α3β4, α3β4α5, α3β4β2) underlie synaptic transmission is to compare the channel kinetics for these nAChRs with the EPSC decay for synaptic currents in the SCG, making the assumption that the decay is dictated largely by channel properties [50] and not by transmitter availability. EPSCs recorded from rat sympathetic neurons decay usually with one time constant in the range of 4–6 ms [51, 52]; Kertser et al. [53] detected a second component of decay at ~20 ms. EPSCs recorded from rat *submandibular* ganglions cells show two phases of decay: ~5 and ~35 ms [54–56]. The decay time constant(s) were found to be weakly voltage dependent, as may be expected if they are explained by channel gating [50]. A decay time constant of ~5 ms for the EPSC matches well the burst durations, but not the single channel open times, of α3β4-containing nAChRs. Ciuraszkiewicz et al. [34] speculate that the EPSC decay in rat SCG could be explained by the α3β4 form, which accounts for 50 % of the nAChRs, given that these have burst times in the appropriate range. The α3β4 nAChRs may therefore be clustered at synaptic sites (Fig. 8.1a). The α3β4β2 species may also be clustered at synaptic sites, since these have similar kinetics; however, the α3β4α5 version is not likely to be clustered at synaptic contacts (Fig. 8.1a), since they display longer openings than can be explained in the decay of EPSCs [34].

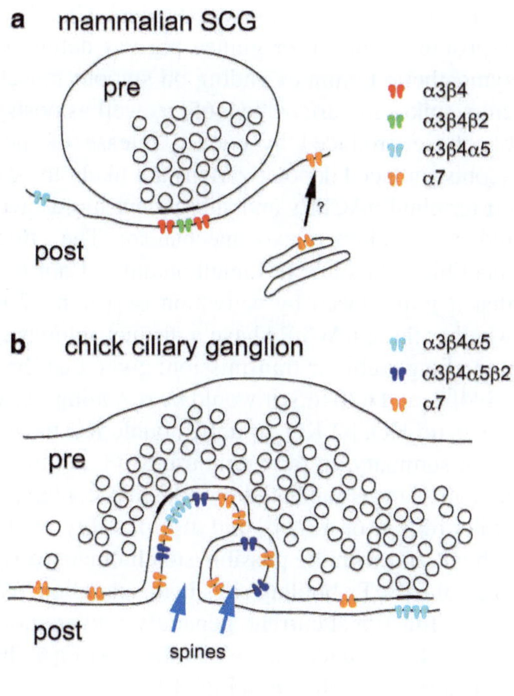

Fig. 8.1 A speculative model of the distribution of distinct nAChR types on autonomic neurons in the neonatal mammalian superior cervical ganglion (SCG), (**a**), and the embryonic chicken ciliary ganglion, (**b**). Legends to the *right* indicate nAChR types. *Arrow* and *question mark* in **a** represent uncertainty about whether α7-nAChRs are transported to the cell surface in SCG neurons. The representation of a "mat" of somatic spines in **b** is highly simplified, for illustration purposes. For additional details refer to text

Rassadi et al. [57] studied ganglionic transmission in neonatal mice in which the α3 gene had been deleted [58]. They found that loss of the α3 gene produced a complete loss of sensitivity of dissociated SCG neurons to ACh (similar results were reported for β2–β4 double knockout mice [59]) and that ganglionic transmission was not detectable in α3 KO mice [57, 60]. These results are supported by Caffrey et al. [44], who engineered a knock-in mouse whose α3 gene was sensitive to α-BuTx and in which ganglionic transmission was completely blocked by α-BuTx. David et al. [41] examined ganglionic transmission in α5β4 double knockout animals and reported no effect on the size of the postganglionic compound action potential. Since the double β2β4 KO has no nAChR currents [59], this finding implies that expression of either β2 and or β4, with α3, is sufficient to support suprathreshold synaptic function in the ganglion. It would be interesting to examine the sensitivity of transmission in these KO animals to α-conotoxin MII and α-conotoxin Au1B, which display selectivity for nAChRs with α3/β2 interfaces or α3/β4 interfaces, respectively [61, 62]. Wang et al. [63] found only subtle phenotypic changes in α5 KO mice, which suggests that α5 isn't required for autonomic function; this again raises questions about how mammalian autonomic neurons utilize α5-containing nAChR oligomers normally. One possibility is that these nAChRs are targeted to non postsynaptic sites (Fig. 8.1a).

Kristufek et al. [43] and Fischer et al. [64] explored differences between somatic nAChRs on sympathetic ganglion cells and "pre-terminal" nAChRs whose activation should elicit TTX-insensitive release of preloaded ^{3}H-NE. Clear differences were detected in agonist potencies between the two nAChR types, but these results have yet to suggest whether nAChRs with distinct subunit composition are targeted to the cell body vs. the axon terminal. Cunnane, Brain, and colleagues found that exposure of mouse or guinea pig vas deferens, which contains a dense network sympathetic terminals ending on smooth muscle, to nicotinic agonists elicits calcium spikes in varicosities [65] as well as postsynaptic depolarizations that resemble those produced by evoked release of the transmitter ATP [66, 67]. These agonist-induced depolarizations are likely to be caused by direct action of agonists on terminal nAChRs and subsequent multi-quantal release of ATP via a calcium-induced calcium release mechanism. The effect was blocked by the nonspecific nicotinic antagonist hexamethonium but not by α-BuTx or MLA, which suggests that it is produced by activation of a non-α7 nAChR. It remains to be explored whether these nAChRs have a distinct subunit composition compared to those that underlie ganglionic transmission; given that the three candidate nAChRs are α3β4, α3β4β2, and α3β4α5, it would be revealing to examine agonist effects in vas deferens in α5 KO, β2 KO, or α5β2 double KO mice.

In summary, in the rat/mouse SCG and in other paravertebral ganglia synaptic transmission is mediated by nAChRs containing both the α3 and the β4 subunit: either $\alpha3_2\beta4_3$ or $\alpha3_3\beta4_2$, and also possibly by a receptor containing as well the β2 subunit and with the possible stoichiometry $\alpha3_2\beta4_2\beta2$ (Fig. 8.1a). Despite the presence of α-BuTx binding sites in the ganglion, α7-nAChRs do not account for much or any functional current, generally, and are not involved in synaptic transmission. A speculative model of the distribution of nAChR types at the neonatal mammalian SCG synapse is shown in Fig. 8.1a.

3 Chick Ciliary Ganglion

The functional role that nAChRs play in ganglionic synaptic transmission is best characterized in the chick ciliary ganglion, a parasympathetic ganglion. At embryonic day (E) 14–15 ciliary neurons, which innervate twitch muscles in the iris, lack dendrites and are innervated by a single preganglionic axon that forms a calyciform ending similar to that made by the glutamatergic calyx of Held [68, 69]. In the adult chicken, the synapse is a mixed chemical/electrical synapse with a more conventional, bouton-containing presynaptic arborization; only the embryonic synapse has been functionally characterized.

Neurons in the embryonic chick ciliary ganglion neurons express the same five nAChR subunit mRNAs that are consistently seen in rat/mice neonatal SCG neurons: α3, α5, α7, β2, and β4 [70]. These subunits associate into two major groups of pentameric nAChRs: those containing α7 subunits and no other known subunits (α7-nAChRs) and heteromeric nAChRs containing the α3, α5, and β4 subunits, and sometimes the β2 subunit as well (α3*-nAChRs; [71–73]). (The complexity of nAChR expression in the chick ciliary ganglion is yet increased by the discovery of a minor species that is recognized by both α-BuTx and by mAb 35 but that contains no known members of the nAChR gene family [74]). The composition of the α3*-nAChRs in chick is similar to that in mammalian SCG in that both have α3 and β4 subunits. In the chick ciliary ganglion all α3*-nAChRs have α5 as well, while in the mammalian SCG only some of them do. In both the chick and mammalian preparations, some of the α3β4-containing nAChRs have β2 and some do not. The major distinction between the two systems is the prominent functional role of α7-nAChRs in the chick; however, this difference may relate to the time of analysis (1 week before hatching in chick vs. 1–3 weeks after birth in rat/mice).

The chick ciliary ganglion was the first site where activation of native α7 nAChRs were shown to produce increases in intracellular calcium [75], cation currents in response to fast application of acetylcholine [76], and fast synaptic currents in response to nerve-released transmitter [77, 78]. Blocking α7 nAChRs reduces peak synaptic current by more than 50 % [77, 78] and reduces the ability of the ganglion cell to follow one-to-one at elevated frequencies of preganglionic stimulation [79]. The consequences of blocking α7-nAChRs on ganglionic transmission is less pronounced at E18/19 than at E13/14 [79], which raises the possibility that α7-nAChRs play an important role in transmission only during early development of the chick ciliary ganglion. The role that these nAChRs play generally in development is covered elsewhere in this volume by Berg.

At a stage when α7-nAChRs underlie most of the synaptic current in the ciliary ganglion (E14/15), a combination of light and electron microscopic evidence reveals that α7-nAChRs are not concentrated at synaptic sites: rather, they are found on collections of somatic spines known as spine mats (Fig. 8.1b) [80–83]. The high calcium permeability of α7-nAChRs allowed Shoop et al. [84] to demonstrate that preganglionic stimulation elevates calcium levels in spines. Only ~10 % of synaptic contacts in the ganglion are located on spine mats [85], and the disparity between the sites of most transmitter release and the importance of α7-nAChRs in generating synaptic current at these embryonic synapses led Coggan et al. [86] to postulate, based on MCell modeling, that ACh is released ectopically onto spines from sites that have no evident presynaptic specializations. This proposal has generated considerable interest; there are a few well-documented examples of ectopic transmission between nerve terminals and glial cells (e.g., cerebellar climbing fiber to Bergmann glial cell transmission: [87, 88]), but little direct evidence for ectopic transmission between neurons. In studying vesicle recycling at calyciform release sites in the ciliary ganglion, Nguyen and Sargent [85] compared synaptic contacts on the somatic surface with those on spine mats and found no differences between the number of vesicles, the distance between vesicles and the presynaptic membrane at active zones, in the number of synaptic vesicles that could be filled by stimulation in the presence of tracer (HRP), or in the number that could subsequently be released upon re-stimulation. Conventional release sites thus do function at spine mats [85]; and it is difficult to rule out the possibility that release from these sites is sufficient to explain the presence of α7-nAChR currents in the EPSC.

In the ciliary ganglion α3-containing nAChRs, recognized by mAb 35 or by neuronal-bungarotoxin (aka bungarotoxin 3.1, toxin F, κ-bungarotoxin) are located in clusters at synaptic contacts but they are also found distributed diffusely at spine mats: the sites where α7-nAChRs are located [82, 83, 89–91]. Thus, ACh release onto these sites might be expected to activate both α7-nAChRs and α3*-nAChRs, while release on the somatic surface of the ganglion, where ~90 % of release occurs, might only activate α3*-nAChRs. To examine the kinetic characteristics of quantal responses, Sargent [92] looked at mEPSCs produced by delayed release in strontium and compared the kinetics of mEPSC populations under native conditions with that observed when either α7-nAChRs or α3*-nAChRs were blocked with α-BuTx or α-conotoxin-MII, respectively. Native populations of mEPSCs contained very few mEPSCs with the slow kinetics expected of α3*-nAChRs; rather they contained populations of fast α7-nAChR-like mEPSCs and populations of mixed mEPSCs: ones with intermediate rise and decay times. These events could be explained by release onto spines, where both receptors are present; what is unresolved is why populations of slowly decaying mEPSCs, expected from release at somatic sites, were not detected. The immunochemical work on α3*-nAChR subunit composition by Berg and colleagues suggests that there are two populations of α3*-nAChRs: those containing and those lacking the β2 subunit [72, 73]. For the α3β4α5 receptors, the stoichiometry is presumably $\alpha3_2\beta4_2\alpha5$, and there should thus be two α3/β4 agonist binding pockets, either of which should be recognized by α-conotoxin-Au1B [62]. For the α3β4α5β2 receptors, the stoichiometry is presumably $\alpha3_2\beta4\alpha5\beta2$, and there should be two different agonist binding pockets, α3/β2 and α3/β4; the first of these would bind α-conotoxin-MII [61] and the second α-conotoxin-Au1B. Thus, the simpler of the two forms of the α3*-nAChR should be blocked by α-conotoxin-Au1B and the more complex form by both it and α-conotoxin-MII, assuming that two binding events are needed to open channels readily [93]. Nai et al. [94] found that both conotoxins blocked virtually all current elicited from acutely dissociated ganglion cells by nicotinic agonists: moreover, more than 90 % of the α-BuTx-resistant synaptic current is blocked by α-conotoxin-MII [92]. This suggests that the $\alpha3_2\beta4\alpha5\beta2$ form of the α3*-nAChR in the ciliary ganglion, which may be outnumbered by the $\alpha3_2\beta4_2\alpha5$ form [72], nonetheless contributes more of the α3*-nAChR-mediated currents, principally because they have a larger P_{open} [94]. Interestingly, this same receptor, with both an α3/β4 and an α3/β2 interface, has been proposed by Quick et al. [95] to underlie a major functional nAChR class in medial habenula neurons. In the ciliary ganglion, the absence of detectable populations of slow mEPSCs [92] may result if the functionally "quiet" α3*-nAChR, lacking β2, is targeted to the PSD at synaptic contacts while the α3*-nAChR containing β2 accounts for the diffusely distributed α3*-nAChR on spines (Fig. 8.1b). While the benefit of such an arrangement to the organism is not immediately evident, it is worth recalling that these synapses are embryonic and undergo subsequent remodeling.

In the chick ciliary ganglion, whole cell recordings from the large calyciform terminal demonstrate the presence of presynaptic α7-nAChRs [96]. These nAChRs, while blocked functionally by α-BuTx, desensitize slowly [96] and may represent a novel α7-containing nAChRs, possibly heteromeric [46]. Rogers and Sargent [97] used optical methods and calcium indicator dyes to demonstrate that these presynaptic receptors could be activated by ACh released from the calyx following orthograde electrical stimulation. The consequences of this "back activation" of presynaptic nAChRs is to elevate calcium levels in the terminal, but the effects of this on subsequent ganglionic transmission are not known.

In summary, in the embryonic chicken ciliary ganglion, both α3*-nAChRs and α7-nAChRs are involved in synaptic transmission. The principal functional α3*-nAChR likely has the stoichiometry $\alpha3_2\beta4\alpha5\beta2$; curiously, the functionally "quiet" $\alpha3_2\beta4_2\alpha5$ is likely targeted to synaptic sites (Fig. 8.1b). α7-nAChRs, which are located on spines and which participate in synaptic signaling (Fig. 8.1b), may be important only at embryonic stages of transmission. A speculative model of the distribution of nAChR types at the chick calyciform synapse is shown in Fig. 8.1b.

4 The Enteric Nervous System

The enteric nervous system includes numerous interconnected ganglia in both the myenteric plexus (between the longitudinal and circular muscle layers) and the submucosal plexus (internal to the circular muscle layer). Each set of ganglia contains motor neurons, sensory neurons, and interneurons. The cells of the myenteric plexus principally control contractility of the muscle layers, while those in the submucosal layer regulate secretions and water movement across the mucosa. nAChRs are expressed in nearly all neurons in both plexuses, but, unlike the situation elsewhere in the autonomic nervous system, they are not the sole ionotropic receptors in play (reviewed by [98, 99]). Circuits within the myenteric plexus underlie reflexes that are either ascending (orally projecting) or descending (anally projecting). Many of the fast EPSPs at orally projecting synapses, be they sensory-interneuron, interneuron-interneuron, or interneuron-motor, are mediated by nAChRs (data primarily from guinea-pig and reviewed in [100]). Many of these connections also result in activation of metabotropic receptors. For anally-projecting reflexes, the situation is more complex, and generally 5HT and/or P2X receptors rather than nAChRs mediate fast EPSPs. In the submucosal plexus, secretomotor neurons receive fast EPSPs mediated by both nAChRs and by P2X receptors (reviewed in [100]). There have been no systematic pharmacological studies of the nAChR subunits involved in nicotinic transmission within the enteric nervous system, but virtually all classes of neurons express immunoreactivity for mAb 35 [101, 102], which is thought to recognize α3 and α5.

Acknowledgements I thank the NIH for their generous support of the work from my laboratory that is cited here (NS 24207, MH 54251, MH 68690).

References

1. Winston N, Vernino S. Recent advances in autoimmune autonomic ganglionopathy. Curr Opin Neurol. 2010;23(5):514–8.
2. Pintérová M, Kuneš J, Zicha J. Altered neural and vascular mechanisms in hypertension. Physiol Res. 2011;60(3):381–402.
3. Krishnaswamy A, Cooper E. Reactive oxygen species inactivate neuronal nicotinic acetylcholine receptors through a highly conserved cysteine near the intracellular mouth of the channel: implications for diseases that involve oxidative stress. J Physiol (Lond). 2012;590 (Pt 1):39–47.
4. Improgo MRD, Scofield MD, Tapper AR, Gardner PD. The nicotinic acetylcholine receptor CHRNA5/A3/B4 gene cluster: dual role in nicotine addiction and lung cancer. Prog Neurobiol. 2010;92(2):212–26.
5. Langley JN. The autonomic nervous system. Cambridge: Heffer; 1921. Available at: http://archive.org/details/autonomicnervous01languoft. Accessed 1 Aug 2013.
6. McLachlan EM. Transmission of signals through sympathetic ganglia—modulation, integration or simply distribution? Acta Physiol Scand. 2003;177(3):227–35.
7. Ivanov A, Purves D. Ongoing electrical activity of superior cervical ganglion cells in mammals of different size. J Comp Neurol. 1989;284(3):398–404.
8. McLachlan EM, Davies PJ, Häbler HJ, Jamieson J. On-going and reflex synaptic events in rat superior cervical ganglion cells. J Physiol. 1997;501(Pt 1):165–81.
9. McLachlan EM, Habler H-J, Jamieson J, Davies PJ. Analysis of the periodicity of synaptic events in neurones in the superior cervical ganglion of anaesthetized rats. J Physiol. 1998; 511(Pt 2):461–78.
10. Bratton B, Davies P, Jänig W, McAllen R. Ganglionic transmission in a vasomotor pathway studied in vivo. J Physiol (Lond). 2010;588(Pt 9):1647–59.
11. Purves D, Rubin E, Snider WD, Lichtman J. Relation of animal size to convergence, divergence, and neuronal number in peripheral sympathetic pathways. J Neurosci. 1986;6(1):158–63.
12. Karila P, Horn JP. Secondary nicotinic synapses on sympathetic B neurons and their putative role in ganglionic amplification of activity. J Neurosci. 2000;20(3):908–18.
13. Rimmer K, Horn JP. Weak and straddling secondary nicotinic synapses can drive firing in rat sympathetic neurons and thereby contribute to ganglionic amplification. Front Neurol. 2010;1:130.
14. Birks RI, Isacoff EY. Burst-patterned stimulation promotes nicotinic transmission in isolated perfused rat sympathetic ganglia. J Physiol. 1988;402:515–32.

15. Briggs CA, Brown TH, McAfee DA. Neurophysiology and pharmacology of long-term potentiation in the rat sympathetic ganglion. J Physiol. 1985;359:503–21.
16. Hamilton SE, Loose MD, Qi M, et al. Disruption of the m1 receptor gene ablates muscarinic receptor-dependent M current regulation and seizure activity in mice. Proc Natl Acad Sci U S A. 1997;94(24):13311–6.
17. Tompkins JD, Ardell JL, Hoover DB, Parsons RL. Neurally released pituitary adenylate cyclase-activating polypeptide enhances guinea pig intrinsic cardiac neurone excitability. J Physiol (Lond). 2007;582(Pt 1):87–93.
18. Pugh PC, Jayakar SS, Margiotta JF. PACAP/PAC1R signaling modulates acetylcholine release at neuronal nicotinic synapses. Mol Cell Neurosci. 2010;43(2):244–57.
19. Valenta DC, Downing JE, Role LW. Peptide modulation of ACh receptor desensitization controls neurotransmitter release from chicken sympathetic neurons. J Neurophysiol. 1993; 69(3):928–42.
20. Ebbesson SO. Quantitative studies of superior cervical sympathetic ganglia in a variety of primates including man. I. The ratio of preganglionic fibers to ganglionic neurons. J Morphol. 1968;124(1):117–32.
21. Ebbesson SO. Quantitative studies of superior cervical sympathetic ganglia in variety of primates including man. II. Neuronal packing density. J Morphol. 1968;124(2):181–6.
22. Langley JN. On the reaction of cells and of nerve-endings to certain poisons, chiefly as regards the reaction of striated muscle to nicotine and to curari. J Physiol (Lond). 1905; 33(4–5):374–413.
23. Mandelzys A, De Koninck P, Cooper E. Agonist and toxin sensitivities of ACh-evoked currents on neurons expressing multiple nicotinic ACh receptor subunits. J Neurophysiol. 1995; 74(3):1212–21.
24. Zoli M, Le Novère N, Hill Jr JA, Changeux JP. Developmental regulation of nicotinic ACh receptor subunit mRNAs in the rat central and peripheral nervous systems. J Neurosci. 1995;15(3 Pt 1):1912–39.
25. Scholze P, Ciuraszkiewicz A, Groessl F, et al. alpha4beta2 nicotinic acetylcholine receptors in the early postnatal mouse superior cervical ganglion. Dev Neurobiol. 2011;71(5):390–9.
26. Rust G, Burgunder JM, Lauterburg TE, Cachelin AB. Expression of neuronal nicotinic acetylcholine receptor subunit genes in the rat autonomic nervous system. Eur J Neurosci. 1994;6(3):478–85.
27. Listerud M, Brussaard AB, Devay P, Colman DR, Role LW. Functional contribution of neuronal AChR subunits revealed by antisense oligonucleotides. Science. 1991;254(5037): 1518–21.
28. Luetje CW, Patrick J. Both alpha- and beta-subunits contribute to the agonist sensitivity of neuronal nicotinic acetylcholine receptors. J Neurosci. 1991;11(3):837–45.
29. Covernton PJ, Kojima H, Sivilotti LG, Gibb AJ, Colquhoun D. Comparison of neuronal nicotinic receptors in rat sympathetic neurones with subunit pairs expressed in Xenopus oocytes. J Physiol (Lond). 1994;481(Pt 1):27–34.
30. Sivilotti LG, McNeil DK, Lewis TM, et al. Recombinant nicotinic receptors, expressed in Xenopus oocytes, do not resemble native rat sympathetic ganglion receptors in single-channel behaviour. J Physiol (Lond). 1997;500(Pt 1):123–38.
31. Lewis TM, Harkness PC, Sivilotti LG, Colquhoun D, Millar NS. The ion channel properties of a rat recombinant neuronal nicotinic receptor are dependent on the host cell type. J Physiol (Lond). 1997;505(Pt 2):299–306.
32. Krashia P, Moroni M, Broadbent S, et al. Human α3β4 neuronal nicotinic receptors show different stoichiometry if they are expressed in Xenopus oocytes or mammalian HEK293 cells. PLoS One. 2010;5(10):e13611.
33. Nelson ME, Kuryatov A, Choi CH, Zhou Y, Lindstrom J. Alternate stoichiometries of alpha-4beta2 nicotinic acetylcholine receptors. Mol Pharmacol. 2003;63(2):332–41.
34. Ciuraszkiewicz A, Schreibmayer W, Platzer D, et al. Single-channel properties of α3β4, α3β45 and α3β4β2 nicotinic acetylcholine receptors in mice lacking specific nicotinic acetylcholine receptor subunits. J Physiol (Lond). 2013;591(Pt 13):3271–88.

35. Mathie A, Cull-Candy SG, Colquhoun D. Conductance and kinetic properties of single nicotinic acetylcholine receptor channels in rat sympathetic neurones. J Physiol (Lond). 1991;439:717–50.
36. Nelson ME, Lindstrom J. Single channel properties of human alpha3 AChRs: impact of beta2, beta4 and alpha5 subunits. J Physiol (Lond). 1999;516(Pt 3):657–78.
37. Nelson ME, Wang F, Kuryatov A, et al. Functional properties of human nicotinic AChRs expressed by IMR-32 neuroblastoma cells resemble those of alpha3beta4 AChRs expressed in permanently transfected HEK cells. J Gen Physiol. 2001;118(5):563–82.
38. Boulter J, O'Shea-Greenfield A, Duvoisin RM, et al. Alpha 3, alpha 5, and beta 4: three members of the rat neuronal nicotinic acetylcholine receptor-related gene family form a gene cluster. J Biol Chem. 1990;265(8):4472–82.
39. Liu Q, Melnikova IN, Hu M, Gardner PD. Cell type-specific activation of neuronal nicotinic acetylcholine receptor subunit genes by Sox10. J Neurosci. 1999;19(22):9747–55.
40. Mao D, Yasuda RP, Fan H, Wolfe BB, Kellar KJ. Heterogeneity of nicotinic cholinergic receptors in rat superior cervical and nodose ganglia. Mol Pharmacol. 2006;70(5):1693–9.
41. David R, Ciuraszkiewicz A, Simeone X, et al. Biochemical and functional properties of distinct nicotinic acetylcholine receptors in the superior cervical ganglion of mice with targeted deletions of nAChR subunit genes. Eur J Neurosci. 2010;31(6):978–93.
42. Trouslard J, Marsh SJ, Brown DA. Calcium entry through nicotinic receptor channels and calcium channels in cultured rat superior cervical ganglion cells. J Physiol (Lond). 1993; 468:53–71.
43. Kristufek D, Stocker E, Boehm S, Huck S. Somatic and prejunctional nicotinic receptors in cultured rat sympathetic neurones show different agonist profiles. J Physiol (Lond). 1999; 516(Pt 3):739–56.
44. Caffery PM, Krishnaswamy A, Sanders T, et al. Engineering neuronal nicotinic acetylcholine receptors with functional sensitivity to alpha-bungarotoxin: a novel alpha3-knock-in mouse. Eur J Neurosci. 2009;30(11):2064–76.
45. Cuevas J, Roth AL, Berg DK. Two distinct classes of functional α7-containing nicotinic receptor on rat superior cervical ganglion neurons. J Physiol (Lond). 2000;525(Pt 3): 735–46.
46. Yu CR, Role LW. Functional contribution of the alpha7 subunit to multiple subtypes of nicotinic receptors in embryonic chick sympathetic neurones. J Physiol (Lond). 1998;509 (Pt 3):651–65.
47. Smolen AJ. Specific binding of alpha-bungarotoxin to synaptic membranes in rat sympathetic ganglion: computer best-fit analysis of electron microscope radioautographs. Brain Res. 1983;289(1–2):177–88.
48. De Koninck P, Cooper E. Differential regulation of neuronal nicotinic ACh receptor subunit genes in cultured neonatal rat sympathetic neurons: specific induction of alpha 7 by membrane depolarization through a Ca^{2+}/calmodulin-dependent kinase pathway. J Neurosci. 1995;15(12):7966–78.
49. Del Signore A, Gotti C, Rizzo A, Moretti M, Paggi P. Nicotinic acetylcholine receptor subtypes in the rat sympathetic ganglion: pharmacological characterization, subcellular distribution and effect of pre- and postganglionic nerve crush. J Neuropathol Exp Neurol. 2004; 63(2):138–50.
50. Magleby KL, Stevens CF. A quantitative description of end-plate currents. J Physiol (Lond). 1972;223(1):173–97.
51. Hirst GD, McLachlan EM. Post-natal development of ganglia in the lower lumbar sympathetic chain of the rat. J Physiol (Lond). 1984;349:119–34.
52. Sacchi O, Rossi ML, Canella R, Fesce R. Synaptic and somatic effects of axotomy in the intact, innervated rat sympathetic neuron. J Neurophysiol. 2006;95(5):2832–44.
53. Kertser S, Bobryshev A, Voitenko S, et al. Dimensions of neuronal nicotinic acetylcholine receptor channel as estimated from the analysis of the channel-blocking effects. J Membr Biol. 1998;163(2):111–8.

54. Rang HP. The characteristics of synaptic currents and responses to acetylcholine of rat submandibular ganglion cells. J Physiol (Lond). 1981;311:23–55.
55. Yawo H. Rectification of synaptic and acetylcholine currents in the mouse submandibular ganglion cells. J Physiol (Lond). 1989;417:307–22.
56. Callister RJ, Walmsley B. Amplitude and time course of evoked and spontaneous synaptic currents in rat submandibular ganglion cells. J Physiol (Lond). 1996;490(Pt 1):149–57.
57. Rassadi S, Krishnaswamy A, Pié B, et al. A null mutation for the alpha3 nicotinic acetylcholine (ACh) receptor gene abolishes fast synaptic activity in sympathetic ganglia and reveals that ACh output from developing preganglionic terminals is regulated in an activity-dependent retrograde manner. J Neurosci. 2005;25(37):8555–66.
58. Xu W, Gelber S, Orr-Urtreger A, et al. Megacystis, mydriasis, and ion channel defect in mice lacking the alpha3 neuronal nicotinic acetylcholine receptor. Proc Natl Acad Sci U S A. 1999;96(10):5746–51.
59. Xu W, Orr-Urtreger A, Nigro F, et al. Multiorgan autonomic dysfunction in mice lacking the beta2 and the beta4 subunits of neuronal nicotinic acetylcholine receptors. J Neurosci. 1999;19(21):9298–305.
60. Krishnaswamy A, Cooper E. An activity-dependent retrograde signal induces the expression of the high-affinity choline transporter in cholinergic neurons. Neuron. 2009;61(2):272–86.
61. Cartier GE, Yoshikami D, Gray WR, et al. A new alpha-conotoxin which targets alpha3beta2 nicotinic acetylcholine receptors. J Biol Chem. 1996;271(13):7522–8.
62. Luo S, Kulak JM, Cartier GE, et al. alpha-conotoxin AuIB selectively blocks alpha3 beta4 nicotinic acetylcholine receptors and nicotine-evoked norepinephrine release. J Neurosci. 1998;18(21):8571–9.
63. Wang N, Orr-Urtreger A, Chapman J, et al. Autonomic function in mice lacking alpha5 neuronal nicotinic acetylcholine receptor subunit. J Physiol (Lond). 2002;542(Pt 2):347–54.
64. Fischer H, Orr-Urtreger A, Role LW, Huck S. Selective deletion of the alpha5 subunit differentially affects somatic-dendritic versus axonally targeted nicotinic ACh receptors in mouse. J Physiol (Lond). 2005;563(Pt 1):119–37.
65. Brain KL, Trout SJ, Jackson VM, Dass N, Cunnane TC. Nicotine induces calcium spikes in single nerve terminal varicosities: a role for intracellular calcium stores. Neuroscience. 2001;106(2):395–403.
66. Williams DJ, Brain KL, Cunnane TC. The effect of epibatidine on spontaneous and evoked neurotransmitter release in the mouse and guinea pig isolated vas deferens. Br J Pharmacol. 2007;150(7):906–12.
67. Williams DJ, Sidaway P, Cunnane TC, Brain KL. Mechanisms involved in nicotinic acetylcholine receptor-induced neurotransmitter release from sympathetic nerve terminals in the mouse vas deferens. PLoS One. 2011;6(12):e29209.
68. Paysan J, Conroy WG, Coggan JS, Berg DK. The neurofilament infrastructure of a developing presynaptic calyx. J Comp Neurol. 2000;425(2):284–94.
69. Sätzler K, Söhl LF, Bollmann JH, et al. Three-dimensional reconstruction of a calyx of Held and its postsynaptic principal neuron in the medial nucleus of the trapezoid body. J Neurosci. 2002;22(24):10567–79.
70. Corriveau RA, Berg DK. Coexpression of multiple acetylcholine receptor genes in neurons: quantification of transcripts during development. J Neurosci. 1993;13(6):2662–71.
71. Conroy WG, Vernallis AB, Berg DK. The alpha 5 gene product assembles with multiple acetylcholine receptor subunits to form distinctive receptor subtypes in brain. Neuron. 1992;9(4):679–91.
72. Vernallis AB, Conroy WG, Berg DK. Neurons assemble acetylcholine receptors with as many as three kinds of subunits while maintaining subunit segregation among receptor subtypes. Neuron. 1993;10(3):451–64.
73. Conroy WG, Berg DK. Neurons can maintain multiple classes of nicotinic acetylcholine receptors distinguished by different subunit compositions. J Biol Chem. 1995;270(9):4424–31.

74. Pugh PC, Corriveau RA, Conroy WG, Berg DK. Novel subpopulation of neuronal acetylcholine receptors among those binding alpha-bungarotoxin. Mol Pharmacol. 1995; 47(4):717–25.
75. Vijayaraghavan S, Pugh PC, Zhang ZW, Rathouz MM, Berg DK. Nicotinic receptors that bind alpha-bungarotoxin on neurons raise intracellular free Ca^{2+}. Neuron. 1992;8(2): 353–62.
76. Zhang ZW, Vijayaraghavan S, Berg DK. Neuronal acetylcholine receptors that bind alpha-bungarotoxin with high affinity function as ligand-gated ion channels. Neuron. 1994;12(1):167–77.
77. Zhang ZW, Coggan JS, Berg DK. Synaptic currents generated by neuronal acetylcholine receptors sensitive to alpha-bungarotoxin. Neuron. 1996;17(6):1231–40.
78. Ullian EM, McIntosh JM, Sargent PB. Rapid synaptic transmission in the avian ciliary ganglion is mediated by two distinct classes of nicotinic receptors. J Neurosci. 1997;17(19): 7210–9.
79. Chang KT, Berg DK. Nicotinic acetylcholine receptors containing alpha7 subunits are required for reliable synaptic transmission in situ. J Neurosci. 1999;19(10):3701–10.
80. Jacob MH, Berg DK. The ultrastructural localization of alpha-bungarotoxin binding sites in relation to synapses on chick ciliary ganglion neurons. J Neurosci. 1983;3(2):260–71.
81. Loring RH, Dahm LM, Zigmond RE. Localization of alpha-bungarotoxin binding sites in the ciliary ganglion of the embryonic chick: an autoradiographic study at the light and electron microscopic level. Neuroscience. 1985;14(2):645–60.
82. Horch HL, Sargent PB. Perisynaptic surface distribution of multiple classes of nicotinic acetylcholine receptors on neurons in the chicken ciliary ganglion. J Neurosci. 1995;15(12): 7778–95.
83. Shoop RD, Martone ME, Yamada N, Ellisman MH, Berg DK. Neuronal acetylcholine receptors with alpha7 subunits are concentrated on somatic spines for synaptic signaling in embryonic chick ciliary ganglia. J Neurosci. 1999;19(2):692–704.
84. Shoop RD, Chang KT, Ellisman MH, Berg DK. Synaptically driven calcium transients via nicotinic receptors on somatic spines. J Neurosci. 2001;21(3):771–81.
85. Nguyen D, Sargent PB. Synaptic vesicle recycling at two classes of release sites in giant nerve terminals of the embryonic chicken ciliary ganglion. J Comp Neurol. 2002;448(2): 128–37.
86. Coggan JS, Bartol TM, Esquenazi E, et al. Evidence for ectopic neurotransmission at a neuronal synapse. Science. 2005;309(5733):446–51.
87. Matsui K, Jahr CE. Ectopic release of synaptic vesicles. Neuron. 2003;40(6):1173–83.
88. Matsui K, Jahr CE, Rubio ME. High-concentration rapid transients of glutamate mediate neural-glial communication via ectopic release. J Neurosci. 2005;25(33):7538–47.
89. Jacob MH, Berg DK, Lindstrom JM. Shared antigenic determinant between the electrophorus acetylcholine receptor and a synaptic component on chicken ciliary ganglion neurons. Proc Natl Acad Sci U S A. 1984;81(10):3223–7.
90. Loring RH, Zigmond RE. Ultrastructural distribution of ^{125}I-toxin F binding sites on chick ciliary neurons: synaptic localization of a toxin that blocks ganglionic nicotinic receptors. J Neurosci. 1987;7(7):2153–62.
91. Williams BM, Temburni MK, Levey MS, et al. The long internal loop of the alpha 3 subunit targets nAChRs to subdomains within individual synapses on neurons in vivo. Nat Neurosci. 1998;1(7):557–62.
92. Sargent PB. Nicotinic receptors concentrated in the subsynaptic membrane do not contribute significantly to synaptic currents at an embryonic synapse in the chicken ciliary ganglion. J Neurosci. 2009;29(12):3749–59.
93. Margiotta JF, Berg DK, Dionne VE. The properties and regulation of functional acetylcholine receptors on chick ciliary ganglion neurons. J Neurosci. 1987;7(11):3612–22.
94. Nai Q, McIntosh JM, Margiotta JF. Relating neuronal nicotinic acetylcholine receptor subtypes defined by subunit composition and channel function. Mol Pharmacol. 2003;63(2): 311–24.

95. Quick MW, Ceballos RM, Kasten M, McIntosh JM, Lester RA. Alpha3beta4 subunit-containing nicotinic receptors dominate function in rat medial habenula neurons. Neuropharmacology. 1999;38(6):769–83.
96. Coggan JS, Paysan J, Conroy WG, Berg DK. Direct recording of nicotinic responses in presynaptic nerve terminals. J Neurosci. 1997;17(15):5798–806.
97. Rogers M, Sargent PB. Rapid activation of presynaptic nicotinic acetylcholine receptors by nerve-released transmitter. Eur J Neurosci. 2003;18(11):2946–56.
98. Galligan JJ, LePard KJ, Schneider DA, Zhou X. Multiple mechanisms of fast excitatory synaptic transmission in the enteric nervous system. J Auton Nerv Syst. 2000;81(1–3):97–103.
99. Galligan JJ, North RA. Pharmacology and function of nicotinic acetylcholine and P2X receptors in the enteric nervous system. Neurogastroenterol Motil. 2004;16 Suppl 1:64–70.
100. Gwynne RM, Bornstein JC. Synaptic transmission at functionally identified synapses in the enteric nervous system: roles for both ionotropic and metabotropic receptors. Curr Neuropharmacol. 2007;5(1):1–17.
101. Kirchgessner AL, Liu MT. Immunohistochemical localization of nicotinic acetylcholine receptors in the guinea pig bowel and pancreas. J Comp Neurol. 1998;390(4):497–514.
102. Obaid AL, Koyano T, Lindstrom J, Sakai T, Salzberg BM. Spatiotemporal patterns of activity in an intact mammalian network with single-cell resolution: optical studies of nicotinic activity in an enteric plexus. J Neurosci. 1999;19(8):3073–93.

Chapter 9
Nicotinic Receptors in the Spinal Cord

Boris Lamotte d'Incamps and Philippe Ascher

Abstract The best known nicotinic receptors (nAChRs) of the spinal cord are the postsynaptic receptors of the Renshaw cells and the presynaptic receptors of the dorsal horn, but pre- and postsynaptic nAChRs are found all over the spinal cord. The subunit composition of the spinal nAChRs is very diverse: α4β2* and α7 nAChRs are the most frequent subtypes, but evidence exists for the presence of α3*, α6*, and β4* receptors. Many neurons bear multiple subtypes of nAChRs: homomeric and heteromeric nAChRs, heteromeric nAChRs associating different subunits, and heteromeric receptors associating the same subunits with distinct stoichiometries. The various nAChRs show some differences in their kinetics and in their ion selectivities but these differences do not match the diversity of their molecular forms. The complete determination of the subunit compositions and of the functional properties of spinal nAChRs is likely to require a better identification of individual neurons (a particularly difficult task in the case of the spinal cord), and recordings from identified neuronal pairs.

Keywords Spinal cord • Motoneuron • Renshaw cell • Presynaptic receptors • Postsynaptic receptors

1 Introduction

The nicotinic receptors (nAChRs) of the Renshaw cells (RCs) were the first nicotinic receptors identified in the CNS. After Renshaw [1] provided evidence that a specific group of inhibitory interneurons (INs) (which now bear his name) is excited

B. Lamotte d'Incamps, Ph.D.
Centre de Neurophysique, Physiologie, Pathologie, CNRS UMR 8119, Université Paris Descartes, 45 rue des Saints Pères, 75006 Paris, France

P. Ascher, Ph.D. (✉)
Laboratoire de Physiologie cérébrale, CNRS UMR 8118, Université Paris Descartes, 45 rue des Saints Pères, 75006 Paris, France
e-mail: philippe.ascher@parisdescartes.fr

R.A.J. Lester (ed.), *Nicotinic Receptors*, The Receptors 26,
DOI 10.1007/978-1-4939-1167-7_9

by collaterals of the motoneuron (MN) axons, Eccles et al. [2] showed that the excitation of the RCs was nearly completely blocked by di-hydro-beta-erythroidine (DHßE), a known antagonist of ganglionic nAChRs. The "central" nAChRs identified on the RCs were clearly postsynaptic, and thus resembled the other nAChRs known at that time—the postsynaptic "peripheral" nAChRs of autonomic ganglia and skeletal muscle. However, in the following years, when "presynaptic" nAChRs were characterized both in the peripheral and the central nervous systems (see [3]), they were found to be particularly abundant in the dorsal horn of the spinal cord. In the following review we will use the postsynaptic nAChRs of the RCs and the presynaptic nAChRs of the dorsal horn as models of the two main neuronal types of nAChRs, but we will also show that pre- and postsynaptic nAChRs are found all over the spinal cord. We will discuss the diversity of these receptors and the persistent uncertainties about their subunit composition and their functional role, and try to outline how these questions could be approached in the future.

2 The Localization of Cholinergic Fibers and Cholinergic Neurons

The cholinergic neurons in the spinal cord were first identified by the presence of AChE, but it was then realized that AChE is present in some non-cholinergic neurons. A more specific labeling was provided by antibodies against choline acetyltransferase (ChAT) and against the vesicular ACh transporter (VAChT) [4–11], and more recently by the use of mice expressing eGFP under the control of the promoters of these proteins [9, 12].

There is no strong evidence for descending cholinergic pathways. Four intrinsic cholinergic neuron types have been described (MNs, dorsal horn neurons, central canal neurons, and partition cells) to which one should add the efferent preganglionic autonomic neurons [4] and possibly afferent neurons since dorsal root ganglia (DRGs) have been shown to contain a splice variant of the ChAT found in the CNS (see [8, 11]) (Fig. 9.1).

The somatic MNs are located in lamina VIII and IX [5, 6]. Their axons have collaterals, most of which terminate on RCs and some on MNs [13–15].

The cholinergic neurons in the dorsal horn (DHI, Fig. 9.1) are mostly found in lamina III and IV [4, 6, 16]. Ribeiro da Silva and Cuello [17] provided the first evidence that in these neurons ChAT is presynaptic to sensory fibers. Mesnage et al. [12] took advantage of transgenic mice expressing eGFP under the control of the ChAT promoter to show that, despite their low numbers (about 24 per segment), these neurons can collect information over a long range and innervate the superficial dorsal horn densely. Pawlowski et al. [18] showed that these neurons are also present in Primates.

In the region between the ventral and the dorsal horns three groups of cholinergic neurons are intermingled: *preganglionic neurons, central canal neurons, and partition neurons*.

Preganglionic neurons are found at the thoracolumbar levels (sympathetic) and at the lumbosacral levels (parasympathetic), in both the intermediolateral and the

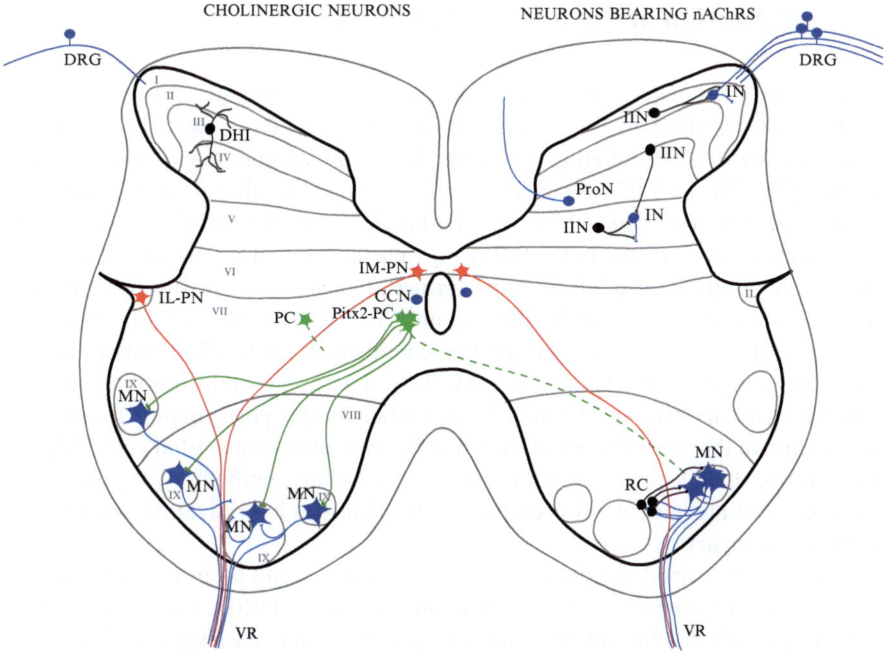

Fig. 9.1 Cholinergic neurons and neurons bearing cholinergic receptors in the spinal cord. The main types of cholinergic neurons are indicated on the *left side* of the drawing whereas the distribution of neurons bearing nAChRs is illustrated on the *right. DRG* dorsal root ganglion neurons, *DHI* dorsal horn interneurons, *IIN* inhibitory interneurons, *IN* dorsal horn interneurons (excitatory or inhibitory), *MN* motoneurons, CCN central canal neurons, *PC* partition cells, among which those expressing Pitx2 project C-boutons on the MNs, *PN* preganglionic neurons from the intermediolateralis (IL-PN) and the intermediomedialis columns (IM-PN), *ProN* projection neurons, *RC* Renshaw cells, *VR* ventral root. The lamina are numbered according to Rexed [93]

intermediomedial columns (IL-PN, IM-PN, Fig. 9.1). They send their axons into the ventral roots, in contrast with central canal neurons and partition neurons, the axons of which remain in the cord but can establish contacts a few segments away [7].

Central canal neurons encircle the central canal in lamina X and in the intermediomedial nucleus along the medial border of lamina VII [4, 6, 7]. Central canal cells receive ChAT immunoreactive fibers and in the thoracic spinal cord they form the beginning of a chain of ChAT immunoreactive neurons which projects across the intermediate gray to the sympathetic preganglionic neurons.

Partition neurons (PC, Fig. 9.1) are distributed between the central canal and the lateral edge of the grey matter, thereby delineating the limit between dorsal and ventral horns. The fact that some of them express the gene Pitx2 [9, 10] (Pitx-2-PC, Fig. 9.1) has allowed to follow their axons and to show that they establish direct connections to motor neurons through C-boutons [9].

Many cholinergic neurons of laminae III–V also contain GABA [12, 16, 19]. Ventral horn MNs in neonates release both ACh and glutamate [20–22]. In adult rats, the presence of VGLUT2 has been reported in some cholinergic terminals which did not appear to be MN axon collaterals [23].

3 Localization of the nAChR Subunits

The first detailed maps of spinal nAChRs used autoradiography of α-bungarotoxin (α-Bgt) which binds to homomeric α7 and α9 nAChRs as well as to α9–α10 nAChRs. Most of the labeling was found in the dorsal horn (see [24]) and in the DRGs [25, 26]. In 1987, Swanson et al. [27] were the first to use an antibody against a purified nAChR which was not labeled by α-Bgt and thus was likely to be a heteromeric nAChR. In the following years a detailed localization of the subunits of heteromeric nAChRs was performed, first based on in situ hybridization and immunochemistry for individual subunits (α2, α3, α4, and β2 [28], α5 [29], α4 [30], α2 [10, 31, 32]), and then immunochemistry for multiple subunits in conjunction with other markers (e.g., [33, 34]). The presynaptic immunochemical labeling was massive in the dorsal horn whereas the postsynaptic localization dominated in the ventral horn. α3, α5, and β2 were also seen in glial cells [33, 34]. Despite the doubts concerning the selectivity of many antibodies [35] there was, overall, good agreement between the results of in situ hybridization and those of immunochemistry.

RT-PCR confirmed the presence of α3, α4, α5, β2, and β4 in the spinal parenchyma [33] and of nearly all the "central" subunits in the DRGs, from α2 to α10 and β2 to β4 [26, 33, 36–38]. RT-PCR, however, rarely went to the single cell level, with the notable exception of the study of Cordero-Erausquin et al. [39] which showed the dominance of α4α6β2 in the dorsal horn inhibitory interneurons, and of α3β2α7 in the dorsal horn excitatory interneurons and projection neurons.

The distribution of the main neurons having been shown to bear nAChRs is summarized in Fig. 9.1 (right).

4 Pharmacological and Physiological Studies of Postsynaptic nAChRs

The characterization of postsynaptic nAChRs has been based on the analysis of electrical responses elicited either by the application of nicotinic agonists or by synaptic release of ACh.

4.1 The Postsynaptic nAChRs of the Renshaw Cell

The pharmacological characterization of the nAChRs of the RC did not go very far in the 1960s due to the limitations of the techniques available at that time, and to the coexistence on the RC of nicotinic and muscarinic receptors [40]. The sensitivity of the RC activation to DHßE [2] indicated the presence of heteromeric nAChRs but later studies showed that, at least in neonatal mice, methyllycaconitine (MLA) eliminates a component of the fast excitatory postsynaptic current (EPSC) due to homomeric nAChRs [22]. During a repetitive stimulation the fast EPSC depresses rapidly, whereas the heteromeric slower EPSC component builds up [41]. Pharmacological or genetic inactivation of acetylcholinesterase (AChE) does not affect the fast EPSC [42] whereas it greatly prolongs the slow EPSC [41].

The subunit composition of the RC heteromeric nAChRs was first proposed to be α4β2 by Dourado and Sargent [43] who observed labeling of the RC by antibodies against α4 and β2. However Ishii et al. [31] noted that α2 mRNA expression was sharply localized in ventral horn cells which could be RCs. This identification was recently confirmed [32]. Taking advantage of previous studies on recombinant nAChRs showing that the different subunit combinations have different sensitivities to DHßE, Lamotte d'Incamps and Ascher [44] compared the IC50s of DHßE on the non-α7 EPSCs in WT mice and in mice lacking one nAChR subunit. They observed a decreased sensitivity to DHßE after the deletion of β2, and an increased sensitivity to DHßE after deletion of β4, a subunit which had not been identified in the ventral horn in previous studies. Overall the results suggested the presence of α4β2* and α2β4* nAChRs. However these two subunit associations have been reported to have similar affinities for ACh and thus their coexistence did not explain the presence of two components in the decay of the (heteromeric) EPSC: a fast component decaying with a time constant of 10 ms and a slow component decaying with a time constant of about 100 ms [41]. It had been proposed that the fast component is due to subsynaptic receptors and the slow one to extrasynaptic receptors [41]. The analysis of the effects of DHßE on the two components suggested that the subsynaptic and extrasynaptic receptors may correspond to receptors having the same subunit composition but different stoichiometries and different affinities for ACh [44]. Overall the results suggest a triple diversity for the nAChRs of RCs as they indicate the presence of homomeric and heteromeric nAChRs, of heteromeric nAChRs made of different subunits, and of heteromeric receptors made of the same subunits with distinct stoichiometries.

4.2 *The Postsynaptic nAChRs of MNs*

Contacts of MN axon collaterals with MNs dendrites have been described in adult cats [13, 15], suggesting the presence of MN-MN synapses. Dourado and Sargent [43] reported α4 immunoreactivity in ventral MNs in young rats (P6-P10) and Khan et al. [33] reported labeling with antibodies against α3, α4, α5, and β2 in MNs of adult rats and rabbits. Ogier et al. [45] measured epibatidine binding in MNs and found that it was much stronger and wider in the young than in the adult. In neonatal mice, ACh induced in ventral MNs a current which was blocked by 1 μM DHßE [45].

Comparable data are lacking for adult MNs, and until now nobody has reported synaptic nicotinic currents on MNs. The cholinergic partition cells regulate MN activity [9, 10] through the activation of muscarinic receptors.

4.3 Postsynaptic nAChRs of the Sympathetic Preganglionic Neurons

Labeling by α-Bgt [46] and by antibodies against heteromeric nAChRs subunits [33] has been observed in the central canal and in lamina X. In the region dorsal to the central canal, Bordey et al. [47], Bradaia and Trouslard [48], and Bradaia et al. [49] recorded neurons that they initially assumed were SPNs and in later studies identified as such by antidromic stimulation of the ventral roots. Both DMPP and choline induced currents blocked respectively by DHßE (1 μM) and α-Bgt. Curiously, however, the EPSCs evoked by local electrical stimulation were completely blocked by α-Bgt (50 nM) and insensitive to DHßE (1 μM). For additional information on nAChRs of the autonomic nervous system refer to Chap. 8

4.4 Postsynaptic nAChRs in the Dorsal Horn

The presence of postsynaptic nAChRs on the soma and dendrites of neurons in the dorsal horn was suggested by the early reports of α-Bgt binding on large multipolar cells in lamina IV or V (e.g., [46]). Later studies with antibodies also showed the presence of postsynaptic heteromeric nAChRs [33]. The RT-PCR studies of Cordero-Erausquin et al. [39] suggested the possible presence of α7 and α3β2 on excitatory INs and projection neurons, and of α7 and α6β2 on inhibitory INs.

The pharmacological analysis of the effects of nicotine on dorsal horn neurons [39, 50–54] has given heterogeneous results. In acutely isolated dorsal horn neurons, Genzen and McGehee [52] reported currents induced by choline and blocked by MLA, as well as currents induced by epibatidine and blocked by DHßE (1 μM), with a limited overlap. In contrast, in slices no evidence was found for the activation of α7 nAChRs and two different pictures emerged for heteromeric nAChRs: in substantia gelatinosa (SG) the currents induced by nicotine persisted in DHßE at 5 μM, suggesting that they were mediated by non-α4β2 heteromeric nAChRs [53], whereas in lamina V all neurons were excited by RJR-2403 (an α4β2 agonist) suggesting "pure" α4β2 receptors [54].

The presence of a variety of postsynaptic nAChRs in the dorsal horn INs is thus well established, but the characterization of the receptors and of the corresponding EPSCs is likely to remain incomplete until one can identify and stimulate the presynaptic neurons.

5 Pharmacological and Physiological Studies of the Presynaptic Nicotinic Receptors

5.1 Presynaptic nAChRs of the Dorsal Spinal Cord

5.1.1 Electrophysiological Recordings in Slices

In the dorsal horn ACh or nicotine increased the frequency of spontaneous synaptic excitatory and inhibitory currents (sEPSCs and sIPSCS). In most cases the effects were also observed after addition of TTX, i.e., on miniature synaptic currents (mEPSCs and mIPSCs), and were therefore attributed to "presynaptic" rather than "preterminal" [55] nAChRs, with one exception [54].

The increase in frequency of sEPSCs and mEPSCs induced by nicotine [52, 53] was nearly eliminated by MLA [52] and was not mimicked by an α4β2 partial agonist [56]. Nicotine also increased some of the EPSCs evoked by stimulation of the DR entry zone [52].

The increase in the frequency of sIPSCs and mIPSCs induced by nicotine was observed in the substantia gelatinosa (SG) of both neonatal [51] and adult rats [53]. In neonates the effect was blocked by DHßE at 0.3 μM [51] but in adult rats it persisted in the presence of DHßE at 5 μM [53], suggesting that part of the effect in the adult was due to a "non-α4β2" nAChR. The presence of a non-α4β2 nAChR was also proposed by other authors [57, 58]. This non-α4β2 nAChR was suggested to be α3β4* by Takeda et al. [53] because cytisine, considered a good agonist of α3β4 receptors, mimicked the effects of nicotine. But this receptor could contain an α6 subunit, since Cordero-Erausquin et al. [39] showed that inhibitory INs of the dorsal horn express α4, α6, and β2.

DHßE decreased the frequency of mIPSCs and sIPSCs (but did not affect the frequency of sEPSCs) in the SG of adult mice [59]. This indicates that there is a "*tonic*" *activation of heteromeric nAChRs* of inhibitory INs. Such a tonic activation was also proposed by Cordero-Erausquin and Changeux [59] to explain the observation that DHßE increased the basal release of serotonin from the pathway descending from the raphe to the spinal cord.

Presynaptic α7 nAChRs were not clearly identified in inhibitory INs of the SG [58], but mixed (α7 and non-α7) presynaptic effects of nicotine on inhibitory INs were seen in lamina V [54].

5.1.2 Electrophysiological Recordings from DRG Neurons

DRG neurons bear α3–α10 and β2–β4 nAChR subunits [25, 26, 36–38, 52] and it is usually assumed that the receptors are on both the soma and on axons, as first shown for α7 nAChRs [46]. Not surprisingly, recordings from the somas of DRG neurons in mice and rats revealed a great variety of functional nAChRs. Genzen et al. [36] in neonates identified four types of responses among which two were proposed to be α7 and α3β4, one α4β2 like, and the fourth atypical. Fucile et al. [60] found three types of responses (fast, slow, and mixed) which could be attributed to α7 and α*β* receptors in isolation or in combination. They attributed the slow response to α3* nAChRs. In neonates the α7 response was very commonly seen; the slow response was nearly absent. The inverse was seen in adults. Rau et al. [61] identified the same three types of nicotinic currents, and concluded that they were all expressed in nociceptive neurons. They also suggested that the heteromeric nAChRs were α3*, with a subgroup possibly associating α5. Finally Hone et al. [38], who analyzed the effects of a set of conotoxins, added to the list of candidates α6β4* receptors. There is thus a consensus that the dominant receptors on the DRGs are α7 and α3*, but α4β2 and α6β4* as well as α9 and α9–α10 are also likely present in some cells.

5.1.3 Intrathecal Administration of Nicotinic Ligands

The effects of nicotinic ligands on behavior have been intensively studied but are difficult to use to characterize spinal nAChRs because nAChRs are present all through the CNS and some of the major effects of drugs administered i.v. or i.p. involve the activation of descending systems (e.g., [62]). However the effects of intrathecal injections of nicotinic ligands have given useful information on the spinal nAChRs.

Intrathecal injections of nicotinic agonists [63–65] were found to have antinociceptive effects. Conversely intrathecal injection of antagonists of heteromeric nAChRs induced thermal and mechanical hyperalgesia and allodynia [63–69] which indicates the presence of a tonic activation of nAChRs. The involvement of α4β2 nAChRs was suggested by the fact that epibatidine was more effective than cytisine [63] and by the strong effect of A-85380, considered to be a selective α4β2 agonist [64]. The data of Marubio et al. [37] supported the role of α4β2 nAChRs, but the persistence of some nicotine antinociceptive effects in ß2-KO mice and α4-KO mice also suggested a role of non-α4β2 receptors. The pro-nociceptive effects of α-CTX-MII (an antagonist of α3β2*:α6β2*) led Young et al. [67] to propose that these non-α4β2 receptors were α3β2*. Yalcin et al. [68] compared the antinociceptive effects of two antagonists, one selective for α4β2 (NDNI) and the other for α3β2*:α6β2* (α-CTX-MII). Both were active in WT mice but not in ß2 KO mice, leading to the suggestion that the effects in WT were mediated by two ß2* receptors, which are thus likely to be α4β2* and α3β2*. These experiments are internally consistent, but not easy to reconcile with the electrophysiological data on DRG neurons which privileged α3β4* subtypes over α3β2* [36, 60, 61]. Young et al. [67] suggested that the α-CTX-MII sensitive receptors were on C-fibers but there was little reduction by the toxin of the eEPSC evoked in lamina I–II by stimulation of the VRs [69].

Intrathecal injections of nicotinic agonists also produce nociceptive and cardiovascular effects [34, 64]. In the model proposed by Khan et al. [34] the "nocifensive" response involves the release of EAAs from excitatory INs, release of EAAs and SP from primary afferents, and excitation of ascending projection neurons.

Little evidence was found for the involvement of α7 nAChRs in any of the experiments described above. This contrasts with the anatomical data which have shown a massive presence of α7 on DRGs, and with the data indicating that activation of α7 nAChRs reduces inflammatory pain (e.g., [70]). Many of these α7 antinociceptive effects may be linked to α7 receptors present on microglial cells and macrophages (see [71] for recent references) but intrathecal injection of MLA blocked some antinociceptive effects of α7 positive allosteric modulators, which suggests a possible role of spinal α7 nAChRs [72].

5.1.4 AChRs and Primary Afferent Depolarization (PAD)

Hochman et al. [73] and Shreckengost et al. [74] have suggested that nAChRs are involved in the PAD. Their main observation is that the PAD is reduced both by α-Bgt and by (+)-tubocurarine [73]. Among the possible interpretations of this observation, one is that the GABAergic INs classically assumed to produce the PAD can be excited by ACh released either by local cholinergic INs or by the primary afferents [73]. Another possibility [74] is that ACh released by DR afferents directly produces a PAD involving nAChRs sensitive to bicuculline (like α7 receptors) [75] which would constitute presynaptic autoreceptors.

5.2 Presynaptic nAChRs in Lamina X

In neurons situated in lamina X and presumed to be SPNs, DMPP increased the frequency of glutamatergic sEPSCs [47], of glycinergic mIPSCs [76], and of GABAergic mIPSCs [77]. For the glycinergic mIPSCs the effect was blocked by (+)-tubocurarine and hexamethonium, reduced by DHßE (10 μM), and reduced by MLA. For the GABAergic mIPSCs the effect was blocked by DHßE (1 μM) and little affected by MLA. Surprisingly, in the case of the glutamatergic and GABAergic inputs the cholinergic agonists increased the amplitude of electrically evoked synaptic currents [47, 77] whereas in the case of the glycinergic inputs the IPSC was reduced [76].

5.3 Presynaptic nAChRs in the Ventral Spinal Cord

Khan et al. [33] observed presynaptic labeling in the ventral spinal cord for α3, α4, α5, and β2, in particular on the C-boutons, which suggests the possible presence of autoreceptors since C-boutons are the terminals of cholinergic partition cells [9]. Varicose terminals labeled by α-Bgt have been seen in the ventral horn [78], but have not been attributed to identified neurons.

5.3.1 Spinal nAChRs During Development

The study of the early development of cholinergic spinal neurons has made great progress recently, in particular for MNs and for the partition cells, but the origin of the dorsal horn INs remains unknown [79]. The co-release of ACh and other transmitters (see Sect. 2) may be more frequent in embryos, as shown in particular in Xenopus embryos [80] or tadpoles [81].

The principal types of spinal nAChRs are already present in embryos, but the ratio of the various subunits changes during development [45, 53, 60, 82, 83].

ACh released from embryonic MNs is thought to be responsible for the early muscle activity that in turn regulates MN survival. Keiger et al. [82] determined the developmental expression profile of nAChRs subunits in the chick lumbar MNs and showed that subunits α1, α4, α7, α8, and β2 are regulated during naturally occurring MN cell death.

MN growth cones bear nAChRs [84] which are likely to orient them during development. In embryonic zebrafish transient application of nicotine alters axonal path finding by secondary MNs [85]. ACh is also likely involved in the survival of MNs: in cultured spinal MNs, activation of α7 nAChRs rescued the neurons from a programmed cell death induced by trophic factor deprivation [86]. In mice, Myers et al. [87] deleted ChAT and showed that cholinergic input is required during embryonic development for proper assembly of spinal locomotor circuits. Diphtheria toxin ablation of the cells which express the α7 subunit [88] produces defects of neural tube closure (spina bifida) which could be reduced by a choline supplemented diet.

6 Conclusions

6.1 Postsynaptic and Presynaptic nAChRs

The dorsoventral asymmetry in the anatomical distribution of pre- and postsynaptic nAChRs in the spinal cord has led to a view in which, in the dorsal horn, small groups of cholinergic neurons distribute their axons over very wide target regions and activate mostly presynaptic nAChRs, while in the ventral horn the MN axon collaterals activate postsynaptic nAChRs and mediate a strict point-to-point transmission. The postsynaptic nAChRs in the ventral horn would thus resemble functionally those of the peripheral nervous system while those of the dorsal horn would resemble the presynaptic nAChRs which dominate in the higher centers of the CNS (e.g., [3]).

Although the dorsoventral asymmetry is undoubtedly present, the frequency and the functional importance of postsynaptic nAChRs in the dorsal horn of the spinal cord may have been underestimated. Inward currents induced by postsynaptic nAChRs have been observed in many dorsal horn neurons [47, 52, 77] which suggests that they may mediate nicotinic EPSCs. The fact that such EPSCs have not been described could stem from the fact that observing such EPSCs requires that one activates selectively the presynaptic neurons. This is possible in the ventral horn because one can use antidromic stimulation of the VRs; it is more problematic in the case of the dorsal horn, where the sparse cholinergic interneurons were difficult to identify until recently [12]. A similar problem is encountered in the higher centers of the CNS where postsynaptic nAChRs and spontaneous nicotinic EPSCs have been described in many neurons, but evoked nicotinic EPSCS have only been observed by combining optogenetic tools and specific molecular markers (e.g., [89, 90]).

Identifying presynaptic cholinergic neurons and stimulating them selectively is also likely to improve the understanding of the role of presynaptic nAChRs. Until now nearly all electrophysiological studies of presynaptic ACh effects have relied on changes of frequency of spontaneous or miniature EPSCs or IPSCs, and assumed that the direction of these changes predicts the changes in evoked PSCs. This hypothesis is plausible but not completely certain, and has only been checked in a few examples [47, 52, 77]. When the two changes were not in the same direction, preterminal nAChRs were invoked [76] but this hypothesis has rarely been submitted to a thorough analysis. It would be very important to analyze the presynaptic cholinergic modulations of release in conditions in which one can control the frequency and the duration of the firing of the presynaptic cholinergic neurons.

6.2 The Subunit Composition of Native nAChRs

One of the main difficulties in establishing the subunit composition of native nAChRs stems from the fact that the presence of a single receptor type seems the exception rather than the rule. A large fraction of spinal neurons seem to bear both a homomeric and at least one heteromeric nAChR [22, 89, 90] and one can suspect that, in a number of cases in which the α7 nAChRs were not detected, they were present but missed either because one had applied α7 antagonists like bicuculline [75] or strychnine [91] (to eliminate inhibitory synaptic currents) or because the agonist application was too slow to avoid the fast desensitization [92]. In addition the presence of more than one heteromeric nAChR seems extremely common. It is particularly striking in the case of the DRGs [25, 26, 36–38, 52] and of the RCs and involves both the presence of multiple α or β subunits and the coexistence at a single synapse of stoichiometric variants of the same receptor assembly [44].

Although a complete identification of the nAChR subtypes in a single neuron has not yet been obtained, some preferred associations of subunits have been identified: α4β2* seem nearly ubiquitous, α3*β2* are abundant in dorsal horn projection neurons and in C fibers, α3*β4* in DRG nociceptors, α6* in dorsal horn INs and in DRGs, α2β4* in RCs, and α9–α10 in primary afferents. α5 may participate in many assemblies. Strong functional differences (in the kinetics of conductance changes and in the ion selectivity of the synaptic channels) have been demonstrated, in particular between α7 and α*β* receptors, and between stoichiometric variants of a given subunit assembly.

The characterization of nicotinic receptors in the nervous system has until now depended mostly on pharmacological compounds activating, potentiating, or blocking specific nAChRs subtypes. More recently genetic tools were introduced which permit the deletion of specific subunits. Overall, the identification of the receptors remains very partial: there are very few subtype-specific compounds, and the number of functional differences identified between subtypes does not match the known combinatorial diversity of the various α*β* assemblies. The diversity of subtypes on a given cell might have been overestimated by the fact that most studies were done on poorly identified neurons or on structures possibly containing heterogeneous neuronal populations.

The identification of individual neurons thus appears to be a priority for future studies, both for the characterization of nAChR subtypes and for the understanding of their functional role. This task appears more difficult in the case of the spinal cord than in laminar structures such as the cerebral cortex or the cerebellum, but it is no less essential. From this point of view the work of those who are attempting to find molecular markers for spinal neurons [79] appears as a major enterprise which could completely renovate the understanding of the functions of nAChRs in the spinal cord.

Funding Acknowledgement Our work is supported by the CNRS and the association française contre les myopathies (AFM).

References

1. Renshaw B. Central effects of centripetal impulses in axons of spinal ventral roots. J Neurophysiol. 1946;9:191–204.
2. Eccles JC, Fatt P, Koketsu K. Cholinergic and inhibitory synapses in a pathway from motor-axon collaterals to motoneurones. J Physiol. 1954;126:524–62.
3. Role LW, Berg DK. Nicotinic receptors in the development and modulation of CNS synapses. Neuron. 1996;16:1077–85.
4. Barber RP, Phelps PE, Houser CR, Crawford GD, Salvaterra PM, Vaughn JE. The morphology and distribution of neurons containing choline acetyltransferase in the adult rat spinal cord: an immunocytochemical study. J Comp Neurol. 1984;229:329–46.
5. Phelps PE, Barber RP, Houser CR, Crawford GD, Salvaterra PM, Vaughn JE. Postnatal development of neurons containing choline acetyltransferase in rat spinal cord: an immunocytochemical study. J Comp Neurol. 1984;229:347–61.
6. Borges LF, Iversen SD. Topography of choline acetyltransferase immunoreactive neurons and fibers in the rat spinal cord. Brain Res. 1986;362:140–8.
7. Sherriff FE, Henderson Z. A cholinergic propriospinal innervation of the rat spinal cord. Brain Res. 1994;634:150–4.
8. Matsumoto M, Xie W, Inoue M, Ueda H. Evidence for the tonic inhibition of spinal pain by nicotinic cholinergic transmission through primary afferents. Mol Pain. 2007;3:41.
9. Zagoraiou L, Akay T, Martin JF, Brownstone RM, Jessell TM, Miles GB. A cluster of cholinergic premotor interneurons modulates mouse locomotor activity. Neuron. 2009;64:645–62.
10. Enjin A, Rabe N, Nakanishi ST, Vallstedt A, Gezelius H, Memic F, et al. Identification of novel spinal cholinergic genetic subtypes disclose Chodl and Pitx2 as markers for fast motor neurons and partition cells. J Comp Neurol. 2010;518:2284–304.
11. Bellier J-P, Kimura H. Peripheral type of choline acetyltransferase: biological and evolutionary implications for novel mechanisms in cholinergic system. J Chem Neuroanat. 2011;42:225–35.

12. Mesnage B, Gaillard S, Godin AG, Rodeau J-L, Hammer M, Von Engelhardt J, et al. Morphological and functional characterization of cholinergic interneurons in the dorsal horn of the mouse spinal cord. J Comp Neurol. 2011;519:3139–58.
13. Cullheim S, Kellerth J-O, Conradi S. Evidence for direct synaptic interconnections between cat spinal alpha-motoneurons via the recurrent axon collaterals: a morphological study using intracellular injection of horseradish peroxidase. Brain Res. 1977;132:1–10.
14. Alvarez FJ, Dewey DE, McMillin P, Fyffe RE. Distribution of cholinergic contacts on Renshaw cells in the rat spinal cord: a light microscopic study. J Physiol. 1999;515(Pt3):787–97.
15. Lagerbäck P-A, Ronnevi L-O, Cullheim S, Kellerth J-O. An ultrastructural study of the synaptic contacts of alpha-motoneurone axon collaterals. I Contacts in Lamina IX and with identified alpha-motoneurone dendrites in lamina VII. Brain Res. 1981;207:247–66.
16. Olave MJ, Puri N, Kerr R, Maxwell DJ. Myelinated and unmyelinated primary afferent axons form contacts with cholinergic interneurons in the spinal dorsal horn. Exp Brain Res. 2002;145:448–56.
17. Ribeiro-Da-Silva A, Cuello Claudio A. Choline Acetyltransferase-Immunoreactive Profiles Are Presynaptic to Primary Sensory Fibers in the Rat Superficial Dorsal Horn. J Comp Neurol. 1990;295:370–84.
18. Pawlowski SA, Gaillard S, Ghorayeb I, Ribeiro-da-Silva A, Schlichter R, Cordero-Erausquin M. A novel population of cholinergic neurons in the macaque spinal dorsal horn of potential clinical relevance for pain therapy. J Neurosci. 2013;33:3727–37.
19. Todd AJ, Spike RC. The localization of classical transmitters and neuropeptides within neurons in laminae I-III of the mammalian spinal dorsal horn. Prog Neurobiol. 1993;41:609–45.
20. Nishimaru H, Restrepo CE, Ryge J, Yanagawa Y, Kiehn O. Mammalian motor neurons corelease glutamate and acetylcholine at central synapses. Proc Natl Acad Sci U S A. 2005;102:5245–9.
21. Mentis GZ, Alvarez FJ, Bonnot A, Richards DS, Gonzalez-Forero D, Zerda R, et al. Noncholinergic excitatory actions of motoneurons in the neonatal mammalian spinal cord. Proc Natl Acad Sci U S A. 2005;102:7344–9.
22. Lamotte d'Incamps B, Ascher P. Four excitatory postsynaptic ionotropic receptors coactivated at the motoneuron-Renshaw cell synapse. J Neurosci. 2008;28:14121–31.
23. Liu TT, Bannatyne B, Jankowska E, Maxwell DJ. Cholinergic terminals in the ventral horn of adult rat and cat: evidence that glutamate is a cotransmitter at putative interneuron synapses but not at central synapses of motoneurons. Neuroscience. 2009;161:111–22.
24. Coggeshall RE, Carlton SM. Receptor localization in the mammalian dorsal horn and primary afferent neurons. Brain Res Brain Res Rev. 1997;24:28–66.
25. Haberberger RV, Bernardini N, Kress M, Hartmann P, Lips KS, Kummer W. Nicotinic acetylcholine receptor subtypes in nociceptive dorsal root ganglion neurons of the adult rat. Auton Neurosci Basic Clin. 2004;113:32–42.
26. Lips KS, Pfeil U, Kummer W. Coexpression of alpha 9 and alpha 10 nicotinic acetylcholine receptors in rat dorsal root ganglion neurons. Neuroscience. 2002;115:1–5.
27. Swanson LW, Simmons DM, Whiting PJ, Lindstrom J. Immunohistochemical localization of neuronal nicotinic receptors in the rodent central nervous system. J Neurosci. 1987;7:3334–42.
28. Wada E, Wada K, Boulter J, Deneris E, Heinemann S, Patrick J, et al. Distribution of alpha 2, alpha 3, alpha 4, and beta 2 neuronal nicotinic receptor subunit mRNAs in the central nervous system: a hybridization histochemical study in the rat. J Comp Neurol. 1989;284:314–35.
29. Wada E, McKinnon D, Heinemann S, Patrick J, Swanson LW. The distribution of mRNA encoded by a new member of the neuronal nicotinic acetylcholine receptor gene family (alpha 5) in the rat central nervous system. Brain Res. 1990;526:45–53.
30. Dineley-Miller K, Patrick J. Gene transcripts for the nicotinic acetylcholine receptor subunit, beta 4, are distributed in multiple areas of the rat central nervous system. Brain Res Mol Brain Res. 1992;16:339–44.
31. Ishii K, Wong JK, Sumikawa K. Comparison of alpha 2 nicotinic acetylcholine receptor subunit mRNA expression in the central nervous system of rats and mice. J Comp Neurol. 2005;493:241–60.

32. Gezelius H. Studies of spinal motor control networks in genetically modified mouse models. Acta Universitatis Upsaliensis 2009, ISBN 1651-6206, pp 47.
33. Khan I, Osaka H, Stanislaus S, Calvo RM, Deerinck T, Yaksh TL, et al. Nicotinic acetylcholine receptor distribution in relation to spinal neurotransmission pathways. J Comp Neurol. 2003;467:44–59.
34. Khan IM, Wart CV, Singletary EA, Stanislaus S, Deerinck T, Yaksh TL, et al. Elimination of rat spinal substance P receptor bearing neurons dissociates cardiovascular and nocifensive responses to nicotinic agonists. Neuropharmacology. 2008;54:269–79.
35. Moser N, Mechawar N, Jones I, Gochberg-Sarver A, Orr-Urtreger A, Plomann M, et al. Evaluating the suitability of nicotinic acetylcholine receptor antibodies for standard immunodetection procedures. J Neurochem. 2007;102:479–92.
36. Genzen JR, Van Cleve W, McGehee DS. Dorsal root ganglion neurons express multiple nicotinic acetylcholine receptor subtypes. J Neurophysiol. 2001;86:1773–82.
37. Marubio LM, Del Mar A-JM, Cordero-Erausquin M, Léna C, Le Novère N, De Kerchove d'Exaerde A. Reduced antinociception in mice lacking neuronal nicotinic receptor subunits. Nature. 1999;398:805–10.
38. Hone AJ, Meyer EL, McIntyre M, McIntosh JM. Nicotinic acetylcholine receptors in dorsal root ganglion neurons include the α6β4* subtype. FASEB J. 2012;26:917–26.
39. Cordero-Erausquin M, Pons S, Faure P, Changeux J-P. Nicotine differentially activates inhibitory and excitatory neurons in the dorsal spinal cord. Pain. 2004;109:308–18.
40. Curtis DR, Ryall RW. Nicotinic and muscarinic receptors of Renshaw cells. Nature. 1964;203:652–3.
41. Lamotte d' Incamps B, Krejci E, Ascher P. Mechanisms shaping the slow nicotinic synaptic current at the motoneuron-Renshaw cell synapse. J Neurosci. 2012;32:8413–23.
42. Farar V, Mohr F, Lamotte d'Incamps B, Legrand M, Cendelin M, Leroy J, et al. Near-complete adaptation of the PRiMA knockout to the lack of central acetylcholinesterase. J Neurochem. 2012;122:1065–80.
43. Dourado M, Sargent PB. Properties of nicotinic receptors underlying Renshaw cell excitation by alpha-motor neurons in neonatal rat spinal cord. J Neurophysiol. 2002;87:3117–25.
44. Lamotte d'Incamps B, Ascher P. Subunit composition and kinetics of the Renshaw cell heteromeric nicotinic receptors. Biochem Pharmacol. 2013;86:1114–21.
45. Ogier R, Liu X, Tribollet E, Bertrand D, Raggenbass M. Identified spinal motoneurons of young rats possess nicotinic acetylcholine receptors of the heteromeric family. Eur J Neurosci. 2004;20:2591–7.
46. Ninkovic M, Hunt SP. Alpha-bungarotoxin binding sites on sensory neurones and their axonal transport in sensory afferents. Brain Res. 1983;272:57–69.
47. Bordey A, Feltz P, Trouslard J. Nicotinic actions on neurones of the central autonomic area in rat spinal cord slices. J Physiol. 1996;497(Pt 1):175–87.
48. Bradaïa A, Trouslard J. Fast synaptic transmission mediated by alpha-bungarotoxin-sensitive nicotinic acetylcholine receptors in lamina X neurones of neonatal rat spinal cord. J Physiol. 2002;544:727–39.
49. Bradaïa A, Seddik R, Schlichter R, Trouslard J. The rat spinal cord slice: Its use in generating pharmacological evidence for cholinergic transmission using the alpha7 subtype of nicotinic receptors in the central autonomic nucleus. J Pharmacol Toxicol Methods. 2005;51:243–52.
50. Urban L, Willetts J, Murase K, Randić M. Cholinergic effects on spinal dorsal horn neurons in vitro: an intracellular study. Brain Res. 1989;500:12–20.
51. Kiyosawa A, Katsurabayashi S, Akaike N, Pang ZP. Nicotine facilitates glycine release in the rat spinal dorsal horn. J Physiol. 2001;536:101–10.
52. Genzen JR, McGehee DS. Short- and long-term enhancement of excitatory transmission in the spinal cord dorsal horn by nicotinic acetylcholine receptors. Proc Natl Acad Sci U S A. 2003;100:6807–12.
53. Takeda D, Nakatsuka T, Papke R, Gu JG. Modulation of inhibitory synaptic activity by a non-alpha4beta2, non-alpha7 subtype of nicotinic receptors in the substantia gelatinosa of adult rat spinal cord. Pain. 2003;101:13–23.

54. Takeda D, Nakatsuka T, Gu JG, Yoshida M. The activation of nicotinic acetylcholine receptors enhances the inhibitory synaptic transmission in the deep dorsal horn neurons of the adult rat spinal cord. Mol Pain. 2007;3:26.
55. Léna C, Changeux JP, Mulle C. Evidence for "preterminal" nicotinic receptors on GABAergic axons in the rat interpeduncular nucleus. J Neurosci. 1993;13:2680–8.
56. Cheng L-Z, Han L, Fan J, Huang L-T, Peng L-C, Wang Y. Enhanced inhibitory synaptic transmission in the spinal dorsal horn mediates antinociceptive effects of TC-2559. Mol Pain. 2011;7:56.
57. Genzen JR, McGehee DS. Nicotinic modulation of GABAergic synaptic transmission in the spinal cord dorsal horn. Brain Res. 2005;1031:229–37.
58. Gao B, Hierl M, Clarkin K, Juan T, Nguyen H, van der Valk M, et al. Pharmacological effects of nonselective and subtype-selective nicotinic acetylcholine receptor agonists in animal models of persistent pain. Pain. 2010;149:33–49.
59. Cordero-Erausquin M, Changeux JP. Tonic nicotinic modulation of serotoninergic transmission in the spinal cord. Proc Natl Acad Sci U S A. 2001;98:2803–7.
60. Fucile S, Sucapane A, Eusebi F. Ca2+ permeability of nicotinic acetylcholine receptors from rat dorsal root ganglion neurones. J Physiol. 2005;565:219–28.
61. Rau KK, Johnson RD, Cooper BY. Nicotinic AChR in subclassified capsaicin-sensitive and -insensitive nociceptors of the rat DRG. J Neurophysiol. 2005;93:1358–71.
62. Iwamoto ET, Marion L. Adrenergic, serotonergic and cholinergic components of nicotinic antinociception in rats. J Pharmacol Exp Ther. 1993;265:777–89.
63. Damaj MI, Fei-Yin M, Dukat M, Glassco W, Glennon RA, Martin BR. Antinociceptive responses to nicotinic acetylcholine receptor ligands after systemic and intrathecal administration in mice. J Pharmacol Exp Ther. 1998;284:1058–65.
64. Khan IM, Stanislaus S, Zhang L, Taylor P, Yaksh TL. A-85380 and epibatidine each interact with disparate spinal nicotinic receptor subtypes to achieve analgesia and nociception. J Pharmacol Exp Ther. 2001;297:230–9.
65. Rashid MH, Ueda H. Neuropathy-specific analgesic action of intrathecal nicotinic agonists and its spinal GABA-mediated mechanism. Brain Res. 2002;953:53–62.
66. Rashid MH, Furue H, Yoshimura M, Ueda H. Tonic inhibitory role of alpha4beta2 subtype of nicotinic acetylcholine receptors on nociceptive transmission in the spinal cord in mice. Pain. 2006;125:125–35.
67. Young T, Wittenauer S, McIntosh JM, Vincler M. Spinal alpha3beta2* nicotinic acetylcholine receptors tonically inhibit the transmission of nociceptive mechanical stimuli. Brain Res. 2008;1229:118–24.
68. Yalcin I, Charlet A, Cordero-Erausquin M, Tessier L-H, Picciotto MR, Schlichter R, et al. Nociceptive thresholds are controlled through spinal β2-subunit-containing nicotinic acetylcholine receptors. Pain. 2011;152:2131–7.
69. Napier IA, Klimis H, Rycroft BK, Jin AH, Alewood PF, Motin L, et al. Intrathecal α-conotoxins Vc1.1, AuIB and MII acting on distinct nicotinic receptor subtypes reverse signs of neuropathic pain. Neuropharmacology. 2012;62:2202–7.
70. Rowley TJ, McKinstry A, Greenidge E, Smith W, Flood P. Antinociceptive and anti-inflammatory effects of choline in a mouse model of postoperative pain. Br J Anaesth. 2010;105:201–7.
71. Loram LC, Taylor FR, Strand KA, Maier SF, Speake JD, Jordan KG, et al. Systemic administration of an alpha-7 nicotinic acetylcholine agonist reverses neuropathic pain in male Sprague Dawley rats. J Pain. 2012;13:1162–71.
72. Freitas K, Carroll FI, Damaj MI. The antinociceptive effects of nicotinic receptors α7-positive allosteric modulators in murine acute and tonic pain models. J Pharmacol Exp Ther. 2013;344:264–75.
73. Hochman S, Shreckengost J, Kimura H, Quevedo J. Presynaptic inhibition of primary afferents by depolarization: observations supporting nontraditional mechanisms. Ann N Y Acad Sci. 2010;1198:140–52.

74. Shreckengost J, Calvo J, Quevedo J, Hochman S. Bicuculline-sensitive primary afferent depolarization remains after greatly restricting synaptic transmission in the mammalian spinal cord. J Neurosci. 2010;30:5283–8.
75. Demuro A, Palma E, Eusebi F, Miledi R. Inhibition of nicotinic acetylcholine receptors by bicuculline. Neuropharmacology. 2001;41:854–61.
76. Bradaïa A, Trouslard J. Nicotinic receptors regulate the release of glycine onto lamina X neurones of the rat spinal cord. Neuropharmacology. 2002;43:1044–54.
77. Seddik R, Schlichter R, Trouslard J. Modulation of GABAergic synaptic transmission by terminal nicotinic acetylcholine receptors in the central autonomic nucleus of the neonatal rat spinal cord. Neuropharmacology. 2006;51:77–89.
78. Shelukhina IV, Kryukova EV, Lips KS, Tsetlin VI, Kummer W. Presence of alpha7 nicotinic acetylcholine receptors on dorsal root ganglion neurons proved using knockout mice and selective alpha-neurotoxins in histochemistry. J Neurochem. 2009;109:1087–95.
79. Alaynick WA, Jessell TM, Pfaff SL. SnapShot: spinal cord development. Cell. 2011;146: 178–178.e1.
80. Borodinsky LN, Root CM, Cronin JA, Sann SB, Gu X, Spitzer NC. Activity-dependent homeostatic specification of transmitter expression in embryonic neurons. Nature. 2004;429: 523–30.
81. Li W-C, Soffe SR, Roberts A. Glutamate and acetylcholine corelease at developing synapses. Proc Natl Acad Sci U S A. 2004;101:15488–93.
82. Keiger CJH, Prevette D, Conroy WG, Oppenheim RW. Developmental expression of nicotinic receptors in the chick and human spinal cord. J Comp Neurol. 2003;455:86–99.
83. Zoli M, Le Novère N, Hill JA, Changeux JP. Developmental regulation of nicotinic ACh receptor subunit mRNAs in the rat central and peripheral nervous systems. J Neurosci. 1995;15:1912–39.
84. Zheng JQ, Felder M, Connor JA, Poo MM. Turning of nerve growth cones induced by neurotransmitters. Nature. 1994;368:140–4.
85. Menelaou E, Svoboda KR. Secondary motoneurons in juvenile and adult zebrafish: axonal pathfinding errors caused by embryonic nicotine exposure. J Comp Neurol. 2009;512: 305–22.
86. Messi ML, Renganathan M, Grigorenko E, Delbono O. Activation of alpha7 nicotinic acetylcholine receptor promotes survival of spinal cord motoneurons. FEBS Lett. 1997;411:32–8.
87. Myers CP, Lewcock JW, Hanson MG, Gosgnach S, Aimone JB, Gage FH, et al. Cholinergic input is required during embryonic development to mediate proper assembly of spinal locomotor circuits. Neuron. 2005;46:37–49.
88. Rogers SW, Tvrdik P, Capecchi MR, Gahring LC. Prenatal ablation of nicotinic receptor alpha7 cell lineages produces lumbosacral spina bifida the severity of which is modified by choline and nicotine exposure. Am J Med Genet A. 2012;158A:1135–44.
89. Ren J, Qin C, Hu F, Tan J, Qiu L, Zhao S, et al. Habenula "cholinergic" neurons co-release glutamate and acetylcholine and activate postsynaptic neurons via distinct transmission modes. Neuron. 2011;69:445–52.
90. Bennett C, Arroyo S, Berns D, Hestrin S. Mechanisms generating dual-component nicotinic EPSCs in cortical interneurons. J Neurosci. 2012;32:17287–96.
91. Matsubayashi H, Alkondon M, Pereira EF, Swanson KL, Albuquerque EX. Strychnine: a potent competitive antagonist of alpha-bungarotoxin-sensitive nicotinic acetylcholine receptors in rat hippocampal neurons. J Pharmacol Exp Ther. 1998;284:904–13.
92. Bouzat C, Bartos M, Corradi J, Sine SM. The interface between extracellular and transmembrane domains of homomeric Cys-loop receptors governs open-channel lifetime and rate of desensitization. J Neurosci. 2008;28:7808–19.
93. Rexed B. A cytoarchitectonic atlas of the spinal cord in the cat. J Comp Neurol. 1954;100: 297–379.

Chapter 10
Slow Synaptic Transmission in the Central Nervous System

A. Rory McQuiston

Abstract New types of nicotinic connections with much slower kinetics than previously described have recently been discovered in several regions of the central nervous system (CNS). Slow nicotinic responses have been reported in the interpeduncular nucleus, hippocampus, neocortex, striatum, spinal cord, and thalamus. Studies have suggested that α4β2 containing nicotinic receptors may be the receptor subtype mediating the slow nicotinic synaptic events. These slow nicotinic excitatory postsynaptic potentials (EPSPs) appear to be the predominant nicotinic synaptic event in regions of the CNS (hippocampus and neocortex) that previously were thought to be dominated by α7 nicotinic receptors. Depending on the region of the CNS, slow nicotinic events may be mediated by classical synaptic transmission or volume transmission. Although the slow nicotinic EPSPs may vary subtly in kinetics and possible mechanisms of transmission, all known neuronal types that respond with slow nicotinic EPSPs are inhibitory. Thus slow nicotinic EPSPs may play an important role in controlling local and global network function through feedforward and feedback inhibition.

Keywords α4β2 nicotinic receptor • Volume transmission • Synaptic transmission • Optogenetics • Inhibitory interneuron • Acetylcholine

1 Introduction

During the late nineteenth and early twentieth century there was much debate on the mechanisms by which the nervous system processed and relayed information among different neural structures. It was not until 1921 that Otto Loewi showed that electrical activation of the vagus nerve resulted in the liberation of a chemical substance that produced a slowing of the heartbeat [1]. This substance was subsequently

A.R. McQuiston, Ph.D. (✉)
Department of Anatomy and Neurobiology, Virginia Commonwealth University, Box 980709, Richmond, VA 23298, USA
e-mail: amcquiston@vcu.edu

R.A.J. Lester (ed.), *Nicotinic Receptors*, The Receptors 26,
DOI 10.1007/978-1-4939-1167-7_10

identified as acetylcholine—the first known neurotransmitter [2]. Although this first demonstration of neurotransmitter action was modulatory in nature, subsequent studies of the action of acetylcholine at the neuromuscular junction were undertaken to investigate the general principles of synaptic transmission [3–5]. This early work indicated that fast responses could also be produced by synaptically released acetylcholine onto nicotinic receptors. In this chapter, we will examine synaptically mediated nicotinic responses in the central nervous system (CNS) and in particular nicotinic postsynaptic responses that are more than an order of magnitude slower than those observed at the neuromuscular junction.

Much is known about the synaptic activation of nicotinic receptors in the periphery. At the neuromuscular junction, both stimulated and spontaneous nicotinic postsynaptic responses have rapid kinetics with 1–2 ms rise times and only 3–4 ms half decay times [4, 5]. In autonomic ganglia, fast nicotinic synaptic transmission also occurs by the release of acetylcholine from preganglionic fibers. The kinetics of the nicotinic excitatory postsynaptic responses (EPSPs) in autonomic ganglia are somewhat slower with rise times of approximately 10–15 ms and decay times to half maximal amplitude of approximately 20–30 ms [6–8]. Both of these types of nicotinic synaptic transmission are capable of exciting their postsynaptic neurons on the time course of milliseconds and relay temporally important information regarding an organism's need to respond to sensory or visceral information.

Although there is an abundance of nicotinic receptors in the CNS, much less is known about their physiological activation and function at the synaptic level in the brain. Until recently, measurements of fast nicotinic excitatory synaptic events had remained elusive and had been detected in only a few areas of the CNS that receive significant cholinergic innervation and express different subsets of nicotinic receptors [9–16]. All of these studies demonstrated synaptic responses that were inhibited by α7 nicotinic receptors antagonists and had similar kinetics to the neuromuscular junction. However, despite the wide distribution of α4β2 subunit containing receptors (α4β2*) throughout the brain, none of these studies observed a nicotinic synaptic event that could be blocked by inhibition of these receptors. Because of the limited evidence for synaptic activation of several nicotinic receptor subtypes in the brain, it had been suggested that cholinergic activation of nicotinic receptors in the CNS was mainly modulatory rather than synaptic [17].

2 Technological Issues with Eliciting the Release of Acetylcholine in Brain Tissue

Measuring the impact of endogenously released acetylcholine has been limited by the available techniques used to stimulate release and record responses in intact nervous tissue. Most studies have used extracellular stimulating electrodes to excite fiber tracts while recording synaptic responses using intracellular electrophysiological techniques from select neurons [9–12, 15, 16]. However, these methods are limited when evaluating cholinergic synaptic transmission because of the diffuse projection of these axons in many areas of innervation. The stimulating electrode has to be positioned in close proximity to activate the cholinergic axon(s) that innervates the recorded neuron. Thus extracellular activation of cholinergic axons can be technically challenging and require many trials to detect a connection between presynaptic axons and postsynaptic neurons. Furthermore, because extracellular stimulation in brain tissue results in the release of other neurotransmitters, pharmacological methods (nonspecific receptor antagonists) are required to isolate cholinergic responses in individual neurons.

Recently, a transgenic mouse approach has been used to improve the success in finding cholinergic synaptic connections in mouse brain slices [12]. By expressing fluorescent proteins in cholinergic neurons, cholinergic axons can be identified near postsynaptic neurons in slices of brain tissue. A small glass pipette can then be placed next to the cholinergic axon for local stimulation and postsynaptic nicotinic responses recorded from adjacent neurons using whole cell patch clamp methods. While these approaches have provided important insights into which types of neurons respond to acetylcholine release by producing nicotinic EPSPs, these techniques are limited to individual cellular responses and provide little information on overall network function. Thus more rigorous and extensive studies of cholinergic synaptic transmission in the mammalian CNS have awaited a method to selectively activate cholinergic elements in intact neural tissue.

3 Optogenetics and Nicotinic Synaptic Transmission in the Mammalian Central Nervous System

The development of optogenetics during the first decade of the twenty-first century [18–22] permitted the selective control of specific neuronal subtypes in intact nervous tissue. This was done by expressing microbial opsins, light-gated ion channels or pumps, in select neuron populations. The expression of different microbial opsins (optogenetic proteins) produces depolarizations or hyperpolarizations through the illumination of specific wavelengths of light in cells expressing the opsin. Thus, neurons could be either excited to produce action potentials [18, 21] or prevented from producing action potentials [19, 20, 22] on a millisecond timescale.

To control acetylcholine release in the central nervous system, two strategies have been used to express the excitatory optogenetic proteins channelrhodopsin or oChIEF in cholinergic neurons. The first strategy used a transgenic mouse model in which channelrhodopsin expression was controlled by the choline acetyltransferase (Chat) promoter allowing cell-type-specific (i.e., cholinergic) activation via flashes of blue light [23, 24]. This animal model was used to investigate cholinergic inputs from the medial habenula to the interpeduncular nucleus (IPN) [23]. However, in this transgenic model, channelrhodopsin expression in cholinergic neurons in brain regions other than the habenula was low and light-driven acetylcholine release directly from axon terminals in other regions of the brain has not be observed or reported. Therefore, to examine nicotinic EPSPs in other regions of the CNS, a second strategy was used that combined adeno-associated viral (AAV) vectors and Crerecombinase-dependent expression of channelrhodopsin or oChIEF in cholinergic neurons

[25–30]. The coding sequence for the optogenetic protein was inserted into an AAV in reverse orientation and flanked by two pairs of incompatible lox sites [31, 32]. After viral infection, Crerecombinase *flips* the coding sequence and *excises* two incompatible lox (*FlEx*) sites locking the coding sequence in the correct orientation and allowing transcription and translation to proceed [33]. Using this strategy, investigators have been able to study the effect of acetylcholine release on neural network properties both in vivo and in brain slices [25, 26, 28–30]. Interestingly, some optogenetic studies have suggested that acetylcholine release may primarily produce slow α4β2*nicotinic receptor-mediated EPSPs in CNS neurons rather than the fast α7 EPSPs that were previously observed in electrical stimulation studies [25–27].

3.1 Slow Nicotinic Excitatory Postsynaptic Potentials in the Central Nervous System

As previously outlined above, the first study to investigate cholinergic synaptic transmission in the CNS using optogenetics examined inputs from the medial habenula to the interpeduncular nucleus (IPN) [23]. Surprisingly, the primary response in IPN neurons was a fast glutamatergic EPSP. However, following blockade of the ionotropic glutamatergic responses, small slow depolarizing current responses could be recorded from IPN neurons that were inhibited by the nonselective nicotinic receptor antagonists hexamethonium and mecamylamine. This slow nicotinic response could only be observed when habenular axon terminals were stimulated for prolonged periods of time at relatively high frequencies of blue light flashes (20 s at >20 Hz). The nicotinic currents did not display excitatory postsynaptic current-like waveforms and instead appeared as a slow ramp of postsynaptic inward current over 20 s. The time for the nicotinic current to decay following stimulation was approximately 5 s. Despite the requirement for prolonged activation of habenular axon terminals, the slow nicotinic depolarization appeared to contribute to action potential firing in IPN neurons ex vivo. The specific receptor subtypes mediating the slow nicotinic EPSP were not explored with subtype-specific antagonists and could have involved α2–6 and β2–4 subunits based on anatomical and expression studies [34, 35]. Interestingly, colocalization studies suggest that both glutamate and acetylcholine are found within the same habenular terminals. Thus, these studies suggest that cholinergic and glutamatergic synaptic transmission at habenular-interpeduncular synapses occur through varying synaptic mechanisms and over different temporal scales [23].

Optogenetics has also been used to study cholinergic synaptic transmission in the hippocampus [26, 29, 30]. In these experiments, excitatory optogenetic proteins oChIEF or channelrhodopsin were expressed in cholinergic neurons of the medial septum/diagonal band of Broca (MS/DBB) complex using AAV-mediated Cre-dependent expression in genetically modified mice. This permitted the use of brief blue light flashes to elicit the release of acetylcholine from MS/DBB terminals throughout hippocampal CA1. Studies that focused on nicotinic synaptic transmission in hippocampal CA1 produced unexpected findings [26, 29]. Instead of observing

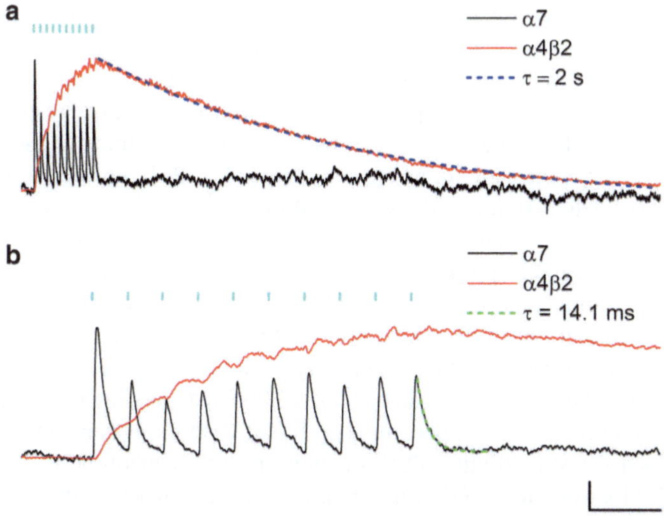

Fig. 10.1 Acetylcholine release produces α4β2* nicotinic receptor-mediated excitatory responses that have much slower kinetics than α7 nicotinic receptor-mediated synaptic responses in hippocampal CA1 interneurons. (**a**) Optogenetic ACh release (*blue lines*, ten blue light flashes (1 ms) at 50 ms intervals) activated α4β2* nicotinic receptors (*red trace*) that produced a much slower rise time and decay time constant (*dotted blue line*) relative to α7 synaptic responses (*black trace*) recorded from different CA1 interneurons. (**b**) Magnification (5×) of traces in (**a**) illustrates that the decay time constant of α7 responses (*green dotted line*) is >100-fold faster than α4β2* responses (*blue dotted line* in (**a**)). Scale bars: Vertical α7 = 0.5 mV, α4β2* = 3 mV; horizontal (**a**) = 500 ms, (**b**) = 100 ms. A.R. McQuiston unpublished observations

α7-mediated nicotinic EPSPs, slow nicotinic EPSPs not affected by α7 nicotinic receptor antagonists were detected in hippocampal CA1 interneurons [26]. These slow synaptic events had slow rise (33 ms) and decay time constants (138 ms) that were much slower than α7 nicotinic excitatory responses observed in electrical stimulation studies (Fig. 10.1) [9, 11]. The optogenetically stimulated nicotinic EPSPs were mediated by α4β2* nicotinic receptors, had small amplitudes (mean approximately 3 mV), and rarely produced action potentials [26]. The interneurons displaying α4β2* nicotinic EPSPs primarily had dendrites confined to the distal dendritic layers of hippocampal CA1 pyramidal neurons. Consistent with anatomical data, voltage-sensitive dye imaging showed that the slow α4β2* nicotinic responses were largest in the stratum lacunosum-moleculare of CA1 [26] where inputs from the entorhinal cortex and nucleus reuniens of the thalamus project. Importantly, the voltage-sensitive dye responses were specifically and completely blocked by α4β2* receptor antagonists (100 nM dihydro-beta-erythroidine DHβE) but not by other subtype-selective antagonists. Because the voltage-sensitive dyes bind to all membranes in the hippocampal slice, the membranes of all cells and processes in the slice will contribute to the voltage-sensitive dye signal. Therefore, any membrane capable of responding to acetylcholine release with a nicotinic receptor-mediated depolarization would contribute to the signal. Because the voltage-sensitive dye nicotinic signal was completely blocked by α4β2* nicotinic receptor antagonists and not α7 receptor antagonist, the most prevalent nicotinic receptor subtype that produces membrane depolarizations in hippocampal CA1 are α4β2* receptors that produce slow excitatory responses.

Optogenetic methods have also been used to study nicotinic synaptic transmission in the primary sensorimotor neocortex of mice [25, 27]. Slow nicotinic synaptic responses were detected in specific subsets of inhibitory interneurons of neocortical layers 1 and 2/3. Every interneuron examined in layer 1 and all interneurons in layer 2/3 that either had a late spiking electrophysiological phenotype or expressed Chat produced a nicotinic synaptic response [25]. In contrast, fast spiking interneurons never displayed nicotinic synaptic responses. Similar to findings in the hippocampus, interneurons in the neocortex displayed nicotinic synaptic events with slow rise times (35 ms) and decay time constants (176–234 ms, depending on the interneuron subtype). In contrast to the hippocampus, 50 % of Chat-expressing and late spiking interneurons had fast nicotinic EPSPs in addition to slow nicotinic EPSPs. The fast nicotinic EPSPs were mediated by α7 nicotinic receptors whereas the slow nicotinic EPSPs were inhibited by 500 nM DHβE suggesting that they may be mediated by α4β2* receptors. Importantly, the neocortical interneuron nicotinic EPSPs were frequently capable of producing barrages of action potentials [25, 27] unlike nicotinic responses in the hippocampus [26]. Therefore, like the hippocampus, specific subsets of interneurons primarily produce slow nicotinic EPSPs in response to acetylcholine release possibly via the activation of α4β2* nicotinic receptors.

Another region of the brain in which optogenetic methods were used to study nicotinic synaptic transmission is the dorsal striatum [28]. Using AAV-mediated and Cre-dependent expression of channelrhodopsin, striatal cholinergic interneurons were shown to produce slow nicotinic EPSPs in inhibitory interneurons of the striatum. One particular inhibitory interneuron subtype that received nicotinic synaptic inputs was the NPY-expressing neurogliaform cell. Striatal neurogliaform interneurons function by producing slow $GABA_A$-mediated IPSPs that suppress firing of projection neurons of the striatum, the medium spiny neurons [36]. The nicotinic EPSPs in the neurogliaform interneurons were slower than EPSPs in the periphery but faster than those measured in the IPN, hippocampus, and neocortex (rise time 17 ms and decay time constant 60 ms). However, like the hippocampus and neocortex, the nicotinic EPSPs could be inhibited by the selective antagonist DHβE at 100 nM, suggesting that the EPSP may be mediated by α4β2* receptors. Finally, single nicotinic EPSPs generated by a population of cholinergic interneurons were in some cases sufficient to generate action potentials in postsynaptic neurogliaform neurons and in turn produce large slow IPSPs in medium spiny neurons. Therefore, like the IPN, hippocampus, and neocortex, nicotinic EPSPs in the striatum may preferentially activate inhibitory neurons.

Following optogenetic studies that identified slow nicotinic synaptic transmission in the CNS, studies using electrical stimulation have isolated slow nicotinic EPSPs in the spinal cord and thalamus [37–39]. In neonatal mouse spinal cord, Renshaw cells receive multicomponent EPSCs with biphasic decay time constants

Table 10.1 Kinetics of slow synaptic nicotinic responses in the CNS

Region of CNS	Rise time (ms)	Decay time constant (ms)	Citation
Medial habenula	ND	>5,000 (vc)	[23]
Hippocampus	33 (cc)	138 (cc)	[26]
Neocortex	35 (vc)	190 (vc)	[27]
Striatum	17 (cc)	60 (cc)	[28]
Spinal cord	13 (vc)	132 (vc)	[38]
Thalamus	11 (vc)	124 (vc)	[39]

Kinetics recorded in voltage clamp (vc) or current clamp (cc)
ND not determined

[38]. Using minimal stimulation, slow nicotinic EPSCs could be isolated with rise times of 13 ms and decay time constants of 132 ms that could be blocked by high concentrations of DHβE [38]. Subsequent studies using nicotinic receptor knockout mice have suggested that the slow nicotinic EPSP in Renshaw cells was mediated by α4β2* and/or α2β4* receptor subtypes [37]. Thus, neurons in the spinal cord also have slow nicotinic EPSPs similar in decay kinetics to those observed in the hippocampus and neocortex but with rise times that were more similar to the striatum.

Slow nicotinic responses were also observed in neurons of the thalamic reticular nucleus (TRN) elicited by electrical stimulation and pharmacological isolation [39]. The kinetics of the thalamic nicotinic excitatory responses (rise time 11 ms and decay time constant 124 ms) were similar to the nicotinic responses observed in the spinal cord [38]. These slow nicotinic synaptic events were inhibited by 300 nM DHβE suggesting that they may be mediated by α4β2* nicotinic receptors similar to the hippocampus, neocortex, striatum, and spinal cord [25, 26, 28, 38]. Interestingly, the nicotinic EPSPs were capable of producing either single or bursts of action potentials depending on the resting membrane potential of the cell [39].

Although there are caveats in comparing voltage clamp data to current clamp data, it appears that the properties of slow nicotinic synaptic transmission in the CNS are not uniform between different cell types and regions of the CNS (Table 10.1). All regions displaying slow nicotinic synaptic responses had slower decay time constants than those measured either at the neuromuscular junction or in autonomic ganglia [5–8]. However, the rise times of the slow nicotinic events in the CNS can be divided into two broad categories. The first group had rise times shorter than 20 ms (striatum, spinal cord, and thalamus) and had kinetics similar to those measured in autonomic ganglia (10–15 ms) [6–8]. The second group had rise times longer than 30 ms (medial habenula, hippocampus, and cortex) and had kinetics more than twofold slower than nicotinic EPSPs measured in the periphery. Therefore, although most slow nicotinic EPSPs in the CNS may be mediated by α4β2* nicotinic receptors, it is likely that acetylcholine release that produces slow nicotinic EPSPs utilizes different transmission methods in different cell types and different regions of the CNS.

3.2 *Mechanisms Responsible for Kinetics of Slow Nicotinic Excitatory Postsynaptic Potentials*

A common feature of all the slow nicotinic EPSPs described in the previous section is their slow decay rate. However, the different rise times of the slow nicotinic EPSPs observed in different cell types and regions of the brain suggest that the slow nicotinic EPSPs may be a result of varying methods of chemical transmission. Some possible explanations include synaptic transmission versus volume transmission, the types of postsynaptic receptors, and the possible presence of accessory proteins.

Although striatal interneurons, TRN neurons, and Renshaw cells have relatively long nicotinic EPSP decay time constants, their rise times are not significantly different than those that occur in autonomic ganglia. Because autonomic preganglionic axons form traditional synapses with postsynaptic specializations, it is probable that cholinergic terminals in the striatum, TRN, and spinal cord release acetylcholine at a classical synapse where nicotinic receptors reside either across the synaptic cleft or perisynaptically from the release site [40]. However, it remains unclear what mediates the long decay rates of the nicotinic EPSPs. It is well known that nicotinic receptors have different kinetics of response [17, 38]. Although $\alpha4\beta2^*$ receptors likely mediate the slow nicotinic EPSPs in the striatum, TRN, and spinal cord, the precise makeup of the postsynaptic receptors remains unknown. Furthermore, it is unclear whether postsynaptic nicotinic receptor kinetics can be modulated by unknown accessory proteins as has been demonstrated for ionotropic glutamate receptors [41, 42]. If such accessory proteins do exist, they could dramatically slow the rise time and decay rates of nicotinic synaptic events.

Slow nicotinic EPSPs with rise times longer than 30 ms have been described in the IPN [23], hippocampus [26], and neocortex [25, 27]. To date there has been no description of a classical synapse that activates a ligand-gated ion channel with such slow rise time kinetics. Perhaps the most parsimonious explanation for the slow rise time is the requirement for acetylcholine to diffuse a significant distance to bind to nicotinic receptors. This process, called *volume transmission*, can either occur from synaptic spillover activating extrasynaptic receptors or release from synaptic varicosities that are not adjacent to a postsynaptic specialization [43]. Anatomical evidence supports the hypothesis that slow nicotinic synaptic transmission in the hippocampus and neocortex is mediated by volume transmission. Studies in rodent have suggested that only 7 % of cholinergic varicosities in the hippocampus and 15 % in the neocortex are opposite postsynaptic membrane specializations [44, 45]. However, others have reported that 66 % of cholinergic terminals in the neocortex form classical synapses with postsynaptic partners [46]. Furthermore, estimating the number of varicosities that form classical synapses is often underestimated when using electron microscopy with single ultrathin sections. This may result in the postsynaptic specialization being absent from that particular section. Therefore, the extent to which cholinergic terminals in the hippocampus and neocortex form classical synaptic connections remains unclear.

Recent physiological studies have added support for volume transmission as the mechanism mediating slow nicotinic EPSPs in the neocortex [27]. More specifically, inhibition of acetylcholinesterases (resulting in an increase in extracellular acetylcholine) has been shown to increase the amplitude and decay time constant of slow nicotinic responses in the neocortex without affecting the kinetics of the fast α7 responses [27]. In contrast, application of exogenous acetylcholinesterase (decreasing extracellular acetylcholine) decreased the amplitude and decay rate of neocortical slow nicotinic EPSCs, again without affecting the α7 response [27]. Together these data suggest that volume transmission is likely the method used for slow nicotinic EPSPs in the neocortex. What is unknown is whether volume transmission in the cortex (and possibly the hippocampus) occurs via synaptic spillover to extrasynaptic receptors or by terminal release of acetylcholine into the extracellular space lacking postsynaptic specializations.

Unfortunately, there are no studies to indicate the method by which slow nicotinic responses are transmitted in the IPN. However, it has been shown that acetylcholine and glutamate are both localized to the same synaptic terminals in the IPN. When these terminals are stimulated, IPN neurons respond with fast glutamatergic postsynaptic responses [23] consistent with a classical synaptic connection. However, slow nicotinic responses in the IPN are likely mediated by volume transmission considering their very slow rise times and decay time constants. Therefore, if acetylcholine is released from the same terminals as glutamate then the volume transmitted nicotinic responses must occur via synaptic spillover to extrasynaptic receptors.

In summary, the methods by which slow nicotinic responses are transmitted in the different regions of the CNS remains incompletely understood. Some nicotinic connections may transmit through classical synaptic connections whereas others likely work through volume transmission.

3.3 *Function of Slow Nicotinic Receptors in the Central Nervous System*

The function of slow nicotinic synaptic transmission in the CNS depends on the region and cell types that respond to acetylcholine release. However, all neurons that produce slow nicotinic responses inhibit their downstream target cells through the release of γ-aminobutyric acid (GABA) (medial habenula, hippocampus, neocortex, striatum, thalamus) or glycine (spinal cord). Thus, the ultimate effect of slow nicotinic EPSPs in various regions in the CNS may be to either inhibit or disinhibit local or global network activity.

In the IPN, the neurons responding with slow nicotinic excitatory events were not anatomically identified [23]. However, the IPN has a relatively large proportion of GABAergic neurons [47] that have been shown anatomically to receive cholinergic input [48]. In turn, the IPN projects to a number of limbic structures (e.g., hippocampus and entorhinal cortex) and neuromodulatory regions (ventral tegmental nucleus and raphe nuclei) suggesting that the slow nicotinic EPSPs in the IPN may play a role in controlling affective states and learning and memory [49].

In the hippocampus, no interneuron subtype was identified from post hoc anatomical reconstruction of neurons that displayed slow α4β2* nicotinic EPSPs [26].

However, more recent studies from our lab have shown that 78 % of interneurons that express vasoactive intestinal peptide (VIP) respond with slow α4β2* nicotinic EPSPs (unpublished observations). Furthermore, anatomical reconstruction of these interneurons showed morphologies consistent with interneuron-selective interneurons that exclusively innervate other inhibitory interneurons in CA1 [50]. Thus, slow nicotinic EPSPs may act to disinhibit the hippocampal CA1 network. Unpublished data from our lab supports this assertion because synaptic activation of α4β2* nicotinic receptors resulted in barrages of inhibitory postsynaptic currents (IPSCs) in interneurons but not in pyramidal neurons. However, other studies have suggested that hippocampal CA1 pyramidal neurons are directly activated by small fast α4β2* receptor-mediated synaptic responses [29], and synaptic activation of nicotinic receptors produced a barrage of IPSCs in pyramidal neurons [30]. Both of these studies did not electrophysiologically characterize the neurons from which the recordings were obtained nor were the neurons anatomically identified by post hoc reconstruction. Thus, this leaves open the possibility that the neurons with direct and indirect responses to synaptically activated α4β2* receptors were interneurons and not pyramidal cells. Therefore, the synaptic activation of α4β2* nicotinic receptors on pyramidal neurons and interneurons that innervate pyramidal cells requires further study.

Inhibitory interneurons also appear to be the primary target for synaptic activation of slow nicotinic responses in the neocortex [25, 27]. However, unlike the hippocampus, slow nicotinic responses could be observed in different subtypes of interneurons. In particular, slow nicotinic responses were observed in interneurons that innervated pyramidal neurons and other interneurons. However, a subclass of fast spiking interneurons was not among the subtype that displayed slow nicotinic EPSPs. Because fast spiking interneurons primarily inhibit the somata of pyramidal neurons, slow nicotinic responses may have a larger influence on integration within pyramidal cell dendritic trees and general disinhibition of the network in neocortical layers 1 through 3. However, these hypotheses remain to be tested. Furthermore, the presence or absence of slow nicotinic responses has not yet been investigated in deeper neocortical layers 4 through 6. Therefore, the effect of slow nicotinic EPSPs on neocortical function is likely to be more complicated.

The function of slow nicotinic EPSPs in the dorsal striatum is understood better than any other region of the CNS [28]. Local cholinergic interneurons of the striatum innervate inhibitory interneurons that can be excited by slow nicotinic EPSPs. These inhibitory interneurons in turn potently inhibit the principal cells of the striatum, the medium spiny neurons. In vivo studies have shown that when an animal is presented with an unexpected reward, or an expected reward is absent following its predictive cue, cholinergic interneurons briefly stop firing action potentials and often produce a burst of excitation following the pause in activity [51]. The rebound excitation following the pause of cholinergic interneuron firing can cause a disynaptic inhibition of medium spiny stellate cells that may be important for processing information of important valence [28].

Renshaw cells in the gray matter of the spinal cord are well known to be involved in controlling alpha motor neuron activity. Renshaw cells receive axon collaterals from motor neurons of the same motor unit and thus feedback to control motor unit activity. Motor neurons likely excite Renshaw cells in part through slow nicotinic EPSPs [37, 38] and thus aid in coordinating muscle contraction.

The TRN receives cholinergic input from tegmental nuclei of the brainstem. Stimulation of cholinergic inputs to the TRN produced slow nicotinic EPSPs that were capable of exciting postsynaptic TRN projection neurons [39]. TRN projection neurons are GABAergic and play crucial roles in generating behaviorally relevant rhythms such as sleep spindles [52]. Indeed, rhythmic activation of slow nicotinic EPSPs can entrain TRN neurons to rhythmically burst [39].

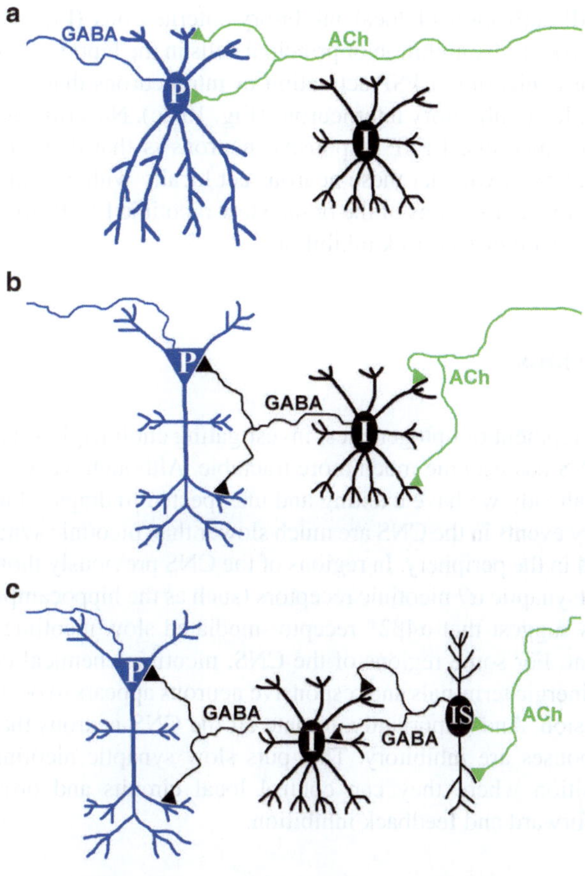

Fig. 10.2 Slow nicotinic excitatory postsynaptic potentials (EPSPs) affect central nervous system network activity through activation of different types of inhibitory neurons. (**a**) Cholinergic afferents (green axon) produce slow nicotinic EPSPs and activate GABAergic principal neurons (P, *blue* neuron) in the reticular nucleus of the thalamus and possibly the interpeduncular nucleus. (**b**) Cholinergic afferents or local cholinergic interneurons (*green* axons) produce slow nicotinic EPSPs in inhibitory GABAergic interneurons (I, *black* neuron) that in turn suppress principal neuron activity (P, *blue* neuron) of the neocortex and dorsal striatum. (**c**) Cholinergic afferents (*green* axons) produce slow nicotinic EPSPs in interneuron-selective interneurons (IS, *black* neuron) that suppress other inhibitory interneurons (I, *black* neuron) resulting in disinhibition of principal neurons (P, *blue* neuron) of the neocortex and hippocampus

Therefore, slow nicotinic EPSPs in TRN neurons may contribute to the generation of behaviorally relevant rhythms in the thalamus.

In summary, depending on the CNS region, slow nicotinic EPSPs can either have effects primarily on the local circuitry (hippocampus, neocortex, striatum, and spinal cord) or have a broader impact on behavioral states within the CNS (IPN and TRN). Global effects with broader impact occur through direct slow nicotinic EPSP activation of GABAergic projection or principal neurons of the TRN and possibly the IPN (Fig. 10.2a). In contrast, slow nicotinic EPSPs can affect local CNS circuitry through different mechanisms. One mechanism involves inhibiting the output and processing in principal cells of the neocortex and striatum through slow nicotinic EPSP activation of local inhibitory interneurons (Fig. 10.2b). Another mechanism involves disinhibition of principal cells in the hippocampus and neocortex through slow nicotinic EPSP activation of interneurons that are specialized to suppress other local inhibitory interneurons (Fig. 10.2c). Nevertheless, the common feature of slow nicotinic EPSP responsive neurons is that they are inhibitory in nature. Regardless of whether these neurons act locally within a single CNS structure or project to other regions of the brain, slow nicotinic EPSPs may act primarily to elicit feedforward or feedback inhibition.

4 Conclusions

With the development of optogenetics, investigating cholinergic synaptic transmission in the CNS has become much more tractable. Although we are just beginning such studies, already we have exciting and unexpected findings. Many of the nicotinic excitatory events in the CNS are much slower than nicotinic synapses that have been observed in the periphery. In regions of the CNS previously thought to primarily utilize postsynaptic α7 nicotinic receptors (such as the hippocampus and cortex), recent studies suggest that α4β2* receptor-mediated slow nicotinic responses are more prevalent. For some regions of the CNS, nicotinic chemical communication between cholinergic terminals and responsive neurons appears to occur through volume transmission. And importantly, to date all the CNS neurons that display slow nicotinic responses are inhibitory. This puts slow synaptic nicotinic EPSPs in a powerful position where they can control local circuits and broader networks through feedforward and feedback inhibition.

References

1. Loewi O. Uber humorale ubertragbarkeit der herznervvenwirkung I. Mitteilung. 2013;2013:239–42.
2. Dale HH, Dudley HW. The presence of histamine and acetylcholine in the spleen of the ox and the horse. J Physiol. 1929;68(2):97–123.

3. del Castillo J, del Katz B. Quantal components of the end-plate potential. J Physiol. 1954;124:560–73.
4. Fatt P, Katz B. Spontaneous subthreshold activity at motor nerve endings. J Physiol. 1952;117:109–28.
5. Fatt P, Katz B. An analysis of the end-plate potential recorded with an intracellular electrode. J Physiol. 1951;115:320–70.
6. Blackman JG, Ginsborg BL, Ray C. Synaptic transmission in the sympathetic ganglion of the frog. J Physiol. 1963;167:355–73.
7. Blackman JG, Ginsborg BL, Ray C. Spontaneous synaptic activity in sympathetic ganglion cells of the frog. J Physiol. 1963;167:389–401.
8. Nishi S, Koketsu K. Electrical properties and activities of single sympathetic neurons in frogs. J Cell Comp Physiol. 1960;55:15–30.
9. Alkondon M, Pereira EF, Albuquerque EX. Alpha-bungarotoxin- and methyllycaconitine-sensitive nicotinic receptors mediate fast synaptic transmission in interneurons of rat hippocampal slices. Brain Res. 1998;810:257–63.
10. Chu ZG, Zhou FM, Hablitz JJ. Nicotinic acetylcholine receptor-mediated synaptic potentials in rat neocortex. Brain Res. 2000;887:399–405.
11. Frazier CJ, Buhler AV, Weiner JL, Dunwiddie TV. Synaptic potentials mediated via alpha-bungarotoxin-sensitive nicotinic acetylcholine receptors in rat hippocampal interneurons. J Neurosci. 1998;18:8228–35.
12. Grybko MJ, Hahm ET, Perrine W, Parnes JA, Chick WS, Sharma G, Finger TE, Vijayaraghavan S. A transgenic mouse model reveals fast nicotinic transmission in hippocampal pyramidal neurons. Eur J Neurosci. 2011;33:1786–98.
13. Hatton GI, Yang QZ. Synaptic potentials mediated by alpha 7 nicotinic acetylcholine receptors in supraoptic nucleus. J Neurosci. 2002;22:29–37.
14. Leao RN, Mikulovic S, Leao KE, Munguba H, Gezelius H, Enjin A, Patra K, Eriksson A, Loew LM, Tort AB, Kullander K. OLM interneurons differentially modulate CA3 and entorhinal inputs to hippocampal CA1 neurons. Nat Neurosci. 2012;15:1524–30.
15. Roerig B, Nelson DA, Katz LC. Fast synaptic signaling by nicotinic acetylcholine and serotonin 5-HT3 receptors in developing visual cortex. J Neurosci. 1997;17:8353–62.
16. Stone TW. Kynurenic acid blocks nicotinic synaptic transmission to hippocampal interneurons in young rats. Eur J Neurosci. 2007;25:2656–65.
17. Dani JA, Bertrand D. Nicotinic acetylcholine receptors and nicotinic cholinergic mechanisms of the central nervous system. Annu Rev Pharmacol Toxicol. 2007;47:699–729.
18. Boyden ES, Zhang F, Bamberg E, Nagel G, Deisseroth K. Millisecond-timescale, genetically targeted optical control of neural activity. Nat Neurosci. 2005;8:1263–8.
19. Chow BY, Han X, Dobry AS, Qian X, Chuong AS, Li M, Henninger MA, Belfort GM, Lin Y, Monahan PE, Boyden ES. High-performance genetically targetable optical neural silencing by light-driven proton pumps. Nature. 2010;463:98–102.
20. Han X, Boyden ES. Multiple-color optical activation, silencing, and desynchronization of neural activity, with single-spike temporal resolution. PLoS One. 2007;2:e299.
21. Lin JY, Lin MZ, Steinbach P, Tsien RY. Characterization of engineered channelrhodopsin variants with improved properties and kinetics. Biophys J. 2009;96:1803–14.
22. Zhang F, Wang LP, Brauner M, Liewald JF, Kay K, Watzke N, Wood PG, Bamberg E, Nagel G, Gottschalk A, Deisseroth K. Multimodal fast optical interrogation of neural circuitry. Nature. 2007;446:633–9.
23. Ren J, Qin C, Hu F, Tan J, Qiu L, Zhao S, Feng G, Luo M. Habenula "cholinergic" neurons co-release glutamate and acetylcholine and activate postsynaptic neurons via distinct transmission modes. Neuron. 2011;69:445–52.
24. Zhao S, Ting JT, Atallah HE, Qiu L, Tan J, Gloss B, Augustine GJ, Deisseroth K, Luo M, Graybiel AM, Feng G. Cell type-specific channelrhodopsin-2 transgenic mice for optogenetic dissection of neural circuitry function. Nat Methods. 2011;8:745–52.

25. Arroyo S, Bennett C, Aziz D, Brown SP, Hestrin S. Prolonged disynaptic inhibition in the cortex mediated by slow, non-alpha7 nicotinic excitation of a specific subset of cortical interneurons. J Neurosci. 2012;32:3859–64.
26. Bell KA, Shim H, Chen CK, McQuiston AR. Nicotinic excitatory postsynaptic potentials in hippocampal CA1 interneurons are predominantly mediated by nicotinic receptors that contain alpha4 and beta2 subunits. Neuropharmacology. 2011;61:1379–88.
27. Bennett C, Arroyo S, Berns D, Hestrin S. Mechanisms generating dual-component nicotinic EPSCs in cortical interneurons. J Neurosci. 2012;32:17287–96.
28. English DF, Ibanez-Sandoval O, Stark E, Tecuapetla F, Buzsaki G, Deisseroth K, Tepper JM, Koos T. GABAergic circuits mediate the reinforcement-related signals of striatal cholinergic interneurons. Nat Neurosci. 2012;15:123–30.
29. Gu Z, Yakel JL. Timing-dependent septal cholinergic induction of dynamic hippocampal synaptic plasticity. Neuron. 2011;71:155–65.
30. Nagode DA, Tang AH, Karson MA, Klugmann M, Alger BE. Optogenetic release of ACh induces rhythmic bursts of perisomatic IPSCs in hippocampus. PLoS One. 2011;6:e27691.
31. Atasoy D, Aponte Y, Su HH, Sternson SM. A FLEX switch targets Channelrhodopsin-2 to multiple cell types for imaging and long-range circuit mapping. J Neurosci. 2008;28: 7025–30.
32. Schnutgen F, Doerflinger N, Calleja C, Wendling O, Chambon P, Ghyselinck NB. A directional strategy for monitoring Cre-mediated recombination at the cellular level in the mouse. Nat Biotechnol. 2003;21:562–5.
33. Rossi J, Balthasar N, Olson D, Scott M, Berglund E, Lee CE, Choi MJ, Lauzon D, Lowell BB, Elmquist JK. Melanocortin-4 receptors expressed by cholinergic neurons regulate energy balance and glucose homeostasis. Cell Metab. 2011;13:195–204.
34. Grady SR, Moretti M, Zoli M, Marks MJ, Zanardi A, Pucci L, Clementi F, Gotti C. Rodent habenulo-interpeduncular pathway expresses a large variety of uncommon nAChR subtypes, but only the alpha3beta4* and alpha3beta3beta4* subtypes mediate acetylcholine release. J Neurosci. 2009;29:2272–82.
35. Perry DC, Xiao Y, Nguyen HN, Musachio JL, Davila-Garcia MI, Kellar KJ. Measuring nicotinic receptors with characteristics of alpha4beta2, alpha3beta2 and alpha3beta4 subtypes in rat tissues by autoradiography. J Neurochem. 2002;82:468–81.
36. Ibanez-Sandoval O, Tecuapetla F, Unal B, Shah F, Koos T, Tepper JM. A novel functionally distinct subtype of striatal neuropeptide Y interneuron. J Neurosci. 2011;31:16757–69.
37. d'Incamps BL, Ascher P. Subunit composition and kinetics of the Renshaw cell heteromeric nicotinic receptors. Biochem Pharmacol. 2013;86(8):1114–21.
38. Lamotte B, Krejci E, Ascher P. Mechanisms shaping the slow nicotinic synaptic current at the motoneuron-renshaw cell synapse. J Neurosci. 2012;32:8413–23.
39. Sun YG, Pita-Almenar JD, Wu CS, Renger JJ, Uebele VN, Lu HC, Beierlein M. Biphasic cholinergic synaptic transmission controls action potential activity in thalamic reticular nucleus neurons. J Neurosci. 2013;33:2048–59.
40. Coggan JS, Bartol TM, Esquenazi E, Stiles JR, Lamont S, Martone ME, Berg DK, Ellisman MH, Sejnowski TJ. Evidence for ectopic neurotransmission at a neuronal synapse. Science. 2005;309:446–51.
41. Copits BA, Swanson GT. Dancing partners at the synapse: auxiliary subunits that shape kainate receptor function. Nat Rev Neurosci. 2012;13:675–86.
42. Straub C, Tomita S. The regulation of glutamate receptor trafficking and function by TARPs and other transmembrane auxiliary subunits. Curr Opin Neurobiol. 2012;22:488–95.
43. Sykova E, Vargova L. Extrasynaptic transmission and the diffusion parameters of the extracellular space. Neurochem Int. 2008;52:5–13.
44. Umbriaco D, Watkins KC, Descarries L, Cozzari C, Hartman BK. Ultrastructural and morphometric features of the acetylcholine innervation in adult rat parietal cortex: an electron microscopic study in serial sections. J Comp Neurol. 1994;348:351–73.

45. Umbriaco D, Garcia S, Beaulieu C, Descarries L. Relational features of acetylcholine, noradrenaline, serotonin and GABA axon terminals in the stratum radiatum of adult rat hippocampus (CA1). Hippocampus. 1995;5:605–20.
46. Turrini P, Casu MA, Wong TP, De KY, Ribeiro-da-Silva A, Cuello AC. Cholinergic nerve terminals establish classical synapses in the rat cerebral cortex: synaptic pattern and age-related atrophy. Neuroscience. 2001;105:277–85.
47. Ferraguti F, Zoli M, Aronsson M, Agnati LF, Goldstein M, Filer D, Fuxe K. Distribution of glutamic acid decarboxylase messenger RNA-containing nerve cell populations of the male rat brain. J Chem Neuroanat. 1990;3:377–96.
48. Kawaja MD, Flumerfelt BA, Hrycyshyn AW. Simultaneous demonstration of choline acetyltransferase and glutamic acid decarboxylase immunoreactivity in the rat interpeduncular nucleus. J Chem Neuroanat. 1990;3:165–77.
49. Klemm WR. Habenular and interpeduncularis nuclei: shared components in multiple-function networks. Med Sci Monit. 2004;10:RA261–73.
50. Acsady L, Gorcs TJ, Freund TF. Different populations of vasoactive intestinal polypeptide-immunoreactive interneurons are specialized to control pyramidal cells or interneurons in the hippocampus. Neuroscience. 1996;73:317–34.
51. Apicella P. Leading tonically active neurons of the striatum from reward detection to context recognition. Trends Neurosci. 2007;30:299–306.
52. Fuentealba P, Steriade M. The reticular nucleus revisited: intrinsic and network properties of a thalamic pacemaker. Prog Neurobiol. 2005;75:125–41.

Chapter 11
The Effects of Nicotine on Learning and Memory

Thomas J. Gould

Abstract Acetylcholine is involved in cognitive processes, and the ability of nicotine to modulate nicotinic acetylcholinergic receptor function contributes to the effects of tobacco on learning and other cognitive processes. The capacity of nicotine to alter learning and cognition may facilitate the development and maintenance of nicotine addiction. Nicotine may enhance the formation of strong but yet maladaptive drug-stimuli associations that can trigger cravings and drug-seeking behavior upon reexposure to the stimuli. In addition, deficits in cognition during periods of abstinence may contribute to smoking relapse. Nicotine may initially alter cell-signaling cascades involved in learning and synaptic plasticity to facilitate cognition, but continued use may lead to compensatory adaptations in nicotinic receptor function, such as receptor desensitization and upregulation that contribute to tolerance and withdrawal deficits in cognition. The effects of nicotine on cognition are influenced by the brain regions involved in the cognitive tasks and by both genetics and developmental stage.

Keywords Acetylcholine • Learning • Memory • Hippocampus • Plasticity • Addiction • Nicotine • Smoking

1 Introduction

Acetylcholine is attributed as the first neurotransmitter identified, largely through the research of Henry Dale and Otto Loewi [1, 2]. Their pioneering work focused on the ability of acetylcholine, originally called Vagusstoff (or vagus stuff) by Loewi and later identified to be the same acetyl derivative of choline that Dale was researching, to modulate cardiovascular function [3]. Since these studies revolutionize physiology and pharmacology and garnered Dale and Loewi the 1936 Nobel Prize in Physiology or Medicine, acetylcholine has been identified to be involved in multiple processes including cognition.

T.J. Gould, Ph.D. (✉)
Department of Psychology and Neuroscience, Temple University,
1701 North 13th Street, Weiss Hall, Philadelphia, PA 19122, USA
e-mail: tgould@temple.edu

R.A.J. Lester (ed.), *Nicotinic Receptors*, The Receptors 26,
DOI 10.1007/978-1-4939-1167-7_11

Numerous studies have identified a wide variety of cognitive processes that acetylcholine is essentially involved with or that are modulated by acetylcholine [4–6]. Some of those processes include spatial learning and working memory [7–9]. In fact, the role of acetylcholine in cognition is so important that Perry and colleagues [10] suggested that acetylcholine may be the basis of consciousness. Acetylcholine binds to two broad classes of receptors, muscarinic and nicotinic. Both receptors are involved in cognitive processes but differ across multiple dimensions. Muscarinic acetylcholinergic receptors are metabotropic receptors that are critically involved in cognitive processes such as long-term memory formation and retrieval [11–13], whereas nicotinic acetylcholinergic receptors (nAChRs) are ionotropic receptors that may be critically involved in some cognitive processes but may more often modulate cognitive processes [14]. While acetylcholine is involved in numerous cognitive processes, it is beyond the scope of this review to cover all of them and instead the review will focus on the role of nAChRs in learning and memory.

Even though nAChRs are largely ubiquitous throughout the central nervous system, they differ in location on neural processes and subtype localization. Nicotinic acetylcholinergic receptors are pentameric ligand-gated receptors composed of α2, α3, α4, α5, α6, α7, α8, α9, α10, β2, β3, or β4 subunits (discussed in greater detail in other chapters of this book). The receptors can be either homomeric or heteromeric with α7 nAChRs and α4β2* nAChRs (* designates potential additional subunit) two of the most common types of nAChRs [15–20]. Both α7 and α4β2* nAChRs are commonly found in areas involved in learning and memory but can differ in the type of neuron they are localized to and location of the nAChR on the neuron (e.g., presynaptic versus postsynaptic location). For example, in the hippocampus, an area involved in declarative learning and memory processes [21, 22], α7 nAChRs and α4β2*nAChRs differentially modulate GABAergic input to CA1 [23], and another study found that CA1 interneurons differed in the types of nAChRs they express [24]. These studies suggest different cellular locations for different nAChR subtypes.

Presynaptic and postsynaptic nAChRs may have vastly different effects on learning and memory. Presynaptic nAChRs have been shown to facilitate the release of a wide range of neurotransmitters, which include acetylcholine, serotonin, dopamine, norepinephrine, GABA, and glutamate [25–32]. The ability of nAChRs to regulate neurotransmitter release may contribute to the ability of nicotine and acetylcholine to modulate learning processes. For example, nAChR-stimulated release of glutamate or norepinephrine during learning processes that involve glutamate and/or norepinephrine could lead to greater synaptic activation and associated plasticity resulting in a stronger memory. Postsynaptic nAChRs could also facilitate ongoing processes or could directly activate cell-signaling cascades involved in learning and memory. For example, NMDA receptors are involved in both learning and synaptic plasticity as measured by long-term potentiation (LTP), and this is thought to occur by NMDA receptor-mediated gating of calcium influx [33, 34]. Similar to NMDA receptors, nAChRs can also gate calcium and can contribute to internal calcium release [35, 36]. Thus, nAChR-mediated calcium influx may facilitate calcium cell-signaling processes involved in learning and synaptic plasticity or directly activate them. In support, behavioral studies have shown that during learning, NMDA receptors and nAChRs mediate similar processes [37, 38], and LTP studies have shown that nicotine can directly induce LTP [39–41].

Learning is not a singular process but instead multiple types of learning exist that involve different brain regions and cell-signaling processes. Therefore, the assumption cannot be made that nicotinic agonists will similarly affect all types of learning. In addition, patterns of nicotine administration may produce variability in the effects of nicotine on learning, with acute effects of nicotinic agonists not necessarily identical to the effects of chronic nicotine on learning and withdrawal effects potentially being different than both acute and chronic effects. Thus, the following sections will review the effects of nicotine on various types of learning, whether those effects change as nicotine administration transitions from acute to chronic to withdrawal from chronic nicotine treatment, and when possible the underlying neural substrates.

Multiple studies suggest that learning processes that involve the hippocampus may be sensitive to the effects of nicotinic agonists [42–44]. The hippocampus is well suited to facilitate the formation and long-term storage of associations that involve multiple information streams and the processing of temporal information. The hippocampus receives input from multiple cortical regions via the entorhinal cortex and projects back to the cortex. In the hippocampus, information flows from the dentate gyrus to CA3 and then to CA1, which send efferents to the entorhinal cortex [45, 46]. This trisynaptic circuitry is strengthened via direct entorhinal cortical afferents to each subdivision. The trisynaptic circuitry, with redundant entorhinal cortical input, may allow the hippocampus to produce long-term spatial and contextual memories. The hippocampus also contains reverberating circuits in the CA3 region that may allow the hippocampus to maintain memory traces of stimuli after the stimuli are no longer present [47]. The primary focus of this chapter will be on the effects of nicotine on learning processes that involve the hippocampus, such as contextual learning, spatial learning, and working memory. Other learning processes will also be examined, often as comparisons with hippocampus-dependent processes. Themes reoccurring throughout the chapter include: (1) the effects of nicotine on learning are not universal; (2) the effects vary as administration duration varies such that, although not always, acute nicotine produces enhancement whereas withdrawal from chronic nicotine produces deficits; (3) the effects are influenced by multiple factors that include age and genetics; and (4) in general there are limited effects on baseline learning associated with nAChR antagonists and nAChR subunit KO mice. The majority of the review will focus on laboratory animal studies followed by a summary of human studies.

2 Fear Conditioning

Fear conditioning is a form of Pavlovian classical conditioning in which an aversive stimulus (unconditioned stimulus, US) is presented either during or after presentation of an initially neutral stimulus (conditioned stimulus, CS). If the CS becomes associated with the US, the CS will evoke a conditioned response (CR) that is often similar to the unconditioned response (UR) evoked by the US. In fear conditioning, an auditory stimulus such as a tone or white noise is often the CS and a mild foot-shock is often the US. In addition to forming an association between the auditory CS and the foot-shock US (cue conditioning), subjects can also form an association between the training context and the foot-shock US. Thus, during fear conditioning, two types of learning occur, cue and context fear conditioning, and these types of learning involve different brain substrates. Context fear conditioning involves the hippocampus and the amygdala (among other brain regions), whereas for cue fear conditioning, the hippocampus is not critically involved but the amygdala is [48–50]. While standard cue fear conditioning does not critically involve the hippocampus, cue fear conditioning can be changed to a form of learning that is hippocampus dependent, trace fear conditioning. In trace fear conditioning, a temporal delay is inserted between the offset of the auditory CS and the onset of the US. This delay necessitates that a memory trace of the CS must remain viable in the brain in order to form an association with the US. The hippocampus and frontal cortex may be involved in maintenance of the memory trace, a process that has similarities to working memory [51, 52]. The effects of a drug on these three forms of fear conditioning can suggest what brain areas the drug may be altering. For example, if a drug enhances cue, context, and trace fear conditioning, the drug may be acting in the amygdala, but if a drug only enhances context and trace fear conditioning, the drug may be acting in the hippocampus.

Acute nicotine has been consistently shown to enhance context and trace fear conditioning, but not cue fear conditioning in mice [44, 53–56]; however, one study in Wistar rats reported nicotine-associated deficits in context fear conditioning [57], while another study in Sprague-Dawley found that nicotine enhanced context memory reconsolidation [58]. Because both trace and context fear conditioning, but not cue fear conditioning, involve the hippocampus, this suggests that nicotine is acting in the hippocampus to enhance learning, and direct drug infusion studies have confirmed this. Context fear conditioning but not cue fear conditioning was enhanced by nicotine infusion into the hippocampus [59], and interestingly, this effect was specific for the dorsal hippocampus. Whereas infusion of nicotine into the dorsal hippocampus enhanced context fear conditioning, infusion into the ventral hippocampus disrupted context fear conditioning [60]. This suggests that the effects of nicotine in the dorsal and ventral hippocampus are in competition for behavioral outcome and that the dorsal hippocampus and ventral hippocampus are discrete brain regions and not divisions of a functionally homogenous hippocampus. Increasing evidence indicates that the dorsal and ventral hippocampus are functionally, physiologically, and even genetically distinct areas; for review, see [61]. The dorsal hippocampus may be more involved in contextual and spatial information processing while the ventral hippocampus may be more involved in anxiety [61].

Direct drug infusion studies suggest that while trace fear conditioning may be more similar to context fear conditioning than cue fear conditioning in regard to the brain regions involved, trace fear conditioning is distinct from context fear conditioning. Similar to context fear conditioning, infusion of acute nicotine into the dorsal hippocampus enhanced trace fear conditioning and infusion into the ventral disrupted trace fear conditioning [62]. Trace fear conditioning and context fear conditioning, however, were differently affected by infusion of nicotine into the medial prefrontal cortex; nicotine infusion enhanced trace but not context or cue fear conditioning. These results are in line with numerous studies demonstrating a critical involvement of the frontal cortex in trace fear conditioning [63, 64].

Although the hippocampus and the frontal cortex are rich in both of the two predominant nAChRs of the CNS, α7 and α4β2*, the effects of acute nicotine on context and trace fear conditioning may be largely mediated by α4β2* nAChRs. Both pharmacological and genetic studies suggest that α4β2* nAChRs are essential for the acute effects of nicotine on context fear conditioning but are largely nonessential for context or cue fear conditioning in the absence of nicotine. Alone, the general nAChR antagonist mecamylamine had no effect on context or cue fear conditioning but did block nicotine enhancement of context fear conditioning [44, 54]. Similarly, systemic administration of dihydro-beta-erythroidine (DHβE, an antagonist of high-affinity nAChRs with highest selectivity for α4β2 nAChRs [65]) did not disrupt baseline context or cue fear conditioning but did prevent enhancement of context fear conditioning by acute systemic nicotine; the α7 nAChR antagonist methyllycaconitine (MLA) [66, 67] had no effect [68].

Studies with nAChR subunit-selective knockout (KO) mice also suggest that nAChR may in large not be critical for contextual fear conditioning but may modulate it. The α7 nAChR does not appear to be essentially involved in context or cue fear conditioning or the enhancement of contextual fear conditioning by nicotine as neither learning nor the effects of nicotine on learning were altered in α7 KO mice [69–71]. Knockout studies suggest that neither the β3 nAChR subunit nor the β4 nAChR subunit is not involved in contextual fear conditioning or the enhancement of contextual fear conditioning by nicotine [71]. However, another study found a sex difference in the effects of the β4 KO on cue fear conditioning with male β4 KO showing a deficit [72]. The α2 nAChR subunit overall does not appear to be involved in context fear conditioning or nicotine effects on it, but female α2 KO mice may have lower levels of cue fear conditioning, especially when compared to wild-type mice given with nicotine [73]. The β2 nAChR subunit is critically involved in the enhancement of context fear conditioning by acute nicotine [69, 71], but in young mice, this subunit may not be critically involved in context or cue fear conditioning in the absence of drug [69, 74], though another study found decreased context fear conditioning in β2 KO mice [71]. Age may increase sensitivity to the effects of altered β2 nAChR subunit function on learning as aged β2 KO mice had deficits in both context and cue fear conditioning [74]. Overall, these studies strongly suggest that β2-containing nAChRs are critically involved in the enhancement of contextual fear conditioning by acute nicotine and that nAChRs may not be essential for contextual or cue fear conditioning in most cases, but age may contribute to variability in the effects of the β2 subunit on fear conditioning. The KO studies demonstrate how genetic differences can lead to phenotypic differences; this is further highlighted by a study that demonstrated the influence of background genotype of inbred mice on the effects of acute nicotine and nicotine withdrawal on context fear conditioning [75], which suggests that comparisons across KO studies should consider the genetic background and the number of backcrosses of the KO mice (i.e., the background genetic variability). In addition, even within the same genetic background, the effects of nicotine change with age as young C57BL/6 mice were more sensitive to the effects of acute nicotine on context fear conditioning than adult C57BL/6 mice [76].

Fewer studies have examined the nAChR receptor subtypes involved in trace fear conditioning, but those studies largely point to a similar involvement of β2 nAChRs in the enhancement of trace fear conditioning by acute nicotine. Studies in KO mice suggest that α2, α7, and β2 nAChR subunits are not needed for normal trace fear conditioning [69, 73] but that the β2 subunit is critically involved in the enhancement of trace fear conditioning by acute nicotine [69]. The effects of direct infusion of the high-affinity nAChR antagonist DHβE and the α7 nAChR antagonist MLA suggest a more complex pattern of nAChR involvement in trace fear conditioning in the absence of nicotine. Infusion of DHβE into the dorsal hippocampus disrupted trace, but not context, fear conditioning, MLA infusion had no effect on either type of fear conditioning, and infusion of either drug into the ventral hippocampus had no effect [62]. Infusion of either DHβE or MLA into the medial prefrontal cortex enhanced trace but not context fear conditioning. These results demonstrate that trace and context fear conditioning are different neurobiological processes involving different brain regions and nAChRs and that nicotine may enhance trace fear conditioning through activation nAChRs in the dorsal hippocampus and desensitizing nAChRs in the medial prefrontal cortex. Differences in results between the antagonist infusion study, which showed involvement of nAChRs in trace fear conditioning, and the KO study, which suggested nAChRs are not critically involved in trace fear conditioning, could be due to differences between transiently disrupting function with an antagonist and permanently knocking-out function and/or selectivity of antagonists versus genetic knockout techniques.

The genetic and pharmacology studies largely suggest that nicotine is modulating hippocampus-dependent context learning and that nAChRs are not necessary for learning to occur under most circumstances. Studies examining the cell-signaling cascades involved in the nicotine enhancement of context fear conditioning support this. Both nAChR and NMDA receptors can gate calcium influx, which is critically involved in synaptic plasticity and learning [34, 77, 78]. Studies have shown that during context fear conditioning, NMDA receptors and nAChRs mediate similar processes and may interact functionally [38, 79]. This may in part explain why many studies have not found a critical involvement of nAChRs in context fear conditioning; that is, disrupting nAChR function could have minimal effect on learning because NMDA receptors can compensate, but if NDMA receptor function is compromised, as may occur with aging [80], deficits might emerge with decrements in nAChR function.

While nicotine may interact with NMDA receptor-mediated processes to enhance context fear conditioning, it also appears that nicotine activates cell-signaling pathways that normally are not involved in learning and that these pathways are critically involved in the enhancement of the hippocampus-dependent learning by nicotine. Acute nicotine enhancement of context fear conditioning produced a different pattern of gene expression in the hippocampus than nicotine administration alone or context fear conditioning without nicotine. Nicotine and learning interacted to increase expression of *Jnk1* in the hippocampus. This change in *Jnk1* expression was mediated by β2-containing nAChRs and involved CREB binding to the *Jnk1* promoter region in the hippocampus [81, 82]. Calcium-associated cell-signaling cascades can activate CREB [83], and thus, it is possible that increased intracellular calcium concentrations mediated by nAChRs led to CREB activation of *Jnk1*. Inhibition of the JNK1 protein during the consolidation phase of context fear conditioning disrupted the nicotine enhancement of the hippocampus-dependent learning [81]. JNK1 phosphorylates microtubule-associated proteins [84] and JNK1 activates transcription factors, such as the JUN family, ATF-2, and ELK-1 [85, 86], processes that could modify and strengthen synaptic signaling.

The effects of nicotine on context and trace fear conditioning change with chronic treatment and subsequent withdrawal of chronic treatment. Chronic nicotine administration produces tolerance for the cognitive enhancing effects of nicotine on trace and context fear conditioning in mice and cessation of treatment results in deficits in both context and trace fear conditioning but not cue fear conditioning [87–91]. The effects of nicotine withdrawal on learning are sensitive to age with younger mice expressing less withdrawal deficits in learning than older mice [76]. Similar to acute nicotine, the withdrawal effects involve β2-containing nicotinic receptors in the hippocampus. Cessation of chronic nicotine infused directly into the dorsal hippocampus produced deficits in context fear conditioning, but similar infusions into the cortex or thalamus had no effect [92]. In addition, this study also found that acute infusion of DHβE, but not MLA, into the dorsal hippocampus of mice treated chronically with systemic nicotine precipitated withdrawal deficits in context fear conditioning. In another study, α7 KO mice showed normal nicotine withdrawal deficits in context fear conditioning, but withdrawal effects in context fear conditioning were absent in β2 KO mice, suggesting that chronic nicotine has to act on β2-containing receptors for withdrawal deficits in hippocampus-dependent learning to emerge [93]. Similarly, for trace fear conditioning, systemic DHβE precipitated withdrawal deficits in trace fear conditioning, and spontaneous nicotine withdrawal did not produce trace fear conditioning deficits in β2 KO mice; MLA did not precipitate withdrawal deficits in learning [90].

The neurobiological change underlying the withdrawal deficit in hippocampus-dependent learning is unknown, but chronic nicotine is associated with both desensitization and upregulation of nAChRs [94–96] and this may contribute to withdrawal deficits. A study that examined the duration of nicotine withdrawal-associated deficits in context fear conditioning and the duration of nAChR upregulation in the cortex, hippocampus, and cerebellum found that the duration of the withdrawal deficits in learning paralleled the duration of high-affinity nAChR upregulation in the hippocampus but not the cortex or cerebellum [97]. In addition, lines of mice that did not develop withdrawal deficits in hippocampus-dependent learning also did not show nAChR upregulation in the hippocampus [98]. This study also further demonstrated differences in the effects of nicotine between the dorsal and the ventral hippocampus with upregulation of nAChRs found in the dorsal hippocampus of mice that showed withdrawal deficits, but no change in ventral hippocampal nAChRs was seen in the same mice. Finally, in young mice that did not show withdrawal deficits in hippocampus-dependent learning, hippocampal nAChR upregulation was not observed [76]. Thus, nAChR upregulation in the hippocampus may be necessary for learning-related withdrawal deficits to emerge. It has been theorized that during nicotine withdrawal, upregulated nAChRs resensitize and that this increase in cholinergic tone contributes to withdrawal deficits [99]. In support, a persistent increase in CA1 pyramidal cell excitability was recorded after withdrawal from chronic nicotine [100], and a recent study suggests that during withdrawal, there may be increased sensitivity to the effects of nicotine on learning [101]. This may explain why varenicline and bupropion are able to ameliorate nicotine withdrawal deficits in hippocampus-dependent learning [102, 103]. Both varenicline, a partial $\alpha 4\beta 2$ nAChR agonist [104], and bupropion, a norepinephrine reuptake inhibitor that is also a $\alpha 4\beta 2$ nAChR antagonist [105], could dampen cholinergic signaling [106].

3 Object Recognition

Object recognition is a form of incidental or exploratory learning in which mice and rats will demonstrate memory for a previously encountered object by decreased exploration of the familiar object and increased exploration of the novel object. Two types of object recognition are commonly used, novel object recognition and spatial object recognition. In novel object recognition, subjects explore an arena that contains two unique objects; at a later time point, the subjects are returned to the arena but one of the original objects is replaced by a novel object. If subjects remember the original objects, they should explore the novel object more. Spatial object recognition is similar to novel object recognition with two differences: the arena is surrounded by external spatial cues and on test day the original objects remain but one is moved to a new spatial location. If subjects remember the spatial position of the objects, they should explore the displaced object more. Spatial object recognition critically involves the hippocampus whereas novel object recognition does not [107, 108]. Similar to results from fear conditioning studies, acute nicotine enhanced the hippocampus-dependent spatial object recognition; however, nicotine also disrupted the hippocampus-independent novel object learning [109]. The same study found that withdrawal from chronic nicotine disrupted spatial but not novel object recognition. These results further support the contention that acute nicotine preferentially modulates hippocampus-dependent learning. However, in cases where there are deficits in baseline novel object recognition learning, nicotine is able to switch the conditions from a non-learning situation to a learning situation [110, 111]. This may occur through recruitment of the hippocampus as direct infusion of nicotine into the perirhinal cortex and hippocampus changed a training procedure that did not produce object recognition to one that resulted in learning [112].

4 Spatial Learning: Morris Water Maze and Barnes Maze

Spatial learning requires subjects to integrate information about environmental stimuli and the relationship between themselves, the environmental cues, and the behavioral goal or object. The necessity to form multimodal associations engages the hippocampus, and in fact the hippocampus has cells that are specifically tuned to spatial locations [113, 114]. In addition to spatial object recognition, other tests of spatial learning exist that include the Morris water maze and the Barnes maze. In the Morris water maze, rodents must find a platform submerged in opaque water; environmental cues surround the pool and can be used to form efficient paths to the platform over trials [115]. Similarly, the goal of the Barnes maze is to find an escape from the maze. The Barnes maze consists of a circular platform with evenly spaced holes along the circumference with one of those holes providing an escape pathway. The maze, which is usually white and in a well-lit environment, takes advantage of rodents' aversion to open and exposed areas as subjects will use external spatial cues to find and remember the exit location across trials [116].

Limited studies have examined nicotine and nAChR function using the Barnes maze. Those that did found that male β4 KO mice tended to use spatial strategies less often to solve the Barnes maze [72], and another study reported no disruption in the Barnes maze in α7 KO mice [117]. Similarly, Morris water maze performance was normal in α7 KO mice [70]. Results from studies of the effects of acute nicotine on Morris water maze learning are variable. Two studies using different time courses for drug administration found enhanced learning in the Morris water maze with acute nicotine administration [118, 119]; however, one study found no effect [120] and another found a deficit in learning with acute nicotine administration [121]. These studies are difficult to compare because of differences across studies in the number of doses used, the doses used in single-dose studies, and the species and strains tested; all of these factors can influence results.

Multiple studies have examined the effects of chronic nicotine on Morris water maze learning and have also reported varying results. A study comparing young and aged rats reported that once a day injections of nicotine for 3 days improved acquisition in the aged rats and memory retention in the young rats [122]. Another study found that twice a day injections of nicotine for 10 days enhanced learning the Morris water maze in rats [118], and a study in mice similarly found that once daily injections of nicotine starting 5 days before training and continuing for 4 days during training enhanced learning [121]. However, a study using minipumps to chronically deliver nicotine for 10 days found a deficit in acquisition and retention in rats [123]. An important difference exists between the Scerri and colleagues study that found a learning deficit and the other three studies that found enhancement; the three studies that found enhancement used repeated injections to administer nicotine, whereas in the Scerri study, nicotine was continually administered. This is an issue because the half-life of nicotine in rodents is substantially shorter than in humans [124], and thus, it is not clear whether a single or even twice daily injection of nicotine in rodents models human chronic tobacco use or reflects more acute use. Thus, the studies that found enhancement of Morris water maze learning may be more reflective of the effects of acute nicotine on spatial learning.

5 Radial Arm Maze

The 8-arm radial maze is a spatial learning task that assesses both reference memory and working memory [125]. Reference memory is tested when a subset of arms is consistently baited across sessions and rodents must remember the location of the baited arms across sessions. Working memory is equated to the ability of rodents to remember within a session which arms they have already extracted the food reward from. Studies suggest that working memory may depend more on the hippocampus than reference memory [126, 127]. Nicotine modulates working memory in rodents but the effects can vary. Multiple studies have shown that acute nicotine enhances working memory [37, 128–132]. However, one study found that age may mediate some of the effects of acute nicotine on working memory as enhancement was seen in aged but not young rats [133] and two studies found no effect on working memory [134, 135].

With chronic nicotine treatment, the majority of studies in rats report enhanced working memory. Some of these studies suggest that the effects of chronic nicotine on working memory may be long-lasting. Two studies found that chronic nicotine treatment enhanced working memory and that this enhancement remained up to 2 weeks after cessation of nicotine treatment [136, 137]. The enhancement of working memory by chronic treatment may be affected by age as enhancement was seen in young but not aged rats [133]. Another study examined working memory after cessation of chronic treatment and also found enhanced working memory [138]. However, two studies found that enhancement of working memory was only present during chronic nicotine treatment [139, 140] and two studies found no effect of chronic treatment [141, 142]. These studies were conducted in Sprague-Dawley rats and suggest that at least for this strain of rat, withdrawal from chronic nicotine does not disrupt spatial working memory.

Multiple studies have examined the neural substrates underlying the effects of nicotine on working memory. Both α7 and α4β2* nAChRs may be involved in working memory but α4β2* nAChRs may play a more significant role. Infusion of the α7 nAChR antagonist MLA or the α4β2* nAChR antagonist DHβE into the ventral hippocampus both disrupted working memory [143], but in another study, chronic infusion of DHβE, but not MLA, into the ventral hippocampus disrupted working memory [144]. In addition, systemic administration of an α7 agonist improved working memory in primates tested at intermediate doses but disrupted working memory at doses above and below [145]. The effects of chronic nicotine on working memory may be through high-affinity nAChRs as systemic chronic nicotine reversed deficits in working memory associate with infusion of DHβE, but not MLA, into the ventral hippocampus [141, 142]. Thus, similar to other hippocampus-dependent learning processes [146, 147], nicotine may modulate working memory through α4β2 nAChRs, but in contrast, both α7 and α4β2 nAChRs may be critically involved in working memory in the absence of nicotine.

The ability of nicotine to modulate working memory may occur through interacting with processes mediated by NMDA receptor. Nicotine reversed the deficits in both working and reference memory induced by the NMDA receptor antagonist dizocilpine [37]. Similarly, acute nicotine ameliorated the deficits in working memory produced by ketamine, another NMDA receptor antagonist [148]. Further demonstrating a link between NMDA receptors and nAChRs, dizocilpine increased α7 nAChR binding in the hippocampus but chronic nicotine reversed this effect [149]. Nicotine may also have an indirect effect on NMDA receptor function during working memory by modulating the noradrenergic system. The α2-noradrenergic antagonist idazoxan blocked the ability of nicotine to reverse dizocilpine-induced deficits in working and reference memory [150]. In sum, these studies strengthen the convention that nAChR and NMDA receptor processes may interact to support and modulate learning processes.

6 Other Cognitive Processes

While the major focus of this chapter is on hippocampus-dependent learning, it should not be interpreted as indicating that nicotine only affects hippocampus-mediated cognitive processes. The following section will briefly review some of the other cognitive processes nicotine can modulate. Attention is an important and adaptive cognitive process that can increase awareness of vital environmental stimuli and increase learning. Multiple studies have examined the effects of nicotine on attention, many of them using the 5-choice serial reaction time task (5CSRT). The 5CSRT is an operant task used to assess attention and impulsivity. Commonly, there are five ports each with a light. Rodents must attend to the ports to see which one is illuminated and then perform a nose-poke or lever press at that port. Stimuli parameters such as the duration the port is lit or the addition of a distracter stimulus can be modified to tax attention and delays between port light offset, and permissible response can be added to assess impulsivity [151, 152]. Mice with the α7 gene knockout had deficits in the 5CSRT [153, 154] and α5 KO mice had reduced accuracy [155]. Acute and chronic nicotine treatment improves performance in the 5CSRT, and cessation of chronic nicotine was associated with deficits in attention [156–161]. The ability of nicotine to enhance performance in the 5CSRT may be mediated by α4β2* nAChRs as DHβE blocked the effects of nicotine on the task [162]. The effects of chronic nicotine on the performance of 5CSRT appear to be mediated by the prelimbic area of the prefrontal cortex and not the dorsal hippocampus as nicotine infusion into the former but not the latter improved attention [163]. In addition, other cognitive processes that involve the prefrontal cortex may be modulated by nicotine. Tests that measure cognitive flexibility were enhanced by acute nicotine [164] and disrupted with chronic nicotine treatment [165], and tasks that require response inhibition were disrupted by deletion of the α7 gene [166] and enhanced with acute but not chronic nicotine treatment [167].

7 Human Cognition

The ability of nicotine to alter cognitive processes in smokers may substantially contribute to nicotine addiction. This may occur through multiple mechanisms. In humans, nicotine has been shown to enhance cognitive processes; for review, see [168, 169]. The ability of nicotine to enhance cognitive processes could contribute to addiction through facilitating the development of strong yet maladaptive drug-cue and drug-context associations that lead to cravings and drug-seeking behavior [170]. In support, smokers showed large activation of the hippocampus and cravings when exposed to smoking-related cues; this activation was absent in nonsmokers [171–173]. In addition, a major symptom of abstinence from tobacco products is detrimental changes in cognition; for review, see [174]. Attention and working memory are two of the most commonly examined cognitive processes in abstinent smokers. Multiple studies have shown that nicotine withdrawal is associated with disrupted attention [175–178]. Similarly, working memory deficits, including spatial working memory, emerge during abstinence [179–184]. These symptoms may be particularly important as changes in working memory during periods of abstinence were associated with relapse to smoking [185]. As in rodent studies, chronic nicotine exposure was associated with upregulated hippocampal nAChRs in the postmortem brains of smokers; the degree of upregulation was correlated with the average number of packs smoked per day [186]. Upregulated β2-containing nAChRs in the brains of smokers were maintained for at least 1 week post cessation of smoking and were associated with levels of craving [187, 188]. Thus, the maintained upregulation of high-affinity β2-containing nAChRs may be an important contributing factor to nicotine withdrawal deficits in cognition.

8 Summary

Nicotine initially has procognitive effects. These effects may support continued nicotine use in an attempt to maintain the beneficial effects of nicotine on cognition while also facilitating the development of maladaptive drug-cue and drug-context associations that can later trigger cravings and drug-seeking behavior. With continued nicotine administration, adaptations occur in brain function that can result in tolerance to the procognitive effects of nicotine and produce withdrawal deficits in cognition during abstinence. The neural changes responsible for withdrawal deficits in cognition are not well understood, but studies in both humans and rodents suggest nAChR upregulation may be a contributing factor. While not all cognitive processes are modulated by nicotine administration, learning and other cognitive tasks that involve the frontal cortex and the hippocampus may be particularly susceptible. The effects of nicotine on learning and other cognitive tasks are influenced by both genetics and the age of the subject. Because changes in cognition are the hallmark of nicotine addiction, understanding how nicotine modulates these processes should facilitate development of more effective treatments for this addiction.

References

1. Dale HH, Dudley HW. The presence of histamine and acetylcholine in the spleen of the ox and the horse. J Physiol. 1929;68(2):97–123.
2. Loewi O. On the antagonism between pressor and depressor agents in the frog's heart. J Pharmacol Exp Ther. 1949;96(3):295–304.
3. Fishman MC. Sir Henry Hallett Dale and acetylcholine story. Yale J Biol Med. 1972; 45(2):104–18.
4. Hasselmo ME. Neuromodulation: acetylcholine and memory consolidation. Trends Cogn Sci. 1999;3(9):351–9.
5. Sarter M, Parikh V, Howe WM. nAChR agonist-induced cognition enhancement: integration of cognitive and neuronal mechanisms. Biochem Pharmacol. 2009;78(7):658–67.
6. Hasselmo ME, Sarter M. Modes and models of forebrain cholinergic neuromodulation of cognition. Neuropsychopharmacology. 2011;36(1):52–73.
7. Fadda F, Cocco S, Stancampiano R. Hippocampal acetylcholine release correlates with spatial learning performance in freely moving rats. Neuroreport. 2000;11(10):2265–9.
8. Hodges H, Allen Y, Kershaw T, Lantos PL, Gray JA, Sinden J. Effects of cholinergic-rich neural grafts on radial maze performance of rats after excitotoxic lesions of the forebrain cholinergic projection system—I. Amelioration of cognitive deficits by transplants into cortex and hippocampus but not into basal forebrain. Neuroscience. 1991;45(3):587–607.
9. Arendt T, Allen Y, Sinden J, et al. Cholinergic-rich brain transplants reverse alcohol-induced memory deficits. Nature. 1988;332(6163):448–50.
10. Perry E, Walker M, Grace J, Perry R. Acetylcholine in mind: a neurotransmitter correlate of consciousness? Trends Neurosci. 1999;22(6):273–80.
11. Feiro O, Gould TJ. The interactive effects of nicotinic and muscarinic cholinergic receptor inhibition on fear conditioning in young and aged C57BL/6 mice. Pharmacol Biochem Behav. 2005;80(2):251–62.
12. Gravius A, Barberi C, Schäfer D, Schmidt WJ, Danysz W. The role of group I metabotropic glutamate receptors in acquisition and expression of contextual and auditory fear conditioning in rats—a comparison. Neuropharmacology. 2006;51(7–8):1146–55.
13. Deutsch JA. The cholinergic synapse and the site of memory. Science. 1971;174(4011): 788–94.
14. Mansvelder H, Aerde K, Couey J, Brussaard A. Nicotinic modulation of neuronal networks: from receptors to cognition. Psychopharmacology (Berl). 2006;184(3–4):292–305.
15. Orr-Urtreger A, Seldin MF, Baldini A, Beaudet al. Cloning and mapping of the mouse alpha 7-neuronal nicotinic acetylcholine receptor. Genomics. 1995;26(2):399–402.
16. Orr-Urtreger A, Goldner FM, Saeki M, et al. Mice deficient in the alpha7 neuronal nicotinic acetylcholine receptor lack alpha-bungarotoxin binding sites and hippocampal fast nicotinic currents. J Neurosci. 1997;17(23):9165–71.
17. Wonnacott S. α bungarotoxin binds to low-affinity nicotine binding sites in rat brain. J Neurochem. 1986;47(6):1706–12.
18. Marks MJ, Collins AC. Characterization of nicotine binding in mouse brain and comparison with the binding of alpha-bungarotoxin and quinuclidinyl benzilate. Mol Pharmacol. 1982;22(3):554–64.
19. Marks MJ, Stitzel JA, Romm E, Wehner JM, Collins AC. Nicotinic binding sites in rat and mouse brain: comparison of acetylcholine, nicotine, and alpha-bungarotoxin. Mol Pharmacol. 1986;30(5):427–36.
20. Perry DC, Xiao Y, Nguyen HN, Musachio JL, Davila-Garcia MI, Kellar KJ. Measuring nicotinic receptors with characteristics of alpha4beta2, alpha3beta2 and alpha3beta4 subtypes in rat tissues by autoradiography. J Neurochem. 2002;82(3):468–81.
21. Penfield DW, Milner B. Memory deficit produced by bilateral lesions in the hippocampal zone. AMA Arch Neurol Psychiatry. 1958;79(5):475–97.

22. Eichenbaum H. The hippocampus and mechanisms of declarative memory. Behav Brain Res. 1999;103(2):123–33.
23. Alkondon M, Albuquerque EX. Nicotinic acetylcholine receptor alpha7 and alpha4beta2 subtypes differentially control GABAergic input to CA1 neurons in rat hippocampus. J Neurophysiol. 2001;86(6):3043–55.
24. McQuiston AR, Madison DV. Nicotinic receptor activation excites distinct subtypes of interneurons in the rat hippocampus. J Neurosci. 1999;19(8):2887–96.
25. Fu Y, Matta SG, James TJ, Sharp BM. Nicotine-induced norepinephrine release in the rat amygdala and hippocampus is mediated through brainstem nicotinic cholinergic receptors. J Pharmacol Exp Ther. 1998;284(3):1188–96.
26. Freund RK, Jungschaffer DA, Collins AC, Wehner JM. Evidence for modulation of GABAergic neurotransmission by nicotine. Brain Res. 1988;453(1–2):215–20.
27. Radcliffe KA, Fisher JL, Gray R, Dani JA. Nicotinic modulation of glutamate and GABA synaptic transmission of hippocampal neurons. Ann N Y Acad Sci. 1999;868:591–610.
28. Rapier C, Lunt GG, Wonnacott S. Nicotinic modulation of [3H]dopamine release from striatal synaptosomes: pharmacological characterisation. J Neurochem. 1990;54(3):937–45.
29. Grady S, Marks MJ, Wonnacott S, Collins AC. Characterization of nicotinic receptor-mediated [H-3] dopamine release from synaptosomes prepared from mouse striatum. J Neurochem. 1992;59(3):848–56.
30. Rowell PP, Winkler DL. Nicotinic stimulation of [3H]acetylcholine release from mouse cerebral cortical synaptosomes. J Neurochem. 1984;43(6):1593–8.
31. Ribeiro EB, Bettiker RL, Bogdanov M, Wurtman RJ. Effects of systemic nicotine on serotonin release in rat brain. Brain Res. 1993;621(2):311–8.
32. Westfall TC, Grant H, Perry H. Release of dopamine and 5-hydroxytryptamine from rat striatal slices following activation of nicotinic cholinergic receptors. Gen Pharmacol. 1983;14(3):321–5.
33. Nicoll RA, Malenka RC. Expression mechanisms underlying NMDA receptor-dependent long-term potentiation. Ann N Y Acad Sci. 1999;868:515–25.
34. Lynch G, Kessler M, Arai A, Larson J. The nature and causes of hippocampal long-term potentiation. Prog Brain Res. 1990;83:233–50.
35. McKay BE, Placzek AN, Dani JA. Regulation of synaptic transmission and plasticity by neuronal nicotinic acetylcholine receptors. Biochem Pharmacol. 2007;74(8):1120–33.
36. Karadsheh MS, Shah MS, Tang X, Macdonald RL, Stitzel JA. Functional characterization of mouse alpha4beta2 nicotinic acetylcholine receptors stably expressed in HEK293T cells. J Neurochem. 2004;91(5):1138–50.
37. Levin ED, Bettegowda C, Weaver T, Christopher NC. Nicotine-dizocilpine interactions and working and reference memory performance of rats in the radial-arm maze. Pharmacol Biochem Behav. 1998;61(3):335–40.
38. André JM, Leach PT, Gould TJ. Nicotine ameliorates NMDA receptor antagonist-induced deficits in contextual fear conditioning through high-affinity nicotinic acetylcholine receptors in the hippocampus. Neuropharmacology. 2011;60(4):617–25.
39. Matsuyama S, Matsumoto A, Enomoto T, Nishizaki T. Activation of nicotinic acetylcholine receptors induces long-term potentiation in vivo in the intact mouse dentate gyrus. Eur J Neurosci. 2000;12(10):3741–7.
40. He J, Deng CY, Chen RZ, Zhu XN, Yu JP. Long-term potentiation induced by nicotine in CA1 region of hippocampal slice is Ca(2+)-dependent. Acta Pharmacol Sin. 2000;21(5):429–32.
41. Matsuyama S, Matsumoto A. Epibatidine induces long-term potentiation (LTP) via activation of alpha4beta2 nicotinic acetylcholine receptors (nAChRs) in vivo in the intact mouse dentate gyrus: both alpha7 and alpha4beta2 nAChRs essential to nicotinic LTP. J Pharmacol Sci. 2003;93(2):180–7.
42. Levin ED. Nicotinic receptor subtypes and cognitive function. J Neurobiol. 2002;53(4):633–40.

43. Kenney JW, Wilkinson DS, Gould TJ. The enhancement of contextual fear conditioning by ABT-418. Behav Pharmacol. 2010;21(3):246–9.
44. Gould TJ, Wehner JM. Nicotine enhancement of contextual fear conditioning. Behav Brain Res. 1999;102(1–2):31–9.
45. Andersen P, Bland BH, Dudar JD. Organization of the hippocampal output. Exp Brain Res. 1973;17(2):152–68.
46. Bartesaghi R, Migliore M, Gessi T. Input–output relations in the entorhinal cortex–dentate–hippocampal system: evidence for a non-linear transfer of signals. Neuroscience. 2006;142(1):247–65.
47. Bennett MR, Gibson WG, Robinson J. Dynamics of the CA3 pyramidal neuron autoassociative memory network in the hippocampus. Philos Trans R Soc Lond B Biol Sci. 1994;343(1304):167–87.
48. Logue SF, Paylor R, Wehner JM. Hippocampal lesions cause learning deficits in inbred mice in the Morris water maze and conditioned-fear task. Behav Neurosci. 1997;111:104–13.
49. Kim JJ, Rison RA, Fanselow MS. Effects of amygdala, hippocampus, and periaqueductal gray lesions on short- and long-term contextual fear. Behav Neurosci. 1993;107(6):1093–8.
50. Phillips RG, Ledoux JE. Differential contribution of amygdala and hippocampus to cued and contextual fear conditioning. Behav Neurosci. 1992;106(2):274–85.
51. Gilmartin MR, Miyawaki H, Helmstetter FJ, Diba K. Prefrontal activity links nonoverlapping events in memory. J Neurosci. 2013;33(26):10910–4.
52. McEchron MD, Bouwmeester H, Tseng W, Weiss C, Disterhoft JF. Hippocampectomy disrupts auditory trace fear conditioning and contextual fear conditioning in the rat. Hippocampus. 1998;8(6):638–46.
53. Gould TJ. Nicotine produces a within subject enhancement of contextual fear conditioning in C57BL/6 mice independent of sex. Integr Physiol Behav Sci. 2003;38:124–32.
54. Gould TJ, Higgins JS. Nicotine enhances contextual fear conditioning in C57BL/6J mice at 1 and 7 days post-training. Neurobiol Learn Mem. 2003;80:147–57.
55. Gould TJ, Feiro O, Moore D. Nicotine enhancement of trace cued fear conditioning but not delay cued fear conditioning in C57BL/6J mice. Behav Brain Res. 2004;155:167–73.
56. Davis JA, Porter J, Gould TJ. Nicotine enhances both foreground and background contextual fear conditioning. Neurosci Lett. 2006;394(3):202–5.
57. Szyndler J, Sienkiewicz-Jarosz H, Maciejak P, et al. The anxiolytic-like effect of nicotine undergoes rapid tolerance in a model of contextual fear conditioning in rats. Pharmacol Biochem Behav. 2001;69(3–4):511–8.
58. Tian S, Huang F, Li P, et al. Nicotine enhances contextual fear memory reconsolidation in rats. Neurosci Lett. 2011;487(3):368–71.
59. Davis JA, Kenney JW, Gould TJ. Hippocampal alpha4beta2 nicotinic acetylcholine receptor involvement in the enhancing effect of acute nicotine on contextual fear conditioning. J Neurosci. 2007;27(40):10870–7.
60. Kenney JW, Raybuck JD, Gould TJ. Nicotinic receptors in the dorsal and ventral hippocampus differentially modulate contextual fear conditioning. Hippocampus. 2012;22(8):1681–90.
61. Fanselow MS, Dong HW. Are the dorsal and ventral hippocampus functionally distinct structures? Neuron. 2010;65(1):7–19.
62. Raybuck JD, Gould TJ. The role of nicotinic acetylcholine receptors in the medial prefrontal cortex and hippocampus in trace fear conditioning. Neurobiol Learn Mem. 2010;94(3):353–63.
63. Knight DC, Cheng DT, Smith CN, Stein EA, Helmstetter FJ. Neural substrates mediating human delay and trace fear conditioning. J Neurosci. 2004;24(1):218–28.
64. Runyan JD, Moore AN, Dash PK. A role for prefrontal cortex in memory storage for trace fear conditioning. J Neurosci. 2004;24(6):1288–95.
65. Harvey SC, Maddox FN, Luetje CW. Multiple determinants of dihydro-beta-erythroidine sensitivity on rat neuronal nicotinic receptor alpha subunits. J Neurochem. 1996;67(5):1953–9.

66. Turek JW, Kang CH, Campbell JE, Arneric SP, Sullivan JP. A sensitive technique for the detection of the alpha 7 neuronal nicotinic acetylcholine receptor antagonist, methyllycaconitine, in rat plasma and brain. J Neurosci Methods. 1995;61(1–2):113–8.
67. Palma E, Bertrand S, Binzoni T, Bertrand D. Neuronal nicotinic alpha 7 receptor expressed in Xenopus oocytes presents five putative binding sites for methyllycaconitine. J Physiol (Lond). 1996;491(Pt 1):151–61.
68. Davis JA, Gould TJ. The effects of DHBE and MLA on nicotine-induced enhancement of contextual fear conditioning in C57BL/6 mice. Psychopharmacology (Berl). 2006;184(3–4):345–52.
69. Davis J, Gould T. Beta2 subunit-containing nicotinic receptors mediate the enhancing effect of nicotine on trace cued fear conditioning in C57BL/6 mice. Psychopharmacology (Berl). 2007;190(3):343–52.
70. Paylor R, Nguyen M, Crawley JN, Patrick J, Beaudet A, Orr-Urtreger A. Alpha 7 nicotinic receptor subunits are not necessary for hippocampal-dependent learning or sensorimotor gating: a behavioral characterization of Acra7-deficient mice. Learn Mem. 1998;5:302–16.
71. Wehner JM, Keller JJ, Keller AB, et al. Role of neuronal nicotinic receptors in the effects of nicotine and ethanol on contextual fear conditioning. Neuroscience. 2004;129(1):11–24.
72. Semenova S, Contet C, Roberts AJ, Markou A. Mice lacking the β4 subunit of the nicotinic acetylcholine receptor show memory deficits, altered anxiety- and depression-like behavior, and diminished nicotine-induced analgesia. Nicotine Tob Res. 2012;14(11):1346–55.
73. Lotfipour S, Byun JS, Leach P, et al. Targeted deletion of the mouse alpha2 nicotinic acetylcholine receptor subunit gene (Chrna2) potentiates nicotine-modulated behaviors. J Neurosci. 2013;33(18):7728–41.
74. Caldarone BJ, Duman CH, Picciotto MR. Fear conditioning and latent inhibition in mice lacking the high affinity subclass of nicotinic acetylcholine receptors in the brain. Neuropharmacology. 2000;39(13):2779–84.
75. Portugal GS, Wilkinson DS, Kenney JW, Sullivan C, Gould TJ. Strain-dependent effects of acute, chronic, and withdrawal from chronic nicotine on fear conditioning. Behav Genet. 2012;42(1):133–50.
76. Portugal GS, Wilkinson DS, Turner JR, Blendy JA, Gould TJ. Developmental effects of acute, chronic, and withdrawal from chronic nicotine on fear conditioning. Neurobiol Learn Mem. 2012;97(4):482–94.
77. Gould TJ, McCarthy MM, Keith RA. MK-801 disrupts acquisition of contextual fear conditioning but enhances memory consolidation of cued fear conditioning. Behav Pharmacol. 2002;13(4):287–94.
78. Kim JJ, DeCola JP, Landeira-Fernandez J, Fanselow MS. N-methyl-D-aspartate receptor antagonist APV blocks acquisition but not expression of fear conditioning. Behav Neurosci. 1991;105(1):126–33.
79. Gould TJ, Lewis MC. Coantagonism of glutamate receptors and nicotinic acetylcholinergic receptors disrupts fear conditioning and latent inhibition of fear conditioning. Learn Mem. 2005;12(4):389–98.
80. Magnusson KR. Aging of glutamate receptors: correlations between binding and spatial memory performance in mice. Mech Ageing Dev. 1998;104(3):227–48.
81. Kenney JW, Florian C, Portugal GS, Abel T, Gould TJ. Involvement of hippocampal jun-N terminal kinase pathway in the enhancement of learning and memory by nicotine. Neuropsychopharmacology. 2010;35(2):483–92.
82. Kenney JW, Poole RL, Adoff MD, Logue SF, Gould TJ. Learning and nicotine interact to increase CREB phosphorylation at the jnk1 promoter in the hippocampus. PLoS One. 2012;7(6):e39939.
83. Silva AJ, Kogan JH, Frankland PW, Kida S. CREB and memory. Annu Rev Neurosci. 1998;21:127–48.
84. Bjorkblom B, Ostman N, Hongisto V, et al. Constitutively active cytoplasmic c-Jun N-terminal kinase 1 is a dominant regulator of dendritic architecture: role of microtubule-associated protein 2 as an effector. J Neurosci. 2005;25(27):6350–61.

85. Bogoyevitch MA, Kobe B. Uses for JNK: the many and varied substrates of the c-Jun N-terminal kinases. Microbiol Mol Biol Rev. 2006;70(4):1061–95.
86. Gupta S, Barrett T, Whitmarsh AJ, et al. Selective interaction of JNK protein kinase isoforms with transcription factors. EMBO J. 1996;15(11):2760–70.
87. Davis JA, James JR, Siegel SJ, Gould TJ. Withdrawal from chronic nicotine administration impairs contextual fear conditioning in C57BL/6 mice. J Neurosci. 2005;25:8708–13.
88. André JM, Gulick D, Portugal GS, Gould TJ. Nicotine withdrawal disrupts both foreground and background contextual fear conditioning but not pre-pulse inhibition of the acoustic startle response in C57BL/6 mice. Behav Brain Res. 2008;190(2):174–81.
89. Portugal GS, Gould TJ. Nicotine withdrawal disrupts new contextual learning. Pharmacol Biochem Behav. 2009;92(1):117–23.
90. Raybuck JD, Gould TJ. Nicotine withdrawal-induced deficits in trace fear conditioning in C57BL/6 mice—a role for high-affinity β2 subunit-containing nicotinic acetylcholine receptors. Eur J Neurosci. 2009;29(2):377–87.
91. Portugal G, Wilkinson D, Kenney J, Sullivan C, Gould T. Strain-dependent effects of acute, chronic, and withdrawal from chronic nicotine on fear conditioning. Behav Genet. 2012; 42:133–50.
92. Davis JA, Gould TJ. Hippocampal nAChRs mediate nicotine withdrawal-related learning deficits. Eur Neuropsychopharmacol. 2009;19(8):551–61.
93. Portugal GS, Kenney JW, Gould TJ. β2 containing acetylcholine receptors mediate nicotine withdrawal deficits in learning. Neurobiol Learn Mem. 2008;89(2):106–13.
94. Gentry CL, Lukas RJ. Regulation of nicotinic acetylcholine receptor numbers and function by chronic nicotine exposure. Curr Drug Targets CNS Neurol Disord. 2002;1(4):359–85.
95. Marks MJ, Burch JB, Collins AC. Effects of chronic nicotine infusion on tolerance development and nicotinic receptors. J Pharmacol Exp Ther. 1983;226(3):817–25.
96. Schwartz RD, Kellar KJ. Nicotinic cholinergic receptor binding sites in the brain: regulation in vivo. Science. 1983;220(4593):214–6.
97. Gould TJ, Portugal GS, Andre JM, et al. The duration of nicotine withdrawal-associated deficits in contextual fear conditioning parallels changes in hippocampal high affinity nicotinic acetylcholine receptor upregulation. Neuropharmacology. 2012;62(5–6):2118–25.
98. Wilkinson DS, Turner JR, Blendy JA, Gould TJ. Genetic background influences the effects of withdrawal from chronic nicotine on learning and high-affinity nicotinic acetylcholine receptor binding in the dorsal and ventral hippocampus. Psychopharmacology (Berl). 2013;225(1): 201–8.
99. Dani JA, Heinemann S. Molecular and cellular aspects of nicotine abuse. Neuron. 1996;16(5): 905–8.
100. Penton RE, Quick MW, Lester RAJ. Short- and long-lasting consequences of in vivo nicotine treatment on hippocampal excitability. J Neurosci. 2011;31(7):2584–94.
101. Wilkinson DS, Gould TJ. Withdrawal from chronic nicotine and subsequent sensitivity to nicotine challenge on contextual learning. Behav Brain Res. 2013;250:58–61.
102. Raybuck JD, Portugal GS, Lerman C, Gould TJ. Varenicline ameliorates nicotine withdrawal-induced learning deficits in C57BL/6 mice. Behav Neurosci. 2008;122(5):1166–71.
103. Portugal GS, Gould TJ. Bupropion dose-dependently reverses nicotine withdrawal deficits in contextual fear conditioning. Pharmacol Biochem Behav. 2007;88(2):179–87.
104. Coe JW, Brooks PR, Vetelino MG, et al. Varenicline: an alpha4beta2 nicotinic receptor partial agonist for smoking cessation. J Med Chem. 2005;48(10):3474–7.
105. Slemmer JE, Martin BR, Damaj MI. Bupropion is a nicotinic antagonist. J Pharmacol Exp Ther. 2000;295(1):321–7.
106. Papke RL, Trocmé-Thibierge C, Guendisch D, Al Rubaiy SAA, Bloom SA. Electrophysiological perspectives on the therapeutic use of nicotinic acetylcholine receptor partial agonists. J Pharmacol Exp Ther. 2011;337(2):367–79.
107. Oliveira AMM, Hawk JD, Abel T, Havekes R. Post-training reversible inactivation of the hippocampus enhances novel object recognition memory. Learn Mem. 2010;17(3):155–60.

108. Barker GRI, Warburton EC. When is the hippocampus involved in recognition memory? J Neurosci. 2011;31(29):10721–31.
109. Kenney JW, Adoff MD, Wilkinson DS, Gould TJ. The effects of acute, chronic, and withdrawal from chronic nicotine on novel and spatial object recognition in male C57BL/6J mice. Psychopharmacology (Berl). 2011;217:353–65.
110. Puma C, Deschaux O, Molimard R, Bizot JC. Nicotine improves memory in an object recognition task in rats. Eur Neuropsychopharmacol. 1999;9(4):323–7.
111. Jacklin DL, Goel A, Clementino KJ, Hall AWM, Talpos JC, Winters BD. Severe cross-modal object recognition deficits in rats treated sub-chronically with NMDA receptor antagonists are reversed by systemic nicotine: implications for abnormal multisensory integration in schizophrenia. Neuropsychopharmacology. 2012;37(10):2322–31.
112. Melichercik AM, Elliott KS, Bianchi C, Ernst SM, Winters BD. Nicotinic receptor activation in perirhinal cortex and hippocampus enhances object memory in rats. Neuropharmacology. 2012;62(5–6):2096–105.
113. Shapiro ML, Eichenbaum H. Hippocampus as a memory map: synaptic plasticity and memory encoding by hippocampal neurons. Hippocampus. 1999;9(4):365–84.
114. O'Keefe J, Dostrovsky J. The hippocampus as a spatial map. Preliminary evidence from unit activity in the freely-moving rat. Brain Res. 1971;34(1):171–5.
115. Morris RG, Garrud P, Rawlins JN, O'Keefe J. Place navigation impaired in rats with hippocampal lesions. Nature. 1982;297(5868):681–3.
116. Barnes CA. Memory deficits associated with senescence: a neurophysiological and behavioral study in the rat. J Comp Physiol Psychol. 1979;93(1):74–104.
117. Azzopardi E, Typlt M, Jenkins B, Schmid S. Sensorimotor gating and spatial learning in α7-nicotinic receptor knockout mice. Genes Brain Behav. 2013;12(4):414–23.
118. Abdulla FA, Bradbury E, Calaminici MR, et al. Relationship between up-regulation of nicotine binding sites in rat brain and delayed cognitive enhancement observed after chronic or acute nicotinic receptor stimulation. Psychopharmacology (Berl). 1996;124(4):323–31.
119. Sharifzadeh M, Tavasoli M, Naghdi N, Ghanbari A, Amini M, Roghani A. Post-training intrahippocampal infusion of nicotine prevents spatial memory retention deficits induced by the cyclo-oxygenase-2-specific inhibitor celecoxib in rats. J Neurochem. 2005;95(4):1078–90.
120. Attaway CM, Compton DM, Turner MD. The effects of nicotine on learning and memory: a neuropsychological assessment in young and senescent fischer 344 rats. Physiol Behav. 1999;67(3):421–31.
121. Bernal MC, Vicens P, Carrasco MC, Redolat R. Effects of nicotine on spatial learning in C57BL mice. Behav Pharmacol. 1999;10(3):333–6.
122. Socci DJ, Sanberg PR, Arendash GW. Nicotine enhances Morris water maze performance of young and aged rats. Neurobiol Aging. 1995;16(5):857–60.
123. Scerri C, Stewart C, Breen K, Balfour D. The effects of chronic nicotine on spatial learning and bromodeoxyuridine incorporation into the dentate gyrus of the rat. Psychopharmacology (Berl). 2006;184(3–4):540–6.
124. Matta SG, Balfour DJ, Benowitz NL, et al. Guidelines on nicotine dose selection for in vivo research. Psychopharmacology (Berl). 2007;190(3):269–319.
125. Meck WH, Church RM, Olton DS. Hippocampus, time, and memory. Behav Neurosci. 1984;98(1):3–22.
126. Davis HP, Baranowski JR, Pulsinelli WA, Volpe BT. Retention of reference memory following ischemic hippocampal damage. Physiol Behav. 1987;39(6):783–6.
127. Olton DS, Feustle WA. Hippocampal function required for nonspatial working memory. Exp Brain Res. 1981;41(3–4):380–9.
128. Levin ED. Chronic haloperidol administration does not block acute nicotine-induced improvements in radial-arm maze performance in the rat. Pharmacol Biochem Behav. 1997;58(4):899–902.
129. Addy N, Levin ED. Nicotine interactions with haloperidol, clozapine and risperidone and working memory function in rats. Neuropsychopharmacology. 2002;27(4):534–41.

130. Levin E, Icenogle L, Farzad A. Ketanserin attenuates nicotine-induced working memory improvement in rats. Pharmacol Biochem Behav. 2005;82(2):289–92.
131. Levin ED, Sledge D, Baruah A, Addy NA. Ventral hippocampal NMDA blockade and nicotinic effects on memory function. Brain Res Bull. 2003;61(5):489–95.
132. Levin ED, Kaplan S, Boardman A. Acute nicotine interactions with nicotinic and muscarinic antagonists: working and reference memory effects in the 16-arm radial maze. Behav Pharmacol. 1997;8(2–3):236–42.
133. Levin ED, Torry D. Acute and chronic nicotine effects on working memory in aged rats. Psychopharmacology (Berl). 1996;123(1):88–97.
134. Arthur D, Levin ED. Chronic inhibition of alpha4beta2 nicotinic receptors in the ventral hippocampus of rats: impacts on memory and nicotine response. Psychopharmacology (Berl). 2002;160(2):140–5.
135. Kholdebarin E, Caldwell DP, Blackwelder WP, Kao M, Christopher NC, Levin ED. Interaction of nicotinic and histamine H3 systems in the radial-arm maze repeated acquisition task. Eur J Pharmacol. 2007;569(1–2):64–9.
136. Levin ED, Christopher NC, Briggs SJ, Rose JE. Chronic nicotine reverses working memory deficits caused by lesions of the fimbria or medial basalocortical projection. Brain Res Cogn Brain Res. 1993;1(3):137–43.
137. Levin ED, Lee C, Rose JE, et al. Chronic nicotine and withdrawal effects on radial-arm maze performance in rats. Behav Neural Biol. 1990;53(2):269–76.
138. Levin ED, Briggs SJ, Christopher NC, Rose JE. Persistence of chronic nicotine-induced cognitive facilitation. Behav Neural Biol. 1992;58(2):152–8.
139. Levin ED, Briggs SJ, Christopher NC, Rose JE. Chronic nicotinic stimulation and blockade effects on working memory. Behav Pharmacol. 1993;4(2):179–82.
140. Levin ED, Christopher NC, Weaver T, Moore J, Brucato F. Ventral hippocampal ibotenic acid lesions block chronic nicotine-induced spatial working memory improvement in rats. Cogn Brain Res. 1999;7(3):405–10.
141. Bettany JH, Levin ED. Ventral hippocampal alpha 7 nicotinic receptor blockade and chronic nicotine effects on memory performance in the radial-arm maze. Pharmacol Biochem Behav. 2001;70(4):467–74.
142. Bancroft A, Levin ED. Ventral hippocampal alpha4beta2 nicotinic receptors and chronic nicotine effects on memory. Neuropharmacology. 2000;39(13):2770–8.
143. Levin ED, Bradley A, Addy N, Sigurani N. Hippocampal alpha 7 and alpha 4 beta 2 nicotinic receptors and working memory. Neuroscience. 2002;109(4):757–65.
144. Pocivavsek A, Icenogle L, Levin ED. Ventral hippocampal alpha7 and alpha4beta2 nicotinic receptor blockade and clozapine effects on memory in female rats. Psychopharmacology (Berl). 2006;188(4):597–604.
145. Castner SA, Smagin GN, Piser TM, et al. Immediate and sustained improvements in working memory after selective stimulation of α7 nicotinic acetylcholine receptors. Biol Psychiatry. 2011;69(1):12–8.
146. Kenney JW, Gould TJ. Modulation of hippocampus-dependent learning and synaptic plasticity by nicotine. Mol Neurobiol. 2008;38(1):101–21.
147. Portugal GS, Gould TJ. Genetic variability in nicotinic acetylcholine receptors and nicotine addiction: converging evidence from human and animal research. Behav Brain Res. 2008;193(1):1–16.
148. Rushforth SL, Steckler T, Shoaib M. Nicotine improves working memory span capacity in rats following sub-chronic ketamine exposure. Neuropsychopharmacology. 2011;36(13):2774–81.
149. Levin ED, Tizabi Y, Rezvani AH, Caldwell DP, Petro A, Getachew B. Chronic nicotine and dizocilpine effects on regionally specific nicotinic and NMDA glutamate receptor binding. Brain Res. 2005;1041(2):132–42.
150. Timofeeva OA, Levin ED. Idazoxan blocks the nicotine-induced reversal of the memory impairment caused by the NMDA glutamate receptor antagonist dizocilpine. Pharmacol Biochem Behav. 2008;90(3):372–81.

151. Carli M, Robbins TW, Evenden JL, Everitt BJ. Effects of lesions to ascending noradrenergic neurones on performance of a 5-choice serial reaction task in rats; implications for theories of dorsal noradrenergic bundle function based on selective attention and arousal. Behav Brain Res. 1983;9(3):361–80.
152. Cole BJ, Robbins TW. Amphetamine impairs the discriminative performance of rats with dorsal noradrenergic bundle lesions on a 5-choice serial reaction time task: new evidence for central dopaminergic-noradrenergic interactions. Psychopharmacology (Berl). 1987;91(4): 458–66.
153. Hoyle E, Genn RF, Fernandes C, Stolerman IP. Impaired performance of alpha7 nicotinic receptor knockout mice in the five-choice serial reaction time task. Psychopharmacology (Berl). 2006;189(2):211–23.
154. Young JW, Crawford N, Kelly JS, et al. Impaired attention is central to the cognitive deficits observed in alpha 7 deficient mice. Eur Neuropsychopharmacol. 2007;17(2):145–55.
155. Bailey CDC, De Biasi M, Fletcher PJ, Lambe EK. The nicotinic acetylcholine receptor α5 subunit plays a key role in attention circuitry and accuracy. J Neurosci. 2010;30(27): 9241–52.
156. Stolerman IP, Naylor CG, Mesdaghinia A, Morris HV. The duration of nicotine-induced attentional enhancement in the five-choice serial reaction time task: lack of long-lasting cognitive improvement. Behav Pharmacol. 2009;20(8):742–54.
157. Young JW, Finlayson K, Spratt C, et al. Nicotine improves sustained attention in mice: evidence for involvement of the alpha7 nicotinic acetylcholine receptor. Neuropsychopharmacology. 2004;29(5):891–900.
158. Semenova S, Stolerman IP, Markou A. Chronic nicotine administration improves attention while nicotine withdrawal induces performance deficits in the 5-choice serial reaction time task in rats. Pharmacol Biochem Behav. 2007;87(3):360–8.
159. Shoaib M, Bizarro L. Deficits in a sustained attention task following nicotine withdrawal in rats. Psychopharmacology (Berl). 2005;178(2):211–22.
160. Amitai N, Markou A. Chronic nicotine improves cognitive performance in a test of attention but does not attenuate cognitive disruption induced by repeated phencyclidine administration. Psychopharmacology (Berl). 2009;202(1–3):275–86.
161. Grottick AJ, Higgins GA. Effect of subtype selective nicotinic compounds on attention as assessed by the five-choice serial reaction time task. Behav Brain Res. 2000;117(1–2): 197–208.
162. Blondel A, Sanger DJ, Moser PC. Characterization of the effects of nicotine in the five-choice serial reaction time task in rats: antagonist studies [in process citation]. Psychopharmacology (Berl). 2000;149(3):293–305.
163. Hahn B, Shoaib M, Stolerman IP. Involvement of the prefrontal cortex but not the dorsal hippocampus in the attention-enhancing effects of nicotine in rats. Psychopharmacology (Berl). 2003;168(3):271–9.
164. Allison C, Shoaib M. Nicotine improves performance in an attentional set shifting task in rats. Neuropharmacology. 2013;64:314–20.
165. Ortega LA, Tracy BA, Gould TJ, Parikh V. Effects of chronic low- and high-dose nicotine on cognitive flexibility in C57BL/6J mice. Behav Brain Res. 2013;238:134–45.
166. Keller JJ, Keller AB, Bowers BJ, Wehner JM. Performance of α7 nicotinic receptor null mutants is impaired in appetitive learning measured in a signaled nose poke task. Behav Brain Res. 2005;162(1):143–52.
167. Leach PT, Cordero KA, Gould TJ. The effects of acute nicotine, chronic nicotine, and withdrawal from chronic nicotine on performance of a cued appetitive response. Behav Neurosci. 2013;127(2):303–10.
168. Heishman SJ. Behavioral and cognitive effects of smoking: relationship to nicotine addiction. Nicotine Tob Res. 1999;1 Suppl 2:S143–7.
169. Heishman S, Kleykamp B, Singleton E. Meta-analysis of the acute effects of nicotine and smoking on human performance. Psychopharmacology (Berl). 2010;210(4):453–69.

170. Gould TJ. Nicotine and hippocampus-dependent learning: implications for addiction. Mol Neurobiol. 2006;34(2):93–107.
171. Due DL, Huettel SA, Hall WG, Rubin DC. Activation in mesolimbic and visuospatial neural circuits elicited by smoking cues: evidence from functional magnetic resonance imaging. Am J Psychiatry. 2002;159(6):954–60.
172. Franklin TR, Wang Z, Wang J, et al. Limbic activation to cigarette smoking cues independent of nicotine withdrawal: a perfusion fMRI study. Neuropsychopharmacology. 2007;32(11): 2301–9.
173. McClernon FJ, Kozink RV, Rose JE. Individual differences in nicotine dependence, withdrawal symptoms, and sex predict transient fMRI-BOLD responses to smoking cues. Neuropsychopharmacology. 2008;33(9):2148–57.
174. Ashare RL, Falcone M, Lerman C. Cognitive function during nicotine withdrawal: Implications for nicotine dependence treatment. Neuropharmacology. 2013;76(Pt B):581–91.
175. Hendricks PS, Ditre JW, Drobes DJ, Brandon TH. The early time course of smoking withdrawal effects. Psychopharmacology (Berl). 2006;187(3):385–96.
176. Hughes JR, Keenan RM, Yellin A. Effect of tobacco withdrawal on sustained attention. Addict Behav. 1989;14(5):577–80.
177. Jacobsen LK, Slotkin TA, Westerveld M, Mencl WE, Pugh KR. Visuospatial memory deficits emerging during nicotine withdrawal in adolescents with prenatal exposure to active maternal smoking. Neuropsychopharmacology. 2005;31(7):1550–61.
178. Rhodes J, Hawk L, Ashare R, Schlienz N, Mahoney M. The effects of varenicline on attention and inhibitory control among treatment-seeking smokers. Psychopharmacology (Berl). 2012;223(2):131–8.
179. Park S, Knopick C, McGurk S, Meltzer HY. Nicotine impairs spatial working memory while leaving spatial attention intact. Neuropsychopharmacology. 2000;22(2):200–9.
180. Jacobsen LK, Krystal JH, Mencl WE, Westerveld M, Frost SJ, Pugh KR. Effects of smoking and smoking abstinence on cognition in adolescent tobacco smokers. Biol Psychiatry. 2005;57(1):56–66.
181. Xu J, Mendrek A, Cohen MS, et al. Effects of acute smoking on brain activity vary with abstinence in smokers performing the N-back task: a preliminary study. Psychiatry Res. 2006;148(2–3):103–9.
182. Mendrek A, Monterosso J, Simon SL, et al. Working memory in cigarette smokers: comparison to non-smokers and effects of abstinence. Addict Behav. 2006;31(5):833–44.
183. Jacobsen LK, Mencl WE, Constable RT, Westerveld M, Pugh KR. Impact of smoking abstinence on working memory neurocircuitry in adolescent daily tobacco smokers. Psychopharmacology (Berl). 2007;193(4):557–66.
184. Patterson F, Jepson C, Strasser AA, et al. Varenicline improves mood and cognition during smoking abstinence. Biol Psychiatry. 2009;65(2):144–9.
185. Patterson F, Jepson C, Loughead J, et al. Working memory deficits predict short-term smoking resumption following brief abstinence. Drug Alcohol Depend. 2010;106(1):61–4.
186. Breese CR, Marks MJ, Logel J, et al. Effect of smoking history on [H-3]nicotine binding in human postmortem brain. J Pharmacol Exp Ther. 1997;282(1):7–13.
187. Staley JK, Krishnan-Sarin S, Cosgrove KP, et al. Human tobacco smokers in early abstinence have higher levels of β2* nicotinic acetylcholine receptors than nonsmokers. J Neurosci. 2006;26(34):8707–14.
188. Cosgrove KP, Batis J, Bois F, et al. β2-nicotinic acetylcholine receptor availability during acute and prolonged abstinence from tobacco smoking. Arch Gen Psychiatry. 2009; 66(6):666–76.

Chapter 12
Nicotinic Receptors as Targets for Novel Analgesics and Anti-inflammatory Drugs

M. Imad Damaj, Kelen Freitas, Deniz Bagdas, and Pamela Flood

Abstract Nicotine and nicotinic receptors have been explored for the past three decades as a strategy for pain control. These receptors are widely expressed throughout the central and peripheral nervous system as well as immune cells. Despite encouraging results with many selective α4β2* agonists in animal models of pain, human studies showed a narrow therapeutic window between analgesic efficacy and toxicity is associated with the use of these agonists as analgesics. α4β2 positive allosteric modulators are being developed with the aim to increase the potency or therapeutic window of these agonists. However, several recent developments have potentially opened new windows of opportunity in the use of nicotinic agents for analgesia. Accumulating evidences suggest that α7 agonists and positive allosteric modulators hold a lot of promise in the treatment of chronic inflammatory pain conditions. In addition, recent animal studies suggest the therapeutic potential of ligands acting at other subtypes of nicotinic receptors. The current review will attempt to highlight these recent developments and outline some important findings that demonstrate further potential for the development of nicotinic ligands as novel analgesics.

Keywords Nicotine • ABT-594 • Choline • Chronic pain • α4β2 neuronal nicotinic acetylcholine • α7 neuronal nicotinic acetylcholine • Neuropathic pain • Inflammatory pain

---

M.I. Damaj, Ph.D. (✉) • K. Freitas
Department of Pharmacology & Toxicology, Virginia Commonwealth University School of Medicine, 1220 East Broad Street, Molecular Medical Research Building, Richmond, VA 23298, USA
e-mail: mdamaj@vcu.edu

D. Bagdas, Ph.D.
Experimental Animals Breeding and Research Center, Faculty of Medicine, Uludag University, Bursa, Turkey

P. Flood, M.D.
Department of Anesthesia and Perioperative Care, University of California at San Francisco, San Francisco, CA, USA

R.A.J. Lester (ed.), *Nicotinic Receptors*, The Receptors 26,
DOI 10.1007/978-1-4939-1167-7_12

Abbreviations

nAChRs	Nicotinic acetylcholine receptors
CNS	Central nervous system
PAM	Positive allosteric modulator

1 Introduction

Current pharmacological treatment approaches to pain include classical analgesics such as opiates and nonsteroidal anti-inflammatory drugs (NSAID) and other agents such as tricyclic antidepressants, local anesthetics, and anticonvulsants. Although there have been improvements to these analgesic mainstays in terms of formulation, and selectivity, there have been few analgesic drugs which have made a significant impact on analgesic treatment. Many patients suffer acute pain after surgery and persistent pain syndromes (e.g., arthritis pain, migraine, cancer pain, and neuropathic pains) represent a significant unmet need in terms of analgesic drug therapy. Thus, NSAIDs exhibit only little if any efficacy for many types of persistent pain, and the long-term use of opioids in patients with nonmalignant pains, for example, remains controversial because of tolerance and sensitization. Moreover, both classes of analgesic drugs upon long-term use produce serious side effects. Accordingly, safer and more efficacious analgesic agents are needed. A target currently under development for analgesic treatment are the nicotinic acetylcholine receptors (nAChRs), which have been found to play a role in modulating pain transmission in the central nervous system (CNS). As originally demonstrated by Davis et al. [1], nicotine and its congeners have been shown to reduce responses to noxious stimuli in experimental animals using a variety of nociceptive tests. Nicotine produces analgesia in humans but its use as a therapeutic is limited by side effects. Nonetheless, interest in a nicotinic cholinergic approach to pain control has been rekindled by the discoveries of epibatidine, a novel nicotinic agonist isolated from the skin of *Edpipedobates tricolor*, an Ecuadoran frog [2] and its synthetic analog ABT-594 [3], both of which exhibit antinociceptive activity with a potency at least two orders of magnitude greater than as well as an efficacy equal to that of morphine. Furthermore, rapid advances in the understanding of the molecular biology, physiology, and biochemistry of nAChRs, which have occurred over the last decade, have led to the creation of a vast array of novel and more selective nicotinic ligands that was essential in exploring the role of these receptors in pain and inflammation.

This chapter (1) reviews the distribution of nicotinic cholinergic receptors along the neuraxis, emphasizing the identities and locations of those that appear to be most important in terms of pain and analgesia; (2) reviews the effects of agonist, partial agonist, and allosteric modulators of α4β2* neuronal nicotinic acetylcholine receptors in animal models of pain; (3) discusses the new findings of agonists and allosteric modulators of α7 neuronal nicotinic acetylcholine pain and inflammatory pain models; (4) describes the emergence of new nicotinic agents acting on other nicotinic receptor subtypes; and (5) reviews the clinical development of nicotine and nicotinic agonists, with an eye toward the prospects for the ultimate development of one or more clinically viable nicotinic analgesic drugs.

2 Nicotinic Receptor Expression in Pain Pathways

Nicotinic receptors are expressed in many pain pathways including primary afferents [4], spinal cord excitatory and inhibitory interneurons and projecting neurons [5, 6], and many areas important for pain perception such as the midbrain [7], the medial habenula [8, 9], the medulla [10], the nucleus raphe magnus [11], the thalamus, the pedunculopontine tegmental nucleus [12], and the spinal cord [13–16]. However, the various nicotinic subtypes are not homogenously distributed between these different peripheral and central regions. Moreover, individual neurons can express multiple subtypes of nAChRs. The most common CNS subtype α4β2* (asterisk indicates assembly with other nAChR subunits) is found in thalamus, dorsal raphe nucleus, nucleus raphe magnus, and locus coeruleus [17–21]. In addition, α3, α5, α6, β3, and β4 subunits are expressed in the locus coeruleus small cells while the larger cells projecting to the hippocampus expressed only α4, α6, β2, and β3 subunits [22]. Many of these areas are thought to play a major role in descending monoaminergic inhibitory pain pathway. α4β2* subtypes are also present in the midbrain periaqueductal gray (PAG), a major component of the descending inhibitory pathway [23]. Multiple subtypes of nAChRs are additionally expressed in the spinal dorsal horn, which is the first site in the CNS where somatosensory information is processed and integrated. For instance, multiple subtypes of nAChRs (α4β2* and α5) are expressed on both inhibitory and excitatory interneurons in the spinal dorsal horn [24].

3 α4β2 nAChRs as Targets

As the most widely distributed receptor subtype in the brain, the α4β2* heteromeric nAChR has been strongly implicated in antinociception and has been a focus of pain modulation research for more than a decade [25–27]. Studies have utilized various preclinical models and techniques to establish α4β2 nAChRs as a target for pain management in animal and human studies. These studies support potential therapeutic benefit of modulating the activity of CNS α4β2 nAChRs to treat chronic pain. Recent studies showed that α4β2 nAChRs exist in alternate stoichiometries: (α4β2)2β2 has high sensitivity to activation by ACh (HS α4β2 nAChRs) and (α4β2)2α4 has lower ACh sensitivity (LS α4β2 nAChRs). Functional and biochemical studies suggest that both HS and LS α4β2 nAChRs coexist in the brain. In addition, α4β2 nAChRs can co-assemble with other nicotinic subunits such as α5, α6, and β3. This section will overview recent findings with respect to these specifics and define the current status of α4β2 nAChRs as targets for analgesia.

3.1 α4β2 nAChRs Agonism

The antinociceptive action of α4β2 agonists has been evidenced by the effects of several compounds that interact with the receptor with various levels of efficacy and selectivity. Selective agonists such as Sazetidine-A [28, 29], ABT-594 [30], NS3956 [31], A-366833 [32, 33], A-85380 [34], RJR-2403 (metanicotine) [35], and ABT-894 [36] have been shown to produce antinociceptive effects in rodents. Newer α4β2-selective agonists have also been found to have antinociceptive properties, such as NS3956, which has shown a strong alleviation of formalin-induced pain [31], and A-366833, which has dose dependently attenuated mechanical hyperalgesia in a complete Freund's adjuvant-induced inflammatory pain model in rats. A-366833 also produced significant antihyperalgesic effects in partial sciatic nerve ligation, chronic constriction injury, and spinal nerve ligation models in rats as well as analgesic effects in writhing pain models in mice and spinal nerve ligation and formalin pain models in rats [32, 33]. In an interesting series of studies with various nicotinic agonists tested in rat formalin and complete Freund's adjuvant models of pain coupled with in vitro functional evaluation, Gao et al. [37] concluded that the activation of α4β2 nAChRs is necessary but not sufficient to produce analgesia in these pain models. They suggested that activation of an α3 containing nAChR might be an additional requirement for an antinociceptive effect in these tests. This was also shown with varenicline in the mouse formalin test [29]. In addition, the efficacy in preclinical pain models of some partial agonists of α4β2* (varenicline [29, 37], sazetidine [28, 29, 38], and TC-2559 [39]) but not others (ABT-089 [40], TC-1734 [37], and A-424274 [41]) raises concerns about the role of selective α4β2* agonists in pain modulation.

It is possible that the activity of these various agonists at the different subtypes of α4β2* nAChRs could explain the discrepancies reported above. Indeed, it is well known that nicotinic subunits such as α5, α6, and β3 can associate with α4β2 nAChRs to form functional nicotinic receptors. Therefore, various agonists may have different sensitivity toward the different α4β2 nAChRs subtypes such as α4β2α5* or α4β2α6* subtypes. The involvement of α5 containing nAChRs receptor in pain modulation was shown by studies using α5 knockout mice in acute pain tests [42] and knockdown of α5 subunit in the rat with spinal nerve ligation [43]. Recently, testing a set of compounds with various activities to different forms and subtypes of human α4β2 nAChRs (LS, HS, and α4β2α5), Zhang et al. [36] showed that α4β2α5 subtypes played an important role in the analgesic activity of the compounds in the formalin test. Currently, the role of nAChRs containing α6 and β3 subunits is unclear. Finally, the desensitization of α4β2 receptors may drive part of the antinociceptive outcome of nicotinic agonists. For example, Zhang et al. [36] reported that compounds that more potently desensitize α4β2 nAChRs display better scores of analgesia in the formalin test. This raises the possibility that α4β2* nicotinic antagonists may exert some analgesic properties on their own. However, many studies including our own showed that selective and nonselective α4β2* nicotinic antagonists caused an increase or no change in pain behaviors in chronic pain rodent models [28, 34, 49, 55, 56, 105, 106].

The proof-of-concept for α4β2 agonists in the treatment of chronic pain, namely diabetic peripheral neuropathic pain, was established with the relatively selective α4β2* agonist, ABT-594. ABT-594's ability to produce antinociception has been particularly successful in preclinical models of pain, such as the formalin injection, chronic constriction injury, and spinal nerve ligation [41, 44–46]. ABT-594 was also found to exert potent antinociception in rat disease models, such as using cyclophosphamide to induce bladder inflammation, or cystitis [47]. In a rat chemotherapy-induced pain model, ABT-594 was found to have potent anti-allodynic effects [48]. Even inactive doses of ABT-594 were found to potentiate gabapentin-mediated antinociception in the rat formalin test [45]. Distinct from epibatidine, which produces coinciding motor and antihyperalgesic effects, ABT-594 has been found to produce distinct motor and antihyperalgesic effects in models of persistent inflammatory and neuropathic pain, supporting the possibility that ABT-594 could have an improved therapeutic window compared to epibatidine [49]. In a randomized, double-blind, placebo-controlled phase II clinical trials, ABT-594 has produced promising results in patients with diabetic peripheral neuropathic pain. ABT-594 improved pain, though adverse effects were frequently reported [50, 51]. These side effects are probably mediated by activation of α3β4* nAChRs peripheral receptors. Indeed, ABT-594 is functionally only 2.4 times more selective for α4β2 over α3β4 nicotinic subtypes [30].

ABT-894 a more highly selective α4β2 agonist that has also demonstrated efficacy in preclinical animal models of neuropathic pain [33] was tested in humans. While ABT-894 was well tolerated, and no significant safety issues were identified, it failed however to improve pain levels in patients with diabetic peripheral neuropathic pain [52], suggesting that defining a therapeutic window for neuropathic pain by selectively targeting the α4β2 nAChR subtypes remains a major challenge and thus a better understanding of the therapeutic profile of this receptor subtype in humans is needed. Bertrand and colleagues recently argued that if indeed α4β2 receptor activation by the above compounds is needed for an analgesic effect to manifest, then relatively high therapeutic doses to achieve brain concentrations necessary for robust activation of α4β2 nAChRs are required. Since the "functional" selectivity of most α4β2 nAChRs agonists is relatively modest, then these high brain concentrations will possibly interact with non-α4β2 nAChRs resulting in a narrow therapeutic index [53]. This could partially explain why the development of α4β2 agonists for pain has been so challenging.

3.2 *α4β2 nAChR Positive Allosteric Modulators*

Targeting the modulation of α4β2 nAChRs to produce analgesia in animal models is not limited to agonists. Allosteric modulators, namely positive allosteric modulators, have been shown to produce or enhance analgesia on their own or in concert with an agonist in various rodent models of pain. For example, the analgesic effect of α4β2 agonist ABT-594 has been shown to be augmented by the positive allosteric modulator (PAM) NS9283 [54] in the carrageenan-induced thermal hyperalgesia test and a model of postoperative pain [41, 44]. It was also found that PAM NS9283 did not potentiate the α4β2 agonist ABT-594's adverse effects [44]. Similarly, the PAM NS9283 was also found to dramatically potentiate the α4β2 nAChR selective agonist NS3956 without potentiating adverse effects [31]. Clinical research investigating the effects of NS9283 in humans would help clarify the therapeutic profile and potential of the PAM NS9283. A greater understanding of how and where α4β2 nAChR agonists as well as PAMs act could better inform the transition from preclinical to clinical studies of the analgesic efficacy of these compounds.

3.3 α4β2 nAChR Sites of Action

As described above, α4β2 nAChRs are found throughout the central nervous system, and preclinical studies have found that α4β2 agonists predominantly act spinally or, more often, supraspinally. The agonist ABT-594 has been shown to be largely centrally acting [47, 55]. It was also found that ABT-594's analgesic effects may involve the activation of the nucleus raphe magnus, suggesting a supraspinal contribution of ABT-594's mechanism [17]. However, other groups have argued that the spinal cord is a key site where the molecular action of α4β2 nAChRs produces analgesia [37]. Whole-cell patch-clamp experiments in spinal cord preparations from adult mice showed that the α4β2 nAChR subtype could tonically inhibit nociceptive transmission through presynaptic facilitation of inhibitory neurotransmission in the substantia gelatinosa [56]. However, there is also some evidence for the involvement of peripheral α4β2 nAChRs. The α4β2 nAChR antagonist chlorisondamine was shown to partially block ABT-594's effect and have no effect on agonist A-85380's antinociceptive effects when administered intraperitoneally while it blocked ABT-594's and A-85380's effects completely when administered intracerebroventricularly [34, 48, 57]. Accordingly, A-85380 was found to act at sites both within and outside the nucleus raphe magnus, potentially suggesting central sites of action for α4β2 agonist A-85380 [58]. Evidence supports a supraspinal, or at least a central, action of compounds that activate the α4β2 nAChR, though peripheral sites are probably involved to a lesser degree. A better understanding of the mechanism of these compounds may explain this discrepancy.

4 α7 nAChRs as Targets

The α7 subtype of homomeric nAChRs is a well-characterized member of the ligand-gated ion channel superfamily [59]. nAChR α7 subtypes are distinguished by their high calcium permeability and their rapid desensitization during agonist stimulation [60] compared to other nAChR subtypes. This α7 nAChR subtype is ubiquitously expressed in both the central and peripheral nervous system [59].

The involvement of α7 nAChRs in cognition and their therapeutic potential for cognitive disorders has been extensively described ([61]; see also Chap. 11). Ever since it has been shown by Pedigo et al. [62] that the activation of receptors by ACh modifies nociception, the nicotinic cholinergic system has been known to play an important role in pain transmission. Numerous studies suggest that activation of α7 nAChR subtypes by an endogenous cholinergic tone or α7 nAChR agonists has a potential role in pain management [60, 63, 64].

These receptors are present in supraspinal and spinal pain-transmission pathways [65, 66]. Autoradiographic analyses revealed that α7 nAChR binding sites were numerous within the substantia gelatinosa (equivalent to Rexed's lamina II) in rat [67] and human [68] spinal cord, and in the rat, these sites were reduced following dorsal rhizotomy [69]. They are also expressed on immune and nonimmune cytokine-producing cells, such as macrophages microglia and keratinocytes [66]. α7 nAChRs are expressed on macrophages which are key immune cells involved in the initiation, maintenance, and resolution of inflammation [66, 71]. In addition, other types of immune cells such as T-cells, B-cells, microglia, and monocytes dendritic cells express α7 nAChR subtypes. It is possible that ACh is closely associated with controlling immune cell functions, attenuation of pro-inflammatory cytokines production, and inhibition of the inflammatory process via activation of α7 nAChRs [70, 71]. This neurophysiological mechanism reduces inflammation by decreasing cytokine synthesis via release of ACh in organs of the reticulo-endothelial system, such as the lungs, spleen, liver, kidneys, and gastrointestinal tract [72]. It has been revealed that α7 nAChRs are implicated in modulating tumor necrosis factor, interleukin-1, interleukin-6, interleukin-18, high mobility group box 1, and some other pro-inflammatory cytokines without affecting the anti-inflammatory cytokine interleukin-10 [66, 70, 73]. A study with α7 subunit knockout mice has demonstrated a critical role for the α7 nAChRs as a peripheral component in cholinergic anti-inflammatory pathway [65]. Hence, these receptors present an alternative therapeutic approach for modulation of inflammation-based pain syndromes [71, 72].

Taken together, these studies suggest that α7 nAChRs represent a significant potential for inflammation-related pain. α7 nAChRs have been seen a new target for inflammation and inflammatory pain and α7 receptor agonists are more efficient than ACh at inhibiting the inflammatory signaling [71, 72]. Studies have shown that α7 nAChR agonists such as choline, CDP-choline, compound B, JN403, PHA-543613, and AR-R17779 were found to exhibit anti-inflammatory effects in various inflammation and pain models in rodents [60, 63, 74–80]. Furthermore, targeting α7 nAChRs in the treatment of chronic neuropathic pain associated with inflammation and nerve injury was used in several preclinical studies [77, 80–82]. Importantly, data indicate a long-lasting neuroprotective effect of nAChR activation, involving mainly α7nAChR subtypes. For example, repeated administration of a selective α7 agonist is able to decrease allodynia in a chronic neuropathic rat model and reverse signs of neuroinflammation and neurodegeneration (macrophagic infiltrate, decrease in axon compactness and diameter together with a significant loss of myelin sheaths) [82]. Modulation of the inflammatory response with α7 nicotinic agonists might open new strategies to cure diseases with important inflammatory component such as ulcerative colitis, sepsis, acute pancreatitis, and asthma.

In addition to agonists, allosteric modulators of α7 nAChRs also afford new target area for rational drug design and discovery [59]. α7 nAChR-selective allosteric modulators could modulate the activity of endogenous ACh in cholinergic neurotransmission without directly activating α7 nAChRs. The α7 nAChR PAMs have typically been divided into two types based on their electrophysiological properties. Type I PAMs increase agonist response amplitudes with little or no effect on desensitization, whereas type II PAMs increase agonist response amplitudes and decrease rate of desensitization [59]. Furthermore, in contrast to α7 agonists, upregulation of α7 nAChRs in the brain does not occur with the α7 nAChR PAMs in vivo [83]. In recent studies, α7 nAChR-selective PAMs have been reported to have anti-inflammatory, antinociceptive, antihyperalgesic, and anti-allodynic effects in experimental pain models [80, 84, 85]. In rats, PNU-120596, a type II α7 nAChRs PAM, significantly reduced mechanical hyperalgesia and weight-bearing deficits in the carrageenan inflammatory test through a reduction of TNF-α and IL-6 within the hind paw edema [79]. In recent studies, it has been shown that PNU-120596 reduced nociception dose dependently [80], while also enhancing synergistically the effects of α7 receptor agonists such as choline to elicit antinociceptive effects in the formalin pain model in mice [85]. Furthermore, tolerance to PNU-120596's antinociceptive effects did not develop after sub-chronic treatment in mice. Additionally, it has been found that the activation of spinal extracellular signal-regulated kinase-1/2 pathways is the likely mechanism of the antinociceptive effect of PNU-120596 in the formalin test. PNU-120596 also reduced paw edema and thermal hyperalgesia induced by intraplantar injection of carrageenan in mice [80]. Finally, the neuropathic pain model of chronic constriction injury (CCI model), a dose-dependent antihyperalgesic and anti-allodynic effect of PNU-120596 for up to 6 h, was observed after systemic administration in mice. However NS-1738, a type I α7 nAChRs PAM, failed to have any effect in the CCI model. This is in contrast to rodent models of cognition and memory, where both type I and type II PAMs for the α7 nAChRs showed cognitive enhancement. Interestingly, the long-acting effects of α7 nAChRs PAMs in both cognitive and pain models were in discordance with their pharmacokinetic profile in rodents, which suggests the involvement of post-receptor signaling mechanisms and raises the possibility that α7 nAChRs PAMs may actually exert their behavioral and pharmacological activity in an ion channel-independent signaling process [86]. Further studies will surely increase our understanding of the similarities and differences in the analgesic and anti-inflammatory properties of type I and type II α7 receptor PAMs.

The recent discovery and characterization of α7 nAChR PAMs in animal models of pain and inflammation has created new opportunities for targeting these receptors as anti-inflammatory and analgesic agents. Further behavioral studies that measure the effects of allosteric modulators for nAChRs will certainly be a necessary part of preclinical studies of these compounds, and such studies will generate important information about these compounds that can influence their further clinical development.

5 Other Nicotinic Receptor Subtypes

The α9 and α10 nicotinic subunits have also become of interest as novel targets for treatment of chronic pain and inflammation [87]. Gene transcripts for α9α10 have been identified in diverse yet limited numbers of tissues such as auditory system, but importantly not in the brain. These subunits are also expressed in immune cells that are involved in the inflammatory response. Interestingly nicotine and other nicotinic agonists such as cytisine and epibatine are antagonists of α9 and α9α10 nAChRs. The α-conotoxins RgIA and Vc1.1. are selective antagonists of α9α10 nAChRs and were found to be effective in rat models of neuropathic and chronic inflammatory pain [87–90], an effect that is possibly mediated via immunological mechanisms. Recently, non-peptide, small molecule antagonists of α9α10 nAChRs were reported to be active in rat models of neuropathic and tonic inflammatory pain [91]. Additionally, a lead analog, ZZ1-61c, was found to be effective in reversing mechanical allodynia in a rat model of neuropathy induced by administration of vincristine [92]. At the doses tested, ZZ1-61c did not cause motor dysfunction or muscular weakness.

6 Analgesic Effects of Nicotine and Other Nicotinic Agonists: Clinical Studies

Nicotine, the prototypical broad spectrum nicotinic antagonist, has been known to have antinociceptive properties in animal models since the 1970s [93]. Nicotine was tested in a human volunteer trial and found to have analgesic properties in response to a cold pressor stimulus in abstinent smokers and long-term ex-smokers [94]. The analgesic response to transdermal nicotine may be modality or route dependent, however, because analgesic activity was not found using a pressure stimulus in a trial of female nonsmokers [95].

Intranasal nicotine (3 mg) was first tested for analgesic activity in a clinical setting in a double-blind randomized control trial in 20 nonsmoking women emerging from anesthesia after gynecological surgery [96]. Visual analog scores for pain were significantly lower 1 h after surgery in the patients treated with nicotine (VAS 7.6 ± 1.4 vs. 5.3 ± 1.6). The reduction in reported pain was in spite of a reduction by half in patient-controlled administration of morphine in the patients treated with nicotine. More recently opioid sparing was demonstrated with the use of intranasal nicotine in a larger trial of female nonsmokers [97]. Several trials have documented efficacy of nicotine delivered by a transcutaneous route for postoperative pain in male nonsmokers having prostatectomy [98], female nonsmokers undergoing gynecological surgery [99], and a mixed gender group of nonsmokers after third molar surgery [100]. Prolonged exposure to nicotine is well known to result in desensitization, inactivation, and internalization of many subtypes of nicotinic receptors. As such, it is perhaps not surprising that transcutaneous nicotine was not effective however, as an analgesic adjuvant in smokers after gynecological surgery [101]. All of the above noted clinical trials that demonstrated analgesic efficacy in the groups treated with nicotine also found an increase in the incidence of nausea. Cardiovascular complications were not observed in clinical trials.

The diversity in the neuronal nicotinic receptors was suspected on the basis of pharmacological studies and was demonstrated with cloning [102] and expression of cDNA [103] for different rat nicotinic acetylcholine receptor subunits. As discussed above, receptors comprised of different nicotinic acetylcholine subunits have been found to be expressed in peripheral and central neurons that contribute to pain pathways. With nausea as a dose-limiting side effect of nicotine itself, the search for clinically useful nicotinic analgesic drugs has focused on subtype-specific nicotinic agonists, partial agonists, and positive allosteric modulators (PAM), many of which have antinociceptive and anti-inflammatory effects in animal models, *described in detail above* (*Sect. 4*). Of the subtype-selective nicotinic agonists, only ABT-594 and ABT-894 which are selective for α4β2 subunit containing nicotinic acetylcholine receptors have been tested in clinical trials. The less selective ABT-594 had dose-dependent efficacy in patients with painful diabetic neuropathy in two trials, although the patients reported side effects including nausea, vomiting, dizziness, headache, and abnormal dreams [50, 51]. As discussed in Sect. 3.1, surprisingly, the more α4β2* selective ABT-894 was not effective as an analgesic in patients with painful diabetic neuropathy [52]. The authors concluded that the therapeutic index may not be sufficient for clinical use of α4β2 selective agonists as analgesics. The naturally occurring α7 nAChR selective agonist choline has efficacy in preclinical models of acute and inflammatory pain. Only one clinical trial has evaluated choline supplementation for acute postoperative pain. Unfortunately, oral perioperative dosing of 20 g of choline did not achieve elevated plasma choline concentration or a reduction in postoperative pain or opioid utilization [104]. The reduced oral absorption compared to that found in previous studies of nutritional supplementation with choline may have been a result of decreased bowel function in the perioperative period.

7 Conclusion

Activation of neuronal nicotinic acetylcholine receptors has clear antinociceptive properties in preclinical models and the broad-spectrum agonist, nicotine, has efficacy as an analgesic adjuvant in the treatment of postoperative pain. The moderately selective α4β2* agonist ABT-594 has efficacy for the treatment of painful diabetic neuropathy. All of the above agonists have in common side effects of nausea and vomiting. It is likely that this side effect is not mediated by α4β2* subtype-selective receptors as treatment with the more α4β2* selective ABT-894 was not associated with excess nausea or vomiting not analgesic efficacy. Future trials with positive allosteric modulators of α4β2* nicotinic receptors hold promise. Selective activation of α7 containing nicotinic receptors is particularly interesting, as they seem to also have anti-inflammatory activity in preclinical models. The one clinical trial with choline, an α7 subtype-selective nicotinic agonist, failed at its primary endpoint, a reduction in postoperative pain; however a significant increase in plasma choline concentrations was not achieved with oral perioperative dosing. As such activation of α7 containing nicotinic receptors remains a viable approach to the treatment of pain and inflammation. Antagonism of α9α10 nAChRs by α-conotoxins RgIA and Vc1.1 was effective in a rodent model of neuropathic and chronic inflammatory pain although to date no α9α10 nAChRs have been identified in the brain. An important challenge facing the development of new nicotinic analgesics candidates is the limitations of the current animal models of pain. They appear to have worked well for mechanistic studies, but poorly as a basis for selecting new analgesic candidates. Most of the animal models of pain rely on detecting a change in the threshold or response to an applied stimulus or injury and the absence of verbal communication in animals is undoubtedly an obstacle to the evaluation of pain. Also, the neurobiology of nociceptive systems differs between species and this limits the extrapolation of findings from animal studies to man. New models that directly measure the affective component of pain are currently being developed. It will be important to test nicotinic ligands in these models since they will ultimately provide better markers to the natural disease state.

The diversity in subunit composition, cellular localization, and pharmacology of the nicotinic acetylcholine receptor family has made it a complicated target for analgesic drug development. However, the potency and selectivity of some of the pharmacological agents available hold forth the promise of important and much needed additions to our pharmacological armamentarium against pain. The reader is referred to Chaps. 5, 9, and 21 for further discussion of the involvement of nAChRs in pain transmission.

Acknowledgements Participants in this research review are supported by the National Institutes of Health (NIH) and The Scientific and Technological Research Council of Turkey (TUBITAK).

References

1. Davis L, Pollock LJ, Stone TT. Visceral pain. Surg Gynecol Obstet. 1932;55:418–26.
2. Spande TF, Garraffo HM, Edwards MW, Yeh HJC, Pannell LK, Daly JW. Epibatidine: a novel (chloropyridyl) azabicycloheptane with potent analgesic activity from an Ecuadoran poison frog. J Am Chem Soc. 1992;114(9):3475–8.
3. Bannon AW, Decker MW, Holladay MW, Curzon P, Donnelly-Roberts D, Puttfarcken PS, et al. Broad-spectrum, non-opioid analgesic activity by selective modulation of neuronal nicotinic acetylcholine receptors. Science. 1998;279(5347):77–81.
4. Khan I, Osaka H, Stanislaus S, Calvo RM, Deerinck T, Yaksh TL, et al. Nicotinic acetylcholine receptor distribution in relation to spinal neurotransmission pathways. J Comp Neurol. 2003;467(1):44–59.
5. Cordero-Erausquin M, Pons S, Faure P, Changeux JP. Nicotine differentially activates inhibitory and excitatory neurons in the dorsal spinal cord. Pain. 2004;109(3):308–18.
6. Genzen JR, McGehee DS. Nicotinic modulation of GABAergic synaptic transmission in the spinal cord dorsal horn. Brain Res. 2005;1031(2):229–37.
7. Mattila MJ, Ahtee L, Saarnivaara L. The analgesic and sedative effects of nicotine in white mice, rabbits and golden hamsters. Ann Med Exp Biol Fenn. 1968;46(1):78–84.

8. Plenge P, Mellerup ET, Wörtwein G. Characterization of epibatidine binding to medial habenula: potential role in analgesia. J Pharmacol Exp Ther. 2002;302(2):759–65.
9. Turner JR, Kellar KJ. Nicotinic cholinergic receptors in the rat cerebellum: multiple heteromeric subtypes. J Neurosci. 2005;25(40):9258–65.
10. Iwamoto ET, Marion L. Adrenergic, serotonergic and cholinergic components of nicotinic antinociception in rats. J Pharmacol Exp Ther. 1993;265(2):777–89.
11. Iwamoto ET. Characterization of the antinociception induced by nicotine in the pedunculopontine tegmental nucleus and the nucleus raphe magnus. J Pharmacol Exp Ther. 1991;257(1):120–33.
12. Iwamoto ET. Antinociception after nicotine administration into the mesopontine tegmentum of rats: evidence for muscarinic actions. J Pharmacol Exp Ther. 1989;251(2):412–21.
13. Aceto MD, Bagley RS, Dewey WL, Fu TC, Martin BR. The spinal cord as a major site for the antinociceptive action of nicotine in the rat. Neuropharmacology. 1986;25(9):1031–6.
14. Christensen MK, Smith DF. Antinociceptive effects of the stereoisomers of nicotine given intrathecally in spinal rats. J Neural Transm Gen Sect. 1990;80(3):189–94.
15. Khan IM, Yaksh TL, Taylor P. Epibatidine binding sites and activity in the spinal cord. Brain Res. 1997;753(2):269–82.
16. Damaj MI, Fei-Yin M, Dukat M, Glassco W, Glennon RA, Martin BR. Antinociceptive responses to nicotinic acetylcholine receptor ligands after systemic and intrathecal administration in mice. J Pharmacol Exp Ther. 1998;284(3):1058–65.
17. Bitner RS, Nikkel AL, Curzon P, Arneric SP, Bannon AW, Decker MW. Role of the nucleus raphe magnus in antinociception produced by ABT-594: immediate early gene responses possibly linked to neuronal nicotinic acetylcholine receptors on serotonergic neurons. J Neurosci. 1998;18(14):5426–32.
18. Cucchiaro G, Commons KG. Alpha 4 nicotinic acetylcholine receptor subunit links cholinergic to brainstem monoaminergic neurotransmission. Synapse. 2003;49(3):195–205.
19. Cucchiaro G, Chaijale N, Commons KG. The dorsal raphe nucleus as a site of action of the antinociceptive and behavioral effects of the alpha4 nicotinic receptor agonist epibatidine. J Pharmacol Exp Ther. 2005;313(1):389–94.
20. Galindo-Charles L, Hernandez-Lopez S, Galarraga E, Tapia D, Bargas J, Garduño J, et al. Serotoninergic dorsal raphe neurons possess functional postsynaptic nicotinic acetylcholine receptors. Synapse. 2008;62(8):601–15.
21. Gotti C, Zoli M, Clementi F. Brain nicotinic acetylcholine receptors: native subtypes and their relevance. Trends Pharmacol Sci. 2006;27(9):482–91.
22. Léna C, de Kerchove D'Exaerde A, Cordero-Erausquin M, Le Novère N, del Mar Arroyo-Jimenez M, Changeux JP. Diversity and distribution of nicotinic acetylcholine receptors in the locus ceruleus neurons. Proc Natl Acad Sci U S A. 1999;96(21):12126–31.
23. Nakamura M, Jang IS. Presynaptic nicotinic acetylcholine receptors enhance GABAergic synaptic transmission in rat periaqueductal gray neurons. Eur J Pharmacol. 2010;640(1–3): 178–84.
24. Cordero-Erausquin M, Changeux JP. Tonic nicotinic modulation of serotoninergic transmission in the spinal cord. Proc Natl Acad Sci U S A. 2001;98(5):2803–7.
25. Marubio LM, del Mar Arroyo-Jimenez M, Cordero-Erausquin M, Léna C, Le Novère N, de Kerchove d'Exaerde A, et al. Reduced antinociception in mice lacking neuronal nicotinic receptor subunits. Nature. 1999;398(6730):805–10.
26. Damaj MI, Fonck C, Marks MJ, Deshpande P, Labarca C, Lester HA, et al. Genetic approaches identify differential roles for alpha4beta2* nicotinic receptors in acute models of antinociception in mice. J Pharmacol Exp Ther. 2007;321(3):1161–9.
27. Nirogi R, Goura V, Abraham R, Jayarajan P. α4β2* neuronal nicotinic receptor ligands (agonist, partial agonist and positive allosteric modulators) as therapeutic prospects for pain. Eur J Pharmacol. 2013;712(1–3):22–9.
28. Cucchiaro G, Xiao Y, Gonzalez-Sulser A, Kellar KJ. Analgesic effects of Sazetidine-A, a new nicotinic cholinergic drug. Anesthesiology. 2008;109(3):512–9.

29. AlSharari SD, Carroll FI, McIntosh JM, Damaj MI. The antinociceptive effects of nicotinic partial agonists varenicline and sazetidine-A in murine acute and tonic pain models. J Pharmacol Exp Ther. 2012;342(3):742–9.
30. Donnelly-Roberts DL, Puttfarcken PS, Kuntzweiler TA, Briggs CA, Anderson DJ, Campbell JE, et al. ABT-594 [(R)-5-(2-azetidinylmethoxy)-2-chloropyridine]: a novel, orally effective analgesic acting via neuronal nicotinic acetylcholine receptors: I. In vitro characterization. J Pharmacol Exp Ther. 1998;285(2):777–86.
31. Rode F, Munro G, Holst D, Nielsen EØ, Troelsen KB, Timmermann DB, Rønn LC, Grunnet M. Positive allosteric modulation of α4β2 nAChR agonist induced behaviour. Brain Res. 2012;1458:67–75.
32. Nirogi R, Jabaris SL, Jayarajan P, Abraham R, Shanmuganathan D, Rasheed MA, et al. Antinociceptive activity of α4β2* neuronal nicotinic receptor agonist A-366833 in experimental models of neuropathic and inflammatory pain. Eur J Pharmacol. 2011;668(1–2):155–62.
33. Ji J, Bunnelle WH, Anderson DJ, Faltynek C, Dyhring T, Ahring PK, et al. A-366833: a novel nicotinonitrile-substituted 3,6-diazabicyclo[3.2.0]-heptane alpha4beta2 nicotinic acetylcholine receptor selective agonist: synthesis, analgesic efficacy and tolerability profile in animal models. Biochem Pharmacol. 2007;74(8):1253–62.
34. Rueter LE, Kohlhaas KL, Curzon P, Surowy CS, Meyer MD. Peripheral and central sites of action for A-85380 in the spinal nerve ligation model of neuropathic pain. Pain. 2003;103(3): 269–76.
35. Damaj MI, Glassco W, Aceto MD, Martin BR. Antinociceptive and pharmacological effects of metanicotine, a selective nicotinic agonist. J Pharmacol Exp Ther. 1999;291(1):390–8.
36. Zhang J, Xiao YD, Jordan KG, Hammond PS, Van Dyke KM, Mazurov AA, et al. Analgesic effects mediated by neuronal nicotinic acetylcholine receptor agonists: correlation with desensitization of α4β2* receptors. Eur J Pharm Sci. 2012;47(5):813–23.
37. Gao B, Hierl M, Clarkin K, Juan T, Nguyen H, Valk MV, et al. Pharmacological effects of nonselective and subtype-selective nicotinic acetylcholine receptor agonists in animal models of persistent pain. Pain. 2010;149(1):33–49.
38. Zwart R, Carbone AL, Moroni M, Bermudez I, Mogg AJ, Folly EA, et al. Sazetidine-A is a potent and selective agonist at native and recombinant alpha4 beta2 nicotinic acetylcholine receptors. Mol Pharmacol. 2008;73(6):1838–43.
39. Cheng LZ, Han L, Fan J, Huang LT, Peng LC, Wang Y. Enhanced inhibitory synaptic transmission in the spinal dorsal horn mediates antinociceptive effects of TC-2559. Mol Pain. 2011;7:56.
40. Decker MW, Meyer MD, Sullivan JP. The therapeutic potential of nicotinic acetylcholine receptor agonists for pain control. Expert Opin Investig Drugs. 2001;10(10):1819–30.
41. Lee CH, Zhu C, Malysz J, Campbell T, Shaughnessy T, Honore P, et al. α4β2 neuronal nicotinic receptor positive allosteric modulation: an approach for improving the therapeutic index of α4β2 nAChR agonists in pain. Biochem Pharmacol. 2011;82(8):959–66.
42. Jackson KJ, Marks MJ, Vann RE, Chen X, Gamage TF, Warner JA, et al. Role of alpha5 nicotinic acetylcholine receptors in pharmacological and behavioral effects of nicotine in mice. J Pharmacol Exp Ther. 2010;334(1):137–46.
43. Vincler MA, Eisenach JC. Knock down of the alpha 5 nicotinic acetylcholine receptor in spinal nerve-ligated rats alleviates mechanical allodynia. Pharmacol Biochem Behav. 2005;80(1):135–43.
44. Zhu CZ, Chin CL, Rustay NR, Zhong C, Mikusa J, Chandran P, et al. Potentiation of analgesic efficacy but not side effects: co-administration of an α4β2 neuronal nicotinic acetylcholine receptor agonist and its positive allosteric modulator in experimental models of pain in rats. Biochem Pharmacol. 2011;82(8):967–76.
45. Munro G, Dyhr H, Grunnet M. Selective potentiation of gabapentin-mediated antinociception in the rat formalin test by the nicotinic acetylcholine receptor agonist ABT-594. Neuropharmacology. 2010;59(3):208–17.
46. Boyce S, Webb JK, Shepheard SL, Russell MG, Hill RG, Rupniak NM. Analgesic and toxic effects of ABT-594 resemble epibatidine and nicotine in rats. Pain. 2000;85(3):443–50.

47. Joshi SK, Mikusa JP, Weaver B, Honore P. Morphine and ABT-594 (a nicotinic acetylcholine agonist) exert centrally mediated antinociception in the rat cyclophosphamide cystitis model of visceral pain. J Pain. 2008;9(2):146–56.
48. Lynch III JJ, Wade CL, Mikusa JP, Decker MW, Honore P. ABT-594 (a nicotinic acetylcholine agonist): anti-allodynia in a rat chemotherapy-induced pain model. Eur J Pharmacol. 2005;509(1):43–8.
49. Kesingland AC, Gentry CT, Panesar MS, Bowes MA, Vernier JM, Cube R, Walker K, Urban L. Analgesic profile of the nicotinic acetylcholine receptor agonists, (+)-epibatidine and ABT-594 in models of persistent inflammatory and neuropathic pain. Pain. 2000; 86(1–2):113–8.
50. Dutta S, Hosmane BS, Awni WM. Population analyses of efficacy and safety of ABT-594 in subjects with diabetic peripheral neuropathic pain. AAPS J. 2012;14(2):168–75.
51. Rowbotham MC, Duan WR, Thomas J, Nothaft W, Backonja MM. A randomized, double-blind, placebo-controlled trial evaluating the efficacy and safety of ABT-594 in patients with diabetic peripheral neuropathic pain. Pain. 2009;146(3):245–52.
52. Rowbotham MC, Arslanian A, Nothaft W, Duan WR, Best AE, Pritchett Y, et al. Efficacy and safety of the α4β2 neuronal nicotinic receptor agonist ABT-894 in patients with diabetic peripheral neuropathic pain. Pain. 2012;153(4):862–8.
53. Hurst R, Rollema H, Bertrand D. Nicotinic acetylcholine receptors: from basic science to therapeutics. Pharmacol Ther. 2013;137(1):22–54.
54. Grupe M, Jensen AA, Ahring PK, Christensen JK, Grunnet M. Unravelling the mechanism of action of NS9283, a positive allosteric modulator of (α4)3(β2)2 nicotinic ACh receptors. Br J Pharmacol. 2013;168(8):2000–10.
55. Decker MW, Bannon AW, Buckley MJ, Kim DJ, Holladay MW, Ryther KB, et al. Antinociceptive effects of the novel neuronal nicotinic acetylcholine receptor agonist, ABT-594, in mice. Eur J Pharmacol. 1998;346(1):23–33.
56. Rashid MH, Furue H, Yoshimura M, Ueda H. Tonic inhibitory role of alpha4beta2 subtype of nicotinic acetylcholine receptors on nociceptive transmission in the spinal cord in mice. Pain. 2006;125(1–2):125–35.
57. Rueter LE, Donnelly-Roberts DL, Curzon P, Briggs CA, Anderson DJ, Bitner RS. A-85380: a pharmacological probe for the preclinical and clinical investigation of the alphabeta neuronal nicotinic acetylcholine receptor. CNS Drug Rev. 2006;12(2):100–12.
58. Curzon P, Nikkel AL, Bannon AW, Arneric SP, Decker MW. Differences between the antinociceptive effects of the cholinergic channel activators A-85380 and (+/–)-epibatidine in rats. J Pharmacol Exp Ther. 1998;287(3):847–53.
59. Sharma G, Vijayaraghavan S. Nicotinic receptors containing the alpha7 subunit: a model for rational drug design. Curr Med Chem. 2008;15(28):2921–32.
60. Feuerbach D, Lingenhoehl K, Olpe HR, Vassout A, Gentsch C, Chaperon F, et al. The selective nicotinic acetylcholine receptor alpha7 agonist JN403 is active in animal models of cognition, sensory gating, epilepsy and pain. Neuropharmacology. 2009;56(1):254–63.
61. Leiser SC, Bowlby MR, Comery TA, Dunlop J. A cog in cognition: how the alpha 7 nicotinic acetylcholine receptor is geared towards improving cognitive deficits. Pharmacol Ther. 2009;122:302–11.
62. Pedigo NW, Dewey WL, Harris LS. Determination and characterization of the antinociceptive activity of intraventricularly administered acetylcholine in mice. J Pharmacol Exp Ther. 1975;193(3):845–52.
63. Damaj MI, Meyer EM, Martin BR. The antinociceptive effects of alpha7 nicotinic agonists in an acute pain model. Neuropharmacology. 2000;39(13):2785–91.
64. Wang Y, Su DM, Wang RH, Liu Y, Wang H. Antinociceptive effects of choline against acute and inflammatory pain. Neuroscience. 2005;132(1):49–56.
65. Wang H, Yu M, Ochani M, Amella CA, Tanovic M, Susarla S, et al. Nicotinic acetylcholine receptor alpha7 subunit is an essential regulator of inflammation. Nature. 2003;421(6921): 384–8.
66. Pavlov VA, Tracey KJ. Neural regulators of innate immune responses and inflammation. Cell Mol Life Sci. 2004;61(18):2322–31.

67. Hunt S, Schmidt J. Some observations on the binding patterns of alpha-bungarotoxin in the central nervous system of the rat. Brain Res. 1978;157(2):213–32.
68. Gillberg PG, Aquilonius SM. Cholinergic, opioid and glycine receptor binding sites localized in human spinal cord by in vitro autoradiography. Changes in amyotrophic lateral sclerosis. Acta Neurol Scand. 1985;72(3):299–306.
69. Gillberg PG, Wiksten B. Effects of spinal cord lesions and rhizotomies on cholinergic and opiate receptor binding sites in rat spinal cord. Acta Physiol Scand. 1986;126(4):575–82.
70. Pavlov VA, Tracey KJ. The cholinergic anti-inflammatory pathway. Brain Behav Immun. 2005;19(6):493–9.
71. de Jonge WJ, Ulloa L. The alpha7 nicotinic acetylcholine receptor as a pharmacological target for inflammation. Br J Pharmacol. 2007;151(7):915–29.
72. Bencherif M, Lippiello PM, Lucas R, Marrero MB. Alpha7 nicotinic receptors as novel therapeutic targets for inflammation-based diseases. Cell Mol Life Sci. 2011;68(6):931–49.
73. Czura CJ, Friedman SG, Tracey KJ. Neural inhibition of inflammation: the cholinergic anti-inflammatory pathway. J Endotoxin Res. 2003;9(6):409–13.
74. Medhurst SJ, Hatcher JP, Hille CJ, Bingham S, Clayton NM, Billinton A, et al. Activation of the alpha7-nicotinic acetylcholine receptor reverses complete Freund adjuvant-induced mechanical hyperalgesia in the rat via a central site of action. J Pain. 2008;9(7):580–7.
75. van Maanen MA, Lebre MC, van der Poll T, LaRosa GJ, Elbaum D, Vervoordeldonk MJ, et al. Stimulation of nicotinic acetylcholine receptors attenuates collagen-induced arthritis in mice. Arthritis Rheum. 2009;60(1):114–22.
76. Gurun MS, Parker R, Eisenach JC, Vincler M. The effect of peripherally administered CDP-choline in an acute inflammatory pain model: the role of alpha7 nicotinic acetylcholine receptor. Anesth Analg. 2009;108(5):1680–7.
77. Bagdas D, Sonat FA, Hamurtekin E, Sonal S, Gurun MS. The antihyperalgesic effect of cytidine-5′-diphosphate-choline in neuropathic and inflammatory pain models. Behav Pharmacol. 2011;22(5–6):589–98.
78. Marrero MB, Bencherif M, Lippiello PM, Lucas R. Application of alpha7 nicotinic acetylcholine receptor agonists in inflammatory diseases: an overview. Pharm Res. 2011;28(2):413–6.
79. Munro G, Hansen R, Erichsen H, Timmermann D, Christensen J, Hansen H. The α7 nicotinic ACh receptor agonist compound B and positive allosteric modulator PNU-120596 both alleviate inflammatory hyperalgesia and cytokine release in the rat. Br J Pharmacol. 2012;167(2):421–35.
80. Freitas K, Ghosh S, Ivy Carroll F, Lichtman AH, Imad Damaj M. Effects of alpha 7 positive allosteric modulators in murine inflammatory and chronic neuropathic pain models. Neuropharmacology. 2013;65:156–64.
81. Loram LC, Taylor FR, Strand KA, Maier SF, Speake JD, Jordan KG, et al. Systemic administration of an alpha-7 nicotinic acetylcholine agonist reverses neuropathic pain in male Sprague Dawley rats. J Pain. 2012;13(12):1162–71.
82. Pacini A, Di Cesare Mannelli L, Bonaccini L, Ronzoni S, Bartolini A, Ghelardini C. Protective effect of alpha7 nAChR: behavioural and morphological features on neuropathy. Pain. 2010;150(3):542–9.
83. Christensen DZ, Mikkelsen JD, Hansen HH, Thomsen MS. Repeated administration of alpha7 nicotinic acetylcholine receptor (nAChR) agonists, but not positive allosteric modulators, increases alpha7 nAChR levels in the brain. J Neurochem. 2010;114(4):1205–16.
84. Freitas K, Carroll FI, Damaj MI. The antinociceptive effects of nicotinic receptors α7-positive allosteric modulators in murine acute and tonic pain models. J Pharmacol Exp Ther. 2013;344(1):264–75.
85. Freitas K, Negus S, Carroll F, Damaj M. In vivo pharmacological interactions between a type II positive allosteric modulator of α7 nicotinic ACh receptors and nicotinic agonists in a murine tonic pain model. Br J Pharmacol. 2013;169(3):567–79.
86. Peng C, Kimbrell MR, Tian C, Pack TF, Crooks PA, Fifer EK, et al. Multiple modes of α7 nAChR non-competitive antagonism of control agonist-evoked and allosterically enhanced currents. Mol Pharmacol. 2013;84(3):459–75 doi:10.1124/mol.113.086462.

87. Vincler M, Wittenauer S, Parker R, Ellison M, Olivera BM, McIntosh JM. Molecular mechanism for analgesia involving specific antagonism of alpha9alpha10 nicotinic acetylcholine receptors. Proc Natl Acad Sci U S A. 2006;103(47):17880–4.
88. Satkunanathan N, Livett B, Gayler K, Sandall D, Down J, Khalil Z. Alpha-conotoxin Vc1.1 alleviates neuropathic pain and accelerates functional recovery of injured neurones. Brain Res. 2005;1059(2):149–58.
89. Nevin ST, Clark RJ, Klimis H, Christie MJ, Craik DJ, Adams DJ. Are alpha9alpha10 nicotinic acetylcholine receptors a pain target for alpha-conotoxins? Mol Pharmacol. 2007;72(6):1406–10.
90. McIntosh JM, Absalom N, Chebib M, Elgoyhen AB, Vincler M. Alpha9 nicotinic acetylcholine receptors and the treatment of pain. Biochem Pharmacol. 2009;78(7):693–702.
91. Zheng G, Zhang Z, Dowell C, Wala E, Dwoskin LP, Holtman JR, et al. Discovery of non-peptide, small molecule antagonists of α9α10 nicotinic acetylcholine receptors as novel analgesics for the treatment of neuropathic and tonic inflammatory pain. Bioorg Med Chem Lett. 2011;21(8):2476–9.
92. Wala EP, Crooks PA, McIntosh JM, Holtman Jr JR. Novel small molecule α9α10 nicotinic receptor antagonist prevents and reverses chemotherapy-evoked neuropathic pain in rats. Anesth Analg. 2012;115(3):713–20.
93. Sahley TL, Berntson GG. Antinociceptive effects of central and systemic administrations of nicotine in the rat. Psychopharmacology (Berl). 1979;65(3):279–83.
94. Fertig JB, Pomerleau OF, Sanders B. Nicotine-produced antinociception in minimally deprived smokers and ex-smokers. Addict Behav. 1986;11(3):239–48.
95. Fukada T, Iwakiri H, Ozaki M. A randomised double-blind crossover trial of the potential analgesic effect of a transdermal nicotine patch in non-smokers based on objective and subjective assessment. Eur J Anaesthesiol. 2011;28(8):592–6.
96. Flood P, Daniel D. Intranasal nicotine for postoperative pain treatment. Anesthesiology. 2004;101(6):1417–21.
97. Jankowski CJ, Weingarten TN, Martin DP, Whalen FX, Gebhart JB, Liedl LM, et al. Randomised trial of intranasal nicotine and postoperative pain, nausea and vomiting in non-smoking women. Eur J Anaesthesiol. 2011;28(8):585–91.
98. Habib AS, White WD, El Gasim MA, Saleh G, Polascik TJ, Moul JW, et al. Transdermal nicotine for analgesia after radical retropubic prostatectomy. Anesth Analg. 2008;107(3):999–1004.
99. Hong D, Conell-Price J, Cheng S, Flood P. Transdermal nicotine patch for postoperative pain management: a pilot dose-ranging study. Anesth Analg. 2008;107(3):1005–10.
100. Yagoubian B, Akkara J, Afzali P, Alfi DM, Olson L, Conell-Price J, Yeh J, Eisig SB, Flood P. Nicotine nasal spray as an adjuvant analgesic for third molar surgery. J Oral Maxillofac Surg. 2011;69(5):1316–9.
101. Olson LC, Hong D, Conell-Price JS, Cheng S, Flood P. A transdermal nicotine patch is not effective for postoperative pain management in smokers: a pilot dose-ranging study. Anesth Analg. 2009;109(6):1987–91.
102. Boulter J, Evans K, Goldman D, Martin G, Treco D, Heinemann S, et al. Isolation of a cDNA clone coding for a possible neural nicotinic acetylcholine receptor alpha-subunit. Nature. 1986;319(6052):368–74.
103. Boulter J, Connolly J, Deneris E, Goldman D, Heinemann S, Patrick J. Functional expression of two neuronal nicotinic acetylcholine receptors from cDNA clones identifies a gene family. Proc Natl Acad Sci U S A. 1987;84(21):7763–7.
104. Sidhu N, Davies S, Nadarajah A, Rivera J, Whittington R, Mercier RJ, Virag L, Wang S, Flood P. Oral choline supplementation for postoperative pain. Br J Anaesth. 2013;111(2):249–55.
105. Hama A, Menzaghi F. Antagonist of nicotinic acetylcholine receptors (nAChR) enhances formalin-induced nociception in rats: tonic role of nAChRs in the control of pain following injury. Brain Res. 2001;888(1):102–6.
106. Ueda M, Iida Y, Tominaga A, Yoneyama T, Ogawa M, Magata Y, Nishimura H, Kuge Y, Saji H. Nicotinic acetylcholine receptors expressed in the ventral posterolateral thalamic nucleus play an important role in anti-allodynic effects. Br J Pharmacol. 2010;159(6):1201–10.

Chapter 13
Nicotinic Acetylcholine Receptors and the Roles of the Alpha7 Subunit

Michael Paolini, Mariella De Biasi, and John A. Dani

Abstract Nicotinic acetylcholine receptors (nAChR) are members of the Cys-loop superfamily of ligand-gated ion channels that includes glycine, γ-aminobutyric acid ($GABA_A$), and serotonin receptor channels. The members of the family are defined by a similar pentameric structure with five membrane-spanning subunits surrounding a central water-filled, cation-selective pore. The nAChRs are further divided into muscle and neuronal types. Muscle nAChRs comprise α1, β1, γ, and δ or ε subunits in a 2:1:1:1 stoichiometric ratio. Neuronal type nAChRs are composed of differing combinations of α and β subunits, with nine genes encoding α subunits (α2–10) and three encoding β subunits (β2–4). This review focuses on the α7 subunit, which was cloned from the chicken in 1990 and from the rat in 1993. α7 has received waning and waxing attention as its involvement in diseases, including Alzheimer's disease and lung cancer, has been defined and redefined. For example, recent reports provide increasing evidence for α7's involvement in the pathogenesis of Alzheimer's disease, suggesting that further work is warranted to understand the roles of the α7 subunit normally and in pathophysiology.

Keywords Nicotine • nAChR • Presynaptic • Synaptic • Calcium

M. Paolini, M.D.
Department of Neuroscience, Baylor College of Medicine, Houston, TX, USA

M. De Biasi, Ph.D.
Department of Psychiatry, Perelman School of Medicine, University of Pennsylvania, Philadelphia, PA, USA

J.A. Dani, Ph.D. (✉)
Department of Neuroscience, Perelman School of Medicine, University of Pennsylvania, 415 Curie Boulevard, Clinical Research Building, Philadelphia, PA, USA
e-mail: Johndani@upenn.edu

R.A.J. Lester (ed.), *Nicotinic Receptors*, The Receptors 26,
DOI 10.1007/978-1-4939-1167-7_13

1 Structure

The structure of Cys-loop receptors, and nAChRs in particular, has been extensively reviewed [1–5]. Therefore, we will focus on a basic overview and the structural characteristics of the α7 subunit. The α7–α9 subunits of the nAChR can form homomeric channels [6–10], but all the other subunits form heteromeric channels commonly with $(\alpha^*)_3(\beta^*)_2$ stoichiometry. Each nAChR subunit has a large N-terminal extracellular domain, four transmembrane regions (TM1–TM4), and a large intracellular loop between TM3 and TM4. The N- and C-terminals of nAChRs are extracellular, as is the loop between TM2 and TM3. TM2 of each subunit makes up much of the lining of the channel pore.

Currently, the highest resolution for an intact nAChR is at 4 Å accomplished with electron microscopy examining the *Torpedo* nAChR [11]. Despite ample supply of muscle-type and electric organ nAChRs, full 3-D crystallization attempts have not been completely successful. Unfortunately, heterologous expression of nAChRs does not produce the receptor in quantities sufficient for crystallization [2]. A breakthrough in the structural understanding of nAChRs occurred when the crystal structure was obtained for a soluble acetylcholine-binding protein (AChBP) from snails [12]. Although it does not function as an ion channel and does not contain a transmembrane region, the AChBP is a functional and structural homologue of the ligand-binding domain of Cys-loop receptors and shares high homology (24 % sequence identity) with the N-terminal ligand-binding domain of the α7 subunit. The study of this structure and that of two other AChBPs [13, 14] coupled with structure-function and mutational analyses [15, 16] have provided detailed information for the ligand-binding properties and for the potential gating mechanisms of nAChRs. Agonists bind at the interface between an α subunit and an adjacent subunit. Therefore, a heteromeric $(\alpha 4)_2(\beta 2)_3$ receptor contains two binding sites. Although α7 subunits are capable of forming heteromeric receptors [17], the more common homomeric α7 receptors have five potential binding sites [18–20].

The N-terminal of the α7 nAChR subunit, as with other nAChR subunits, provides the extracellular ligand-binding domain [21]. The extracellular ligand-binding domain consists of six loops: three on the principal side of the α subunit (loops a–c) and three on the adjacent subunit (loops d–f) (Fig. 13.1a, b). Since α7 is expressed as a homomeric receptor, the interface of these two sets of loops is between identical subunits. Two important loops in the N-terminal extracellular domain are the c-loop and the Cys-loop (Fig. 13.1a, b). The c-loop is the loop between β9 and 10. The cys-loop, for which the superfamily is identified, is found in the β6-7 loop. In the crystal structure of the α1 subunit extracellular domain, the Cys-loop is a 13 amino acid sequence linked by a cysteine disulfide bond. It is located at the bottom of the beta-barrel, lying in close proximity to the TM2–TM3 loop, which is extracellular in a functional receptor (Fig. 13.1a). When an agonist, such as nicotine or ACh, enters the binding site, the c-loop moves and caps the ligand [22] (Fig. 13.1c, d). Ligand binding also has been shown to require a series of aromatic residues whose architectural structure is shared by all members of the Cys-loop family of channels [23].

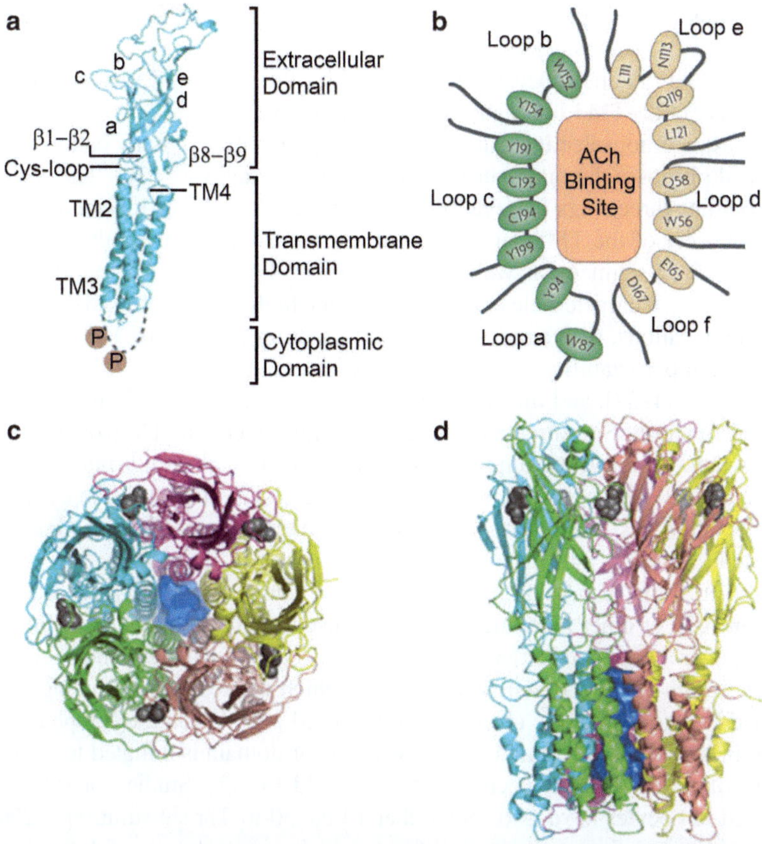

Fig. 13.1 Model structures of the α7 nAChR subunit and receptor. (**a**) The folded structure for one α7 subunit is represented, including the extracellular loops, the transmembrane alpha helixes, and the cytoplasmic phosphorylation sites. (**b**) A schematic representation of the ACh-binding site. The binding site is composed of amino-acid residues that make up loops a, b, and c (the principal component from 1 subunit) and loops d, e, and f (the complementary component from another subunit). (**c**) A top view looking down on the nAChR with its central water-filled pore showing five nicotine molecules (*dark grey*) in binding sites. (**d**) A side view of the nAChR. The model structures were obtained by comparative modeling by Taly et al., 2009, based on homology to the ACh-binding protein from *Erwinia chrysanthemi*. The panels were adapted with permission from Taly A. et al., Nat Rev Drug Discov, 2009. 8:733-50

The TM2 segment lines the ion channel along the axis of symmetry so that each one of the five nAChR subunits is arranged like the staves in a barrel. The channel contributes to three fundamental aspects of nAChR function: gating (i.e., opening and closing), ion permeation, and ion selectivity. The gate, which is thought to be located in the middle–upper part of the channel, is allosterically coupled with the agonist binding [21]. It comprises three rings of hydrophobic residues that prevent passage of permeant ions when the channel is in the closed state [11, 24]. The analysis of bacterial protein homologue to nAChRs has suggested that channel opening is produced by the concerted tilting of the TM2 helices, the TM2–TM3 loop, and the TM3 segment [25].

nAChRs are nonselective cationic channels permeable to many cations, and for biological purposes the main permeants are Na^+, K^+, and Ca^{2+} [26]. The selectivity filter contains three rings of negatively charged residues and is located at the cytoplasmic border of the TM2 segment, with key contributions from the cytoplasmic end of the TM2 segment, as well as from the loop linking the TM1 and TM2 segments [27–31]. Characteristic of α7 nAChR is the high permeability to divalent over monovalent cations, with a ratio of calcium to sodium of about 10:1 [8, 32, 33]. The high calcium permeability of the α7 nAChR is similar to that of NMDA glutamate receptor [32, 34–37], and this defining feature is critical to the α7 subunit's role in neuronal function. Site-directed mutagenesis experiments in TM2 or neighboring residues have revealed the amino acids that modulate Ca^{2+} influx into the channel [32, 36, 38, 39]. For example, when a glutamate residue found in the cytoplasmic portion of TM2 is mutated to an alanine, calcium permeability is significantly decreased [32, 40].

Desensitization of nAChRs arises most commonly after prolonged exposure to sufficient ACh or other nicotinic ligands to cause a decrease in the activation of the nAChRs over a time course ranging from milliseconds to minutes. The phenomenon leads to a high-affinity, desensitized state in which the receptor's binding sites are occupied by agonist but the channel pore is closed [41]. When a tryptophan residue located in the β2 subunit strand of the extracellular domain is mutated to an alanine, the rate of desensitization decreases by >30-fold [3, 42]. Studies of α7 nAChRs expressed in oocytes have also shown that a Leu250-to-Thr substitution (L250T) in the channel domain increases agonist affinity and decreases the rate of desensitization, creating a gain-of-function α7 nAChR [43–45]. Hippocampal neurons from mice homozygous (T/T) for the α7-L250T nAChR "knock-in" mutation have properties consistent with those observed for α7-L250T nAChRs in oocytes [45]. Homozygosity for the L250T mutation leads to perinatal death while partial gain of function in heterozygous (+/T) mice is compatible with normal growth and life-span.

2 Assembly and Trafficking

The assembly and trafficking of nAChRs are tightly regulated processes that are critically important for a properly functioning cholinergic system. This regulation can be demonstrated by the difficulty with heterologous expression of the α7 nAChRs in cell lines that do not endogenously express α7 [46]. Although the nicotinic receptor was the first ion channel to be cloned and characterized, the progress was relatively slow toward understanding the mechanisms responsible for the trafficking of the receptor to the plasma membrane. Most of the initial work was conducted on muscle-type nAChRs and showed that nAChRs are folded and assembled in the rough endoplasmic reticulum (ER) [47–51]. The process of folding and pentameric assembly is a prerequisite for exit from the ER [52]. Unassembled or incompletely assembled subunits are retained in the ER and are ultimately degraded through the process of ER-associated degradation (ERAD). The receptors are subsequently transported to the Golgi apparatus where they are segregated into vesicles for delivery and insertion into the plasma membrane [53].

Throughout the process, a number of signals have been discovered that are necessary for proper delivery of the fully assembled receptor to the cell surface. An α-helix located at the N-terminal is necessary for expression of the receptor in the membrane of *Xenopus* oocytes [54]. Fully assembled receptors mask an ER retention signal contained within the TM1 domain. Although nAChR subunits capable of forming homomeric channels do not contain this motif, the TM1 region is still important for export from the ER [55]. The C-terminal domain is also important for the export process as α7-5HT3A chimeras are retained in the ER without this domain [56].

Correct folding of individual subunits allows for the large intracellular loop (TM3–TM4) to be exposed to the cytoplasm. While all nAChR subunits are highly homologous, the TM3–TM4 loop shows the highest degree of divergence [57], suggesting its involvement in the determination of some nAChR subtype-specific properties. The large intracellular loop contains a signal targeting the assembled receptor for coat protein II (COPII)-mediated ER to Golgi transport [58]. The intracellular loop also contains a signal that targets the receptor for COPI-mediated Golgi to ER retrograde trafficking [59]. This signal must be masked in order for transport to continue out of the Golgi and to the cell surface. In α7* nAChRs, the large intracellular loop is important for targeting the receptor to its proper location in vivo. In normal chick ciliary ganglia, α7 homopentamers are targeted to perisynaptic sites and are excluded from synaptic locations whereas α3* nAChRs can be expressed at the synapse. However, α7 chimeras containing the large intracellular loop of the α3 subunit are efficiently targeted to the synapse in vivo [60].

The large intracellular loop is also a binding site for adaptor proteins/chaperones that influence receptor trafficking and degradation [61, 62]. The most well-characterized molecular chaperone for α7 nAChRs is RIC-3. RIC-3 is a single-pass transmembrane protein found in the endoplasmic reticulum that interacts with unfolded and folded α7 to facilitate its assembly and trafficking to the cell surface [63–66]. In cultured mammalian cell lines that do not endogenously express α7, transfection of α7 cDNA leads to rapid degradation of protein and no functional channels at the cell surface [67]. However, co-transfection with RIC-3 is sufficient to detect functional homomeric α7 nAChRs [64, 68]. The nature of this facilitation is dependent on RIC-3 expression levels. At low levels, RIC-3 indeed increases surface expression of α7; however, at higher levels assembled α7 receptors are retained in the ER. In cultured neurons, this retention process is associated with trafficking of α7 to dendrites along the ER-restricted pathway and preventing transport to axons [69]. Thus, RIC-3 assists in the folding, assembly of pentamers, and localization of α7 nAChRs.

A common property of neurotransmitter receptors is high-density clustering at a controlled location in the cell. α7 nAChRs are maintained in clusters at the cell surface through interactions with various scaffolding proteins and cytoskeletal elements [70]. In the spines of chick ciliary ganglion neurons, α7 forms discrete clusters.

Upon depolymerization of actin filaments, these clusters dissipate and α7 becomes undetectable [71]. Among the proteins that interact with α7 and may be important for clustering are rapsyn and PDZ-domain proteins such as PICK1 [46, 72, 73]. Postsynaptic density (PSD)-95, another PDZ-domain-containing protein, is a scaffolding protein that is important for the postsynaptic localization of NMDA receptors [74]. α7 nAChRs are found in close proximity to NMDA receptors postsynaptically [75], and PSD-95 can be detected after immunoprecipitation of α7 [76]. Therefore, scaffolding by PSD-95-containing complexes may contribute for the postsynaptic localization of α7.

Posttranslational modifications are also critical for the proper function of α7 nAChRs. Palmitoylation is the covalent attachment of the fatty acid palmitate to cysteine residues and is required for the formation of α-BTX-binding properties of α7 [69, 77]. This modification occurs in the ER leading to highly palmitoylated receptors that undergo a pruning process as the receptor travels to the membrane. Although glycosylation does not affect surface expression of the receptor, expression of glycosylation-deficient α7 causes changes in whole-cell currents [78, 79].

Another factor regulating the surface expression of nAChRs is long-term repeated exposure to agonist, as commonly occurs with nicotine during tobacco use. The usual response to overstimulation by an agonist is downregulation of that receptor. On the contrary, chronic nicotine exposure upregulates nAChRs in a subtype-specific manner. Although high-affinity β2-containing nAChRs are most potently upregulated, in some locations of the brain α7 nAChRs also increase [80–82]. A variety of mechanisms have been proposed to explain the phenomenon. Chronic nicotine induces long-term desensitization of a significant proportion of nAChRs, particularly of the high-affinity subtypes. Therefore, a homeostatic response to desensitization is to increase nAChR levels to make up for the decreased availability of receptors [83, 84]. This homeostatic reaction occurs because nAChRs that reach the surface remain there longer when chronically exposed to nicotine owing to decreased turnover rates of receptors [85]. Nicotine also acts as a pharmacological chaperone for its own receptor, stabilizing nAChRs during assembly and trafficking them to the surface [86]. Increased levels of chaperones can account for increased surface expression of nAChRs. Upregulation is also aided by a decreased receptor degradation rate. Nicotine inhibits proteasomal activity [82] and by doing so increases the stability of many synaptic proteins, including nAChR subunits [87].

3 Brain α7* nAChRs

Nicotinic acetylcholine receptors are expressed throughout the brain and can be found on neurons as well as astrocytes, microglia, and vascular endothelial cells [88–90]. α7 homomers and α4β2 heteromers represent the two major subtypes of nAChRs found in the mammalian brain [88]. High levels of the α7 mRNA transcript are found via in situ hybridization in the olfactory lobes, cerebral cortex, hippocampus, hypothalamus, and amygdala [8]. α7 can also be detected with α-bungarotoxin (α-BTX), a toxin that binds to α1 nAChR subunits at the neuromuscular junction as well as neuronal subunits α7–α10 [91–93]. In the rat brain, α-BTX-binding sites contain primarily α7 subunits [94], and their distribution is similar to that of α7 mRNA in the adult rat brain [95]. However, α-BTX also binds to and inhibits $GABA_A$ receptors containing the $GABA_A$-β3 subunit [92], but the extent to which this binding is reflected in vivo has yet to be determined. Immunohistochemistry is also commonly used to detect α7* nAChRs. However, because commercially available antibodies may not always be specific [96, 97] a combination of techniques is advisable for the detection of brain nAChRs, especially in the mouse. Despite these limitations, the overall expression pattern of α7 has been extensively examined [88] and is summarized in Table 13.1 for the human brain.

The α7 nAChR subunit is detected at pre- and postsynaptic locations, as well as dendrites, axons, and somas of neurons [40, 99–103]. Like other nAChRs, α7* nAChRs are commonly localized to the presynaptic membrane, where a basic function is to modulate the release of neurotransmitters [103]. As stated earlier, the calcium permeability through α7 homomers relative to that of sodium (P_{Ca}/P_{Na}) is estimated near 10, which is comparable to that of NMDA receptors [6, 8, 32, 104]. As influx of Ca^{2+} into the presynaptic terminal is a critical driver of neurotransmitter vesicle fusion and consequent release [105, 106], α7* nAChRs are positioned to influence the efficacy of synaptic transmission. As action potentials reach the synaptic terminal, voltage-gated calcium channels open to provide increased intracellular calcium, and temporal summation of repeated action potentials can lead to calcium-induced calcium release, all of which produce neurotransmitter release [107]. Activation of presynaptic α7* nAChRs in concomitance with the arrival of an action potential contributes to cell membrane depolarization, thereby lowering the threshold for neurotransmitter release. For example, the frequency of miniature excitatory postsynaptic potentials induced by glutamatergic neurons from the hippocampus or the medial habenula is increased after inducing nicotinic currents in those cells [101, 103, 108]. In the CA3 region of the hippocampus, exposure to nicotine in concentrations achievable by smoking leads to increases in the intracellular calcium concentration in mossy fiber presynaptic terminals. This effect is blocked by α7 nAChR antagonists such as methyllycaconitine (MLA) and α-BTX, indicating that presynaptic α7 nAChRs are at least partly responsible [103]. Therefore, activation of α7 nAChRs found presynaptically may be sufficient to induce neurotransmitter release even without the appearance of action potentials. As presynaptic nAChRs make the release of neurotransmitter more likely, a stimulus reaching presynaptic terminals is more likely to induce a postsynaptic depolarization and, thus, long-term potentiation (LTP) [109]. Additionally, given that some synapses in the CA1 region express the α7 nAChR [40, 75], nicotinic activation in that region can also enhance the induction of long-term potentiation, depotentiation, or long-term depression [110–112].

Although not as common in the mammalian brain, α7* nAChRs are found in postsynaptic membranes and agonists can induce fast synaptic transmission at these sites. α7 nAChRs are most abundant in the hippocampus and neocortex [113], especially on GABAergic interneurons [114], where they may mediate cholinergic synaptic

Table 13.1 Localization of CHRNA7 mRNA expression [88, 98]

Peripheral	
Epithelial cells	*Immune cells*
• Airway	• Monocytes
– Surface epithelium	• Eosinophils
– Alveolar type 2 cells	• Macrophage
• Skin	• Lymphocytes
– Keratinocytes	– T and B cells
– Pilosebaceous unit	*Mesenchymal cells*
– Myoepithelial cells of sweat glands	• Fibroblasts
– Melanocytes	– Lung
– Urothelium	• Tenocytes (rat)
Endothelial cells	• Adipocytes (rat)
• Aorta and Pulmonary Vessels	• Smooth muscle fibres
	– Lung
	– Vasculature
	Mesothelial cells
	• Mesothelioma cells
Central	
Neocortex	*Basal ganglia*
Entorhinal cortex	• Expression in the striatum
Hippocampus	*Thalamus*
• Interneurons in striatum oriens and striatum radiatum, pyramidal cells in CA2/3, CA4 > CA1, hilar cells	• Reticular thalamic nuclei
– Moderate/strong expression	– High expression
• Dentate granular cells, subiculum, presubiculum	• LGN, MD, VA
– Weak/moderate expression	– Lower expression
	Brainstem
	• Midbrain
	– Minimal and controversial
	• Pons
	• Medulla
	– Strong expression
	Cerebellum
	• Large ovoid cells, astrocytes, endothelial cells

input [115]. The α7-mediated activation of such interneurons can result in either inhibition or disinhibition of pyramidal neurons [116–118]. α7-Mediated postsynaptic nAChR activity on pyramidal neurons is capable of boosting the impact of a weak electrical stimulation on the Schaffer collateral pathway, thereby influencing synaptic plasticity [112].

The study of single channels from outside-out patches pulled from somas of rat hippocampal neurons demonstrated channels with α7 characteristics that are inhibited by MLA [119]. To very finely localize functional α7 nAChRs, laser photolysis of α-carboxy-2-nitrobenzyl (CNB)-caged carbachol can be used to release the agonist in a very small, controlled area (~6–7 μM) and for a short period of time (~100 μs) to avoid the effects of desensitization. This technique helped to map functional somatodendritic α7 nAChRs in rat hippocampal CA1 and substantia nigra pars reticulata neurons [120, 121]. To further elucidate the localization of α7 nAChRs, the C-terminus of α7 was tagged with either hemagglutinin epitope or GFP. Interestingly, immunostaining of cultured hippocampal neurons demonstrated that α7 was almost entirely localized to the somatodendritic compartments, with obvious caveat of the effects of such protein modifications. Additionally, a 48-residue motif in the large intracellular loop of the α7 receptor led to dendritic localization when fused to proteins with normally ubiquitous distribution [122].

Given that acetylcholine half-life and the distance it travels from its release site depend on the distribution and density of acetylcholinesterase (AChE), the importance of extra-synaptic α7 expression is not well understood. The acetylcholine breakdown product, choline, acts at higher concentrations as a specific agonist for α7 nAChRs [123, 124]. Choline is an essential nutrient that can be found in membrane phospholipids as well as being a key component of acetylcholine. Within synapses, it is liberated from acetylcholine by acetylcholinesterase and diffuses some distance away before being eventually transported back into the cell. Choline can be found in cerebral spinal fluid at concentrations of 4–12 μM [125]. Studies utilizing electron microscopy have determined that a significant proportion of nAChRs are localized to positions within the neuron away from any synapse, including along axons and somas of neurons [118, 120, 126]. It follows that nAChRs, α7 in particular, may function through diffuse, or volume, transmission of endogenous agonists (acetylcholine and choline), rather than solely through synaptic action of released neurotransmitters [40, 109, 123, 124, 127]. Located outside the synapse, these receptors may influence neuronal function by altering the threshold for action potentials or direct the travel of action potentials toward certain branches on an axon via local membrane depolarizations [40]. Additionally, through calcium-mediated signal transduction, activation of these channels likely has other downstream consequences involving regulation of transcription and other cellular processes [128]. Axonal expression of α7 nAChRs has also been shown to modulate presynaptic NMDAR expression and presynaptic and postsynaptic maturation of glutamatergic synapses [129]. Hippocampal α7* nAChRs have also been shown to exert a profound effect on adult hippocampal neurogenesis [130].

4 Peripheral and Other α7* Receptors

Neuronal nAChRs are also expressed in the peripheral nervous system, where autonomic nervous system nAChRs mediate fast ganglionic transmission and other functions [131]. Although α3, β2, α5, and β4 subunits are the predominate subtypes, α7 nAChRs are known to be expressed in parasympathetic and sympathetic ganglionic neurons and are physiologically important [132, 133]. Sympathetic neurons isolated from rat superior cervical ganglia express two classes of functional α7 nAChRs. One class rapidly desensitizes, recognizes choline as an agonist, and is blocked almost irreversibly by α-BTX. The other class desensitizes much more slowly, does not respond to choline, and is blocked by α-BTX in a readily reversible manner [134, 135]. These two types of currents reflect the activation of two splicing variants of the α7 gene: one with extremely rapid desensitization kinetics and low amplitude (α7-1), and a second with relatively slow desensitization kinetics and larger current amplitudes (α7-2) [135, 136]. Autonomic α7* nAChRs provide a relatively small contribution to fast ganglionic transmission, which is mainly mediated by α3* nAChRs [133, 137]. In addition, α7* nAChRs at extrasynaptic locations might monitor and respond to the extracellular choline released by hydrolysis of ACh [123, 124], providing an additional level of modulation to ganglionic transmission.

α7* nAChRs are also found in dorsal root ganglia [138–141], where they influence the processing of sensory information [142, 143]. α7* nAChRs have a prominent role in pain processing. The α7-receptor antagonists MLA and α-BTX inhibit nicotine-induced nociception [144], suggesting an analgesic effect associated with α7 nAChR activation. In mice, the α7 nAChR selective agonist choline produces analgesia in the late phase of the formalin test, an assay associated with an inflammatory response [145]. α7 nAChR agonists can also reverse hypersensitivity in an acute inflammatory pain model [144]. Although the analgesic effects of α7 nAChR stimulation seem to be centrally mediated [144], the receptors might also influence inflammatory pain by reducing inflammation. The anti-inflammatory effects of α7 nAChRs result in part from the activation of the vagus nerve which can produce an anti-inflammatory state reflected by decreased levels of pro-inflammatory cytokines such as TNF-α, IL-1, and HMGB1 [146]. The spleen is a required component of the cholinergic anti-inflammatory pathway [147, 148]. It is innervated by the splenic nerve arising from the celiac-superior mesenteric plexus [149–151], and delivers primarily catecholaminergic afferents [152, 153]. α7* nAChRs expressed in the celiac ganglion may contribute to the effects of vagal nerve stimulation by influencing ganglionic transmission and controlling the amount of norepinephrine released from splenic nerve endings. The norepinephrine released from the splenic nerve binds to adrenergic receptors on lymphocytes. Lymphocytes produce acetylcholine that then binds α7 nAChRs expressed on macrophages to mitigate the inflammatory response [147]. Vagus nerve activity has been found to be decreased in patients suffering from rheumatoid arthritis, and α7-specific agonists are being pursued as a possible therapeutic treatment for this condition [154]. In summary, stimulation of α7 nAChRs leads to an anti-inflammatory state and associated mechanism has been implicated in the pathophysiology of a wide variety of inflammatory diseases such as sepsis, pancreatitis, rheumatoid arthritis, and inflammatory bowel disease [147].

As an indication of the ubiquity of cholinergic signaling, it is interesting to note that ACh is also found and synthesized in a wide variety of life forms that do not possess a mammalian-type nervous system, including archaea, bacteria, plants, fungi, and animals [155]. This observation suggests that cholinergic signaling developed as a very early form of intercellular communication and that it could be expected to hold important functions outside of the nervous system. Morris was the first to report non-neuronal synthesis of acetylcholine in the placenta [156]. Choline acetyltransferase (ChAT), a key component of the ACh synthetic pathway, has subsequently been shown to be expressed in a wide variety of tissues including endothelial cells [157], glial cells [158, 159], immune cells [160], and epithelial cells [161–164]. α7 nAChRs can be detected in many types of cells including epithelial, endothelial, and mesenchymal cell types [98], where they may control various aspects of cellular function via paracrine and/or autocrine mechanisms [160].

5 α7 Knockout and Gain-of-Function Mice

Despite the many roles played by the α7 nAChR in the mammalian brain, behavioral phenotypes have been difficult to find in α7 knockout mice. These mice were first described in 1997 [165] to have properties expected of a null mutation: little or no α-BTX binding in the brain and a lack of kinetically fast nicotinic currents in the hippocampus. These mice were then subjected to a full battery of neurologic and behavioral tests [166]. The α7 knockout mice showed no easily detectable differences in any motor or sensory responses compared with wild-type mice. They performed similarly in the initial tests of learning and memory, such as the Morris water maze and conditioned fear paradigms. Nor did these mice show differences in sensorimotor gating or spatial learning. The only difference found was that α7 null mice spent more time in the center of an open field test of anxiety-related behavior. This result may indicate that α7 null mice experience less baseline anxiety. However, the authors caution that this finding is not supported by further tests of anxiety-related behavior (e.g., light/dark exploration test).

The baroreflex is the autonomic neural mechanism to control blood pressure. In response to increased blood pressure, baroreceptors found in the carotid artery are activated and lead to an increase in parasympathetic tone and decreased sympathetic outflow. The overall result is decreased blood pressure secondary to decreases in heart rate, cardiac contractility, and vascular resistance. The opposite response is observed with a decrease in blood pressure [167]. α7-deficient mice demonstrate a decreased baroreflex in response to a drop in blood pressure. Wild-type mice administered sodium nitroprusside displayed a 48 % increase in heart rate, whereas that of α7-deficient mice only increased 21 % [168].

In another study [169], α7 knockout mice and wild-type mice were exposed to nicotine chronically in their drinking water. Then, they were observed as they were allowed to go spontaneously into nicotine withdrawal. α7 knockout mice showed significantly less withdrawal-induced hyperalgesia, but were otherwise similar to wild-type animals, including the presentation of somatic signs of withdrawal. However, when precipitated by mecamylamine, nicotine withdrawal does lead to decreased somatic signs in α7 null mice compared to wild-type mice [170]. These mice were observed after 2 weeks of treatment with subcutaneous nicotine via micro-osmotic pumps and acute precipitation of withdrawal, which may account for the discrepancy in the two studies. A subsequent report using a similar method [171] did not confirm the role of α7 in somatic signs of nicotine withdrawal, but did show decreased hyperalgesia in α7 null mice. Interestingly, MLA-induced nicotine withdrawal was shown to lead to identical increases in somatic signs between wild-type and α7 null mice [170]. This result was found using MLA concentrations that had been previously assumed to be α7 specific. This same study demonstrated that α7 null mice have wild-type levels of nicotine tolerance and basal anxiety levels.

Tonic-clonic seizures are observed when high doses of nicotine are administered to mice, and the sensitivity to this effect is strain dependent [172]. Many studies indicate that α7 nAChRs are involved in the presentation of seizures. Increased α-BTX binding in the hippocampus is correlated with an increased sensitivity to nicotine-induced seizure behavior [172–176]. MLA, an α7-specific antagonist, significantly inhibits the convulsive effects of nicotine when administered either peripherally or centrally [173]. However, despite the evidence for the role of α7 in nicotine-induced seizure, α7 null mice that have little or no α-BTX binding in the brain demonstrate a similar sensitivity to nicotine as wild-type mice [177].

To further confuse the situation, specific mutations within α7 can alter seizure sensitivity in mice. The TM2 region of nAChRs contains a highly conserved hydrophobic leucine residue that if mutated to threonine (L250T) or other polar amino acids leads to an interesting gain-of-function phenotype for α7 nAChRs. These channels demonstrate higher affinity to acetylcholine and decreased desensitization. In oocytes, dihydro-β-erythroidine, (+)-tubocurarine, and hexamethonium were shown to inhibit the response of wild-type α7 receptors to acetylcholine. All three of these compounds, in the absence of acetylcholine, led to channel opening when the L250T-mutant α7 subunits were expressed in oocytes [43, 44]. A transgenic mouse line was created expressing this gain-of-function mutation in order to further elucidate its role in vivo. Mice homozygous for the mutation (T/T) die within 24 h after birth. Mice expressing one copy of the mutant L250T allele, but no wild-type allele, i.e., α7 (–/T), also demonstrate the lethal phenotype. However, with one wild-type copy of the gene (+/T), these mice survive and are largely normal as determined by a battery of behavioral tests [45, 178, 179]. These surviving mice demonstrated differences in their response to nicotine. They displayed two stereotypic movements soon after injection of nicotine: "head bobbing" and a crisscross forepaw tapping. In addition, these mice were twice as sensitive to nicotine-induced seizures [178]. Nicotinic currents elicited by 1 mM acetylcholine from three different hippocampal cell types (interneurons, pyramidal, and granule cells) were severalfold larger in the +/T slices than in +/+ slices [112]. Also, these currents were slower to desensitize [178]. These results are consistent with previous findings from mutant α7 expressed in oocytes.

Nicotinic receptors have also been shown to be important components of the fertilization process. Human sperm expresses α3, α5, α7, α9, and β4 nAChR subunit mRNA. When a sperm reaches the zona pellucida of an unfertilized egg, the sperm undergoes the acrosomal reaction. During this reaction, a secretory vesicle, the acrosome, located at the anterior head of the sperm releases its contents toward the egg, which allow the sperm to penetrate the egg. Sperm from α7 −/− mice do not undergo the zona pellucida-initiated acrosomal reaction and display a 25 % reduction of *in vitro* fertilization as compared to wild-type sperm [180]. Calcium influx is a critical initiating event for the acrosomal reaction. Acetylcholine causes an increase in intracellular calcium in sperm and can also induce the acrosomal reaction in wild-type sperm. However, ACh is unable to do so in α7-deficient sperm [181]. Additionally, sperm from α7-deficient mice demonstrate impaired motility. Therefore, drugs targeting the α7 nAChR may have a role to aid in cases of infertility.

6 Pathophysiology

A number of human pathologies involve dysfunctions in cholinergic signaling. Because nAChRs are widely distributed in the peripheral and central nervous system, it is not surprising that they have involvement in diverse neuronal diseases. For example, a principal feature of the pathogenesis of Alzheimer's disease is the loss of nAChRs and of cholinergic neurons, particularly from the basal forebrain nuclei. Also, inhibition of acetylcholine esterase is a treatment for the cognitive symptoms of AD [182–186]. Furthermore, β-amyloid is known to interact with the α7 nAChR, suggesting that it may have a role in the pathogenesis possibly through astrocytic mechanisms [187] (see Chap. 19 by Dineley).

Patients with schizophrenia are known to have a disproportionately high smoking rate, which may reflect an inappropriate form of self-medication [186]. Schizophrenics have decreased numbers of nAChRs and of the α7 subunit in particular. This finding implies that overall cholinergic signaling is also diminished [186, 188]. As with schizophrenia, adults who have attention-deficit hyperactivity disorder (ADHD) as children are more likely to be smokers. In addition, they demonstrate a decreased performance in tests of attention that can be improved with administration of nicotinic agonists [186] (see Chap. 20 by Leonard).

Lung cancer is the leading cause of cancer death in the world and most cases are attributable to smoking. However, it is important to note that smoking not only increases the risk of lung cancer, but also many other types of cancer including mouth, laryngeal, bladder, esophageal, and stomach cancers [189]. Nicotine and, more importantly, its active metabolites may act via various routes, but they also bind to nAChRs throughout the body and activate many molecular cascades that may favor the development of neoplasm. α7 activation induces indirect initiation of cell proliferation, an effect kept in check by α4β2 activity. Importantly, α7 nAChRs have a lower affinity for nicotine than α4β2 nAChRs, resulting in significant α4β2 receptor desensitization with no profound change in the sensitivity of α7 receptors [190, 191]. Recent work has implicated α7 as a potential chemotherapeutic target for small and non-small-cell lung cancer, mesothelioma, gastric, colon, oral, esophageal, and pancreatic cancer [190].

The preceding was only a brief introduction to the role of the α7 nAChR subunit in the pathophysiology of human disease inside and outside of the central nervous system. This receptor is likely involved in many other pathologic states. For example, nicotinic signaling through α7 receptors is also implicated in pro-angiogenic activity, which may be an important target for many diseases with a neovascular component, such as cardiovascular disease, cancer, and macular degeneration [192]. Also, in human neoplasms, epithelial-to-mesenchymal transitions (EMT) are known to be an important step in the pathogenesis of a malignant phenotype. Nicotine, through α7 signaling, leads to increased migration and gene expression changes consistent with the epithelial-mesenchymal transition in cancer cell lines [193, 194].

Despite the widespread expression and seemingly important physiology related to the α7 nAChR subunit, it is interesting that the phenotype associated with the knockout mouse is relatively mild. The reasons for this apparent discrepancy are many. Compensatory upregulation of a complementary subunit could replace the function of α7. Additionally, it is possible that α7 functions as one of several redundant pathways to accomplish the same goal and we are only able to tease out a small portion of the effect. Alternatively, loss of function of this subunit may be a less important cause of pathophysiology than a gain of function. The L250T transgenic mice show considerably more dramatic phenotypes as described above. Recent work has indicated that increased dosage of the Chrna7 gene in humans can be linked to cognitive impairment and neuropsychiatric disease [195]. In any case, the evidence is overwhelming that α7 is an important contributor to normal human physiology as well as disease states and, therefore, should receive increased attention in the future. Drugs designed to target this receptor may find utility in such wide-ranging disease states as cancer, schizophrenia (see also Chap. 20), infertility, and Alzheimer's disease (see also Chap. 19).

References

1. Chen L. In pursuit of the high-resolution structure of nicotinic acetylcholine receptors. J Physiol. 2010;588(Pt 4):557–64.
2. Wells GB. Structural answers and persistent questions about how nicotinic receptors work. Front Biosci. 2008;13:5479–510.
3. Yakel JL. Gating of nicotinic ACh receptors: latest insights into ligand binding and function. J Physiol. 2010;588(Pt 4):597–602.
4. Tsetlin V, Hucho F. Nicotinic acetylcholine receptors at atomic resolution. Curr Opin Pharmacol. 2009;9(3):306–10.
5. Kalamida D, et al. Muscle and neuronal nicotinic acetylcholine receptors. Structure, function and pathogenicity. FEBS J. 2007;274(15):3799–845.

6. Couturier S, et al. A neuronal nicotinic acetylcholine receptor subunit (α7) is developmentally regulated and forms a homo-oligomeric channel blocked by α-BgTx. Neuron. 1990;5:847–56.
7. Schoepfer R, et al. Brain alpha-bungarotoxin binding protein cDNAs and MAbs reveal subtypes of this branch of the ligand-gated ion channel gene superfamily. Neuron. 1990;5(1): 35–48.
8. Seguela P, et al. Molecular cloning, functional properties, and distribution of rat brain α7: a nicotinic cation channel highly permeable to calcium. J Neurosci. 1993;13:596–604.
9. Corringer PJ, et al. Identification of a new component of the agonist binding site of the nicotinic alpha 7 homooligomeric receptor. J Biol Chem. 1995;270(20):11749–52.
10. Elgoyen AB, et al. α9: an acetylcholine receptor with novel pharmacological properties expressed in rat cochlear hair cells. Cell. 1994;79:705–15.
11. Unwin N. Refined structure of the nicotinic acetylcholine receptor at 4A resolution. J Mol Biol. 2005;346(4):967–89.
12. Brejc K, et al. Crystal structure of an ACh-binding protein reveals the ligand-binding domain of nicotinic receptors. Nature. 2001;411(6835):269–76.
13. Celie PH, et al. Crystal structure of acetylcholine-binding protein from Bulinus truncatus reveals the conserved structural scaffold and sites of variation in nicotinic acetylcholine receptors. J Biol Chem. 2005;280(28):26457–66.
14. Hansen SB, et al. Structures of Aplysia AChBP complexes with nicotinic agonists and antagonists reveal distinctive binding interfaces and conformations. EMBO J. 2005;24(20):3635–46.
15. Sine SM. The nicotinic receptor ligand binding domain. J Neurobiol. 2002;53(4):431–46.
16. Brams M, et al. Crystal structures of a cysteine-modified mutant in loop D of acetylcholine binding protein. J Biol Chem. 2010;286(6):4420–8.
17. Murray TA, et al. Alpha7beta2 nicotinic acetylcholine receptors assemble, function, and are activated primarily via their alpha7-alpha7 interfaces. Mol Pharmacol. 2012;81(2):175–88.
18. Rucktooa P, Smit AB, Sixma TK. Insight in nAChR subtype selectivity from AChBP crystal structures. Biochem Pharmacol. 2009;78(7):777–87.
19. Galzi JL, et al. Functional significance of aromatic amino acids from three peptide loops of the alpha 7 neuronal nicotinic receptor site investigated by site-directed mutagenesis. FEBS Lett. 1991;294(3):198–202.
20. Williams DK, et al. The effective opening of nicotinic acetylcholine receptors with single agonist binding sites. J Gen Physiol. 2011;137(4):369–84.
21. Taly A, et al. Nicotinic receptors: allosteric transitions and therapeutic targets in the nervous system. Nat Rev Drug Discov. 2009;8(9):733–50.
22. Celie PH, et al. Nicotine and carbamylcholine binding to nicotinic acetylcholine receptors as studied in AChBP crystal structures. Neuron. 2004;41(6):907–14.
23. Galzi JL, Changeux JP. Neurotransmitter-gated ion channels as unconventional allosteric proteins. Curr Opin Struct Biol. 1994;4(4):554–65.
24. Hilf RJ, Dutzler R. X-ray structure of a prokaryotic pentameric ligand-gated ion channel. Nature. 2008;452(7185):375–9.
25. Popot JL, et al. Interaction of the acetylcholine (nicotinic) receptor protein from Torpedo marmorata electric organ with monolayers of pure lipids. Eur J Biochem. 1978;85(1): 27–42.
26. Dani JA. Open channel structure and ion binding sites of the nicotinic acetylcholine receptor channel. J Neurosci. 1989;9(3):884–92.
27. Gunthorpe MJ, Lummis SC. Conversion of the ion selectivity of the 5-HT(3a) receptor from cationic to anionic reveals a conserved feature of the ligand-gated ion channel superfamily. J Biol Chem. 2001;276(24):10977–83.
28. Imoto K, et al. Rings of negatively charged amino acids determine the acetylcholine receptor channel conductance. Nature. 1988;335(6191):645–8.
29. Konno T, et al. Rings of anionic amino acids as structural determinants of ion selectivity in the acetylcholine receptor channel. Proc Biol Sci. 1991;244(1310):69–79.

30. Galzi JL, et al. Mutations in the channel domain of a neuronal nicotinic receptor convert ion selectivity from cationic to anionic. Nature. 1992;359(6395):500–5.
31. Corringer PJ, et al. Mutational analysis of the charge selectivity filter of the alpha7 nicotinic acetylcholine receptor. Neuron. 1999;22(4):831–43.
32. Bertrand D, et al. Mutations at two distinct sites within the channel domain M2 alter calcium permeability of neuronal alpha 7 nicotinic receptor. Proc Natl Acad Sci U S A. 1993;90(15): 6971–5.
33. Vernino S, et al. Calcium modulation and high calcium permeability of neuronal nicotinic acetylcholine receptors. Neuron. 1992;8(1):127–34.
34. Mayer ML, Westbrook GL. Permeation and block of N-methyl-D-aspartic acid receptor channels by divalent cations in mouse cultured central neurones. J Physiol. 1987;394:501–27.
35. Fucile S. Ca2+ permeability of nicotinic acetylcholine receptors. Cell Calcium. 2004;35(1): 1–8.
36. Fucile S, et al. Fractional Ca(2+) current through human neuronal alpha7 nicotinic acetylcholine receptors. Cell Calcium. 2003;34(2):205–9.
37. Rogers M, Dani JA. Comparison of quantitative calcium flux through NMDA, ATP, and ACh receptor channels. Biophys J. 1995;68(2):501–6.
38. Rogers M, et al. Calcium flux through predominantly independent purinergic ATP and nicotinic acetylcholine receptors. J Neurophysiol. 1997;77(3):1407–17.
39. Fucile S, et al. Human neuronal threonine-for-leucine-248 alpha 7 mutant nicotinic acetylcholine receptors are highly Ca2+ permeable. Proc Natl Acad Sci U S A. 2000;97(7): 3643–8.
40. Dani JA, Bertrand D. Nicotinic acetylcholine receptors and nicotinic cholinergic mechanisms of the central nervous system. Annu Rev Pharmacol Toxicol. 2007;47:699–729.
41. Changeux JP, Edelstein SJ. Nicotinic acetylcholine receptors: from molecular biology to cognition. New York: Odile Jacob; 2005.
42. Gay EA, et al. Aromatic residues at position 55 of rat alpha7 nicotinic acetylcholine receptors are critical for maintaining rapid desensitization. J Physiol. 2008;586(4):1105–15.
43. Revah F, et al. Mutations in the channel domain alter desensitization of a neuronal nicotinic receptor. Nature. 1991;353(6347):846–9.
44. Bertrand D, et al. Unconventional pharmacology of a neuronal nicotinic receptor mutated in the channel domain. Proc Natl Acad Sci U S A. 1992;89(4):1261–5.
45. Orr-Urtreger A, et al. Mice homozygous for the L250T mutation in the alpha7 nicotinic acetylcholine receptor show increased neuronal apoptosis and die within 1 day of birth. J Neurochem. 2000;74(5):2154–66.
46. Millar NS, Harkness PC. Assembly and trafficking of nicotinic acetylcholine receptors (review). Mol Membr Biol. 2008;25(4):279–92.
47. Gelman MS, et al. Role of the endoplasmic reticulum chaperone calnexin in subunit folding and assembly of nicotinic acetylcholine receptors. J Biol Chem. 1995;270(25):15085–92.
48. Keller SH, Lindstrom J, Taylor P. Involvement of the chaperone protein calnexin and the acetylcholine receptor beta -subunit in the assembly and cell surface expression of the receptor. J Biol Chem. 1996;271(37):22871–7.
49. Chang W, Gelman MS, Prives JM. Calnexin-dependent enhancement of nicotinic acetylcholine receptor assembly and surface expression. J Biol Chem. 1997;272(46):28925–32.
50. Jeanclos EM, et al. The chaperone protein 14-3-3eta interacts with the nicotinic acetylcholine receptor alpha 4 subunit. Evidence for a dynamic role in subunit stabilization. J Biol Chem. 2001;276(30):28281–90.
51. Wanamaker CP, Green WN. Endoplasmic reticulum chaperones stabilize nicotinic receptor subunits and regulate receptor assembly. J Biol Chem. 2007;282(43):31113–23.
52. Christianson JC, Green WN. Regulation of nicotinic receptor expression by the ubiquitin-proteasome system. EMBO J. 2004;23(21):4156–65.
53. Wanamaker CP, Christianson JC, Green WN. Regulation of nicotinic acetylcholine receptor assembly. Ann N Y Acad Sci. 2003;998:66–80.

54. Castillo M, et al. Role of the N-terminal alpha-helix in biogenesis of alpha7 nicotinic receptors. J Neurochem. 2009;108(6):1399–409.
55. Wang JM, et al. A transmembrane motif governs the surface trafficking of nicotinic acetylcholine receptors. Nat Neurosci. 2002;5(10):963–70.
56. Pons S, et al. Critical role of the C-terminal segment in the maturation and export to the cell surface of the homopentameric alpha 7-5HT3A receptor. Eur J Neurosci. 2004;20(8):2022–30.
57. Temburni MK, Blitzblau RC, Jacob MH. Receptor targeting and heterogeneity at interneuronal nicotinic cholinergic synapses in vivo. J Physiol. 2000;525(Pt 1):21–9.
58. Ren XQ, et al. Structural determinants of alpha4beta2 nicotinic acetylcholine receptor trafficking. J Neurosci. 2005;25(28):6676–86.
59. Keller SH, et al. Adjacent basic amino acid residues recognized by the COP I complex and ubiquitination govern endoplasmic reticulum to cell surface trafficking of the nicotinic acetylcholine receptor alpha-Subunit. J Biol Chem. 2001;276(21):18384–91.
60. Williams BM, et al. The long internal loop of the alpha 3 subunit targets nAChRs to subdomains within individual synapses on neurons in vivo. Nat Neurosci. 1998;1(7):557–62.
61. Rezvani K, et al. UBXD4, a UBX-containing protein, regulates the cell surface number and stability of alpha3-containing nicotinic acetylcholine receptors. J Neurosci. 2009;29(21):6883–96.
62. Ficklin MB, Zhao S, Feng G. Ubiquilin-1 regulates nicotine-induced up-regulation of neuronal nicotinic acetylcholine receptors. J Biol Chem. 2005;280(40):34088–95.
63. Wang Y, et al. Mouse RIC-3, an endoplasmic reticulum chaperone, promotes assembly of the alpha7 acetylcholine receptor through a cytoplasmic coiled-coil domain. J Neurosci. 2009;29(40):12625–35.
64. Lansdell SJ, et al. RIC-3 enhances functional expression of multiple nicotinic acetylcholine receptor subtypes in mammalian cells. Mol Pharmacol. 2005;68(5):1431–8.
65. Halevi S, et al. The C. elegans ric-3 gene is required for maturation of nicotinic acetylcholine receptors. EMBO J. 2002;21(5):1012–20.
66. Halevi S, et al. Conservation within the RIC-3 gene family. Effectors of mammalian nicotinic acetylcholine receptor expression. J Biol Chem. 2003;278(36):34411–7.
67. Kassner PD, Berg DK. Differences in the fate of neuronal acetylcholine receptor protein expressed in neurons and stably transfected cells. J Neurobiol. 1997;33(7):968–82.
68. Williams ME, et al. Ric-3 promotes functional expression of the nicotinic acetylcholine receptor alpha7 subunit in mammalian cells. J Biol Chem. 2005;280(2):1257–63.
69. Alexander JK, et al. Palmitoylation of nicotinic acetylcholine receptors. J Mol Neurosci. 2010;40(1–2):12–20.
70. Huh KH, Fuhrer C. Clustering of nicotinic acetylcholine receptors: from the neuromuscular junction to interneuronal synapses. Mol Neurobiol. 2002;25(1):79–112.
71. Shoop RD, Yamada N, Berg DK. Cytoskeletal links of neuronal acetylcholine receptors containing alpha 7 subunits. J Neurosci. 2000;20(11):4021–9.
72. Baer K, et al. PICK1 interacts with alpha7 neuronal nicotinic acetylcholine receptors and controls their clustering. Mol Cell Neurosci. 2007;35(2):339–55.
73. Berg DK, et al. Nicotinic signal transduction machinery. J Mol Neurosci. 2006;30(1–2):149–52.
74. Sheng M. Molecular organization of the postsynaptic specialization. Proc Natl Acad Sci U S A. 2001;98(13):7058–61.
75. Fabian-Fine R, et al. Ultrastructural distribution of the alpha7 nicotinic acetylcholine receptor subunit in rat hippocampus. J Neurosci. 2001;21(20):7993–8003.
76. Farias GG, et al. Wnt-7a induces presynaptic colocalization of alpha 7-nicotinic acetylcholine receptors and adenomatous polyposis coli in hippocampal neurons. J Neurosci. 2007;27(20):5313–25.
77. Drisdel RC, Manzana E, Green WN. The role of palmitoylation in functional expression of nicotinic alpha7 receptors. J Neurosci. 2004;24(46):10502–10.

78. Chen D, Dang H, Patrick JW. Contributions of N-linked glycosylation to the expression of a functional alpha7-nicotinic receptor in Xenopus oocytes. J Neurochem. 1998;70(1):349–57.
79. Dellisanti CD, et al. Crystal structure of the extracellular domain of nAChR alpha1 bound to alpha-bungarotoxin at 1.94 A resolution. Nat Neurosci. 2007;10(8):953–62.
80. Peng X, et al. Chronic nicotine treatment up-regulates alpha3 and alpha7 acetylcholine receptor subtypes expressed by the human neuroblastoma cell line SH-SY5Y. Mol Pharmacol. 1997;51(5):776–84.
81. Gaimarri A, et al. Regulation of neuronal nicotinic receptor traffic and expression. Brain Res Rev. 2007;55(1):134–43.
82. Rezvani K, et al. Nicotine regulates multiple synaptic proteins by inhibiting proteasomal activity. J Neurosci. 2007;27(39):10508–19.
83. Fenster CP, et al. Desensitization of nicotinic receptors in the central nervous system. Ann N Y Acad Sci. 1999;868:620–3.
84. Dani JA, Heinemann S. Molecular and cellular aspects of nicotine abuse. Neuron. 1996;16(5):905–8.
85. Peng X, et al. Nicotine-induced increase in neuronal nicotinic receptors results from a decrease in the rate of receptor turnover. Mol Pharmacol. 1994;46(3):523–30.
86. Kuryatov A, et al. Nicotine acts as a pharmacological chaperone to up-regulate human alpha-4beta2 acetylcholine receptors. Mol Pharmacol. 2005;68(6):1839–51.
87. Rezvani K, Teng Y, De Biasi M. The ubiquitin-proteasome system regulates the stability of neuronal nicotinic acetylcholine receptors. J Mol Neurosci. 2010;40(1–2):177–84.
88. Wevers A. Localisation of pre- and postsynaptic cholinergic markers in the human brain. Behav Brain Res. 2010;221(2):341–55.
89. Cooke JP, Ghebremariam YT. Endothelial nicotinic acetylcholine receptors and angiogenesis. Trends Cardiovasc Med. 2008;18(7):247–53.
90. Conejero-Goldberg C, Davies P, Ulloa L. Alpha7 nicotinic acetylcholine receptor: a link between inflammation and neurodegeneration. Neurosci Biobehav Rev. 2008;32(4):693–706.
91. Gotti C, Clementi F. Neuronal nicotinic receptors: from structure to pathology. Prog Neurobiol. 2004;74(6):363–96.
92. McCann CM, et al. The cholinergic antagonist alpha-bungarotoxin also binds and blocks a subset of GABA receptors. Proc Natl Acad Sci U S A. 2006;103(13):5149–54.
93. Colquhoun LM, Patrick JW. Pharmacology of neuronal nicotinic acetylcholine receptor subtypes. Adv Pharmacol. 1997;39:191–220.
94. Chen D, Patrick JW. The alpha-bungarotoxin-binding nicotinic acetylcholine receptor from rat brain contains only the alpha7 subunit. J Biol Chem. 1997;272(38):24024–9.
95. Clarke PB, et al. Nicotinic binding in rat brain: autoradiographic comparison of [3H]acetylcholine, [3H]nicotine, and [125I]-alpha-bungarotoxin. J Neurosci. 1985;5(5):1307–15.
96. Moser N, et al. Evaluating the suitability of nicotinic acetylcholine receptor antibodies for standard immunodetection procedures. J Neurochem. 2007;102(2):479–92.
97. Herber DL, et al. Biochemical and histochemical evidence of nonspecific binding of alpha7nAChR antibodies to mouse brain tissue. J Histochem Cytochem. 2004;52(10):1367–76.
98. Wessler I, Kirkpatrick CJ. Acetylcholine beyond neurons: the non-neuronal cholinergic system in humans. Br J Pharmacol. 2008;154(8):1558–71.
99. Girod R, et al. Facilitation of glutamatergic neurotransmission by presynaptic nicotinic acetylcholine receptors. Neuropharmacology. 2000;39(13):2715–25.
100. Aramakis VB, Metherate R. Nicotine selectively enhances NMDA receptor-mediated synaptic transmission during postnatal development in sensory neocortex. J Neurosci. 1998;18(20):8485–95.
101. McGehee DS, et al. Nicotine enhancement of fast excitatory synaptic transmission in CNS by presynaptic receptors. Science. 1995;269:1692–6.

102. De Filippi G, Baldwinson T, Sher E. Nicotinic receptor modulation of neurotransmitter release in the cerebellum. Prog Brain Res. 2005;148:307–20.
103. Gray R, et al. Hippocampal synaptic transmission enhanced by low concentration of nicotine. Nature. 1996;383:713–6.
104. Castro NG, Albuquerque EX. Alpha-Bungarotoxin-sensitive hippocampal nicotinic receptor channel has a high calcium permeability. Biophys J. 1995;68(2):516–24.
105. Pang ZP, Sudhof TC. Cell biology of Ca2+-triggered exocytosis. Curr Opin Cell Biol. 2010;22(4):496–505.
106. Cheng Q, Yakel JL. Presynaptic alpha7 nicotinic acetylcholine receptors enhance hippocampal mossy fiber glutamatergic transmission via PKA activation. J Neurosci. 2014;34(1):124–33.
107. Vernino S, et al. Quantitative measurement of calcium flux through muscle and neuronal nicotinic acetylcholine receptors. J Neurosci. 1994;14(9):5514–24.
108. Radcliffe KA, Dani JA. Nicotinic stimulation produces multiple forms of increased glutamatergic synaptic transmission. J Neurosci. 1998;18(18):7075–83.
109. Placzek AN, Zhang TA, Dani JA. Nicotinic mechanisms influencing synaptic plasticity in the hippocampus. Acta Pharmacol Sin. 2009;30(6):752–60.
110. Fujii S, Sumikawa K. Nicotine accelerates reversal of long-term potentiation and enhances long-term depression in the rat hippocampal CA1 region. Brain Res. 2001;894(2):340–6.
111. Fujii S, Sumikawa K. Acute and chronic nicotine exposure reverse age-related declines in the induction of long-term potentiation in the rat hippocampus. Brain Res. 2001;894(2):347–53.
112. Ji D, Lape R, Dani JA. Timing and location of nicotinic activity enhances or depresses hippocampal synaptic plasticity. Neuron. 2001;31(1):131–41.
113. Burli T, et al. Single particle tracking of alpha7 nicotinic AChR in hippocampal neurons reveals regulated confinement at glutamatergic and GABAergic perisynaptic sites. PLoS One. 2010;5(7):e11507.
114. Freedman R, et al. Alpha-bungarotoxin binding to hippocampal interneurons: immunocytochemical characterization and effects on growth factor expression. J Neurosci. 1993;13(5):1965–75.
115. Frazier CJ, et al. Synaptic potentials mediated via alpha-bungarotoxin-sensitive nicotinic acetylcholine receptors in rat hippocampal interneurons. J Neurosci. 1998;18(20):8228–35.
116. Ji D, Dani JA. Inhibition and disinhibition of pyramidal neurons by activation of nicotinic receptors on hippocampal interneurons. J Neurophysiol. 2000;83(5):2682–90.
117. Buhler AV, Dunwiddie TV. Alpha7 nicotinic acetylcholine receptors on GABAergic interneurons evoke dendritic and somatic inhibition of hippocampal neurons. J Neurophysiol. 2002;87(1):548–57.
118. Frazier CJ, Strowbridge BW, Papke RL. Nicotinic receptors on local circuit neurons in dentate gyrus: a potential role in regulation of granule cell excitability. J Neurophysiol. 2003;89(6):3018–28.
119. Castro NG, Albuquerque EX. Brief-lifetime, fast-inactivating ion channels account for the alpha-bungarotoxin-sensitive nicotinic response in hippocampal neurons. Neurosci Lett. 1993;164(1–2):137–40.
120. Poisik OV, et al. Functional alpha7-containing nicotinic acetylcholine receptors localize to cell bodies and proximal dendrites in the rat substantia nigra pars reticulata. J Physiol. 2008;586(5):1365–78.
121. Khiroug L, et al. Functional mapping and Ca2+ regulation of nicotinic acetylcholine receptor channels in rat hippocampal CA1 neurons. J Neurosci. 2003;23(27):9024–31.
122. Xu J, Zhu Y, Heinemann SF. Identification of sequence motifs that target neuronal nicotinic receptors to dendrites and axons. J Neurosci. 2006;26(38):9780–93.
123. Papke RL, Bencherif M, Lippiello P. An evaluation of neuronal nicotinic acetylcholine receptor activation by quaternary nitrogen compounds indicates that choline is selective for the alpha 7 subtype. Neurosci Lett. 1996;213(3):201–4.

124. Alkondon M, et al. Choline is a selective agonist of alpha7 nicotinic acetylcholine receptors in the rat brain neurons. Eur J Neurosci. 1997;9(12):2734–42.
125. Klein J, et al. Uptake and metabolism of choline by rat brain after acute choline administration. J Neurochem. 1992;58(3):870–6.
126. Klein RC, Yakel JL. Functional somato-dendritic alpha7-containing nicotinic acetylcholine receptors in the rat basolateral amygdala complex. J Physiol. 2006;576(Pt 3):865–72.
127. Descarries L, Gisiger V, Steriade M. Diffuse transmission by acetylcholine in the CNS. Prog Neurobiol. 1997;53(5):603–25.
128. Berg DK, Conroy WG. Nicotinic alpha 7 receptors: synaptic options and downstream signaling in neurons. J Neurobiol. 2002;53(4):512–23.
129. Lin H, et al. Axonal alpha7 nicotinic ACh receptors modulate presynaptic NMDA receptor expression and structural plasticity of glutamatergic presynaptic boutons. Proc Natl Acad Sci U S A. 2010;107(38):16661–6.
130. Campbell NR, et al. Endogenous signaling through alpha7-containing nicotinic receptors promotes maturation and integration of adult-born neurons in the hippocampus. J Neurosci. 2010;30(26):8734–44.
131. De Biasi M. Nicotinic mechanisms in the autonomic control of organ systems. J Neurobiol. 2002;53(4):568–79.
132. Stokes C, Papke RL. Use of an alpha3beta4 nicotinic acetylcholine receptor subunit concatamer to characterize ganglionic receptor subtypes with specific subunit composition reveals species-specific pharmacologic properties. Neuropharmacology. 2012;63(4):538–46.
133. David R, et al. Biochemical and functional properties of distinct nicotinic acetylcholine receptors in the superior cervical ganglion of mice with targeted deletions of nAChR subunit genes. Eur J Neurosci. 2010;31(6):978–93.
134. Cuevas J, Berg DK. Mammalian nicotinic receptors with alpha7 subunits that slowly desensitize and rapidly recover from alpha-bungarotoxin blockade. J Neurosci. 1998;18(24):10335–44.
135. Cuevas J, Roth AL, Berg DK. Two distinct classes of functional 7-containing nicotinic receptor on rat superior cervical ganglion neurons. J Physiol. 2000;525(Pt 3):735–46.
136. Severance EG, et al. The alpha7 nicotinic acetylcholine receptor subunit exists in two isoforms that contribute to functional ligand-gated ion channels. Mol Pharmacol. 2004;66(3):420–9.
137. Rassadi S, et al. A null mutation for the alpha3 nicotinic acetylcholine (ACh) receptor gene abolishes fast synaptic activity in sympathetic ganglia and reveals that ACh output from developing preganglionic terminals is regulated in an activity-dependent retrograde manner. J Neurosci. 2005;25(37):8555–66.
138. Genzen JR, Van Cleve W, McGehee DS. Dorsal root ganglion neurons express multiple nicotinic acetylcholine receptor subtypes. J Neurophysiol. 2001;86(4):1773–82.
139. Haberberger RV, et al. Nicotinic acetylcholine receptor subtypes in nociceptive dorsal root ganglion neurons of the adult rat. Auton Neurosci. 2004;113(1–2):32–42.
140. Lips KS, Pfeil U, Kummer W. Coexpression of alpha 9 and alpha 10 nicotinic acetylcholine receptors in rat dorsal root ganglion neurons. Neuroscience. 2002;115(1):1–5.
141. Shelukhina IV, et al. Presence of alpha7 nicotinic acetylcholine receptors on dorsal root ganglion neurons proved using knockout mice and selective alpha-neurotoxins in histochemistry. J Neurochem. 2009;109(4):1087–95.
142. Abdrakhmanova GR, et al. {Alpha}7-nAChR-mediated suppression of hyperexcitability of colonic dorsal root ganglia neurons in experimental colitis. Am J Physiol Gastrointest Liver Physiol. 2010;299(3):G761–8.
143. Fucile S, Sucapane A, Eusebi F. Ca2+ permeability of nicotinic acetylcholine receptors from rat dorsal root ganglion neurones. J Physiol. 2005;565(Pt 1):219–28.
144. Medhurst SJ, et al. Activation of the alpha7-nicotinic acetylcholine receptor reverses complete Freund adjuvant-induced mechanical hyperalgesia in the rat via a central site of action. J Pain. 2008;9(7):580–7.

145. Wang Y, et al. Antinociceptive effects of choline against acute and inflammatory pain. Neuroscience. 2005;132(1):49–56.
146. Ulloa L. The vagus nerve and the nicotinic anti-inflammatory pathway. Nat Rev Drug Discov. 2005;4(8):673–84.
147. Rosas-Ballina M, Tracey KJ. Cholinergic control of inflammation. J Intern Med. 2009;265(6):663–79.
148. Huston JM, et al. Splenectomy protects against sepsis lethality and reduces serum HMGB1 levels. J Immunol. 2008;181(5):3535–9.
149. Bellinger DL, et al. Origin of noradrenergic innervation of the spleen in rats. Brain Behav Immun. 1989;3(4):291–311.
150. Nance DM, Burns J. Innervation of the spleen in the rat: evidence for absence of afferent innervation. Brain Behav Immun. 1989;3(4):281–90.
151. Cano G, et al. Characterization of the central nervous system innervation of the rat spleen using viral transneuronal tracing. J Comp Neurol. 2001;439(1):1–18.
152. Klein RL, et al. Opioid peptides and noradrenaline co-exist in large dense-cored vesicles from sympathetic nerve. Neuroscience. 1982;7(9):2255–61.
153. Felten DL, et al. Noradrenergic sympathetic innervation of the spleen: I. Nerve fibers associate with lymphocytes and macrophages in specific compartments of the splenic white pulp. J Neurosci Res. 1987;18(1):28–36. 118-21.
154. Bruchfeld A, et al. Whole blood cytokine attenuation by cholinergic agonists ex vivo and relationship to vagus nerve activity in rheumatoid arthritis. J Intern Med. 2010;268(1): 94–101.
155. Kawashima K, et al. Ubiquitous expression of acetylcholine and its biological functions in life forms without nervous systems. Life Sci. 2007;80(24–25):2206–9.
156. Morris D. The choline acetyltransferase of human placenta. Biochem J. 1966;98(3):754–62.
157. Parnavelas JG, Kelly W, Burnstock G. Ultrastructural localization of choline acetyltransferase in vascular endothelial cells in rat brain. Nature. 1985;316(6030):724–5.
158. Lan CT, et al. Ultrastructural localization of acetylcholinesterase and choline acetyltransferase in oligodendrocytes, glioblasts and vascular endothelial cells in the external cuneate nucleus of the gerbil. Anat Embryol (Berl). 1996;194(2):177–85.
159. Wessler I, et al. Mammalian glial cells in culture synthesize acetylcholine. Naunyn Schmiedebergs Arch Pharmacol. 1997;356(5):694–7.
160. Kawashima K, et al. Reconciling neuronally and nonneuronally derived acetylcholine in the regulation of immune function. Ann N Y Acad Sci. 2012;1261:7–17.
161. Zia S, et al. Receptor-mediated inhibition of keratinocyte migration by nicotine involves modulations of calcium influx and intracellular concentration. J Pharmacol Exp Ther. 2000;293(3):973–81.
162. Nguyen VT, et al. Choline acetyltransferase, acetylcholinesterase, and nicotinic acetylcholine receptors of human gingival and esophageal epithelia. J Dent Res. 2000;79(4):939–49.
163. Klapproth H, et al. Non-neuronal acetylcholine, a signalling molecule synthesized by surface cells of rat and man. Naunyn Schmiedebergs Arch Pharmacol. 1997;355(4):515–23.
164. Reinheimer T, et al. Acetylcholine in isolated airways of rat, guinea pig, and human: species differences in role of airway mucosa. Am J Physiol. 1996;270(5 Pt 1):L722–8.
165. Orr-Urtreger A, et al. Mice deficient in the alpha7 neuronal nicotinic acetylcholine receptor lack alpha-bungarotoxin binding sites and hippocampal fast nicotinic currents. J Neurosci. 1997;17(23):9165–71.
166. Paylor R, et al. α7 nicotinic receptor subunits are not necessary for hippocampal-dependent learning or sensimotor gating: a behavioral characterization of Acra7-deficient mice. Learn Memory. 1998;5(4–5):302–16.
167. Sykora M, et al. Baroreflex: a new therapeutic target in human stroke? Stroke. 2009;40(12):e678–82.
168. Franceschini D, et al. Altered baroreflex in α7 deficient mice. Behav Brain Res. 2000;113:3–10.

169. Grabus SD, Martin BR, Imad Damaj M. Nicotine physical dependence in the mouse: involvement of the alpha7 nicotinic receptor subtype. Eur J Pharmacol. 2005;515(1–3):90–3.
170. Salas R, et al. Decreased withdrawal symptoms but normal tolerance to nicotine in mice null for the [alpha]7 nicotinic acetylcholine receptor subunit. Neuropharmacology. 2007;53(7): 863–9.
171. Jackson KJ, et al. Differential role of nicotinic acetylcholine receptor subunits in physical and affective nicotine withdrawal signs. J Pharmacol Exp Ther. 2008;325(1):302–12.
172. Miner LL, Collins AC. Strain comparison of nicotine-induced seizure sensitivity and nicotinic receptors. Pharmacol Biochem Behav. 1989;33(2):469–75.
173. Damaj MI, et al. Pharmacological characterization of nicotine-induced seizures in mice. J Pharmacol Exp Ther. 1999;291(3):1284–91.
174. Miner LL, Marks MJ, Collins AC. Genetic analysis of nicotine-induced seizures and hippocampal nicotinic receptors in the mouse. J Pharmacol Exp Ther. 1986;239(3):853–60.
175. Miner LL, Marks MJ, Collins AC. Classical genetic analysis of nicotine-induced seizures and nicotinic receptors. J Pharmacol Exp Ther. 1984;231(3):545–54.
176. Miner LL, Marks MJ, Collins AC. Relationship between nicotine-induced seizures and hippocampal nicotinic receptors. Life Sci. 1985;37(1):75–83.
177. Franceschini D, et al. Absence of alpha7-containing neuronal nicotinic acetylcholine receptors does not prevent nicotine-induced seizures. Brain Res Mol Brain Res. 2002;98(1–2): 29–40.
178. Broide RS, et al. Increased sensitivity to nicotine-induced seizures in mice expressing the L250T alpha7 nicotinic acetylcholine receptor mutation. Mol Pharmacol. 2002;61(3): 695–705.
179. Broide R, Orr-Urtreger A, Patrick JW. Normal apoptosis levels in mice expressing one α7 nicotinic receptor null and one L250T mutant allele. Neuroreport. 2001;12(8):1643–8.
180. Meizel S, Son JH. Studies of sperm from mutant mice suggesting that two neurotransmitter receptors are important to the zona pellucida-initiated acrosome reaction. Mol Reprod Dev. 2005;72(2):250–8.
181. Bray C, et al. Mice deficient in CHRNA7, a subunit of the nicotinic acetylcholine receptor, produce sperm with impaired motility. Biol Reprod. 2005;73(4):807–14.
182. Wang HY, et al. beta-Amyloid(1-42) binds to alpha7 nicotinic acetylcholine receptor with high affinity. Implications for Alzheimer's disease pathology. J Biol Chem. 2000;275(8): 5626–32.
183. Wang HY, et al. Amyloid peptide Abeta(1-42) binds selectively and with picomolar affinity to alpha7 nicotinic acetylcholine receptors. J Neurochem. 2000;75(3):1155–61.
184. Liu Q, Kawai H, Berg DK. beta-Amyloid peptide blocks the response of alpha 7-containing nicotinic receptors on hippocampal neurons. Proc Natl Acad Sci U S A. 2001;98(8): 4734–9.
185. Barrantes FJ, Borroni V, Valles S. Neuronal nicotinic acetylcholine receptor-cholesterol crosstalk in Alzheimer's disease. FEBS Lett. 2010;584(9):1856–63.
186. Levin ED, McClernon FJ, Rezvani AH. Nicotinic effects on cognitive function: behavioral characterization, pharmacological specification, and anatomic localization. Psychopharmacology (Berl). 2006;184(3–4):523–39.
187. Talantova M, et al. Abeta induces astrocytic glutamate release, extrasynaptic NMDA receptor activation, and synaptic loss. Proc Natl Acad Sci U S A. 2013;110(27):E2518–27.
188. Leiser SC, et al. A cog in cognition: how the alpha 7 nicotinic acetylcholine receptor is geared towards improving cognitive deficits. Pharmacol Ther. 2009;122(3):302–11.
189. Hecht SS. Cigarette smoking and lung cancer: chemical mechanisms and approaches to prevention. Lancet Oncol. 2002;3(8):461–9.
190. Schuller HM. Is cancer triggered by altered signalling of nicotinic acetylcholine receptors? Nat Rev Cancer. 2009;9(3):195–205.
191. Kawai H, Berg DK. Nicotinic acetylcholine receptors containing alpha 7 subunits on rat cortical neurons do not undergo long-lasting inactivation even when up-regulated by chronic nicotine exposure. J Neurochem. 2001;78(6):1367–78.

192. Egleton RD, Brown KC, Dasgupta P. Angiogenic activity of nicotinic acetylcholine receptors: implications in tobacco-related vascular diseases. Pharmacol Ther. 2009;121(2):205–23.
193. Wei PL, et al. Tobacco-specific carcinogen enhances colon cancer cell migration through alpha7-nicotinic acetylcholine receptor. Ann Surg. 2009;249(6):978–85.
194. Dasgupta P, et al. Nicotine induces cell proliferation, invasion and epithelial-mesenchymal transition in a variety of human cancer cell lines. Int J Cancer. 2009;124(1):36–45.
195. Soler-Alfonso C, et al. CHRNA7 triplication associated with cognitive impairment and neuropsychiatric phenotypes in a three-generation pedigree. Eur J Hum Genet. 2014;doi:10.1038/ejhg.2013.302.

Chapter 14
Role of Central Serotonin Receptors in Nicotine Addiction

Massimo Pierucci, Stephanie Chambers, Lucy Partridge, Philippe De Deurwaerdère, and Giuseppe Di Giovanni

Abstract Regulation of normal or abnormal behaviour is critically controlled by the central serotonergic systems. Recent evidence has suggested that serotonin (5-HT) neurotransmission dysfunction contributes to a variety of pathological conditions, including depression, anxiety, schizophrenia and Parkinson's disorders. There is also a great amount of evidence indicating that 5-HT signalling may affect the reinforcing properties of drugs of abuse by the interaction and modulation of dopamine (DA) function. This chapter is focused on one of the more addictive drugs, nicotine. It is widely recognised that the effects of nicotine are strongly associated with the stimulatory action it exhibits on mesolimbic DAergic function. We outline the role of 5-HT and its plethora of receptors, focusing on 5-HT_2 subtypes with relation to their involvement in the neurobiology of nicotine addiction. We also explore the novel pharmacological approaches using 5-HT agents for the treatment of nicotine dependence. Compelling evidence shows that 5-HT_{2C} receptor agonists may be possible therapeutic targets for smoking cessation, although further investigation is required.

M. Pierucci, B.Sc. • G. Di Giovanni, Ph.D. (✉)
Faculty of Medicine, Department of Physiology & Biochemistry, Malta University, Tal-Qroqq, Msida MSD 2080, Malta
e-mail: giuseppe.digiovanni@um.edu.mt; massimo.pierucci@um.edu.mt

S. Chambers
Department of Medical Sciences, Exeter Medical School, St. Luke's Campus, Exeter, EX1 2LU, Devon, UK
e-mail: sc505@exeter.ac.uk

L. Partridge
Department of Bioscience, Cardiff School of Biosciences, Sir Martin Evans Building, 2 Museum Avenue, Cardiff CF10 3AX, South Glamorgan, UK
e-mail: PartridgeLM@cardiff.ac.uk

P. De Deurwaerdère, Ph.D.
Centre National de la Recherche Scientifique, Institut des Maladies Neurodégénératives (UMR CNRS 5293), Universitè de Bordeaux, 146 rue Léo Saignat, Bordeaux Cedex 33076, France
e-mail: deurwaer@u-bordeaux.fr

R.A.J. Lester (ed.), *Nicotinic Receptors*, The Receptors 26,
DOI 10.1007/978-1-4939-1167-7_14

Keywords Serotonin receptors • Nicotine addiction • Serotonergic system • Neurotransmission • Parkinson's • 5-HT • 5-HT receptors • Nicotine • DRN • MRN • Serotonergic projections • Smoking cessation treatment • Serotonergic drugs

1 Introduction

There are a large number of scientific publications on serotonin (5-HT)-containing neurons in the brain and their role in normal and pathological behaviour [1, 2]. Disturbances in 5-HT function have been linked to many psychiatric and behavioural conditions, particularly anxiety and mood disorders [3]. This is not surprising, due to the complexity of 5-HT function, as it has a multitude of receptors and widespread 5-HT neuronal innervation and is coupled with diverse signalling pathways [4, 5].

Rather than exhibiting direct effects on cell bodies, the effects of 5-HT are usually indirect, involving the modulation of complex neuronal circuitry. This modulation is generally subtle as it is dependent on a number of other neurotransmitter baseline activities and their activation. The increased availability of 5-HT receptor knockout mice (KO) and molecular understanding of the 5-HT receptors have encouraged a great amount of new research in recent years [5, 6].

Dysfunction of 5-HT neurotransmission contributes to the pathophysiology of a variety of neuropsychiatric disorders such as schizophrenia, depression and drug abuse [2, 7, 8]. Therefore, developing selective and specific pharmacological agents for 5-HT receptor subtypes has given researchers the opportunity to gain insight into the roles these receptors play in various disorders of the brain and clinicians' new therapies with better efficacy and fewer side effects [9–13].

This chapter focuses on the functional role of 5-HT and its receptors in nicotine addiction. It reviews the anatomy of the 5-HT system and explores the distribution of 5-HT receptors, and the physiological and pharmacological aspects of these receptors in the central nervous system (CNS). Previously published experimental data is reviewed showing the relationship between agents specifically targeted to 5-HT_2 receptors and the effect of nicotine. Consideration of the potential use of these agents for treatment of cessation of tobacco smoking and subsequently nicotine addiction is also included.

2 Serotonergic Systems

5-HT has been suggested to have both excitatory and inhibitory roles in the nervous system; therefore there is great difficulty in elucidating the precise nature of modulation [14]. The reason for this difficulty in classification may be due to the number of different roles exhibited by the 5-HT receptor subtypes in association with different neurotransmitter systems [14–16]. Due to the significance of 5-HT receptor control on dopamine (DA) activity and the relationship between this and pathophysiology of specific DA-related diseases such as schizophrenia, depression, Parkinson's disease (PD) and drug abuse, this area has received a great amount of attention [17].

The distribution of 5-HT in organisms gives rise to the description of two 5-HT subsystems, those being central and peripheral [18]. 5-HT is found in many different tissues such as the heart, lungs, blood vessels, platelets and pancreatic tissue. The amount of 5-HT present in the CNS is relatively low. The peripheral subsystem contains most of the 5-HT (90 %), which is around 10 mg in the human body, and is found mainly in mucosal enterochromaffin cells of the gastrointestinal (GI) tract [19, 20]. Overall, 5-HT expression of all neurons in the enteric nervous system (ENS) reaches 2–20 % [21, 22].

Research into 5-HT began in the ENS to accentuate the high concentrations of the neurotransmitter in this region, more specifically in the GI tract. Between 1937 and 1940, the Italian researcher Vittorio Erspamer identified a substance they called *enteramine*, a "gut-stimulating factor" found in the mucosa of the intestine, later found to be 5-HT [23]. In the late 1940s a research group led by Page also purified the same substance from beef blood, highlighting that it had vasoconstrictive activity. Page named this as "*… serotonin, which indicates that its source is serum and its activity is one of causing constriction …*" [24]. Twarog and Page [25] performed a sensitive bioassay which used extracts of mammalian brain and detected 5-HT. Brodie and colleagues [26] suggested that 5-HT may serve as a neurotransmitter in the CNS, which was later supported further by the work of Costa and Aprison [27] who discovered that it was in fact present in the human brain. This resulted in one of the most pivotal discoveries in science and the birth of a new branch of neuroscience [28].

The serotonergic system has a topographical organisation corresponding to the functional and anatomical properties that they exhibit [29], and is one of the mammalian brain's diffusively organised projective systems. The majority of serotonergic neurons reside in the brainstem, being particularly present in the raphe nuclei, part of the reticular formation. The 5-HT cells in this region are multipolar, with size and orientation being extremely different to other complex axonal systems in the CNS. They send projections virtually to all the CNS areas from limbic structures, basal ganglia, cerebral cortex and brainstem to the spinal cord grey matter. Upon the discovery that the brainstem contained these 5-HT cell groups, Dahlstrom and Fuxe [30] produced a system to code these groups based on the rat brainstem. These morphologically heterogeneous cells have been divided into nine nuclei groups from B1 (the most caudal cell cluster) to B9 (Fig. 14.1). These can then be combined into two major groups, the caudal and rostral serotonergic groups. The caudal group, also referred to as the inferior group, is situated in the medulla and consists of three separate nuclei. These nuclei project to the grey matter of the spinal cord and are called the nucleus raphe magnus (NRM, group B5), the nucleus raphe obscures (NRO, groups B1, B2 and B3) and the nucleus raphe pallidus (NRP, group B4). The rostral group, also known as the superior group, is positioned in the pons/mesencephalon, and consists of two main nuclei: the first being the dorsal raphe nucleus (DRN, groups B6 and B7), which in the human brain is thought to contain about 235,000 neurons [31], and the second being the median raphe nucleus (MRN, cell group B8).

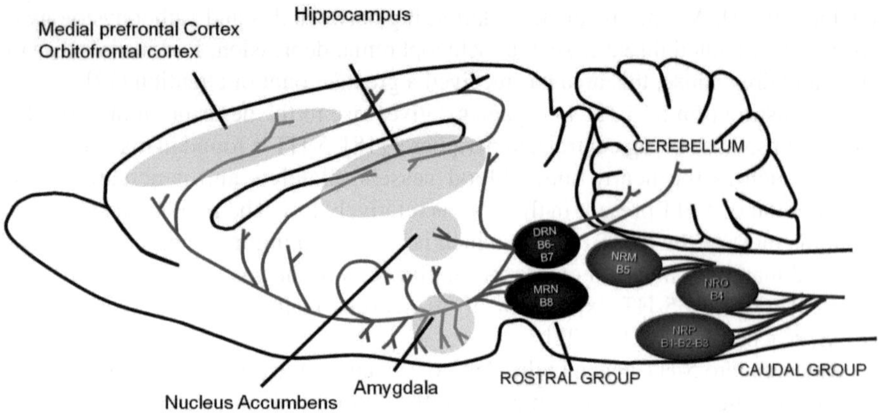

Fig. 14.1 Midsagittal view of the rat brainstem with serotonin-immunoreactive cell body groups. The *ovals* encompass the two major subdivisions of the brain serotonergic system. *DRN* dorsal raphe nucleus, *MRN* medial raphe nucleus, *NRM* nucleus raphe magnus, *NRO* nucleus raphe obscurus. Cell groups B1 to B9 according to the terminology of Dahlström and Fuxe [176]

2.1 *Ascending Serotonergic Projections*

The DRN and MRN are the main subdivisions of the raphe nucleus, providing innervation to nearly all structures of the CNS [32–37]. The primary targets of 5-HT projections lie in the forebrain and spinal cord, with the ascending projection collaterals reaching many different regions of the brain such as the cerebral cortex, basal ganglia, limbic system and diencephalon. These projections primarily travel in the medial forebrain bundle from which many axons extend toward other fibre pathways in order to reach their target areas. Innervation patterns vary greatly between the different areas of the forebrain and are extremely specific. Fibre density in the cortex varies between layers; in primates the cortex receives a particularly dense 5-HT innervation in layer IV [35, 38]. The striatum also has a very dense innervation, along with the hypothalamus, septal area and specific areas of the thalamus.

Most areas of the brain contain an overlapping innervation coming from the DRN and MRN, apart from the dorsal hippocampus, the suprachiasmatic nucleus, the olfactory bulb and the medial septum nucleus, which receive 5-HT innervation from only the MRN. On the other hand, the corpus striatum, the globus pallidus, the lateral septum nucleus, the amygdala and most of the prefrontal cortex (PFC) are all preferentially innervated by the DRN with the substantia nigra pars reticulata (SNr) receiving the largest DRN 5-HT innervation in the CNS [39, 40]. The ventral part of the hippocampus, the nucleus accumbens (NAcc) and various nuclei in the thalamus receive innervation from both the DRN and MRN [32]. Moreover, there are extensive serotonergic connections between the two main serotonergic nuclei of the brainstem [41].

It is noteworthy that all the brain areas involved in drug addiction [33, 34, 37], such as the ventral tegmental area (VTA) and substantia nigra (SN), including their terminal fields, receive projections from 5-HT containing cell bodies in the raphe nuclei (Fig. 14.1) [34, 35, 42, 43], confirming the implication of 5-HT in drug abuse.

2.2 Descending Serotonergic Projections

Descending serotonergic projections to the spinal cord originate from the medullary raphe nuclear complex, forming a network of fibres with a strong density of 5-HT axons throughout the grey matter [44]. More specifically, the dorsal horn, ventral horn motor nuclei and thoracic cord intermediolateral column have extremely dense serotonergic innervation. Additional regions receiving strong 5-HT input include the central grey matter and the ependyma of the central canal (containing a 5-HT nerve plexus). As highlighted above, the main origin of 5-HT afferents is the raphe nuclei, although other minor projections are also present. These are shown to originate from B9 cells scattered in the pontomesencephalic reticular area of the brain and also from cells in the medial longitudinal fasciculus [44, 45].

3 5-HT Receptors

The diversity of effects produced by 5-HT when administered in the brain is due to the large variety of receptors to which it binds. These are classified into seven main classes, those being 5-HT_1 to 5-HT_7. These are then further divided into 14 different receptor types, classified by the means of pharmacological profiling, cDNA-deduced primary sequencing and signal transduction methods [4, 5, 16, 46]. Most 5-HT receptors are metabotropic G-protein-coupled receptors (GPCRs), excluding the ionotropic 5-HT_3 receptors. The 5-HT GPCRs act through intracellular signalling pathways in order to depolarise or hyperpolarise their host cell [16]. 5-HT receptors may have synergistic or opposing activity. For example, in the administration of cocaine, blocking the 5-HT_{2A} receptors attenuated some of the effects of cocaine, whereas blocking the 5-HT_{2C} receptors caused an enhancement in some of the behavioural effects induced by cocaine [47–49]. The behavioural aspects linked to impulsivity are also mediated by the opposing actions of these two 5-HT_2 receptor subtypes [49]. It is of interest to note that the same 5-HT receptor subtypes located in different neuronal regions can act synergistically; that is, the 5-HT_{1A} receptors that are located both on raphe cell bodies and postsynaptically in limbic regions control impulse flow through the excitability of 5-HT raphe neurons and the postsynaptic neurons [50].

4 Serotonergic Involvement in the CNS Effects of Nicotine

It has been suggested in several studies that 5-HT is one of the more important neurotransmitter systems involved in the reinforcing aspects of drugs of abuse [51–54]. The physiological and pharmacological effects of nicotine are complex. Interactions between nicotinic acetylcholine receptors (nAChRs), gamma-aminobutyric acid (GABA), glutamate (GLU) and DA in the VTA are involved in the reinforcing effects of nicotine (for reviews, see [55, 56] and Chap. 15). For instance, the activation of mesolimbic DA neurons is essential in the behavioural effects of nicotine, including locomotor activity and reinforcement [57–59].

5-HT hypofunction is a defining factor in depression [60]. A number of animal and human studies suggest that changes in 5-HT transmission can contribute to the symptoms of nicotine withdrawal such as anxiety and depression [61] and an increased risk of suicide and self-harm [62]. Consistently, nicotine has been shown to manifest antidepressant effect on the user [63]; indeed prevalence of smoking is much higher amongst people suffering with depression [64].

Previous studies into the molecular properties of nAChRs in the DRN have shown the presence of α4, α7 and β2 subunits [65–68]. The α7 subunit was found to be highly expressed on DRN neurons with a large diameter of 15–25 μm and colocalised with tryptophan hydroxylase [69]. It was also found on neurons with a smaller diameter of 5–10 μm that are usually GABAergic [66]. The cholinergic input to the serotonergic DRN cells is sustained by the pedunculopontine tegmentum (PPTg) neurons [70]. In vitro electrophysiology studies have shown that the neuronal activity of 5-HT neurons is influenced by nicotine. Most 5-HT neurons of the DRN increase their action potential firing rate when nicotine is administered, which results in an overall increase in 5-HT output to its target tissues [71–73]. The excitability of these DRN neurons is also indirectly stimulated by their synaptic drive, which is influenced by nicotinic presynaptic heteroreceptors. Indeed, the effects of nicotine include the enhancement of excitatory glutamatergic input to DRN-NAcc projection neurons, and the positive or negative modulation of inhibitory GABAergic input to the same subset. Nonetheless, in vivo systemic nicotine has been shown to decrease neuronal firing in the majority of DRN neurons, increasing it in only a small subset of neurons [74, 75]. When nicotine was applied locally to the DRN it evoked 5-HT release [76], while systemic nicotine produced only a small and restricted increase in 5-HT [77]. Thus, the 5-HT firing inhibition may be a secondary effect, depending on 5-HT release from the somatic dendrites. Some in vivo data are in conflict with in vitro results. It is important to consider that 5-HT neurons are strongly impacted by the arousal state (sleep-waking) [78]; therefore there is a possibility that the effects in the DRN neurons, upon nicotine administration in vivo, may be influenced by factors such as anaesthesia and the time of day that the experiment takes place. Moreover, acute exposure of DRN neurons to nicotine may therefore cause direct depolarisation and indirect synaptic modulation. Given the excitatory effects of nicotine on DRN neurons, it is likely that 5-HT contributes to the rewarding effects of nicotine.

A number of the behavioural effects induced by nicotine may be mediated by the 5-HT neurotransmitter system [61]. In this context, Summers and colleagues [79] showed that nicotine, as well as the nicotinic agonist RJR-2403, significantly increased cortical release of 5-HT in rats. Since there has been no observation of nicotinic binding sites being present on 5-HT axon terminals in the cortex [80], the increased release of 5-HT might be due to the DRN projections to the cortex that express nAChRs in their soma [79].

In addition, [^{3}H]5-HT release from striatal synaptosomes is increased by ACh, nicotine and the nicotinic receptor agonists epibatidine and cytosine; this effect can be inhibited by administration of the non-competitive nicotinic receptor antagonist mecamylamine [81]. The release of 5-HT into striatal slices due to nicotine exposure showed an increase over a number of days; for example a significant enhancement after 10 days of nicotine treatment has been described [82]. Acute and chronic nicotine administration increased striatal 5-HT release in freely moving rats only when they were exposed to stress, likely by stimulating presynaptic nicotinic receptors in the striatum [83]. Nicotinic receptor agonists such as 1,1-dimethyl-4-phenylpiperazinium (DMPP), lobeline and nicotine have been shown to increase [^{3}H]5-HT release in hippocampal slices in the rat; however cytisine, epibatidine and nicotine had no effect in others [84, 85]. Reuben and Clarke [81] also found supporting evidence that nicotine had no effect on 5-HT release from the cerebral cortex or hippocampal synaptosomes. Furthermore, there is conflicting evidence on the effects of chronic treatment with nicotine and the concentration of 5-HT in the dorsal hippocampus, with both decreases [86] and increases [87] being described. The duration of treatment is likely to be a factor in the effect of nicotine on the brain's 5-HT system. Although, in one study, 5-HT levels in the hypothalamus increased following both acute and chronic nicotinic administration, rates of 5-HT synthesis were not affected [61].

A number of pieces of research suggest that the release of 5-HT in the spinal cord is both directly and indirectly controlled through multiple nAChR populations. One nAChR population, located on 5-HT terminals, might have an excitatory effect on 5-HT outflow. Conversely, a second nAChR population, expressed on GABAergic interneurons and tonically activated by endogenous ACh, inhibits 5-HT release in the spinal grey matter [61].

Nicotine is the main alkaloid found in nicotiana plants (95–97 %) and is largely responsible for the deleterious effects of tobacco including addiction. Two factors have to be taken into account when considering the mode of which nicotine is administered. First of all, a number of minor alkaloids are found in the plant (cotinine, anabasine, nornicotine, tabagisine, myosmine). These minor tobacco alkaloids have similar structural and pharmacological activity compared to nicotine, although they are generally less potent [88, 89]. Nornicotine and cotinine also play a role as major metabolites of nicotine [89]. Intravenous infusion of nicotine, combined with five minor alkaloids found in tobacco smoke (anabasine, nornicotine, anatabine, cotinine and myosmine), increased locomotor activity and behavioural sensitisation following self-administration in rats [90]. A tobacco extract specifically containing nicotine and these alkaloids was more potent to enhancing striatal DA release in freely moving rats compared to pure nicotine. The potentiation was not observed in the nucleus accumbens [91]. These experiments, using an extract, confirm that the striatum plays an important role in the psychoactive properties of nicotine and/or tobacco (see below). Interestingly, the extract of alkaloids from the plant is more potent in inhibiting DRN neuronal firing rate compared to pure nicotine [75].

Second, smoking tobacco generates a very high number of molecules (>4,000) that are absorbed which can also modulate the psychoactive properties of nicotine. It has been reported for instance that some of these compounds block the monoamine oxidases A and B (MAOA and MAOB), mitochondrial enzymes involved in the degradation of monoamines. These compounds may be relevant to tobacco addiction because their co-administration in rats has been shown to dramatically enhance the DA effects of nicotine and the motivation to self-administer nicotine [92]. The blockade of MAO-A and MAO-B, by limiting the degradation of 5-HT at the terminals, could favour 5-HT output and reinforce the dichotomy between the ability of nicotine to reduce DRN neuronal firing rate and to enhance 5-HT release at some terminals. However, this remains to be investigated.

On the whole, there is good supporting evidence from a number of sources to conclude that nicotine elicits an increase in 5-HT neuronal activity and an overall release of 5-HT in several target tissues in the brain including the DRN [76].

5 5-HT Receptors and the Rewarding Properties of Nicotine

In regard to the involvement of the various 5-HT receptor subtypes in nicotine use and dependence, very little work has been done to attribute the effects of administered ligands on these receptors, to the behavioural effects of nicotine. The 5-HT$_{1A}$, 5-HT$_{2A}$, 5-HT$_{2C}$, 5-HT$_{3}$, 5-HT$_{4}$ and 5-HT$_{6}$ subtypes are considered to be the most likely targets involved in nicotine use and dependence. In this chapter we focus on the 5-HT$_{2}$ receptor class.

5.1 5-HT$_{2A}$ Receptors

5-HT$_{2A}$ receptors are the predominant receptor subtypes in cortical areas, but are also present in DA-rich areas such as the NAcc, striatum, VTA and SN [93–95]. Such receptors are situated mainly postsynaptically; for instance 5-HT$_{2A}$ receptors are found on pyramidal neurons, as well as GABAergic interneurons in the PFC [96, 97]. 5-HT$_{2A}$ receptors can also be found on DAergic and non-DAergic cells of the VTA and SN [97–99]. 5-HT$_{2A}$ receptors are therefore seen, due to their anatomical distribution, as a good candidate for the modulation of DA-mediated functions.

A number of behavioural effects of nicotine are altered by 1-(2,5-dimethoxy-4-iodophenyl)-2-aminopropane (DOI) administration. For example, a study showed that DOI did not alter the acute stimulant action of nicotine, yet it did inhibit the development of sensitisation to the locomotor stimulant effect when nicotine was administered repeatedly in rats [100]. When given to mice, DOI was shown to avert the initial locomotion suppression induced by nicotine [101]. DOI administration mitigated the discriminative stimulus properties of nicotine, effects that were inhibited by 5-HT_{2A} receptor antagonists, but not by 5-HT_{2C} receptor antagonists [101, 102]. These results provide evidence that 5-HT_{2A} receptor stimulation may oppose some of the behavioural effects of nicotine. Drug discrimination studies have shown a relationship between DOI and lysergic acid diethylamide (LSD) [103], with 5-HT_{2A} receptor stimulation by LSD and other drugs being recognised to induce hallucinations [104]. Even with this evidence, and although DOI can reduce some behavioural effects of nicotine, it is difficult to apply this to the treatment of nicotine dependence in humans with 5-HT_{2A} receptor agonists. 5-HT_{2A} receptor blockade inhibits the extracellular increase of DA when induced by amphetamine and cocaine [105, 106]. This effect can also be seen on a behavioural level, in which the selective 5-HT_{2A} receptor antagonist M100907 attenuates locomotor stimulant effects of cocaine, amphetamine and other such psychomotor stimulants [47, 48, 107]. The 5-HT_{2A} receptors resident in the VTA appear to be integral in modulating psychostimulant-induced behaviours mediated by the mesocorticoaccumbens circuit. Indeed local infusion of M100907 to the VTA blocked amphetamine- and cocaine-induced locomotion [47, 106]. Consistently, it has recently been shown that vulnerability of adult male rats to hyperactivity, induced by cocaine, is enhanced following virally mediated overexpression of 5-HT_{2A} receptors in the VTA [108]. Due to this evidence, and the relationship between the VTA and reinforcing effects of nicotine, it may be expected that 5-HT_{2A} receptor antagonists would modify some of the behavioural effects observed following the administration of nicotine. However, phasic but not tonic activation of 5-HT_{2A} receptors appear to have inhibitory influence on the nicotine cue, since their pharmacological stimulation by DOI attenuates the discriminative stimulus effects of nicotine, while their blockade by M100907 was ineffective [102]. Recently, Levin and co-authors [109] showed that the 5-$HT_{2A/2C}$ receptor antagonist, ketanserin, significantly decreased nicotine self-administration. This effect is likely to be due to 5-HT_{2C} receptor blockade since, unlike ketanserin, the selective 5-HT_{2A} receptor antagonist M100907 did not alter nicotine self-administration on the fixed ratio schedule 5 (FR5) or progressive ratio schedules of reinforcement [110]. The reinstatement of nicotine-seeking behaviour elicited by cues previously associated with self-administered nicotine or by priming injections of nicotine were both reduced by activation of 5-HT_{2C} receptors and by blockade of 5-HT_{2A} receptors.

Collectively, these results show that while a 5-HT_{2C} receptor agonist and a 5-HT_{2A} receptor antagonist have differing effects on nicotine self-administration, they both reduce the reinstatement of nicotine-seeking behaviour [110]. This evidence shows that the reinforcing, psychomotor stimulant and discriminative effects of nicotine are not affected by 5-HT_{2A} receptor blockade, but that 5-HT_{2A} receptors may have a role in inducing nicotine-seeking behaviour. However, since M100907 reduced the reinstatement of cocaine-seeking behaviour, induced by cocaine or cocaine-associated cues [110], it can be inferred that its effects are not nicotine specific.

5.2 5-HT_{2C} Receptors

5-HT has an influential role on the effects of many drugs of addiction and of particular interest is its action via the 5-HT_{2C} receptor subtype in modulating the effects of nicotine [1, 51, 54, 111]. 5-HT_{2C} receptors are located on DA cell bodies and GABAergic interneurons in the VTA area, observed by mapping studies [112, 113]. Furthermore, there is evidence that shows how central DA function is influenced by 5-HT_{2C} receptors [10, 114–116], whereas blockage of the 5-HT_{2A} receptor has consistently no effect on reinforcing the discriminative or the psychomotor stimulatory effects of nicotine. Administration of RO 60-0175, a selective 5-HT_{2C} receptor agonist [117], in in vivo electrophysiological and neurochemical studies, induced a decrease in both the burst firing activity and basal firing rate of VTA DA neurons and decreased both striatal and accumbal DA release [118–121]. Moreover, the administration of the selective 5-$HT_{2C/2B}$ receptor antagonist, SB 206553 [122], and the selective 5-HT_{2C} receptor antagonist, SB 242084 [123], both showed the opposite effect on DA neuronal activity and DA release [115, 121, 124, 125]. Additionally, the use of RO 60-0175 to stimulate 5-HT_{2C} receptors in the VTA caused suppression of the stress-stimulated DA outflow in the rat PFC [126], whereas SB 242084 was found to potentiate it [126]. Alongside this, the phencyclidine-induced increase in accumbal DA release was enhanced by SB 242084 pretreatment [127]. Collectively, this evidence suggests that these receptors play a role in mediating the evoked DA release. In line with these studies, the increase in morphine-induced DA release in the NAcc was inhibited by stimulation of 5-HT_{2C} receptors in freely moving rats [96]. Furthermore, DA release in the NAcc and corpus striatum, in addition to the DA neuronal firing rate in the VTA and the substantia nigra pars compacta (SNc) induced by morphine, was enhanced by SB 206553 administration [128]. The effects of these agonists were blocked by SB 242084. The pharmacological stimulation of the 5-HT_{2C} receptor with RO 60-0175 and WAY163909 (another agonist of the 5-HT_{2C} receptor) has been shown to attenuate the discriminative stimulus effects induced by nicotine [102, 129]. SB 242084 alone induced some nicotine-induced responses (~30 %), although it did not alter the nicotine cue. Overall, these studies show that pharmacological stimulation of the 5-HT_{2C} receptors causes an attenuation of the discriminative stimulus response, although the tonic activation of these receptors does not influence the subjective effects of nicotine.

Investigations carried out by independent groups into the role of a 5-HT_{2C} receptor agonist, RO 60-0175, in nicotine self-administration, showed that on a fixed-ratio/60-s timeout (FR5/ TO-60s) and FR5 schedule, RO 60-0175 caused a diminished response rate for nicotine [110, 130], and this effect was blocked by SB 242084 [110]. Priming injections of nicotine were administered to induce reinstatement of the response; this response was reduced by 5-HT_{2C} receptor agonism by RO 60-0175 and lorcaserin (a relatively selective 5-HT_{2C} receptor agonist which has received the US Food and Drug Administration (FDA) approval as an anti-obesity drug) and by 5-HT_{2A} antagonism using M100907 [110, 131]. The observed diminished response rate for nicotine occurred over a similar range of doses to that observed for cocaine self-administration [132], which was also reduced by 5-HT_{2C} manipulation [110, 132–134]. Moreover, the stimulation of 5-HT_{2C} receptors by RO 60-0175 reduced the previous nicotine-induced self-administration and nicotine-induced hyperactivity [130, 135]. In addition, a reduction in nicotine-induced self-administration and hyperactivity was observed with ketanserin [109] and lorcaserin [136].

Mixed results have been obtained from studies on nicotine-induced conditioned place preference involving the effects of 5-HT_{2C} receptor agonists such as RO 60-0175, WAY161503 and WAY163909. In mice, RO 60-0175 blocked both nicotine-induced place preference and the conditioned place preference observed with tetrahydrocannabinol (THC) [137]. However, in rats, nicotine-induced place preference was not blocked by WAY161503, even at doses which induced locomotion [138]. It is important to consider that the comparison of these studies is difficult, as it is unclear whether these discrepancies are due to the different 5-HT_{2C} receptor agonists or the different species on which they were carried out.

Interestingly RO 60-0175 was also able to block sensitisation to nicotine (occurring after repeated exposure to the drug) as well as reduce nicotine-induced hyperlocomotion and the operant response for nicotine [130]. Two separate studies using 5-HT_{2C} agonists found that the stimulation of locomotion by nicotine in rats with prior nicotine exposure was blocked by both WAY161503 [138] and WAY163909 [139], and furthermore that the administration of SB 242084 reinstated this effect of nicotine.

Two studies by Esposito and colleagues [140, 141] were conducted to investigate whether DA is involved in the behavioural effects of the interaction of RO 60-0175 and nicotine-induced hyperlocomotion and reward. At 1 and 3 mg/kg doses of RO 60-0175, the increase in DA release induced by the administration of acute nicotine was prevented in the corpus striatum but not in the NAcc [140]. However, enhanced DA release, induced by chronic nicotine administration, was prevented both in the corpus striatum and NAcc by the same doses of RO 60-0175 [140]. The selective 5-HT_{2C} receptor antagonists, SB 242084 and SB243213, both inhibited the effects of RO 60-0175 [140]. At the doses used, it is probable that both SB 242084 and SB243213 specifically block 5-HT_{2C} receptors in the brain. Therefore, it can be inferred that blockage of the hyperlocomotive and rewarding effects of nicotine by RO 60-0175 [132] are likely to be due to its ability to inhibit mesolimbic DA function. Differences in the effects of acute and repeated exposure to nicotine on the central 5-HT system may explain why RO 60-0175 had varying effects on DA release in the NAcc, although further investigations are needed to elucidate this point. The efficacy of the 5-HT_{2C} agonists could be indirect by acting against DA transmission within the basal ganglia. For instance, RO 60-0175 (1 mg/kg) has been shown to enhance the activity of GABAergic neurons of the VTA neurons [142] while DA transmission may favour their inhibition [143, 144]. Also, RO 60-0175 (1 mg/kg) enhanced the phasic and excitatory impact of the prefrontal but not the motor cortex stimulation on the activity of GABAergic neuron of the SNr [145]. The efficacy of the 5-HT_{2C} agonists could be more directly related to changes of DA indices especially because 5-HT_{2C} receptors control well an increase in DA release associated with an increase in DA neuron firing rate [96, 128, 146]. Indeed, Pierucci et al. [141] tested the hypothesis that the effects seen in the NAcc and striatum with stimulation of 5-HT_{2C} receptors and nicotine were due to changes in DA neuronal activity (see Figs. 14.2, 14.3 and 14.4). RO 60-0175 pretreatment prevented the enhancement in DA neuronal firing rate elicited by acute nicotine administration in the SNc, in both drug-naive and chronically treated rats, but was devoid of any significant effect in the VTA. Moreover, RO 60-0175 significantly reduced the stimulatory effect on the VTA DA neurons, induced by acute nicotine, both in drug-naive and chronically treated rats. However, after repeated administration of nicotine there were no changes observed in the nicotine-induced excitation of DA neurones in the VTA (see Figs. 14.3 and 14.4). Thus it seems that the chronic administration of nicotine does not potentiate the nicotine effect previously observed, via increasing the firing activity of mesolimbic DAergic neurons, but via another mechanism. Moreover, in rats previously treated with chronic nicotine, administration of acute nicotine caused a significant increase in the burst firing activity of DA neurons in the VTA but not in the SNc [141]. It has been postulated that after repeated nicotine administration, tolerance to the stimulatory effect of nicotine occurs in the nigrostriatal but not in the mesolimbic DAergic system, a hypothesis also supported by the presented data.

In further agreement, studies have recently revealed that there is a reduced 5-HT turnover and an increase in 5-HT_{2C} receptor sensitivity after the repeated administration of nicotine. However the up-regulation of 5-HT_{2C} receptors is only observed following repeated treatment with this alkaloid. Furthermore, the expression of nicotine-induced locomotor sensitisation is counteracted by repeated administration of citalopram, a selective 5-HT re-uptake inhibitor [147]; therefore it can be hypothesised that the development and expression of sensitisation to mesolimbic DA system, by repeated administration of nicotine, can be prevented by the activation of the 5-HT system. This effect could be beneficial to smoking cessation therapies as the activation of the 5-HT system could lead to the extinction of nicotine-induced rewarding effect on the DAergic system. The activation of the 5-HT_{2C} receptors was previously associated with the effect of stimulating the GABAergic neurons which negatively impact on DA-containing neurons in the SNc and the VTA [1]; therefore it is speculated that the inhibitory action of RO 60-0175 on nicotine-induced DA release [140] and in the neuronal activity changes [141] might be also partially mediating this effect in a similar manner. Collectively, these results show that

Fig. 14.2 (continued) nicotine administration on single DAergic neurons' firing rates compared to controls (saline, 100 μl i.v., at *arrows*), while the graphs *below* report the cumulative dose-response curves showing the mean percentage changes (±S.E.M.) in firing rate after either nicotine or saline administration. *Arrows* indicate the time of nicotine or saline injections; apomorphine (Apo), a D1/D2 receptor agonist, was injected (10–30 μg/kg, i.v.) at the end of most of the experiments to confirm the DAergic identity of recorded neurons (adapted from [141] with permission). (**b**) Time course of nicotine (1 mg/kg, i.p.)-induced effect on DA terminal release in both the striatum and nucleus accumbens. Changes in DA extracellular levels were assessed using in vivo microdialysis on freely moving animals. The *arrows* indicate the time of injection of either nicotine or saline (adapted from [140] with permission)

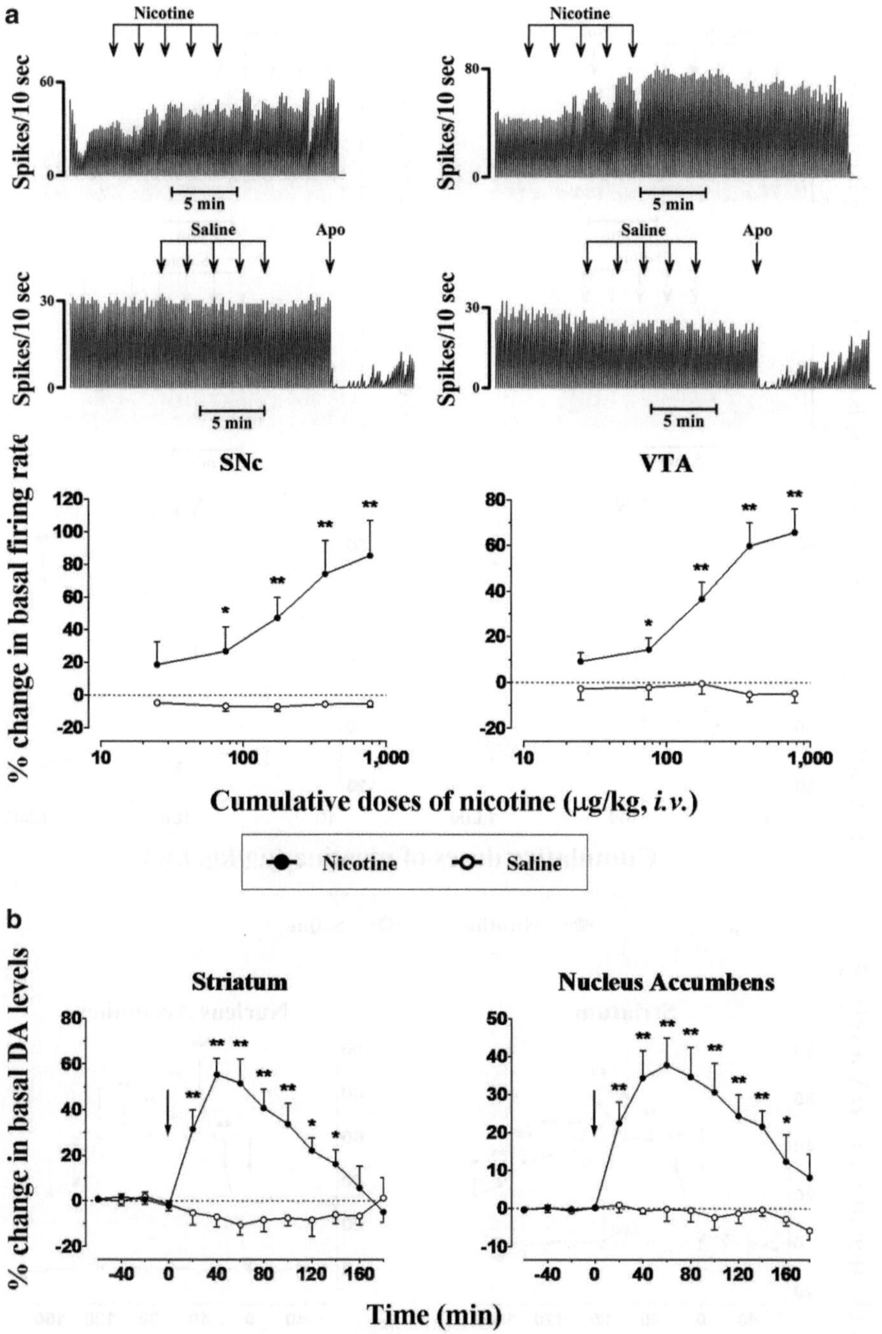

Fig. 14.2 Systemic administration of nicotine increases the activity of the both nigro-striatal and meso-corticolimbic DAergic systems of drug-naive rats in vivo. (**a**) The injection of cumulative doses of nicotine (25–775 µg/kg, i.v.) increased the firing rate of DA neurons recorded using in vivo single-unit extracellular recording technique. The rate meters on *top* show the effect of

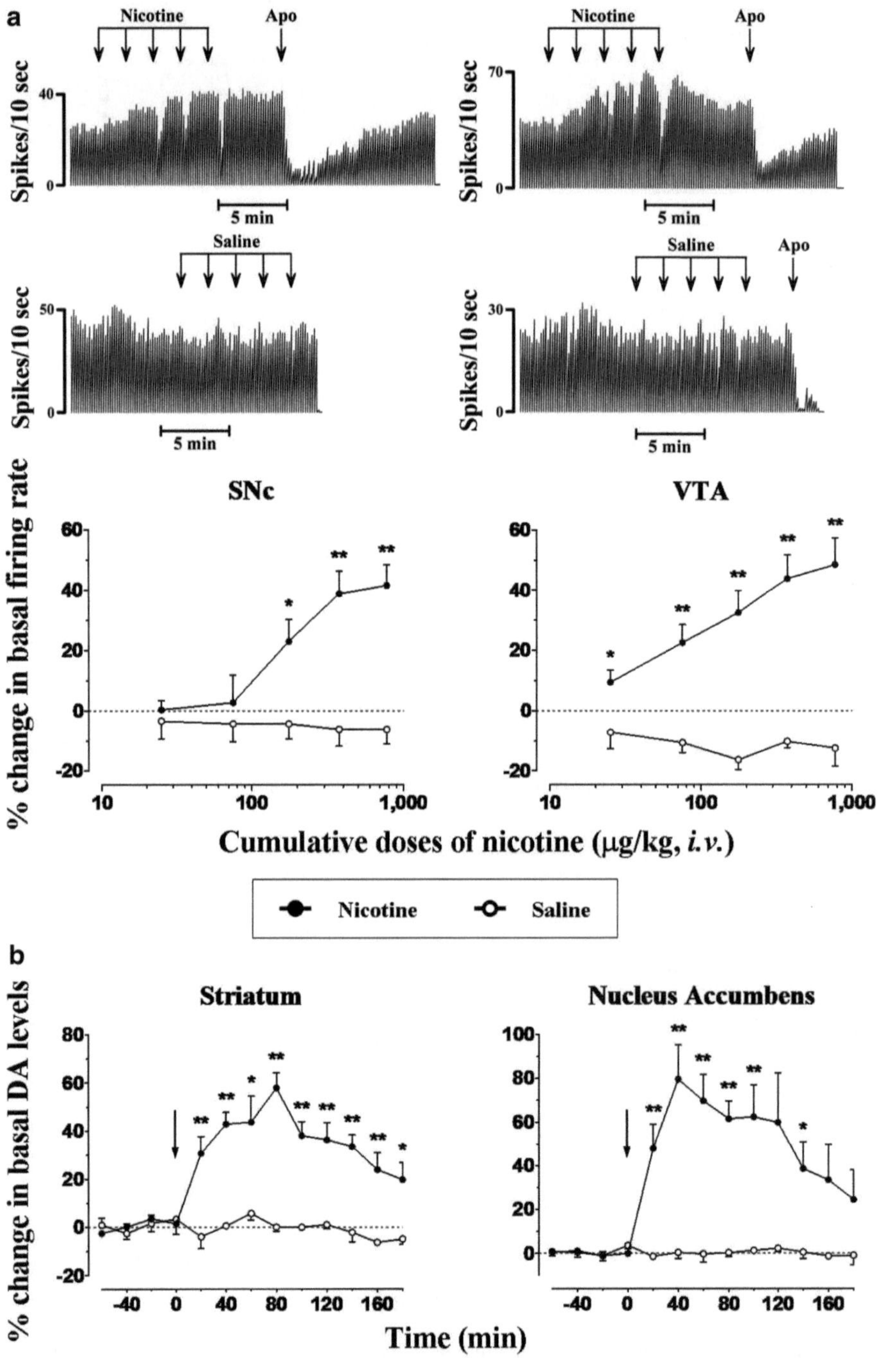

Fig. 14.3 Systemic administration of nicotine increases the activity of both nigro-striatal and meso-corticolimbic DAergic systems in vivo of rats chronically treated with nicotine (1 mg/kg, i.p.) for 10 consecutive days. (**a**) The injection of cumulative doses of nicotine (25–775 μg/kg, i.v.)

5-HT_{2C} receptor agonists reduce the stimulatory, discriminative and perhaps reinforcing effects of nicotine as well as reduce the impact of nicotine on electrophysiological and neurochemical indices of DA function (Fig. 14.4).

The cessation of chronic nicotine use can produce aversive effects, including both motivational and somatic withdrawal symptoms such as nicotine craving, anxiety, reduced concentration, irritability, depressed mood and an increased appetite [148]. These withdrawal effects are considered to be the main factors in the maintenance of the tobacco habit in human smokers. The role of 5-HT and the 5-HT_{2C} receptor in nicotine withdrawal and its negative effects is gradually being elucidated; recently it was shown that both RO 60-0175 and M100907 reversed the resultant depressive-like behaviour from nicotine withdrawal [149]. Conversely, preliminary results show that RO 60-0175 and lorcaserin do not reduce mecamylamine-precipitated somatic signs of withdrawal in nicotine-dependent rats [131]. 5-HT_{2C} drugs could be potentially beneficial to relieve the aversive symptoms induced by nicotine cessation when combined with adjunct smoking cessation therapies.

6 Smoking Cessation Treatment with the Use of Serotonergic Drugs

In the USA and Europe only approximately 6 % of those who quit smoking succeed in abstaining, despite 70 % of the population trying to quit at least once; therefore relapse is deemed to be the limiting factor in successful smoking cessation [150]. Half of relapses occur within 2 days of smoking cessation, and the majority occur within 3 months of quitting [151]. Environmental and sensory stimuli associated with smoking can act as conditioning stimuli which re-enforces smoking behaviour in humans [152]; in both active and former smokers the motivation to smoke can be increased on exposure to such cues (i.e. produce cue-induced craving) [153, 154]. All these factors are important considerations when generating the overall picture of smoking addiction.

Smoking is estimated to reduce the overall life expectancy of an individual by 8 years and reduce the amount of "healthy" years by 12 [150]; therefore the development of beneficial therapies to facilitate smoking cessation and to reduce relapse

Fig. 14.3 (continued) increased the firing rate of DA neurons recorded using in vivo single-unit extracellular recording technique. The rate meters on *top* show the effect of nicotine administration on single DAergic neurons' firing rates compared to controls (saline, 100 μl i.v., at *arrows*), while the graphs *below* report the cumulative dose-response curves showing the mean percentage changes (±S.E.M.) in firing rate after either nicotine or saline administration. *Arrows* indicate the time of nicotine or saline injections; apomorphine (Apo), a D1/D2 receptor agonist, was injected (10–30 μg/kg, i.v.) at the end of most of the experiments to confirm the DAergic identity of recorded neurons (adapted from [141] with permission). (**b**) Time course of nicotine (1 mg/kg, i.p.)-induced effect on DA terminal release in both the striatum and nucleus accumbens. Changes in DA extracellular levels were assessed using in vivo microdialysis on freely moving animals. The *arrows* indicate the time of injection of either nicotine or saline (adapted from [140] with permission)

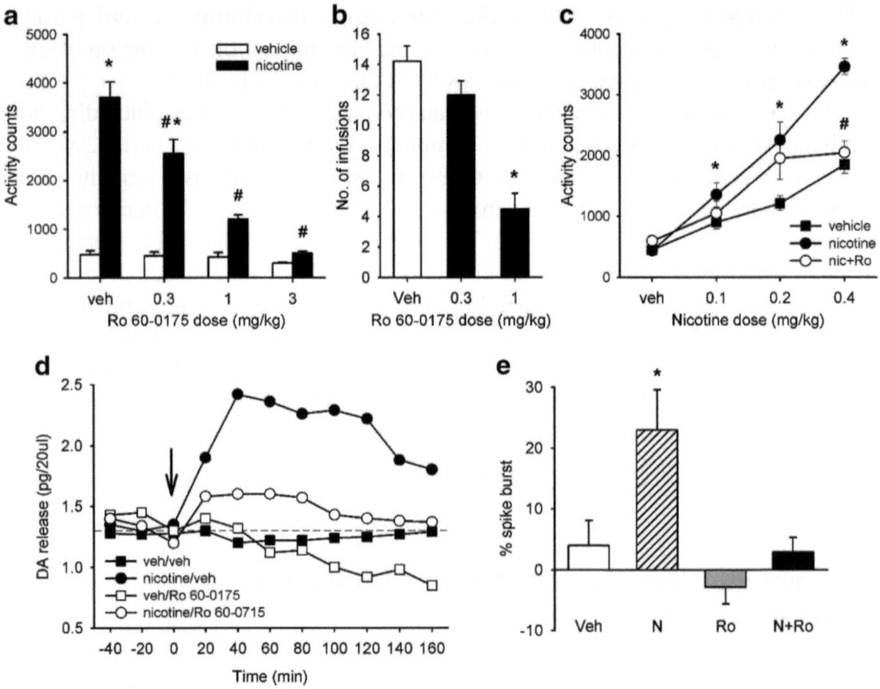

Fig. 14.4 Summary of evidence supporting a potential role for 5-HT$_{2C}$ receptor agonists as treatments for smoking cessation. (**a**) Effect of RO 60-0175 on hyperactivity induced by nicotine (0.4 mg/kg s.c), in rats previously sensitised to nicotine (ten daily injections of nicotine 0.4 mg/kg s.c). $^{\#}p<0.05$ vs. vehicle/vehicle pretreatment, $^{*}p<0.05$ vs. vehicle/nicotine pretreatment. (**b**) Effect of RO 60-0175 on nicotine self-administration. Nicotine (0.03 mg/infusion) was available for 1 h each day, under an FR5TO 1-min schedule of reinforcement. (**c**) Effect of chronic treatment with vehicle, nicotine (0.4 mg/kg s.c) or nicotine (0.4 mg/kg s.c)+RO 60-0175 (1 mg/kg s.c) on locomotor activity produced by nicotine (0–0.4 mg/kg s.c). Prior nicotine exposure resulted in an enhanced locomotor response to nicotine ($^{*}p<0.05$) that was blocked by RO 60-0175 ($^{\#}p<0.05$). (**d**) Time course of the effect of acute nicotine (1 mg/kg i.p.) on extracellular dopamine levels in the nucleus accumbens measured by microdialysis. Rats were treated for 10 days with nicotine (1 mg/kg i.p.) prior to the microdialysis experiment. Rats were treated with either vehicle or RO 60-0175 (1 mg/kg i.p) 20 min before nicotine (1 mg/kg i.p) or vehicle (indicated by *arrow*). Nicotine increased extracellular dopamine release in the nucleus accumbens. This effect was significantly attenuated by RO 60-0175 pretreatment. (**e**) Effect of nicotine (i.v.) and RO 60-0175 (0.1 mg/kg i.v.) on the firing pattern of VTA dopamine neurons in rats treated for 10 days with nicotine (1 mg/kg i.p.). The data represent the mean ± SEM difference between the percentages of spikes occurring in bursts during baseline vs. post-drug periods. The data show that nicotine increases burst firing of VTA dopamine neurons, and that this effect is blocked by RO 60-0175. Figure reproduced from [53] with permission

is of great interest. Nicotine replacement therapy (NRT) and the non-nicotine-based therapies are the two types of pharmacological interventions that have been approved by the US FDA for smoking cessation. NRT involves the replacement of nicotine with other nicotine-based formulations, which are less harmful than tobacco, in the form of chewing gum, transdermal patches or inhalers [155]. Non-nicotine-based therapy is represented mostly by bupropion, an atypical antidepressant [156, 157], or varenicline, the α4β2* nAChR partial agonist [158]. Comparing these therapies, NRTs have proved to be the least effective for smoking cessation, with varenicline being the most effective, followed by bupropion [159]. The increased efficacy of varenicline is likely to be due to its ability to both aid smoking cessation and prevent relapse; varenicline partially activates the α4β2* nAChRs (also activated by smoking); therefore when paired with smoking cessation it effectively mimics the effects of smoking and, additionally, if a relapse occurs varenicline prevents the full activation of α4β2* nAChRs [160].

Bupropion was the first non-NRT to be used for smoking cessation, although it later appeared to be a non-competitive antagonist of several nAChRs. The sustained-release bupropion formulation, the phenylaminoketone atypical antidepressant agent, was approved in the USA in 1997 by the FDA. It is thought that the ability of bupropion to block DA and noradrenaline (NA) reuptake, although modestly, accounts for its ability to treat nicotine dependence, but its exact mechanism of action is still unclear [156]. Moreover, the clinical efficacy of bupropion as a smoking cessation aid may be due to its antagonistic activity at nAChRs [161]. Threohydroxybupropion and hydroxybupropion are the major metabolites of bupropion, and they inhibit DA and NA reuptake to the same or lesser extent as bupropion [156]. Bupropion and hydroxybupropion administered in a dose-dependent manner were found to reduce the firing rate of noradrenergic neurons in the locus coeruleus, similar to tricyclic antidepressants (TCAs); the α_2-adrenergic antagonist yohimbine reversed this action [162]. In vitro, bupropion diminished the tonic inhibition of DA neurones via a reduction in the GABAergic transmission to these neurones, and also greatly reduced the effects of nicotine on DA neuron excitability [163]. The antidepressant action of bupropion may be aided, in the absence of nicotine, by these increases in DA neuron excitability during treatment [163]. Overall, this suggests that the antidepressant effect of bupropion may be partially due to a DAergic or a noradrenergic counterpart. Additionally, in the DRN, bupropion caused a sustained increase in the basal firing activity of 5-HT neurons [164], but currently the significance of the mechanism of action of bupropion in this latter effect, in a clinical setting, is unknown.

Varenicline, its trade name being Chantix in the USA and Champix in Canada, Europe and other countries, was developed by Pfizer and introduced as a novel smoking cessation aid with a high efficacy for the α4β2 nAChR, in 2006 [158, 160]. Varenicline affects monoaminergic neurotransmission by modulating release of DA, NA and 5-HT. Varenicline selectively binds to α4β2* nAChRs with high affinity, similar to nicotine-evoked DA release in the rat NAcc due to its interaction with α4β2* nAChRs in the VTA [158, 160, 165]. Varenicline competitively blocks nicotine from binding to the nAChRs, and thus stimulates the mesolimbic DA system.

In the rat, only high doses of varenicline (10 mg/kg) produce increases in PFC DA and NA release [166], whereas 1 mg/kg produces no significant changes in extracellular levels of 5-HT, NA and DA [167]. The effect of a 10 mg/kg dose, that is associated with very high brain concentrations of >1 μM, may be due to the interaction of varenicline with several other nAChR subtypes other than $\alpha4\beta2$ nAChRs. Varenicline has a very low affinity for 5-HT receptors and the 5-HT transporter; its lack of interaction with central 5-HT receptors and its inability to block 5-HT reuptake mean it is unable to increase 5-HT; thus its *in vitro* properties explain why it has little effect on cortical monoamine release. Varenicline is also a full agonist of 5-HT_{3A} receptors; however, despite having a modest affinity for these receptors, the brain concentrations of therapeutic unbound varenicline are hypothesised to be insufficient to activate central 5-HT_{3A} receptors. Moreover varenicline does not inhibit MAO-A, the enzyme which metabolises 5-HT, and does not bind to DA receptors that modulate 5-HT release. Additionally, when varenicline is combined with one or both antidepressant drugs it did not cause significant effects on the increased neurotransmitter levels, induced by sertraline and/or clorgyline [167]. Despite the findings that varenicline is more effective than bupropion and NRT in recent meta-analyses [159, 168], there was an observed increase in neuropsychiatric symptoms in those taking varenicline, causing the FDA to issue an alert [169]. These neuropsychiatric symptoms include behavioural changes, agitation, depressed mood, hostility and suicidal thoughts. This has led to a large clinical trial to reassess the safety of varenicline, evaluating the neuropsychiatric adverse events; this is expected to be complete by 2017. Nevertheless, a recent prospective cohort study showed that there is no evidence of an increased risk of suicidal behaviour in patients prescribed varenicline or bupropion compared with those prescribed nicotine replacement therapy [62]. These findings should be reassuring for users and prescribers of smoking cessation medicines.

Antidepressants are the most common drug, in the class of non-nicotine-based drugs, to be effective in aiding smoking cessation. TCAs which inhibit NA and 5-HT such as nortriptyline [170–172] and doxepin [173] have been proposed to be of benefit in aiding smoking cessation, in combination with behavioural treatment.

Due to the primary role of 5-HT in the modulation of smoking behaviour and nicotine reward, selective serotonin reuptake inhibitors (SSRIs) have been studied as a safer and better tolerated antidepressant to TCAs. Indeed, anticholinergic activity and frequent lethality in overdose are associated with TCAs; thus these significant aversive side effects mean other substances are being targeted for the beneficial therapeutic use [61, 111]. Fluoxetine, a prototypical SSRI, was shown to have equivocal results to TCAs and may relieve most of the adverse withdrawal symptoms experienced by smokers with a history of depression [174]. The 5-HT_{1A} receptor is involved in anxiolytic activity; therefore buspirone, a 5-HT_{1A} receptor partial agonist, has been studied for its use as an anti-anxiolytic in order to treat nicotine addiction. Buspirone administration stimulates the 5-HT_{1A} receptors leading to a reduction in presynaptic release of 5-HT, which appears to mediate anxiolytic effects. Unfortunately buspirone was shown to have little effect in aiding smoking cessation, shown by a placebo-controlled clinical trial [175]. 5-HT_{2C} receptor agonists could also potentially be utilised for the treatment of smoking cessation [49, 111, 129]. Weight gain is viewed as a contributory factor in preventing smoking cessation; therefore the role of 5-HT_{2C} receptor agonists in decreasing feeding behaviour and thus preventing weight gain could be of particular interest in beneficially treating nicotine dependence [176] (Table 14.1).

Table 14.1 Effects of varenicline, bupropion, RO 60-0175 and lorcaserin against various nicotine-motivated behaviours in the rodent (adapted from [111] with permission)

	Varenicline	Bupropion	RO 60-0175	Lorcaserin
Nicotine self-administration	↓	↔	↓	↓
Nicotine-induced hyperlocomotion	↓	↔	↓	↓
Nicotine interoceptive cue	↑/↓	↔	↓	↓
Reinstatement of nicotine-seeking behaviour[a]	↓	↔	↓	↓
Nicotine-conditioned place preference	↓	↑	NT	NT
Nicotine-precipitated W/D[b] (somatic)	NT	↓	↔	↔
Nicotine-precipitated W/D (affective)	NT	↓	↓	NT

↓, Reduced response; ↑, increased response; ↔, no reliable effect; ↑/↓, mixed effects reported; NT = not tested

[a]In the case of varenicline, bupropion and lorcaserin, each was assessed against reinstatement produced by a combination of nicotine prime and cues. RO 60-0175 was shown to be effective against reinstatement induced by either a nicotine prime or cues

[b]*W/D* withdrawal/dependence

Doses of lorcaserin that reduce feeding behaviour and behavioural effects induced by nicotine in rats [131] result in plasma levels similar to those observed in people taking a therapeutic dose of lorcaserin for obesity [111], thus suggesting that dose regimens used in obesity treatment would be appropriate for testing lorcaserin in a smoking-cessation trial. However one disadvantage of using this drug is its potential interaction with 5-HT_{2A} receptors, thought to induce hallucinogenic effects. Moreover, it is as of yet unknown to what extent lorcaserin, along with other 5-HT_{2C} agonists, occupies and activates the 5-HT_{2C} receptors at therapeutic exposure. However, due to lorcaserin being the first 5-HT_{2C} receptor agonist with clinical approval, it is a preferable and viable method for the application of testing animal evidence in humans [111].

7 Conclusions

Nicotine exerts a wide range of effects on multiple neurotransmitter systems including ACh, GLU, DA, NA, 5-HT and GABA by acting on nAChRs; it is considered to be an extraordinary psychotropic drug as it is difficult to find another with such a pleiotropic action. The 5-HT pathway is implicated in nicotine dependence and may influence smoking cessation; the 5-HT_{2C} receptor seems to be the most likely receptor to be involved.

The pharmacotherapy of nicotine addiction is still considered to be unsatisfactory, despite a large amount of data being produced on the neurobiological basis of withdrawal, action and dependence of nicotine. An increasing body of evidence has indicated that 5-HT_{2C} neurotransmission is a critical neurological substrate in the process of smoking cessation or its role in the suppression of nicotine withdrawal in smokers. 5-HT_{2C} receptor agonists could be useful in treating not only the wide range of behaviours produced by nicotine addiction, but also other drugs of abuse, because of the widespread influence of these receptors on behaviour. In conclusion, a more complete preclinical evaluation of the 5-HT_{2C} receptor role in nicotine addiction will ultimately allow educated, proof-of-concept trials to test the concept that selective 5-HT_{2C} agonists such as lorcaserin may be useful as adjunctive therapy.

Acknowledgments This work was supported by University of Malta funding scheme (GDG) and EU COST Action CM1103 "Structure-based drug design for diagnosis and treatment of neurological diseases: dissecting and modulating complex function in the monoaminergic systems of the brain" (GDG and PDD). SC. and LP. were supported by Erasmus placement scholarships.

References

1. Di Giovanni G, Esposito E, Di Matteo V. 5-HT2C receptors in the pathophysiology of CNS disease. The receptors. New York: Springer; 2011. p. 1–557.
2. Di Giovanni G, Di Matteo V, Esposito E. Serotonin–dopamine interaction: experimental evidence and therapeutic relevance. Progress in brain research, vol. 172. Amsterdam: Elsevier; 2008. p. 1–665.
3. Müller CP, Jacobs BL. Handbook of behavioral neurobiology of serotonin: handbook of behavioral neuroscience. 1st ed. Massachusetts: Academic Press; 2010. p. 836.
4. Hannon J, Hoyer D. Molecular biology of 5-HT receptors. Behav Brain Res. 2008;195: 198–213.
5. Hoyer D, Hannon JP, Martin GR. Molecular, pharmacological and functional diversity of 5-HT receptors. Pharmacol Biochem Behav. 2002;71(4):533–54.
6. Bonasera SJ, Tecott LH. Mouse models of serotonin receptor function: toward a genetic dissection of serotonin systems. Pharmacol Ther. 2000;88(2):133–42.
7. Koob GF. Drugs of abuse: anatomy, pharmacology and function of reward pathways. Trends Pharmacol Sci. 1992;13(5):177–84.
8. Brown AS, Gershon S. Dopamine and depression. J Neural Transm Gen Sect. 1993;91(2–3): 75–109.
9. Jenck F, et al. The role of 5-HT2C receptors in affective disorders. Expert Opin Investig Drugs. 1998;7(10):1587–99.
10. Di Matteo V, et al. Role of 5-HT2C receptors in the control of central dopamine function. Trends Pharmacol Sci. 2001;22(5):229–32.
11. Higgins GA, Fletcher PJ. Serotonin and drug reward: focus on 5-HT2C receptors. Eur J Pharmacol. 2003;480(1–3):151–62.
12. Giorgetti M, Tecott LH. Contributions of 5-HT(2C) receptors to multiple actions of central serotonin systems. Eur J Pharmacol. 2004;488(1–3):1–9.
13. Alex KD, Pehek EA. Pharmacologic mechanisms of serotonergic regulation of dopamine neurotransmission. Pharmacol Ther. 2007;113(2):296–320.
14. Di Giovanni G, Esposito E, Di Matteo V. 5-HT2C receptors in the pathophysiology of CNS disease. New York: Humana Press; 2011. p. 560.
15. Hoyer D, et al. International Union of Pharmacology classification of receptors for 5-hydroxytryptamine (Serotonin). Pharmacol Rev. 1994;46(2):157–203.
16. Barnes NM, Sharp T. A review of central 5-HT receptors and their function. Neuropharmacology. 1999;38(8):1083–152.
17. Di Giovanni G, Esposito E, Di Matteo V. Role of serotonin in central dopamine dysfunction. CNS Neurosci Ther. 2010;16(3):179–94.
18. Murphy DL, et al. How the serotonin story is being rewritten by new gene-based discoveries principally related to SLC6A4, the serotonin transporter gene, which functions to influence all cellular serotonin systems. Neuropharmacology. 2008;55(6):932–60.

19. Thompson JH. Serotonin and the alimentary tract. Res Commun Chem Pathol Pharmacol. 1971;2(4):687–781.
20. Cirillo C, Vanden Berghe P, Tack J. Role of serotonin in gastrointestinal physiology and pathology. Minerva Endocrinol. 2011;36(4):311–24.
21. Feldberg W, Toh CC. Distribution of 5-hydroxytryptamine (serotonin, enteramine) in the wall of the digestive tract. J Physiol. 1953;119:352–62.
22. Costa M, et al. Neurons with 5-hydroxytryptamine-like immunoreactivity in the enteric nervous system: their visualization and reactions to drug treatment. Neuroscience. 1982;7:351–63.
23. Erspamer V, Asero B. Identification of enteramine, the specific hormone of the enterochromaffin cell system, as 5-hydroxytryptamine. Nature. 1952;169(4306):800–1.
24. Rapport MM, Green AA, Page IH. Serum vasoconstrictor (serotonin).4. Isolation and characterization. J Biol Chem. 1948;176(3):1243–51.
25. Twarog BM, Page IH. Serotonin content of some mammalian tissues and urine and a method for its determination. Am J Physiol. 1953;175(1):157–61.
26. Brodie BB, Pletscher A, Shore PA. Evidence that serotonin has a role in brain function. Science. 1955;122(3177):968.
27. Costa E, Aprison MH. Studies on the 5-hydroxytryptamine (serotonin) content in human brain. J Nerv Ment Dis. 1958;126(3):289–93.
28. Whitaker-Azmita PM. Serotonin and brain development: role in human developmental diseases. Brain Res Bull. 2001;56(5):479–85.
29. Abrams JK, et al. Anatomic and functional topography of the dorsal raphe nucleus. Ann N Y Acad Sci. 2004;1018:46–57.
30. Dahlström A, Fuxe K. Evidence for the existence of monoamine-containing neurons in the central nervous system I Demonstration of monoamines in the cell bodies of brain stem neurons. Acta Physiol Scand Suppl. 1964;(Suppl. 232):1–55.
31. Baker KG, Halliday GM, Tork I. Cytoarchitecture of the human dorsal raphe nucleus. J Comp Neurol. 1990;301(2):147–61.
32. Azmitia EC, Segal M. An autoradiographic analysis of the differential ascending projections of the dorsal and median raphe nuclei in the rat. J Comp Neurol. 1978;179(3): 641–67.
33. van der Kooy D, Hattori T. Dorsal raphe cells with collateral projections to the caudate-putamen and substantia nigra: a fluorescent retrograde double labeling study in the rat. Brain Res. 1980;186(1):1–7.
34. Herve D, et al. Serotonin axon terminals in the ventral tegmental area of the rat: fine structure and synaptic input to dopaminergic neurons. Brain Res. 1987;435(1–2):71–83.
35. Van Bockstaele EJ, Biswas A, Pickel VM. Topography of serotonin neurons in the dorsal raphe nucleus that send axon collaterals to the rat prefrontal cortex and nucleus accumbens. Brain Res. 1993;624(1–2):188–98.
36. Van Bockstaele EJ, Cestari DM, Pickel VM. Synaptic structure and connectivity of serotonin terminals in the ventral tegmental area: potential sites for modulation of mesolimbic dopamine neurons. Brain Res. 1994;647(2):307–22.
37. Moukhles H, et al. Quantitative and morphometric data indicate precise cellular interactions between serotonin terminals and postsynaptic targets in rat substantia nigra. Neuroscience. 1997;76(4):1159–71.
38. Hillegaart V. Functional topography of brain serotonergic pathways in the rat. Acta Physiol Scand Suppl. 1991;598:1–54.
39. Corvaja N, Doucet G, Bolam JP. Ultrastructure and synaptic targets of the raphe-nigral projection in the rat. Neuroscience. 1993;55(2):417–27.
40. Di Matteo V, et al. Serotonin control of central dopaminergic function: focus on in vivo microdialysis studies. Prog Brain Res. 2008;172:7–44.
41. Jacobs BL, Azmitia EC. Structure and function of the brain serotonin system. Physiol Rev. 1992;72(1):165–229.
42. Phillipson OT. Afferent projections to the ventral tegmental area of Tsai and interfascicular nucleus: a horseradish peroxidase study in the rat. J Comp Neurol. 1979;187(1):117–43.

43. Esposito E, Di Matteo V, Di Giovanni G. Serotonin-dopamine interaction: an overview. Prog Brain Res. 2008;172:3–6.
44. Bowker RM, et al. Organization of descending serotonergic projections to the spinal cord. Prog Brain Res. 1982;57:239–65.
45. Bowker RM. The relationship between descending serotonin projections and ascending projections in the nucleus raphe magnus: a double labeling study. Neurosci Lett. 1986;70(3):348–53.
46. McMahon LL, Yoon KW, Chiappinelli VA. Nicotinic receptor activation facilitates gabaergic neurotransmission in the avian lateral spiriform nucleus. Neuroscience. 1994;59(3):689–98.
47. Fletcher PJ, Grottick AJ, Higgins GA. Differential effects of the 5-HT(2A) receptor antagonist M100907 and the 5-HT(2C) receptor antagonist SB242084 on cocaine-induced locomotor activity, cocaine self-administration and cocaine-induced reinstatement of responding. Neuropsychopharmacology. 2002;27(4):576–86.
48. Fletcher A, et al. Opposing effects of 5-HT2A and 5-HT2C receptor antagonists in the rat and mouse on premature responding in the five-choice serial reaction time test. Psychopharmacology. 2007;195:223–34.
49. Hjorth S, Magnusson T. The 5-HT 1A receptor agonist, 8-OH-DPAT, preferentially activates cell body 5-HT autoreceptors in rat brain in vivo. Naunyn Schmiedebergs Arch Pharmacol. 1988;338(5):463–71.
50. Di Giovanni G, et al. Central serotonin2C receptor: from physiology to pathology. Curr Top Med Chem. 2006;6(18):1909–25.
51. Di Giovanni G, et al. Serotonin involvement in the basal ganglia pathophysiology: could the 5-HT2C receptor be a new target for therapeutic strategies? Curr Med Chem. 2006;13(25):3069–81.
52. Fletcher PJ, Le AD, Higgins GA. Serotonin receptors as potential targets for modulation of nicotine use and dependence. Prog Brain Res. 2008;172:361–83.
53. Fletcher A, Higgins GA. Serotonin and reward-related behaviour: focus on 5-HT2C receptors. In: Di Giovanni G, Esposito E, Di Matteo V, editors. 5-HT2C receptors in the pathophysiology of CNS disease. New York: Springer; 2011. p. 293–324.
54. Di Matteo V, Esposito E, Di Giovanni G. Neurodegenerative disorders: from molecules to man (part 1). CNS Neurol Disord Drug Targets. 2007;6(6):375–6.
55. Markou A. Neurobiology of nicotine dependence. Philos Trans R Soc Biol Sci. 2008;363(1507):3159–68.
56. Clarke PBS, et al. Evidence that mesolimbic dopaminergic activation underlies the locomotor stimulant action of nicotine in rats. Journal of Pharmacology and Experimental Therapeutics. 1988;246(2):701–8.
57. Corrigall WA, Coen KM, Adamson KL. Self-administered nicotine activates the mesolimbic dopamine system through the ventral tegmental area. Brain Res. 1994;653(1–2):278–84.
58. Di Chiara G. Role of dopamine in the behavioural actions of nicotine related to addiction. Eur J Pharmacol. 2000;393(1–3):295–314.
59. Coppen A. The biochemistry of affective disorders. Br J Psychiatry. 1967;113(504):1237–64.
60. Seth P, et al. Nicotinic–serotonergic interactions in brain and behaviour. Pharmacol Biochem Behav. 2002;71(4):795–805.
61. Thomas KH, et al. Smoking cessation treatment and risk of depression, suicide, and self harm in the Clinical Practice Research Datalink: prospective cohort study. BMJ. 2013;347:f5704.
62. Balfour DJ, Ridley DL. The effects of nicotine on neural pathways implicated in depression: a factor in nicotine addiction? Pharmacol Biochem Behav. 2000;66(1):79–85.
63. Quattrocki E, Baird A, Yurgelun-Todd D. Biological aspects of the link between smoking and depression. Harv Rev Psychiatry. 2000;8(3):99–110.
64. Bitner RS, et al. Reduced nicotinic receptor-mediated antinociception following in vivo antisense knock-down in rat. Brain Res. 2000;871(1):66–74.
65. Bitner RS, Nikkel AL. Alpha-7 nicotinic receptor expression by two distinct cell types in the dorsal raphe nucleus and locus coeruleus of rat. Brain Res. 2002;938(1–2):45–54.
66. Cucchiaro G, Chaijale N, Commons KG. The dorsal raphe nucleus as a site of action of the antinociceptive and behavioral effects of the alpha4 nicotinic receptor agonist epibatidine. J Pharmacol Exp Ther. 2005;313(1):389–94.

67. Cucchiaro G, Commons KG. Alpha 4 nicotinic acetylcholine receptor subunit links cholinergic to brainstem monoaminergic neurotransmission. Synapse. 2003;49(3):195–205.
68. Enkhbaatar P, et al. The inhibition of inducible nitric oxide synthase in ovine sepsis model. Shock. 2006;25(5):522–7.
69. Galindo-Charles L, et al. Serotoninergic dorsal raphe neurons possess functional postsynaptic nicotinic acetylcholine receptors. Synapse. 2008;62(8):601–15.
70. Li X, et al. Presynaptic nicotinic receptors facilitate monoaminergic transmission. J Neurosci. 1998;18(5):1904–12.
71. Mihailescu S, Guzman-Marin R, Drucker-Colin R. Nicotine stimulation of dorsal raphe neurons: effects on laterodorsal and pedunculopontine neurons. Eur Neuropsychopharmacol. 2001;11(5):359–66.
72. Chang B, et al. Nicotinic excitation of serotonergic projections from dorsal raphe to the nucleus accumbens. J Neurophysiol. 2011;106(2):801–8.
73. Engberg G, et al. Nicotine inhibits firing activity of dorsal raphe 5-HT neurones in vivo. Naunyn Schmiedebergs Arch Pharmacol. 2000;362(1):41–5.
74. Touiki K, et al. Effects of tobacco and cigarette smoke extracts on serotonergic raphe neurons in the rat. Neuroreport. 2007;18(9):925–9.
75. Ma Z, et al. Effects on serotonin of (−)nicotine and dimethylphenylpiperazinium in the dorsal raphe and nucleus accumbens of freely behaving rats. Neuroscience. 2005;135(3):949–58.
76. Ribeiro EB, et al. Effects of systemic nicotine on serotonin release in rat brain. Brain Res. 1993;621(2):311–8.
77. Ranade SP, Mainen ZF. Transient firing of dorsal raphe neurons encodes diverse and specific sensory, motor, and reward events. J Neurophysiol. 2009;102(5):3026–37.
78. Summers KL, Lippiello P, Giacobini E. A microdialysis study of the effects of the nicotinic agonist RJR-2403 on cortical release of acetylcholine and biogenic amines. Neurochem Res. 1996;21(10):1181–6.
79. Schwartz RD, Lehmann J, Kellar KJ. Presynaptic nicotinic cholinergic receptors labeled by [3H]acetylcholine on catecholamine and serotonin axons in brain. J Neurochem. 1984;42(5):1495–8.
80. Reuben M, Clarke PB. Nicotine-evoked [3H]5-hydroxytryptamine release from rat striatal synaptosomes. Neuropharmacology. 2000;39(2):290–9.
81. Yu ZJ, Wecker L. Chronic nicotine administration differentially affects neurotransmitter release from rat striatal slices. J Neurochem. 1994;63(1):186–94.
82. Takahashi H, et al. Nicotine increases stress-induced serotonin release by stimulating nicotinic acetylcholine receptor in rat striatum. Synapse. 1998;28(3):212–9.
83. Lendvai B, et al. Differential mechanisms involved in the effect of nicotinic agonists DMPP and lobeline to release [3H]5-HT from rat hippocampal slices. Neuropharmacology. 1996;35(12):1769–77.
84. Kenny PJ, File SE, Neal MJ. Evidence for a complex influence of nicotinic acetylcholine receptors on hippocampal serotonin release. J Neurochem. 2000;75(6):2409–14.
85. Benwell ME, Balfour DJ. Effects of nicotine administration and its withdrawal on plasma corticosterone and brain 5-hydroxyindoles. Psychopharmacology (Berl). 1979;63(1):7–11.
86. Takada Y, et al. Changes in the central and peripheral serotonergic system in rats exposed to water-immersion restrained stress and nicotine administration. Neurosci Res. 1995;23(3):305–11.
87. Matta SG, et al. Guidelines on nicotine dose selection for in vivo research. Psychopharmacology (Berl). 2007;190(3):269–319.
88. Crooks PA, Dwoskin LP. Contribution of CNS nicotine metabolites to the neuropharmacological effects of nicotine and tobacco smoking. Biochem Pharmacol. 1997;54(7):743–53.
89. Clemens KJ, et al. The addition of five minor tobacco alkaloids increases nicotine-induced hyperactivity, sensitization and intravenous self-administration in rats. Int J Neuropsychopharmacol. 2009;12(10):1355–66.
90. Khalki H, et al. A tobacco extract containing alkaloids induces distinct effects compared to pure nicotine on dopamine release in the rat. Neurosci Lett. 2013;544:85–8.

91. Guillem K, et al. Monoamine oxidase inhibition dramatically increases the motivation to self-administer nicotine in rats. J Neurosci. 2005;25(38):8593–600.
92. Pazos A, Cortes R, Palacios JM. Quantitative autoradiographic mapping of serotonin receptors in the rat-brain. 2. Serotonin-2 receptors. Brain Res. 1985;346(2):231–49.
93. Pompeiano M, Palacios JM, Mengod G. Distribution of the serotonin 5-Ht2 receptor family messenger-Rnas—comparison between 5-Ht(2a) and 5-Ht(2c) receptors. Mol Brain Res. 1994;23(1–2):163–78.
94. Bubser M, et al. Distribution of serotonin 5-HT(2A) receptors in afferents of the rat striatum. Synapse. 2001;39(4):297–304.
95. Willins DL, Meltzer HY. Serotonin 5-HT2C agonists selectively inhibit morphine-induced dopamine efflux in the nucleus accumbens. Brain Res. 1998;781(1–2):291–9.
96. Cornea-Hébert V, et al. Cellular and subcellular distribution of the serotonin 5-HT_{2A} receptor in the central nervous system of adult rat. J Comp Neurol. 1999;409(2):187–209.
97. Doherty MD, Pickel VM. Ultrastructural localization of the serotonin 2A receptor in dopaminergic neurons in the ventral tegmental area. Brain Res. 2000;864(2):176–85.
98. Nocjar C, Roth BL, Pehek EA. Localization of 5-HT(2A) receptors on dopamine cells in subnuclei of the midbrain A10 cell group. Neuroscience. 2002;111(1):163–76.
99. Olausson P, et al. Effects of 5-HT1A and 5-HT2 receptor agonists on the behavioral and neurochemical consequences of repeated nicotine treatment. Eur J Pharmacol. 2001; 420(1):45–54.
100. Batman AM, Munzar P, Beardsley PM. Attenuation of nicotine's discriminative stimulus effects in rats and its locomotor activity effects in mice by serotonergic 5-HT2A/2C receptor agonists. Psychopharmacology (Berl). 2005;179(2):393–401.
101. Zaniewska M, et al. Effects of the serotonin 5-HT2A and 5-HT2C receptor ligands on the discriminative stimulus effects of nicotine in rats. Eur J Pharmacol. 2007;571(2–3):156–65.
102. Arnt J. Characterization of the discriminative stimulus properties induced by 5-HT1 and 5-HT2 agonists in rats. Pharmacol Toxicol. 1989;64(2):165–72.
103. Nichols DE. Hallucinogens. Pharmacol Ther. 2004;101(2):131–81.
104. Porras G, et al. 5-HT2A and 5-HT2C/2B receptor subtypes modulate dopamine release induced in vivo by amphetamine and morphine in both the rat nucleus accumbens and striatum. Neuropsychopharmacology. 2002;26(3):311–24.
105. Auclair A, et al. Role of serotonin 2A receptors in the D-amphetamine-induced release of dopamine: comparison with previous data on alpha1b-adrenergic receptors. J Neurochem. 2004;91(2):318–26.
106. O'Neill MF, Heron-Maxwell CL, Shaw G. 5-HT2 receptor antagonism reduces hyperactivity induced by amphetamine, cocaine, and MK-801 but not D1 agonist C-APB. Pharmacol Biochem Behav. 1999;63(2):237–43.
107. Herin DV, et al. Elevated expression of serotonin 5-HT(2A) receptors in the rat ventral tegmental area enhances vulnerability to the behavioral effects of cocaine. Front Psychiatry. 2013;4:2.
108. Levin ED, et al. Ketanserin, a 5-HT2 receptor antagonist, decreases nicotine self-administration in rats. Eur J Pharmacol. 2008;600(1–3):93–7.
109. Fletcher PJ, et al. Effects of the 5-HT2C receptor agonist Ro60-0175 and the 5-HT2A receptor antagonist M100907 on nicotine self-administration and reinstatement. Neuropharmacology. 2012;62(7):2288–98.
110. Higgins GA, Sellers EM, Fletcher PJ. From obesity to substance abuse: therapeutic opportunities for 5-HT2C receptor agonists. Trends Pharmacol Sci. 2013;34(10):560–70.
111. Bubar MJ, et al. Validation of a selective serotonin 5-HT2C receptor antibody for utilization in fluorescence immunohistochemistry studies. Brain Res. 2005;1063(2):105–13.
112. Bubar MJ, Cunningham KA. Distribution of serotonin 5-HT2C receptors in the ventral tegmental area. Neuroscience. 2007;146(1):286–97.
113. Prisco S, Pagannone S, Esposito E. Serotonin-dopamine interaction in the rat ventral tegmental area: an electrophysiological study in vivo. J Pharmacol Exp Ther. 1994;271(1):83–90.
114. Di Giovanni G, et al. Selective blockade of serotonin-2C/2B receptors enhances mesolimbic and mesostriatal dopaminergic function: a combined in vivo electrophysiological and microdialysis study. Neuroscience. 1999;91(2):587–97.

115. De Deurwaerdere P, et al. Multiple controls exerted by 5-HT2C receptors upon basal ganglia function: from physiology to pathophysiology. Exp Brain Res. 2013;230(4):477–511.
116. Martin JR, et al. 5-HT2C receptor agonists: pharmacological characteristics and therapeutic potential. J Pharmacol Exp Ther. 1998;286(2):913–24.
117. Millan MJ, Dekeyne A, Gobert A. Serotonin (5-HT)2C receptors tonically inhibit dopamine (DA) and noradrenaline (NA), but not 5-HT, release in the frontal cortex in vivo. Neuropharmacology. 1998;37(7):953–5.
118. Di Matteo V, et al. SB 242084, a selective serotonin2C receptor antagonist, increases dopaminergic transmission in the mesolimbic system. Neuropharmacology. 1999;38(8):1195–205.
119. Di Matteo V, et al. Biochemical and electrophysiological evidence that RO 60-0175 inhibits mesolimbic dopaminergic function through serotonin(2C) receptors. Brain Res. 2000;865(1):85–90.
120. Gobert A, et al. Serotonin(2C) receptors tonically suppress the activity of mesocortical dopaminergic and adrenergic, but not serotonergic, pathways: a combined dialysis and electrophysiological analysis in the rat. Synapse. 2000;36(3):205–21.
121. Kennett GA, et al. In vitro and in vivo profile of SB 206553, a potent 5-HT2C/5-HT2B receptor antagonist with anxiolytic-like properties. Br J Pharmacol. 1996;117(3):427–34.
122. Kennett GA, et al. SB 242084, a selective and brain penetrant 5-HT2C receptor antagonist. Neuropharmacology. 1997;36(4–5):609–20.
123. Di Matteo V, et al. Selective blockade of serotonin2C/2B receptors enhances dopamine release in the rat nucleus accumbens. Neuropharmacology. 1998;37(2):265–72.
124. De Deurwaerdere P, et al. Constitutive activity of the serotonin2C receptor inhibits in vivo dopamine release in the rat striatum and nucleus accumbens. J Neurosci. 2004;24(13):3235–41.
125. Pozzi L, et al. Stimulation of 5-hydroxytryptamine (5-HT(2C)) receptors in the ventrotegmental area inhibits stress-induced but not basal dopamine release in the rat prefrontal cortex. J Neurochem. 2002;82(1):93–100.
126. Hutson PH, et al. Activation of mesolimbic dopamine function by phencyclidine is enhanced by 5-HT(2C/2B) receptor antagonists: neurochemical and behavioural studies. Neuropharmacology. 2000;39(12):2318–28.
127. Porras G, et al. Central serotonin4 receptors selectively regulate the impulse-dependent exocytosis of dopamine in the rat striatum: in vivo studies with morphine, amphetamine and cocaine. Neuropharmacology. 2002;43(7):1099–109.
128. Quarta D, Naylor GG, Stolerman IP. The serotonin 2C receptor agonist Ro-60-0175 attenuates effects of nicotine in the five-choice serial reaction time task and in drug discrimination. Psychopharmacology (Berl). 2007;193:391–402.
129. Grottick AJ, Corrigall WA, Higgins GA. Activation of 5-HT(2C) receptors reduces the locomotor and rewarding effects of nicotine. Psychopharmacology (Berl). 2001;157(3):292–8.
130. Higgins GA, et al. Evaluation of chemically diverse 5-HT(2)c receptor agonists on behaviours motivated by food and nicotine and on side effect profiles. Psychopharmacology (Berl). 2012;226(3):475–90.
131. Grottick AJ, Fletcher PJ, Higgins GA. Studies to investigate the role of 5-HT(2C) receptors on cocaine- and food-maintained behavior. J Pharmacol Exp Ther. 2000;295(3):1183–91.
132. Neisewander JL, Acosta JI. Stimulation of 5-HT2C receptors attenuates cue and cocaine-primed reinstatement of cocaine-seeking behavior in rats. Behav Pharmacol. 2007;18(8):791–800.
133. Pockros LA, et al. Blockade of 5-HT2A receptors in the medial prefrontal cortex attenuates reinstatement of cue-elicited cocaine-seeking behavior in rats. Psychopharmacology (Berl). 2010;213(2–3):307–20.
134. Tomkins DM, et al. An investigation of the role of 5-HT(2C) receptors in modifying ethanol self-administration behaviour. Pharmacol Biochem Behav. 2002;71(4):735–44.
135. Levin ED, et al. Lorcaserin, a 5-HT2C agonist, decreases nicotine self-administration in female rats. J Pharmacol Exp Ther. 2011;338(3):890–6.
136. Ji SP, et al. Disruption of PTEN coupling with 5-HT2C receptors suppresses behavioral responses induced by drugs of abuse. Nat Med. 2006;12(3):324–9.

137. Hayes DJ, Mosher TM, Greenshaw AJ. Differential effects of 5-HT2C receptor activation by WAY 161503 on nicotine-induced place conditioning and locomotor activity in rats. Behav Brain Res. 2009;197(2):323–30.
138. Zaniewska M, McCreary AC, Filip M. Interactions of serotonin (5-HT)2 receptor-targeting ligands and nicotine: locomotor activity studies in rats. Synapse. 2009;63(8):653–61.
139. Di Matteo V, Pierucci M, Esposito E. Selective stimulation of serotonin2C receptors blocks the enhancement of striatal and accumbal dopamine release induced by nicotine administration. J Neurochem. 2004;89(2):418–29.
140. Pierucci M, Di Matteo V, Esposito E. Stimulation of serotonin2C receptors blocks the hyperactivation of midbrain dopamine neurons induced by nicotine administration. J Pharmacol Exp Ther. 2004;309(1):109–18.
141. Di Giovanni G, et al. m-Chlorophenylpiperazine excites non-dopaminergic neurons in the rat substantia nigra and ventral tegmental area by activating serotonin-2C receptors. Neuroscience. 2001;103(1):111–6.
142. Chevalier G, et al. Disinhibition as a basic process in the expression of striatal functions. I. The striato-nigral influence on tecto-spinal/tecto-diencephalic neurons. Brain Res. 1985;334(2):215–26.
143. Maurice N, et al. Relationships between the prefrontal cortex and the basal ganglia in the rat: physiology of the cortico-nigral circuits. J Neurosci. 1999;19(11):4674–81.
144. Beyeler A, et al. Stimulation of serotonin2C receptors elicits abnormal oral movements by acting on pathways other than the sensorimotor one in the rat basal ganglia. Neuroscience. 2010;169(1):158–70.
145. Navailles S, et al. In vivo evidence that 5-HT2C receptor antagonist but not agonist modulates cocaine-induced dopamine outflow in the rat nucleus accumbens and striatum. Neuropsychopharmacology. 2004;29(2):319–26.
146. Olausson P, Engel JA, Soderpalm B. Behavioral sensitization to nicotine is associated with behavioral disinhibition; counteraction by citalopram. Psychopharmacology (Berl). 1999; 142(2):111–9.
147. Kenny PJ, Markou A. Neurobiology of the nicotine withdrawal syndrome. Pharmacol Biochem Behav. 2001;70(4):531–49.
148. Zaniewska M, et al. Effects of serotonin (5-HT)2 receptor ligands on depression-like behavior during nicotine withdrawal. Neuropharmacology. 2010;58(7):1140–6.
149. Anderson JE, et al. Treating tobacco use and dependence: an evidence-based clinical practice guideline for tobacco cessation. Chest. 2002;121(3):932–41.
150. Jain A. Treating nicotine addiction. BMJ. 2003;327(7428):1394–5.
151. Shahan TA, et al. Comparing the reinforcing efficacy of nicotine containing and de-nicotinized cigarettes: a behavioral economic analysis. Psychopharmacology (Berl). 1999;147(2):210–6.
152. Mucha RF, Geier A, Pauli P. Modulation of craving by cues having differential overlap with pharmacological effect: evidence for cue approach in smokers and social drinkers. Psychopharmacology (Berl). 1999;147(3):306–13.
153. Dols M, et al. Smokers can learn to influence their urge to smoke. Addict Behav. 2000; 25(1):103–8.
154. Cryan JF, et al. Non-nicotinic neuropharmacological strategies for nicotine dependence: beyond bupropion. Drug Discov Today. 2003;8(22):1025–34.
155. Ascher JA, et al. Bupropion: a review of its mechanism of antidepressant activity. J Clin Psychiatry. 1995;56(9):395–401.
156. Rigotti NA. Clinical practice. Treatment of tobacco use and dependence. N Engl J Med. 2002;346(7):506–12.
157. Coe JW, et al. Varenicline: an alpha4beta2 nicotinic receptor partial agonist for smoking cessation. J Med Chem. 2005;48(10):3474–7.
158. Eisenberg MJ, et al. Pharmacotherapies for smoking cessation: a meta-analysis of randomized controlled trials. CMAJ. 2008;179(2):135–44.
159. Rollema H, et al. Rationale, pharmacology and clinical efficacy of partial agonists of alpha-4beta2 nACh receptors for smoking cessation. Trends Pharmacol Sci. 2007;28(7):316–25.

160. Slemmer JE, Martin BR, Damaj MI. Bupropion is a nicotinic antagonist. J Pharmacol Exp Ther. 2000;295(1):321–7.
161. Cooper BR, et al. Evidence that the acute behavioral and electrophysiological effects of bupropion (Wellbutrin) are mediated by a noradrenergic mechanism. Neuropsychopharmacology. 1994;11(2):133–41.
162. Mansvelder HD, et al. Bupropion inhibits the cellular effects of nicotine in the ventral tegmental area. Biochem Pharmacol. 2007;74(8):1283–91.
163. Dong J, Blier P. Modification of norepinephrine and serotonin, but not dopamine, neuron firing by sustained bupropion treatment. Psychopharmacology (Berl). 2001;155(1):52–7.
164. Reperant C, et al. Effect of the [alpha]4[beta]2* nicotinic acetylcholine receptor partial agonist varenicline on dopamine release in [beta]2 knock-out mice with selective re-expression of the [beta]2 subunit in the ventral tegmental area. Neuropharmacology. 2010;58(2): 346–50.
165. Rollema H, et al. Preclinical pharmacology of the alpha4beta2 nAChR partial agonist varenicline related to effects on reward, mood and cognition. Biochem Pharmacol. 2009; 78(7):813–24.
166. Rollema H, et al. Effect of co-administration of varenicline and antidepressants on extracellular monoamine concentrations in rat prefrontal cortex. Neurochem Int. 2011;58(1):78–84.
167. Mills EJ, et al. Efficacy of pharmacotherapies for short-term smoking abstinance: a systematic review and meta-analysis. Harm Reduct J. 2009;6:25.
168. US Food and Drug Administration CfDEaR. Varenicline (marketed as Chantix) information. 2008. Washington (DC): US Department of Health and Human Services. www.fda.gov/CDER/Drug/infopage/varenicline/default.htm. Accessed 12 Dec 2010.
169. Hall SM, et al. Nortriptyline and cognitive-behavioral therapy in the treatment of cigarette smoking. Arch Gen Psychiatry. 1998;55(8):683–90.
170. Prochazka AV, et al. A randomized trial of nortriptyline for smoking cessation. Arch Intern Med. 1998;158(18):2035–9.
171. Hall SM, et al. Psychological intervention and antidepressant treatment in smoking cessation. Arch Gen Psychiatry. 2002;59(10):930–6.
172. Edwards NB, et al. Doxepin as an adjunct to smoking cessation: a double-blind pilot study. Am J Psychiatry. 1989;146(3):373–6.
173. Dalack GW, et al. Mood, major depression, and fluoxetine response in cigarette smokers. Am J Psychiatry. 1995;152(3):398–403.
174. Schneider NG, et al. Efficacy of buspirone in smoking cessation: a placebo-controlled trial. Clin Pharmacol Ther. 1996;60(5):568–75.
175. Farley AC, et al. Interventions for preventing weight gain after smoking cessation. Cochrane Database Syst Rev. 2012;1, CD006219.
176. Dahlstrom A, Fuxe K. Localization of monoamines in the lower brain stem. Experientia. 1964;20(7):398–9.

Chapter 15
Neuronal Nicotinic Acetylcholine Receptors in Reward and Addiction

Linzy M. Hendrickson and Andrew R. Tapper

Abstract Drugs of abuse stimulate the pleasure centers of the brain to initiate addiction. During the beginning stages of addiction, the rewarding or reinforcing properties of abused drugs drive intake. However, as addiction develops drug intake is more likely to be dominated by negative reinforcement. The main reward center of the brain is the mesolimbic pathway which consists of dopaminergic neurons originating in the ventral tegmental area that project to the nucleus accumbens. Most, if not all, abused drugs stimulate this circuit resulting in increased release of the neurotransmitter, dopamine, in the nucleus accumbens, a phenomenon intimately associated with reward and reinforcement. Neuronal nAChRs are robustly expressed within the microcircuitry of this reward pathway. Drugs of abuse such as nicotine and alcohol directly interact with nAChRs expressed within the mesolimbic circuit to affect drug reward sensitivity, whereas with other drugs of abuse such as the psychostimulants and opioids, nAChRs play a more indirect, modulatory role on drug reward. In this chapter, the expression and function of nAChRs in the reinforcing/rewarding properties of drugs of abuse are explored.

Keywords Dopamine • Reinforcement • Alcohol • Nicotine • Psychostimulants • Opioids

1 Introduction

Species that learned to respond to natural rewards (such as when and where they could obtain food, have the opportunity to mate) ensured their survival. Achieving these goals function as rewards [1]. Consequently, many neural substrates that modulate reward systems are conserved across species from *Drosophila*, mice,

L.M. Hendrickson, Ph.D. • A.R. Tapper, Ph.D. (✉)
Department of Psychiatry, Brudnick Neuropsychiatric Research Institute, University of Massachusetts Medical School, 303 Belmont Street, Worcester, MA 01604, USA
e-mail: Andrew.tapper@umassmed.edu; lmhendrickson@gmail.com

R.A.J. Lester (ed.), *Nicotinic Receptors*, The Receptors 26,
DOI 10.1007/978-1-4939-1167-7_15

and rats to humans and include conserved circuitry, neurotransmitters, receptors, signaling molecules, and transcription factors [2]. Not surprisingly, this endogenous system can be exogenously altered via drugs that have potential to become abused. We now know that responses to natural rewards and addictive drugs have many similarities and shared pathways within the central nervous system (CNS). For example, studies in rats have shown a cross-sensitization between the natural reward sugar and the drug amphetamine [3]. In addition, a recent study found similar neuroadaptations in reward circuitry between chronic exposure of abused drugs and high-energy palatable food [4].

A common effect of natural rewards and most drugs of abuse is an enhancement of activity in the mesolimbic dopamine (DA) system (discussed in more detail below), leading to an increase of DA release in the nucleus accumbens (NAc) [5–7]. While it is widely accepted that the epicenter of reward stimuli processing within the brain, whether natural or drug, is the mesolimbic DA circuitry, much controversy exists regarding the precise role of DA in modulating goal-directed behavior. Mesolimbic DA is critical for a variety of physiological and affective behaviors such as movement, motivation, reward, learning, arousal, attention, and emotion [8]. Indeed, each of these individual behavioral components is necessary for the outward, measurable behavior of reward (i.e., an organism must locate a reward, pay attention, learn where to find it, like it, and have a desire to return to it).

Most of what is known regarding the underlying circuitry and molecular underpinnings of reward in addiction stems from pharmacological and genetic manipulations in rodent models. How does one measure the rewarding properties of drugs in animal models of dependence? The rewarding properties of drugs of abuse are typically measured via operant self-administration and/or conditioned place preference assays (CPP). In the former assay, an animal learns to self-administer a drug by pressing an active lever or nose poke that delivers a fixed dose to the animal by way of intravenous catheter, cannula to the brain, or, in the case of ethanol, a consumable liquid [9]. If a drug is reinforcing, then the animal will press on the active lever to self-administer the drug while ignoring a second inactive lever which yields no drug. In the CPP assay, an animal prefers a chamber where it received drug over the chamber where it received vehicle (i.e., the drug conditions a place preference as a measure of reward [10]).

Current theories on drug addiction suggest that the acute, rewarding properties of abused drugs drive intake during the initial stages of dependence; whereas drug intake in later stages is motivated by negative reinforcement (i.e., drugs are taken to predominantly alleviate negative affective states precipitated by withdrawal) [11]. This chapter focuses on nAChRs in the acute rewarding properties of drugs of abuse, while chapter 18 will focus on nAChRs in negative reinforcement, aversion, and withdrawal. It is important to point out that the circuitry underlying positive reinforcement (i.e., reward) and negative reinforcement (i.e., aversion) likely interact. However, the most well-studied circuit in the context of reward, addiction, and nAChRs is the mesolimbic pathway.

2 The Mesolimbic DA Pathway

It is widely accepted that the mesolimbic DA system plays a central role in modulating the rewarding effects of drugs of abuse [12, 13]. Olds and Milner first identified this pathway in 1954. Using brain stimulation reward (BSR) they discovered that rats returned to the same region of a testing apparatus where they had received electrical stimulation to the septal area of the brain [14]. Upon further examination using mapping and lesion studies, it was determined that the most sensitive sites in the brain (i.e., lowest stimulation threshold) were along the medial forebrain bundle (MFB) which connects the ventral tegmental area (VTA) to the basal forebrain [14–16]. Next, using pharmacology, studies showed that DAergic receptor blockade attenuated brain stimulation reward [17, 18], suggesting that specific neurotransmitter systems were involved in reward mechanisms [19].

Flash-forward almost 60 years and what was once commonly referred to as the "reward circuit" is now known as the mesolimbic DA pathway. This pathway consists of DAergic neurons whose cell bodies originate in the ventral tegmental area (VTA), a region of the midbrain, and project to regions of the limbic system including the NAc, amygdala, and hippocampus among other regions. An additional DAergic pathway, the mesocortical pathway, also originates in the VTA and project to regions of the prefrontal cortex. These pathways are shown in a simplified diagram in Fig. 15.1.

3 The Ventral Tegmental Area

The VTA is known to at least partially mediate the rewarding effects of nicotine, opiates, psychostimulants, ethanol, and cannabinoids [20]. For example, rats and mice will self-administer opiates [21], cannabinoids [22], cocaine [23], nicotine [24], or ethanol [25, 26] directly into the VTA. Additionally, intravenous nicotine self-administration is attenuated by either selective lesions of VTA DAergic neurons in rats [27] or a local VTA infusion of a nicotinic receptor antagonist [28]. The VTA is located in the midbrain, medial to the substantia nigra and ventral to the red nucleus [29]. It is referred to as an "area" and not considered to be a "nucleus" because the cryoarchitecture of the region is not well defined such that the boundaries of the VTA are determined by its neighboring structures [20, 30]. Within the VTA are two main cell populations, the A10 DAergic projection neurons, which comprise ~60 % of cells in this region [31], as well as local GABAergic interneurons [32, 33]. Although data are emerging indicating that different subpopulations of neurons within the VTA exist including DAergic neurons that also co-release glutamate, GABAergic projection neurons, and a small number of purely glutamatergic neurons [34, 35], the expression and function of nAChRs in these neuronal subpopulations as they relate to reward are unknown. The VTA receives inputs from regions throughout the CNS [36] including glutamatergic projections from the

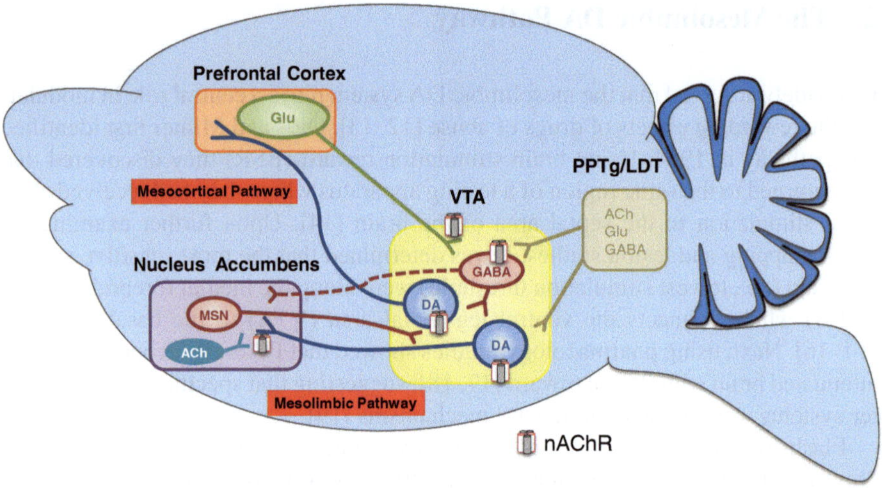

Fig. 15.1 Neuronal nAChR expression in the mesolimbic and mesocortical pathways. A sagittal rodent brain section depicting a simplified circuit diagram of the mesolimbic and mesocortical pathways is shown. The VTA (*yellow box*) consists of DAergic neurons projecting to the NAc (*purple box*) and prefrontal cortex (*orange box*). VTA GABAergic neurons provide local inhibition within the VTA and also project to the NAc. Glutamatergic neurons provide excitatory input into the VTA. Cholinergic, GABAergic, and glutamatergic VTA inputs also stem from laterodorsal tegmental (LTD) and pedunculopontine (PPTg) afferents. Drugs of abuse ultimately increase release of DA into the NAc to affect medium spiny projection neuron (MSN) activity. DA release at DAergic neuron presynaptic terminals is modulated by endogenous ACh provided by large aspiny cholinergic interneurons. Location of nAChR expression within the mesolimbic and mesocortical circuitry is indicated by the receptor icons

prefrontal cortex [37], as well as glutamatergic, cholinergic, and GABAergic projections from two groups of mesopontine tegmental area neurons, the pedunculopontine tegmental nucleus (PPTg) and the laterodorsal tegmental nucleus (LDT) [38–40]. Other regions that project to the VTA include the NAc, amygdala, ventral pallidum, superior colliculus, and lateral hypothalamus [30]. Additionally, the lateral habenula, a small nucleus that is a part of the epithalamus, has been shown to project to and stimulate midbrain areas that inhibit the release of DA from the VTA and substantia nigra pars compacta [41–43].

Projections from the VTA are primarily to the ventromedial striatum including the NAc shell and core as well as smaller projections to the prefrontal cortex (PFC), hippocampus, entorhinal cortex, and lateral septal areas [30]. Furthermore, studies using retrograde markers have shown that distinct groups of neurons originating in the VTA project to specific forebrain regions [44, 45]. Projections to the NAc contain the largest proportion of DA neurons, with 65–85 % being DAergic, while the PFC projections are only 30–40 % DAergic [31, 45]. The remaining component of VTA afferents to the NAc and PFC contain GABAergic neurons [32]. Although the VTA consists of two predominant neuronal subtypes, there is mounting evidence that this brain structure is not homogenous but can be divided into discrete subregions including anterior (aVTA), posterior (pVTA), and tail (tVTA) [20, 46–48]. Recent data indicate that the aVTA and pVTA project to distinct regions of the ventral striatum and are differentially responsive to various drugs of abuse suggesting functional heterogeneity [22, 49–52]. For example, rats will self-administer nicotine and ethanol directly in the pVTA but not the aVTA although the mechanistic basis of this regional selectivity is unknown [49].

4 The Nucleus Accumbens

For decades, the NAc has been a main focus of mesolimbic DA in studies of natural and drug reward [8]. It is located in the ventromedial striatum and is primarily composed of GABAergic medium spiny neurons (~95 %) and to a lesser extent a variety of interneurons (1–2 %) including cholinergic, fast-spiking GABAergic and low-threshold spiking. Two distinct regions of the NAc have been described, the core and shell, based on differences in functions and anatomical connectivity [53, 54]. Additionally, studies have shown that the response to extracellular DA release of these two regions differs. For example, it has been shown that the DA release induced by a food reward is rapidly habituated in the shell, but not the core [55]. Another study showed differential NAc shell and core Fos immunolabeling (a marker of neuronal activation) of cholinergic interneurons after cocaine self-administration [56]. These and other data suggest the possibility that the shell may act to modulate the initiation of drug-seeking behavior by mediating the hedonic states associated with reward [57, 58] while the core may modulate acquisition and maintenance of drug seeking [59].

The extracellular DA concentration in the NAc is regulated by two main factors: (1) the rate of release of DA from DAergic neurons that originate in the VTA and (2) dopamine uptake through dopamine transporters located in perisynaptic areas [60]. DAergic neurons of the VTA are known to be the main input source of extracellular DA in the NAc. Under normal conditions, the action potential (AP) firing rate of DAergic neurons is tonic with spike activity at 1–5 Hz [61]. However, when an unexpected presentation of a primary reward or a reward-predicting stimulus occurs, the firing rate increases to 2–10 APs at 10–30 Hz [62, 63].

4.1 Neuronal nAChR Expression in Reward Circuitry

Neuronal nicotinic acetylcholine receptors (nAChRs) are ligand-gated cation channels that, under normal conditions, are activated by the endogenous neurotransmitter, acetylcholine (ACh) [64, 65]. Eleven mammalian genes encoding nAChR subunits have been identified (α2–α7, α9–α10, β2–β4) and five subunits coassemble to form a functional receptor [64, 66]. The majority of nAChRs with high affinity for agonist are heteromeric consisting of two or three alpha subunits coassembled with two or three beta subunits while a subset of low-affinity receptors are homomeric, consisting of predominantly α7 subunits [64]. The subunit composition of the receptor determines the biophysical and pharmacological properties of each receptor subtype. Given the large number of nAChR subunits, the potential for a vast array of nAChR subtypes exists.

Multiple studies have examined nAChR expression and function within the VTA [67–72]. Klink et al. compared nAChR expression and function in DAergic and GABAergic neurons between the VTA and substantia nigra pars compacta (SNc). Utilizing β2, α4, and α7 KO mice in combination with nAChR antagonists, they concluded that most DAergic neurons express nAChRs containing α4, α5, α6, β2, and β3 subunits while most GABAergic neurons express nAChRs containing α4 and β2 subunits [67]. Using a similar strategy, Wooltorton et al. determined that α7 expression was more prevalent in VTA neurons than SNc neurons while nAChRs containing the β2 subunit (denoted β2*) are prevalent in DAergic and non-DAergic neurons throughout both brain regions [72]. The α6 nAChR subunit is predominantly expressed in DAergic neurons (although it may also be expressed in GABAergic terminal boutons) and can coassemble with β2, β3, and α4 subunits [70, 71, 73–76]. Using immunoprecipitation approaches in ventral midbrain, Gotti et al. deduced that at least five distinct nAChR subtypes were expressed in DAergic neurons at the level of soma/dendrites including α4β2, α2α4β2, α4α5β2, α4β2β3, and α4α6β2β3 nAChRs [77]. Within the NAc, the majority of nAChRs are expressed in DAergic presynaptic terminals where they modulate the probability of DA release by endogenous ACh and DAergic neuron firing frequency [78, 79]. DAergic neuron terminal nAChRs consist of α4β2, α4α5β2, α4β2β3, α4α6β2β3, and α6β2β3 subtypes [77]. Of these subtypes, α4α6β2β3 appears to dominate control of DA release at least in the NAc core [80].

4.2 *Nicotinic Receptor Subtypes Involved in Nicotine Reward/Reinforcement*

Smoking is the primary cause of preventable mortality in the world [81]. When volatized, nicotine, the addictive component of tobacco smoke, is absorbed into the bloodstream via the lungs and rapidly, on the order of seconds, crosses the blood-brain barrier [65]. Although nAChRs are expressed throughout the CNS, nicotine-induced activation of the mesocorticolimbic reward circuitry likely initiates addiction [66]. Indeed, pharmacological blockade of DA receptors or destruction of DA neurons or lesioning of the NAC reduces nicotine self-administration [27, 82]. Within this pathway, nicotine ultimately drives activity of DAergic neurons originating in the VTA resulting in increased DA release in the NAc and prefrontal cortex (PFC) [83]. More recently, nicotine has been found to also increase DA release in the hippocampus where it facilitates memory formation of nicotine reward [84].

With the great diversity of potential nAChR subunit combinations possible in nAChR subtypes within the VTA, a major goal of nicotine dependence research is to identify nAChR subunit combinations that are critical for the rewarding properties of nicotine. The majority of insights into reward circuitry nAChRs in reward and reinforcement stems from pharmacological and genetic studies in rodent models. Infusion of the nonspecific nAChR antagonist, mecamylamine, into the VTA reduces self-administration of nicotine in rodents while also blocking nicotine-mediated increases in NAc DA [49, 85]. In addition, the β2*-selective antagonist dihydro-β-erythroidine (dhβe) also reduces nicotine self-administration in rats when infused into the VTA [28]. Finally, infusion of the α6β2-selective antagonist, α-conotoxin MII, into the VTA or NAc reduces nicotine self-administration [77, 86].

Because of the limited nAChR subtype selectivity of most pharmacological agents, a more direct approach to address nAChR subunit composition in nicotine reward is through the use of genetically engineered mouse models. To date, several studies have utilized traditional knockout mice, which do not express a given nAChR subunit, or mice that express "gain-of-function" receptors that harbor a mutated subunit hypersensitive to nicotine, to examine the role of individual nAChR subunits in nicotine reward and reinforcement [87, 88]. Mice that do not express the β2 subunit fail to maintain nicotine self-administration indicating that nAChRs containing β2 are necessary for nicotine reinforcement [89]. These knockout mice also do not condition a place preference to nicotine consistent with a critical role for β2* nAChRs in nicotine reward [90]. In addition, mice that express a single-point mutation in the gene encoding the α4 subunit (a leucine residue mutated to an alanine residue in the pore forming transmembrane domain of the α4 subunit) that renders α4* nAChRs supersensitive to agonist condition a place preference to nicotine at sub-reward-threshold doses indicating that selective activation of α4* nAChRs is sufficient for nicotine reward [91]. In addition, mice harboring a distinct mutation within the α4 subunit also resulting in nicotine-hypersensitive α4* nAChRs self-administer nicotine at lower doses [92] than mice with non-mutated receptors. Knockout mice that do not express β2, α4, or α6* nAChRs fail to self-administer nicotine but nicotine intake can be rescued via viral mediated expression of these subunits in the VTA, indicating that expression of nAChRs specifically in the VTA is sufficient to support nicotine reinforcement [93, 94]. Thus, the emerging consensus across laboratories, based on a combination of pharmacology and mouse genetics, is that expression of α4β2* and α6* nAChRs in the VTA is necessary and sufficient for nicotine reward and reinforcement.

The identification of α4β2* nAChRs as critical for nicotine reward has led to rational design of small-molecule compounds to target these receptors in an effort to facilitate smoking cessation. The most successful smoking cessation aid to date is varenicline. Varenicline was designed as a high-affinity partial agonist at α4β2* nAChRs [95]. Studies in rodent midbrain slices indicate that varenicline activates α4β2* nAChRs in the mesolimbic circuitry modestly increasing DA release in the NAc while blocking further stimulation by the full agonist, nicotine [96]. In doing so, it is hypothesized that, in smokers, varenicline will alleviate affective withdrawal symptoms through increasing mesolimbic DA stimulation but also block the pleasurable effects of nicotine achieved through smoking.

4.3 Mechanisms of VTA DAergic Neuron Activation by Nicotine

VTA DAergic neurons fire tonically and also fire bursts [97, 98]. Recent studies using optogenetics to precisely depolarize DAergic neurons through light activation of the cationic ion channel, channelrhodopsin, indicate that bursting, but not tonic, DAergic neuron firing is sufficient to condition a place preference [99]. Conversely optogenetic activation of VTA GABAergic neurons *alone* inhibits DAergic neurons and signal aversion [100]. Acutely, nicotine elicits both an increase in baseline DAergic neuron firing frequency and an increase in burst firing that can persist up to an hour after a single bolus of nicotine [101, 102]. Previous studies indicate that nicotine can directly activate DAergic neurons in rodent midbrain slices [103, 104] and neuronal α4β2* nAChR subunits are critical for this effect. Indeed, nicotine fails to condition a place preference in mice that do not express α4* nAChRs selectively in DAergic neurons [105]. However, how VTA GABAergic neurons, which make up as many as half the neurons in the VTA [106] and also robustly express α4β2* nAChRs [67, 68, 70, 89, 91], contribute to shaping nicotine responses in DAergic neurons is emerging. In rat midbrain slices, nicotine may desensitize α4β2* nAChRs on GABAergic neurons, thereby disinhibiting DAergic neurons, increasing their activation [107]. In addition, blood nicotine concentrations achieved by smoking rapidly and persistently desensitize a portion of nAChRs on both DAergic and GABAergic neurons [102, 107]. Low-affinity α7 nAChRs, which are expressed on glutamatergic terminals that innervate the VTA, may rapidly recover from desensitization and drive glutamate release, thereby allowing for persistent activation of DAergic neurons by nicotine [107]. This is consistent with previous data indicating that glutamate release into the VTA is critical for nicotine reinforcement [108]. More recently, Tolu et al. found that nicotine, at least acutely, activates both DAergic and GABAergic VTA neurons in vivo [109]. Using viral mediated gene delivery to selectively re-express β2 nAChR subunits in VTA DAergic neurons of β2 KO mice was insufficient to restore nicotine self-administration and nicotine-mediated DA release in NAc. Surprisingly, β2 expression in both VTA DAergic *and* GABAergic neurons was required for rescue of nicotine self-administration. Remarkably, β2 expression in GABAergic neurons was critical for nicotine-mediated burst firing of DAergic neurons. These data indicate that nicotine activation of GABAergic interneurons in concert with activation of DAergic neurons may shape the firing pattern of DAergic neurons and modulate nicotine reward and reinforcement. Finally, recent studies have identified a unique nAChR subtype in VTA DAergic neurons consisting of both α4 and α6 subunits. These α4α6* nAChRs remain active with prolonged exposure to nicotine, and cause persistent depolarization of DAergic neurons [110, 111]. This persistent activation leads to changes in NMDA/AMPA receptor expression which may underlie sensitization to repeated nicotine exposure and enhance nicotine reward over time [111].

5 Neuronal nAChRs in Alcohol Reward

Alcohol abuse is the third largest cause of preventable mortality in the world [112]. As with nicotine, the rewarding or reinforcing properties of alcohol are associated with an increase in DA release in the NAc [113–117]. Ethanol-induced release of DA is critical for the onset and maintenance of dependence [118–121].

Multiple mechanisms underlying alcohol-mediated activation of VTA DAergic neurons have been proposed including modulation of intrinsic ion channels within these neurons, as well as alcohol-mediated alterations in synaptic input, both excitatory and inhibitory [122–128]. However, cholinergic signaling through nAChRs also contributes to NAc DA release and ethanol reinforcement [129–132]. For example, in rats, ethanol-mediated DA elevation in the NAc is inhibited by systemic or VTA but not NAc infusion of the noncompetitive, nonselective, nAChR antagonist, mecamylamine [130, 131, 133–136]. Blocking midbrain nAChRs via mecamylamine also decreases ethanol consumption and sensitization in rats. In addition, patients administered mecamylamine report reduced pleasurable effects of alcoholic beverages [137].

As discussed above, neuronal nAChR subtypes are expressed throughout the VTA in both DAergic neurons projecting to the NAc and in local GABAergic interneurons [67, 72]. How does ethanol interact with these receptors? Systemic ethanol has been shown to increase ACh concentrations in the VTA, presumably activating nAChRs in this area [135]. In addition, ethanol can directly modulate nAChR activity depending on the subtype of nicotinic receptor expressed [138–140]. In ventral midbrain slices containing the VTA, acetylcholine-induced activation of DAergic neurons is potentiated by ethanol and blocked by mecamylamine. In addition, the effects of ethanol on VTA DAergic neuron activity is reduced in α4 KO mice and enhanced in gain-of-function α4 knock-in mice [141]. Finally, potentiation is also blocked by an α6* nAChR-selective antagonist and reduced in α6 KO mice [142]. Thus, α4, α6, and/or α4α6* nAChRs may contribute to activation of VTA DAergic neurons by ethanol.

5.1 *What Are the nAChR Subtypes Involved in Ethanol Reward and Reinforcement?*

Identifying the nAChR subtype(s) that may underlie ethanol reward and consumption is necessary as they may represent therapeutic targets to reduce alcohol consumption. This endeavor is complicated by the fact that ethanol physiological and behavioral effects involve additional non-cholinergic mechanisms. In an effort to tease out individual nAChR subunits in *ethanol*-related behaviors, several studies have utilized pharmacology. As mentioned above, the nonspecific nAChR antagonist, mecamylamine, when injected systemically or locally within the VTA blocks ethanol consumption [132, 143, 144]. Alcohol consumption and alcohol-mediated DA release in the NAc are resistant to dhβe [133–135, 145–149]. In addition, the α7-selective antagonist, methyllycaconitine (MLA), does not affect alcohol-mediated behaviors precluding a role for homomeric α7 nAChRs [133, 144, 150]. On the other hand, the α3β2*, β3*, and α6* subtype-selective antagonist, α-conotoxin MII, does inhibit ethanol consumption and DA release in the NAc [151, 152]. Importantly, recent data indicate that approximately half of α-conotoxin MII-sensitive nAChRs in the striatum contain the α4 subunit [74, 153] and deletion of β2* nAChRs nearly abolishes α-conotoxin MII binding in the VTA [68]. Varenicline, an α4β2 partial agonist clinically approved as a smoking cessation therapeutic [95, 154–156], can reduce both ethanol intake and seeking in rats [155] and acute alcohol consumption in mice [157]. However, at high concentrations, varenicline is also a partial agonist at α6β2* nAChRs, a full agonist at α3β4 and α7 nAChRs, as well as at 5-hydroxytryptophan-3 receptors, which may also explain some of its effects on alcohol consumption [158–161]. Sazetidine-A, an α4β2* nAChR-selective "desensitizer," can also reduce alcohol consumption in rats [162]. Cytisine, a partial agonist that preferentially activates high-affinity β2* nAChRs at low doses but also is a full β4* nAChR agonist at high doses, also reduces alcohol consumption [163–165]. Novel partial agonists targeting α3β4* nAChRs reduce alcohol consumption and seeking in rats [166]. However, infusion of the α3β4* nAChR antagonist 18-methoxycoronaridine into the VTA fails to reduce alcohol consumption [167] consistent with data indicating low expression of β4* nAChRs in VTA DAergic neurons [76, 77].

Behavioral studies in genetically engineered mice have also been used to glean information on nAChR subtypes that are involved in alcohol consumption. To date, mice that do not express α6, α4, α7, β2, or β3 subunits have been evaluated in a two-bottle alcohol consumption assay. α6, β2, and β3 nAChR subunit KO mice consume and prefer alcohol similarly to WT controls [157, 168, 169], whereas α7 KO mice consume less alcohol at high concentrations [157]. In addition, α4 KO mice consume acutely less alcohol in a binge-drinking assay compared to WT littermates and are less sensitive to ethanol reward as measured in the CPP assay. In contrast, ethanol conditions a place preference at low doses in gain-of-function α4 knock-in mice (i.e., mice that are hypersensitive to acetylcholine) compared to WT mice [141]. Similarly, mice expressing gain-of-function α6* nAChRs consume more ethanol than WT mice and are sensitive to ethanol reward at sub-reward-threshold doses [170]. Thus, consistent with a potential role in activation of VTA DAergic neurons by ethanol, α4 and/or α6 or α4α6* nAChRs within the VTA may be inherently critical for the rewarding properties of ethanol, although additional experiments are needed to identify the precise brain region and circuitry where these nAChRs are expressed.

6 Neuronal nAChRs in Psychostimulant Reward

Whereas nicotine and ethanol interact with nAChRs directly to modulate function of the mesolimbic reward circuitry, the interaction between nAChRs and psychostimulant is likely indirect occurring at the circuit level. Indeed, psychostimulants such as cocaine and amphetamine bind to the dopamine transporter (DAT), which, under basal conditions, takes up DA at the synaptic cleft from the presynaptic side where it can be recycled to help terminate DA receptor signaling [171]. Cocaine blocks DAT while amphetamine reverses transport resulting in increased NAc DA and reward. Neuronal nAChRs modulate the rewarding and reinforcing properties of psychostimulants. Nicotine preexposure potentiates self-administration of low doses of cocaine in rats and augments conditioned place preference in mice [172, 173], whereas mecamylamine reduces cocaine self-administration in rats and reduces low-dose cocaine place preference in mice [173–175]. Neuronal nAChRs that influence psychostimulant reward are likely expressed at DAergic presynaptic terminals where they modulate DA release through cholinergic input from large aspiny cholinergic interneurons within the NAc. Cholinergic neuron activity, and hence cholinergic signaling, is critical for cocaine reward as the drug fails to condition a place preference if these interneurons are silenced [176]. Supporting a role for NAc DAergic presynaptic terminal nAChRs on cocaine reinforcement, infusion of mecamylamine or dhβe and MLA into the NAc reduces DA release elicited by an i.p. injection of cocaine in rats [177]. While the precise nAChR subtype involved in cocaine reward has not been fully elucidated, they most likely contain the β2 subunit, as β2 KO show reduced CPP in response to low doses of cocaine [173].

7 Neuronal nAChRs in Opioid Reward

Morphine and commonly abused prescription opioids are opioid receptor agonists. Like the psychostimulants, opioids do not interact with nAChRs directly. However, they do indirectly stimulate VTA DAergic neurons in the mesolimbic pathway by binding to and activating mu opioid receptors on VTA GABAergic interneurons and reducing interneuron activity [178]. Infusion of nicotine in the VTA potentiates morphine-conditioned place preference, whereas infusion of mecamylamine into the VTA inhibits morphine CPP suggesting a role for VTA nAChRs in opioid reward [179]. In addition, dhβe or MLA blocks drug priming-induced reinstatement of morphine CPP [180]. However, few studies have directly examined the role of nAChRs in the mesolimbic pathway in opioid reward. Thus, further studies to identify the mechanism of action of nAChRs in opioid reward are needed.

8 Conclusions

Although neuronal nAChRs are expressed throughout the CNS, most studies examining the role of nAChRs in drug reward have focused on the DAergic mesolimbic reward circuitry. Indeed, nAChRs are robustly expressed within the mesolimbic circuitry in multiple neuronal subpopulation including DAergic projection neurons and GABAergic interneuron among others. Direct stimulation of α4β2, α6, and/or α4α6* nAChRs within the VTA by nicotine underlies the acute rewarding properties of the drug. Neuronal nAChRs containing the α4 and/or α6 subunit also contribute to alcohol reward. Ethanol potentiates the response to ACh at these receptors. In addition, ethanol may enhance release of ACh in the VTA to activate DAergic neurons in this pathway through indirect nAChR activation. Emerging evidence indicates that nAChRs within the mesolimbic pathway may also modestly affect psychostimulant and opioid reward through modulation of DA release in the NAc. Identification of nAChR subtypes involved in drug reward may provide novel molecular targets for therapeutics designed to help treat drug addiction.

References

1. Hyman SE, Malenka RC, Nestler EJ. Neural mechanisms of addiction: the role of reward-related learning and memory. Annu Rev Neurosci. 2006;29:565–98.
2. Kelley AE, Berridge KC. The neuroscience of natural rewards: relevance to addictive drugs. J Neurosci. 2002;22(9):3306–11.
3. Avena NM, Hoebel BG. A diet promoting sugar dependency causes behavioral cross-sensitization to a low dose of amphetamine. Neuroscience. 2003;122(1):17–20.
4. Johnson PM, Kenny PJ. Dopamine D2 receptors in addiction-like reward dysfunction and compulsive eating in obese rats. Nat Neurosci. 2010;13(5):635–41.
5. Di Chiara G, Imperato A. Preferential stimulation of dopamine release in the nucleus accumbens by opiates, alcohol, and barbiturates: studies with transcerebral dialysis in freely moving rats. Ann N Y Acad Sci. 1986;473:367–81.
6. Koob GF, Bloom FE. Cellular and molecular mechanisms of drug dependence. Science. 1988;242(4879):715–23.
7. Wise RA. Drug-activation of brain reward pathways. Drug Alcohol Depend. 1998; 51(1–2):13–22.
8. Gonzales RA, Job MO, Doyon WM. The role of mesolimbic dopamine in the development and maintenance of ethanol reinforcement. Pharmacol Ther. 2004;103(2):121–46.
9. Fowler CD, Kenny PJ. Intravenous nicotine self-administration and cue-induced reinstatement in mice: effects of nicotine dose, rate of drug infusion and prior instrumental training. Neuropharmacology. 2011;61(4):687–98.
10. Cunningham CL, Gremel CM, Groblewski PA. Drug-induced conditioned place preference and aversion in mice. Nat Protoc. 2006;1(4):1662–70.
11. Koob GF, Le Moal M. Review. Neurobiological mechanisms for opponent motivational processes in addiction. Philos Trans R Soc Lond B Biol Sci. 2008;363(1507):3113–23.
12. Koob GF. Drugs of abuse: anatomy, pharmacology and function of reward pathways. Trends Pharmacol Sci. 1992;13(5):177–84.
13. Wise RA, Bozarth MA. A psychomotor stimulant theory of addiction. Psychol Rev. 1987; 94(4):469–92.
14. Olds J, Milner P. Positive reinforcement produced by electrical stimulation of septal area and other regions of rat brain. J Comp Physiol Psychol. 1954;47(6):419–27.
15. Corbett D, Wise RA. Intracranial self-stimulation in relation to the ascending dopaminergic systems of the midbrain: a moveable electrode mapping study. Brain Res. 1980;185(1):1–15.
16. Wise RA. Intracranial self-stimulation: mapping against the lateral boundaries of the dopaminergic cells of the substantia nigra. Brain Res. 1981;213(1):190–4.
17. Wise RA. Catecholamine theories of reward: a critical review. Brain Res. 1978;152(2): 215–47.
18. Liebman JM. Discriminating between reward and performance: a critical review of intracranial self-stimulation methodology. Neurosci Biobehav Rev. 1983;7(1):45–72.

19. Wise RA. The role of reward pathways in the development of drug dependence. Pharmacol Ther. 1987;35(1–2):227–63.
20. Ikemoto S. Dopamine reward circuitry: two projection systems from the ventral midbrain to the nucleus accumbens-olfactory tubercle complex. Brain Res Rev. 2007;56(1):27–78.
21. Bozarth MA, Wise RA. Intracranial self-administration of morphine into the ventral tegmental area in rats. Life Sci. 1981;28(5):551–5.
22. Zangen A, et al. Two brain sites for cannabinoid reward. J Neurosci. 2006;26(18):4901–7.
23. Rodd ZA, et al. Intracranial self-administration of cocaine within the posterior ventral tegmental area of Wistar rats: evidence for involvement of serotonin-3 receptors and dopamine neurons. J Pharmacol Exp Ther. 2005;313(1):134–45.
24. David V, et al. Reinforcing effects of nicotine microinjections into the ventral tegmental area of mice: dependence on cholinergic nicotinic and dopaminergic D1 receptors. Neuropharmacology. 2006;50(8):1030–40.
25. Gatto GJ, et al. Ethanol self-infusion into the ventral tegmental area by alcohol-preferring rats. Alcohol. 1994;11(6):557–64.
26. Rodd-Henricks ZA, et al. Regional heterogeneity for the intracranial self-administration of ethanol within the ventral tegmental area of female Wistar rats. Psychopharmacology (Berl). 2000;149(3):217–24.
27. Corrigall WA, et al. The mesolimbic dopaminergic system is implicated in the reinforcing effects of nicotine. Psychopharmacology (Berl). 1992;107(2–3):285–9.
28. Corrigall WA, Coen KM, Adamson KL. Self-administered nicotine activates the mesolimbic dopamine system through the ventral tegmental area. Brain Res. 1994;653(1–2):278–84.
29. Paxinos G, Franklin KBJ. The mouse brain in stereotaxic coordinates. 2nd ed. San Diego: Academic; 2000. p. 296.
30. Fields HL, et al. Ventral tegmental area neurons in learned appetitive behavior and positive reinforcement. Annu Rev Neurosci. 2007;30:289–316.
31. Swanson LW. The projections of the ventral tegmental area and adjacent regions: a combined fluorescent retrograde tracer and immunofluorescence study in the rat. Brain Res Bull. 1982;9(1–6):321–53.
32. Carr DB, Sesack SR. GABA-containing neurons in the rat ventral tegmental area project to the prefrontal cortex. Synapse. 2000;38(2):114–23.
33. Margolis EB, et al. The ventral tegmental area revisited: is there an electrophysiological marker for dopaminergic neurons? J Physiol. 2006;577(Pt 3):907–24.
34. Stuber GD, et al. Dopaminergic terminals in the nucleus accumbens but not the dorsal striatum corelease glutamate. J Neurosci. 2010;30(24):8229–33.
35. Hnasko TS, et al. Vesicular glutamate transport promotes dopamine storage and glutamate corelease in vivo. Neuron. 2010;65(5):643–56.
36. Geisler S, Zahm DS. Afferents of the ventral tegmental area in the rat-anatomical substratum for integrative functions. J Comp Neurol. 2005;490(3):270–94.
37. Sesack SR, Pickel VM. Prefrontal cortical efferents in the rat synapse on unlabeled neuronal targets of catecholamine terminals in the nucleus accumbens septi and on dopamine neurons in the ventral tegmental area. J Comp Neurol. 1992;320(2):145–60.
38. Cornwall J, Cooper JD, Phillipson OT. Afferent and efferent connections of the laterodorsal tegmental nucleus in the rat. Brain Res Bull. 1990;25(2):271–84.
39. Oakman SA, et al. Distribution of pontomesencephalic cholinergic neurons projecting to substantia nigra differs significantly from those projecting to ventral tegmental area. J Neurosci. 1995;15(9):5859–69.
40. Semba K, Fibiger HC. Afferent connections of the laterodorsal and the pedunculopontine tegmental nuclei in the rat: a retro- and antero-grade transport and immunohistochemical study. J Comp Neurol. 1992;323(3):387–410.
41. Herkenham M, Nauta WJ. Efferent connections of the habenular nuclei in the rat. J Comp Neurol. 1979;187(1):19–47.
42. Ji H, Shepard PD. Lateral habenula stimulation inhibits rat midbrain dopamine neurons through a GABA(A) receptor-mediated mechanism. J Neurosci. 2007;27(26):6923–30.

43. Matsumoto M, Hikosaka O. Lateral habenula as a source of negative reward signals in dopamine neurons. Nature. 2007;447(7148):1111–5.
44. Margolis EB, et al. Kappa opioids selectively control dopaminergic neurons projecting to the prefrontal cortex. Proc Natl Acad Sci U S A. 2006;103(8):2938–42.
45. Fallon JH, et al. Neurons in the ventral tegmentum have separate populations projecting to telencephalon and inferior olive, are histochemically different, and may receive direct visual input. Brain Res. 1984;321(2):332–6.
46. Ikemoto S, Murphy JM, McBride WJ. Regional differences within the rat ventral tegmental area for muscimol self-infusions. Pharmacol Biochem Behav. 1998;61(1):87–92.
47. Perrotti LI, et al. DeltaFosB accumulates in a GABAergic cell population in the posterior tail of the ventral tegmental area after psychostimulant treatment. Eur J Neurosci. 2005; 21(10):2817–24.
48. Kaufling J, et al. Afferents to the GABAergic tail of the ventral tegmental area in the rat. J Comp Neurol. 2009;513(6):597–621.
49. Ikemoto S, Qin M, Liu ZH. Primary reinforcing effects of nicotine are triggered from multiple regions both inside and outside the ventral tegmental area. J Neurosci. 2006;26(3):723–30.
50. Boehm 2nd SL, et al. Ventral tegmental area region governs GABA(B) receptor modulation of ethanol-stimulated activity in mice. Neuroscience. 2002;115(1):185–200.
51. Ericson M, et al. Nicotinic acetylcholine receptors in the anterior, but not posterior, ventral tegmental area mediate ethanol-induced elevation of accumbal dopamine levels. J Pharmacol Exp Ther. 2008;326(1):76–82.
52. Shabat-Simon M, et al. Dissociation between rewarding and psychomotor effects of opiates: differential roles for glutamate receptors within anterior and posterior portions of the ventral tegmental area. J Neurosci. 2008;28(34):8406–16.
53. Heimer L, et al. Specificity in the projection patterns of accumbal core and shell in the rat. Neuroscience. 1991;41(1):89–125.
54. Zahm DS. Functional-anatomical implications of the nucleus accumbens core and shell subterritories. Ann N Y Acad Sci. 1999;877:113–28.
55. Bassareo V, De Luca MA, Di Chiara G. Differential expression of motivational stimulus properties by dopamine in nucleus Accumbens shell versus core and prefrontal cortex. J Neurosci. 2002;22(11):4709–19.
56. Berlanga ML, et al. Cholinergic interneurons of the nucleus accumbens and dorsal striatum are activated by the self-administration of cocaine. Neuroscience. 2003;120(4):1149–56.
57. Pecina S, Berridge KC. Opioid site in nucleus accumbens shell mediates eating and hedonic 'liking' for food: map based on microinjection Fos plumes. Brain Res. 2000;863(1–2):71–86.
58. Rodd-Henricks ZA, et al. Cocaine is self-administered into the shell but not the core of the nucleus accumbens of Wistar rats. J Pharmacol Exp Ther. 2002;303(3):1216–26.
59. Ito R, Robbins TW, Everitt BJ. Differential control over cocaine-seeking behavior by nucleus accumbens core and shell. Nat Neurosci. 2004;7(4):389–97.
60. Nirenberg MJ, et al. The dopamine transporter: comparative ultrastructure of dopaminergic axons in limbic and motor compartments of the nucleus accumbens. J Neurosci. 1997; 17(18):6899–907.
61. Grace AA, Bunney BS. The control of firing pattern in nigral dopamine neurons: single spike firing. J Neurosci. 1984;4(11):2866–76.
62. Pan WX, et al. Dopamine cells respond to predicted events during classical conditioning: evidence for eligibility traces in the reward-learning network. J Neurosci. 2005;25(26):6235–42.
63. Schultz W. Predictive reward signal of dopamine neurons. J Neurophysiol. 1998;80(1):1–27.
64. Albuquerque EX, et al. Mammalian nicotinic acetylcholine receptors: from structure to function. Physiol Rev. 2009;89(1):73–120.
65. Tapper AR, Nashmi R, Lester HA. Neuronal nicotinic acetylcholine receptors and nicotine dependence. In: Cell biology of addiction. Cold Spring Harbor: Cold Spring Harbor Laboratory Press; 2006. p. 179–91.
66. Laviolette SR, van der Kooy D. The neurobiology of nicotine addiction: bridging the gap from molecules to behaviour. Nat Rev Neurosci. 2004;5(1):55–65.

67. Klink R, et al. Molecular and physiological diversity of nicotinic acetylcholine receptors in the midbrain dopaminergic nuclei. J Neurosci. 2001;21(5):1452–63.
68. Marubio LM, et al. Effects of nicotine in the dopaminergic system of mice lacking the alpha4 subunit of neuronal nicotinic acetylcholine receptors. Eur J Neurosci. 2003;17(7):1329–37.
69. Champtiaux N, Changeux JP. Knockout and knock-in mice to investigate the role of nicotinic receptors in the central nervous system. Prog Brain Res. 2004;145:235–51.
70. Champtiaux N, et al. Subunit composition of functional nicotinic receptors in dopaminergic neurons investigated with knock-out mice. J Neurosci. 2003;23(21):7820–9.
71. Champtiaux N, et al. Distribution and pharmacology of alpha 6-containing nicotinic acetylcholine receptors analyzed with mutant mice. J Neurosci. 2002;22(4):1208–17.
72. Wooltorton JR, et al. Differential desensitization and distribution of nicotinic acetylcholine receptor subtypes in midbrain dopamine areas. J Neurosci. 2003;23(8):3176–85.
73. Drenan RM, et al. In vivo activation of midbrain dopamine neurons via sensitized, high-affinity alpha6* nicotinic acetylcholine receptors. Neuron. 2008;60(1):123–36.
74. Salminen O, et al. Pharmacology of alpha-conotoxin MII-sensitive subtypes of nicotinic acetylcholine receptors isolated by breeding of null mutant mice. Mol Pharmacol. 2007; 71(6):1563–71.
75. Salminen O, et al. Subunit composition and pharmacology of two classes of striatal presynaptic nicotinic acetylcholine receptors mediating dopamine release in mice. Mol Pharmacol. 2004;65(6):1526–35.
76. Zhao-Shea R, et al. Nicotine-mediated activation of dopaminergic neurons in distinct regions of the ventral tegmental area. Neuropsychopharmacology. 2011;36(5):1021–32.
77. Gotti C, et al. Nicotinic acetylcholine receptors in the mesolimbic pathway: primary role of ventral tegmental area alpha6beta2* receptors in mediating systemic nicotine effects on dopamine release, locomotion, and reinforcement. J Neurosci. 2010;30(15):5311–25.
78. Zhou FM, Liang Y, Dani JA. Endogenous nicotinic cholinergic activity regulates dopamine release in the striatum. Nat Neurosci. 2001;4(12):1224–9.
79. Zhang H, Sulzer D. Frequency-dependent modulation of dopamine release by nicotine. Nat Neurosci. 2004;7(6):581–2.
80. Exley R, et al. Distinct contributions of nicotinic acetylcholine receptor subunit {alpha}4 and subunit {alpha}6 to the reinforcing effects of nicotine. Proc Natl Acad Sci U S A. 2011;108(18):7577–82.
81. Harris DS, Anthenelli RM. Expanding treatment of tobacco dependence. Curr Psychiatry Rep. 2005;7(5):344–51.
82. Corrigall WA, Coen KM. Selective dopamine antagonists reduce nicotine self-administration. Psychopharmacology (Berl). 1991;104(2):171–6.
83. Dani JA. Roles of dopamine signaling in nicotine addiction. Mol Psychiatry. 2003; 8(3):255–6.
84. Tang J, Dani JA. Dopamine enables in vivo synaptic plasticity associated with the addictive drug nicotine. Neuron. 2009;63(5):673–82.
85. Nisell M, Nomikos GG, Svensson TH. Systemic nicotine-induced dopamine release in the rat nucleus accumbens is regulated by nicotinic receptors in the ventral tegmental area. Synapse. 1994;16(1):36–44.
86. Brunzell DH, et al. Alpha-conotoxin MII-sensitive nicotinic acetylcholine receptors in the nucleus accumbens shell regulate progressive ratio responding maintained by nicotine. Neuropsychopharmacology. 2010;35(3):665–73.
87. Drenan RM, Lester HA. Insights into the neurobiology of the nicotinic cholinergic system and nicotine addiction from mice expressing nicotinic receptors harboring gain-of-function mutations. Pharmacol Rev. 2012;64(4):869–79.
88. Champtiaux N, Changeux JP. Knock-out and knock-in mice to investigate the role of nicotinic receptors in the central nervous system. Curr Drug Target CNS Neurol Disord. 2002;1(4):319–30.
89. Picciotto MR, et al. Acetylcholine receptors containing the beta2 subunit are involved in the reinforcing properties of nicotine. Nature. 1998;391(6663):173–7.

90. Walters CL, et al. The beta2 but not alpha7 subunit of the nicotinic acetylcholine receptor is required for nicotine-conditioned place preference in mice. Psychopharmacology (Berl). 2006;184(3–4):339–44.
91. Tapper AR, et al. Nicotine activation of alpha4* receptors: sufficient for reward, tolerance, and sensitization. Science. 2004;306(5698):1029–32.
92. Cahir E, et al. The necessity of alpha4* nicotinic receptors in nicotine-driven behaviors: dissociation between reinforcing and motor effects of nicotine. Neuropsychopharmacology. 2011;36(7):1505–17.
93. Maskos U, et al. Nicotine reinforcement and cognition restored by targeted expression of nicotinic receptors. Nature. 2005;436(7047):103–7.
94. Pons S, et al. Crucial role of alpha4 and alpha6 nicotinic acetylcholine receptor subunits from ventral tegmental area in systemic nicotine self-administration. J Neurosci. 2008;28(47): 12318–27.
95. Gonzales D, et al. Varenicline, an alpha4beta2 nicotinic acetylcholine receptor partial agonist, vs sustained-release bupropion and placebo for smoking cessation: a randomized controlled trial. JAMA. 2006;296(1):47–55.
96. Rollema H, et al. Pharmacological profile of the alpha4beta2 nicotinic acetylcholine receptor partial agonist varenicline, an effective smoking cessation aid. Neuropharmacology. 2007;52(3):985–94.
97. Li W, Doyon WM, Dani JA. Acute in vivo nicotine administration enhances synchrony among dopamine neurons. Biochem Pharmacol. 2011;82(8):977–83.
98. De Biasi M, Dani JA. Reward, addiction, withdrawal to nicotine. Annu Rev Neurosci. 2011;34:105–30.
99. Tsai HC, et al. Phasic firing in dopaminergic neurons is sufficient for behavioral conditioning. Science. 2009;324(5930):1080–4.
100. Tan KR, et al. GABA neurons of the VTA drive conditioned place aversion. Neuron. 2012;73(6):1173–83.
101. Mansvelder HD, et al. Cholinergic modulation of dopaminergic reward areas: upstream and downstream targets of nicotine addiction. Eur J Pharmacol. 2003;480(1–3):117–23.
102. Mansvelder HD, McGehee DS. Cellular and synaptic mechanisms of nicotine addiction. J Neurobiol. 2002;53(4):606–17.
103. Fisher JL, Pidoplichko VI, Dani JA. Nicotine modifies the activity of ventral tegmental area dopaminergic neurons and hippocampal GABAergic neurons. J Physiol Paris. 1998;92(3–4):209–13.
104. Pidoplichko VI, et al. Nicotine activates and desensitizes midbrain dopamine neurons. Nature. 1997;390(6658):401–4.
105. McGranahan TM, et al. alpha4beta2 nicotinic acetylcholine receptors on dopaminergic neurons mediate nicotine reward and anxiety relief. J Neurosci. 2011;31(30):10891–902.
106. Nashmi R, et al. Chronic nicotine cell specifically upregulates functional alpha 4* nicotinic receptors: basis for both tolerance in midbrain and enhanced long-term potentiation in perforant path. J Neurosci. 2007;27(31):8202–18.
107. Mansvelder HD, Keath JR, McGehee DS. Synaptic mechanisms underlie nicotine-induced excitability of brain reward areas. Neuron. 2002;33(6):905–19.
108. Kenny PJ, et al. NMDA receptors regulate nicotine-enhanced brain reward function and intravenous nicotine self-administration: role of the ventral tegmental area and central nucleus of the amygdala. Neuropsychopharmacology. 2009;34(2):266–81.
109. Tolu S, et al. Co-activation of VTA DA and GABA neurons mediates nicotine reinforcement. Mol Psychiatry. 2013;18(3):382–93.
110. Liu L, et al. Nicotine persistently activates ventral tegmental area dopaminergic neurons via nicotinic acetylcholine receptors containing alpha4 and alpha6 subunits. Mol Pharmacol. 2012;81(4):541–8.
111. Engle SE, et al. alpha4alpha6beta2* nAChR activation on VTA DA neurons is sufficient to stimulate a depolarizing conductance and enhance surface AMPA receptor function. Mol Pharmacol. 2013;84(3):393–406.

112. W.H.O. Global status report on alcohol. Geneva: World Health Organization; 2011.
113. Samson HH, et al. Alcohol self-administration: role of mesolimbic dopamine. Ann N Y Acad Sci. 1992;654:242–53.
114. Diana M, Rossetti ZL, Gessa G. Rewarding and aversive effects of ethanol: interplay of GABA, glutamate and dopamine. Alcohol Alcohol Suppl. 1993;2:315–9.
115. Lanca AJ. Reduction of voluntary alcohol intake in the rat by modulation of the dopaminergic mesolimbic system: transplantation of ventral mesencephalic cell suspensions. Neuroscience. 1994;58(2):359–69.
116. Lewis MJ, June HL. Neurobehavioral studies of ethanol reward and activation. Alcohol. 1990;7(3):213–9.
117. Weiss F, et al. Oral alcohol self-administration stimulates dopamine release in the rat nucleus accumbens: genetic and motivational determinants. J Pharmacol Exp Ther. 1993; 267(1):250–8.
118. Rassnick S, Stinus L, Koob GF. The effects of 6-hydroxydopamine lesions of the nucleus accumbens and the mesolimbic dopamine system on oral self-administration of ethanol in the rat. Brain Res. 1993;623(1):16–24.
119. Ikemoto S, et al. 6-OHDA-lesions of the nucleus accumbens disrupt the acquisition but not the maintenance of ethanol consumption in the alcohol-preferring P line of rats. Alcohol Clin Exp Res. 1997;21(6):1042–6.
120. Kiianmaa K. Decreased intoxicating effect of ethanol in rats after 6-hydroxydopamine-induced degeneration of ascending dopamine pathways. Pharmacol Biochem Behav. 1978;9(3):391–3.
121. Koob GF, Weiss F. Pharmacology of drug self-administration. Alcohol. 1990;7(3):193–7.
122. Okamoto T, Harnett MT, Morikawa H. Hyperpolarization-activated cation current (Ih) is an ethanol target in midbrain dopamine neurons of mice. J Neurophysiol. 2006;95(2):619–26.
123. Rodd ZA, et al. Serotonin-3 receptors in the posterior ventral tegmental area regulate ethanol self-administration of alcohol-preferring (P) rats. Alcohol. 2010;44(3):245–55.
124. Xiao C, Ye JH. Ethanol dually modulates GABAergic synaptic transmission onto dopaminergic neurons in ventral tegmental area: role of mu-opioid receptors. Neuroscience. 2008; 153(1):240–8.
125. Xiao C, et al. Ethanol facilitates glutamatergic transmission to dopamine neurons in the ventral tegmental area. Neuropsychopharmacology. 2009;34(2):307–18.
126. Guan Y, et al. GABAergic actions mediate opposite ethanol effects on dopaminergic neurons in the anterior and posterior ventral tegmental area. J Pharmacol Exp Ther. 2012; 341(1):33–42.
127. Theile JW, et al. GABAergic transmission modulates ethanol excitation of ventral tegmental area dopamine neurons. Neuroscience. 2011;172:94–103.
128. Job MO, et al. Mu (mu) opioid receptor regulation of ethanol-induced dopamine response in the ventral striatum: evidence of genotype specific sexual dimorphic epistasis. Biol Psychiatry. 2007;62(6):627–34.
129. Soderpalm B, et al. Nicotinic mechanisms involved in the dopamine activating and reinforcing properties of ethanol. Behav Brain Res. 2000;113(1–2):85–96.
130. Blomqvist O, et al. The mesolimbic dopamine-activating properties of ethanol are antagonized by mecamylamine. Eur J Pharmacol. 1993;249(2):207–13.
131. Blomqvist O, et al. Voluntary ethanol intake in the rat: effects of nicotinic acetylcholine receptor blockade or subchronic nicotine treatment. Eur J Pharmacol. 1996;314(3):257–67.
132. Blomqvist O, Soderpalm B, Engel JA. Ethanol-induced locomotor activity: involvement of central nicotinic acetylcholine receptors? Brain Res Bull. 1992;29(2):173–8.
133. Larsson A, et al. Role of different nicotinic acetylcholine receptors in mediating behavioral and neurochemical effects of ethanol in mice. Alcohol. 2002;28(3):157–67.
134. Le AD, et al. Involvement of nicotinic receptors in alcohol self-administration. Alcohol Clin Exp Res. 2000;24(2):155–63.
135. Ericson M, et al. Ethanol elevates accumbal dopamine levels via indirect activation of ventral tegmental nicotinic acetylcholine receptors. Eur J Pharmacol. 2003;467(1–3):85–93.

136. Blomqvist O, et al. Accumbal dopamine overflow after ethanol: localization of the antagonizing effect of mecamylamine. Eur J Pharmacol. 1997;334(2–3):149–56.
137. Chi H, de Wit H. Mecamylamine attenuates the subjective stimulant-like effects of alcohol in social drinkers. Alcohol Clin Exp Res. 2003;27(5):780–6.
138. Forman SA, Zhou Q. Nicotinic receptor pore mutations create a sensitive inhibitory site for ethanol. Alcohol Clin Exp Res. 2000;24(9):1363–8.
139. Zhou QL, Zhou Q, Forman SA. The n-alcohol site in the nicotinic receptor pore is a hydrophobic patch. Biochemistry. 2000;39(48):14920–6.
140. Zuo Y, et al. Alcohol modulation of neuronal nicotinic acetylcholine receptors is alpha subunit dependent. Alcohol Clin Exp Res. 2002;26(6):779–84.
141. Liu L, et al. Nicotinic acetylcholine receptors containing the alpha4 subunit modulate alcohol reward. Biol Psychiatry. 2013;73(8):738–46.
142. Liu L, et al. Nicotinic acetylcholine receptors containing the alpha6 subunit contribute to ethanol activation of ventral tegmental area dopaminergic neurons. Biochem Pharmacol. 2013;86(8):1194–200.
143. Ericson M, et al. Voluntary ethanol intake in the rat and the associated accumbal dopamine overflow are blocked by ventral tegmental mecamylamine. Eur J Pharmacol. 1998; 358(3):189–96.
144. Hendrickson LM, Zhao-Shea R, Tapper AR. Modulation of ethanol drinking-in-the-dark by mecamylamine and nicotinic acetylcholine receptor agonists in C57BL/6J mice. Psychopharmacology (Berl). 2009;204(4):563–72.
145. Harvey SC, Luetje CW. Determinants of competitive antagonist sensitivity on neuronal nicotinic receptor beta subunits. J Neurosci. 1996;16(12):3798–806.
146. Harvey SC, Maddox FN, Luetje CW. Multiple determinants of dihydro-beta-erythroidine sensitivity on rat neuronal nicotinic receptor alpha subunits. J Neurochem. 1996;67(5): 1953–9.
147. Kamens HM, Phillips TJ. A role for neuronal nicotinic acetylcholine receptors in ethanol-induced stimulation, but not cocaine- or methamphetamine-induced stimulation. Psychopharmacology (Berl). 2008;196(3):377–87.
148. Lof E, et al. Nicotinic acetylcholine receptors in the ventral tegmental area mediate the dopamine activating and reinforcing properties of ethanol cues. Psychopharmacology (Berl). 2007;195(3):333–43.
149. Moroni M, et al. alpha4beta2 nicotinic receptors with high and low acetylcholine sensitivity: pharmacology, stoichiometry, and sensitivity to long-term exposure to nicotine. Mol Pharmacol. 2006;70(2):755–68.
150. Kuzmin A, et al. Effects of subunit selective nACh receptors on operant ethanol self-administration and relapse-like ethanol-drinking behavior. Psychopharmacology (Berl). 2009;203(1):99–108.
151. Jerlhag E, et al. Role of the subunit composition of central nicotinic acetylcholine receptors for the stimulatory and dopamine-enhancing effects of ethanol. Alcohol Alcohol. 2006; 41(5):486–93.
152. Larsson A, et al. Is an alpha-conotoxin MII-sensitive mechanism involved in the neurochemical, stimulatory, and rewarding effects of ethanol? Alcohol. 2004;34(2–3):239–50.
153. Grady SR, et al. The subtypes of nicotinic acetylcholine receptors on dopaminergic terminals of mouse striatum. Biochem Pharmacol. 2007;74(8):1235–46.
154. Coe JW, et al. Varenicline: an alpha4beta2 nicotinic receptor partial agonist for smoking cessation. J Med Chem. 2005;48(10):3474–7.
155. Steensland P, et al. Varenicline, an alpha4beta2 nicotinic acetylcholine receptor partial agonist, selectively decreases ethanol consumption and seeking. Proc Natl Acad Sci U S A. 2007; 104(30):12518–23.
156. Tonstad S, et al. Effect of maintenance therapy with varenicline on smoking cessation: a randomized controlled trial. JAMA. 2006;296(1):64–71.
157. Kamens HM, Andersen J, Picciotto MR. Modulation of ethanol consumption by genetic and pharmacological manipulation of nicotinic acetylcholine receptors in mice. Psychopharmacology (Berl). 2010;208(4):613–26.

158. Mihalak KB, Carroll FI, Luetje CW. Varenicline is a partial agonist at alpha4beta2 and a full agonist at alpha7 neuronal nicotinic receptors. Mol Pharmacol. 2006;70(3):801–5.
159. Lummis SC, et al. Varenicline is a potent agonist of the human 5-hydroxytryptamine3 receptor. J Pharmacol Exp Ther. 2011;339(1):125–31.
160. Bordia T, et al. Varenicline is a potent partial agonist at alpha6beta2* nicotinic acetylcholine receptors in rat and monkey striatum. J Pharmacol Exp Ther. 2012;342(2):327–34.
161. Papke RL, Wecker L, Stitzel JA. Activation and inhibition of mouse muscle and neuronal nicotinic acetylcholine receptors expressed in Xenopus oocytes. J Pharmacol Exp Ther. 2010;333(2):501–18.
162. Rezvani AH, et al. Effects of sazetidine-A, a selective alpha4beta2 nicotinic acetylcholine receptor desensitizing agent on alcohol and nicotine self-administration in selectively bred alcohol-preferring (P) rats. Psychopharmacology (Berl). 2010;211(2):161–74.
163. Sajja RK, Rahman S. Neuronal nicotinic receptor ligands modulate chronic nicotine-induced ethanol consumption in C57BL/6J mice. Pharmacol Biochem Behav. 2012;102(1):36–43.
164. Sajja RK, Rahman S. Lobeline and cytisine reduce voluntary ethanol drinking behavior in male C57BL/6J mice. Prog Neuropsychopharmacol Biol Psychiatry. 2011;35(1):257–64.
165. Bell RL, et al. Nicotinic receptor ligands reduce ethanol intake by high alcohol-drinking HAD-2 rats. Alcohol. 2009;43(8):581–92.
166. Chatterjee S, et al. Partial agonists of the alpha3beta4* neuronal nicotinic acetylcholine receptor reduce ethanol consumption and seeking in rats. Neuropsychopharmacology. 2011;36(3):603–15.
167. Carnicella S, et al. Noribogaine, but not 18-MC, exhibits similar actions as ibogaine on GDNF expression and ethanol self-administration. Addict Biol. 2010;15(4):424–33.
168. Kamens HM, et al. The alpha6 nicotinic acetylcholine receptor subunit influences ethanol-induced sedation. Alcohol. 2012;46(5):463–71.
169. Dawson A, Miles MF, Damaj MI. The beta2 nicotinic acetylcholine receptor subunit differentially influences ethanol behavioral effects in the mouse. Alcohol. 2013;47(2):85–94.
170. Powers MS, et al. Nicotinic acetylcholine receptors containing alpha6 subunits contribute to alcohol reward-related behaviors. Genes Brain Behav. 2013;12(5):543–53.
171. Williams JM, Galli A. The dopamine transporter: a vigilant border control for psychostimulant action. Handb Exp Pharmacol. 2006;175:215–32.
172. Horger BA, Giles MK, Schenk S. Preexposure to amphetamine and nicotine predisposes rats to self-administer a low dose of cocaine. Psychopharmacology (Berl). 1992;107(2–3):271–6.
173. Zachariou V, et al. Nicotine receptor inactivation decreases sensitivity to cocaine. Neuropsychopharmacology. 2001;24(5):576–89.
174. Hansen ST, Mark GP. The nicotinic acetylcholine receptor antagonist mecamylamine prevents escalation of cocaine self-administration in rats with extended daily access. Psychopharmacology (Berl). 2007;194(1):53–61.
175. Levin ED, et al. The nicotinic antagonist mecamylamine preferentially inhibits cocaine vs. food self-administration in rats. Physiol Behav. 2000;71(5):565–70.
176. Witten IB, et al. Cholinergic interneurons control local circuit activity and cocaine conditioning. Science. 2010;330(6011):1677–81.
177. Zanetti L, Picciotto MR, Zoli M. Differential effects of nicotinic antagonists perfused into the nucleus accumbens or the ventral tegmental area on cocaine-induced dopamine release in the nucleus accumbens of mice. Psychopharmacology (Berl). 2007;190(2):189–99.
178. Kreek MJ, et al. Opiate and cocaine addiction: from bench to clinic and back to the bench. Curr Opin Pharmacol. 2009;9(1):74–80.
179. Rezayof A, et al. Morphine-induced place preference: involvement of cholinergic receptors of the ventral tegmental area. Eur J Pharmacol. 2007;562(1–2):92–102.
180. Feng B, et al. Blocking alpha4beta2 and alpha7 nicotinic acetylcholine receptors inhibits the reinstatement of morphine-induced CPP by drug priming in mice. Behav Brain Res. 2011;220(1):100–5.

Chapter 16
Genetic Contributions of the α5 Nicotinic Receptor Subunit to Smoking Behavior

Laura J. Bierut and Jerry A. Stitzel

Abstract Cigarette smoking and tobacco use remain a significant global health problem that kills nearly 6 million people every year [1]. Differences in heaviness of smoking are influenced by genetic variation, and compelling evidence from large-scale, genome-wide association meta-analyses identifies the chromosome 15 region, which contains the α5, α3, and β4 nicotinic receptor subunit gene cluster (*CHRNA5*, *CHRNA3*, *CHRNB4*), as the locus that most strongly contributes to heaviness of smoking. Further dissection of this region through human genetic, functional, and animal studies links differences in smoking behaviors to the α5 nicotinic receptor subunit. The same genetic variants in the α5 nicotinic receptor subunit gene that increase the risk of heavy smoking also play a role in failed smoking cessation. Importantly, an interplay exists between these high-risk genetic variants in the α5 nicotinic receptor subunit and pharmacologic treatment so that those at highest genetic risk for failed smoking cessation respond most favorably to pharmacologic treatment to aid cessation. Hopefully this knowledge will improve smoking cessation efforts and reduce the health burdens associated with cigarette smoking.

Keywords Smoking • Nicotine dependence • Nicotinic receptor genes • Genetics • Addiction

1 Introduction

The reduction of cigarette smoking over the last three decades from 33 % of the population in 1980 to 19 % in 2010 is one of the most important public health successes in the USA. This reduction occurred through restrictions on indoor smoking

L.J. Bierut, M.D. (✉)
Department of Psychiatry, Washington University School of Medicine,
660 South Euclid Avenue, Campus Box 8134, St. Louis, MO 63110, USA
e-mail: Laura@wustl.edu

J.A. Stitzel, Ph.D.
Institute for Behavioral Genetics, University of Colorado Boulder,
1480 30th Street, Boulder, CO 80303, USA
e-mail: stitzel@colorado.edu

R.A.J. Lester (ed.), *Nicotinic Receptors*, The Receptors 26,
DOI 10.1007/978-1-4939-1167-7_16

and increases in taxation of cigarettes, along with robust educational campaigns to educate the public about the adverse health consequences of smoking. These policy changes and educational efforts decreased initiation of smoking among teens and increased smoking cessation by established smokers.

Though this decrease in smoking represents a public health success, cigarette smoking remains a significant health concern. In the USA nearly a fifth of adults are current smokers [2] despite a majority (68 %) reporting wanting to quit [3]. A large number of people with smoking-related illnesses continue to smoke. Of those with chronic bronchitis, 41 % smoke, as do 29 % of those with coronary heart disease and 49 % of those with emphysema [4]. These high rates of ongoing cigarette smoking in the presence of a smoking-related illness starkly illustrate the difficulty of changing smoking behaviors. Ongoing public health concerns associated with cigarette smoking motivate studies to understand biologic factors, including genetic risk factors, that drive smoking behaviors, and the goal is to gain knowledge that can be used to reduce smoking and facilitate cessation. This chapter reviews the contributions of genetic variation in α5 nicotinic receptor subunit gene as an example of how human, animal, and other laboratory studies inform our knowledge about the development of heavy smoking and smoking cessation.

2 The Modern Genetic World

In parallel with the dramatic reduction of smoking, the past decade has witnessed a revolution in genetic technologies. Millions of genetic variants (or single-nucleotide polymorphisms, SNPs) can be queried in genome-wide association studies, and it is feasible to test tens to hundreds of thousands of individuals. This depth of inquiry into human genomes represents a tremendous expansion of scale of study in the past 10 years. These new technologies facilitate genetic investigations into many complex diseases and have resulted in thousands of new genetic discoveries [5, 6].

This modern genetic technology has been applied to the study of smoking behaviors. Cigarette smoking clusters in families and strong evidence from twin studies demonstrate that this familial clustering of smoking is due to both environmental and genetic factors. Environmental factors appear to have a stronger effect on smoking initiation, whereas genetic factors play a larger role in the transition from regular cigarette use to the development of heavy smoking and nicotine dependence [7]. Estimated heritability for heavy smoking and nicotine dependence is at least 50 % [8], and this high heritability motivates genetic studies of smoking behavior.

Number of cigarettes smoked per day, which is highly correlated with nicotine dependence, has been measured in many studies because of the strong contribution of smoking to medical illnesses. Individual studies can search for genetic contributions to heaviness of smoking, and through meta-analysis, genetic results are combined across studies. Meta-analysis is a powerful technique and allows the examination of large numbers of individuals so that new genetic contributions to diseases can be discovered.

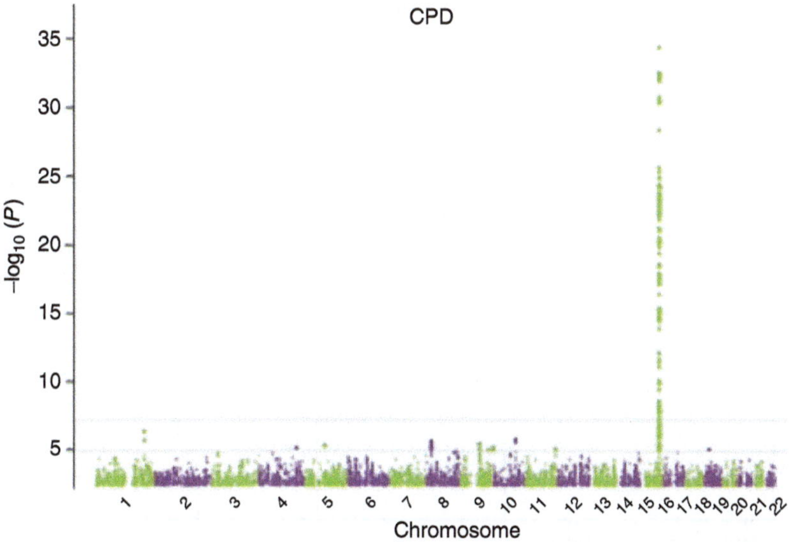

Fig.16.1 Genome-wide association results for the TAG Consortium. Plot shows SNPs plotted along the *x*-axis according to their position on each chromosome and *y*-axis shows -log10 p value. The peak on chromosome 15 represents the strong association with cigarettes per day and encompasses the *CHRNA5-CHRNA3-CHRNB4* gene cluster. Reprinted by permission from Macmillan Publishers Ltd: Nature Genetics. Tobacco and Genetics Consortium. Genome-wide meta-analyses identify multiple loci associated with smoking behavior. Nat Genet. 2010;42(5):441-7. Copyright 2010

Compelling genetic evidence is provided by several large-scale, genome-wide association meta-analyses that variation in nicotinic receptor subunit genes contributes to the heaviness of smoking and nicotine dependence [9–12]. These meta-analyses of over 80,000 individuals of European ancestry identify the chromosomal region 15q24, which contains the α5, α3, and β4 nicotinic receptor subunit gene cluster (*CHRNA5*, *CHRNA3*, *CHRNB4*), as the genetic region that most strongly contributes to heaviness of smoking (refer to Chap. 3 for additional information on regulation of this gene cluster). This finding is best illustrated by Figure 16.1. In this chromosomal region, the variant rs16969968 is strongly associated with smoking behavior ($p = 5.57 \times 10^{-72}$) [12]. Further examination of this chromosome 15 region demonstrates that there are at least two distinct genetic risk variants in this region that independently contribute to differences in smoking behavior [12, 13].

3 Dissecting Genetic Findings

When a genetic association is found, it represents not only association with the tested variants, but all genetic variants (tested and untested) that are highly correlated. Figure 16.2 shows the genetic correlation of polymorphisms with rs16969968 in

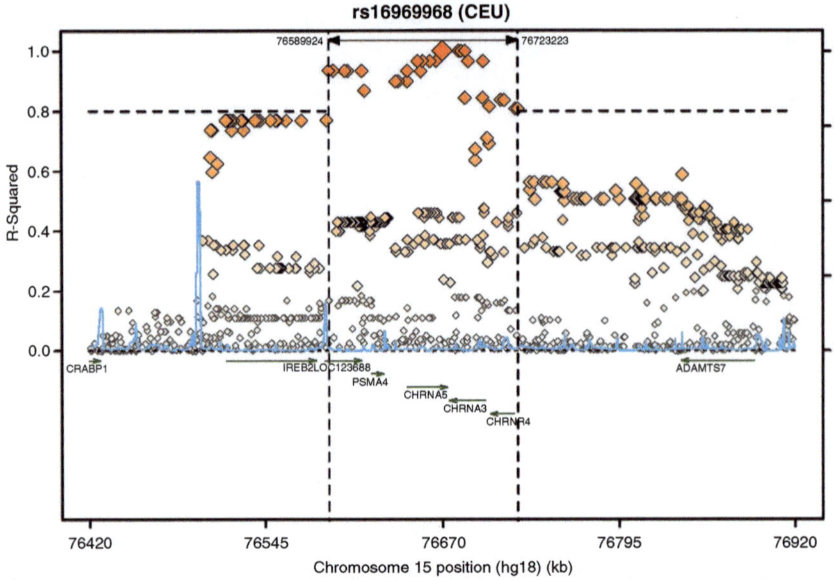

Fig.16.2 Genetic architecture of the chromosome 15 region associated with cigarettes per day in European ancestry populations. Plot shows SNPs plotted along the *x*-axis according to their position in the chromosome 15 region and genes are shown below the *x*-axis. *Y*-axis shows correlation of SNPs with rs16969968. SNPs highly correlated (r2>0.9) extend across many genes. Figure generated through SNAP. Johnson AD, Handsaker RE, Pulit S, Nizzari MM, O'Donnell CJ, de Bakker, PIW. SNAP: A web-based tool for identification and annotation of proxy SNPs using HapMap. Bioinformatics. 2008;24(24):2938-9

the chromosome 15 region. Many genes are implicated in this association with rs16969968, and identifying the variant or variants that cause functional changes requires further investigation. One approach to refine an association signal is population contrast mapping, which works on the hypothesis that important biological mechanisms underlying disease are shared across human populations. Because the genetic architecture of chromosome 15q24 varies across populations, identifying a consistent association across the diverse world populations points to a variant more likely to have functional relevance [14–16].

Cross-population genetic association studies of this chromosome 15q24 region identify variation in *CHRNA5* as an important influence to heaviness of smoking. An international collaborative study examined the association between variants in the *CHRNA5-CHRNA3-CHRNB4* region and smoking quantity in a meta-analysis of over 22,000 smokers of European, Asian, and African descent [17]. Despite diverse genetic backgrounds across these populations, the variant rs16969968 is associated with smoking behavior across all populations. The frequency of the "A" risk allele of rs16969968 differs dramatically across human populations. It is common in populations of European and Middle Eastern descent (approximately 35 % frequency) and is less common in populations of African,

Asian, or American origin (less than 5 %). This cross-population finding points to *CHRNA5* as a key gene associated with heaviness of smoking and nicotine dependence and provides evidence that rs16969968 is most likely a causative functional variant in this region.

4 From Genetic Association to Function

Once a genetic association is found and a putative functional variant is identified, it is important to move beyond genetic association into biological function. The most biologically credible variant associated with nicotine dependence in the chromosome 15q24 region is rs16969968, a polymorphism that causes an amino acid change from aspartic acid to asparagine (Asp398Asn) in the α5 nicotinic receptor subunit. The aspartic acid at position 398 in *CHRNA5* in humans occurs at a residue that is highly conserved across species. This variant is located in the large cytoplasmic loop between transmembrane domains three and four. Frogs, chickens, rodents, cattle, and nonhuman primates possess an aspartic acid residue at this location. In humans, the amino acid may be either an aspartic acid, the predominant residue at this position, or asparagine.

5 α5 D398N Alters Maximal Response of $(\alpha4\beta2)_2\alpha5$ Receptors to Nicotine Agonists

Evidence that this amino acid change is functionally relevant is supported by *in vitro* experiments demonstrating that $(\alpha4\beta2)_2\alpha5$ nicotinic receptors with the aspartic acid variant (D398) exhibit a greater maximal response to a nicotinic agonist than $(\alpha4\beta2)_2\alpha5$ nicotinic receptors with the asparagine amino acid substitution (N398) [18]. Because the allele that codes for asparagine is associated with increased risk for developing nicotine dependence, and nicotinic receptors containing the α5 subunit with this amino acid (N398) exhibit decreased function *in vitro*, reduced function of $(\alpha4\beta2)_2\alpha5$ nicotinic receptors is associated with an elevated risk for developing nicotine dependence. Other nicotinic receptors $(\alpha3\beta4)_2\alpha5$ demonstrate a similar decrease in function with this α5 amino acid change [19, 20]. Further studies show that $(\alpha4\beta2)_2\alpha5$ nicotinic acetylcholine receptors that contain an α5 subunit protein with the asparagine substitution have lower calcium permeability and increased short-term desensitization compared to receptors that contain the α5 protein with aspartic acid in that position [21]. This combined evidence of high conservation across species and biological changes in receptor function nominates this amino acid change in the α5 nicotinic receptor as a causative biological factor that alters the risk of developing heavy smoking and nicotine dependence. The mechanism of action is hypothesized to be reduced function of the receptor with the asparagine containing α5 subunit protein.

6 Noncoding Variant(s) Result in a Two- to Threefold Difference in mRNA Expression

Genome-wide association meta-analyses demonstrate that a second, distinct variant independently contributes to the risk of heavy smoking and nicotine dependence, and laboratory evidence points to a different biological mechanism underlying this association. In brain tissue (frontal cortex), expression levels of *CHRNA5* mRNA vary across samples, and a strong association between *CHRNA5* mRNA expression levels and a distinct group of SNPs has been identified [22]. Fine mapping studies using allele-specific gene expression in tissue from European and African American brain narrow down the most likely locus containing the functional alleles for expression regulation to a region upstream of the transcriptional start site of *CHRNA5* [23]. These putative functional variants alter mRNA expression of *CHRNA5* and contribute to the risk of developing nicotine dependence in the human association studies.

Low expression of *CHRNA5* mRNA in the frontal cortex is associated with a decreased risk of developing nicotine dependence. At this time, it is difficult to explain why reduced function of the α5 nicotinic acetylcholine receptor subunit leads to increased risk for developing nicotine dependence while low expression of α5 nicotinic acetylcholine receptor subunit mRNA leads to protection. However, it is important to note that the expression analysis was performed with mRNA isolated from frontal cortex. Currently, little is known about the role of cortical α5 nicotinic acetylcholine receptor subunits in nicotine dependence. It would be of interest to determine the relationship between the *CHRNA5* mRNA expression-associated SNPS and α5 nicotinic acetylcholine receptor subunit mRNA expression in brain areas where the α5 nicotinic acetylcholine receptor subunit is currently thought to contribute to nicotine-seeking behavior.

In summary, through *in vitro* studies, two distinct biological mechanisms, an amino acid substitution in the α5 nicotinic receptor subunit protein that changes receptor conductance and an alteration of mRNA expression of *CHRNA5*, have been identified. In humans, the genetic risk for nicotine dependence is conferred by the genetic variation associated with each of these two mechanisms [10, 24]. A next step to improve our understanding of the biological processes influenced by this region has been undertaken in animal studies.

7 Animal Studies of Nicotine Use and the α5 Nicotinic Receptor Subunit

The amount of nicotine consumed reflects a balance between reinforcing and aversive properties of nicotine. At lower doses, nicotine use can stimulate reinforcing properties, which leads to increased intake [25]. At higher levels of nicotine use, aversive effects become more prominent, which limits further intake [26, 27].

The α5 nicotinic receptor subunit plays a role in the balance between the reinforcing and aversive properties related to nicotine use.

Studies with Chrna5 knockout mice suggest a potentially complex role of the α5 nicotinic acetylcholine subunit in nicotine-seeking behavior. Fowler et al. [28] reported that at lower nicotine concentrations both wild-type and α5 subunit knockout mice have similar consumption patterns. However, at higher doses of nicotine, differences in nicotine intake are seen between wild-type and α5 knockout mice. Wild-type mice decrease intake at higher doses of nicotine, consistent with stimulation of aversive pathways related to high levels of nicotine. However, α5 subunit knockout mice consume greater amounts of nicotine at high doses, which suggests an attenuation of the aversive effects of nicotine with the loss of the α5 nicotinic receptor subunit. A dose-dependent Chrna5 genotype effect on nicotine reward also was reported by Jackson et al. [29]. Similar to what Fowler et al. [28] described, Jackson et al. [29] found that α5 knockout mice exhibited comparable reward to nicotine at low nicotine doses. However, at high nicotine doses, α5 knockout mice continued to find nicotine rewarding, whereas wild-type animals did not. In a recent study, Fowler et al. [30] suggested that the role of α5 in nicotine aversion is not through a general aversion to noxious stimuli, but more specifically, through high-dose nicotine inhibition of the reward pathway. However, this interpretation seems to be somewhat at odds with a study by Salas et al. [31] in which α5 nicotinic acetylcholine receptor subunit knockout mice were found to exhibit significantly reduced sensitivity to several adverse responses to nicotine.

Further experiments using mice with the α5 nicotinic acetylcholine receptor subunit knocked out highlight the role of the medial habenula in the control of nicotine intake. Expression of α5 nicotinic acetylcholine receptor subunit is concentrated in the habenulo-interpeduncular pathway in mice [28]. The medial habenula projects almost exclusively to the interpeduncular pathway, and high-dose nicotine activates this pathway [32, 33]. Reintroduction of α5 nicotinic acetylcholine receptor subunits into the medial habenula of α5 knockout mice restores the nicotine consumption patterns of wild-type mice [28]. These findings demonstrate that the α5 nicotinic acetylcholine receptor subunit expression in the habenulo-interpeduncular neurocircuit is a key component controlling nicotine intake at high nicotine doses. Interestingly, the habenulo-interpeduncular pathway also has been implicated in α5 nicotinic acetylcholine receptor subunit-mediated withdrawal following chronic nicotine exposure [34, 35].

A new study suggests a role for ventral tegmental area-expressed α5 nicotinic acetylcholine receptor subunits in nicotine reinforcement. Similar to what Fowler et al. reported, Morel et al. [36] found that α5 nicotinic acetylcholine receptor subunit knockout mice continued nicotine self-administration at high nicotine doses whereas wild-type animals ceased their nicotine intake. However, in contrast to the Fowler et al. report, the Morel et al. study found that deletion of the α5 nicotinic acetylcholine receptor subunit resulted in decreased nicotine self-administration at low nicotine concentrations. Reintroduction of the α5 nicotinic acetylcholine receptor subunit into the ventral tegmental area restored low-dose nicotine self-administration to control levels and also reduced nicotine intake at high nicotine doses back to wild-type levels. Intriguingly, re-expression of the α5 nicotinic acetylcholine receptor subunit possessing the variation associated with risk for nicotine dependence in humans (N398) neither restored low-dose nicotine self-administration nor reduced high-dose nicotine intake. Taken together, the studies with α5 nicotinic acetylcholine receptor subunit knockout mice paint a complex picture in which the α5 nicotinic acetylcholine receptor subunit contributes to both the reinforcing and aversive effects of nicotine depending upon the brain structure in which they are expressed. It is important to note that the α5 nicotinic acetylcholine receptor subunit is expressed in several other brain areas including the cortex, hippocampus, and raphe nuclei to name a few [31, 37, 38]. The relevance of α5 nicotinic acetylcholine receptor subunit expression in these regions remains to be determined.

8 The *Chrna5* D398N Mouse

In addition to a knockout mouse model, a knock-in mouse has been engineered to possess the human rs16969968 variant, which results in the asparagine amino acid substitution in the α5 nicotine receptor subunit. This mouse is the first that has been engineered to possess a human polymorphism associated with nicotine dependence, and this novel resource will permit us to further explore molecular, neurobiological, and behavioral mechanisms through which this important genetic variant alters risk for nicotine dependence.

9 Convergence of Findings

In sum, human genetic, functional, and animal studies link differences in smoking behaviors to the α5 nicotinic receptor subunit and to specific genetic variants. In turn, these variants are linked to basic cellular mechanisms and to specific brain regions. These studies converge on a conceptual framework that balances the reinforcing and aversive effects of nicotine use and involves the medial habenula, ventral tegmental area, and perhaps other brain regions in which the α5 nicotinic acetylcholine receptor subunit is expressed. Individuals who smoke more heavily are more likely to have variation in the α5 nicotinic acetylcholine receptor subunit that results in an amino acid change from aspartic acid to asparagine. The incorporation of this high-genetic-risk α5 subunit (asparagine containing) into nicotinic acetylcholine receptors decreases receptor function compared to receptors containing the low-genetic-risk α5 subunit. In contrast, lowered mRNA expression of *CHRNA5* in the frontal cortex is associated with decreased risk of heavy smoking. Mice with the α5 subunit gene knocked out show increased nicotine intake at high nicotine concentrations, consistent with a loss of the aversive effects related to nicotine. Reintroduction of the wild-type α5 subunit into either the medial habenula or ventral tegmental area of knockout mice restores wild-type patterns of nicotine consumption. In contrast, re-expression of the risk form of the α5 nicotinic acetylcholine receptor subunit into the ventral tegmental area neither restores nicotine consumption at low nicotine concentrations nor reduces nicotine intake at high nicotine concentrations. These convergent findings support a model of the habenulo-interpeduncular pathway as well as the reward pathway as key neurocircuits that act through the α5 nicotinic acetylcholine receptor subunit as an inhibitory pathway that limits nicotine intake. Within this circuit the α5 subunit plays an important role in nicotinic acetylcholine receptor function, and reduced receptor function decreases the inhibitory signaling that in turn limits nicotine intake. This model is consistent with the finding that individuals who carry risk alleles for nicotine dependence are less sensitive to the aversive effects of nicotine and smoke more heavily.

10 Smoking Cessation

The goal of genetic discovery and functional investigation is to understand the biological basis of nicotine dependence and then to translate this knowledge into improved treatment for smoking cessation. This translation process is under way and it is clear that genetic variants in the α5 nicotinic receptor subunit gene, which increase the risk of heavy smoking and nicotine dependence, play a role in failed smoking cessation.

The first study to clearly show an association of the chromosome 15q24 region with smoking cessation was an examination of women who smoked during pregnancy [39]. Women with high-risk genetic variants in *CHRNA5* were less likely to quit smoking during pregnancy compared to women with the low-risk genetic variants. These results have been confirmed in a second, independent sample of pregnant women who smoked [40]. In both studies, smoking cessation was undertaken in a naturalistic setting of a normal pregnancy, focused on this specific time period of pregnancy, and most likely occurred without pharmacologic intervention. This finding implies that the strong genetic influence of variation in *CHRNA5* on smoking cessation is seen during a period when environmental and social influences maximally encourage smoking cessation. Several other lines of evidence support this conclusion. Individuals with the high-risk variants in a general population setting quit smoking at an older age [41]. In smoking cessation treatment trials, failed smoking cessation is associated with the same high-risk genetic risk variants that predict a greater likelihood of developing nicotine dependence [41, 42]. Though smoking cessation is a complex behavior that is clearly influenced by many environmental factors, these different studies indicate that variants in the α5 nicotinic receptor subunit gene that predispose to heavy smoking and nicotine dependence also contribute to failed smoking cessation.

Importantly, new evidence demonstrates an interplay between these high-risk genetic variants in the α5 nicotinic receptor subunit and pharmacologic treatment. Those at highest genetic risk for failed smoking cessation respond most favorably to pharmacologic treatment to aid cessation [41, 42]. That this genetic risk for unsuccessful smoking cessation can be modified by pharmacologic treatment is great news for the field of smoking cessation and represents a first step in the goal of personalized medicine based on a person's genetic makeup.

11 Future Directions

This review highlights results from the strongest region associated with nicotine dependence, the genomic area encompassing the α5-α3-β4 nicotinic receptor subunit gene cluster. New evidence suggests that both common and less frequent genetic variants in the chromosome 15 region containing the α5-α3-β4 nicotinic receptor subunit gene cluster contribute to the development of nicotine dependence [43]. For example, less common (frequency 1 %-5 %) and rare (frequency < 1 %) variation in the α3 and β4 nicotinic receptor subunit genes is associated with an altered risk for nicotine dependence and cigarette consumption [43]. These less common variants also alter the protein structure of the α3 and β4 nicotinic receptor subunits. These results demonstrate that we must delve more deeply into the regions of association. Variation in other nicotinic receptor subunits, α6 and β3 on chromosome 8, is also associated with heaviness of smoking. Further investigation into these genes is under way and it is likely that many more genetic contributions await discovery.

12 Summary

Smoking remains one of the major contributors to disease in the USA. The economic burden of smoking is estimated at nearly $100 billion in direct medical expenses and a similar amount in lost productivity [44]. Even with these significant adverse health consequences, a large number of teenagers initiate cigarette smoking every year, with many 12th graders reporting ever smoking (38.1 %), smoking in the past 30 days (16.3 %), and daily smoking (8.5 %) [45]. Once smoking behaviors are established, cessation is a challenge and it may take years before smoking cessation is successful. Hopefully with our increased knowledge about the biology of how the nicotinic receptors contribute to persistent smoking behaviors, we will be able to improve smoking cessation efforts and reduce the burdens associated with cigarette smoking. The reader is referred to Chap. 17 for further discussions on gene linkage and smoking.

References

1. World Health Organization [Internet]. WHO report on the global tobacco epidemic, 2011: Warnings about the dangers of tobacco. 2011 [cited 2011 Dec 7]. Available from: http://whqlibdoc.who.int/publications/2011/9789240687813_eng.pdf.
2. Centers for Disease Control. Current cigarette smoking among adults - United States, 2011. Morbidity & Mortality Weekly Report. 2012;61(44):889–94.

3. Centers for Disease Control. Quitting smoking among adults – United States, 2001-2010. Morbidity & Mortality Weekly Report. 2011;60(44):1513–9.
4. Centers for Disease Control. Cigarette smoking among adults - United States, 2006. Morbidity & Mortality Weekly Report. 2007;56(44):1157–61.
5. Hindorff LA, MacArthur J, Morales J, Junkins HA, Hall PN, Klemm AK, et al. [Internet]. A catolog of published genome-wide association studies. 2013 [cited 2013 Mar 13]. Available from: www.genome.gov/gwastudies.
6. Hindorff LA, Sethupathy P, Junkins HA, Ramos EM, Mehta JP, Collins FS, et al. Potential etiologic and functional implications of genome-wide association loci for human diseases and traits. Proc Natl Acad Sci U S A. 2009;106(23):9362–7.
7. Vink JM, Willemsen G, Boomsma DI. Heritability of smoking initiation and nicotine dependence. Behav Genet. 2005;35(4):397–406.
8. Li MD. The genetics of nicotine dependence. Curr Psychiatry Rep. 2006;8(2):158–64.
9. Liu JZ, Tozzi F, Waterworth DM, Pillai SG, Muglia P, Middleton L, et al. Meta-analysis and imputation refines the association of 15q25 with smoking quantity. Nat Genet. 2010;42(5):436–40.
10. Saccone NL, Culverhouse RC, Schwantes-An TH, Cannon DS, Chen X, Cichon S, et al. Multiple independent loci at chromosome 15q25.1 affect smoking quantity: a meta-analysis and comparison with lung cancer and COPD. PLoS Genet. 2010;6(8):e1001053.
11. Thorgeirsson TE, Gudbjartsson DF, Surakka I, Vink JM, Amin N, Geller F, et al. Sequence variants at CHRNB3-CHRNA6 and CYP2A6 affect smoking behavior. Nat Genet. 2010; 42(5):448–53.
12. Tobacco and Genetics Consortium. Genome-wide meta-analyses identify multiple loci associated with smoking behavior. Nat Genet. 2010;42(5):441–7.
13. Saccone NL, Schwantes-An TH, Wang JC, Grucza RA, Breslau N, Hatsukami D, et al. Multiple cholinergic nicotinic receptor genes affect nicotine dependence risk in African and European Americans. Genes Brain Behav. 2010;9(7):741–50.
14. Rotimi CN, Jorde LB. Ancestry and disease in the age of genomic medicine. N Engl J Med. 2010;363(16):1551–8.
15. Saccone NL, Saccone SF, Goate AM, Grucza RA, Hinrichs AL, Rice JP, et al. In search of causal variants: refining disease association signals using cross-population contrasts. BMC Genet. 2008;9:58.
16. Zaitlen N, Pasaniuc B, Gur T, Ziv E, Halperin E. Leveraging genetic variability across populations for the identification of causal variants. Am J Hum Genet. 2010;86(1):23–33.
17. Chen LS, Saccone NL, Culverhouse RC, Bracci PM, Chen CH, Dueker N, et al. Smoking and genetic risk variation across populations of European, Asian, and African American ancestry–a meta-analysis of chromosome 15q25. Genet Epidemiol. 2012;36(4):340–51.
18. Bierut LJ, Stitzel JA, Wang JC, Hinrichs AL, Grucza RA, Xuei X, et al. Variants in nicotinic receptors and risk for nicotine dependence. Am J Psychiatry. 2008;165(9):1163–71.
19. Tammimaki A, Herder P, Li P, Esch C, Laughlin JR, Akk G, et al. Impact of human D398N single nucleotide polymorphism on intracellular calcium response mediated by alpha3beta4alpha5 nicotinic acetylcholine receptors. Neuropharmacology. 2012;63(6):1002–11.
20. George AA, Lucero LM, Damaj MI, Lukas RJ, Chen X, Whiteaker P. Function of human alpha3beta4alpha5 nicotinic acetylcholine receptors is reduced by the alpha5(D398N) variant. J Biol Chem. 2012;287(30):25151–62.
21. Kuryatov A, Berrettini W, Lindstrom J. Acetylcholine receptor (AChR) alpha5 subunit variant associated with risk for nicotine dependence and lung cancer reduces (alpha4beta2)alpha5 AChR function. Mol Pharmacol. 2011;79(1):119–25.
22. Wang JC, Cruchaga C, Saccone NL, Bertelsen S, Liu P, Budde JP, et al. Risk for nicotine dependence and lung cancer is conferred by mRNA expression levels and amino acid change in CHRNA5. Hum Mol Genet. 2009;18(16):3125–35.
23. Smith RM, Alachkar H, Papp AC, Wang D, Mash DC, Wang JC, et al. Nicotinic alpha5 receptor subunit mRNA expression is associated with distant 5' upstream polymorphisms. Eur J Hum Genet. 2011;19(1):76–83.

24. Wang JC, Bierut LJ, Goate AM. Variants weakly correlated with CHRNA5 D398N polymorphism should be considered in transcriptional deregulation at the 15q25 locus associated with lung cancer risk. Clin Cancer Res. 2009;15(17):5599.
25. Lynch WJ, Carroll ME. Regulation of drug intake. Exp Clin Psychopharmacol. 2001; 9(2):131–43.
26. Henningfield JE, Goldberg SR. Nicotine as a reinforcer in human subjects and laboratory animals. Pharmacol Biochem Behav. 1983;19(6):989–92.
27. Lynch WJ, Carroll ME. Regulation of intravenously self-administered nicotine in rats. Exp Clin Psychopharmacol. 1999;7(3):198–207.
28. Fowler CD, Lu Q, Johnson PM, Marks MJ, Kenny PJ. Habenular alpha5 nicotinic receptor subunit signalling controls nicotine intake. Nature. 2011;471(7340):597–601.
29. Jackson KJ, Marks MJ, Vann RE, Chen X, Gamage TF, Warner JA, et al. Role of alpha5 nicotinic acetylcholine receptors in pharmacological and behavioral effects of nicotine in mice. J Pharmacol Exp Ther. 2010;334(1):137–46.
30. Fowler CD, Tuesta L, Kenny PJ. Role of alpha5* nicotinic acetylcholine receptors in the effects of acute and chronic nicotine treatment on brain reward function in mice. Psychopharmacology (Berl). Epub 2013 Aug 20.
31. Salas R, Orr-Urtreger A, Broide RS, Beaudet A, Paylor R, De Biasi M. The nicotinic acetylcholine receptor subunit alpha 5 mediates short-term effects of nicotine in vivo. Mol Pharmacol. 2003;63(5):1059–66.
32. Herkenham M, Nauta WJ. Efferent connections of the habenular nuclei in the rat. J Comp Neurol. 1979;187(1):19–47.
33. London ED, Connolly RJ, Szikszay M, Wamsley JK, Dam M. Effects of nicotine on local cerebral glucose utilization in the rat. J Neurosci. 1988;8(10):3920–8.
34. Jackson KJ, Martin BR, Changeux JP, Damaj MI. Differential role of nicotinic acetylcholine receptor subunits in physical and affective nicotine withdrawal signs. J Pharmacol Exp Ther. 2008;325(1):302–12.
35. Salas R, Sturm R, Boulter J, De Biasi M. Nicotinic receptors in the habenulo-interpeduncular system are necessary for nicotine withdrawal in mice. J Neurosci. 2009;29(10):3014–8.
36. Morel C, Fattore L, Pons S, Hay YA, Marti F, Lambolez B, et al. Nicotine consumption is regulated by a human polymorphism in dopamine neurons. Mol Psychiatry Epub. 2013. doi:10.1038/mp.2013.158.
37. Marks MJ, Pauly JR, Gross SD, Deneris ES, Hermans-Borgmeyer I, Heinemann SF, et al. Nicotine binding and nicotinic receptor subunit RNA after chronic nicotine treatment. J Neurosci. 1992;12(7):2765–84.
38. Wada E, McKinnon D, Heinemann S, Patrick J, Swanson LW. The distribution of mRNA encoded by a new member of the neuronal nicotinic acetylcholine receptor gene family (alpha 5) in the rat central nervous system. Brain research. 1990;526(1):45–53.
39. Freathy RM, Ring SM, Shields B, Galobardes B, Knight B, Weedon MN, et al. A common genetic variant in the 15q24 nicotinic acetylcholine receptor gene cluster (CHRNA5-CHRNA3-CHRNB4) is associated with a reduced ability of women to quit smoking in pregnancy. Hum Mol Genet. 2009;18(15):2922–7.
40. Thorgeirsson TE, Stefansson K. Commentary: gene-environment interactions and smoking-related cancers. Int J Epidemiol. 2010;39(2):577–9.
41. Chen LS, Baker TB, Piper ME, Breslau N, Cannon DS, Doheny KF, et al. Interplay of genetic risk factors (CHRNA5-CHRNA3-CHRNB4) and cessation treatments in smoking cessation success. Am J Psychiatry. 2012;169(7):735–42.
42. Bergen AW, Javitz HS, Krasnow R, Nishita D, Michel M, Conti DV, et al. Nicotinic acetylcholine receptor variation and response to smoking cessation therapies. Pharmacogenet Genomics. 2013;23(2):94–103.
43. Haller G, Druley T, Vallania FL, Mitra RD, Li P, Akk G, et al. Rare missense variants in CHRNB4 are associated with reduced risk of nicotine dependence. Hum Mol Genet. 2012;21(3):647–55.

44. Centers for Disease Control. Smoking-attributable mortality, years of potential life lost, and productivity losses--United States, 2000-2004. Morbidity & Mortality Weekly Report. 2008; 57(45):1226-8.
45. Johnston LD, O'Malley PM, Bachman JG, Schulenberg JE [Internet]. Teen smoking continues to decline in 2013. Ann Arbor, MI: University of Michigan News Service; 2013 [cited 2013 Dec 20]. Available from: http://www.monitoringthefuture.org.

Chapter 17
Smoking-Related Genes and Functional Consequences

Ines Ibañez-Tallon and Jessica L. Ables

Abstract As the leading preventable cause of cancer and death, nicotine use and dependence have been the subject of a multitude of genetic studies in the past decade, ranging from candidate gene studies to genome-wide association studies (GWAS) to prospective studies. The genetics of nicotine addiction, smoking, and cancer are multifactorial, as would be expected from a complex behavior such as cigarette smoking. The combined heritability based on twin studies is estimated at 50–75 % (Li, Am J Med Sci 326(4): 168–73, 2003; Hall et al. Tob Control 11(2): 119–24, 2002; Lessov et al. Psychol Med 34(5): 865–79, 2004.; Lessov-Schlaggar et al. Int J Epidemiol 35(5): 1278–85, 2006; Maes et al. Psychol Med 34(7): 1251–61, 2004), with a large number of genes contributing to a small amount of risk individually. Some genes contribute to the quantity of nicotine used, while another set is associated with the duration of nicotine use, and yet another set is linked with severity of nicotine dependence. Associated "risky" genes comprise genes encoding for nicotinic acetylcholine receptor (nAChR) subunits, but also include genes like the bitter taste receptor. Still others have been associated with initiation and with success, or lack thereof, in cessation. In this chapter we review the current findings in the genetics of smoking, focusing on those studies that have linked nicotinic acetylcholine receptors (nAChR) to nicotine addiction, and further discuss how mutations in these receptors alter their function.

Keywords SNP • GWAS • Alpha5 • nAChR • Addiction • Mouse models

1 Human Genetic Studies

The genes associated with an increased risk to develop nicotine dependence fall into three broad categories: nicotinic receptors, nicotine-metabolizing enzymes, and neurotransmitter systems. Table 17.1 summarizes the surprisingly large number of genes that have been associated with nicotine use/dependence. It is important to

I. Ibañez-Tallon, Ph.D. (✉) • J.L. Ables, M.D., Ph.D.
Laboratory of Molecular Biology, The Rockefeller University, 1230 York Avenue, New York, NY 10065, USA
e-mail: iibanez@rockefeller.edu

R.A.J. Lester (ed.), *Nicotinic Receptors*, The Receptors 26,
DOI 10.1007/978-1-4939-1167-7_17

Table 17.1 Genes associated with nicotine behaviors in humans

Gene	Allele/SNP	Population features	Associated feature	Studies
CHRNB3/ CHRNA6	rs 6474412-T	31,266 Subjects of European descent	Quantity (CPD)	Genome-wide association meta analysis [15]
CYP2A6	rs 4105144	31,266 Subjects of European descent	Quantity (CPD)	Genome-wide association meta analysis [15]
CHRNB3/ CHRNA6	rs 1051730	31,266 Subjects of European descent	Quantity (CPD)	Genome-wide association meta analysis [15]
CHRNA3	rs 8034191 (rs 1051730 synonymous variant)	4,435 Cases, 7,272 controls of European descent		GWAS [9]
CHRNA5	rs 16969968	4,435 Cases, 7,272 controls of European descent		GWAS [9]
CHRNA3	rs 1051730, rs 8192475	301(MT1)+276(MT2)=577	Quantity (CO in expired air)	Clinical trial [25]
		MT1: 78.4 % Caucasian, 9.6 % Hispanic, 5.7 % Asian, 3 % African-American, 1 % Native American, <0.3 % Pacific Islander, 2 % other		
		MT2: 76.3 % Caucasian, 8.5 % Hispanic, 5.9 % Asian, 1.1 % Pacific Islander, 3.7 % African-American, 4.1 % other		
CHRNA5	rs16969968, proxy rs1051730	33,348 (heavy and light smokers with age onset information), European descent (25 and older)	Quantity (CPD)	Meta-analysis [16]
CHRNB4	rs 12914008	301(MT1)+276(MT2)=577	Quantity (CO in expired air)	Clinical trial [25]
		MT1: 78.4 % Caucasian, 9.6 % Hispanic, 5.7 % Asian, 3 % African-American, 1 % Native American, <0.3 % Pacific Islander, 2 % other		
		MT2: 76.3 % Caucasian, 8.5 % Hispanic, 5.9 % Asian, 1.1 % Pacific Islander, 3.7 % African-American, 4.1 % other		
CHRNA5	rs 1051730	13,945 Icelandic smokers	Quantity (CPD)	GWAS [8]

CHAT	rs 4838547, rs 11101202, rs 867687	472 Smokers of European descent	Quantity (HSI)	System-based genomic association analysis [42]
CHAT	Haplotype association	2,037 Subjects of European-American (EA) and African-American (AA) descent	Quantity (CPD, HIS, FTND)	SNP-based association analysis [43]
CHAT	rs 3793790 in AA sample	2,037 Subjects of European-American (EA) and African-American (AA) descent	Quantity (CPD, HIS, FTND)	SNP-based association analysis [43]
CHAT	rs 11101202 in EA sample	2,037 Subjects of European-American (EA) and African-American (AA) descent	Quantity (CPD, FTND)	SNP-based association analysis [43]
CHAT	rs 4838547, rs 11101202, rs 867687 in EA sample	2,037 Subjects of European-American (EA) and African-American (AA) descent	Quantity (HSI)	SNP-based association analysis [43]
OPRM1	rs9479757-rs2075572-rs10485057	688 Caucasian subjects of lifetime smokers and nonsmokers of European ancestry	Fagerström Tolerance Questionnaire (FTQ)	Clinical study [54]
OPRM1	A118G SNP	30 Smokers of each OPRM1 genotype (A/A vs. A/G or G/G)	Cigarette choice paradigm	Within-subject, double-blind human laboratory study [60]
OPRM1	rs1074287, rs6912029, rs1799971, rs12209447, rs510769, rs3798676, rs553202, rs7748401, rs495491, rs10457090, rs589046, rs3778152, rs563649	366 Taiwanese methadone maintenance treatment patients	Quantity (cotinine)	Clinical study [56]
OPRM1	rs1799971	598 Subjects of European descent	Quantity (exhaled CO, cotinine)	Clinical study [61]
OPRM1	A118G SNP	19 Male tobacco smokers	PET brain imaging study using raclopride	Clinical study [57]
OPRM1	A118G SNP	Twenty-two smokers prescreened for genotype (12 A/A, 10 */G)	PET brain imaging	Clinical study [58]
OPRM1	chr6q26 gene loci	Male 204, female 215 consisting of experimenters, current and former smokers	Rasch-modeled nicotine withdrawal score	Clinical study [59]

note though that the interpretation of genetic studies might be affected by some limitations. First, sample size and composition are critically important for genetic studies, particularly for genome-wide association studies (GWAS). A small sample size may yield false positives or false negatives, particularly when the effect size is small for any given locus, as is the case in even the most strongly associated gene *CHRNA5* (see below). Similarly, a population that is genetically divergent has too much inherent diversity to allow detection of a weakly linked single-nucleotide polymorphism (SNP). For this reason, the vast majority of studies to date have been performed in subjects of European descent. These two problems have been overcome most recently by performing meta-analyses on all available published datasets and expanding datasets to include subjects of African and Asian descent. Second, the phenotypes assessed critically affect the outcome of the study [1]. Nicotine-related phenotypes include cigarettes per day (CPD), dependence via the Fagerström Nicotine Test of Dependence (FNTD) or the DSM-IV, heaviness of smoking index (HSI), smoking initiation and cessation, peripheral blood levels of nicotine or its metabolite cotinine, cognition, cancer, and peripheral artery disease. While the dependent measure(s) in each of the studies discussed below varies, the most replicable and strongest associations are with CPD [2] or nicotine/cotinine levels [3], indicating that the more relevant variants in the associated genes predominantly affect how much nicotine a person consumes. Most of the studies find relatively weak associations and small effects of risk genes on CPD, but it could be the case that the best phenotype was simply not assessed. For example, a recent meta-analysis found that *CHRNB3* is more strongly associated with nicotine dependence than with CPD [4]. Third, many of the studies are limited by retrospective reporting and subjective measures such as self-reported CPD. Even the measures of nicotine dependence suffer from a subjective component. More objective measures, such as nicotine or cotinine levels, are more highly correlated with SNPs, although not with all of the genes, as we shall see below. Lastly, the definition of "control" subjects varies widely among the studies. Some studies compare never-smokers to ever-smokers, while others compare smokers who are nicotine dependent to smokers who are not, and yet others compare light smokers (also called "chippers") to heavy smokers. In the following sections, we try to highlight the findings of each study as well as the pertinent design information, such as population composition and phenotype(s) examined. This information is also included in Table 17.1.

1.1 Nicotinic Acetylcholine Receptors

In this section we highlight the major associations between nAChRs themselves and nicotine-related diseases. By far, the strongest and most replicable evidence exists for a role of the *CHRNA5-A3-B4* gene cluster (Fig. 17.1). For additional information on the regulation of this gene cluster refer to Chap. 3. Two SNPs in particular, rs16969968 in *CHRNA5* and rs1051730 in *CHRNA3*, have been linked to

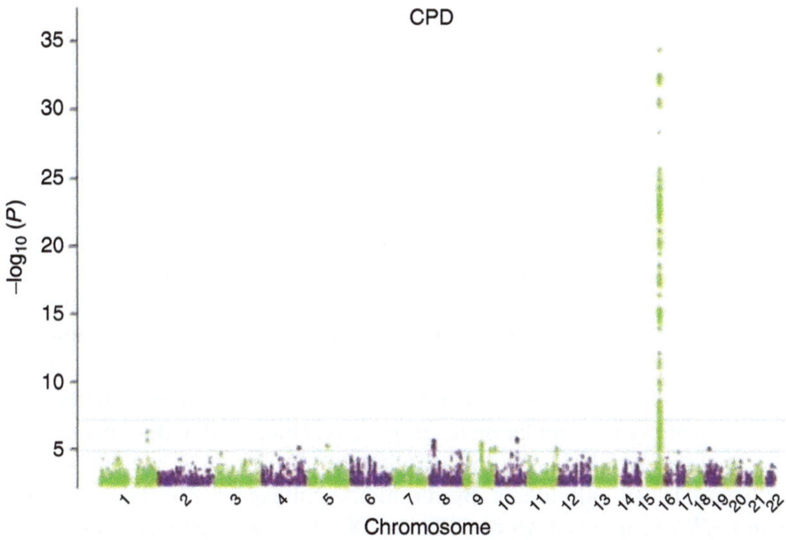

Fig. 17.1 Chromosome 15 contains the strongest genetic contribution to the risk of developing nicotine dependence. Genome-wide association results for cigarettes per day indicating significance of association of all SNPs in the TAG Consortium meta-analysis for cigarettes per day. Figure courtesy of TAG Consortium (2010). Figure used with permission from [72]

smoking-related behaviors in multiple studies. For authoritative reviews on the associations between CHRNA5 and CHRNA3 and nicotine, the reader is directed to [5].

The first evidence linking the *CHRNA5-A3-B4* gene cluster to nicotine addiction came from a candidate gene study [6] and a parallel GWAS [7] of 1,929 subjects of European descent. They used the FTND to assess dependence and set a threshold at 4, dividing their population into controls (i.e., non-dependent smokers) and nicotine-dependent cases. In the candidate gene study, Saccone and colleagues looked at SNPs across 300 candidate genes and found that two SNPs, one in the 3′-untranslated region of *CHRNA3* (rs578776) and a non-synonymous SNP in *CHRNA5* (rs16969968), were strongly associated with nicotine dependence. In fact, individuals with two copies of the A risk allele of *CHRNA5* were twice as likely to be nicotine dependent. SNP rs1051730, a synonymous SNP in *CHRNA3*, was also associated with nicotine dependence, although this SNP was in strong linkage disequilibrium with rs16969968, suggesting that it served as a surrogate mark for the SNP in *CHRNA5*. Interestingly, the *CHRNA5-A3-B4* cluster was not among the top hits in this first GWAS, although none of the SNPs identified remained significantly associated after correcting for multiple comparisons, suggesting that the study was underpowered. A year later, Thorgeirsson and colleagues found that the same SNP rs1051730 in *CHRNA3* was significantly associated with smoking quantity and nicotine dependence in a GWAS of 15,771 individuals of European descent [8]. Each copy of the T risk allele increased the amount smoked per day by one cigarette. Further, they found that frequency of the variant increased with addiction severity and carriers were less likely to quit smoking. Interestingly, frequency of the variant in low-quantity (i.e., non-nicotine dependent) smokers was significantly less than in nonsmokers. They also found an association between the variant and lung cancer and peripheral artery disease as well, but it is impossible to say whether or not this is due to increased smoking quantity or a direct risk. A separate GWAS reported that the same SNPs, rs1051730 in *CHRNA3* and rs16969968 in *CHRNA5*, were significantly associated with lung cancer in a GWAS of 11,707 individuals from Central Europe or countries with predominantly European background [9]. They found that the increased risk of lung cancer was seen for former and current smokers, as well as never-smokers, and that there is no association between these SNPs and other tobacco-associated cancers such as oral cavity, larynx, pharynx, and esophagus. In contrast, a more recent study of Asian women found no association between *CHRNA5* and lung cancer in never-smokers [10]. Highlighting the importance of controlling for population heterogeneity, Hung et al. found that this particular risk haplotype (rs1051730 and rs16969968) is rare in Asian individuals and absent from African individuals. Berrettini and colleagues also described the same SNPs in *CHRNA3* in a GWAS of CPD in 7,000 European subjects, suggesting that a common European haplotype confers risk for nicotine addiction [11]. In order to address the risk in multiple populations, a recent meta-analysis of European, Asian, and African populations found that only SNP rs16969968 in *CHRNA5* is associated with nicotine dependence in all populations, while other SNPs are population dependent [12]. This gene cluster has since been replicated in multiple associations of nicotine addiction (mostly in Europeans) that exceed our ability to include in this chapter, such as nicotine dependence [13, 14] and cotinine levels [3]. The reader is encouraged to look for the most comprehensive and up-to-date collection of genetic nicotine studies on PubMed itself, as the field is rapidly adding new findings.

Fewer studies have examined the *CHRNA5-A3-B4* gene cluster in relation to initiation and cessation, and with mixed results. In 2010, Thorgeirsson et al. again identified the same SNP rs1051730 in *CHRNA5-A3-B4* in a GWAS meta-analysis of CPD (31,266 subjects) and initiation (46,481 subjects) with replication in another 9,040 subjects, all of European descent [15]. The SNP was strongly linked to CPD, with each copy of the allele conferring an increase of 0.8 CPD, but it was not linked to initiation. A meta-analysis of 43 studies (33,348 individuals) found that the non-synonymous SNP rs16969968 in *CHRNA5* was associated with higher risk of heavy smoking in early-onset smokers, but that the allele itself was not responsible for the age of smoking onset [16]. On the other hand, a study of 200 individuals found that the SNP rs16969968 in *CHRNA5* was significantly associated with "enhanced pleasurable responses" to a person's first cigarette [17], as did a candidate gene study of early-onset nicotine addiction [18]. Another study found that SNP rs8040868 in *CHRNA5-A3-B4* is associated with more "externalizing" or risky behavior, including experimenting with drugs and nicotine [19]. The data for cessation seem to be more encouraging, as certain SNPs predict whether or not an individual will have success with a particular treatment. A community-based cross-sectional study (5,216 subjects) and randomized cessation trial (1,073 subjects) found that presence of SNP rs16969968 was associated with increased cessation failure when abstinence was the mode of cessation, but subjects with this SNP were three times more likely to quit if given pharmacotherapy [20]. The same findings were replicated with SNP rs1051730 [21]. Another study found that presence of the *CHRNA5* SNP was associated with reduced ability to quit during pregnancy [22]. An association between SNP rs871058 in *CHRNA5* and treatment response to bupropion was found in a placebo-controlled clinical trial [23], but this finding was not replicated [24]. The non-synonymous SNP rs8192475 (R37H) in *CHRNA3* has been associated with increased withdrawal symptoms and craving over time in a study of 276 smokers receiving both bupropion and transdermal nicotine [25]. The same study found that SNP rs680244 in *CHRNA5* predicted lack of abstinence at week 52.

It is interesting to note that fewer studies have linked *CHRNB4* itself with smoking phenotypes despite the preponderance evidence linking the *CHRNA5-A3-B4* gene cluster to nicotine-related diseases. While SNPs have been nominally associated with CPD [26], they failed to reach significance in meta-analysis [15, 27, 28]. Even in the less stringent pathway meta-analysis of CPD, SNPs in *CHRNB4* were only nominally significant in two of the three studies included, while it was highly significant in the third study [29]. However, variants located upstream of *CHRNB4* have been found to affect the age at which individuals transition to daily and habitual smoking [30], independently of other SNPs identified in the gene cluster and of other smoking-related phenotypes. Pooled sequencing of the coding regions and flanking sequence followed by comparison of rare missense variants at conserved sites of the *CHRNA5*, *CHRNA3*, *CHRNB4*, *CHRNA6*, and *CHRNB3* genes in African-American and European-American nicotine-dependent smokers and smokers without symptoms of dependence revealed that missense variants in *CHRNB4* (T375I and T91I) and in *CHRNA3* (R37H) are associated with lower risk for nicotine dependence and fewer CPD [31]. SNP rs12914008 in *CHRNB4* is also associated with less abstinence over time [25], suggesting that *CHRNB4* does play a role in some aspects of nicotine-related disease, but not CPD.

Like the *CHRNA5-A3-B4* gene cluster, *CHRNB3* and *CHRNA6* form a gene cluster on chromosome 8p11. The first evidence linking the *CHRNB3-A6* gene cluster to nicotine addiction derived from the same parallel candidate gene study and GWAS linking the *CHRNA5-A3-B4* gene cluster [6, 7]. Saccone and colleagues identified two SNPs in the coding region of *CHRNB3*, a synonymous SNP rs4593 and a non-synonymous SNP rs4952, as well as an SNP rs13277254. Thorgeirsson and colleagues found that the CHRNB3-A6 cluster was significantly linked to CPD but not to smoking initiation [15]. SNP rs6474412 in the 5′ region of *CHRNB3* and SNP rs13280604 reached genome-wide significance and were in linkage disequilibrium with the SNP rs13277254 identified by Saccone and Bierut in 2007. SNP rs6474412 is also part of a group of correlated SNPs including those identified in the coding regions of *CHRNB3* [6]. However, the effect size was particularly small, with the risk alleles conferring an increase of only 0.3 CPD [15]. As we discussed previously, a more recent study found that *CHRNB3* is linked to nicotine dependence, and not to CPD [4]. Indeed, the studies that first identified *CHRNB3-A6* examined nicotine dependence, not CPD [32, 33]. Nicotine dependence was not examined in the study by Thorgeirsson [15], and perhaps they would have found a larger effect size if they had examined that phenotype. Conversely, the sample size examined by Rice and colleagues was small (3,365, including both European- and African-Americans), perhaps explaining why they did not observe a significant association between *CHRNB3* and CPD [4].

While the most replicable genetic associations are for the two gene clusters we have already discussed, almost all of the nAChRs have been linking to some aspect of nicotine addiction. *CHRNA7*, *CHRAN10*, *CHRNA4*, *CHRNB1*, and a novel locus containing the *CHRND-CHRNG* were identified by Saccone and colleagues [34, 35]. Carriers of SNP rs2072661 in the 3′ UTR of *CHRNB2* have substantially decreased odds of quitting, due to decreased success with bupropion, shorter time to relapse, and more severe withdrawal symptoms [24].

1.2 Other Genes (Non-nicotinic)

Nicotinic receptors are not the only genes that have been associated with smoking or nicotine. In this section we briefly highlight the genes that comprise the two other broad categories: nicotine-metabolizing enzymes and neurotransmitter systems.

1.2.1 Metabolism

Nicotine is metabolized by a variety of enzymes, including the notoriously polymorphic cytochrome P450 enzymes, CYP2A6, CYP2A13, and CYP2B6 [36, 37], as well as the related flavin-containing monooxygenase FMO3 [38]. Similar to the case for many prescription drugs, variants in these enzymes affect the clearance of nicotine, especially CYP2A6 which is responsible for converting ~80 % of nicotine to cotinine [37]. Some mutations increase clearance of nicotine and these have been linked to more CPD, relapse, and craving [39, 40]. Other mutations decrease nicotine clearance, resulting in more sustained levels of nicotine and the major metabolite cotinine in the blood. This, in turn, leads to lower levels of nicotine consumption and perhaps less withdrawal [41].

1.2.2 Cholinergic System

nAChRs, which bind nicotine, are obvious players in nicotine addiction, serving as the point of action for the drug itself. However, in the absence of nicotine, these receptors bind the endogenous ligand acetylcholine. As nicotine would primarily affect the cholinergic system, researchers have focused their efforts to link other genes in this system to nicotine use and dependence. Choline acetyltransferase (ChAT) is the enzyme that synthesizes acetylcholine from acetyl co-A and choline. In two recent studies, SNPs in the gene encoding this enzyme were also linked to several measures of nicotine dependence, namely smoking quantity, heaviness of smoking, smoking cessation, and FTND. The initial study was a prospective study of 472 treatment-seeking smokers of European descent looking at smoking cessation success [42]. The second study extended and replicated the ChAT findings in the first study in a larger cohort (2,037) of both African-Americans and European-Americans [43]. While individual SNPs in the ChAT gene were only nominally associated with any measure of nicotine dependence, several SNP haplotypes (or combinations of four different SNPs) were significantly associated. This finding needs to be extended and replicated in a larger cohort. Using less stringent pathway analysis, most of the cholinergic genes, including the muscarinic acetylcholine receptor genes, have been associated with smoking quantity [29]. It is important to consider, however, that this is a relatively new method for linking pathways, or multiple gene families, to a particular disease and the findings must be followed up by more traditional GWAS.

1.2.3 Other Neurotransmitters

The cholinergic system modulates multiple other neurotransmitters in the brain fine-tuning activity. So it is not surprising that other neurotransmitter systems, or even that all of the major neurotransmitter systems, have also been linked directly or indirectly to cholinergic neurotransmission and nicotine addiction. For example, the metabotropic glutamate receptor genes *GRM7* and *GRM8* have been linked to both nicotine dependence [44] and smoking initiation [45]. *GRM7* was linked to depression in heavy smoking families [46]. These findings were replicated in a pathway analysis that found that SNP rs1555764 conferred protection in terms of CPD while SNPs rs963843 and rs1018854 in *GRM8* increased CPD [29]. Likewise, multiple SNPs in genes of the serotonergic system have been linked to CPD, HIS, and FTND in both African- and European-Americans [47]. SNPs in the genes that encode GABA receptors have also been linked to nicotine addiction [48–50], as have SNPs in *CB1*, which encodes the cannabinoid receptor [51].

The mu opioid receptor is well characterized for regulating addiction phenotypes [52], is required for nicotine reinforcement in rodents [53], and has been linked to nicotine initiation and dependence [54]. The non-synonymous SNP rs1799971 (A118G polymorphism) in *OPRM1*, the gene encoding the mu opioid receptor, leads to alterations in the rewarding effect of nicotine [55]. Several studies have looked at the effect of this allele on nicotine consumption and cessation. In a Taiwanese cohort, carriers of SNP rs1799971 had lower cotinine levels, suggesting that they found nicotine more rewarding and thus consumed less nicotine to achieve the same response as non-carriers [56]. PET imaging of male smokers found that indeed the presence of the A118G polymorphism led to more dopamine release in the striatum of male smokers [57] and higher binding potential and greater reward in the thalamus, cortex, and amygdala [58]. Presence of this SNP has also been linked with increased severity of withdrawal and relapse [59]. In women, however, the presence of the variant allele seems to lead to decreased rewarding effects of nicotine [60].

Given that nicotine is addictive, it seems fairly obvious that mutations in the dopamine (DA) receptor might be linked to nicotine dependence and/or cessation success. A large number of studies have examined the association of known mutations in *DRD2* that encodes one of the DA receptors, such as the reduced expression Taq1A1 allele (rs1800497) or the increased function A2 allele, with mixed results. A meta-analysis found no association with smoking behavior [61]; however, presence of the A1 allele generally predicts a better response to nicotine replacement therapy [62–64] while presence of the A2 allele predicts better response to bupropion [63, 65, 66].

Multiple SNPs in the *GALR1* gene of European- and African-American smokers have been associated with self-reported heavy smoking [67], and analysis of treatment-seeking smokers identified a highly significant association of a single *GALR1* SNP, rs2717162, with retrospective reports of tobacco craving during a previous quit attempt [68]. The same SNP was then examined in a several pharmacogenetic clinical trials that assessed smoking cessation and tobacco craving in 1,217 smokers of European ancestry who participated in one of the three smoking cessation clinical trials [63, 69, 70]. Data were pooled across trials and they found that there was a reduction in quitting success and more severe cravings in the presence of at least one minor (C) allele in the bupropion-treated group [71].

Finally, it is worth mentioning that even genes such as *TAS2R1*, which encodes the bitter taste receptor, and the olfactory receptor genes *OR10P1*, *OR52E2*, *OR52J3*, and *OR8D4* can contribute to nicotine dependence by modulating the amount of nicotine consumed, presumably by modifying the taste perception of nicotine [29].

2 Functional Consequences of Genetic Variants

Numerous genetic studies have linked SNPs in specific genes to behavioral differences related to smoking or difficulty quitting, highlighting how human genetic tools have aided in the identification of genetic variants contributing to the addiction cycle. Yet fewer studies have explored the functional consequences of allelic variants. In this context it is important to note that a genetic association characterizes only the first stage in understanding the underlying biology that leads to disease. A genetic association represents not only an association with tested genetic variants, but also an association with untested, highly correlated SNPs that can span across many genes on the same chromosome [72]. Thus once a genetic association is confirmed the challenge is to then understand which of these variants contribute to the biological mechanism underlying the correlation with a disease. Here we present some of the studies, which have employed such a reverse-like genetics strategy to elucidate whether specific SNPs cause variations in expression levels or functional changes to the receptor or the enzyme they encode. In this section we first review the data available on regulation of expression levels by noncoding SNPs, second, on point mutants that affect receptor/enzyme activity, and, third, we expand on in vivo studies that have addressed the functional consequences of these variants in laboratory animals.

2.1 Genetic Variants that Affect Expression Levels

Interestingly, the genes most strongly linked to nicotine dependence are arranged in gene clusters. These include the *CHRNA5-A3-B4*, in chromosome 15q25, the *CHRNA6-B3* chromosome 8p11, and the *CYP2A6- CYP2A7- CYP2B7P1- CYP2B6-CYP2A13* in chromosome 19q13 (Fig. 17.2). The clustering facilitates their coordinated expression by transcriptional or posttranscriptional co-regulation mechanisms.

The fact that a large number of SNPs map to noncoding segments of genes or gene cluster suggests that altered regulation of these genes can contribute to the pathophysiology of tobacco use. Indeed the risk for nicotine dependence seems to stem from at least two separate mechanisms: the variability in the mRNA levels of these genes and functional changes due to non-synonymous amino acid variants [72]. Several nAChR subunits have been found to be upregulated in smokers or differentially expressed in lung tumor cells [73, 74]. For instance the second most frequent genetic association with nicotine dependence in the *CHRNA5-A3-B4* locus is marked by rs880395 [75], and functional studies suggest that this SNP results in altered α5 nicotinic receptor mRNA expression [74, 76, 77]. Variants tagged by rs880395, which are more than 10 kb upstream of *CHRNA5* (Fig. 17.3), result in a 2.5- to 4-fold difference in α5 nicotinic receptor mRNA expression in the brain. High expression of *CHRNA5* mRNA is correlated with an increased risk of heavy smoking and nicotine dependence [77]. In support of this correlation it has been shown that the relative levels of α5 and β4 subunits strongly affect α3β4α5 nAChR currents [78]. Thus inclusion of more α5 competes with β4 and results in lower currents and increased nicotine consumption [78].

In general, gene expression can be modified either pre- or posttranscriptionally. Detailed studies on these mechanisms have been done for the *CHRNA5-A3-B4* but very few studies have analyzed the *CHRNA6-B3* [79]. The studies on the

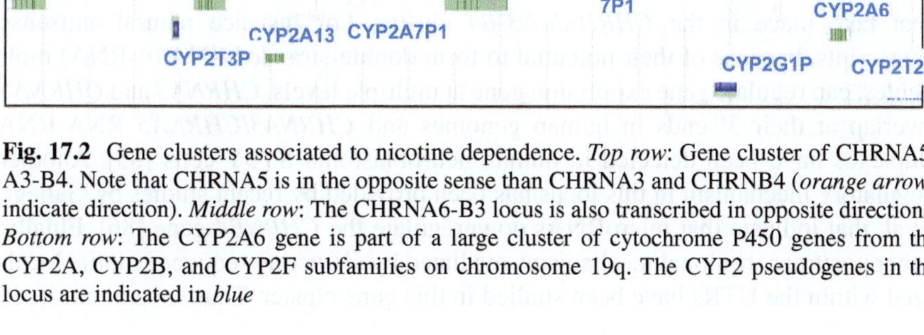

Fig. 17.2 Gene clusters associated to nicotine dependence. *Top row*: Gene cluster of CHRNA5-A3-B4. Note that CHRNA5 is in the opposite sense than CHRNA3 and CHRNB4 (*orange arrows* indicate direction). *Middle row*: The CHRNA6-B3 locus is also transcribed in opposite directions. *Bottom row*: The CYP2A6 gene is part of a large cluster of cytochrome P450 genes from the CYP2A, CYP2B, and CYP2F subfamilies on chromosome 19q. The CYP2 pseudogenes in the locus are indicated in *blue*

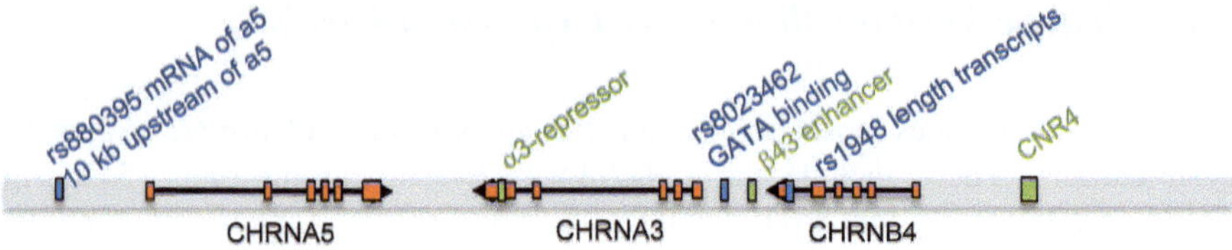

Fig. 17.3 Transcriptional and posttranscriptional regulation of the CHRNA5/A3/B4 gene cluster. Coding regions of the subunits are represented as *orange boxes* (exons) with *arrows* indicating the direction of transcription. Three transcriptional regulatory elements, the intronic α3 repressor in the fifth intron of α3, the 3′ enhancer in the β4 gene, and the conserved coding region (CNR4) are shown (*green boxes*). Allelic variations encoded by three SNPs located in noncoding regions (indicated in *blue*) have been associated to specific functional dysregulation of the expression of the gene cluster

transcriptional control of the CHRNA5/A3/B4 gene cluster indicate that this gene cluster is tightly co-regulated (for review see [80]). The *CHRNA5-A3-B4* genes are co-expressed in many cell types; thus their clustering reflects coordinate regulation. This hypothesis is supported by the fact that the transcriptional activities of the promoter regions of the three genes are regulated by many of the same transcription factors (reviewed in [80]). However, the *CHRNA5-A3-B4* genes are not always co-expressed, suggesting that independent regulation of each gene also occurs. The co-expression and clustering of the *CHRNA5-A3-B4* genes suggest that they may share common regulatory mechanisms in addition to specific regulation of each gene. This is supported by several observations including their lack of classical CAAT and TATA boxes [81]. Instead, the promoters are GC rich and contain several binding sites for the transcription factors, Sp1 and Sp3, which have been proposed to tether the basal transcription machinery to the TATA-less nAChR subunit gene promoters [82]. In addition it has been shown that three transcriptional regulatory elements, β43′, a conserved noncoding region 4 (CNR4), and α3 repressor (Fig. 17.3), play key roles in directing expression of the clustered nAChR genes in a tissue-specific manner with β43′ being important for expression in the adrenal gland and CNR4 being critical for expression in the pineal gland, superior cervical ganglion, and brain [83]. Studies of this intergenic region between *CHRNA3* and *CHRNB4* in the human sequences have shown that GATA transcription factors appeared to bind rs8023462 only when the minor/risk allele was present [84] (Fig. 17.3).

Additional studies have uncovered very specific posttranscriptional mechanisms that take place in the *CHRNA5-A3-B4* cluster. For instance natural antisense transcripts, because of their potential to form double-stranded RNA (dsRNA) molecules, can regulate gene expression gene at multiple levels. *CHRNA3* and *CHRNA5* overlap at their 3′ ends in human genomes and *CHRNA3*/*CHRNA5* RNA-RNA duplexes have been detected in human neuroblastoma SY5Y cells [85]. Another regulatory mechanism in this locus has been provided by recent studies by Gallego et al. that indicate that microRNAs downregulate the *CHRNB4* gene [86]. Finally, other posttranscriptional mechanisms mediated by cis-acting sequence motifs localized within the UTRs have been studied in this gene cluster. Thus several motifs in the 5′ UTR responsible for translational regulation in eukaryotic genes such as upstream open reading frames (uORFs) have been investigated as regulators of protein translation of the cluster [87].

2.2 Genetic Variants that Impact Receptor Activity

Some of the most prevalent SNPs associated with heavy smoking correspond to nucleotide substitutions in coding sequences that change a critical amino acid in the protein sequence. This is the case for the functional variants: D398N in the *CHRNA5* gene and A118G variant in the *OPRM1* gene.

The most common SNP in the *CHRNA5-A3-B4* gene cluster linked to higher risk to develop nicotine dependence is the D398N polymorphism [6, 7 and replicated in numerous studies]. It also corresponds to an Asn-to-Asp substitution. This change is in amino acid 398 of the α5 nAChR subunit. D398 (and the equivalent D397 in the mouse) is located in the amphipathic membrane-associated (MA) stretch (Fig. 17.4a) that is predicted to fold as a curved helix. When the α5 subunit assembles with other four subunits to form a functional α3β4α5 or α4β2α5 pentameric receptor, the α-helices help create an intracellular vestibule in the shape of an inverted cone (Fig. 17.2b) with five portals for the passage of Ca^{2+} and Na^{+} ions. Electrostatic mapping of the vestibule has shown that this is a highly charged domain of the receptor and that the substitution of D398 (Asp negatively charged) to N398 (Asn polar uncharged) at the more distal part of the vestibule has a very strong effect on receptor activity [78].

The D398N corresponds to a "loss-of-function" phenotype. Thus α4β2α5 nAChRs containing the α5 N398 variant exhibited reduced calcium permeability and agonist-evoked intracellular calcium response as well as enhanced short-term desensitization compared to α4β2α5 nAChRs [13, 88]. However, the effect of the D398N polymorphism on the function of α3b4a5 nAChRs has been controversial until recently. We found a similar loss-of-function effect of the D398N polymorphism on the function of α3β4α5 nAChRs [78].

However, two other reports indicated that the variant forms of the α5 subunit did not differentially affect the electrophysiological properties of α3β4α5 nAChRs [88, 89]. A recent study has solved this conflict, demonstrating that the α5 N398 variant indeed affects the function and pharmacology of α3β4α5 nAChRs by showing that at high extracellular Ca^{2+} concentrations the N398 variant exhibits a reduced response to agonists [90]. Thus inclusion of α5 decreases current amplitude of α3β4α5 nAChRs and D398N mutation decreases current amplitude even further [78, 90]. The fact that the 398N loss of function is only detectable at high Ca^{2+} or in the absence of Na^{+} [90] provides further support for the key role of this charged residue for the nAChR intracellular vestibule conductance. Indeed this same domain is a key location in determining the channel conductance of other nAChR, 5HT, and glycine receptors [91–93].

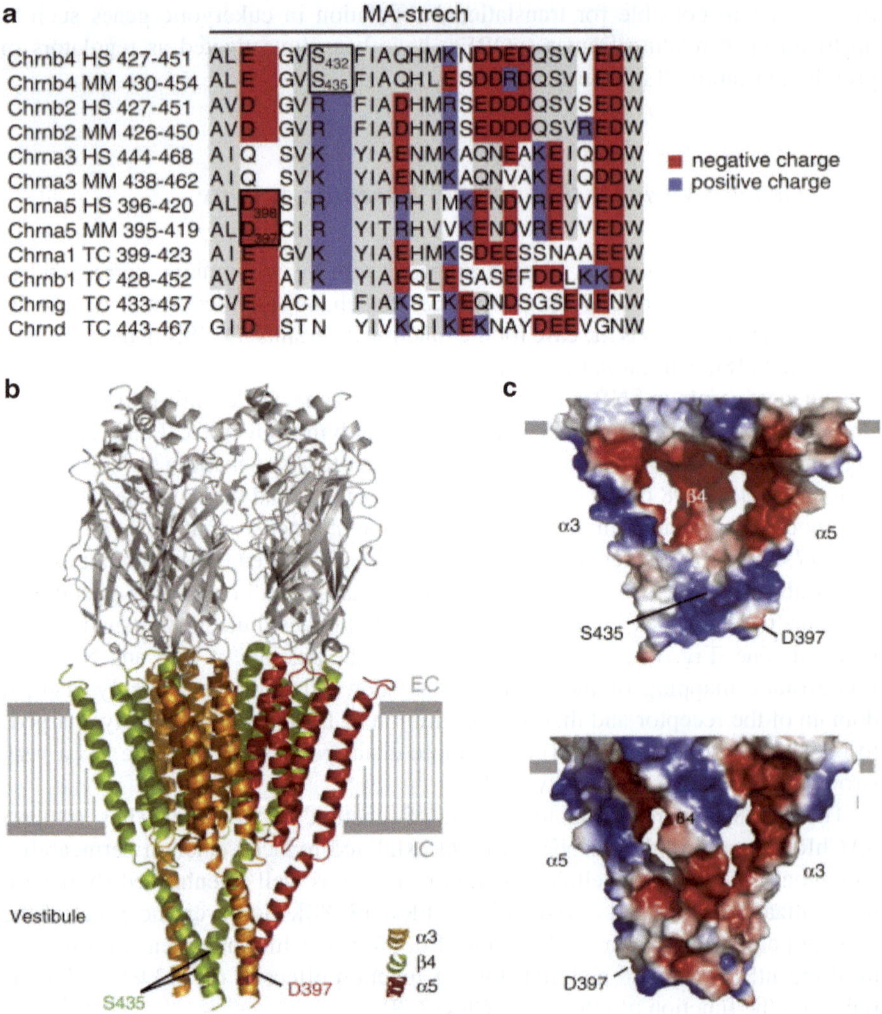

Fig. 17.4 Electrostatic mapping of the intracellular vestibule of the α3β4α5 nAChR complex. (**a**) Alignment of the human (HS), mouse (MM), and torpedo (TC) sequences spanning the MA stretch of the indicated nAChR subunits. The S435 residue in β4 and the D397 residue in α5 are indicated by a *black frame*. (**b**) Model of the 3D structure of α32β42α51 nAChR. Transmembrane domains of each of the subunits are colored as indicated. *EC* extracellular, *IC* intracellular. The S435 and D397 residues are located at the tip of the intracellular vestibule. (**c**) Electrostatic potential surface representations showing frontal (*top*) and top-down (*bottom*) views of the vestibule. Negative and positive charges are marked in *red* and *blue*, respectively. Figure used with permission from [78]

The A118G polymorphism is the most frequent SNP in *OPRM1* and corresponds to a nucleotide substitution at position 118 (A118G) that changes an Asn to Asp in amino acid residue 40 of the receptor (N40D). This Asn is a putative site for N-glycosylation important for protein folding and processing. The variant receptor (Asp40) encoded by the minor (G) allele was initially considered as a "gain-of-function" variant, on the basis of increased affinity of β-endorphins [94]. However subsequent studies suggested that the G allele was correlated with lower mRNA and protein expression levels [95, 96], consistent with studies of knock-in mice homozygous for the equivalent (112G) allele [97]. Recent in vivo investigations in smokers showed that the G allele displays significantly reduced mu opioid receptor-binding potential [58], providing additional support for a "loss-of-function" phenotype of the minor G allele.

3 Behavioral Consequences of Genetic Variants

Animal models, particularly rodent models, have allowed a thorough exploration of the role of each of the nAChR subunits in nicotine addiction [98]. Very few animal models, however, have focused on the effect that a particular SNP has on function at the behavioral level. In this section, we summarize the findings from the few studies that have started exploring the effect of SNPs associated with human nicotine dependence in animal behavioral models.

Contrary to the usual route in modern science, the role of the *CHRNA5-A3-B4* gene cluster was first discovered in human subjects, and only subsequently did the animal models begin to focus on the gene. Cumulative data from rodent models suggest that *CHRNA5* and *CHRNB4*, particularly in the medial habenula (MHb)-interpeduncular (IPN) circuit, mediate the aversive properties of nicotine [78, 99, 100]. For example, animals lacking the α5 nAChR subunit self-administer more nicotine [99] and do not display somatic signs of nicotine withdrawal [100]. Restoration of *CHRNA5* to the MHb leads to normalization of nicotine consumption; that is, the mice reach a plateau in administration, suggesting that it is the aversive aspect that limits intake [99]. Overexpression of the β4 nAChR subunit in the Tabac mouse model leads to increased sensitivity to the aversive properties of nicotine and decreased consumption [78]. Overexpression of the entire human gene cluster in mice likewise leads to increased sensitivity to nicotine with higher activation of the MHb and reduced activation of the VTA [101]. Together these studies suggest that it is the balance of receptor subtypes in the MHb-IPN that limit nicotine intake, with α5 limiting the activity of MHb β4* nAChRs, which are responsible for mediating aversion to nicotine.

Data from rodent models also suggest that α5 plays a role in mediating the rewarding effects of nicotine. Striatal dopamine (DA) transmission is critical to the acquisition and maintenance of drug addiction and is modulated strongly by nicotine acting at heteromeric β2-containing (β2*) nAChRs. DA transmission depends critically upon α4α5β2 nAChRs in the caudate putamen of the dorsal striatum, a region associated with habitual and instrumental responses, but not in the nucleus accumbens, where α4α6β2β3-nAChRs are required [102]. This DA is released from neurons in the ventral tegmental area (VTA) that project to the striatum. Within the VTA, nicotine acts through α4β2* nAChRs on both DA and GABAergic cells, where concerted activity of DA and GABA systems is necessary for the reinforcing actions of nicotine through burst firing of DA neurons [103]. On DA neurons in the

VTA, presence of the α5 subunit increases expression of the α4 subunit by 60 % and strengthens baseline nAChR currents, suggesting that α5 increases expression of α4* nAChRs on the cell surface. Presence of the α5 subunit also blunts the desensitization of nAChRs following nicotine exposure, suggesting that this subunit modulates the rewarding response of nicotine [104].

Only one study to date has examined how SNP rs16969968 in *CHRNA5* affects nicotine intake in mice. Expression of the D398N variant (D397N in mice) in the MHb of Tabac mice is sufficient to abolish the aversion exhibited by these mice to moderate doses of nicotine [78]. This suggests that the D398N variant is more effective at limiting β4-mediated nicotine aversion than the wild-type allele. Perhaps this explains the increase in nicotine consumption seen in smokers with the D398N variant. Unpublished data presented at the annual Society for Neuroscience conference demonstrated that the effects of this SNP in a knock-in mouse model contributed to higher nicotine consumption, as well as lower NAc DA levels, higher DA turnover, higher VTA 5-HT levels, and lower 5-HT turnover [105] (Fig. 17.5).

Though not a mouse model, a human resting-state functional-MRI imaging study found that presence of the risk allele decreases the connectivity strength of the dor-

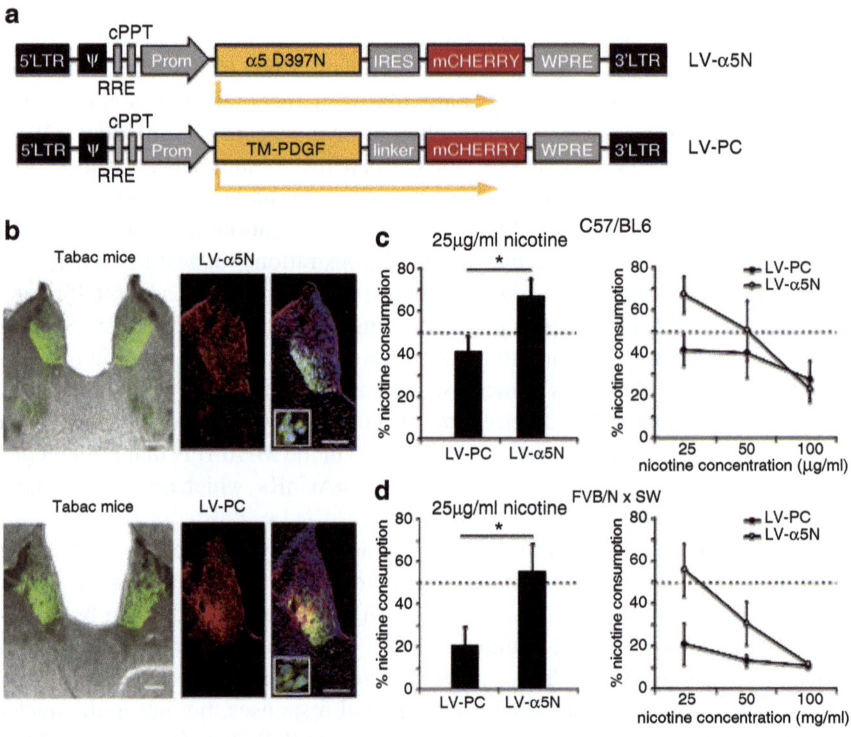

Fig. 17.5 Reversal of nicotine aversion in Tabac mice by lentiviral mediated expression of the a5 D397N variant in the MHb. (**a**) Schematic representation of the lentiviral constructs used. (**b**) Coronal brain section of Tabac mice stereotactically injected in the MHb with the indicated virus. (**c, d**) Two-bottle choice nicotine consumption in Tabac mice after bilateral MHb injection. Figure used with permission from [78]

sal anterior cingulate–ventral striatum/extended amygdala circuit. Importantly, this effect is observed independently in nonsmokers and smokers, although decreased circuit strength distinguishes smokers from nonsmokers and predicts addiction severity in smokers [106].

4 Future and Further Perspectives

It is important to remember that nicotinic receptor function is but a single factor in the complex interaction of genes and environment that contributes to nicotine-related disease susceptibility. One has to smoke or otherwise consume nicotine to develop most of the phenotypes discussed below, and without this crucial exposure there would be no disease. However, many people do smoke, and many try to quit, often with great difficulty. Understanding the cumulative risk of each of these contributors is the first step toward truly personalized medicine. We are on the brink of having genomic sequencing available to the masses. This ability to evaluate the patient genome quickly can then educate us toward cessation therapies that would be most effective for a given "risk genotype." It is also important to remember that in the case of lung cancer, there is a link with the *CHRNA5* independent of smoking, but that does not mean that it is not a gene-environment interaction. Many of the pesticides in use today are derivatives of the original natural pesticide, nicotine. It might be that we are unknowingly exposing ourselves to nicotine in other forms, contributing to cancer development through nAChRs by chronic pesticide exposure.

We must also continue to develop animal models of the variants that are associated with nicotine dependence. While knockout mouse models have been informative for understanding the function of nAChRs, we must now knock-in the SNPs that have been most frequently implicated, such as *CHRNA5*. Future efforts are also needed to expand sample size and diversity in the GWAS association studies, in order to refine the risk contributed by each of the genes that have already been identified, and to understand how the genetic and ethnic background contributes to nicotine use and abuse and the development of cancer. The reader is referred to Chap. 16 for further discussions on gene linkage and smoking.

Acknowledgements The authors would like to thank Samin Shehab for his invaluable assistance in assembling the reference material in Table 17.1.

References

1. Lessov CN, et al. Defining nicotine dependence for genetic research: evidence from Australian twins. Psychol Med. 2004;34(5):865–79.
2. Chen LS, et al. Dissection of the phenotypic and genotypic associations with nicotinic dependence. Nicotine Tob Res. 2012;14(4):425–33.

3. Keskitalo K, et al. Association of serum cotinine level with a cluster of three nicotinic acetylcholine receptor genes (CHRNA3/CHRNA5/CHRNB4) on chromosome 15. Hum Mol Genet. 2009;18(20):4007–12.
4. Rice JP, et al. CHRNB3 is more strongly associated with Fagerstrom test for cigarette dependence-based nicotine dependence than cigarettes per day: phenotype definition changes genome-wide association studies results. Addiction. 2012;107(11):2019–28.
5. Ware JJ, van den Bree M, Munafo MR. From men to mice: CHRNA5/CHRNA3, smoking behavior and disease. Nicotine Tob Res. 2012;14(11):1291–9.
6. Saccone SF, et al. Cholinergic nicotinic receptor genes implicated in a nicotine dependence association study targeting 348 candidate genes with 3713 SNPs. Hum Mol Genet. 2007;16(1):36–49.
7. Bierut LJ, et al. Novel genes identified in a high-density genome wide association study for nicotine dependence. Hum Mol Genet. 2007;16(1):24–35.
8. Thorgeirsson TE, et al. A variant associated with nicotine dependence, lung cancer and peripheral arterial disease. Nature. 2008;452(7187):638–42.
9. Hung RJ, et al. A susceptibility locus for lung cancer maps to nicotinic acetylcholine receptor subunit genes on 15q25. Nature. 2008;452(7187):633–7.
10. Lan Q, et al. Genome-wide association analysis identifies new lung cancer susceptibility loci in never-smoking women in Asia. Nat Genet. 2012;44(12):1330–5.
11. Berrettini W, et al. Alpha-5/alpha-3 nicotinic receptor subunit alleles increase risk for heavy smoking. Mol Psychiatry. 2008;13(4):368–73.
12. Chen LS, et al. Smoking and genetic risk variation across populations of European, Asian, and African American ancestry – a meta-analysis of chromosome 15q25. Genet Epidemiol. 2012;36(4):340–51.
13. Bierut LJ, et al. Variants in nicotinic receptors and risk for nicotine dependence. Am J Psychiatry. 2008;165(9):1163–71.
14. Caporaso N, et al. Genome-wide and candidate gene association study of cigarette smoking behaviors. PLoS One. 2009;4(2):e4653.
15. Thorgeirsson TE, et al. Sequence variants at CHRNB3-CHRNA6 and CYP2A6 affect smoking behavior. Nat Genet. 2010;42(5):448–53.
16. Hartz SM, et al. Increased genetic vulnerability to smoking at CHRNA5 in early-onset smokers. Arch Gen Psychiatry. 2012;69(8):854–60.
17. Sherva R, et al. Association of a single nucleotide polymorphism in neuronal acetylcholine receptor subunit alpha 5 (CHRNA5) with smoking status and with 'pleasurable buzz' during early experimentation with smoking. Addiction. 2008;103(9):1544–52.
18. Weiss RB, et al. A candidate gene approach identifies the CHRNA5-A3-B4 region as a risk factor for age-dependent nicotine addiction. PLoS Genet. 2008;4(7):e1000125.
19. Stephens SH, et al. Externalizing behaviors are associated with SNPs in the CHRNA5/CHRNA3/CHRNB4 gene cluster. Behav Genet. 2012;42(3):402–14.
20. Chen LS, et al. Interplay of genetic risk factors (CHRNA5-CHRNA3-CHRNB4) and cessation treatments in smoking cessation success. Am J Psychiatry. 2012;169(7):735–42.
21. Bergen AW, et al. Nicotinic acetylcholine receptor variation and response to smoking cessation therapies. Pharmacogenet Genomics. 2013;23(2):94–103.
22. Freathy RM, et al. A common genetic variant in the 15q24 nicotinic acetylcholine receptor gene cluster (CHRNA5-CHRNA3-CHRNB4) is associated with a reduced ability of women to quit smoking in pregnancy. Hum Mol Genet. 2009;18(15):2922–7.
23. Heitjan DF, et al. Identification of pharmacogenetic markers in smoking cessation therapy. Am J Med Genet B Neuropsychiatr Genet. 2008;147B(6):712–9.
24. Conti DV, et al. Nicotinic acetylcholine receptor beta2 subunit gene implicated in a systems-based candidate gene study of smoking cessation. Hum Mol Genet. 2008;17(18):2834–48.
25. Sarginson JE, et al. Markers in the 15q24 nicotinic receptor subunit gene cluster (CHRNA5-A3-B4) predict severity of nicotine addiction and response to smoking cessation therapy. Am J Med Genet B Neuropsychiatr Genet. 2011;156B(3):275–84.

26. Saccone NL, et al. The CHRNA5-CHRNA3-CHRNB4 nicotinic receptor subunit gene cluster affects risk for nicotine dependence in African-Americans and in European-Americans. Cancer Res. 2009;69(17):6848–56.
27. Liu JZ, et al. Meta-analysis and imputation refines the association of 15q25 with smoking quantity. Nat Genet. 2010;42(5):436–40.
28. Tobacco and Genetics Consortium. Genome-wide meta-analyses identify multiple loci associated with smoking behavior. Nat Genet. 2010;42(5):441–7.
29. Harari O, et al. Pathway analysis of smoking quantity in multiple GWAS identifies cholinergic and sensory pathways. PLoS One. 2012;7(12):e50913.
30. Kapoor M, et al. Variants located upstream of CHRNB4 on chromosome 15q25.1 are associated with age at onset of daily smoking and habitual smoking. PLoS One. 2012;7(3):e33513.
31. Haller G, et al. Rare missense variants in CHRNB4 are associated with reduced risk of nicotine dependence. Hum Mol Genet. 2012;21(3):647–55.
32. Zeiger JS, et al. The neuronal nicotinic receptor subunit genes (CHRNA6 and CHRNB3) are associated with subjective responses to tobacco. Hum Mol Genet. 2008;17(5):724–34.
33. Hoft NR, et al. Genetic association of the CHRNA6 and CHRNB3 genes with tobacco dependence in a nationally representative sample. Neuropsychopharmacology. 2009;34(3): 698–706.
34. Saccone NL, et al. Multiple distinct risk loci for nicotine dependence identified by dense coverage of the complete family of nicotinic receptor subunit (CHRN) genes. Am J Med Genet B Neuropsychiatr Genet. 2009;150B(4):453–66.
35. Saccone NL, et al. Multiple cholinergic nicotinic receptor genes affect nicotine dependence risk in African and European Americans. Genes Brain Behav. 2010;9(7):741–50.
36. Bloom J, et al. The contribution of common CYP2A6 alleles to variation in nicotine metabolism among European-Americans. Pharmacogenet Genomics. 2011;21(7):403–16.
37. Raunio H, Rahnasto-Rilla M. CYP2A6: genetics, structure, regulation, and function. Drug Metabol Drug Interact. 2012;27(2):73–88.
38. Bloom AJ, et al. Effects upon in-vivo nicotine metabolism reveal functional variation in FMO3 associated with cigarette consumption. Pharmacogenet Genomics. 2013;23(2):62–8.
39. Lerman C, et al. Nicotine metabolite ratio predicts efficacy of transdermal nicotine for smoking cessation. Clin Pharmacol Ther. 2006;79(6):600–8.
40. Malaiyandi V, Sellers EM, Tyndale RF. Implications of CYP2A6 genetic variation for smoking behaviors and nicotine dependence. Clin Pharmacol Ther. 2005;77(3):145–58.
41. Benowitz NL. Pharmacology of nicotine: addiction, smoking-induced disease, and therapeutics. Annu Rev Pharmacol Toxicol. 2009;49:57–71.
42. Ray R, et al. Convergent evidence that choline acetyltransferase gene variation is associated with prospective smoking cessation and nicotine dependence. Neuropsychopharmacology. 2010;35(6):1374–82.
43. Wei J, et al. Replication and extension of association of choline acetyltransferase with nicotine dependence in European and African American smokers. Hum Genet. 2010;127(6): 691–8.
44. Uhl GR, et al. Molecular genetics of nicotine dependence and abstinence: whole genome association using 520,000 SNPs. BMC Genet. 2007;8:10.
45. Vink JM, et al. Genome-wide association study of smoking initiation and current smoking. Am J Hum Genet. 2009;84(3):367–79.
46. Pergadia ML, et al. A 3p26-3p25 genetic linkage finding for DSM-IV major depression in heavy smoking families. Am J Psychiatry. 2011;168(8):848–52.
47. Yang Z, et al. Serotonin transporter and receptor genes significantly impact nicotine dependence through genetic interactions in both European American and African American smokers. Drug Alcohol Depend. 2013;129(3):217–25.
48. Cui WY, et al. Genetics of GABAergic signaling in nicotine and alcohol dependence. Hum Genet. 2012;131(6):843–55.
49. Han S, et al. Meta-analysis of 15 genome-wide linkage scans of smoking behavior. Biol Psychiatry. 2010;67(1):12–9.

50. Li MD, et al. Association and interaction analyses of GABBR1 and GABBR2 with nicotine dependence in European- and African-American populations. PLoS One. 2009;4(9):e7055.
51. Chen X, et al. Cannabinoid receptor 1 gene association with nicotine dependence. Arch Gen Psychiatry. 2008;65(7):816–24.
52. Le Merrer J, et al. Reward processing by the opioid system in the brain. Physiol Rev. 2009;89(4):1379–412.
53. Liu X, Jernigan C. Activation of the opioid mu1, but not delta or kappa, receptors is required for nicotine reinforcement in a rat model of drug self-administration. Prog Neuropsychopharmacol Biol Psychiatry. 2011;35(1):146–53.
54. Zhang L, Kendler KS, Chen X. The mu-opioid receptor gene and smoking initiation and nicotine dependence. Behav Brain Funct. 2006;2:28.
55. Verhagen M, Kleinjan M, Engels RC. A systematic review of the A118G (Asn40Asp) variant of OPRM1 in relation to smoking initiation, nicotine dependence and smoking cessation. Pharmacogenomics. 2012;13(8):917–33.
56. Chen YT, et al. OPRM1 genetic polymorphisms are associated with the plasma nicotine metabolite cotinine concentration in methadone maintenance patients: a cross sectional study. J Hum Genet. 2013;58(2):84–90.
57. Domino EF, et al. Tobacco smoking produces greater striatal dopamine release in G-allele carriers with mu opioid receptor A118G polymorphism. Prog Neuropsychopharmacol Biol Psychiatry. 2012;38(2):236–40.
58. Ray R, et al. Human Mu Opioid Receptor (OPRM1 A118G) polymorphism is associated with brain mu-opioid receptor binding potential in smokers. Proc Natl Acad Sci U S A. 2011;108(22):9268–73.
59. Hardin J, et al. Nicotine withdrawal sensitivity, linkage to chr6q26, and association of OPRM1 SNPs in the SMOking in FAMilies (SMOFAM) sample. Cancer Epidemiol Biomarkers Prev. 2009;18(12):3399–406.
60. Ray R, et al. Association of OPRM1 A118G variant with the relative reinforcing value of nicotine. Psychopharmacology (Berl). 2006;188(3):355–63.
61. Munafo MR, et al. Association of the DRD2 gene Taq1A polymorphism and smoking behavior: a meta-analysis and new data. Nicotine Tob Res. 2009;11(1):64–76.
62. Johnstone EC, et al. Genetic variation in dopaminergic pathways and short-term effectiveness of the nicotine patch. Pharmacogenetics. 2004;14(2):83–90.
63. Lerman C, et al. Role of functional genetic variation in the dopamine D2 receptor (DRD2) in response to bupropion and nicotine replacement therapy for tobacco dependence: results of two randomized clinical trials. Neuropsychopharmacology. 2006;31(1):231–42.
64. Yudkin P, et al. Effectiveness of nicotine patches in relation to genotype in women versus men: randomised controlled trial. BMJ. 2004;328(7446):989–90.
65. David SP, et al. Bupropion efficacy for smoking cessation is influenced by the DRD2 Taq1A polymorphism: analysis of pooled data from two clinical trials. Nicotine Tob Res. 2007;9(12):1251–7.
66. Swan GE, et al. Dopamine receptor DRD2 genotype and smoking cessation outcome following treatment with bupropion SR. Pharmacogenomics J. 2005;5(1):21–9.
67. Jackson KJ, et al. The neuropeptide galanin and variants in the GalR1 gene are associated with nicotine dependence. Neuropsychopharmacology. 2011;36(11):2339–48.
68. Lori A, et al. The galanin receptor 1 gene associates with tobacco craving in smokers seeking cessation treatment. Neuropsychopharmacology. 2011;36(7):1412–20.
69. Lerman C, et al. Individualizing nicotine replacement therapy for the treatment of tobacco dependence: a randomized trial. Ann Intern Med. 2004;140(6):426–33.
70. Schnoll RA, et al. Effectiveness of extended-duration transdermal nicotine therapy: a randomized trial. Ann Intern Med. 2010;152(3):144–51.
71. Gold AB, et al. Pharmacogenetic association of the galanin receptor (GALR1) SNP rs2717162 with smoking cessation. Neuropsychopharmacology. 2012;37(7):1683–8.
72. Bierut LJ. Genetic vulnerability and susceptibility to substance dependence. Neuron. 2011;69(4):618–27.

73. Corringer PJ, Sallette J, Changeux JP. Nicotine enhances intracellular nicotinic receptor maturation: a novel mechanism of neural plasticity? J Physiol Paris. 2006;99(2–3):162–71.
74. Falvella FS, et al. Transcription deregulation at the 15q25 locus in association with lung adenocarcinoma risk. Clin Cancer Res. 2009;15(5):1837–42.
75. Saccone NL, et al. Multiple independent loci at chromosome 15q25.1 affect smoking quantity: a meta-analysis and comparison with lung cancer and COPD. PLoS Genet. 2010;6(8): e1001053.
76. Smith RM, et al. Nicotinic alpha5 receptor subunit mRNA expression is associated with distant 5′ upstream polymorphisms. Eur J Hum Genet. 2011;19(1):76–83.
77. Wang JC, et al. Risk for nicotine dependence and lung cancer is conferred by mRNA expression levels and amino acid change in CHRNA5. Hum Mol Genet. 2009;18(16):3125–35.
78. Frahm S, et al. Aversion to nicotine is regulated by the balanced activity of beta4 and alpha5 nicotinic receptor subunits in the medial habenula. Neuron. 2011;70(3):522–35.
79. Ebihara M, et al. Genomic organization and promoter analysis of the human nicotinic acetylcholine receptor alpha6 subunit (CHNRA6) gene: Alu and other elements direct transcriptional repression. Gene. 2002;298(1):101–8.
80. Improgo MR, et al. The nicotinic acetylcholine receptor CHRNA5/A3/B4 gene cluster: dual role in nicotine addiction and lung cancer. Prog Neurobiol. 2010;92(2):212–26.
81. Boulter J, et al. Alpha 3, alpha 5, and beta 4: three members of the rat neuronal nicotinic acetylcholine receptor-related gene family form a gene cluster. J Biol Chem. 1990;265(8): 4472–82.
82. Pugh BF, Tjian R. Transcription from a TATA-less promoter requires a multisubunit TFIID complex. Genes Dev. 1991;5(11):1935–45.
83. Xu X, Scott MM, Deneris ES. Shared long-range regulatory elements coordinate expression of a gene cluster encoding nicotinic receptor heteromeric subtypes. Mol Cell Biol. 2006;26(15):5636–49.
84. Flora AV, et al. Functional characterization of SNPs in CHRNA3/B4 intergenic region associated with drug behaviors. Brain Res. 2013;1529:1–15.
85. Solda G, et al. In vivo RNA-RNA duplexes from human alpha3 and alpha5 nicotinic receptor subunit mRNAs. Gene. 2005;345(2):155–64.
86. Gallego X, et al. Alternative CHRNB4 3′-UTRs mediate the allelic effects of SNP rs1948 on gene expression. PLoS One. 2013;8(5):e63699.
87. Eggert M, et al. Nicotinic acetylcholine receptor subunits alpha4 and alpha5 associated with smoking behaviour and lung cancer are regulated by upstream open reading frames. PLoS One. 2013;8(7):e66157.
88. Kuryatov A, Berrettini W, Lindstrom J. Acetylcholine receptor (AChR) alpha5 subunit variant associated with risk for nicotine dependence and lung cancer reduces (alpha4beta2)(2) alpha5 AChR function. Mol Pharmacol. 2011;79(1):119–25.
89. Li P, et al. Functional characterization of the alpha5(Asn398) variant associated with risk for nicotine dependence in the alpha3beta4alpha5 nicotinic receptor. Mol Pharmacol. 2011; 80(5):818–27.
90. Tammimaki A, et al. Impact of human D398N single nucleotide polymorphism on intracellular calcium response mediated by alpha3beta4alpha5 nicotinic acetylcholine receptors. Neuropharmacology. 2012;63(6):1002–11.
91. Peters JA, et al. Novel structural determinants of single channel conductance and ion selectivity in 5-hydroxytryptamine type 3 and nicotinic acetylcholine receptors. J Physiol. 2010; 588(Pt 4):587–96.
92. Carland JE, et al. Characterization of the effects of charged residues in the intracellular loop on ion permeation in alpha1 glycine receptor channels. J Biol Chem. 2009;284(4):2023–30.
93. Kelley SP, et al. A cytoplasmic region determines single-channel conductance in 5-HT3 receptors. Nature. 2003;424(6946):321–4.
94. Bond C, et al. Single-nucleotide polymorphism in the human mu opioid receptor gene alters beta-endorphin binding and activity: possible implications for opiate addiction. Proc Natl Acad Sci U S A. 1998;95(16):9608–13.

95. Beyer A, et al. Effect of the A118G polymorphism on binding affinity, potency and agonist-mediated endocytosis, desensitization, and resensitization of the human mu-opioid receptor. J Neurochem. 2004;89(3):553–60.
96. Zhang Y, et al. Allelic expression imbalance of human mu opioid receptor (OPRM1) caused by variant A118G. J Biol Chem. 2005;280(38):32618–24.
97. Mague SD, et al. Mouse model of OPRM1 (A118G) polymorphism has sex-specific effects on drug-mediated behavior. Proc Natl Acad Sci U S A. 2009;106(26):10847–52.
98. Changeux JP. Nicotine addiction and nicotinic receptors: lessons from genetically modified mice. Nat Rev Neurosci. 2010;11(6):389–401.
99. Fowler CD, et al. Habenular alpha5 nicotinic receptor subunit signalling controls nicotine intake. Nature. 2011;471(7340):597–601.
100. Salas R, et al. Nicotinic receptors in the habenulo-interpeduncular system are necessary for nicotine withdrawal in mice. J Neurosci. 2009;29(10):3014–8.
101. Gallego X, et al. Overexpression of the CHRNA5/A3/B4 genomic cluster in mice increases the sensitivity to nicotine and modifies its reinforcing effects. Amino Acids. 2012;43(2):897–909.
102. Exley R, et al. Striatal alpha5 nicotinic receptor subunit regulates dopamine transmission in dorsal striatum. J Neurosci. 2012;32(7):2352–6.
103. Tolu S, et al. Co-activation of VTA DA and GABA neurons mediates nicotine reinforcement. Mol Psychiatry. 2013;18(3):382–93.
104. Chatterjee S, et al. The alpha5 subunit regulates the expression and function of alpha4*-containing neuronal nicotinic acetylcholine receptors in the ventral-tegmental area. PLoS One. 2013;8(7):e68300.
105. Zalud AW, Helfand RS, Tammimaki AE, Nguyen V, Stitzel JA. The nicotinic receptor α5 subunit polymorphism (CHRNA5 D398N) increases nicotine consumption and mitigates withdrawal-related disruptions in mesolimbic neurochemistry. 2012. 532.11/B38.
106. Hong LE, et al. A genetically modulated, intrinsic cingulate circuit supports human nicotine addiction. Proc Natl Acad Sci U S A. 2010;107(30):13509–14.
107. Lessov-Schlaggar CN, et al. Heritability of cigarette smoking and alcohol use in Chinese male twins: the Qingdao twin registry. Int J Epidemiol. 2006;35(5):1278–85.
108. Li MD. The genetics of smoking related behavior: a brief review. Am J Med Sci. 2003;326(4): 168–73.
109. Hall W, Madden P, Lynskey M. The genetics of tobacco use: methods, findings and policy implications. Tob Control. 2002;11(2):119–24.
110. Maes HH, et al. A twin study of genetic and environmental influences on tobacco initiation, regular tobacco use and nicotine dependence. Psychol Med. 2004;34(7):1251–61.

Chapter 18
Nicotinic Acetylcholine Receptors Along the Habenulo-Interpeduncular Pathway: Roles in Nicotine Withdrawal and Other Aversive Aspects

Dang Q. Dao, Ramiro Salas, and Mariella De Biasi

Abstract Addiction to tobacco smoking is a deadly disease that consumes millions of lives each year. However, the neurobiology underlying the disease remains an enigma. One reason for this is the relative complexity of nicotine's effects on the brain, with a multitude of targets throughout many different brain regions, each subserving individual components of the disease. Still, a handful of brain circuits mediate particularly significant roles in the disease. The epithalamic habenulo-interpeduncular (Hb-IPN) pathway participates in the aversive aspects of nicotine dependence, including the aversive experience of nicotine withdrawal. Many hypotheses regarding the exact mechanisms for these behavioral roles exist, but the convergent feature of those hypotheses is that nicotine acts at populations of nicotinic acetylcholine receptors (nAChRs) across the brain, including the Hb-IPN pathway. Of note, the Hb-IPN pathway is one of the brain regions with the highest density of nAChRs, including both heteromeric (e.g., α3β4 and α4β2) and homomeric (i.e., α7) receptors. As nAChR subtypes that subserve multiple aspects of affective and reinforcement behaviors are expressed along this pathway, it is of no surprise that the Hb-IPN pathway participates in similar affective behaviors. This chapter will discuss the roles of nAChRs along the Hb-IPN in aversive nicotine-associated behaviors, as well as touch upon the innate roles of those populations of nAChRs over biology and behavior in healthy animals.

D.Q. Dao, Ph.D.
Department of Neuroscience, Baylor College of Medicine,
One Baylor Plaza, Houston, TX 77030, USA
e-mail: ddao@alumni.bcm.edu

R. Salas, Ph.D.
Department of Psychiatry, Baylor College of Medicine,
One Baylor Plaza, Houston, TX 77030, USA
e-mail: rsalas@bcm.edu

M. De Biasi, Ph.D. (✉)
Department of Psychiatry, Perelman School of Medicine, University of Pennsylvania,
415 Curie Boulevard, Clinical Research Building, Philadelphia, PA 19104, USA
e-mail: marielde@upenn.edu; marielde@mail.med.upenn.edu

R.A.J. Lester (ed.), *Nicotinic Receptors*, The Receptors 26,
DOI 10.1007/978-1-4939-1167-7_18

Keywords Nicotinic acetylcholine receptor • α5 subunit • β4 subunit • Medial habenula • Interpeduncular nucleus • Nicotine • Aversion • Negative reinforcement • Nicotine withdrawal

1 Introduction

Tobacco smoking is the leading cause of preventable death in the world, with estimates of four to five million annual deaths worldwide [1–3]. Containing over 60 identified carcinogenic compounds [4], tobacco smoke is highly carcinogenic, as roughly one third of all cancer-related mortalities in developed countries can be attributed to tobacco use [1]. The fact that rates of tobacco use in developing countries remain high, despite costly antitobacco campaigns, speaks to the global pervasiveness of this health threat [5, 6]. As a consequence, there is strong need to develop therapies to aid smoking cessation by addressing dependence to nicotine, the primary addictive component of tobacco smoke [3, 7–9].

The first and arguably greatest barrier to successful smoking cessation is the collection of withdrawal symptoms that emerges soon after an attempt to quit [10–14]. Comprising both physical and affective symptoms, nicotine withdrawal can be a considerably unpleasant experience, with an onset as early as a few hours following the suspension of nicotine consumption. Successful strategies to develop new treatments for nicotine dependence should incorporate the existing knowledge of nicotine's effects over the neuronal pathways and molecular mechanisms that underlie this disease.

Fortunately, inroads toward understanding the neurobiology of nicotine dependence have been made on many fronts [9, 15, 16]. A significant body of knowledge has already been obtained regarding the biophysical, pharmacological, and cellular properties of nicotinic acetylcholine receptors (nAChRs), which constitute the primary molecular targets of nicotine in the body [7–9, 17]. Progress is also being made toward the definition of the brain circuits that underlie various aspects of nicotine dependence, from reward to withdrawal symptoms [9, 18].

nAChRs are pentameric acetylcholine (ACh)-gated ion channels that exist as homomeric (all α subunits) or heteromeric (α and β subunits) structures [7, 17, 19, 20]. Genes encoding nAChR subunits are found in both vertebrates and invertebrates [7, 21], and sequences among mammals are fairly conserved [22, 23]. Within mammalian genomes, separate genes encode eight distinct α nAChR subunits and three distinct β subunits [7]. Expression in heterologous systems has allowed the study of the contribution of individual subunits to receptor function [24–30]. nAChRs are expressed in almost all brain regions, including the circuits that underlie nicotine's influence over reinforcement, aversion, attention, and learning and memory [7, 9, 18, 31, 32].

In this chapter, we discuss molecular, cellular, circuit, and behavioral facets of nicotine withdrawal and related negative aspects of nicotine dependence. In so doing, we focus primarily on the nAChRs along the habenulo-interpeduncular

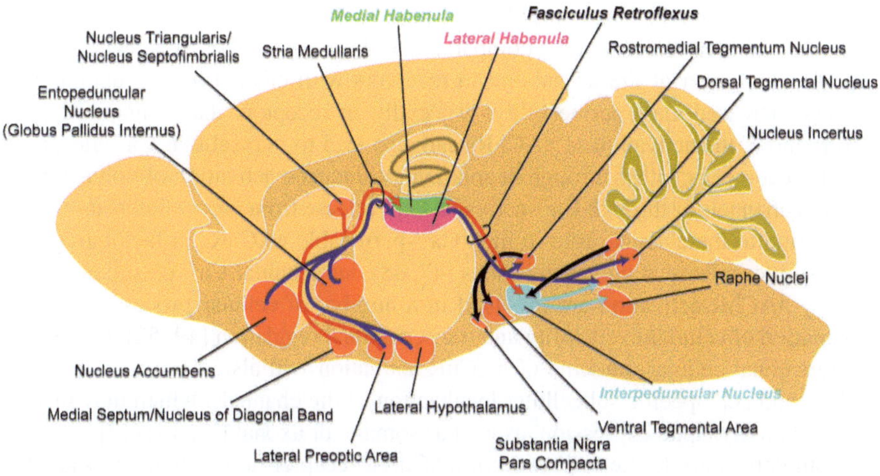

Fig. 18.1 The Hb-IPN pathway bridges forebrain nuclei to those within the mid- and hindbrain. Diagram of a sagittal view of the mouse brain illustrating the anatomical connectivity of the Hb-IPN pathway in mice. The Hb-IPN pathway, principally composed of the medial (*green*) and lateral (*pink*) habenulae, the fasciculus retroflexus, and the interpeduncular nucleus (*sky blue*), bridges various nuclei within the forebrain to mid- and hindbrain nuclei. Afferent projections to and efferent projections from the medial (*red arrows*) and lateral (*purple arrows*) habenulae are displayed

(Hb-IPN) pathway, a circuit with emerging roles in negative reinforcement and aversion (Fig. 18.1) [33–35]. As we enumerate the physiological and behavioral roles of this circuit, we discuss the relevant functional roles of the nAChRs expressed along this pathway. Ultimately, we integrate these topics into a basic framework for the understanding of the function of habenulo-interpeduncular nAChRs in overall dependence to nicotine.

2 Nicotine's Influences on nAChRs and Cell Function

To understand the effects of nicotine on brain circuits and ultimately behavior, it is necessary to first consider its effects at a molecular level. The binding of nicotine to nAChRs occurs at specific sites on the interface between two adjacent subunits [7, 36, 37]. In homomeric nAChRs, such as in α7 nAChRs, binding occurs between any two of the subunits and results in a total of five binding sites. In heteromeric nAChRs, such as α4β2 nAChRs, nicotine binding occurs at the interface between specific α and β subunits, resulting in a total of two binding sites. Nicotine binding activates the nAChR, resulting in the flux of mono- and divalent cations such as Na^+, K^+, and Ca^{2+} across the plasma membrane [7, 38]. The influx of cations (primarily Na^+, but also Ca^{2+}) leads to a membrane depolarization that consequently triggers a variety of intracellular events. In addition, intracellular Ca^{2+} signaling is important in many cellular processes [39–42]. Influenced by the specific subunit composition, many nAChR subtypes have sizable Ca^{2+} permeabilities. For example, α7 nAChRs have Ca^{2+} permeabilities that are comparable to those of NMDA glutamate receptors [43, 44]. Inclusion of the α5 subunit in receptors with other α and β subunits also confers increased Ca^{2+} permeability, especially in receptors that contain β2 [45]. Furthermore, the activation of nAChRs can also lead to intracellular Ca^{2+} elevations via indirect means, either through depolarization-induced activation of voltage-gated calcium channels or through Ca^{2+}-activated Ca^{2+} release from intracellular stores [46]. In addition to the classical role of high Ca^{2+}-permeable nAChRs in facilitating neurotransmitter release at presynaptic sites [47, 48], intracellular Ca^{2+} elevations generated by nAChR activation are involved in a number of cellular processes, such as modulation of cellular excitability and transcriptional regulation [49–52]. Ultimately, the functional cellular outcome of nAChR activation will also depend on the brain region- and cell-specific subcellular localization of the channels, which may include pre- and postsynaptic expression, as well as somatic or axonal locations [7].

Complementary to the activation of nAChRs, receptor desensitization is another important property of nAChRs that determines their overall function and must be considered when examining the acute or chronic effects of nicotine [7, 53–56]. In addition to open and closed conformations, nAChRs can exist in desensitized conformations following exposure to elevated concentrations of ligand. In a desensitized conformation, nAChRs are unable to evoke the response that occurs during open states, despite the presence of bound ligands on receptors. The kinetics of desensitization are characterized by multiple exponential functions and are influenced by receptor composition. α7 nAChRs desensitize quickly, while α4β2 nAChRs desensitize with slower kinetics [57–60]. However, due to their high affinity for nicotine, α4β2 nAChRs desensitize at lower concentrations of nicotine (<0.1 μM), while α7 nAChRs do not efficiently desensitize below a concentration of 1 μM. At concentrations of nicotine typically found in smokers, α4β2 nAChRs in the human brain are nearly fully occupied by nicotine [61]. Given that desensitization occurs at these concentrations or lower, it is likely that most of the brain's α4β2 nAChRs are maintained in a desensitized state during habitual cigarette smoking. nAChR desensitization is regulated by many factors, including intracellular Ca^{2+}. In both heterologous expression systems and neurons isolated from the medial habenula, the level of intracellular Ca^{2+} is inversely proportional to the recovery of nAChRs from desensitization to nicotine [62–64]. It is suggested that the presence of Ca^{2+} stabilizes a desensitized conformation. Since nAChR desensitization is functionally analogous to blockade of those receptors, to some degree, this phenomenon can have significant consequences across many levels of brain function, from molecular and cellular to systems and behavioral.

Finally, one should consider that as a tertiary amine, nicotine exists in both charged and uncharged states. In the nonpolar uncharged form, nicotine becomes membrane permeable and can freely enter the cytosolic space where it interferes with various cellular mechanisms, including actions at the endoplasmic reticulum and/or proteasome complex [65, 66]. Given that nAChR subunits are degraded by the proteasome and proteasome inhibition enhances nAChR assembly within the endoplasmic reticulum [67], nicotine-mediated inhibition of the proteasome complex causes enhanced nAChR plasma membrane expression as well [68, 69].

3 Neuroadaptations During Chronic Nicotine Exposure

The chronic use of nicotine causes multiple neuroadaptations in the brain, demonstrated by many in vitro studies in heterologous expression systems, as well as in vivo studies in rodents [25, 70–74]. The most commonly observed molecular phenomena are alterations in membrane expression of nAChRs that occur in a subtype-, brain region-, and time-dependent fashion [75]. The α4β2 nAChR subtype, in particular, has been shown in vitro to exhibit functional upregulation in response to chronic nicotine treatment, in the form of acetylcholine-induced current increases, in both heterologous expression systems and cultured neurons [72, 76–78]. Furthermore, in vivo nicotine exposure increases these measures for neurons in mouse brain slices [79]. It is suggested that these functional upregulations are due to dual mechanisms [80]. A short-lived switch in the conformation of surface nAChRs from a low-affinity to a higher-affinity state constitutes a rapid response following nicotine exposure. Secondarily, an effect with slower kinetics ensues, increasing surface α4β2 nAChRs via reduced proteasomal degradation of subunits and increased maturation of the receptors [65, 66].

4 Nicotine Withdrawal Syndrome: nAChRs Along the MHb-IPN Pathway Are Critical for the Physical Symptoms of Nicotine Withdrawal

Several physical and affective symptoms emerge during the period of acute nicotine withdrawal that may last for as long as a month following the start of abstinence [9, 14, 81]. These symptoms include a collection of unpleasant and aversive experiences such as intense cravings for nicotine, irritability, anger, anxiety, difficulty concentrating, insomnia, and increased appetite with consequent weight gain [11, 82–84]. Other physical manifestations accompany these behavioral symptoms of nicotine withdrawal, including restlessness, decreased heart rates, fluctuations of hormonal levels, drowsiness, headaches, gastrointestinal disturbances, and reduction in the electroencephalography (EEG) theta band [11, 85–87]. The emergence of these negative and/or aversive symptoms is the result of brain circuits, accustomed to the chronic presence of nicotine, readapting to a new steady state in its absence [9].

Behaviorally, symptoms of nicotine withdrawal can be classified as physical or affective [88, 89]. Physical symptoms of nicotine withdrawal have been successfully simulated in the laboratory to study the neurobiology underlying these behavioral disruptions [90–93]. Normal naïve mice display a number of typical behaviors when idle, including grooming, scratching, and chewing. However, when mice are subjected to withdrawal from nicotine following chronic treatment, the instances of these behaviors increase substantially and are accompanied by the emergence of other behaviors indicative of physical discomfort, including shaking, cage scratching, head nodding, and jumping. Using this behavioral paradigm in combination with mutant mice, the nAChR subunits important for the emergence of these physical symptoms of withdrawal were determined. In wild-type mice chronically treated with nicotine, systemic injection of the broad nAChR antagonist mecamylamine elicited an elevation of the physical signs of nicotine withdrawal over that of control mice chronically treated with a vehicle solution. However, in mice null for the β4 nAChR subunit, this elevation was completely abolished [92]. Along with previously established roles of this subunit in the modulation of anxiety and the anxiolytic properties of nicotine [94], this finding began to build a framework for the function of β4-containing (β4*) nAChRs in aversive and negatively reinforcing behaviors. In vivo, the most common assembly partner of the β4 subunit in neuronal nAChRs is the α3 subunit [9, 14, 27, 95].

Further experiments with additional nAChR subunit mutant mice revealed that physical withdrawal from nicotine also depends on α5, α2, and partially on α7 nAChR subunits [93, 96, 97]. Human genetic studies also identified multiple single-nucleotide polymorphisms in the gene cluster encoding the α5, α3, and β4 nAChR subunits that associate with various aspects of nicotine dependence and tobacco-related diseases [98–106]. The MHb-IPN pathway is among the brain areas with the highest co-expression of α5, α3, α2, and β4 [98, 107–111]. The habenular complex is composed of the medial (MHb) and lateral (LHb) habenular nuclei, with projections traveling along the fasciculus retroflexus, the white matter tract that bridges the habenular nuclei and their projection sites. The interpeduncular nucleus (IPN) is the main projection site for the MHb, while the LHb extends behaviorally important projections to the rostromedial tegmental nucleus (RMTg) in the midbrain. These brain areas are now understood to mediate negative reinforcement, negative prediction errors, negative motivation, and aversion [34, 112–117]. The emerging roles of the LHb were motivation for the investigation of the nAChRs along the MHb-IPN pathway in the nicotine withdrawal syndrome, and behavioral pharmacological experiments indicated that these receptors are indeed important for this behavioral manifestation [93]. In mice chronically treated with nicotine, the nAChR antagonist, mecamylamine, was sufficient to produce nicotine withdrawal behaviors only when microinjected into the Hb or the IPN, but not when microinjected into other brain areas such as the hippocampus or cerebral cortex. Interestingly, experiments using mice bearing an α2 null mutation suggest that the roles of MHb-IPN pathway nAChRs in physical withdrawal are context specific [93, 97]. While this mutation produced decreased physical signs in animals assayed in their home environments, those assayed in novel environments exhibited potentiated physical signs [97]. Altogether, this series of experiments demonstrates the importance of the MHb-IPN pathway to the nicotine withdrawal syndrome.

5 The Affective Symptoms of Nicotine Withdrawal

Affective symptoms accompany the physical symptoms of nicotine withdrawal and have a major role in relapse [89]. These withdrawal symptoms can manifest in animals as anhedonia, conditioned place aversion, anxiety-related behaviors, and conditioned fear [9]. Anhedonia, the inability to experience pleasure from activities that are normally pleasurable, has been modeled in electrical self-stimulation assays [118]. In animals that are trained to press a lever to electrically stimulate reward nuclei, the threshold for continued brain stimulation is viewed as a measure of the rewarding effect of electrical stimulation. Anhedonic animals will exhibit an increase in this threshold, suggesting a decrease in the reward value of the stimulation. Increases in self-stimulation thresholds are observed during both spontaneous withdrawal [119, 120] and withdrawal precipitated by a systemic injection of the nAChR antagonist, mecamylamine [121].

Humans learn negative associations with specific environments, and this is modeled in rodents in the conditioned place aversion (CPA) paradigm, wherein animals try to avoid an environment that was previously paired with a negative stimulus [89]. Successful CPA in chronic nicotine-treated mice was induced by pairing an environment with injections of nAChR or opioid receptor antagonists, such as mecamylamine, dihydro-β-erythroidine, and naloxone [89, 122]. The aversion generated by the induced withdrawal was sufficient to cause animals to associate the aversive experience with a specific environment.

Smokers undergoing nicotine withdrawal may experience extreme anxiety resembling levels experienced by depressed individuals or those with anxiety disorders [123, 124]. Anxiety is routinely analyzed in rodents using the elevated plus maze (EPM) assay [125]. This assay is essentially a four-armed maze elevated above the ground, with two open arms and two closed arms. Mice generally prefer to remain in the closed arms, and the amount of time spent in the closed vs. open arms is considered a measure of the animal's state of anxiety. Multiple investigations have demonstrated that both mice and rats experiencing nicotine withdrawal exhibit increased anxiety-like behavior in this assay [126, 127], mimicking symptoms of withdrawal observed in humans. It is possible that the MHb-IPN pathway also participates in this facet of the nicotine withdrawal syndrome. Mice null for the β4 nAChR subunit, which is densely expressed along the MHb-IPN pathway [111], exhibit modified anxiety-related behavior from wild-type mice [94]. β4 null mice display anxiolytic behavior in the elevated plus and staircase mazes but also display increased anxiety in the social isolation test, suggesting that nAChRs along the MHb-IPN pathway regulate anxiety-related behavior in a nuanced manner, with the output behavior dependent on specific environmental conditions.

A type of learning influenced by nicotine withdrawal is fear conditioning, a hippocampus-dependent form of Pavlovian learning where a conditioned stimulus is associated with an aversive unconditioned stimulus [128]. The conditioned fear assay measures the degree to which an animal is able to display this type of learning. Acutely administered nicotine enhances conditioned fear responses, regardless of whether the context is a foreground or background stimulus [129]. Furthermore, nicotine withdrawal impairs novel contextual fear conditioning but does not affect previously learned conditioned responses [130]. The impaired contextual fear conditioning occurs with or without pairing to an auditory stimulus (i.e., background vs. foreground) [131]. Evidence indicates that withdrawal-mediated deficits in contextual fear conditioning are mediated through β2* nAChRs [130, 132].

6 Other Medial Habenula-Dependent Behaviors Relevant to Nicotine Dependence

More recent studies have complemented the work in nicotine withdrawal to highlight roles for the MHb, in particular, with respect to its functional ties to nAChRs [34, 116, 117, 133]. Contributing further to the involvement of this brain area in aversion-related behaviors, those studies showed that α5* nAChRs along the MHb-IPN pathway mediate the aversive properties of nicotine at high doses, thereby regulating nicotine intake [34]. Using a self-administration paradigm, in which mice chose to intravenously self-administer nicotine over placebo, the authors demonstrated that mice lacking the α5 nAChR subunit will self-administer nicotine at substantially elevated levels compared to wild-type mice. That is, α5 null mice will continue to self-administer nicotine despite reaching a threshold at which wild-type mice would find nicotine to be aversive. They further demonstrated, through focal pharmacological microinjection and lentiviral RNAi knockdown or re-expression of the α5 subunit in the MHb or IPN, that α5* nAChRs in those nuclei are directly involved in the regulation of nicotine intake.

Other investigators used genomic and lentiviral overexpression of the β4 and α5 nAChR subunits, respectively, to further corroborate the role of α5β4* nAChRs in aversion to nicotine and described the functional interplay between these subunits [116]. They demonstrated that the β4 subunit enhances nAChR-mediated currents when overexpressed. Conversely, the α5 subunit competes with β4 to temper its effect, particularly when α5 is a variant (398N α5) that is linked to increased genetic risk of nicotine dependence in humans [99–106]. With β4 overexpression, mice experience reduced nicotine intake and nicotine-associated conditioned place aversion. Furthermore, lentiviral expression of the D398N α5 variant in the MHb alongside β4 overexpression reverted the nicotine intake phenotype to wild-type levels.

Utilizing immunotoxin-mediated ablation of two separate afferents to the MHb, another study dissected their contribution to MHb-dependent behavior [117]. Ablating the inputs from the nucleus triangularis (NT) in the septal area decreased anxiety-related behaviors in the open field, elevated plus maze, and marble burying task, while the analogous lesion of the bed nucleus of the anterior commissure (BAC) spared the performance in these tasks. Conversely, only ablation of the BAC inputs to the MHb caused deficits in the fear conditioning and passive avoidance tasks, indicating that these inputs are vital for proper fear expression.

Lastly, through a genetic approach, another group elaborated the influence of the MHb and IPN in behaviors involving motivational and emotive processes [133]. Neurons in the MHb were genetically ablated in mice using Cre-recombinase-mediated expression of diphtheria toxin A (DTA) in transgenic mice with strong Cre expression in the MHb and very sparse expression in few other brain areas. The expression of DTA induces apoptotic cell death [134] and, in these experiments, resulted in the dramatic loss of Nissl-stained neurons in the MHb. Furthermore, the IPN suffered reductions in acetylcholine concentration and overall volume. Thus, mice bearing the genetic ablation of the MHb (MHb:DTA) showed significant damage to the Hb-IPN pathway, with habenular damage predominantly restricted to the MHb.

As a consequence of the genetic ablation, many behaviors in MHb:DTA mice were severely impaired [133]. Where wild-type mice exhibited habituation to novel environments, habituation was absent in MHb:DTA mice. In the 5-choice serial reaction time task (5-CSRTT), which assesses impulsiveness, compulsiveness, and attention [135, 136], MHb:DTA mice were found to have increased premature responses, which is indicative of impulsive behavior. Compulsive behavior is displayed through perseverative nose-pokes following correct trials, even if reward is delivered only once for the initial correct choice. Also, sensorimotor gating is disrupted, as MHb:DTA mice have impaired acoustic pre-pulse inhibition, while baseline startle responses were unaffected.

To further investigate the MHb's influence over impulsiveness and compulsiveness, delay- and effort-based decision-making tasks were used [133]. MHb:DTA mice are more likely to choose a low-reward choice if a high-reward choice is associated with a delay longer than 10 s or they encounter an obstacle barrier (effort). These results indicate that, as delay and effort increase, mice lacking the MHb will discount reward value more readily than wild-type mice and will select the option that provides the quickest reward.

In the open field arena (OFA) and elevated plus maze (EPM), which assess anxiety, MHb:DTA mice exhibited minor deficits in both tasks. MHb:DTA mice made slightly fewer entries to the center in the OFA and to the open arms of the EPM, together indicating a modest increase in anxiety. This appears in slight contrast to the previous study [117], which found decreased anxiety following ablation of afferent innervation from the NT, excitatory (glutamatergic and ATPergic) inputs into the MHb [137]. Our group demonstrated that mice null for the α5 and β4 nAChR subunits display reduced anxiety-like behavior in the EPM, suggesting a direct role for the nicotinic cholinergic system in these behaviors [94, 138].

To test whether the nAChRs along the MHb-IPN pathway modulate impulsivity and compulsivity, performance in the 5-CSRTT was measured following systemic nicotine administration [133]. In wild-type animals, nicotine administration induced delayed nose-pokes and increased errors due to omission. However, neither of these measures was affected in MHb:DTA mice. Furthermore, while habituation to a novel environment was accelerated by nicotine within and across sessions, neither of these measures was affected by nicotine in MHb:DTA mice. Altogether, this genetic ablation study strengthens the role of MHb-IPN pathway nAChRs in the modulation of these motivational and emotive behaviors.

7 The Potential Role of the MHb-IPN Pathway in Aversion and Negative Reinforcement: Inferences from LHb Studies

The role of the MHb-IPN pathway in basal- and nicotine-related behaviors is consistent with the established influences of the LHb on behavior [117, 139, 140]. Early work hinted at the role of the LHb in the representation of negative motivational value, negative reinforcement, and aversion [140–148]. Physiological studies reported the ability of the LHb (most likely due to its projections caudally, toward the RMTg) to modulate the reward-related centers [149, 150]. In rats, electrical stimulation of the LHb inhibited the firing of DA neurons in the SNc and VTA [151], as well as that of serotonergic neurons in the dorsal and median raphe nuclei [149].

A series of studies in macaques from Hikosaka and colleagues focusing on the behavioral roles of the LHb led to the maturation of our understanding of this modulatory circuit [112, 114, 115, 152, 153]. Since their seminal study, in which they demonstrated that LHb neurons fire in response to negative outcomes, as well as inhibit the firing of SNc DA neurons [112], they have expanded their studies to further clarify the behavioral roles of this nucleus. In cleverly designed experiments that varied the severity of negative outcomes in a task, so as to include punishments and lack of rewards as possible outcomes, LHb neurons fired most robustly in response to the worse-case scenario between the options of the particular task [114]. They further demonstrated that LHb neurons signal reward values derived from both the animal's experience and inference [152]. Additional studies have also implicated the LHb in the representation of memory for reward, signaling of reward prediction errors, and learning of behaviors to avoid unpleasant outcomes [115, 152, 153]. The involvement of the Hb in error signaling during the prediction of rewards was subsequently demonstrated in humans using functional magnetic resonance imaging (fMRI) [35].

8 Dopaminergic Adaptations During Withdrawal

Given the global reach of nicotine within the brain, multiple mechanisms in different brain areas are likely responsible for the behavioral experiences during nicotine withdrawal. For example, the dopaminergic mesolimbic pathway participates in the mechanisms underlying nicotine abstinence manifestations. Principally consisting of dopaminergic projections from the VTA and SNc to the striatum, the mesolimbic pathway is known to influence behaviors associated with reward and motivation [154, 155]. Upon cessation of nicotine intake, the extracellular levels of DA decrease in the nucleus accumbens [156–160]. Consistent with withdrawal as a qualitatively aversive and generally unpleasant experience, this decrease in extracellular accumbal DA is also observed in withdrawal from many other drugs of abuse, such as that from ethanol, morphine, cocaine, and amphetamine [161, 162]. Therefore, common mechanisms and circuits operate to produce similar behavioral states during withdrawal to nicotine and other drugs of abuse. Because the mesolimbic pathway functionally interacts with the habenular circuitry, which is associated with negative reinforcement and aversion, a potential shift in the balance between these two systems could be responsible for the hypodopaminergic state during withdrawal from nicotine. As already mentioned, the LHb sends excitatory projections to the RMTg [140, 163, 164], which, in turn, projects GABAergic efferents to DA neurons in the VTA and SNc [112, 163]. This inhibitory control of RMTg projections onto DA neurons is a substantial modulator of their firing behavior [164].

Notably, α3β4* nAChRs within the MHb modulate the accumbal DA release in response to acute nicotine [165]. As projections from the MHb to LHb are documented [166], modulation of DA release by the MHb might occur via its anatomical connections with the LHb. However, as discussed above, there is robust evidence that the MHb-IPN pathway mediates aspects of nicotine aversion and withdrawal. Data suggest that MHb can affect the activity of the dopaminergic neurons in the ventral tegmental area (VTA) via the IPN [167], but the anatomical underpinning of this phenomenon remains unclear. One possibility through which the MHb-IPN pathway might regulate the activity of VTA DA neuron firing is via connections of the IPN to the laterodorsal tegmentum (LDTg) [168]. The LDTg is a cholinergic nucleus that sends inputs to the VTA that are required for proper bursting activity of DA neurons [169, 170]. Whether and how this circuit participates in the mechanisms of nicotine withdrawal remains to be established.

It should be noted that the hypodopaminergic state during nicotine withdrawal itself likely reflects a combination of many neuroadaptive processes triggered by withdrawal from nicotine. A reduction of striatal DA release certainly contributes and is accompanied by increased protein levels of vesicular monoamine transporter 2 (VMAT2) in the striatum [171]. An important player in DA reuptake, this elevation of VMAT2 is proposed to be a compensatory mechanism to counteract the deficiencies in DA release. Indeed, increased DA uptake into striatal synaptosomes was observed during nicotine withdrawal, as well as an increase in mRNA expression of another key participant in DA reuptake, the DA transporter (DAT), in the SNc and VTA [172]. Furthermore, an increase in DA clearance during nicotine withdrawal has been observed in vivo using microdialysis, corroborating the model of enhanced DA reuptake during withdrawal [158]. Regardless, these alterations in DA reuptake are transient, as the changes during withdrawal return to basal levels by 48 h of abstinence from nicotine [172]. Due to the synchronicity with nicotine withdrawal behavior, these alterations in DA release and reuptake might be a key mechanism underlying nicotine withdrawal symptoms [9].

Interestingly, DA transmission does not uniformly decrease throughout the brain during nicotine withdrawal. In contrast to striatal effects, DA release in the prefrontal cortex (PFC) is heightened during withdrawal to nicotine [159]. The role of this mesocortical innervation is related to the roles of some DA neurons in motivational salience, as those DA neurons signal the beginning of stimuli via phasic bursts, regardless of valence [115]. Since the experience during nicotine withdrawal can be significantly aversive and PFC DA release increases during aversive and stressful situations, the increased PFC DA release during nicotine withdrawal might coordinate the necessary mechanisms for the proper aversive behavioral response [173–177].

9 Conclusion

Addiction to nicotine, similar to other drugs of abuse, likely results from multiple mechanisms that involve interactions among various brain circuits. nAChRs, which are distributed on almost all brain circuits, stimulate responses to nicotine intake that ultimately produce an addicted state under prolonged use of the drug. Here, we discussed a specific brain circuit, the MHb-IPN pathway, that is involved in the nicotine withdrawal syndrome and other aversive aspects of nicotine use. Recent genome-based studies have identified genes encoding the α5, α3, and β4 subunits as important genetic determinants for the risk of nicotine dependence. All three of these subunits are highly expressed along the MHb-IPN pathway, highlighting its importance to overall addiction to nicotine. Resulting from a series of studies, the prevailing functional model of the Hb and its projection pathways is that it regulates dopaminergic and serotonergic function in the midbrain and, consequently, aversion and negative reinforcement. Given that withdrawal is arguably the most aversive and unpleasant experience associated with nicotine dependence, this model is consistent with the circuit's roles in withdrawal behavior. Targeting the nAChRs along the MHb-IPN pathway should be a goal in future approaches at pharmacological treatment of nicotine dependence.

References

1. Vineis P, Alavanja M, Buffler P, et al. Tobacco and cancer: recent epidemiological evidence. J Natl Cancer Inst. 2004;96(2):99–106.
2. Kenfield SA, Stampfer MJ, Rosner BA, Colditz GA. Smoking and smoking cessation in relation to mortality in women. JAMA. 2008;299(17):2037–47.
3. Benowitz NL. Pharmacology of nicotine: addiction, smoking-induced disease, and therapeutics. Annu Rev Pharmacol Toxicol. 2009;49:57–71.
4. Tonini G, D'Onofrio L, Dell'Aquila E, Pezzuto A. New molecular insights in tobacco-induced lung cancer. Future Oncol. 2013;9(5):649–55.
5. Giovino GA, Mirza SA, Samet JM, et al. Tobacco use in 3 billion individuals from 16 countries: an analysis of nationally representative cross-sectional household surveys. Lancet. 2012; 380(9842):668–79.
6. Kabir MA, Goh KL, Khan MH. A cross-country comparison of tobacco consumption among youths from selected South-Asian countries. BMC Public Health. 2013;13(1):379.
7. Dani JA, Bertrand D. Nicotinic acetylcholine receptors and nicotinic cholinergic mechanisms of the central nervous system. Annu Rev Pharmacol Toxicol. 2007;47:699–729.
8. Mansvelder HD, McGehee DS. Cellular and synaptic mechanisms of nicotine addiction. J Neurobiol. 2002;53(4):606–17.
9. De Biasi M, Dani JA. Reward, addiction, withdrawal to nicotine. Annu Rev Neurosci. 2011;34:105–30.
10. Piasecki TM, Jorenby DE, Smith SS, Fiore MC, Baker TB. Smoking withdrawal dynamics: I. Abstinence distress in lapsers and abstainers. J Abnorm Psychol. 2003;112(1):3–13.
11. Hughes JR. Effects of abstinence from tobacco: valid symptoms and time course. Nicotine Tob Res. 2007;9(3):315–27.
12. al'Absi M, Amunrud T, Wittmers LE. Psychophysiological effects of nicotine abstinence and behavioral challenges in habitual smokers. Pharmacol Biochem Behav. 2002;72(3):707–16.

13. Swan GE, Ward MM, Jack LM. Abstinence effects as predictors of 28-day relapse in smokers. Addict Behav. 1996;21(4):481–90.
14. De Biasi M, Salas R. Influence of neuronal nicotinic receptors over nicotine addiction and withdrawal. Exp Biol Med (Maywood). 2008;233(8):917–29.
15. Drenan RM, Lester HA. Insights into the neurobiology of the nicotinic cholinergic system and nicotine addiction from mice expressing nicotinic receptors harboring gain-of-function mutations. Pharmacol Rev. 2012;64(4):869–79.
16. Jasinska AJ, Zorick T, Brody AL, Stein EA. Dual role of nicotine in addiction and cognition: a review of neuroimaging studies in humans. Neuropharmacology. 2014;84C:111–22.
17. Role LW, Berg DK. Nicotinic receptors in the development and modulation of CNS synapses. Neuron. 1996;16(6):1077–85.
18. Picciotto MR, Mineur YS. Molecules and circuits involved in nicotine addiction: the many faces of smoking. Neuropharmacology. 2014;76:543–53.
19. McGehee DS, Role LW. Physiological diversity of nicotinic acetylcholine receptors expressed by vertebrate neurons. Annu Rev Physiol. 1995;57:521–46.
20. Alkondon M, Pereira EF, Barbosa CT, Albuquerque EX. Neuronal nicotinic acetylcholine receptor activation modulates gamma-aminobutyric acid release from CA1 neurons of rat hippocampal slices. J Pharmacol Exp Ther. 1997;283(3):1396–411.
21. Tomizawa M, Casida JE. Structure and diversity of insect nicotinic acetylcholine receptors. Pest Manag Sci. 2001;57(10):914–22.
22. Kent WJ, Sugnet CW, Furey TS, et al. The human genome browser at UCSC. Genome Res. 2002;12(6):996–1006.
23. Meyer LR, Zweig AS, Hinrichs AS, et al. The UCSC Genome Browser database: extensions and updates 2013. Nucleic Acids Res. 2013;41(Database issue):D64–9.
24. Nelson ME, Lindstrom J. Single channel properties of human alpha3 AChRs: impact of beta2, beta4 and alpha5 subunits. J Physiol. 1999;516(Pt 3):657–78.
25. Peng X, Gerzanich V, Anand R, Wang F, Lindstrom J. Chronic nicotine treatment up-regulates alpha3 and alpha7 acetylcholine receptor subtypes expressed by the human neuroblastoma cell line SH-SY5Y. Mol Pharmacol. 1997;51(5):776–84.
26. Tapia L, Kuryatov A, Lindstrom J. Ca2+ permeability of the (alpha4)3(beta2)2 stoichiometry greatly exceeds that of (alpha4)2(beta2)3 human acetylcholine receptors. Mol Pharmacol. 2007;71(3):769–76.
27. Wang F, Gerzanich V, Wells GB, et al. Assembly of human neuronal nicotinic receptor alpha5 subunits with alpha3, beta2, and beta4 subunits. J Biol Chem. 1996;271(30):17656–65.
28. Gerzanich V, Wang F, Kuryatov A, Lindstrom J. alpha 5 Subunit alters desensitization, pharmacology, Ca++ permeability and Ca++ modulation of human neuronal alpha 3 nicotinic receptors. J Pharmacol Exp Ther. 1998;286(1):311–20.
29. Kuryatov A, Berrettini W, Lindstrom J. Acetylcholine receptor (AChR) alpha5 subunit variant associated with risk for nicotine dependence and lung cancer reduces (alpha4beta2)(2) alpha5 AChR function. Mol Pharmacol. 2011;79(1):119–25.
30. Albuquerque EX, Pereira EF, Alkondon M, Rogers SW. Mammalian nicotinic acetylcholine receptors: from structure to function. Physiol Rev. 2009;89(1):73–120.
31. Woolf NJ. Cholinergic systems in mammalian brain and spinal cord. Prog Neurobiol. 1991;37(6):475–524.
32. Miwa JM, Freedman R, Lester HA. Neural systems governed by nicotinic acetylcholine receptors: emerging hypotheses. Neuron. 2011;70(1):20–33.
33. Sutherland RJ. The dorsal diencephalic conduction system: a review of the anatomy and functions of the habenular complex. Neurosci Biobehav Rev. 1982;6(1):1–13.
34. Fowler CD, Lu Q, Johnson PM, Marks MJ, Kenny PJ. Habenular alpha5 nicotinic receptor subunit signalling controls nicotine intake. Nature. 2011;471(7340):597–601.
35. Salas R, Baldwin P, de Biasi M, Montague PR. BOLD responses to negative reward prediction errors in human habenula. Front Hum Neurosci. 2010;4:36.
36. Karlin A. Emerging structure of the nicotinic acetylcholine receptors. Nat Rev Neurosci. 2002;3(2):102–14.

37. Sine SM. The nicotinic receptor ligand binding domain. J Neurobiol. 2002;53(4):431–46.
38. Galzi JL, Devillers-Thiery A, Hussy N, Bertrand S, Changeux JP, Bertrand D. Mutations in the channel domain of a neuronal nicotinic receptor convert ion selectivity from cationic to anionic. Nature. 1992;359(6395):500–5.
39. Berridge MJ, Lipp P, Bootman MD. The versatility and universality of calcium signalling. Nat Rev Mol Cell Biol. 2000;1(1):11–21.
40. Bootman MD, Berridge MJ, Roderick HL. Calcium signalling: more messengers, more channels, more complexity. Curr Biol. 2002;12(16):R563–5.
41. Sudhof TC. Calcium control of neurotransmitter release. Cold Spring Harb Perspect Biol. 2012;4(1):a011353.
42. Ross WN. Understanding calcium waves and sparks in central neurons. Nat Rev Neurosci. 2012;13(3):157–68.
43. Seguela P, Wadiche J, Dineley-Miller K, Dani JA, Patrick JW. Molecular cloning, functional properties, and distribution of rat brain alpha 7: a nicotinic cation channel highly permeable to calcium. J Neurosci. 1993;13(2):596–604.
44. Bertrand D, Galzi JL, Devillers-Thiery A, Bertrand S, Changeux JP. Mutations at two distinct sites within the channel domain M2 alter calcium permeability of neuronal alpha 7 nicotinic receptor. Proc Natl Acad Sci U S A. 1993;90(15):6971–5.
45. Kuryatov A, Gerzanich V, Nelson M, Olale F, Lindstrom J. Mutation causing autosomal dominant nocturnal frontal lobe epilepsy alters Ca2+ permeability, conductance, and gating of human alpha4beta2 nicotinic acetylcholine receptors. J Neurosci. 1997;17(23):9035–47.
46. Shen JX, Yakel JL. Nicotinic acetylcholine receptor-mediated calcium signaling in the nervous system. Acta Pharmacol Sin. 2009;30(6):673–80.
47. McGehee DS, Heath MJ, Gelber S, Devay P, Role LW. Nicotine enhancement of fast excitatory synaptic transmission in CNS by presynaptic receptors. Science. 1995;269(5231):1692–6.
48. Garduno J, Galindo-Charles L, Jimenez-Rodriguez J, et al. Presynaptic alpha4beta2 nicotinic acetylcholine receptors increase glutamate release and serotonin neuron excitability in the dorsal raphe nucleus. J Neurosci. 2012;32(43):15148–57.
49. Griguoli M, Scuri R, Ragozzino D, Cherubini E. Activation of nicotinic acetylcholine receptors enhances a slow calcium-dependent potassium conductance and reduces the firing of stratum oriens interneurons. Eur J Neurosci. 2009;30(6):1011–22.
50. Ji D, Dani JA. Inhibition and disinhibition of pyramidal neurons by activation of nicotinic receptors on hippocampal interneurons. J Neurophysiol. 2000;83(5):2682–90.
51. Hu M, Liu QS, Chang KT, Berg DK. Nicotinic regulation of CREB activation in hippocampal neurons by glutamatergic and nonglutamatergic pathways. Mol Cell Neurosci. 2002;21(4):616–25.
52. McKay BE, Placzek AN, Dani JA. Regulation of synaptic transmission and plasticity by neuronal nicotinic acetylcholine receptors. Biochem Pharmacol. 2007;74(8):1120–33.
53. Dani JA, Radcliffe KA, Pidoplichko VI. Variations in desensitization of nicotinic acetylcholine receptors from hippocampus and midbrain dopamine areas. Eur J Pharmacol. 2000; 393(1–3):31–8.
54. Quick MW, Lester RA. Desensitization of neuronal nicotinic receptors. J Neurobiol. 2002;53(4):457–78.
55. Wooltorton JR, Pidoplichko VI, Broide RS, Dani JA. Differential desensitization and distribution of nicotinic acetylcholine receptor subtypes in midbrain dopamine areas. J Neurosci. 2003;23(8):3176–85.
56. Giniatullin R, Nistri A, Yakel JL. Desensitization of nicotinic ACh receptors: shaping cholinergic signaling. Trends Neurosci. 2005;28(7):371–8.
57. Briggs CA, McKenna DG. Activation and inhibition of the human alpha7 nicotinic acetylcholine receptor by agonists. Neuropharmacology. 1998;37(9):1095–102.
58. Papke RL, Porter Papke JK. Comparative pharmacology of rat and human alpha7 nAChR conducted with net charge analysis. Br J Pharmacol. 2002;137(1):49–61.
59. Fenster CP, Rains MF, Noerager B, Quick MW, Lester RA. Influence of subunit composition on desensitization of neuronal acetylcholine receptors at low concentrations of nicotine. J Neurosci. 1997;17(15):5747–59.

60. Paradiso KG, Steinbach JH. Nicotine is highly effective at producing desensitization of rat alpha4beta2 neuronal nicotinic receptors. J Physiol. 2003;553(Pt 3):857–71.
61. Brody AL, Mandelkern MA, London ED, et al. Cigarette smoking saturates brain alpha 4 beta 2 nicotinic acetylcholine receptors. Arch Gen Psychiatry. 2006;63(8):907–15.
62. Khiroug L, Giniatullin R, Sokolova E, Talantova M, Nistri A. Imaging of intracellular calcium during desensitization of nicotinic acetylcholine receptors of rat chromaffin cells. Br J Pharmacol. 1997;122(7):1323–32.
63. Khiroug L, Sokolova E, Giniatullin R, Afzalov R, Nistri A. Recovery from desensitization of neuronal nicotinic acetylcholine receptors of rat chromaffin cells is modulated by intracellular calcium through distinct second messengers. J Neurosci. 1998;18(7):2458–66.
64. Guo X, Lester RA. Regulation of nicotinic acetylcholine receptor desensitization by Ca2+. J Neurophysiol. 2007;97(1):93–101.
65. Rezvani K, Teng Y, Shim D, De Biasi M. Nicotine regulates multiple synaptic proteins by inhibiting proteasomal activity. J Neurosci. 2007;27(39):10508–19.
66. Lester HA, Xiao C, Srinivasan R, et al. Nicotine is a selective pharmacological chaperone of acetylcholine receptor number and stoichiometry. Implications for drug discovery. AAPS J. 2009;11(1):167–77.
67. Christianson JC, Green WN. Regulation of nicotinic receptor expression by the ubiquitin-proteasome system. EMBO J. 2004;23(21):4156–65.
68. Rezvani K, Teng Y, Pan Y, et al. UBXD4, a UBX-containing protein, regulates the cell surface number and stability of alpha3-containing nicotinic acetylcholine receptors. J Neurosci. 2009;29(21):6883–96.
69. Rezvani K, Teng Y, De Biasi M. The ubiquitin-proteasome system regulates the stability of neuronal nicotinic acetylcholine receptors. J Mol Neurosci. 2010;40(1–2):177–84.
70. Marks MJ, Pauly JR, Gross SD, et al. Nicotine binding and nicotinic receptor subunit RNA after chronic nicotine treatment. J Neurosci. 1992;12(7):2765–84.
71. Pauly JR, Marks MJ, Robinson SF, van de Kamp JL, Collins AC. Chronic nicotine and mecamylamine treatment increase brain nicotinic receptor binding without changing alpha 4 or beta 2 mRNA levels. J Pharmacol Exp Ther. 1996;278(1):361–9.
72. Vallejo YF, Buisson B, Bertrand D, Green WN. Chronic nicotine exposure upregulates nicotinic receptors by a novel mechanism. J Neurosci. 2005;25(23):5563–72.
73. Huang YY, Kandel ER, Levine A. Chronic nicotine exposure induces a long-lasting and pathway-specific facilitation of LTP in the amygdala. Learn Mem. 2008;15(8):603–10.
74. Aydin C, Oztan O, Isgor C. Nicotine-induced anxiety-like behavior in a rat model of the novelty-seeking phenotype is associated with long-lasting neuropeptidergic and neuroplastic adaptations in the amygdala: effects of the cannabinoid receptor 1 antagonist AM251. Neuropharmacology. 2012;63(8):1335–45.
75. Vezina P, McGehee DS, Green WN. Exposure to nicotine and sensitization of nicotine-induced behaviors. Prog Neuropsychopharmacol Biol Psychiatry. 2007;31(8):1625–38.
76. Buisson B, Vallejo YF, Green WN, Bertrand D. The unusual nature of epibatidine responses at the alpha4beta2 nicotinic acetylcholine receptor. Neuropharmacology. 2000;39(13):2561–9.
77. Buisson B, Bertrand D. Chronic exposure to nicotine upregulates the human (alpha)4(beta)2 nicotinic acetylcholine receptor function. J Neurosci. 2001;21(6):1819–29.
78. Nashmi R, Dickinson ME, McKinney S, et al. Assembly of alpha4beta2 nicotinic acetylcholine receptors assessed with functional fluorescently labeled subunits: effects of localization, trafficking, and nicotine-induced upregulation in clonal mammalian cells and in cultured midbrain neurons. J Neurosci. 2003;23(37):11554–67.
79. Alkondon M, Albuquerque EX. Nicotinic receptor subtypes in rat hippocampal slices are differentially sensitive to desensitization and early in vivo functional up-regulation by nicotine and to block by bupropion. J Pharmacol Exp Ther. 2005;313(2):740–50.
80. Govind AP, Walsh H, Green WN. Nicotine-induced upregulation of native neuronal nicotinic receptors is caused by multiple mechanisms. J Neurosci. 2012;32(6):2227–38.
81. Benowitz NL. Nicotine addiction. N Engl J Med. 2010;362(24):2295–303.
82. American Psychiatric Association. Diagnostic and statistical manual of mental disorders. 4th ed. Washington, DC: American Psychiatric Association; 2000. Text Revision.

83. Volkow ND, Fowler JS, Wang GJ, Goldstein RZ. Role of dopamine, the frontal cortex and memory circuits in drug addiction: insight from imaging studies. Neurobiol Learn Mem. 2002;78(3):610–24.
84. Hogle JM, Kaye JT, Curtin JJ. Nicotine withdrawal increases threat-induced anxiety but not fear: neuroadaptation in human addiction. Biol Psychiatry. 2010;68(8):719–25.
85. Buchhalter AR, Fant RV, Henningfield JE. Novel pharmacological approaches for treating tobacco dependence and withdrawal: current status. Drugs. 2008;68(8):1067–88.
86. Yamamoto KI, Domino EF. Cholinergic agonist-antagonist interactions on neocortical and limbic EEG activation. Int J Neuropharmacol. 1967;6(5):357–73.
87. Kadoya C, Domino EF, Matsuoka S. Relationship of electroencephalographic and cardiovascular changes to plasma nicotine levels in tobacco smokers. Clin Pharmacol Ther. 1994; 55(4):370–7.
88. Hughes JR, Gust SW, Skoog K, Keenan RM, Fenwick JW. Symptoms of tobacco withdrawal. A replication and extension. Arch Gen Psychiatry. 1991;48(1):52–9.
89. Kenny PJ, Markou A. Neurobiology of the nicotine withdrawal syndrome. Pharmacol Biochem Behav. 2001;70(4):531–49.
90. Malin DH, Lake JR, Newlin-Maultsby P, et al. Rodent model of nicotine abstinence syndrome. Pharmacol Biochem Behav. 1992;43(3):779–84.
91. Malin DH, Lake JR, Carter VA, et al. The nicotinic antagonist mecamylamine precipitates nicotine abstinence syndrome in the rat. Psychopharmacology (Berl). 1994;115(1–2):180–4.
92. Salas R, Pieri F, De Biasi M. Decreased signs of nicotine withdrawal in mice null for the beta4 nicotinic acetylcholine receptor subunit. J Neurosci. 2004;24(45):10035–9.
93. Salas R, Sturm R, Boulter J, De Biasi M. Nicotinic receptors in the habenulo-interpeduncular system are necessary for nicotine withdrawal in mice. J Neurosci. 2009;29(10):3014–8.
94. Salas R, Orr-Urtreger A, Broide RS, Beaudet A, Paylor R, De Biasi M. The nicotinic acetylcholine receptor subunit alpha 5 mediates short-term effects of nicotine in vivo. Mol Pharmacol. 2003;63(5):1059–66.
95. Flores CM, DeCamp RM, Kilo S, Rogers SW, Hargreaves KM. Neuronal nicotinic receptor expression in sensory neurons of the rat trigeminal ganglion: demonstration of alpha3beta4, a novel subtype in the mammalian nervous system. J Neurosci. 1996;16(24):7892–901.
96. Salas R, Main A, Gangitano D, De Biasi M. Decreased withdrawal symptoms but normal tolerance to nicotine in mice null for the alpha7 nicotinic acetylcholine receptor subunit. Neuropharmacology. 2007;53(7):863–9.
97. Lotfipour S, Byun JS, Leach P, et al. Targeted deletion of the mouse alpha2 nicotinic acetylcholine receptor subunit gene (Chrna2) potentiates nicotine-modulated behaviors. J Neurosci. 2013;33(18):7728–41.
98. Boulter J, O'Shea-Greenfield A, Duvoisin RM, et al. Alpha 3, alpha 5, and beta 4: three members of the rat neuronal nicotinic acetylcholine receptor-related gene family form a gene cluster. J Biol Chem. 1990;265(8):4472–82.
99. Saccone SF, Hinrichs AL, Saccone NL, et al. Cholinergic nicotinic receptor genes implicated in a nicotine dependence association study targeting 348 candidate genes with 3713 SNPs. Hum Mol Genet. 2007;16(1):36–49.
100. Thorgeirsson TE, Geller F, Sulem P, et al. A variant associated with nicotine dependence, lung cancer and peripheral arterial disease. Nature. 2008;452(7187):638–42.
101. Amos CI, Wu X, Broderick P, et al. Genome-wide association scan of tag SNPs identifies a susceptibility locus for lung cancer at 15q25.1. Nat Genet. 2008;40(5):616–22.
102. Bierut LJ, Stitzel JA, Wang JC, et al. Variants in nicotinic receptors and risk for nicotine dependence. Am J Psychiatry. 2008;165(9):1163–71.
103. Stevens VL, Bierut LJ, Talbot JT, et al. Nicotinic receptor gene variants influence susceptibility to heavy smoking. Cancer Epidemiol Biomarkers Prev. 2008;17(12):3517–25.
104. Berrettini W, Yuan X, Tozzi F, et al. Alpha-5/alpha-3 nicotinic receptor subunit alleles increase risk for heavy smoking. Mol Psychiatry. 2008;13(4):368–73.
105. Berrettini WH, Doyle GA. The CHRNA5-A3-B4 gene cluster in nicotine addiction. Mol Psychiatry. 2012;17(9):856–66.

106. Weiss RB, Baker TB, Cannon DS, et al. A candidate gene approach identifies the CHRNA5-A3-B4 region as a risk factor for age-dependent nicotine addiction. PLoS Genet. 2008;4(7):e1000125.
107. Boulter J, Evans K, Goldman D, et al. Isolation of a cDNA clone coding for a possible neural nicotinic acetylcholine receptor alpha-subunit. Nature. 1986;319(6052):368–74.
108. Duvoisin RM, Deneris ES, Patrick J, Heinemann S. The functional diversity of the neuronal nicotinic acetylcholine receptors is increased by a novel subunit: beta 4. Neuron. 1989;3(4):487–96.
109. Wada E, McKinnon D, Heinemann S, Patrick J, Swanson LW. The distribution of mRNA encoded by a new member of the neuronal nicotinic acetylcholine receptor gene family (alpha 5) in the rat central nervous system. Brain Res. 1990;526(1):45–53.
110. Sheffield EB, Quick MW, Lester RA. Nicotinic acetylcholine receptor subunit mRNA expression and channel function in medial habenula neurons. Neuropharmacology. 2000; 39(13):2591–603.
111. Gahring LC, Persiyanov K, Rogers SW. Neuronal and astrocyte expression of nicotinic receptor subunit beta4 in the adult mouse brain. J Comp Neurol. 2004;468(3):322–33.
112. Matsumoto M, Hikosaka O. Lateral habenula as a source of negative reward signals in dopamine neurons. Nature. 2007;447(7148):1111–5.
113. Matsumoto M, Hikosaka O. Negative motivational control of saccadic eye movement by the lateral habenula. Prog Brain Res. 2008;171:399–402.
114. Matsumoto M, Hikosaka O. Representation of negative motivational value in the primate lateral habenula. Nat Neurosci. 2009;12(1):77–84.
115. Bromberg-Martin ES, Hikosaka O. Lateral habenula neurons signal errors in the prediction of reward information. Nat Neurosci. 2011;14(9):1209–16.
116. Frahm S, Slimak MA, Ferrarese L, et al. Aversion to nicotine is regulated by the balanced activity of beta4 and alpha5 nicotinic receptor subunits in the medial habenula. Neuron. 2011;70(3):522–35.
117. Yamaguchi T, Danjo T, Pastan I, Hikida T, Nakanishi S. Distinct roles of segregated transmission of the septo-habenular pathway in anxiety and fear. Neuron. 2013;78(3):537–44.
118. Paterson NE, Markou A. Animal models and treatments for addiction and depression co-morbidity. Neurotox Res. 2007;11(1):1–32.
119. Epping-Jordan MP, Watkins SS, Koob GF, Markou A. Dramatic decreases in brain reward function during nicotine withdrawal. Nature. 1998;393(6680):76–9.
120. Harrison AA, Liem YT, Markou A. Fluoxetine combined with a serotonin-1A receptor antagonist reversed reward deficits observed during nicotine and amphetamine withdrawal in rats. Neuropsychopharmacology. 2001;25(1):55–71.
121. Watkins SS, Stinus L, Koob GF, Markou A. Reward and somatic changes during precipitated nicotine withdrawal in rats: centrally and peripherally mediated effects. J Pharmacol Exp Ther. 2000;292(3):1053–64.
122. Suzuki T, Ise Y, Tsuda M, Maeda J, Misawa M. Mecamylamine-precipitated nicotine-withdrawal aversion in rats. Eur J Pharmacol. 1996;314(3):281–4.
123. Dani JA, Harris RA. Nicotine addiction and comorbidity with alcohol abuse and mental illness. Nat Neurosci. 2005;8(11):1465–70.
124. Pomerleau OF, Pomerleau CS, Mehringer AM, Snedecor SM, Ninowski R, Sen A. Nicotine dependence, depression, and gender: characterizing phenotypes based on withdrawal discomfort, response to smoking, and ability to abstain. Nicotine Tob Res. 2005;7(1):91–102.
125. Pellow S, Chopin P, File SE, Briley M. Validation of open:closed arm entries in an elevated plus-maze as a measure of anxiety in the rat. J Neurosci Methods. 1985;14(3):149–67.
126. Damaj MI, Kao W, Martin BR. Characterization of spontaneous and precipitated nicotine withdrawal in the mouse. J Pharmacol Exp Ther. 2003;307(2):526–34.
127. Irvine EE, Cheeta S, File SE. Tolerance to nicotine's effects in the elevated plus-maze and increased anxiety during withdrawal. Pharmacol Biochem Behav. 2001;68(2):319–25.
128. Maren S, Phan KL, Liberzon I. The contextual brain: implications for fear conditioning, extinction and psychopathology. Nat Rev Neurosci. 2013;14(6):417–28.

129. Coe JW, Brooks PR, Vetelino MG, et al. Varenicline: an alpha4beta2 nicotinic receptor partial agonist for smoking cessation. J Med Chem. 2005;48(10):3474–7.
130. Portugal GS, Gould TJ. Genetic variability in nicotinic acetylcholine receptors and nicotine addiction: converging evidence from human and animal research. Behav Brain Res. 2008;193(1):1–16.
131. Andre JM, Gulick D, Portugal GS, Gould TJ. Nicotine withdrawal disrupts both foreground and background contextual fear conditioning but not pre-pulse inhibition of the acoustic startle response in C57BL/6 mice. Behav Brain Res. 2008;190(2):174–81.
132. Raybuck JD, Gould TJ. Nicotine withdrawal-induced deficits in trace fear conditioning in C57BL/6 mice – a role for high-affinity beta2 subunit-containing nicotinic acetylcholine receptors. Eur J Neurosci. 2009;29(2):377–87.
133. Kobayashi Y, Sano Y, Vannoni E, et al. Genetic dissection of medial habenula-interpeduncular nucleus pathway function in mice. Front Behav Neurosci. 2013;7:17.
134. Kobayakawa K, Kobayakawa R, Matsumoto H, et al. Innate versus learned odour processing in the mouse olfactory bulb. Nature. 2007;450(7169):503–8.
135. Robbins TW. The 5-choice serial reaction time task: behavioural pharmacology and functional neurochemistry. Psychopharmacology (Berl). 2002;163(3–4):362–80.
136. Patel S, Stolerman IP, Asherson P, Sluyter F. Attentional performance of C57BL/6 and DBA/2 mice in the 5-choice serial reaction time task. Behav Brain Res. 2006;170(2):197–203.
137. Qin C, Luo M. Neurochemical phenotypes of the afferent and efferent projections of the mouse medial habenula. Neuroscience. 2009;161(3):827–37.
138. Gangitano D, Salas R, Teng Y, Perez E, De Biasi M. Progesterone modulation of alpha5 nAChR subunits influences anxiety-related behavior during estrus cycle. Genes Brain Behav. 2009;8(4):398–406.
139. Hikosaka O, Sesack SR, Lecourtier L, Shepard PD. Habenula: crossroad between the basal ganglia and the limbic system. J Neurosci. 2008;28(46):11825–9.
140. Hikosaka O. The habenula: from stress evasion to value-based decision-making. Nat Rev Neurosci. 2010;11(7):503–13.
141. Brady JV, Nauta WJ. Subcortical mechanisms in emotional behavior: the duration of affective changes following septal and habenular lesions in the albino rat. J Comp Physiol Psychol. 1955;48(5):412–20.
142. Nielson HC, McIver AH. Cold stress and habenular lesion effects on rat behaviors. J Appl Physiol. 1966;21(2):655–60.
143. Dafny N, Qiao JT. Habenular neuron responses to noxious input are modified by dorsal raphe stimulation. Neurol Res. 1990;12(2):117–21.
144. Cohen SR, Melzack R. The habenula and pain: repeated electrical stimulation produces prolonged analgesia but lesions have no effect on formalin pain or morphine analgesia. Behav Brain Res. 1993;54(2):171–8.
145. Matsumoto N, Yahata F, Kawarada K, Kamata K, Suzuki TA. Tooth pulp stimulation induces c-fos expression in the lateral habenular nucleus of the cat. Neuroreport. 1994;5(17):2397–400.
146. Mahieux G, Benabid AL. Naloxone-reversible analgesia induced by electrical stimulation of the habenula in the rat. Brain Res. 1987;406(1–2):118–29.
147. Lee EH, Huang SL. Role of lateral habenula in the regulation of exploratory behavior and its relationship to stress in rats. Behav Brain Res. 1988;30(3):265–71.
148. Thornton EW, Bradbury GE, Davies C. Increased immobility in an automated forced swimming test following lesion of the habenula in rats: absence of evidence for a contribution from motor impairment. Behav Neurosci. 1990;104(1):37–43.
149. Wang RY, Aghajanian GK. Physiological evidence for habenula as major link between forebrain and midbrain raphe. Science. 1977;197(4298):89–91.
150. Herkenham M, Nauta WJ. Efferent connections of the habenular nuclei in the rat. J Comp Neurol. 1979;187(1):19–47.
151. Christoph GR, Leonzio RJ, Wilcox KS. Stimulation of the lateral habenula inhibits dopamine-containing neurons in the substantia nigra and ventral tegmental area of the rat. J Neurosci. 1986;6(3):613–9.

152. Bromberg-Martin ES, Matsumoto M, Hikosaka O. Dopamine in motivational control: rewarding, aversive, and alerting. Neuron. 2010;68(5):815–34.
153. Matsumoto M, Hikosaka O. Electrical stimulation of the primate lateral habenula suppresses saccadic eye movement through a learning mechanism. PLoS One. 2011;6(10):e26701.
154. Wise RA. Roles for nigrostriatal – not just mesocorticolimbic – dopamine in reward and addiction. Trends Neurosci. 2009;32(10):517–24.
155. Dani JA, De Biasi M. Mesolimbic dopamine and habenulo-interpeduncular pathways in nicotine withdrawal. Cold Spring Harb Perspect Med. 2013;3(6):a012138.
156. Hildebrand BE, Nomikos GG, Hertel P, Schilstrom B, Svensson TH. Reduced dopamine output in the nucleus accumbens but not in the medial prefrontal cortex in rats displaying a mecamylamine-precipitated nicotine withdrawal syndrome. Brain Res. 1998;779(1–2):214–25.
157. Rada P, Jensen K, Hoebel BG. Effects of nicotine and mecamylamine-induced withdrawal on extracellular dopamine and acetylcholine in the rat nucleus accumbens. Psychopharmacology (Berl). 2001;157(1):105–10.
158. Rahman S, Zhang J, Engleman EA, Corrigall WA. Neuroadaptive changes in the mesoaccumbens dopamine system after chronic nicotine self-administration: a microdialysis study. Neuroscience. 2004;129(2):415–24.
159. Carboni E, Bortone L, Giua C, Di Chiara G. Dissociation of physical abstinence signs from changes in extracellular dopamine in the nucleus accumbens and in the prefrontal cortex of nicotine dependent rats. Drug Alcohol Depend. 2000;58(1–2):93–102.
160. Gaddnas H, Piepponen TP, Ahtee L. Mecamylamine decreases accumbal dopamine output in mice treated chronically with nicotine. Neurosci Lett. 2002;330(3):219–22.
161. Weiss F, Markou A, Lorang MT, Koob GF. Basal extracellular dopamine levels in the nucleus accumbens are decreased during cocaine withdrawal after unlimited-access self-administration. Brain Res. 1992;593(2):314–8.
162. Rossetti ZL, Hmaidan Y, Gessa GL. Marked inhibition of mesolimbic dopamine release: a common feature of ethanol, morphine, cocaine and amphetamine abstinence in rats. Eur J Pharmacol. 1992;221(2–3):227–34.
163. Jhou TC, Geisler S, Marinelli M, Degarmo BA, Zahm DS. The mesopontine rostromedial tegmental nucleus: a structure targeted by the lateral habenula that projects to the ventral tegmental area of Tsai and substantia nigra compacta. J Comp Neurol. 2009;513(6):566–96.
164. Jhou TC, Fields HL, Baxter MG, Saper CB, Holland PC. The rostromedial tegmental nucleus (RMTg), a GABAergic afferent to midbrain dopamine neurons, encodes aversive stimuli and inhibits motor responses. Neuron. 2009;61(5):786–800.
165. McCallum SE, Cowe MA, Lewis SW, Glick SD. alpha3beta4 nicotinic acetylcholine receptors in the medial habenula modulate the mesolimbic dopaminergic response to acute nicotine in vivo. Neuropharmacology. 2012;63(3):434–40.
166. Kim U, Chang SY. Dendritic morphology, local circuitry, and intrinsic electrophysiology of neurons in the rat medial and lateral habenular nuclei of the epithalamus. J Comp Neurol. 2005;483(2):236–50.
167. Nishikawa T, Fage D, Scatton B. Evidence for, and nature of, the tonic inhibitory influence of habenulo-interpeduncular pathways upon cerebral dopaminergic transmission in the rat. Brain Res. 1986;373(1–2):324–36.
168. Groenewegen HJ, Ahlenius S, Haber SN, Kowall NW, Nauta WJ. Cytoarchitecture, fiber connections, and some histochemical aspects of the interpeduncular nucleus in the rat. J Comp Neurol. 1986;249(1):65–102.
169. Lodge DJ, Grace AA. The laterodorsal tegmentum is essential for burst firing of ventral tegmental area dopamine neurons. Proc Natl Acad Sci U S A. 2006;103(13):5167–72.
170. Grace AA, Floresco SB, Goto Y, Lodge DJ. Regulation of firing of dopaminergic neurons and control of goal-directed behaviors. Trends Neurosci. 2007;30(5):220–7.
171. Duchemin AM, Zhang H, Neff NH, Hadjiconstantinou M. Increased expression of VMAT2 in dopaminergic neurons during nicotine withdrawal. Neurosci Lett. 2009;467(2):182–6.
172. Hadjiconstantinou M, Neff NH. Nicotine and endogenous opioids: neurochemical and pharmacological evidence. Neuropharmacology. 2011;60(7–8):1209–20.

173. Thierry AM, Tassin JP, Blanc G, Glowinski J. Selective activation of mesocortical DA system by stress. Nature. 1976;263(5574):242–4.
174. Inglis FM, Moghaddam B. Dopaminergic innervation of the amygdala is highly responsive to stress. J Neurochem. 1999;72(3):1088–94.
175. Bradberry CW, Lory JD, Roth RH. The anxiogenic beta-carboline FG 7142 selectively increases dopamine release in rat prefrontal cortex as measured by microdialysis. J Neurochem. 1991;56(3):748–52.
176. Broersen LM, Abbate F, Feenstra MG, de Bruin JP, Heinsbroek RP, Olivier B. Prefrontal dopamine is directly involved in the anxiogenic interoceptive cue of pentylenetetrazol but not in the interoceptive cue of chlordiazepoxide in the rat. Psychopharmacology (Berl). 2000; 149(4):366–76.
177. Kawasaki H, Kaufman O, Damasio H, et al. Single-neuron responses to emotional visual stimuli recorded in human ventral prefrontal cortex. Nat Neurosci. 2001;4(1):15–6.

Chapter 19
Nicotinic Acetylcholine Receptors in Alzheimer's and Parkinson's Disease

Kelly T. Dineley

Abstract In the CNS, nicotinic acetylcholine receptors prominent roles in modulating presynaptic and postsynaptic signaling and have been implicated in several CNS disorders including the two most prominent neurodegenerative diseases: Alzheimer's disease and Parkinson's disease. These neurodegenerative diseases affect scores of millions of persons worldwide with their prevalence increasing as human longevity increases.

This chapter will provide an overview of Alzheimer's disease and Parkinson's disease, the cholinergic system affected in each disorder, and the types of nicotinic acetylcholine receptors affected during disease progression. Finally, a discussion of therapeutic strategies targeting nicotinic acetylcholine receptors is included based upon the most current preclinical and clinical research.

Keywords Neurodegeneration • Cholinergic • Basal forebrain cholinergic system • Hippocampus • Striatum • Substantia nigra • Acetylcholine • Dopamine

1 Alzheimer's Disease

Alzheimer's disease affects an estimated 5.2 million people in the USA and 26.6 million people worldwide; Alzheimer's disease is the leading cause of dementia in elderly people [1]. With the proportion of elderly people in the population increasing steadily, the burden of the disease, both to caregivers and national economies, is expected to become substantially greater over the next 2–3 decades [1]. Alzheimer's disease is a progressive neurodegenerative disorder with survival of 4–8 years between diagnosis and death [1]. Brain regions that are associated with higher mental functions lose cholinergic innervation; thus, cholinergic denervation is most severe in the temporal lobes and the adjacent limbic and paralimbic areas including the hippocampus [2, 3]. These brain regions preferentially accumulate misfolded

K.T. Dineley, B.A., M.S., Ph.D. (✉)
Department of Neurology, University of Texas Medical Branch,
301 University Boulevard, Galveston, TX 77555, USA
e-mail: ktdinele@utmb.edu

R.A.J. Lester (ed.), *Nicotinic Receptors*, The Receptors 26,
DOI 10.1007/978-1-4939-1167-7_19

amyloid-β peptide (Aβ) and tau-containing neurofibrillary tangles [4–6]. The soma of cholinergic projection neurons resides within the nucleus basalis of Meynert and the septal diagonal band complex to provide the major source of cholinergic innervation to the cerebral cortex and hippocampus, respectively [7]. In early Alzheimer's disease, there is impairment in hippocampus-based episodic memory that is improved through enhancement of cholinergic transmission, indicating that compromise of septo-hippocampal connectivity underlies the earliest symptomatology in Alzheimer's disease dementia [8]. The current model for the cholinergic deficit in Alzheimer's disease posits that inappropriate accumulation of misfolded oligomeric aggregates of amyloid-β (Aβ) peptide leads to presynaptic deterioration of the septo-hippocampal pathway first manifest as loss of the cholinergic phenotype, e.g., loss of cholinergic markers and eventually cholinergic neurons from the basal forebrain cholinergic system, and is postulated to result from altered nicotinic acetylcholine receptor function and disruption of the nerve growth factor trophic support system [9–19].

1.1 The Basal Forebrain Cholinergic System

Acetylcholine is an essential neurotransmitter for a variety of attentional, learning, and memory processes [20–22]. The cholinergic basal forebrain, a midbrain complex that most notably includes the medial septum, diagonal band, and nucleus basalis of Meynert, is a major source of acetylcholine and provides the principal cholinergic innervation to the cortex and hippocampus. In presynaptic nerve terminals, choline acetyltransferase synthesizes acetylcholine, and the vesicular acetylcholine transporter is responsible for the transport of acetylcholine into synaptic vesicles for storage until exocytotic release into the synapse [22, 23].

Nerve growth factor (NGF) neurotrophic signaling maintains the cholinergic phenotype by promoting cholinergic neuron survival and expression of cholinergic markers such as choline acetyltransferase and the high-affinity receptor for NGF, TrkA (tyrosine kinase receptor A), through its role in maintaining and promoting the cholinergic phenotype; NGF plays an important role in learning and memory. Lesions of the septo-hippocampal pathway are an avenue by which NGF deprivation can be induced in vivo, and this leads to cholinergic hypofunction expressed as reduced choline acetyltransferase protein and activity accompanied by impaired hippocampus-dependent memory [24, 25] which can be ameliorated through infusion of NGF into the septum and restoration of cholinergic activity [26, 27]. Furthermore, endogenous NGF levels correlate with an animal's capacity for hippocampus-dependent learning [28, 29]. As such, a central concept regarding the maintenance of the cholinergic phenotype in the basal forebrain (i.e., neuronal survival and expression of cholinergic marker genes) requires NGF-mediated signaling through interaction with both high-affinity TrkA and low-affinity p75 neurotrophin receptors.

NGF is a target-derived neurotrophin that is internalized upon binding to its receptor(s) on cholinergic nerve terminals then packaged for retrograde axonal transport to the cholinergic soma and nuclei in the cholinergic basal forebrain [30]. In order to exert its pro-survival and cholinergic phenotype-promoting activities, NGF must be synthesized and secreted in adequate quantity from basal forebrain target regions such as the hippocampus. Mature NGF is generated after cleavage at the carboxy terminus of its monomeric precursor form, pro-NGF [31, 32]. Thus, the secreted neurotrophin is a collective mixture of pro- and mature NGF that will potentially bind TrkA and/or p75 neurotrophin receptors.

Mature NGF binds to the TrkA receptor, while pro-NGF preferentially binds to p75 neurotrophin receptors over TrkA leading to apoptotic death of cells co-expressing both receptors [33]. Although the regulatory mechanisms underlying the retrograde trafficking of NGF-engaged TrkA and p75 neurotrophin receptors are not fully characterized, it is postulated that when NGF bound to its receptor(s) arrives at its presynaptic locus within cholinergic basal forebrain target regions, it activates pro-survival signaling cascades to influence target gene transcription (e.g., ChAT, TrkA; [34].

For example, NGF receptor signaling activates the mitogen-activated protein kinase pathway, which participates in a wide array of biologic functions, including cell survival, synaptic plasticity, and learning and memory [35]. The classic MAPK cascade involves activation of the ribosomal S6 kinases which phosphorylates several transcription factors including cAMP-regulated response element-binding protein (CREB) [35, 36]. A second downstream pathway is the PI3K/Akt pathway that also mediates neurotrophin-mediated survival [37]. As will be discussed below, NGF receptor signaling and α7 nAChR activation share these downstream mediators and support the hypothesis that α7 nAChRs contribute to cholinergic basal forebrain integrity which may be exploited during Alzheimer's disease progression.

1.2 Nicotinic Receptors in the Basal Forebrain Cholinergic System

The nuclei of the basal forebrain primarily contain cholinergic and GABAergic neurons that project to the hippocampus, cortex, and olfactory bulbs, brain regions that are particularly vulnerable to the ravages of early Alzheimer's disease. Nicotinic acetylcholine receptors, primarily α4β2* and α7 subtypes (where the asterisk denotes the possible contribution of an additional α or β subunit), are expressed within the cholinergic forebrain nuclei as well pre- and postsynaptically in the basal forebrain cholinergic neuron targets. For example, in the hippocampus, α4β2* and α7 nicotinic acetylcholine receptors have been localized presynaptically and somatodendritically on GABAergic interneurons as well as principal cells [38], and the cholinergic deficit in Alzheimer's disease is due in part to altered expression of these receptors within the septo-hippocampal pathway [39–42]. Within this circuit, α4β2* nicotinic acetylcholine receptors and α7 nAChRs are important players in hippocampal excitability, the induction of LTP, and learning and memory, as well as mediating the cognitive enhancing effects of in vivo administered nicotine [43–47].

1.3 Alzheimer's Disease and Nicotinic Acetylcholine Receptors

In advanced Alzheimer's disease, there is profound loss of cholinergic markers (e.g., choline acetyltransferase protein) and nicotinic acetylcholine receptors within neocortical areas, with muscarinic acetylcholine receptors left relatively intact [48–54]. Binding studies on brain tissue obtained at autopsy employing radiolabeled nicotinic agonists such as [^{3}H]acetylcholine, [^{3}H]nicotine, and [^{3}H]methyl carbamylcholine at nmol/L concentrations that define high-affinity α4β2* nicotinic acetylcholine receptor binding sites have consistently concluded that 20–50 % of these receptors are lost from a number of neocortical areas and hippocampi of patients with Alzheimer's disease [54–60]. Although follow-up studies utilizing immunoblot and immunohistochemistry have been called into question due to potential antibody technical issues, it is thought that the major contributor to the loss of high-affinity nicotinic acetylcholine receptors in Alzheimer's brain cortical areas is likely due to loss of the α4 subunit [61–68]. The reduction in α4 subunit protein does not appear to be the result of attenuated gene transcription because mRNA levels in Alzheimer's brain samples appear comparable to those of age-matched control subjects [69].

With regard to α7 nicotinic acetylcholine receptors in Alzheimer's disease, the extent of deficits in cerebral cortical areas appears to be more restricted than for α4-containing receptors in terms of magnitude and areas involved. Of note is the observed lack of consistency between individual cases [51]. To summarize several studies utilizing radiolabeled toxin specific for CNS α7 nicotinic acetylcholine receptors, it has been determined that frontal cortex does not exhibit loss of α-bungarotoxin-binding sites [70, 71]; within temporal cortex, entorhinal cortex, and hippocampus, considerable variation has been found leading to reports of both no significant difference and 25–40 % reduction in [^{125}I]α-bungarotoxin-binding sites [69, 70].

While loss of nicotinic acetylcholine receptors at end-stage disease is not actively debated, the story for early phases of Alzheimer's disease is quite different. An early form of Alzheimer's disease has recently been clinically defined and has been termed "mild cognitive impairment due to probable Alzheimer's disease" and is identified as a prodromal phase based on newly defined diagnostic and staging criteria [72]. Studies utilizing postmortem samples from this stage of Alzheimer's disease have concluded that choline acetyltransferase activity is preserved in the neocortex of these patients; in fact, some studies have concluded that choline acetyltransferase activity is actually elevated in the hippocampus and frontal cortex of mild cognitive impairment subjects [73–75]. Mechanisms proposed to underlie this increase include sprouting in response to loss of glutamatergic input and/or resilience of particular cholinergic nuclei to Alzheimer's disease processes [76–78]. An alternative explanation is ligand-induced nicotinic receptor upregulation resulting from an interaction with elevated levels of soluble Aβ during early Alzheimer's disease [79–82].

Along the lines of transient compensatory changes in cholinergic markers within the basal forebrain cholinergic system, one study specifically investigated mRNA expression levels of nicotinic and muscarinic acetylcholine receptor subtypes and choline acetyltransferase in single cells isolated from the cholinergic basal forebrain of postmortem Alzheimer's disease tissue and age-matched controls. No differences in mRNA expression were observed for the other nicotinic acetylcholine receptor subunits, muscarinic acetylcholine receptor subtypes, or choline acetyltransferase [75]. However, cells from Alzheimer's disease basal forebrain exhibited a significant upregulation of α7 nicotinic acetylcholine receptor subunit mRNAs [75]. This increase in α7 nicotinic acetylcholine receptor expression levels within cholinergic basal forebrain neurons was inversely correlated with global cognitive score and with mini-mental state examination performance, neuropsychological tests that are used to help diagnose Alzheimer's disease and other dementias [75]. As such, increased α7 nAChR may be a compensatory attempt to regulate cholinergic tone at the gene transcription level.

Elucidating the molecular mechanisms that underlie the selective vulnerability of cholinergic neurons to Aβ toxicity would greatly advance our capability to treat the Alzheimer's cholinergic deficit. The observation that Aβ preferentially accumulates in neuronal populations that are also enriched for α4β2 and α7 nicotinic acetylcholine receptors may provide an important clue given the mounting evidence that these two abundant nicotinic acetylcholine receptors interact with Aβ [80, 81]. Importantly, α7 nicotinic acetylcholine receptors exhibit an exceptionally high affinity (picomolar range) for Aβ peptides [83], suggesting that these two proteins may interact under normal physiological conditions and evidence is mounting that this interaction may influence synaptic transmission and plasticity in the hippocampus [79, 84, 85]. Regarding Aβ and the α4β2* subtypes of nicotinic receptors, there are indications that basal forebrain cholinergic neurons express such a nicotinic receptor that is responsive to nM Aβ. Neurons acutely isolated from the diagonal band nucleus of the basal forebrain cholinergic region were found to be responsive to Aβ [86]. In these studies, electrophysiological recordings demonstrated that a nicotinic current was blocked by the broad-spectrum nicotinic acetylcholine receptor antagonist mecamylamine but not the α7-selective antagonist methyllycaconitine. These studies indicate that on basal forebrain cholinergic neurons (rather than GABAergic which more abundantly express α7 nicotinic acetylcholine receptors [87]), there is a population of postsynaptically located non-α7 nAChRs that are activated by nM concentrations of Aβ. Recent reports that α7 nicotinic acetylcholine receptors are upregulated on astrocytes, peripheral blood leukocytes, and cortical and hippocampal neurons harvested from the tissue of Alzheimer's patients [19, 69, 88, 89] further suggest that these increases in α7 nAChR protein may be in response to a direct interaction with the increasing burden of Aβ in Alzheimer's disease. Nicotinic acetylcholine receptors upregulate in response to agonist and antagonist exposure through receptor desensitization-mediated upregulation [90–94]. Recent studies have shown that in Alzheimer's disease tissue samples, much of the α7 nAChR protein in brain regions targeted by the cholinergic basal forebrain is associated with Aβ; disruption of this association in postmortem Alzheimer's cortex leads to increased availability of functional α7 nAChRs [18, 19]. These observations suggest that in Alzheimer's disease, α7 nAChRs are likely inactive due to desensitization as a consequence of prolonged association with Aβ peptide [83, 95]. This model of α7 nAChR desensitization is supported by in vitro studies showing that acute exposure of organotypic slice cultures derived from hippocampus to low (picomolar) concentrations of Aβ leads to PI3K activation that results in CREB phosphorylation [95–99]; second messengers that are thought to promote neuronal survival as well as contribute to neuronal plasticity [100–103]. In contrast, intermediate exposure times to Aβ (i.e., minutes to hours) reduced activation of these signaling pathways [96, 99], whereas extended exposure to Aβ (i.e., over the course of days) upregulated α7 nAChRs in hippocampal slice cultures, which is observed in Aβ animal models [82] and for Alzheimer's disease autopsy brain samples [75]. Thus, several have speculated that under normal physiological conditions and early in the disease, α7 nAChR–Aβ interaction likely results in transient α7 nAChR activation and downstream signal transduction cascades that promote neuronal survival and function; however, as Aβ concentration increases, prolonged association with Aβ leads to receptor desensitization and *functional* downregulation [79, 104, 105]. Thus, the observed increase in α7 nAChR expression in human Alzheimer's disease basal forebrain cholinergic system may reflect an α7 nAChR–Aβ interaction that actually contributes to the disease process.

Although relatively uncharacterized in Alzheimer's disease, it is imperative to mention a recently identified nicotinic acetylcholine receptor subtype in basal forebrain cholinergic neurons and interneurons of the hippocampus that exhibit particular sensitivity to Aβ antagonism [106–108]. This new subtype appears to be comprised of α7 and β2 subunits in a heteromeric complex. Not too surprising given previous reports of both subunit mRNA's localized to cholinergic neurons in the basal forebrain [87]. In comparison to homomeric α7 nAChRs, heteromeric α7β2* nAChRs exhibit relatively slow whole-cell current kinetics and are more sensitive to the β2 subunit-containing nAChR selective antagonist, DHβE. This current profile was absent in neurons prepared from β2 nicotinic acetylcholine receptor knockout mice. α7β2* nicotinic acetylcholine receptors were antagonized by 1 nM Aβ, whereas homomeric α7 nicotinic acetylcholine receptors were not, suggesting that the heteromeric receptors are more sensitive to Aβ antagonism. Such selective sensitivity to relatively low concentrations of Aβ suggests that this nicotinic receptor may contribute to the selective sensitivity of the basal forebrain and hippocampus to early Alzheimer's disease pathology.

1.4 Targeting Nicotinic Acetylcholine Receptors in Alzheimer's Disease

Since the discovery that cholinergic basal forebrain neurons degenerate in Alzheimer's patients and treatment with cholinergic antagonists disrupts learning and memory function in humans and rodents, the cholinergic hypothesis of cognitive dysfunction has been under intense investigation [9]. An early observation in Alzheimer's research and a fundamental principle for current treatment strategies is the loss of cholinergic markers within the basal forebrain cholinergic system.

Current understanding of the disease posits that the manifestation of episodic memory impairments during early-stage Alzheimer's disease is thought to be triggered by accumulating Aβ and is associated with synaptic dysfunction and compromised cholinergic neurotransmission between the cholinergic basal forebrain and its targets in the cortex and hippocampus [109–112]. As such, most FDA-approved drugs currently used in Alzheimer's treatment are geared toward boosting acetylcholine-mediated neurotransmission through the use of drugs that block the main enzyme responsible for degrading acetylcholine at the synapse, acetylcholinesterase, with the idea that prolonging the half-life of acetylcholine at the synapse will boost cholinergic signaling [9, 113].

In general, long-term clinical assessments indicate that the main effect of anticholinesterase drugs is symptomatic treatment with limited disease-modifying actions [114]. Since the cholinergic deficit may not represent an early defect in the progression of Alzheimer's disease [73–75], the use of these drugs in the prodromal stages of Alzheimer's disease and for the treatment of cognitive decline in mild-to-moderate Alzheimer's disease should be continued. However, the limited effect of cholinesterase inhibitors on behavioral symptoms in severe Alzheimer's disease poses a significant clinical challenge [115]; loss of nicotinic acetylcholine receptor function and protein may be a significant contributor to perceived therapeutic limitations associated with cholinesterase inhibitor drugs [116, 117].

Nicotine treatment has been shown to improve attention, as well as learning and memory performance in patients with mild-to-moderate Alzheimer's. In fact, α4β2* and α7* nicotinic acetylcholine receptors, either on cholinergic projection neurons or within the neocortex of the basal forebrain cholinergic system, are important for the types of cognitive performance that are impaired in early Alzheimer's disease. An additional bonus from nicotinic acetylcholine agonist therapy is a neuroprotective effect that appears to result following activation of nicotinic acetylcholine receptors. Such activation has been shown to protect neurons from a variety of toxin insults including Aβ through stimulation of the PI3K pathway, presumably through transactivation of src kinase receptors [118–122]. Thus, several preclinical as well as clinical trials have tested subtype-selective agonists and partial agonists that target α4β2 nicotinic receptors [123–133] as well as α7 nicotinic receptors [134–138] in mild-to-moderate Alzheimer's disease. While, in all cases, efficacy was found in preclinical models, trials in subjects diagnosed with Alzheimer's yielded mixed results [139]. While study design and placebo effects may account for some of the variability, it is now evident that many Alzheimer's disease clinical trials likely "fail" due to enrollment of patients at disease stages that preclude any opportunity for disease-modifying effects, thus the emerging consensus that the best treatment strategy for Alzheimer's disease is to treat people during the earliest stages of disease, e.g., prodromal Alzheimer's or mild cognitive impairment due to Alzheimer's disease [140–143]. The current challenge is to reliably diagnose at such early stages using a combination of biomarkers, brain imaging, and cognitive testing [72, 144].

An alternative therapeutic strategy was initially pursued by Servier. S-24795 is an α7 nAChR partial agonist that both inhibits and partially reverses Aβ binding to these receptors [18, 19]. However, it remains to be seen if the dislodged Aβ is then free to interact in alternative but equally deleterious ways. Possibly coincident Aβ immunotherapy would alleviate this potential negative side effect of S-247945 therapy. Another possible strategy, albeit somewhat difficult to envision at the receptor level, would be to develop a compound that is capable of maintaining α7 nAChR neuroprotective signaling capabilities on the one hand and continue to sequester Aβ on the other. Again, this in conjunction with interventions that decrease Aβ levels might prove most efficacious.

An α7 nAChR–Aβ interaction was first described over a decade ago. Initial studies reported seemingly incongruent consequences of this interaction such as receptor antagonism versus activation [80, 81]. As we delve deeper and refine our understanding of this interaction and how it relates to the pathophysiology of Alzheimer's disease progression, it is evident that Aβ effects on α7 nAChRs are quite dynamic and, at minimum, depend upon the concentration and aggregation state of Aβ which may have profound effects on receptor responsiveness which, in turn, may have profound effects on the responsiveness of the basal forebrain cholinergic network.

Assuming that, under normal physiological conditions, Aβ and α7 nAChRs interact and result in receptor activation implies that this interaction may serve a neuroprotective role given that α7 nAChRs couple to neuroprotective signaling cascades (PI3K, etc.). Therefore, it seems unwise to prophylactically block *all* α7 nAChR–Aβ interaction. However, as Alzheimer's disease progresses and soluble Aβ acquires pathological concentrations and conformations, it might be useful to develop ways in which to interrupt specific α7 nAChR–Aβ interactions, especially if this interaction antagonizes receptor function or is involved in accumulating intracellular Aβ [145]. As is being currently pursued, targeting Aβ directly with immunotherapy is one approach that holds promise [140].

Nicotinic acetylcholine receptors comprised of α4β2 or α7 subunits have been strongly implicated in the molecular etiology of Alzheimer's disease. Specifically, these nicotinic receptor subtypes have been implicated in conferring Alzheimer's disease-specific vulnerability of the cholinergic basal forebrain system, the initial presentation of attentional and episodic memory deficits, and the therapeutic efficacy of cholinesterase inhibitors during initial disease progression. Continued design and synthesis of nicotinic receptor ligands with the desired pharmacokinetics and pharmacodynamics to target the desired receptor population in combination with continued elucidation of the disease-stage properties of nicotinic receptors during the prodromal phase of Alzheimer's disease as well as during early progression may ultimately achieve the goal of developing an interventional tool to combat the loss of cholinergic basal forebrain connectivity in this most prevalent of the devastating neurodegenerative disorders.

2 Parkinson's Disease

Parkinson's disease is the second most common neurodegenerative disease affecting 10 % of people over the age of 65 (for review, see [146]. This neurodegenerative movement disorder is characterized by postural instability, bradykinesia, and asymmetric onset of tremor and rigidity that result from degeneration of the nigrostriatal dopaminergic pathway [147, 148]. Although numerous CNS neurotransmitter systems are compromised in Parkinson's disease, including the adrenergic, cholinergic, serotonergic, glutamatergic, and GABAergic systems, the nigrostriatal dopaminergic pathway is most severely affected. Degeneration of these other neurotransmitter systems both centrally and peripherally likely contribute to the nonmotor symptoms associated with Parkinson's disease such as deficits in cognition and memory, depression, affect, circadian patterns, and autonomic function.

Similar to Alzheimer's disease, Parkinson's disease is comprised of a minority (~5 %) of familial cases that arise from mutations that cause either autosomal dominant (SNCA and LRRK2) or autosomal recessive (PARKIN, PINK1, and DJ1) Parkinson's disease [149]. Mutations in *PARK2* and *PINK1* appear to be the most common causes of early-onset (<45 years of age) Parkinson's disease. Mutations in LRRK2, coding for leucine-rich repeat serine/threonine-protein kinase 2, are the most frequent cause of genetic Parkinson's disease, with the most common mutation (G2019S) accounting for 1 % of sporadic and 4 % of familial cases [150]. Furthermore, the penetrance of LRRK2 mutations is age dependent—less than 20 % at age 45 and more than 80 % at age 80. The remainder of Parkinson's disease cases are classified as sporadic and are attributed to a complex interplay between genetic and environmental factors (for review, see [151]. However, similar to Alzheimer's disease, age or the aging process is the most significant contributing factor to the development of Parkinson's disease, and a current challenge in the Parkinson's disease field is to understand how autosomal genes as well as GWAS-identified risk genes and the aging process (including environmental factors) may intermingle to contribute to the disease process [152]. From an environmental perspective, Parkinson's disease risk is most strongly linked to pesticide exposure whereas tobacco use has consistently been linked to a decreased frequency of the disease [153]. Nonetheless, the etiology of Parkinson's disease is currently indeterminate.

Currently, the most effective treatment for Parkinson's disease is dopamine replacement therapy with levodopa, also called L-DOPA, which is the precursor to the neurotransmitters dopamine, norepinephrine (noradrenaline), and epinephrine (adrenaline) collectively known as catecholamines. L-DOPA is sometimes prescribed in combination with dopamine agonists and inhibitors of dopamine turnover as adjunctive therapies [154]. These drugs remain the most effective symptomatic treatment of Parkinson's disease; however, long-term administration of L-DOPA and related dopaminergic therapies is marred by the emergence of abnormal involuntary movements, i.e., L-DOPA-induced dyskinesia, that diminishes its long-term therapeutic efficacy [155]. Moreover, the nonmotor symptoms, especially depression and dementia, are not alleviated with dopamine-directed therapies [156]. In short, pharmacological management of Parkinson's disease is complex and requires continual monitoring and individualization for each patient such as appropriate timing of dopaminergic therapy and adoption of strategies to delay and treat L-DOPA-related motor complications and nonmotor Parkinson's disease-related symptoms. Moreover, as is the case in all neurodegenerative diseases, there is great need for disease-modifying therapies and interventions that can address the spectrum of Parkinson's disease symptoms. Given the extensive anatomical and functional overlap between the nicotinic cholinergic and dopaminergic systems in the nigrostriatal pathway and the epidemiological and pharmacological evidence that nicotinic receptor drugs alleviate some of the nonmotor as well as motor symptoms of Parkinson's disease, the nicotinic acetylcholine receptor system has been proposed as a potential therapeutic strategy in Parkinson's disease. The following sections will discuss these aspects with particular emphasis on nicotinic acetylcholine receptor functional anatomy and nicotinic receptor subtypes affected during disease progression and when targeted in Parkinson's disease preclinical models.

2.1 The Nicotinic Cholinergic and Dopaminergic Systems Involved in Parkinson's Disease

2.1.1 Striatum

While the function of the striatum in motor control is not completely understood, it appears that it is involved in the enabling of practiced motor movements and in gating the initiation of voluntary movements. Thus, voluntary movements are not initiated in the striatum; however, proper functioning of the striatum appears to be necessary in order for the motor cortex to relay the appropriate motor commands to the rest of the circuitry during voluntary movement. The following is a brief discussion of the circuitry and neurotransmitter systems that comprise the striatum and how nicotinic acetylcholine receptors are functionally integrated.

Parkinson's disease has traditionally been considered a motor syndrome that results from degeneration of the dopaminergic afferents from the substantia nigra to the striatal medium spiny neurons. The striatum is a subcortical part of the forebrain that in primates consists of the caudate and putamen separated by a white matter tract termed the internal capsule; in rodents the caudate and putamen are a single merged structure. The dorsal striatum is primarily innervated by dopamine neurons from the substantia nigra pars compacta, with little input from the ventral tegmental area. It is this nigrostriatal pathway that degenerates in Parkinson's disease. The ventral striatum (or nucleus accumbens core), on the other hand, mainly receives dopaminergic input from the ventral tegmental area with a minor component coming from the dorsal substantia nigra. As stated above, the dorsal striatum and the nigrostriatal dopaminergic projections are most severely affected in Parkinson's disease leaving the mesolimbic projection from the ventral tegmental area relatively spared (for review, see [157, 158].

The dopaminergic input to the striatum arrives from the substantia nigra to innervate GABAergic medium spiny neuron dendritic shafts (corticostriatal glutamatergic afferents converge on the dendritic spine heads) and substantially contribute to the output of the striatum via the direct and indirect pathways to the basal ganglia whereby motor function is arbitrated [158–160]. The direct pathway sends information from the striatum directly to the substantia nigra pars reticulata and the entopeduncular nucleus (rodents) or the internal segment of the globus pallidus (primates), then onward to the brainstem to mediate motor movements of the head, neck, and eye or onward to the thalamus and motor cortex to mediate motor movements involving the arms, legs, and trunk. It should be noted that both the striatal and thalamic/brainstem connections are inhibitory such that when the direct pathway striatal neurons fire, they inhibit the activity of their downstream connections (substantia nigra/entopeduncular nucleus in rodents and the internal segment of the globus pallidus in primates) that leads to disinhibition of thalamus/brainstem allowing them to excite the motor cortex and cranial nerves.

The indirect pathway starts with a different set of cells in the striatum that make inhibitory connections to the globus pallidus (the external segment in primates) that make subsequent inhibitory connections within the subthalamic nucleus which in turn make excitatory connections with the substantia nigra pars reticulata and the entopeduncular nucleus (the internal segment of the globus pallidus in primates). Again, take note that when the substantia nigra pars reticulata and entopeduncular nucleus (internal segment of the globus pallidus in primates) cells are active, they inhibit thalamic neurons, thus making the motor cortex/brainstem less active. When the subthalamic neurons are firing, they increase the firing rate of substantia nigra pars reticulata and entopeduncular nucleus neurons, thus increasing the net inhibition on motor cortex/brainstem. Thus, when the indirect pathway striatal neurons are active, they inhibit the globus pallidus neurons, thus disinhibiting the subthalamic neurons. With the subthalamic neurons free to fire, the substantia nigra pars reticulata and entopeduncular nucleus neurons inhibit the thalamus/brainstem, thereby producing a net inhibition on the motor activity.

Within the striatum, in addition to the symmetric dopaminergic inputs from the substantia nigra pars compacta that impinge upon synaptic as well as extrasynaptic dopamine receptors, there are corticostriatal glutamatergic afferents that synapse on the medium spiny neuron dendritic heads to activate metabotropic glutamate receptors to modulate striatal activity. Additionally, there is a widespread cholinergic interneuron system that is tonically active under basal conditions. Since these cholinergic arborizations extensively intermingle with that of the dopaminergic network, there is much cross talk between the two systems within the striatum. These, in addition to a GABAergic interneuron system and serotonergic input from the raphe nucleus, intrinsic activity within the striatum is integrated with the various afferent inputs to determine whether the striatum maintains "resting" state tonic activity or shifts to burst firing.

Tonic striatal activity has been attributed to low dopaminergic afferent activity and high intrinsic pulsatile acetylcholine release. The high level of acetylcholinesterase in the striatum expedites the hydrolysis of acetylcholine, enabling extracellular acetylcholine to reflect its release pattern as well as minimizing receptor desensitization [161]. It is thought that the coordinated reciprocal actions of the dopaminergic afferents and cholinergic interneurons within the striatum, which in turn rely upon substantia nigra and thalamic inputs to the striatum, drive striatal dopaminergic activity [162–164]. However, recent work using simultaneous electrophysiological recordings of striatal channelrhodopsin2-expressing cholinergic interneurons with simultaneous detection of dopamine release in striatal slices suggests that cholinergic neurons facilitate, or potentially drive, dopamine release from dopaminergic axonal inputs [165]. Thus, presynaptic nicotinic acetylcholine receptors on dopaminergic afferents from the substantia nigra potentially synergize with ascending dopaminergic activity [165, 166].

Therefore, the presynaptic location of nicotinic acetylcholine receptors throughout the striatum (except the cholinergic interneuron population which are endowed with inhibitory muscarinic acetylcholine autoreceptors) designates this receptor system as a major influence on striatal output despite being greatly outnumbered by muscarinic acetylcholine receptors which are located presynaptically on afferents and somatodendritically on medium spiny neurons [167–172].

2.1.2 Substantia Nigra

The dopaminergic neurons arising from the substantia nigra pars compacta projecting to the dorsal striatum are the seat of control for modulating motor functions as well as cognitive aspects of motor learning (for review, see [173]. In addition to this major striatal innervation, the substantia nigra sends projections to innervate the cortex and related limbic areas that modulate the principal glutamatergic and GABAergic transmission. Within the substantia nigra (as well as the ventral tegmental area), dopamine is also released; this local release can be modulated by nicotinic acetylcholine receptors located somatodendritically on dopaminergic neurons as well as by nicotinic acetylcholine receptors located presynaptically on GABAergic interneurons as well as GABAergic and glutamatergic afferents. The substantia nigra pars compacta GABAergic interneurons serve to inhibit dopaminergic neuron activation thus contributing to the reciprocal striatal dopamine neuron tonic and burst firing. The switch to dopaminergic neuron burst firing is dependent on glutamatergic innervation of the substantia nigra pars compacta received from the subthalamic nucleus and pedunculopontine nucleus [174] which may also be endowed with presynaptic nicotinic acetylcholine receptors [175]. Again, as is the case in the striatum, nicotinic acetylcholine receptors significantly influence the activity of the substantia nigra pars compacta due to their presynaptic location on glutamatergic and GABAergic afferents as well as somatodendritic location on GABAergic interneurons and dopaminergic neurons.

The acetylcholine that binds nicotinic acetylcholine receptors in the substantia nigra pars compacta comes from the pedunculopontine nucleus located in the pons, the origin for the glutamatergic input to this brain region. While stimulation of the pedunculopontine nucleus elicits burst firing in the substantia nigra pars compacta and dopamine release in the striatum, either nicotinic or glutamatergic receptor broad antagonists applied to the substantia nigra inhibits these processes suggesting that these two receptor systems work in concert to maximize dopaminergic neuron burst firing [176, 177].

2.2 *Nicotinic Acetylcholine Receptor Subtypes in the Nigrostriatum*

Much effort has been expended to decipher the subunit combinations that form functional nicotinic acetylcholine receptors within defined brain regions and circuits as well as their subcellular location. While in situ hybridization provides the first level of information along these lines, genetic deletion of specific subunits and lesion studies (e.g., lesioning the nigrostriatal pathway with 6-hydroxydopamine or 1-methyl-4-phenyl-1,2,3,6-tetrahydropyradine), in combination with pharmacological tools that define receptors at the subunit composition level (most notably α-conotoxin MII; [178, 179]), the subunit composition, and in some cases, the stoichiometry, can be deduced based upon affinity of these subunit-selective compounds. These approaches, in combination with electrophysiological recordings complemented by binding studies and immunoprecipitation assays, have significantly contributed to the identification of nicotinic acetylcholine receptor subtypes and stoichiometry in the striatum and substantia nigra [180].

In the substantia nigra pars compacta, where the dopaminergic cell bodies that innervate the striatum reside, a broad array of nicotinic acetylcholine receptor subunits is expressed. The spectrum expressed by dopamine neurons is diverse and more complex than those expressed by GABAergic neurons. The dopamine neuron cell bodies in the substantia nigra pars compacta exhibit α4β2* and α6β2* nicotinic acetylcholine receptor subtypes where the asterisk denotes the possible contribution of an additional α4 or β2 subunit or α5 in the case of α4β2* and denotes the possible contribution of an α5 or β3 subunit with two each of the α6 and β2 subunits [181, 182]. In addition, α7 nicotinic acetylcholine homopentameric receptors are present on a smaller fraction of these cell bodies [175, 182]. Those expressed by the GABAergic interneurons and afferents exhibit a somewhat simpler expression pattern comprised mainly of α4β2* subunits somatodendritically and presynaptically, respectively. The asterisk in this instance designates the possible participation of an additional α4 or β2 subunit or α3 [175, 182]. Clearly, these populations of nicotinic acetylcholine receptors comprised of discrete subunit combinations are predicted to exhibit unique pharmacological profiles. This, in addition to the discrete localization of certain of these receptor subtypes to GABAergic and dopaminergic neurons, offers an opportunity to target specific substantia nigra functions with nicotinic acetylcholine receptor pharmaceuticals possessing subtype-selective properties.

In the striatum, dopamine terminals express a remarkable diversity of nicotinic acetylcholine receptors, yet it must be noted that α3 and α7 subunits do not contribute to this population of presynaptic nicotinic acetylcholine receptors. Presynaptic dopaminergic nicotinic receptor subtypes include α4β2, α4α5β2, α4α6β2β3, α6β2β3, and α6β2 [183–185]. The alternate forms of the α4β2 subtype have distinct agonist sensitivities depending on whether α4 (lower) or β2 (higher) occupies the non-ligand-binding site receptor subunit position [186, 187]. α6β2* varieties are unique to dopamine presynaptic terminals in the striatum and are lost during progression of

Parkinson's disease [183–185, 188, 189], thus providing a unique pharmaceutical target to boost dopamine release in early Parkinson's disease. Interestingly, the α4β2* varieties, which represent the high-affinity nicotine-binding sites in brain, are resistant to nigrostriatal degeneration, possibly due to the more diverse expression pattern for this receptor type that includes non-dopaminergic neuronal components in the striatum [183, 185, 188, 190, 191], in contrast to the α6- and β3-containing nicotinic acetylcholine receptors which are restricted to catecholaminergic neurons [192, 193]. For example, α4β2* nicotinic acetylcholine receptors are found on the presynaptic terminals of the GABAergic medium spiny neurons and interneurons of the striatum [194]. There is also evidence for functional α7 nicotinic acetylcholine receptors on glutamatergic afferents as well as on a proportion of medium spiny neurons, GABAergic interneurons, and cholinergic interneurons [195–197]. While α7 nicotinic acetylcholine receptors are detectable using electrophysiological techniques [195], the expression level of this receptor subtype in striatum is apparently low based upon binding studies [198]. While much has been revealed regarding the α6β2* nicotinic receptor subtype on dopamine terminals in the striatum, much less is known regarding the subunit composition and cellular localization of the other nicotinic receptor varieties within the striatum, not to mention how they contribute to striatal output so critical to motor movement and the etiology of Parkinson's disease.

The activity of dopamine neurons in the substantia nigra pars compacta that project to the striatum is thought to be driven, in part, by cholinergic input from the pedunculopontine nucleus. As discussed previously, substantia nigra dopamine neuron firing exhibits a tonic pacemaker pattern [199, 200] predominantly in response to these cholinergic inputs. Nicotinic acetylcholine receptors, most likely α6β2* somatodendritically located to dopaminergic neurons within the substantia nigra pars compacta, are considered the primary mediators of acetylcholine effects in addition to promoting responses to glutamatergic afferents as well as facilitating the contribution of presynaptic α7 nicotinic acetylcholine receptors on these afferents. In an analogous fashion to their role in the ventral tegmental area [201, 202], presynaptic α7 nicotinic acetylcholine receptors on glutamatergic afferents may contribute to substantia nigra pars compacta burst firing and synaptic plasticity. Excitability of dopamine neurons within the substantia nigra pars compacta is constrained by GABA afferents and local interneurons [174, 203]; the presence of nicotinic acetylcholine receptors on these elements presents a more complex scenario for nicotinic regulation of dopamine cell activity. For example, nicotine, but not intrinsic acetylcholine release, preferentially desensitizes α4β2* nicotinic acetylcholine receptors on GABA interneurons, relieving this inhibitory influence, thus providing a potential therapeutic strategy for Parkinson's disease. This, in combination with the effect of chronic nicotine on upregulation of α4β2 nicotinic acetylcholine receptors in human tobacco users as well as nonhuman primates and rodents exposed to nicotine [194, 204–210], suggests an explanation as to why tobacco use is protective against Parkinson's disease and nicotine is neuroprotective against nigrostriatal damage in animal models of the disease [211–215].

Given that substantia nigra pars compacta dopamine neuron activity results in dopamine release in the striatum, it is a likely conclusion that the nicotinic acetylcholine receptor populations within this brain region as well as the cholinergic input from the pedunculopontine nucleus are crucial for proper striatal influence over motor movement modulation. However, intrinsic striatal properties that govern sensitivity and efficiency of responses to dopamine release within the striatum are an equally important component of the equation. In this regard, striatal nicotinic acetylcholine receptors are a major determinant of dopamine release probability in the striatum and shaping local dopaminergic responses in this brain region [165, 216]. For example, presynaptic nicotinic acetylcholine receptors resident on dopaminergic afferents, GABAergic interneurons, and corticostriatal glutamatergic afferents may serve to integrate acetylcholine input reciprocally with dopamine neuron firing in order to discriminate between tonic and burst firing of substantia nigra pars compacta dopamine neurons. Nicotinic acetylcholine receptor agents could be exploited to improve the signal-to-noise ratio between tonic and phasic patterns of striatal stimulation with recognizable implications for Parkinson's disease treatment. For additional information on nAChRs in the striatal dopaminergic system refer to Chap. 15.

2.3 Parkinson's Disease and Nicotinic Acetylcholine Receptors

A key pathologic hallmark of Parkinson's disease is loss of midbrain dopaminergic neurons of the substantia nigra pars compacta and their terminals that innervate the dorsal striatum. In addition to the well-known reduction in dopamine and damage to the nigrostriatal dopaminergic pathway whether it is due to Parkinson's disease or parkinsonian animal model lesions, on nicotinic acetylcholine receptor expression in striatum is also significantly affected [52, 53, 55, 188, 190, 191, 217–219]. The $\alpha6\beta2^*$ variety of nicotinic acetylcholine receptors is particularly vulnerable to the ravages of Parkinson's disease and may be due to the observation that this subtype of receptor is restricted to dopaminergic terminals that innervate the striatum [185, 188, 190]. The extent of $\alpha6\beta2^*$ nicotinic acetylcholine receptor loss parallels that of the dopamine transporter, and pharmacological studies using toxins that can discriminate $\alpha6$-containing nicotinic receptors at the stoichiometric level indicate that the $\alpha4\alpha6\beta2\beta3$ receptor subtype may represent a particularly vulnerable set of dopaminergic afferents to the striatum [220]. Since this pattern of loss is observed across rodents, monkeys, and humans, it suggests that this particular nicotinic acetylcholine receptor subtype represents the signature vulnerable neuronal population for Parkinson's disease (nigrostriatal dopaminergic neurons). Furthermore, this receptor variety may also represent a fruitful therapeutic target to maintain dopaminergic afferents if Parkinson's disease could be diagnosed early enough in the disease process.

As mentioned previously, $\alpha4\beta2^*$ nicotinic acetylcholine receptors do not suffer the same fate as the $\alpha6\beta2^*$ nicotinic acetylcholine receptors in Parkinson's disease. In fact, an almost complete dopaminergic lesion leads to only a 30–50 % decline in striatal $\alpha4\beta2^*$ nicotinic receptors. This selective sparing is thought to be due to the more promiscuous distribution of this receptor subtype compared to the $\alpha6\beta2$ subunit-containing nicotinic acetylcholine receptors that are restricted to catecholaminergic neurons by an as yet to be determined mechanism [193]. These studies are further supported by studies on striatal synaptosomal preparations (containing predominantly presynaptic components) which have shown that striatal α4β2* nicotinic receptors are comprised of both α4β2 and α4α5β2 subtypes [221, 222] and that nigrostriatal damage leads to ~90 % loss of α5 subunits whereas α4 declines by only ~50 % [185]. Thus, the vast majority of α4α5β2 nicotinic acetylcholine receptors are located presynaptically on dopaminergic afferents in the striatum, whereas α4β2 nicotinic acetylcholine receptors are more widely distributed.

Although studies on α7 nicotinic acetylcholine receptors following nigrostriatal lesions are sparse and difficult to execute due to the already low expression level of this receptor type in striatum of rodents and primates, such studies have found that α7 nicotinic acetylcholine receptor expression is, for the most part, unaffected by nigrostriatal damage [188, 223]. Akin to the α4β2* receptor, these results with α7 nicotinic acetylcholine receptors have been attributed to their presence on non-dopaminergic terminals, for example, the glutamatergic afferents that arrive from the cortex [197].

Nicotinic acetylcholine receptor changes within the substantia nigra pars compacta resulting from Parkinson's disease and nigrostriatal lesions should also be considered since the main neurodegenerative feature of Parkinson's disease is loss of striatal dopaminergic innervation from the substantia nigra pars compacta. Because the decrease in α6β2* is greater than that of α4β2* nicotinic acetylcholine receptor expression in the substantia nigra pars compacta following nigrostriatal damage [224, 225] and α6β2* loss correlates with dopamine transporter reductions, it is thought that these nicotinic acetylcholine receptors reside somatodendritically on dopaminergic neurons [182]. As is the case in the striatum, the less severely affected α4β2* nicotinic acetylcholine receptor population in the substantia nigra pars compacta likely reflects its presence on GABAergic interneurons and afferents from the substantia nigra pars reticulata in addition to dopaminergic neurons. Based on these observations, nicotinic acetylcholine receptors most affected in preclinical models of Parkinson's disease nigrostriatal damage are those receptors present on dopaminergic neurons that reside in the substantia nigra pars compacta and project to the striatum. Since α6β2* and α4α5β2 are most affected with less dramatic changes in α4β2 and essentially no loss of α7 nicotinic acetylcholine receptors, targeting α6β2* and α4β2* nicotinic acetylcholine receptors that remain on the dopaminergic neurons resident in the substantia nigra pars compacta during early stages of Parkinson's disease may represent a viable nicotinic receptor-targeted therapeutic strategy.

2.4 Targeting Nicotinic Acetylcholine Receptors in Parkinson's Disease

Several lines of evidence suggest that nicotinic acetylcholine receptors represent a valid therapeutic target in Parkinson's disease in spite of the historical perspective that underlying the disorder is degeneration and loss of functionality of the nigrostriatal dopaminergic pathway. The first evidence arose from epidemiological studies that emerged over a half-century ago establishing that tobacco use (smoking duration, intensity, and recentness) is inversely correlated with the incidence of Parkinson's disease without a relationship to mortality risk related to smoking; that is, Parkinson's disease patients appear to die of Parkinson's rather than of smoking-related health issues [226–234]. More recent work implicates this receptor system as a potential therapeutic target in Parkinson's disease as it has been discovered that the acetylcholine and dopamine neurotransmitter systems are functionally intertwined within the nigrostriatal pathway [216, 235, 236]. Further, work in preclinical models suggests that nicotine and nicotinic receptor ligands may directly stimulate dopaminergic transmission to alleviate motor-related symptoms as well as serve a neuroprotective role against nigrostriatal damage [211–215, 237, 238]. Augmented dopamine release via nicotinic acetylcholine receptor activation is postulated to occur mainly through α6α4β2 nicotinic receptors due to the notion that systemic nicotine levels would desensitize α4β2 nicotinic acetylcholine receptors more quickly and at lower concentrations. Although there is no direct evidence of a role for α7 nicotinic acetylcholine receptors in nicotine-mediated neuroprotection, the fact that α7 nicotinic acetylcholine receptors are present somatodendritically on the substantia nigra pars compacta dopaminergic neurons as well as couple to the PI3K/Akt neuroprotective pathway indicates that targeting this receptor class may also benefit preservation of nigrostriatal dopamine neurons. Additional neuroprotective mechanisms are based upon nicotine-induced nicotinic acetylcholine receptor upregulation, effects on receptor stoichiometry, and enhanced nicotinic receptor chaperoning [239, 240].

Additional neuroprotective properties may be provided by non-nicotine components of tobacco smoke that act as monoamine oxidase inhibitors [241, 242] presumably through inhibiting the synthesis of nigrostriatal toxins as well as enhancing the availability of dopamine. Finally, there is some evidence that nicotinic receptor stimulation augments cytochrome p450 expression [243] and ameliorates Parkinson's disease treatment-related side effects, specifically dyskinesias that result from long-term dopamine precursor treatment with L-DOPA as a replacement therapy [244–249]. This side effect has stimulated much debate within the treatment community as how to best initiate and maintain L-DOPA therapy [250, 251]. Although long-term nicotine administration has shown benefit in preclinical models, the subtype and anatomical location of the nicotinic acetylcholine receptors responsible for this effect have only recently been elucidated. Using β2 nicotinic acetylcholine receptor knockout mice in combination with nigrostriatal damage, it was revealed that the β2* nicotinic receptor population plays an essential role in nicotine reduction of L-DOPA-induced dyskinesias [252]. Subsequent studies indicate that α4β2* and α6β2* nicotinic receptor subtypes differentially contribute to this therapeutic benefit, possibly due to respective anatomical location and nicotine pharmacology for each of these receptor subtypes [253]. Furthermore, these nicotinic receptor subtypes, in addition to α7* nicotinic acetylcholine receptors, are implicated in the expression of L-DOPA-induced dyskinesias that also likely involve their anatomical location within the nigrostriatal system and involvement in dopamine transmission

[253–255]. Thus, the intricate subtype receptor profile, anatomical localization, and pharmacologic properties of nigrostriatal nicotinic acetylcholine receptors present a challenging pharmaceutical therapeutic conundrum. Nonetheless, a handful of preclinical trials using FDA-approved and candidate nicotinic receptor ligands have provided some insight to carry the field forward. For example, recent preclinical studies tested the broad-spectrum nicotinic agonist varenicline and 5-iodo-A-85380, a nicotinic acetylcholine receptor agonist that preferentially interacts with the $\alpha4\beta2^*$ and $\alpha6\beta2^*$ subtypes, and showed both to be of benefit [256, 257].

Since preclinical studies indicate that nicotine provides an approximate 30 % protection against nigrostriatal toxin-induced damage if administered prior to delivery of the toxin [258] and since Parkinson's disease motor symptoms typically arise only when nigral dopaminergic neurons are depleted by at least 50 % and striatal dopamine release by greater than 70 % [259, 260], one might consider utilizing a therapeutic strategy targeting nicotinic acetylcholine receptors for nigrostriatal neuroprotection to be provided prior to this therapeutic window of opportunity [261]; again, early diagnosis is the major challenge in developing disease-modifying interventions for neurodegenerative diseases such as Parkinson's. This latter issue may have contributed to the mixed results thus far obtained from several clinical trials [262–271].

In summary, extensive epidemiological evidence supports that long-term smoking is associated with a decreased risk for Parkinson's disease. This association is thought to be due to a neuroprotective role for the nicotine in tobacco acting at nicotinic acetylcholine receptors in the nigrostriatal pathway to preserve the dopaminergic system; thus, agents that target $\alpha4\beta2^*$ and $\alpha6\beta2^*$ nicotinic acetylcholine receptors may be useful for the treatment of Parkinson's disease if administered early enough in its etiology. Targeting nicotinic acetylcholine receptors may also be useful to counteract L-DOPA-induced dyskinesias; for this applicability, again $\alpha4\beta2^*$ and $\alpha6\beta2^*$ receptors appear to be important. Still, much work remains to elucidate the specific composition and location of nicotinic acetylcholine receptors that mediate the beneficial effects of nicotine in Parkinson's disease and models for nigrostriatal damage as well as elucidate the mechanism by which nicotinic receptors may alleviate L-DOPA-induced dyskinesias.

References

1. Thies W, Bleiler L, Alzheimer's A. Alzheimer's disease facts and figures. Alzheimers Dement. 2013;9(2):208–45. PubMed PMID: 23507120.
2. Mesulam MM. Cholinergic circuitry of the human nucleus basalis and its fate in Alzheimer's disease. J Comp Neurol. 2013;521(18):4124–44. PubMed PMID: 23852922.
3. Schliebs R, Arendt T. The cholinergic system in aging and neuronal degeneration. Behav Brain Res. 2011;221(2):555–63. PubMed PMID: 21145918.
4. Auld DS, Kornecook TJ, Bastianetto S, Quirion R. Alzheimer's disease and the basal forebrain cholinergic system: relations to beta-amyloid peptides, cognition, and treatment strategies. Prog Neurobiol. 2002;68(3):209–45. PubMed PMID: 12450488.

5. Mufson EJ, Ginsberg SD, Ikonomovic MD, DeKosky ST. Human cholinergic basal forebrain: chemoanatomy and neurologic dysfunction. J Chem Neuroanat. 2003;26(4):233–42. PubMed PMID: 14729126.
6. Mufson EJ, Counts SE, Perez SE, Ginsberg SD. Cholinergic system during the progression of Alzheimer's disease: therapeutic implications. Expert Rev Neurother. 2008;8(11):1703–18. PubMed PMID: 18986241, Pubmed Central PMCID: 2631573.
7. Everitt BJ, Robbins TW. Central cholinergic systems and cognition. Annu Rev Psychol. 1997;48:649–84. PubMed PMID: 9046571.
8. Riekkinen PJ, Laulumaa V, Sirvio J, Soininen H, Helkala EL. Recent progress in the research of Alzheimer's disease. Med Biol. 1987;65(2–3):83–8. PubMed PMID: 3309490.
9. Bartus RT. On neurodegenerative diseases, models, and treatment strategies: lessons learned and lessons forgotten a generation following the cholinergic hypothesis. Exp Neurol. 2000;163(2):495–529. PubMed PMID: 10833325.
10. Bartus RT, Dean 3rd RL, Beer B, Lippa AS. The cholinergic hypothesis of geriatric memory dysfunction. Science. 1982;217(4558):408–14. PubMed PMID: 7046051.
11. Perry EK, Tomlinson BE, Blessed G, Bergmann K, Gibson PH, Perry RH. Correlation of cholinergic abnormalities with senile plaques and mental test scores in senile dementia. Br Med J. 1978;2(6150):1457–9.
12. Banerjee C, Nyengaard JR, Wevers A, de Vos RA, Jansen Steur EN, Lindstrom J, et al. Cellular expression of alpha7 nicotinic acetylcholine receptor protein in the temporal cortex in Alzheimer's and Parkinson's disease–a stereological approach. Neurobiol Dis. 2000;7(6 Pt B): 666–72. PubMed PMID: 11114264.
13. Mitsis EM, Cosgrove KP, Staley JK, Bois F, Frohlich EB, Tamagnan GD, et al. Age-related decline in nicotinic receptor availability with [(123)I]5-IA-85380 SPECT. Neurobiol Aging. 2009;30(9):1490–7. PubMed PMID: 18242781.
14. Pappas BA, Bayley PJ, Bui BK, Hansen LA, Thal LJ. Choline acetyltransferase activity and cognitive domain scores of Alzheimer's patients. Neurobiol Aging. 2000;21(1):11–7. PubMed PMID: 10794843.
15. Selkoe DJ. Alzheimer's disease is a synaptic failure. Science. 2002;298(5594):789–91. PubMed PMID: 12399581.
16. Ikonomovic M, Wecker L, Abrahamson E, Wuu J, Counts S, Ginsberg S, et al. Cortical alpha7 nicotinic acetylcholine receptor and beta-amyloid levels in early Alzheimer's disease. Arch Neurol. 2009;66(5):646–51.
17. Ellis JR, Villemagne VL, Nathan PJ, Mulligan RS, Gong SJ, Chan JG, et al. Relationship between nicotinic receptors and cognitive function in early Alzheimer's disease: a 2-[18F] fluoro-A-85380 PET study. Neurobiol Learn Mem. 2008;90(2):404–12. PubMed PMID: 18620875.
18. Wang HY, Bakshi K, Shen C, Frankfurt M, Trocme-Thibierge C, Morain P. S 24795 limits beta-amyloid-alpha7 nicotinic receptor interaction and reduces Alzheimer's disease-like pathologies. Biol Psychiatry. 2010;67(6):522–30. PubMed PMID: 19932469.
19. Wang HY, Stucky A, Liu J, Shen C, Trocme-Thibierge C, Morain P. Dissociating beta-amyloid from alpha 7 nicotinic acetylcholine receptor by a novel therapeutic agent, S 24795, normalizes alpha 7 nicotinic acetylcholine and NMDA receptor function in Alzheimer's disease brain. J Neurosci. 2009;29(35):10961–73. PubMed PMID: 19726654.
20. Hasselmo ME, Giocomo LM. Cholinergic modulation of cortical function. J Mol Neurosci. 2006;30(1–2):133–5. PubMed PMID: 17192659.
21. Micheau J, Marighetto A. Acetylcholine and memory: a long, complex and chaotic but still living relationship. Behav Brain Res. 2011;221(2):424–9. PubMed PMID: 21130809.
22. Picciotto MR, Higley MJ, Mineur YS. Acetylcholine as a neuromodulator: cholinergic signaling shapes nervous system function and behavior. Neuron. 2012;76(1):116–29. PubMed PMID: 23040810, Pubmed Central PMCID: 3466476.
23. Prado VF, Roy A, Kolisnyk B, Gros R, Prado MA. Regulation of cholinergic activity by the vesicular acetylcholine transporter. Biochem J. 2013;450(2):265–74. PubMed PMID: 23410039.

24. Nilsson OG, Leanza G, Rosenblad C, Lappi DA, Wiley RG, Bjorklund A. Spatial learning impairments in rats with selective immunolesion of the forebrain cholinergic system. Neuroreport. 1992;3(11):1005–8. PubMed PMID: 1482757.
25. Rossner S, Schliebs R, Hartig W, Bigl V. 192IGG-saporin-induced selective lesion of cholinergic basal forebrain system: neurochemical effects on cholinergic neurotransmission in rat cerebral cortex and hippocampus. Brain Res Bull. 1995;38(4):371–81. PubMed PMID: 8535860.
26. Dekker AJ, Gage FH, Thal LJ. Delayed treatment with nerve growth factor improves acquisition of a spatial task in rats with lesions of the nucleus basalis magnocellularis: evaluation of the involvement of different neurotransmitter systems. Neuroscience. 1992;48(1):111–9. PubMed PMID: 1374860.
27. Janis LS, Glasier MM, Stein DG. Effects of a single intraseptal injection of NGF on spatial learning in the water maze. Physiol Behav. 1997;62(1):69–76. PubMed PMID: 9226344.
28. Woolf NJ, Milov AM, Schweitzer ES, Roghani A. Elevation of nerve growth factor and antisense knockdown of TrkA receptor during contextual memory consolidation. J Neurosci. 2001;21(3):1047–55. PubMed PMID: 11157090.
29. Ruberti F, Capsoni S, Comparini A, Di Daniel E, Franzot J, Gonfloni S, et al. Phenotypic knockout of nerve growth factor in adult transgenic mice reveals severe deficits in basal forebrain cholinergic neurons, cell death in the spleen, and skeletal muscle dystrophy. J Neurosci. 2000;20(7):2589–601. PubMed PMID: 10729339.
30. DiStefano PS, Friedman B, Radziejewski C, Alexander C, Boland P, Schick CM, et al. The neurotrophins BDNF, NT-3, and NGF display distinct patterns of retrograde axonal transport in peripheral and central neurons. Neuron. 1992;8(5):983–93. PubMed PMID: 1375039.
31. Darling TL, Petrides PE, Beguin P, Frey P, Shooter EM, Selby M, et al. The biosynthesis and processing of proteins in the mouse 7S nerve growth factor complex. Cold Spring Harb Symp Quant Biol. 1983;48(Pt 1):427–34. PubMed PMID: 6586361.
32. Edwards RH, Selby MJ, Garcia PD, Rutter WJ. Processing of the native nerve growth factor precursor to form biologically active nerve growth factor. J Biol Chem. 1988;263(14):6810–5. PubMed PMID: 3360808.
33. Lee R, Kermani P, Teng KK, Hempstead BL. Regulation of cell survival by secreted proneurotrophins. Science. 2001;294(5548):1945–8. PubMed PMID: 11729324.
34. Niewiadomska G, Mietelska-Porowska A, Mazurkiewicz M. The cholinergic system, nerve growth factor and the cytoskeleton. Behav Brain Res. 2011;221(2):515–26. PubMed PMID: 20170684.
35. Sweatt JD. The neuronal MAP kinase cascade: a biochemical signal integration system subserving synaptic plasticity and memory. J Neurochem. 2001;76(1):1–10. PubMed PMID: 11145972.
36. Sweatt JD. Mitogen-activated protein kinases in synaptic plasticity and memory. Curr Opin Neurobiol. 2004;14(3):311–7. PubMed PMID: 15194111.
37. Sofroniew MV, Howe CL, Mobley WC. Nerve growth factor signaling, neuroprotection, and neural repair. Annu Rev Neurosci. 2001;24:1217–81. PubMed PMID: 11520933.
38. Yakel JL. Nicotinic ACh receptors in the hippocampus: role in excitability and plasticity. Nicotine Tob Res. 2012;14(11):1249–57. Epub 2012 Apr 3. doi: 10.1093/ntr/nts091.
39. Perry EK, Perry RH, Smith CJ, Dick DJ, Candy JM, Edwardson JA, et al. Nicotinic receptor abnormalities in Alzheimer's and Parkinson's diseases. J Neurol Neurosurg Psychiatry. 1987;50(6):806–9. PubMed PMID: 2956364.
40. Quirion R, Martel JC, Robitaille Y, Etienne P, Wood P, Nair NP, et al. Neurotransmitter and receptor deficits in senile dementia of the Alzheimer type. Can J Neurol Sci. 1986;13 (4 Suppl):503–10. PubMed PMID: 2878714.
41. Fabian-Fine R, Skehel P, Errington ML, Davies HA, Sher E, Stewart MG, et al. Ultrastructural distribution of the alpha7 nicotinic acetylcholine receptor subunit in rat hippocampus. J Neurosci. 2001;21(20):7993–8003. PubMed PMID: 11588172.
42. Hunt SP, Schmidt J. The electron microscopic autoradiographic localization of alpha-bungarotoxin binding sites within the central nervous system of the rat. Brain Res. 1978;142(1):152–9. PubMed PMID: 626914.

43. Gray R, Rajan AS, Radcliffe KA, Yakehiro M, Dani JA. Hippocampal synaptic transmission enhanced by low concentrations of nicotine. Nature. 1996;383(6602):713–6. PubMed PMID: 8878480.
44. Ji D, Lape R, Dani JA. Timing and location of nicotinic activity enhances or depresses hippocampal synaptic plasticity. Neuron. 2001;31(1):131–41. PubMed PMID: 11498056.
45. Dani JA. Nicotinic receptor activity alters synaptic plasticity. ScientificWorldJournal. 2001;1(8):393–5. PubMed PMID: 12806076.
46. Rezvani AH, Bushnell PJ, Levin ED. Effects of nicotine and mecamylamine on choice accuracy in an operant visual signal detection task in female rats. Psychopharmacology (Berl). 2002;164(4):369–75. PubMed PMID: 12457266.
47. Levin ED, Rezvani AH. Nicotinic treatment for cognitive dysfunction. Curr Drug Targets CNS Neurol Disord. 2002;1(4):423–31. PubMed PMID: 12769614.
48. Bowen DM, Allen SJ, Benton JS, Goodhardt MJ, Haan EA, Palmer AM, et al. Biochemical assessment of serotonergic and cholinergic dysfunction and cerebral atrophy in Alzheimer's disease. J Neurochem. 1983;41(1):266–72. PubMed PMID: 6306169.
49. Cummings JL, Benson DF. The role of the nucleus basalis of Meynert in dementia: review and reconsideration. Alzheimer Dis Assoc Disord. 1987;1(3):128–55. PubMed PMID: 3331115.
50. Perry EK, Haroutunian V, Davis KL, Levy R, Lantos P, Eagger S, et al. Neocortical cholinergic activities differentiate Lewy body dementia from classical Alzheimer's disease. Neuroreport. 1994;5(7):747–9. PubMed PMID: 8018843.
51. Court J, Martin-Ruiz C, Piggott M, Spurden D, Griffiths M, Perry E. Nicotinic receptor abnormalities in Alzheimer's disease. Biol Psychiatry. 2001;49(3):175–84. PubMed PMID: 11230868.
52. Aubert I, Araujo DM, Cecyre D, Robitaille Y, Gauthier S, Quirion R. Comparative alterations of nicotinic and muscarinic binding sites in Alzheimer's and Parkinson's diseases. J Neurochem. 1992;58(2):529–41. PubMed PMID: 1729398.
53. Court JA, Piggott MA, Lloyd S, Cookson N, Ballard CG, McKeith IG, et al. Nicotine binding in human striatum: elevation in schizophrenia and reductions in dementia with Lewy bodies, Parkinson's disease and Alzheimer's disease and in relation to neuroleptic medication. Neuroscience. 2000;98(1):79–87. PubMed PMID: 10858614.
54. Martin-Ruiz CM, Court JA, Molnar E, Lee M, Gotti C, Mamalaki A, et al. Alpha4 but not alpha3 and alpha7 nicotinic acetylcholine receptor subunits are lost from the temporal cortex in Alzheimer's disease. J Neurochem. 1999;73(4):1635–40. PubMed PMID: 10501210.
55. Perry EK, Morris CM, Court JA, Cheng A, Fairbairn AF, McKeith IG, et al. Alteration in nicotine binding sites in Parkinson's disease, Lewy body dementia and Alzheimer's disease: possible index of early neuropathology. Neuroscience. 1995;64(2):385–95. PubMed PMID: 7700528.
56. Whitehouse PJ, Martino AM, Wagster MV, Price DL, Mayeux R, Atack JR, et al. Reductions in [3H]nicotinic acetylcholine binding in Alzheimer's disease and Parkinson's disease: an autoradiographic study. Neurology. 1988;38(5):720–3. PubMed PMID: 3362368.
57. Svensson AL, Warpman U, Hellstrom-Lindahl E, Bogdanovic N, Lannfelt L, Nordberg A. Nicotinic receptors, muscarinic receptors and choline acetyltransferase activity in the temporal cortex of Alzheimer patients with differing apolipoprotein E genotypes. Neurosci Lett. 1997;232(1):37–40. PubMed PMID: 9292886.
58. Warpman U, Nordberg A. Epibatidine and ABT 418 reveal selective losses of alpha 4 beta 2 nicotinic receptors in Alzheimer brains. Neuroreport. 1995;6(17):2419–23. PubMed PMID: 8747166.
59. Nordberg A, Winblad B. Reduced number of [3H]nicotine and [3H]acetylcholine binding sites in the frontal cortex of Alzheimer brains. Neurosci Lett. 1986;72(1):115–9. PubMed PMID: 3808458.
60. Pimlott SL, Piggott M, Owens J, Greally E, Court JA, Jaros E, et al. Nicotinic acetylcholine receptor distribution in Alzheimer's disease, dementia with Lewy bodies, Parkinson's disease, and vascular dementia: in vitro binding study using 5-[(125)i]-a-85380. Neuropsychopharmacology. 2004;29(1):108–16. PubMed PMID: 12955099.

61. Guan ZZ, Zhang X, Ravid R, Nordberg A. Decreased protein levels of nicotinic receptor subunits in the hippocampus and temporal cortex of patients with Alzheimer's disease. J Neurochem. 2000;74(1):237–43. PubMed PMID: 10617125.
62. Moser N, Mechawar N, Jones I, Gochberg-Sarver A, Orr-Urtreger A, Plomann M, et al. Evaluating the suitability of nicotinic acetylcholine receptor antibodies for standard immunodetection procedures. J Neurochem. 2007;102(2):479–92. PubMed PMID: 17419810.
63. Schroder H, Giacobini E, Struble RG, Zilles K, Maelicke A. Nicotinic cholinoceptive neurons of the frontal cortex are reduced in Alzheimer's disease. Neurobiol Aging. 1991; 12(3):259–62. PubMed PMID: 1876232.
64. Schroder H, Giacobini E, Struble RG, Zilles K, Maelicke A, Luiten PG, et al. Cellular distribution and expression of cortical acetylcholine receptors in aging and Alzheimer's disease. Ann N Y Acad Sci. 1991;640:189–92. PubMed PMID: 1776738.
65. Herber DL, Severance EG, Cuevas J, Morgan D, Gordon MN. Biochemical and histochemical evidence of nonspecific binding of alpha7nAChR antibodies to mouse brain tissue. J Histochem Cytochem. 2004;52(10):1367–76. PubMed PMID: 15385583.
66. Gotti C, Moretti M, Bohr I, Ziabreva I, Vailati S, Longhi R, et al. Selective nicotinic acetylcholine receptor subunit deficits identified in Alzheimer's disease, Parkinson's disease and dementia with Lewy bodies by immunoprecipitation. Neurobiol Dis. 2006;23(2):481–9. PubMed PMID: 16759874.
67. Engidawork E, Gulesserian T, Balic N, Cairns N, Lubec G. Changes in nicotinic acetylcholine receptor subunits expression in brain of patients with Down syndrome and Alzheimer's disease. J Neural Transm Suppl. 2001;61:211–22. PubMed PMID: 11771745.
68. Burghaus L, Schutz U, Krempel U, de Vos RA, Jansen Steur EN, Wevers A, et al. Quantitative assessment of nicotinic acetylcholine receptor proteins in the cerebral cortex of Alzheimer patients. Brain Res Mol Brain Res. 2000;76(2):385–8. PubMed PMID: 10762715.
69. Hellstrom-Lindahl E, Mousavi M, Zhang X, Ravid R, Nordberg A. Regional distribution of nicotinic receptor subunit mRNAs in human brain: comparison between Alzheimer and normal brain. Brain Res Mol Brain Res. 1999;66(1–2):94–103. PubMed PMID: 10095081.
70. Davies P, Feisullin S. Postmortem stability of alpha-bungarotoxin binding sites in mouse and human brain. Brain Res. 1981;216(2):449–54. PubMed PMID: 7248786.
71. Sugaya K, Giacobini E, Chiappinelli VA. Nicotinic acetylcholine receptor subtypes in human frontal cortex: changes in Alzheimer's disease. J Neurosci Res. 1990;27(3):349–59. PubMed PMID: 2097379.
72. Hyman BT, Phelps CH, Beach TG, Bigio EH, Cairns NJ, Carrillo MC, et al. National Institute on Aging-Alzheimer's Association guidelines for the neuropathologic assessment of Alzheimer's disease. Alzheimers Dement. 2012;8(1):1–13. PubMed PMID: 22265587, Pubmed Central PMCID: 3266529.
73. Davis KL, Mohs RC, Marin D, Purohit DP, Perl DP, Lantz M, et al. Cholinergic markers in elderly patients with early signs of Alzheimer disease. JAMA. 1999;281(15):1401–6. PubMed PMID: 10217056.
74. DeKosky ST, Ikonomovic MD, Styren SD, Beckett L, Wisniewski S, Bennett DA, et al. Upregulation of choline acetyltransferase activity in hippocampus and frontal cortex of elderly subjects with mild cognitive impairment. Ann Neurol. 2002;51(2):145–55. PubMed PMID: 11835370.
75. Counts SE, He B, Che S, Ikonomovic MD, DeKosky ST, Ginsberg SD, et al. Alpha7 nicotinic receptor up-regulation in cholinergic basal forebrain neurons in Alzheimer disease. Arch Neurol. 2007;64(12):1771–6. PubMed PMID: 18071042.
76. Hyman BT, Van Hoesen GW, Damasio AR, Barnes CL. Alzheimer's disease: cell-specific pathology isolates the hippocampal formation. Science. 1984;225(4667):1168–70. PubMed PMID: 6474172.
77. Cotman CW, Matthews DA, Taylor D, Lynch G. Synaptic rearrangement in the dentate gyrus: histochemical evidence of adjustments after lesions in immature and adult rats. Proc Natl Acad Sci U S A. 1973;70(12):3473–7. PubMed PMID: 4519639, Pubmed Central PMCID: 427262.
78. Savaskan NE, Nitsch R. Molecules involved in reactive sprouting in the hippocampus. Rev Neurosci. 2001;12(3):195–215. PubMed PMID: 11560368.

79. Parri HR, Hernandez CM, Dineley KT. Research update: alpha7 nicotinic acetylcholine receptor mechanisms in Alzheimer's disease. Biochem Pharmacol. 2011;82(8):931–42. PubMed PMID: 21763291.
80. Parri RH, Dineley TK. Nicotinic acetylcholine receptor interaction with beta-amyloid: molecular, cellular, and physiological consequences. Curr Alzheimer Res. 2010;7(1):27–39. PubMed PMID: 20205670.
81. Dineley KT. Beta-amyloid peptide–nicotinic acetylcholine receptor interaction: the two faces of health and disease. Front Biosci. 2007;12:5030–8. PubMed PMID: 17569627.
82. Dineley KT, Xia X, Bui D, Sweatt JD, Zheng H. Accelerated plaque accumulation, associative learning deficits, and up-regulation of alpha 7 nicotinic receptor protein in transgenic mice co-expressing mutant human presenilin 1 and amyloid precursor proteins. J Biol Chem. 2002;277(25):22768–80. PubMed PMID: 11912199.
83. Wang HY, Lee DH, D'Andrea MR, Peterson PA, Shank RP, Reitz AB. beta-Amyloid(1-42) binds to alpha7 nicotinic acetylcholine receptor with high affinity. Implications for Alzheimer's disease pathology. J Biol Chem. 2000;275(8):5626–32.
84. Puzzo D, Privitera L, Fa M, Staniszewski A, Hashimoto G, Aziz F, et al. Endogenous amyloid-beta is necessary for hippocampal synaptic plasticity and memory. Ann Neurol. 2011;69(5):819–30. PubMed PMID: 21472769.
85. Puzzo D, Privitera L, Leznik E, Fa M, Staniszewski A, Palmeri A, et al. Picomolar amyloid-beta positively modulates synaptic plasticity and memory in hippocampus. J Neurosci. 2008;28(53):14537–45. PubMed PMID: 19118188, Pubmed Central PMCID: 2673049.
86. Fu W, Jhamandas JH. Beta-amyloid peptide activates non-alpha7 nicotinic acetylcholine receptors in rat basal forebrain neurons. J Neurophysiol. 2003;90(5):3130–6. PubMed PMID: 12890800.
87. Azam L, Winzer-Serhan U, Leslie FM. Co-expression of alpha7 and beta2 nicotinic acetylcholine receptor subunit mRNAs within rat brain cholinergic neurons. Neuroscience. 2003;119(4):965–77. PubMed PMID: 12831856.
88. Teaktong T, Graham A, Court J, Perry R, Jaros E, Johnson M, et al. Alzheimer's disease is associated with a selective increase in alpha7 nicotinic acetylcholine receptor immunoreactivity in astrocytes. Glia. 2003;41(2):207–11. PubMed PMID: 12509811.
89. Chu LW, Ma ES, Lam KK, Chan MF, Lee DH. Increased alpha 7 nicotinic acetylcholine receptor protein levels in Alzheimer's disease patients. Dement Geriatr Cogn Disord. 2005;19(2–3):106–12. PubMed PMID: 15591800.
90. Govind AP, Walsh H, Green WN. Nicotine-induced upregulation of native neuronal nicotinic receptors is caused by multiple mechanisms. J Neurosci. 2012;32(6):2227–38. PubMed PMID: 22323734, Pubmed Central PMCID: 3286518.
91. Govind AP, Vezina P, Green WN. Nicotine-induced upregulation of nicotinic receptors: underlying mechanisms and relevance to nicotine addiction. Biochem Pharmacol. 2009;78(7):756–65. PubMed PMID: 19540212, Pubmed Central PMCID: 2728164.
92. Colombo SF, Mazzo F, Pistillo F, Gotti C. Biogenesis, trafficking and up-regulation of nicotinic ACh receptors. Biochem Pharmacol. 2013;86(8):1063–73. PubMed PMID: 23830821.
93. Marks MJ, McClure-Begley TD, Whiteaker P, Salminen O, Brown RW, Cooper J, et al. Increased nicotinic acetylcholine receptor protein underlies chronic nicotine-induced up-regulation of nicotinic agonist binding sites in mouse brain. J Pharmacol Exp Ther. 2011;337(1):187–200. PubMed PMID: 21228066, Pubmed Central PMCID: 3063733.
94. Gentry CL, Lukas RJ. Regulation of nicotinic acetylcholine receptor numbers and function by chronic nicotine exposure. Curr Drug Targets CNS Neurol Disord. 2002;1(4):359–85. PubMed PMID: 12769610.
95. Dineley KT, Bell KA, Bui D, Sweatt JD. beta -Amyloid peptide activates alpha 7 nicotinic acetylcholine receptors expressed in Xenopus oocytes. J Biol Chem. 2002;277(28):25056–61. PubMed PMID: 11983690.
96. Dineley KT, Westerman M, Bui D, Bell K, Ashe KH, Sweatt JD. Beta-amyloid activates the mitogen-activated protein kinase cascade via hippocampal alpha7 nicotinic acetylcholine receptors: in vitro and in vivo mechanisms related to Alzheimer's disease. J Neurosci. 2001;21(12):4125–33. PubMed PMID: 11404397.

97. Abbott JJ, Howlett DR, Francis PT, Williams RJ. Abeta(1-42) modulation of Akt phosphorylation via alpha7 nAChR and NMDA receptors. Neurobiol Aging. 2008;29(7):992–1001. PubMed PMID: 17292512.
98. Dougherty JJ, Wu J, Nichols RA. Beta-amyloid regulation of presynaptic nicotinic receptors in rat hippocampus and neocortex. J Neurosci. 2003;23(17):6740–7. PubMed PMID: 12890766.
99. Bell KA, O'Riordan KJ, Sweatt JD, Dineley KT. MAPK recruitment by beta-amyloid in organotypic hippocampal slice cultures depends on physical state and exposure time. J Neurochem. 2004;91(2):349–61. PubMed PMID: 15447668.
100. Marambaud P, Dreses-Werringloer U, Vingtdeux V. Calcium signaling in neurodegeneration. Mol Neurodegener. 2009;4:20. PubMed PMID: 19419557, Pubmed Central PMCID: 2689218.
101. Jancic D. Lopez de Armentia M, Valor LM, Olivares R, Barco A. Inhibition of cAMP response element-binding protein reduces neuronal excitability and plasticity, and triggers neurodegeneration. Cereb Cortex. 2009;19(11):2535–47.
102. Baki L, Neve RL, Shao Z, Shioi J, Georgakopoulos A, Robakis NK. Wild-type but not FAD mutant presenilin-1 prevents neuronal degeneration by promoting phosphatidylinositol 3-kinase neuroprotective signaling. J Neurosci. 2008;28(2):483–90. PubMed PMID: 18184791.
103. Abdul HM, Butterfield DA. Involvement of PI3K/PKG/ERK1/2 signaling pathways in cortical neurons to trigger protection by cotreatment of acetyl-L-carnitine and alpha-lipoic acid against HNE-mediated oxidative stress and neurotoxicity: implications for Alzheimer's disease. Free Radic Biol Med. 2007;42(3):371–84. PubMed PMID: 17210450, Pubmed Central PMCID: 1808543.
104. Hernandez CM, Dineley KT. alpha7 nicotinic acetylcholine receptors in Alzheimer's disease: neuroprotective, neurotrophic or both? Curr Drug Targets. 2012;13(5):613–22. PubMed PMID: 22300028.
105. Puzzo D, Arancio O. Amyloid-beta peptide: Dr. Jekyll or Mr. Hyde? J Alzheimers Dis. 2013;33 Suppl 1:S111–20. PubMed PMID: 22735675, Pubmed Central PMCID: 3696497.
106. Liu Q, Kawai H, Berg DK. beta -Amyloid peptide blocks the response of alpha 7-containing nicotinic receptors on hippocampal neurons. Proc Natl Acad Sci U S A. 2001;98(8):4734–9. PubMed PMID: 11274373.
107. Liu Q, Huang Y, Shen J, Steffensen S, Wu J. Functional alpha7beta2 nicotinic acetylcholine receptors expressed in hippocampal interneurons exhibit high sensitivity to pathological level of amyloid beta peptides. BMC Neurosci. 2012;13:155. PubMed PMID: 23272676, Pubmed Central PMCID: 3573893.
108. Liu Q, Huang Y, Xue F, Simard A, DeChon J, Li G, et al. A novel nicotinic acetylcholine receptor subtype in basal forebrain cholinergic neurons with high sensitivity to amyloid peptides. J Neurosci. 2009;29(4):918–29. PubMed PMID: 19176801, Pubmed Central PMCID: 2857410.
109. Hasselmo ME, Wyble BP, Wallenstein GV. Encoding and retrieval of episodic memories: role of cholinergic and GABAergic modulation in the hippocampus. Hippocampus. 1996; 6(6):693–708. PubMed PMID: 9034856.
110. Hasselmo ME. Neuromodulation: acetylcholine and memory consolidation. Trends Cogn Sci. 1999;3(9):351–9. PubMed PMID: 10461198.
111. Chrobak JJ, Hanin I, Schmechel DE, Walsh TJ. AF64A-induced working memory impairment: behavioral, neurochemical and histological correlates. Brain Res. 1988;463(1):107–17. PubMed PMID: 3196899.
112. Olson L, Nordberg A, von Holst H, Backman L, Ebendal T, Alafuzoff I, et al. Nerve growth factor affects 11C-nicotine binding, blood flow, EEG, and verbal episodic memory in an Alzheimer patient (case report). J Neural Transm Park Dis Dement Sect. 1992;4(1):79–95. PubMed PMID: 1540306.
113. Doody RS, Cummings JL, Farlow MR. Reviewing the role of donepezil in the treatment of Alzheimer's disease. Curr Alzheimer Res. 2012;9(7):773–81. PubMed PMID: 22175653.

114. Rogers SL, Doody RS, Pratt RD, Ieni JR. Long-term efficacy and safety of donepezil in the treatment of Alzheimer's disease: final analysis of a US multicentre open-label study. Eur Neuropsychopharmacol. 2000;10(3):195–203. PubMed PMID: 10793322.
115. Gauthier S, Cummings J, Ballard C, Brodaty H, Grossberg G, Robert P, et al. Management of behavioral problems in Alzheimer's disease. Int Psychogeriatr. 2010;22(3):346–72. PubMed PMID: 20096151.
116. Di Santo SG, Prinelli F, Adorni F, Caltagirone C, Musicco M. A meta-analysis of the efficacy of donepezil, rivastigmine, galantamine, and memantine in relation to severity of Alzheimer's disease. J Alzheimers Dis. 2013;35(2):349–61. PubMed PMID: 23411693.
117. Cummings JL. Treatment of Alzheimer's disease: the role of symptomatic agents in an era of disease-modifying therapies. Rev Neurol Dis. 2007;4(2):57–62. PubMed PMID: 17609636.
118. Takada-Takatori Y, Kume T, Sugimoto M, Katsuki H, Sugimoto H, Akaike A. Acetylcholinesterase inhibitors used in treatment of Alzheimer's disease prevent glutamate neurotoxicity via nicotinic acetylcholine receptors and phosphatidylinositol 3-kinase cascade. Neuropharmacology. 2006;51(3):474–86. PubMed PMID: 16762377.
119. Kihara T, Shimohama S, Sawada H, Honda K, Nakamizo T, Shibasaki H, et al. alpha 7 nicotinic receptor transduces signals to phosphatidylinositol 3-kinase to block A beta-amyloid-induced neurotoxicity. J Biol Chem. 2001;276(17):13541–6. PubMed PMID: 11278378.
120. Li XD, Buccafusco JJ. Effect of beta-amyloid peptide 1-42 on the cytoprotective action mediated by alpha7 nicotinic acetylcholine receptors in growth factor-deprived differentiated PC-12 cells. J Pharmacol Exp Ther. 2003;307(2):670–5. PubMed PMID: 12970390.
121. Li XD, Arias E, Jonnala RR, Mruthinti S, Buccafusco JJ. Effect of amyloid peptides on the increase in TrkA receptor expression induced by nicotine in vitro and in vivo. J Mol Neurosci. 2005;27(3):325–36. PubMed PMID: 16280603.
122. Hernandez CM, Terry Jr AV. Repeated nicotine exposure in rats: effects on memory function, cholinergic markers and nerve growth factor. Neuroscience. 2005;130(4):997–1012. PubMed PMID: 15652996.
123. Dunbar GC, Kuchibhatla RV, Lee G, Group T-ACS. A randomized double-blind study comparing 25 and 50 mg TC-1734 (AZD3480) with placebo, in older subjects with age-associated memory impairment. J Psychopharmacol. 2011;25(8):1020–9. PubMed PMID: 20542923.
124. Arneric SP, Sullivan JP, Briggs CA, Donnelly-Roberts D, Anderson DJ, Raszkiewicz JL, et al. (S)-3-methyl-5-(1-methyl-2-pyrrolidinyl) isoxazole (ABT 418): a novel cholinergic ligand with cognition-enhancing and anxiolytic activities: I. In vitro characterization. J Pharmacol Exp Ther. 1994;270(1):310–8.
125. Papke RL, Thinschmidt JS, Moulton BA, Meyer EM, Poirier A. Activation and inhibition of rat neuronal nicotinic receptors by ABT-418. Br J Pharmacol. 1997;120(3):429–38. PubMed PMID: 9031746, Pubmed Central PMCID: 1564486.
126. Bencherif M, Lovette ME, Fowler KW, Arrington S, Reeves L, Caldwell WS, et al. RJR-2403: a nicotinic agonist with CNS selectivity I. In vitro characterization. J Pharmacol Exp Ther. 1996;279(3):1413–21. PubMed PMID: 8968366.
127. Lippiello PM, Bencherif M, Gray JA, Peters S, Grigoryan G, Hodges H, et al. RJR-2403: a nicotinic agonist with CNS selectivity II. In vivo characterization. J Pharmacol Exp Ther. 1996;279(3):1422–9. PubMed PMID: 8968367.
128. Gatto GJ, Bohme GA, Caldwell WS, Letchworth SR, Traina VM, Obinu MC, et al. TC-1734: an orally active neuronal nicotinic acetylcholine receptor modulator with antidepressant, neuroprotective and long-lasting cognitive effects. CNS Drug Rev. 2004;10(2):147–66. PubMed PMID: 15179444.
129. Wilson AL, Langley LK, Monley J, Bauer T, Rottunda S, McFalls E, et al. Nicotine patches in Alzheimer's disease: pilot study on learning, memory, and safety. Pharmacol Biochem Behav. 1995;51(2–3):509–14. PubMed PMID: 7667377.
130. Lopez-Arrieta JM, Rodriguez JL, Sanz F. Nicotine for Alzheimer's disease. Cochrane Database Syst Rev 2000 (2):CD001749. PubMed PMID: 10796667.
131. Zamani MR, Allen YS. Nicotine and its interaction with beta-amyloid protein: a short review. Biol Psychiatry. 2001;49(3):221–32. PubMed PMID: 11230873.

132. Lopez-Arrieta JM, Rodriguez JL, Sanz F. Efficacy and safety of nicotine on Alzheimer's disease patients. Cochrane Database Syst Rev 2001 (2):CD001749. PubMed PMID: 11406005
133. O'Neill MJ, Murray TK, Lakics V, Visanji NP, Duty S. The role of neuronal nicotinic acetylcholine receptors in acute and chronic neurodegeneration. Curr Drug Targets CNS Neurol Disord. 2002;1(4):399–411. PubMed PMID: 12769612.
134. Rezvani AH, Kholdebarin E, Brucato FH, Callahan PM, Lowe DA, Levin ED. Effect of R3487/MEM3454, a novel nicotinic alpha7 receptor partial agonist and 5-HT3 antagonist on sustained attention in rats. Prog Neuropsychopharmacol Biol Psychiatry. 2009;33(2):269–75. PubMed PMID: 19110025.
135. Wallace TL, Callahan PM, Tehim A, Bertrand D, Tombaugh G, Wang S, et al. RG3487, a novel nicotinic alpha7 receptor partial agonist, improves cognition and sensorimotor gating in rodents. J Pharmacol Exp Ther. 2011;336(1):242–53. PubMed PMID: 20959364.
136. Bitner RS, Bunnelle WH, Decker MW, Drescher KU, Kohlhaas KL, Markosyan S, et al. In vivo pharmacological characterization of a novel selective alpha7 neuronal nicotinic acetylcholine receptor agonist ABT-107: preclinical considerations in Alzheimer's disease. J Pharmacol Exp Ther. 2010;334(3):875–86. PubMed PMID: 20504913.
137. Biton B, Bergis OE, Galli F, Nedelec A, Lochead AW, Jegham S, et al. SSR180711, a novel selective alpha7 nicotinic receptor partial agonist: (1) binding and functional profile. Neuropsychopharmacology. 2007;32(1):1–16. PubMed PMID: 17019409.
138. Barak S, Arad M, De Levie A, Black MD, Griebel G, Weiner I. Pro-cognitive and antipsychotic efficacy of the alpha7 nicotinic partial agonist SSR180711 in pharmacological and neurodevelopmental latent inhibition models of schizophrenia. Neuropsychopharmacology. 2009;34(7):1753–63. PubMed PMID: 19158670.
139. Hurst R, Rollema H, Bertrand D. Nicotinic acetylcholine receptors: from basic science to therapeutics. Pharmacol Ther. 2013;137(1):22–54. PubMed PMID: 22925690.
140. Vellas B, Carrillo MC, Sampaio C, Brashear HR, Siemers E, Hampel H, et al. Designing drug trials for Alzheimer's disease: what we have learned from the release of the phase III antibody trials: a report from the EU/US/CTAD Task Force. Alzheimers Dement. 2013;9(4):438–44. PubMed PMID: 23809364.
141. Becker RE, Greig NH. Fire in the ashes: can failed Alzheimer's disease drugs succeed with second chances? Alzheimers Dement. 2013;9(1):50–7. PubMed PMID: 22465172.
142. Becker RE, Greig NH. Increasing the success rate for Alzheimer's disease drug discovery and development. Exp Opin Drug Dis. 2012;7(4):367–70. PubMed PMID: 22439785.
143. Vellas B. Recruitment, retention and other methodological issues related to clinical trials for Alzheimer's disease. J Nutr Health Aging. 2012;16(4):330. PubMed PMID: 22499451.
144. Jack Jr CR, Albert MS, Knopman DS, McKhann GM, Sperling RA, Carrillo MC, et al. Introduction to the recommendations from the National Institute on Aging-Alzheimer's Association workgroups on diagnostic guidelines for Alzheimer's disease. Alzheimers Dement. 2011;7(3):257–62. PubMed PMID: 21514247, Pubmed Central PMCID: 3096735.
145. Nagele RG, D'Andrea MR, Anderson WJ, Wang HY. Intracellular accumulation of beta-amyloid(1-42) in neurons is facilitated by the alpha 7 nicotinic acetylcholine receptor in Alzheimer's disease. Neuroscience. 2002;110(2):199–211. PubMed PMID: 11958863.
146. Dickson DW. Parkinson's disease and parkinsonism: neuropathology. Cold Spring Harb Perspec Med. 2012;2(8):a009258. PubMed PMID: 22908195.
147. Poewe W, Mahlknecht P. The clinical progression of Parkinson's disease. Parkinsonism Relat Disord. 2009;15 Suppl 4:S28–32. PubMed PMID: 20123553.
148. Schapira AH. Etiology and pathogenesis of Parkinson disease. Neurol Clin. 2009;27(3):583–603. v, PubMed PMID: 19555823.
149. Kumar KR, Lohmann K, Klein C. Genetics of Parkinson disease and other movement disorders. Curr Opin Neurol. 2012;25(4):466–74. PubMed PMID: 22772876.
150. Healy DG, Falchi M, O'Sullivan SS, Bonifati V, Durr A, Bressman S, et al. Phenotype, genotype, and worldwide genetic penetrance of LRRK2-associated Parkinson's disease: a case-control study. Lancet Neurol. 2008;7(7):583–90. PubMed PMID: 18539534, Pubmed Central PMCID: 2832754.

151. Houlden H, Singleton AB. The genetics and neuropathology of Parkinson's disease. Acta Neuropathol. 2012;124(3):325–38. PubMed PMID: 22806825, Pubmed Central PMCID: 3589971.
152. Schapira AH, Jenner P. Etiology and pathogenesis of Parkinson's disease. Mov disord. 2011;26(6):1049–55. PubMed PMID: 21626550.
153. Kieburtz K, Wunderle KB. Parkinson's disease: evidence for environmental risk factors. Mov disord. 2013;28(1):8–13. PubMed PMID: 23097348.
154. Ferreira JJ, Katzenschlager R, Bloem BR, Bonuccelli U, Burn D, Deuschl G, et al. Summary of the recommendations of the EFNS/MDS-ES review on therapeutic management of Parkinson's disease. Eur J Neurol. 2013;20(1):5–15. PubMed PMID: 23279439.
155. Huot P, Johnston TH, Koprich JB, Fox SH, Brotchie JM. The pharmacology of L-DOPA-induced dyskinesia in Parkinson's disease. Pharmacol Rev. 2013;65(1):171–222. PubMed PMID: 23319549.
156. Jankovic J. Medical treatment of dystonia. Mov disord. 2013;28(7):1001–12. PubMed PMID: 23893456.
157. Obeso JA, Rodriguez-Oroz MC, Benitez-Temino B, Blesa FJ, Guridi J, Marin C, et al. Functional organization of the basal ganglia: therapeutic implications for Parkinson's disease. Mov disord. 2008;23 Suppl 3:S548–59. PubMed PMID: 18781672.
158. Smith Y, Kieval JZ. Anatomy of the dopamine system in the basal ganglia. Trends Neurosci. 2000;23(10 Suppl):S28–33. PubMed PMID: 11052217.
159. Bolam JP, Hanley JJ, Booth PA, Bevan MD. Synaptic organisation of the basal ganglia. J Anat. 2000;196(Pt 4):527–42. PubMed PMID: 10923985, Pubmed Central PMCID: 1468095.
160. Obeso JA, Lanciego JL. Past, present, and future of the pathophysiological model of the Basal Ganglia. Front Neuroanat. 2011;5:39. PubMed PMID: 21808607, Pubmed Central PMCID: 3136734.
161. Zhou FM, Liang Y, Dani JA. Endogenous nicotinic cholinergic activity regulates dopamine release in the striatum. Nat Neurosci. 2001;4(12):1224–9. PubMed PMID: 11713470.
162. Zhou FM, Wilson CJ, Dani JA. Cholinergic interneuron characteristics and nicotinic properties in the striatum. J Neurobiol. 2002;53(4):590–605. PubMed PMID: 12436423.
163. Ungless MA, Cragg SJ. A choreography of nicotinic receptors directs the dopamine neuron routine. Neuron. 2006;50(6):815–6. PubMed PMID: 16772163.
164. Threlfell S, Cragg SJ. Dopamine signaling in dorsal versus ventral striatum: the dynamic role of cholinergic interneurons. Front Syst Neurosci. 2011;5:11. PubMed PMID: 21427783, Pubmed Central PMCID: 3049415.
165. Threlfell S, Lalic T, Platt NJ, Jennings KA, Deisseroth K, Cragg SJ. Striatal dopamine release is triggered by synchronized activity in cholinergic interneurons. Neuron. 2012;75(1):58–64. PubMed PMID: 22794260.
166. Ding JB, Guzman JN, Peterson JD, Goldberg JA, Surmeier DJ. Thalamic gating of corticostriatal signaling by cholinergic interneurons. Neuron. 2010;67(2):294–307. PubMed PMID: 20670836.
167. Flynn DD, Mash DC. Distinct kinetic binding properties of N-[3H]-methylscopolamine afford differential labeling and localization of M1, M2, and M3 muscarinic receptor subtypes in primate brain. Synapse. 1993;14(4):283–96. PubMed PMID: 8248852.
168. Nastuk MA, Graybiel AM. Autoradiographic localization and biochemical characteristics of M1 and M2 muscarinic binding sites in the striatum of the cat, monkey, and human. J Neurosci. 1988;8(3):1052–62. PubMed PMID: 3346715.
169. Ariano MA, Kenny SL. Striatal muscarinic receptors are associated with substance P and somatostatin containing neurons. Brain Res. 1989;497(1):51–8. PubMed PMID: 2477112.
170. Alcantara AA, Mrzljak L, Jakab RL, Levey AI, Hersch SM, Goldman-Rakic PS. Muscarinic m1 and m2 receptor proteins in local circuit and projection neurons of the primate striatum: anatomical evidence for cholinergic modulation of glutamatergic prefronto-striatal pathways. J Comp Neurol. 2001;434(4):445–60. PubMed PMID: 11343292.
171. Hersch SM, Gutekunst CA, Rees HD, Heilman CJ, Levey AI. Distribution of m1-m4 muscarinic receptor proteins in the rat striatum: light and electron microscopic immunocytochem-

istry using subtype-specific antibodies. J Neurosci. 1994;14(5 Pt 2):3351–63. PubMed PMID: 8182478.
172. Goldberg JA, Ding JB, Surmeier DJ. Muscarinic modulation of striatal function and circuitry. Handb Exp Pharmacol. 2012;208:223–41. PubMed PMID: 22222701.
173. Girault JA, Greengard P. The neurobiology of dopamine signaling. Arch Neurol. 2004; 61(5):641–4. PubMed PMID: 15148138.
174. Lee CR, Tepper JM. Basal ganglia control of substantia nigra dopaminergic neurons. J Neural Transm Suppl. 2009;73:71–90. PubMed PMID: 20411769.
175. Keath JR, Iacoviello MP, Barrett LE, Mansvelder HD, McGehee DS. Differential modulation by nicotine of substantia nigra versus ventral tegmental area dopamine neurons. J Neurophysiol. 2007;98(6):3388–96. PubMed PMID: 17942622.
176. Futami T, Takakusaki K, Kitai ST. Glutamatergic and cholinergic inputs from the pedunculopontine tegmental nucleus to dopamine neurons in the substantia nigra pars compacta. Neurosci Res. 1995;21(4):331–42. PubMed PMID: 7777224.
177. Forster GL, Blaha CD. Pedunculopontine tegmental stimulation evokes striatal dopamine efflux by activation of acetylcholine and glutamate receptors in the midbrain and pons of the rat. Eur J Neurosci. 2003;17(4):751–62. PubMed PMID: 12603265.
178. McIntosh JM, Azam L, Staheli S, Dowell C, Lindstrom JM, Kuryatov A, et al. Analogs of alpha-conotoxin MII are selective for alpha6-containing nicotinic acetylcholine receptors. Mol Pharmacol. 2004;65(4):944–52. PubMed PMID: 15044624.
179. Quik M, McIntosh JM. Striatal alpha6* nicotinic acetylcholine receptors: potential targets for Parkinson's disease therapy. J Pharmacol Exp Ther. 2006;316(2):481–9. PubMed PMID: 16210393.
180. Quik M, Wonnacott S. alpha6beta2* and alpha4beta2* nicotinic acetylcholine receptors as drug targets for Parkinson's disease. Pharmacol Rev. 2011;63(4):938–66. PubMed PMID: 21969327, Pubmed Central PMCID: 3186078.
181. Gotti C, Guiducci S, Tedesco V, Corbioli S, Zanetti L, Moretti M, et al. Nicotinic acetylcholine receptors in the mesolimbic pathway: primary role of ventral tegmental area alpha-6beta2* receptors in mediating systemic nicotine effects on dopamine release, locomotion, and reinforcement. J Neurosci. 2010;30(15):5311–25. PubMed PMID: 20392953.
182. Klink R. de Kerchove d'Exaerde A, Zoli M, Changeux JP. Molecular and physiological diversity of nicotinic acetylcholine receptors in the midbrain dopaminergic nuclei. J Neurosci. 2001;21(5):1452–63.
183. Champtiaux N, Gotti C, Cordero-Erausquin M, David DJ, Przybylski C, Lena C, et al. Subunit composition of functional nicotinic receptors in dopaminergic neurons investigated with knock-out mice. J Neurosci. 2003;23(21):7820–9. PubMed PMID: 12944511.
184. Cui C, Booker TK, Allen RS, Grady SR, Whiteaker P, Marks MJ, et al. The beta3 nicotinic receptor subunit: a component of alpha-conotoxin MII-binding nicotinic acetylcholine receptors that modulate dopamine release and related behaviors. J Neurosci. 2003;23(35):11045–53. PubMed PMID: 14657161.
185. Zoli M, Moretti M, Zanardi A, McIntosh JM, Clementi F, Gotti C. Identification of the nicotinic receptor subtypes expressed on dopaminergic terminals in the rat striatum. J Neurosci. 2002;22(20):8785–9. PubMed PMID: 12388584.
186. Anderson DJ, Malysz J, Gronlien JH, El Kouhen R, Hakerud M, Wetterstrand C, et al. Stimulation of dopamine release by nicotinic acetylcholine receptor ligands in rat brain slices correlates with the profile of high, but not low, sensitivity alpha4beta2 subunit combination. Biochem Pharmacol. 2009;78(7):844–51. PubMed PMID: 19555668.
187. Grady SR, Drenan RM, Breining SR, Yohannes D, Wageman CR, Fedorov NB, et al. Structural differences determine the relative selectivity of nicotinic compounds for native alpha 4 beta 2*-, alpha 6 beta 2*-, alpha 3 beta 4*- and alpha 7-nicotine acetylcholine receptors. Neuropharmacology. 2010;58(7):1054–66. PubMed PMID: 20114055, Pubmed Central PMCID: 2849849.

188. Quik M, Sum JD, Whiteaker P, McCallum SE, Marks MJ, Musachio J, et al. Differential declines in striatal nicotinic receptor subtype function after nigrostriatal damage in mice. Mol Pharmacol. 2003;63(5):1169–79. PubMed PMID: 12695545.
189. Quik M, Vailati S, Bordia T, Kulak JM, Fan H, McIntosh JM, et al. Subunit composition of nicotinic receptors in monkey striatum: effect of treatments with 1-methyl-4-phenyl-1,2,3, 6-tetrahydropyridine or L-DOPA. Mol Pharmacol. 2005;67(1):32–41. PubMed PMID: 15470079.
190. Quik M, Polonskaya Y, Kulak JM, McIntosh JM. Vulnerability of 125I-alpha-conotoxin MII binding sites to nigrostriatal damage in monkey. J Neurosci. 2001;21(15):5494–500. PubMed PMID: 11466420.
191. Kulak JM, Musachio JL, McIntosh JM, Quik M. Declines in different beta2* nicotinic receptor populations in monkey striatum after nigrostriatal damage. J Pharmacol Exp Ther. 2002;303(2):633–9. PubMed PMID: 12388645.
192. Wada E, Wada K, Boulter J, Deneris E, Heinemann S, Patrick J, et al. Distribution of alpha 2, alpha 3, alpha 4, and beta 2 neuronal nicotinic receptor subunit mRNAs in the central nervous system: a hybridization histochemical study in the rat. J Comp Neurol. 1989;284(2):314–35. PubMed PMID: 2754038.
193. Le Novere N, Zoli M, Changeux JP. Neuronal nicotinic receptor alpha 6 subunit mRNA is selectively concentrated in catecholaminergic nuclei of the rat brain. Eur J Neurosci. 1996;8(11):2428–39. PubMed PMID: 8950106.
194. Xiao C, Yang KC, Zhou CY, Jin GZ, Wu J, Ye JH. Nicotine modulates GABAergic transmission to dopaminergic neurons in substantia nigra pars compacta. Acta Pharmacol Sin. 2009;30(6):851–8. PubMed PMID: 19498424.
195. Marchi M, Risso F, Viola C, Cavazzani P, Raiteri M. Direct evidence that release-stimulating alpha7* nicotinic cholinergic receptors are localized on human and rat brain glutamatergic axon terminals. J Neurochem. 2002;80(6):1071–8. PubMed PMID: 11953457.
196. Sharples CG, Kaiser S, Soliakov L, Marks MJ, Collins AC, Washburn M, et al. UB-165: a novel nicotinic agonist with subtype selectivity implicates the alpha4beta2* subtype in the modulation of dopamine release from rat striatal synaptosomes. J Neurosci. 2000;20(8):2783–91. PubMed PMID: 10751429.
197. Kaiser S, Wonnacott S. alpha-bungarotoxin-sensitive nicotinic receptors indirectly modulate [(3)H]dopamine release in rat striatal slices via glutamate release. Mol Pharmacol. 2000;58(2):312–8. PubMed PMID: 10908298.
198. Marks MJ, Stitzel JA, Romm E, Wehner JM, Collins AC. Nicotinic binding sites in rat and mouse brain: comparison of acetylcholine, nicotine, and alpha-bungarotoxin. Mol Pharmacol. 1986;30(5):427–36. PubMed PMID: 3534542.
199. Grace AA, Bunney BS. The control of firing pattern in nigral dopamine neurons: burst firing. J Neurosci. 1984;4(11):2877–90. PubMed PMID: 6150071.
200. Grace AA, Bunney BS. The control of firing pattern in nigral dopamine neurons: single spike firing. J Neurosci. 1984;4(11):2866–76. PubMed PMID: 6150070.
201. Mansvelder HD, McGehee DS. Long-term potentiation of excitatory inputs to brain reward areas by nicotine. Neuron. 2000;27(2):349–57. PubMed PMID: 10985354.
202. Placzek AN, Dani JA. Synaptic plasticity within midbrain dopamine centers contributes to nicotine addiction. Nebr Symp Motiv. 2009;55:5–15. PubMed PMID: 19013936.
203. Misgeld U. Innervation of the substantia nigra. Cell Tissue Res. 2004;318(1):107–14. PubMed PMID: 15338269.
204. Marks MJ, Burch JB, Collins AC. Effects of chronic nicotine infusion on tolerance development and nicotinic receptors. J Pharmacol Exp Ther. 1983;226(3):817–25. PubMed PMID: 6887012.
205. Schwartz RD, Kellar KJ. Nicotinic cholinergic receptor binding sites in the brain: regulation in vivo. Science. 1983;220(4593):214–6. PubMed PMID: 6828889.
206. Benwell ME, Balfour DJ, Anderson JM. Evidence that tobacco smoking increases the density of (-)-[3H]nicotine binding sites in human brain. J Neurochem. 1988;50(4):1243–7. PubMed PMID: 3346676.

207. Perry DC, Davila-Garcia MI, Stockmeier CA, Kellar KJ. Increased nicotinic receptors in brains from smokers: membrane binding and autoradiography studies. J Pharmacol Exp Ther. 1999;289(3):1545–52. PubMed PMID: 10336551.
208. McCallum SE, Parameswaran N, Bordia T, Fan H, Tyndale RF, Langston JW, et al. Increases in alpha4* but not alpha3*/alpha6* nicotinic receptor sites and function in the primate striatum following chronic oral nicotine treatment. J Neurochem. 2006;96(4):1028–41. PubMed PMID: 16412091.
209. Staley JK, Krishnan-Sarin S, Cosgrove KP, Krantzler E, Frohlich E, Perry E, et al. Human tobacco smokers in early abstinence have higher levels of beta2* nicotinic acetylcholine receptors than nonsmokers. J Neurosci. 2006;26(34):8707–14. PubMed PMID: 16928859.
210. Moretti M, Mugnaini M, Tessari M, Zoli M, Gaimarri A, Manfredi I, et al. A comparative study of the effects of the intravenous self-administration or subcutaneous minipump infusion of nicotine on the expression of brain neuronal nicotinic receptor subtypes. Mol Pharmacol. 2010;78(2):287–96. PubMed PMID: 20439469.
211. Belluardo N, Mudo G, Blum M, Fuxe K. Central nicotinic receptors, neurotrophic factors and neuroprotection. Behav Brain Res. 2000;113(1–2):21–34. PubMed PMID: 10942029.
212. O'Neill MJ, Siemers ER. Pharmacological approaches to disease-modifying therapies in Parkinson's disease. Expert Rev Neurother. 2002;2(6):819–34. PubMed PMID: 19810916.
213. Quik M, O'Neill M, Perez XA. Nicotine neuroprotection against nigrostriatal damage: importance of the animal model. Trends Pharmacol Sci. 2007;28(5):229–35. PubMed PMID: 17412429.
214. Picciotto MR, Zoli M. Neuroprotection via nAChRs: the role of nAChRs in neurodegenerative disorders such as Alzheimer's and Parkinson's disease. Front Biosci. 2008;13:492–504. PubMed PMID: 17981563.
215. Bencherif M. Neuronal nicotinic receptors as novel targets for inflammation and neuroprotection: mechanistic considerations and clinical relevance. Acta Pharmacol Sin. 2009;30(6):702–14. PubMed PMID: 19498416.
216. Exley R, Cragg SJ. Presynaptic nicotinic receptors: a dynamic and diverse cholinergic filter of striatal dopamine neurotransmission. Br J Pharmacol. 2008;153 Suppl 1:S283–97. PubMed PMID: 18037926, Pubmed Central PMCID: 2268048.
217. Quik M, Bordia T, Forno L, McIntosh JM. Loss of alpha-conotoxinMII- and A85380-sensitive nicotinic receptors in Parkinson's disease striatum. J Neurochem. 2004;88(3):668–79. PubMed PMID: 14720216.
218. Bohr IJ, Ray MA, McIntosh JM, Chalon S, Guilloteau D, McKeith IG, et al. Cholinergic nicotinic receptor involvement in movement disorders associated with Lewy body diseases. An autoradiography study using [(125)I]alpha-conotoxinMII in the striatum and thalamus. Exp Neurol. 2005;191(2):292–300.
219. Schwartz RD, Lehmann J, Kellar KJ. Presynaptic nicotinic cholinergic receptors labeled by [3H]acetylcholine on catecholamine and serotonin axons in brain. J Neurochem. 1984;42(5):1495–8. PubMed PMID: 6707650.
220. Bordia T, Grady SR, McIntosh JM, Quik M. Nigrostriatal damage preferentially decreases a subpopulation of alpha6beta2* nAChRs in mouse, monkey, and Parkinson's disease striatum. Mol Pharmacol. 2007;72(1):52–61. PubMed PMID: 17409284.
221. Grady SR, Salminen O, McIntosh JM, Marks MJ, Collins AC. Mouse striatal dopamine nerve terminals express alpha4alpha5beta2 and two stoichiometric forms of alpha4beta2*-nicotinic acetylcholine receptors. J Mol Neurosci. 2010;40(1–2):91–5. PubMed PMID: 19693710.
222. McClure-Begley TD, King NM, Collins AC, Stitzel JA, Wehner JM, Butt CM. Acetylcholine-stimulated [3H]GABA release from mouse brain synaptosomes is modulated by alpha4beta2 and alpha4alpha5beta2 nicotinic receptor subtypes. Mol Pharmacol. 2009;75(4):918–26. PubMed PMID: 19139153, Pubmed Central PMCID: 2684932.
223. Quik M, Polonskaya Y, Gillespie A. G KL, Langston JW. Differential alterations in nicotinic receptor alpha6 and beta3 subunit messenger RNAs in monkey substantia nigra after nigrostriatal degeneration. Neuroscience. 2000;100(1):63–72.

224. Quik M, Polonskaya Y, McIntosh JM, Kulak JM. Differential nicotinic receptor expression in monkey basal ganglia: effects of nigrostriatal damage. Neuroscience. 2002;112(3):619–30. PubMed PMID: 12074903.
225. Quik M, Campos C, Parameswaran N, Langston JW, McIntosh JM, Yeluashvili M. Chronic nicotine treatment increases nAChRs and microglial expression in monkey substantia nigra after nigrostriatal damage. J Mol Neurosci. 2010;40(1–2):105–13. PubMed PMID: 19685015, Pubmed Central PMCID: 3133952.
226. Morens DM, Grandinetti A, Reed D, White LR, Ross GW. Cigarette smoking and protection from Parkinson's disease: false association or etiologic clue? Neurology. 1995;45(6):1041–51. PubMed PMID: 7783862.
227. Gorell JM, Rybicki BA, Johnson CC, Peterson EL. Smoking and Parkinson's disease: a dose-response relationship. Neurology. 1999;52(1):115–9. PubMed PMID: 9921857.
228. Hernan MA, Zhang SM, Rueda-deCastro AM, Colditz GA, Speizer FE, Ascherio A. Cigarette smoking and the incidence of Parkinson's disease in two prospective studies. Ann Neurol. 2001;50(6):780–6. PubMed PMID: 11761476.
229. Alves G, Kurz M, Lie SA, Larsen JP. Cigarette smoking in Parkinson's disease: influence on disease progression. Mov disord. 2004;19(9):1087–92. PubMed PMID: 15372603.
230. Allam MF, Campbell MJ, Hofman A, Del Castillo AS, Fernandez-Crehuet NR. Smoking and Parkinson's disease: systematic review of prospective studies. Mov disord. 2004;19(6):614–21. PubMed PMID: 15197698.
231. Thacker EL, O'Reilly EJ, Weisskopf MG, Chen H, Schwarzschild MA, McCullough ML, et al. Temporal relationship between cigarette smoking and risk of Parkinson disease. Neurology. 2007;68(10):764–8. PubMed PMID: 17339584, Pubmed Central PMCID: 2225169.
232. Elbaz A, Moisan F. Update in the epidemiology of Parkinson's disease. Curr Opin Neurol. 2008;21(4):454–60. PubMed PMID: 18607207.
233. Chen H, Huang X, Guo X, Mailman RB, Park Y, Kamel F, et al. Smoking duration, intensity, and risk of Parkinson disease. Neurology. 2010;74(11):878–84. PubMed PMID: 20220126, Pubmed Central PMCID: 2836869.
234. Ritz B, Rhodes SL. After half a century of research on smoking and PD, where do we go now? Neurology. 2010;74(11):870–1. PubMed PMID: 20220125.
235. Calabresi P, Di Filippo M, Ghiglieri V, Picconi B. Molecular mechanisms underlying levodopa-induced dyskinesia. Mov disord. 2008;23 Suppl 3:S570–9. PubMed PMID: 18781674.
236. Barik J, Wonnacott S. Molecular and cellular mechanisms of action of nicotine in the CNS. Handb Exp Pharmacol. 2009;192:173–207. PubMed PMID: 19184650.
237. Quik M, Parameswaran N, McCallum SE, Bordia T, Bao S, McCormack A, et al. Chronic oral nicotine treatment protects against striatal degeneration in MPTP-treated primates. J Neurochem. 2006;98(6):1866–75. PubMed PMID: 16882311.
238. Quik M, Chen L, Parameswaran N, Xie X, Langston JW, McCallum SE. Chronic oral nicotine normalizes dopaminergic function and synaptic plasticity in 1-methyl-4-phenyl-1,2,3,6-tetrahydropyridine-lesioned primates. J Neurosci. 2006;26(17):4681–9. PubMed PMID: 16641249.
239. Srinivasan R, Pantoja R, Moss FJ, Mackey ED, Son CD, Miwa J, et al. Nicotine up-regulates alpha4beta2 nicotinic receptors and ER exit sites via stoichiometry-dependent chaperoning. J Gen Physiol. 2011;137(1):59–79. PubMed PMID: 21187334, Pubmed Central PMCID: 3010053.
240. Xiao C, Nashmi R, McKinney S, Cai H, McIntosh JM, Lester HA. Chronic nicotine selectively enhances alpha4beta2* nicotinic acetylcholine receptors in the nigrostriatal dopamine pathway. J Neurosci. 2009;29(40):12428–39. PubMed PMID: 19812319, Pubmed Central PMCID: 2787412.
241. Castagnoli K, Murugesan T. Tobacco leaf, smoke and smoking, MAO inhibitors, Parkinson's disease and neuroprotection; are there links? Neurotoxicology. 2004;25(1–2):279–91. PubMed PMID: 14697903.

242. Castagnoli KP, Steyn SJ, Petzer JP, Van der Schyf CJ, Castagnoli Jr N. Neuroprotection in the MPTP Parkinsonian C57BL/6 mouse model by a compound isolated from tobacco. Chem Res Toxicol. 2001;14(5):523–7. PubMed PMID: 11368550.
243. Miksys S, Tyndale RF. Nicotine induces brain CYP enzymes: relevance to Parkinson's disease. J Neural Transm Suppl. 2006;70:177–80. PubMed PMID: 17017527.
244. Fahn S. The history of dopamine and levodopa in the treatment of Parkinson's disease. Mov disord. 2008;23 Suppl 3:S497–508. PubMed PMID: 18781671.
245. Lang AE. When and how should treatment be started in Parkinson disease? Neurology. 2009;72(7 Suppl):S39–43. PubMed PMID: 19221313.
246. Poewe W. Treatments for Parkinson disease–past achievements and current clinical needs. Neurology. 2009;72(7 Suppl):S65–73. PubMed PMID: 19221317.
247. Schapira AH, Emre M, Jenner P, Poewe W. Levodopa in the treatment of Parkinson's disease. Eur J Neurol. 2009;16(9):982–9. PubMed PMID: 19538218.
248. Jankovic J, Stacy M. Medical management of levodopa-associated motor complications in patients with Parkinson's disease. CNS Drugs. 2007;21(8):677–92. PubMed PMID: 17630819.
249. Obeso JA, Olanow W. Continuing efforts to obtain continuous delivery of levodopa. Mov disord. 2011;26(12):2149–50. PubMed PMID: 22021155.
250. Worth PF. How to treat Parkinson's disease in 2013. Clin Med. 2013;13(1):93–6. PubMed PMID: 23472504.
251. Jankovic J, Poewe W. Therapies in Parkinson's disease. Curr Opin Neurol. 2012;25(4):433–47. PubMed PMID: 22691758.
252. Huang LZ, Grady SR, Quik M. Nicotine reduces L-DOPA-induced dyskinesias by acting at beta2* nicotinic receptors. J Pharmacol Exp Ther. 2011;338(3):932–41. PubMed PMID: 21665941, Pubmed Central PMCID: 3164339.
253. Quik M, Campos C, Bordia T, Strachan JP, Zhang J, McIntosh JM, et al. Alpha4beta2 nicotinic receptors play a role in the nAChR-mediated decline in L-dopa-induced dyskinesias in parkinsonian rats. Neuropharmacology. 2013;71:191–203. PubMed PMID: 23583932, Pubmed Central PMCID: 3685407.
254. Quik M, Campos C, Grady SR. Multiple CNS nicotinic receptors mediate l-dopa-induced dyskinesias: studies with parkinsonian nicotinic receptor knockout mice. Biochem Pharmacol. 2013;86(8):1153–62. PubMed PMID: 23831952.
255. Quik M, Park KM, Hrachova M, Mallela A, Huang LZ, McIntosh JM, et al. Role for alpha6 nicotinic receptors in l-dopa-induced dyskinesias in parkinsonian mice. Neuropharmacology. 2012;63(3):450–9. PubMed PMID: 22579614, Pubmed Central PMCID: 3726309.
256. Mihalak KB, Carroll FI, Luetje CW. Varenicline is a partial agonist at alpha4beta2 and a full agonist at alpha7 neuronal nicotinic receptors. Mol Pharmacol. 2006;70(3):801–5. PubMed PMID: 16766716.
257. Rollema H, Chambers LK, Coe JW, Glowa J, Hurst RS, Lebel LA, et al. Pharmacological profile of the alpha4beta2 nicotinic acetylcholine receptor partial agonist varenicline, an effective smoking cessation aid. Neuropharmacology. 2007;52(3):985–94. PubMed PMID: 17157884.
258. Huang LZ, Parameswaran N, Bordia T, Michael McIntosh J, Quik M. Nicotine is neuroprotective when administered before but not after nigrostriatal damage in rats and monkeys. J Neurochem. 2009;109(3):826–37. PubMed PMID: 19250334, Pubmed Central PMCID: 2677631.
259. Olanow CW. The scientific basis for the current treatment of Parkinson's disease. Annu Rev Med. 2004;55:41–60. PubMed PMID: 14746509.
260. Samii A, Nutt JG, Ransom BR. Parkinson's disease. Lancet. 2004;363(9423):1783–93. PubMed PMID: 15172778.
261. Huang LZ, Campos C, Ly J, Ivy Carroll F, Quik M. Nicotinic receptor agonists decrease L-dopa-induced dyskinesias most effectively in partially lesioned parkinsonian rats. Neuropharmacology. 2011;60(6):861–8. PubMed PMID: 21232546, Pubmed Central PMCID: 3133531.

262. Ishikawa A, Miyatake T. Effects of smoking in patients with early-onset Parkinson's disease. J Neurol Sci. 1993;117(1–2):28–32. PubMed PMID: 8410063.
263. Kelton MC, Kahn HJ, Conrath CL, Newhouse PA. The effects of nicotine on Parkinson's disease. Brain Cogn. 2000;43(1–3):274–82. PubMed PMID: 10857708.
264. Mitsuoka T, Kaseda Y, Yamashita H, Kohriyama T, Kawakami H, Nakamura S, et al. Effects of nicotine chewing gum on UPDRS score and P300 in early-onset parkinsonism. Hiroshima J Med Sci. 2002;51(1):33–9. PubMed PMID: 11999458.
265. Villafane G, Cesaro P, Rialland A, Baloul S, Azimi S, Bourdet C, et al. Chronic high dose transdermal nicotine in Parkinson's disease: an open trial. Eur J Neurol. 2007;14(12):1313–6. PubMed PMID: 17941858.
266. Lemay S, Chouinard S, Blanchet P, Masson H, Soland V, Beuter A, et al. Lack of efficacy of a nicotine transdermal treatment on motor and cognitive deficits in Parkinson's disease. Prog Neuropsychopharmacol Biol Psychiatry. 2004;28(1):31–9. PubMed PMID: 14687854.
267. Clemens P, Baron JA, Coffey D, Reeves A. The short-term effect of nicotine chewing gum in patients with Parkinson's disease. Psychopharmacology (Berl). 1995;117(2):253–6. PubMed PMID: 7753975.
268. Ebersbach G, Stock M, Muller J, Wenning G, Wissel J, Poewe W. Worsening of motor performance in patients with Parkinson's disease following transdermal nicotine administration. Mov disord. 1999;14(6):1011–3. PubMed PMID: 10584678.
269. Vieregge A, Sieberer M, Jacobs H, Hagenah JM, Vieregge P. Transdermal nicotine in PD: a randomized, double-blind, placebo-controlled study. Neurology. 2001;57(6):1032–5. PubMed PMID: 11571330.
270. Parkinson Study G. Randomized placebo-controlled study of the nicotinic agonist SIB-1508Y in Parkinson disease. Neurology. 2006;66(3):408–10. PubMed PMID: 16476941.
271. Fagerstrom KO, Pomerleau O, Giordani B, Stelson F. Nicotine may relieve symptoms of Parkinson's disease. Psychopharmacology (Berl). 1994;116(1):117–9. PubMed PMID: 7862924.

Chapter 20
Nicotinic Receptors and Mental Illness

Sherry Leonard

Abstract The prevalence of smoking in the mentally ill, particularly in schizophrenic patients, is much higher than in the general population. While smoking demographics are altered in these patients, nicotinic receptors are implicated in the disorder. Nicotine normalizes several sensory processing deficits in schizophrenia, as well as improving cognition and disease symptomatology. Smoking has a large effect on gene expression in human brain, and many genes abnormally expressed in schizophrenic nonsmokers are brought to control levels in schizophrenic smokers. The α7 nicotinic receptor gene, *CHRNA7*, is genetically linked to the disorder in multiple studies. Deletion of *Chrna7* in mice results in several traits found in schizophrenic individuals. The expression of *α7nAChR*s is decreased in postmortem brain of schizophrenic subjects, as measured by α-bungarotoxin binding. Nicotine binding is also decreased in schizophrenic brain, suggesting that high-affinity nicotinic receptor expression is reduced as well. Regulation of the *CHRNA7* gene is complex. Promoter methylation and several transcription factors have been identified that affect transcription. The human *CHRNA7* gene is unusual in that it is partially duplicated. The duplicated sequences are expressed with exons from a second partial duplication, forming a new, chimeric gene, *CHRFAM7A*. The duplicated gene is human specific, not being found in rodents or primates. The duplicated sequences in *CHRFAM7A* are nearly identical to exons 5–10 of the full-length gene, *CHRNA7*. Thus, exons 5–10 cannot be accurately queried for *CHRNA7* in genome-wide association studies. Further, the duplicated gene product, dupα7, is a dominant negative regulator of α7nAChR function, reducing current in response to acetylcholine application. Functional mutations in the *CHRNA7* gene promoter and in *CHRFAM7A* have been identified and are associated with schizophrenia. Several agonists of the α7nAChR have been identified as possible therapeutic drugs, including DMXB-A and choline. Type II allosteric modulators appear to potentiate function of the human heteromeric receptor, containing both *CHRNA7* and *CHRFAM7A* gene products.

Keywords Nicotinic receptor • Smoking • Alpha 7 • Schizophrenia • Gene duplication • *CHRNA7* • *CHRFAM7A*

S. Leonard, Ph.D. (✉)
Department of Psychiatry, University of Colorado, Anschutz Medical Campus,
12800 East 19th Avenue, RC-1, Room 8107, Aurora, CO 81230, USA
e-mail: sherry.leonard@ucdenver.edu

R.A.J. Lester (ed.), *Nicotinic Receptors*, The Receptors 26,
DOI 10.1007/978-1-4939-1167-7_20

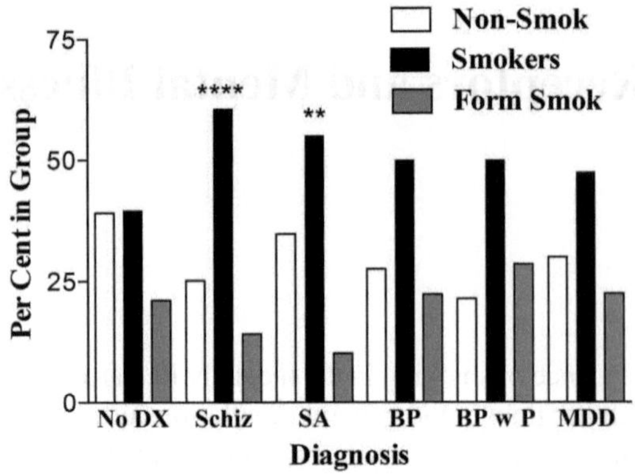

Fig. 20.1 Comparison of smoking prevalence in the mentally ill. Subjects are from the Denver Schizophrenia Center Brain Bank and from locally collected blood samples. Smoking history was reported as nonsmoker (NS), smoker (S), or former smoker (FS). The percent of subjects in each of the smoking history groups is shown for controls (No Dx), schizophrenia (Schiz), schizoaffective disorder (SA), bipolar disorder (BP), bipolar with psychosis (BP w P), and major depressive disorder (MDD). There were significantly more smokers among the schizophrenics than in the control group ($\chi^2 = 31.61$, 1; $P < 0.0001$****). Patients with schizoaffective disorder also had a higher prevalence of smoking ($\chi^2 = 4.497$, 1; $P = 0.0340$**). Although there were also more smokers than nonsmokers in the bipolar and MDD groups, the difference was not statistically significant

1 Introduction

Mental illness, including depression, attention-deficit hyperactivity disorder (ADHD), autism, bipolar disorder, and schizophrenia, occurs in approximately 30 % of our population [1]. This group of people uses tobacco products more frequently than the general population. Somewhere between 50 and 90 % of individuals, depending on the specific mental illness, smoke cigarettes, compared to 20 % of the general population [2]. The use of tobacco products has declined in this country since its connection to cancer etiology, but it has not decreased in individuals who suffer from mental illness. It is estimated that of the total number of smokers in the United States today, 30 % are mentally ill [1, 3]. Further, they are heavier smokers generally and purchase the majority of cigarettes [4]. Utilizing demographic data on subjects from our postmortem brain bank and blood samples, Fig. 20.1 illustrates the use of tobacco across the mental illness spectrum.

Smoking prevalence is highest in schizophrenic patients, most of which use some type of tobacco product [3, 5–9]. Schizophrenics appear to extract more nicotine from each cigarette, as evidenced by higher blood nicotine and cotinine levels [10, 11]. Higher levels are not related to nicotine metabolism differences [12], but may be related to topography. Patients have shortened inter-puff intervals and take more puffs per cigarette [13, 14]. Normally aversive, rapid smoking appears to be a normal pattern in schizophrenia [15].

Although smoking levels are highest in the schizophrenia spectrum disorders, subjects with depression are also heavy users of tobacco products [16]. A recent study of substance use in the mentally ill suggested that schizophrenics were more likely to use tobacco for coping and illness motives, while individuals with depression were more likely to use alcohol for the same reasons [17]. Many years ago, Janowsky suggested that mental illnesses involving mania and affective symptoms might both involve cholinergic function with mania being a hypocholinergic state and depression a hypercholinergic condition [18, 19]. Brain imaging studies show that high-affinity nicotinic receptor occupancy is high in depression [20] and low in schizophrenia [21, 22], consistent with either low levels of acetylcholine or receptors in schizophrenia and high cholinergic function in depression. Supportive of this hypothesis, recent work shows that many antidepressants are AChR blockers [23, 24], and in rats, a decrease in α7nAChR function increases self-administration of nicotine [25]. Thus, while there may be underlying biological etiology for the heavy use of tobacco products in both depression and schizophrenia, the aberrant pathways may be quite different.

2 Why Are Schizophrenics Heavy Smokers?

A biological basis for smoking in schizophrenia could be informative for treatment and drug development. It has been suggested that smoking in these patients may be a form of self-medication [26, 27]. Schizophrenics often describe smoking not only as enjoyable but as a crutch [2, 28].

Nicotine normalizes an auditory sensory processing deficit in schizophrenia, the P50 deficit [29, 30]. This deficit is measured utilizing paired auditory stimuli 500 ms apart. In the normal response, the second stimulus is inhibited; in schizophrenics, this does not occur [31, 32]. Nicotine improves eye-tracking deficits in patients [33–35]. Smoking also improves cognition in schizophrenic subjects [36–40] and decreases psychiatric symptoms [41, 42].

Another reason why smoking might offer relief to schizophrenic patients is the release of dopamine. Dopamine levels in patient brains are high in the mesolimbic pathways and low in the mesocortical pathways [43]. Virtually all effective antipsychotic drugs block D_2 dopamine receptors. Nicotine releases dopamine through a presynaptic mechanism stimulating the release of glutamate and subsequent release of dopamine [44–48]. Smoking may, thus, compensate for low levels of dopamine in these patients.

Smoking cigarettes profoundly changes gene expression in human brain. In a microarray study of mRNA levels in postmortem hippocampus of control and schizophrenic smokers and nonsmokers, we found that smoking in all subjects resulted in expression changes for more than 200 genes, including increased expression of genes in the NMDA postsynaptic density [49]. Interestingly, there were 70 genes that were differentially regulated by smoking in schizophrenia. Expression was either up- or downregulated in schizophrenic nonsmokers compared to control nonsmokers. In schizophrenic smokers, however, expression was more similar to that in control smokers, suggesting that patients may be utilizing smoking to normalize underlying biological deficits.

Smoking cessation in mental illness is usually not successful, particularly in schizophrenic patients [2, 50]. Failure to quit smoking in schizophrenia seems to be associated with a poor performance on cognitive tests that depend on the prefrontal cortex [51].

3 Nicotinic Receptor Genetics in Mental Illness

The heavy use of tobacco products in schizophrenia suggests the involvement of nicotinic receptors that respond to nicotine in cigarettes and other forms of tobacco. Genetic analysis, including linkage and association studies, shows that the gene cluster of *CHRNA5-CHRNA3-CHRNB4* at 15q25 is strongly associated with nicotine dependence [52–54]. A recent meta-analysis of variants in the 15q25 gene cluster was positive for both schizophrenia and bipolar disorder, but none of the three genes was specifically targeted [55]. It is possible, however, that the strong association of this locus to smoking and nicotine addiction could have influenced the study outcome. A comprehensive association study in comorbid psychiatric disorders, including alcohol dependence, cannabis dependence, major depression, panic attack, social phobia, posttraumatic stress disorder, attention-deficit hyperactivity disorder, conduct disorder, and antisocial personality disorder, did not show association at the A5-A3-B4 locus [56].

Somewhat stronger genetic evidence exists for the α7 nicotinic receptor gene, *CHRNA7*, which has been linked to schizophrenia in multiple linkage and association studies [57–64], and also to bipolar disorder [65, 66]. Deletions at 15q13-14, involving the *CHRNA7* gene are rare, but strongly associated with schizophrenia [67, 68]. Such deletions, and sometimes duplications, are also seen in rare cases of developmental delay, mental retardation, autism, seizures, and bipolar disorder [69]. The part of chromosome 15 affected by these chromosomal anomalies involves two large regions of homology originally identified as breakpoint regions (BP) in Prader–Willi syndrome [70, 71]. Deletions at 15q13.3 involve BP4 and BP5, which contain the *CHRNA7* gene cluster. Approximately 20 % of Prader–Willi patients have psychotic disorders [72, 73].

Chrna7 knockout animals exhibit many of the traits seen in schizophrenia. *Chrna7*$^{(-/-)}$ mice have impaired attention [74] and delayed procedural learning, where repeated patterns signal acceptance of a rule [75]. A *Chrna7* knockout generated in the C3H/2j mouse exhibits impaired sensory gating [76] and decreased LTP (manuscript submitted). Pharmacologically decreasing expression of *Chrna7* results in an increase in the motivation for nicotine self-administration [25]. The *CHRNA7* gene has also been associated with smoking in schizophrenia [77, 78].

4 Decreased Expression of Nicotinic Receptors in Postmortem Brain of Schizophrenic Subjects

Both high- and low-affinity nicotinic receptors are decreased in expression in postmortem brain of schizophrenic patients. High-affinity receptors, as measured by [^{3}H]-nicotine binding, are increased in postmortem brain of normal smokers by about 50 % [79–81]. Receptor number is also increased in blood leucocytes of smokers compared to nonsmokers [82]. In both tissues, the receptor number is dose dependent, being correlated with the number of cigarettes smoked per day. In hippocampus from schizophrenic smokers, however, there was 50 % less [^{3}H]-nicotine binding [83]. We also found decreased [^{3}H]-nicotine binding in blood leukocytes from schizophrenic patients (Fig. 20.2). As in postmortem brain, nicotine binding was increased by 50 % in polymorphonuclear cells (PMN) of control smokers. This increase was not seen in PMN from schizophrenic smokers. The specific high-affinity receptor subtypes involved in these changes in schizophrenic brain and periphery have not yet been identified.

The α7nAChR is downregulated by 50 % in schizophrenia, as measured by [^{125}I]-α-bungarotoxin binding, in the hippocampus [84], in the cortex [85, 86], and in the reticular thalamic nucleus [87]. While nicotine abundantly upregulates high-affinity receptors in both the brain and periphery of control subjects, only in very heavy smokers did we find an increase in [^{125}I]-α-bungarotoxin binding [83].

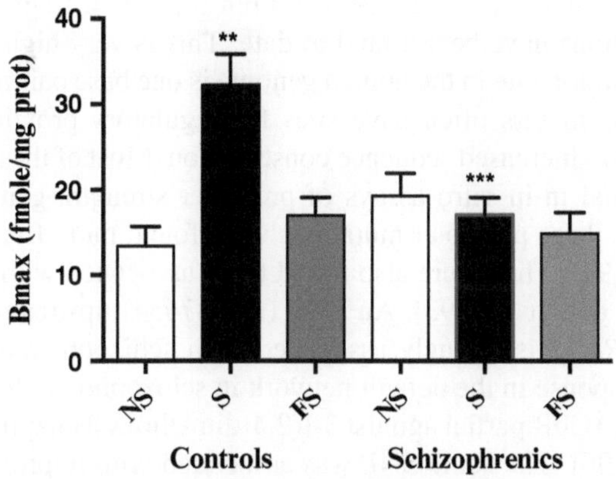

Fig. 20.2 High-affinity nicotinic receptors are not upregulated in schizophrenic smokers. Polymorphonuclear (PMN) cells were isolated from the blood of control and schizophrenic, smokers and nonsmokers (12 in each group), and [^{3}H]-nicotine binding was performed as described [82]. Control subjects upregulate binding by approximately 50 %; there was no increase in schizophrenic smokers. ** $p=0.0022$; *** $p=0.0007$; *NS* nonsmoker, *S* smoker, *FS* former smoker

5 Structure of the Human 15q14 Chromosomal Region

Within this region of human genetic linkage and chromosomal instability on 15q13.3, a further chromosomal rearrangement occurred, which partially duplicated the *CHRNA7* gene [88, 89]. Exons 5–10 were duplicated along with a 3′ cassette of DNA; this duplicon was inserted into a partial duplication of another gene from chromosome 3 and lies 5′ to the full-length *CHRNA7* gene at chromosome 15q13.3. The new chimeric gene, with at least 4 exons from the gene on chromosome 3 (*ULK4*) and exons 5–10 of *CHRNA7*, now named *CHRFAM7A*, is expressed in the human brain and periphery [88]. This is a new duplication and is human specific, not being found in either primates or rodents [90]. The duplicated sequences from the *CHRNA7* gene are 99.9 % identical. This is a major problem for association studies of *CHRNA7*; the entire gene cannot be interrogated because polymorphisms are not mapped in the current genome-wide association studies (GWAS). Only exons 1–4 and the promoter of the *CHRNA7* gene can be properly studied. This has likely resulted in an underestimation of linkage and association of nicotine addiction and mental illness to this gene cluster!

6 Mutations in the CHRNA7/CHRFAM7A Gene Cluster

Several polymorphisms in the regions of *CHRNA7* and *CHRFAM7A* that can be mapped have been associated with schizophrenia. The proximal promoter of *CHRNA7* is approximately 250 bp [91]. Within this small fragment of regulatory control, 21 mutations have been found to date. This is very high, considering that the estimated mutation rate in the human genome is one base pair in ~10^{-8} base pairs [92], and gene promoters often have sites for regulatory protein binding, which would likely lead to increased sequence conservation. Most of these promoter mutations are functional in in vitro assays of promoter strength, generally decreasing function. Further, these promoter mutations were found more frequently in schizophrenic patients [91]. They were also found to be associated with the P50 auditory evoked potential deficit [91, 93]. An SNP (rs3087454) upstream of the proximal promoter in *CHRNA7* is strongly associated with schizophrenia [94] and with a positive fMRI response in the default network in schizophrenia following administration of the α7nAChR partial agonist 3-[(2,4-dimethoxy)benzylidene]-anabaseine (DMXB-A) [95, 96]. The same SNP was associated with improved P50 gating in newborns of mothers treated prenatally with choline, a specific α7AChR agonist [97]. These results suggest SNP rs3087454 as a possible pharmacogenomics target for the evaluation of α7nAChR therapeutic drugs.

Polymorphisms in the duplicated gene, *CHRFAM7A*, are also associated with schizophrenia. A 2 bp deletion in exon 6 was mapped to *CHRFAM7A*, utilizing mRNA [98]. This deletion is evolutionarily new, being found almost exclusively in Caucasian individuals. The 2 bp deletion is associated with schizophrenia [99], and also with the P50 gating deficit [100, 101]. Presence of the 2 bp deletion in *CHRFAM7A* apparently results in a gene inversion. The normal copy of *CHRFAM7A* is in a head-to-head orientation with the full-length *CHRNA7* gene; the copy of *CHRFAM7A* containing the 2 bp deletion in exon 6 is in a tail-to-head orientation with *CHRNA7* [102]. This could have implications for the regulation of the duplicated gene, since the DNA upstream of the promoter would be different in each case.

7 Function of the Duplicated Gene, CHRFAM7A

The *CHRFAM7A* gene differs in copy number. Approximately 10 % of individuals have only one copy and about 5 % have no copies [88, 99]. We have also found subjects with more than two copies of this chimeric gene. Thus, if *CHRFAM7A* were to have a function, the number of copies could be important. The *CHRFAM7A* gene product, dupα7, was found to be a dominant negative regulator of α7nAChR function. When expressed in oocytes, *CHRFAM7A* does not have a function as a cholinergic receptor [103]. As it is missing exons 1–4 of *CHRNA7*, it has no signal peptide and would require co-assembly in the endoplasmic reticulum for surface expression. While this is likely a very inefficient process, it appears to occur. Co-expression of *CHRNA7* and *CHRFAM7A* in oocytes results in a large decrease in current following stimulation with acetylcholine compared to *CHRNA7* expressed alone [104, 105]. Co-expression of the copy of *CHRFAM7A* with the 2 bp deletion in exon 6 and full-length *CHRNA7* resulted in a further decrease in current [104]. Reduced current amplitude was not associated with a decrease in $[^{125}I]$-α-bungarotoxin binding, suggesting a decrease in the number of acetylcholine activatable receptors. This hypothesis was supported by the stimulation of current by the type II allosteric modulator 1-(5-chloro-2,4-dimethoxy-phenyl)-3-(5-methyl-isoxazol-3-yl)-urea, PNU-120596, which binds in the transmembrane segment of the receptor [104]. It is not yet known how dupα7 results in decreased function of the receptor. It is possible that at least some of the α7/dupα7 receptors could be sequestered in the endoplasmic reticulum, resulting in fewer surface receptors. *CHRFAM7A* is transcribed very efficiently but appears to be translated at low levels. However, it was clear from titration experiments that even low levels of expression were sufficient to disrupt function [104, 105].

The *CHRFAM7A* gene is downregulated by both nicotine [105] and by bacterial infection [106], which could result in increased activity of α7nAChRs. Thus, the dupα7 peptide may play a role in important α7nAChR pathways in inflammation.

8 Complex Regulation of CHRNA7 and CHRFAM7A

Like all genes, *CHRNA7* and *CHRFAM7A* are regulated by a complex set of mechanisms. *CHRNA7* is regulated by at least a dozen different pathways, and it is clear that we are far from understanding them all. Some are direct, including mutations in each of the genes and by differences in copy number. Others are indirect, involving influence from regulatory molecules. These regulatory modulators are sometimes also associated with schizophrenia.

As described earlier, both *CHRNA7* and *CHRFAM7A* differ in copy number due either to large segmental deletions or duplications [67, 68, 88]. These are rare but are associated with schizophrenia. The duplications appear more frequently in the autistic spectrum disorders [70]. Deletion does not always involve both genes [67]. This is important as additional copies of *CHRFAM7A*, where the gene product, dupα7, is a dominant regulator of α7nAChR function [104], could affect the level of activatable receptor in the patient.

Transcription of *CHRNA7* is affected directly by promoter mutations, which are unusually common [91]. Multiple other mutations were found in both *CHRNA7* and *CHRFAM7A* by mutation screening of mRNA [107], but most were rare and not associated with schizophrenia. A 2 bp deletion in exon 6 of *CHRFAM7A* is associated with schizophrenia [99], with the P50 deficit [100, 101], and with a more dominantly negative gene product [104]. It is not known whether the 2 bp deletion affects transcription of *CHRFAM7A*. A promoter for *CHRFAM7A* has not been identified.

Regulation of the *CHRNA7* promoter occurs by several other methods. The promoter is methylated in cells that do not express this gene, including in a cell line commonly used in neuroscience research, SHEP1 [108]. In human tissues, methylation was associated with decreased expression. Additionally, the transcription factor AP2α is a potent repressor of *CHRNA7* transcription [109].

CHRNA7 can be regulated by steroids, which affect both transcription and function [110]. Neurosteroids and antidepressive drugs, such as sertraline, are noncompetitive inhibitors of α7nAChR function, likely binding in the ion channel [23, 111]. Glucocorticoid regulation appears to involve the transcription factor *Egr1* [112].

Another gene that is strongly associated with schizophrenia is neuregulin 1 (*NRG1*) [113–115]. An isoform of *NRG1* regulates the expression of α7nAChRs [116, 117]. Further, genetic variants in *NRG1* are associated with mRNA levels of *CHRNA7* in human prefrontal cortex [118], suggesting that these two genes are functionally connected.

nAChRs are modified posttranslationally. Phosphorylation of the large intracellular loop occurs in both alpha 4 [119] and alpha 7 [120], principally by protein kinase A. Lipid moieties may be important in the regulation of surface expression of α7nAChRs. Palmitoylation in the N-terminus occurs in the endoplasmic reticulum during assembly and promotes the formation of functional receptors [121, 122]. Phospholipid abnormalities in plasma membranes isolated from human postmortem brain occur in schizophrenia and potentially could alter nAChR function [123].

CHRNA7 also interacts with various chaperone moieties, which can affect function. *RIC3*, resistance to inhibitors of cholinesterase 3 homolog, promotes surface expression of α7nAChR [124, 125]. Lynx1, a protein that shares characteristics with snake toxins, is an enhancer of cholinergic function, apparently without receptor subtype specificity [126, 127]. And nicotine, being membrane soluble, can bind to intracellular receptors, facilitating migration to the plasma membrane [128].

Recent research suggests that the glutamatergic postsynaptic density is modulated by α7nAChRs [129]. The α7nAChR lies either in or very near the NMDA postsynaptic density [130], where it has a permissive role on NR2B-NMDARs following treatment with α7-specific agonists [129]. Chronic nicotine promotes the formation of an α7nAChR-NMDA(NR2A) complex [131], disruption of which blocks smoking cessation relapse. Both NR2A and NR2B are upregulated in postmortem brain of smokers, as are other genes in the NMDA PSD [49].

There are surely yet to be defined regulators of *CHRNA7* expression and function. A recent study targeted to Rett syndrome and autism suggests that there are long-distance chromatin interactions between the *CHRNA7* gene at 15q13.3 and the Prader–Willi syndrome region at 15q11.2-13.3 that involve epigenetic regulation [132].

9 Nicotinic Cholinergic Drug Development for Schizophrenia

Several approaches have been taken to study the effects of nicotinic cholinergic drugs in schizophrenia. These include effects on sensory processing deficits, on cognition, and on perinatal development. Cognition is a major focus, as it is almost universally affected in schizophrenic patients [133], and is important in other types of mental illness [70, 134]. It was noted early that schizophrenics are heavy smokers [5, 6] and that such heavy use might be a form of self-medication [26, 27]. This has led to the investigation of the nicotinic cholinergic system as a therapeutic target [43, 135]. Some atypical drugs already utilized for schizophrenia affect cholinergic systems. Clozapine, the most efficacious drug developed to date for schizophrenia [136, 137] interacts with multiple receptors, including the 5HT3 receptor where it is an antagonist [138], resulting in massive release of acetylcholine [139]. Olanzapine is even more efficient for acetylcholine release at a dose equivalent to the dose of clozapine, but not at therapeutic doses. Clozapine decreases smoking in schizophrenia [140] and has positive effects on cognitive measures [141], suggesting that the release of acetylcholine might play a role in these improvements. Other 5HT3 antagonists, such as tropisetron, have similar effects on cognition [138].

Nicotine and smoking normalize auditory sensory processing in schizophrenia [29]. Nicotine normalizes the P50 deficit in patients [29] and in relatives [142]. Abnormal gating is also normalized by the 5HT3 antagonists clozapine [143] and tropisetron [144], consistent with a cholinergic effect. Varenicline, a partial agonist–antagonist of nicotinic receptors, efficacious in smoking cessation [145], has also shown some promise for cognitive and antismoking effects in schizophrenia [146, 147]. However, reports of exacerbation of psychosis in a few patients suggest caution [148, 149].

Initial clinical trials of the α7nAChR partial agonist 3-[(2,4-dimethoxy) benzylidene]-anabaseine (DMXB-A) have been encouraging. DMXB-A normalizes the P50 sensory processing deficit and improves cognition, with specific effects on attention [150, 151]. DMXB-A has recently been shown to improve the fMRI default network response in schizophrenic patients [95]. The effect was associated with an SNP in *CHRNA7* (rs3087454), previously associated with schizophrenia [152].

Another α7nAChR-specific agonist, choline, has also shown promise, particularly for perinatal treatment. Choline levels, essential for both membrane and brain development, are variable and decrease during pregnancy [153, 154]. A recent study, which utilized the P50 deficit recorded shortly after birth [155], showed that prenatal administration of choline to the mother resulted in normal sensory processing in the infant and was associated with the same *CHRNA7* SNP (rs3087454) [97]. Thus, prenatal treatment with a dietary supplement shows promise for prevention of sensory deficits associated with schizophrenia in the offspring. Such treatment might also affect normal development of inhibitory processes. *CHRNA7* gene expression regulates the chloride switch during development, controlled by the Cl^- transporters *NKCC1* and *KCC2* [156]. Regulators of *NKCC1* expression are elevated in schizophrenia, which could have serious effects on neuronal differentiation [157]. This potentially might be corrected in utero with perinatal choline.

A complication for the development of drugs acting at α7nChRs is the partial duplication of the *CHRNA7* gene discussed earlier. The product of this chimeric gene, *CHRFAM7A*, appears to assemble with *CHRNA7* subunits, resulting in a dominant negative effect [104]. The *CHRFAM7A* gene is found only in humans, not in primates nor in rodents [90], and, thus, presents a problem for preclinical studies in these animal species. The *CHRFAM7A* gene product is missing amino acids coded by exons 1–4 and does not, therefore, have a signal peptide [88]. However, the membrane and pore structure of a receptor composed of both α7 and dup-α7 subunits could be unchanged. This hypothesis is supported by the potentiation of α7/dup-α7 receptors with a type II modulator, 1-(5-chloro-2,4-dimethoy-phyl)-3-(5-methyl-isoxazol-3-yl)-urea (PNU-120596) [104], which acts in the transmembrane portion of the receptor [158]. It is important, therefore, that drugs targeting human α7nAChRs be evaluated in a tissue model with *CHRFAM7A* expression.

10 Summary and Future Directions

An important future contribution to research and drug development for the α7nAChR will be the generation of rodent and primate models expressing the duplicated gene, *CHRFAM7A*. Without a preclinical model for the human receptor, assembling with both α7 and dupα7 subunits, it will be difficult to optimize drug doses in the development of human therapeutics. This could lead to toxicity or inadequate dosing, obscuring efficacy. Additional contributions could be made in study of the regulation of *CHRFAM7A* expression.

The *CHRNA7*/*CHRFAM7A* gene complex is expressed widely in human tissues and has multiple functions. The development of therapies involving the α7nAChR, and the many other molecules involved in its regulation, is important not only for mental illness but also for many other diseases involving cognitive deficits, such as Alzheimer's disease, and brain development disorders. Additionally, such treatments might be useful in inflammation and cancer.

References

1. NIMH. The numbers count: mental disorders in America. Bethesda, MD: NIH; 2013.
2. Ziedonis D, Hitsman B, Beckham JC, Zvolensky M, Adler LE, Audrain-McGovern J, et al. Tobacco use and cessation in psychiatric disorders: National Institute of Mental Health report. Nic Tob Res. 2008;10:1691–715.
3. Leonard S, Adler LE, Benhammou K, Berger R, Breese CR, Drebing C, et al. Smoking and mental illness. Pharmacol Biochem Behav. 2001;70:561–70.
4. Lasser K, Boyd JW, Woolhandler S, Himmelstein DU, McCormick D, Bor DH. Smoking and mental illness - a population-based prevalence study. JAMA. 2000;284(20):2606–10.
5. Dalack GW, Healy DJ, Meador-Woodruff JH. Nicotine dependence in schizophrenia: clinical phenomena and laboratory findings. Am J Psychiatry. 1998;155:1490–501.
6. de Leon J, Diaz FJ. A meta-analysis of worldwide studies demonstrates an association between schizophrenia and tobacco smoking behaviors. Schiz Res. 2005;76:135–57.
7. Williams JM, Ziedonis D. Addressing tobacco among individuals with a mental illness or an addiction. Add Behav. 2004;29:1067–83.
8. Leonard S, Gault J, Adams C, Breese CR, Rollins Y, Adler LE, et al. Nicotinic receptors, smoking and schizophrenia. Restor Neurol Neurosci. 1998;12:195–201.
9. Aubin HJ, Rollema H, Svensson RH, Winterer G. Smoking, quitting, and psychiatric disease: a review. Neurosci Biobehav Rev. 2012;36:271–84.
10. Williams JM, Gandhi KK, Lu SE, Kumar S, Shen JW, Foulds J, et al. Higher nicotine levels in schizophrenia compared with controls after smoking a single cigarette. Nic Tob Res. 2010;12:855–9.
11. Olincy A, Young DA, Freedman R. Increased levels of the nicotine metabolite cotinine in schizophrenic smokers compared to other smokers. Biol Psychiatry. 1997;42:1–5.
12. Williams JM, Ziedonis DM, Abanyie F, Steinberg ML, Foulds J, Benowitz NL. Increased nicotine and cotinine levels in smokers with schizophrenia and schizoaffective disorder is not a metabolic effect. Schiz Res. 2005;79:323–35.
13. Williams JM, Gandhi KK, Lu SE, Kumar S, Steinberg ML, Cottler B, et al. Shorter interpuff interval is associated with higher nicotine intake in smokers with schizophrenia. Drug Alc Depend. 2011;118:313–9.
14. Tidey JW, Rohsenow DJ, Kaplan GB, Swift RM. Cigarette smoking topography in smokers with schizophrenia and matched non-psychiatric controls. Drug Alc Depend. 2005;80:259–65.
15. Williams JM, Gandhi KK, Lu SE, Steinberg ML, Benowitz NL. Rapid smoking may not be aversive in schizophrenia. Nic Tob Res. 2013;15:262–6.
16. Glassman AH, Helzer JE, Covey LS, Cottler LB, Stetner F, Tipp JE, et al. Smoking, smoking cessation, and major depression. JAMA. 1990;264:1546–9.
17. Thornton LK, Baker AL, Lewin TJ, Kay-Lambkin FJ, Kavanagh D, Richmond R, et al. Reasons for substance use among people with mental disorders. Addict Behav. 2012;37:1134–43.
18. Janowsky DS, El-Yousef MK, David JM, Sekerke HJ. A cholinergic-adrenergic hypothesis of mania and depression. Lancet. 1972;2:632–5.
19. Janowsky DS, El-Yousef MK, David JM. Acetylcholine and depression. Psychosom Med. 1974;36:248–57.
20. Saricicek A, Esterlis I, Maloney KH, Mineur YS, Ruf BM, Muralidharan A, et al. Persistent beta(2)*-nicotinic acetylcholinergic receptor dysfunction in major depressive disorder. Am J Psychiatry. 2012;169:851–9.
21. Brasic JR, Cascella N, Kumar A, Zhou Y, Hilton J, Raymont V, et al. Positron emission tomography experience with 2-[^{18}F]fluoro-3-(2(s)-azetidinylmethoxy)pyridine (2-[^{18}F]fa) in the living human brain of smokers with paranoid schizophrenia. Synapse. 2012;66:352–68.
22. D'Souza DC, Esterlis I, Carbuto M, Krasenics M, Seibyl J, Bois F, et al. Lower beta(2)*-nicotinic acetylcholine receptor availability in smokers with schizophrenia. Am J Psychiatry. 2012;169:326–34.

23. Fryer JD, Lukas RJ. Antidepressants noncompetitively inhibit nicotinic acetylcholine receptor function. J Neurochem. 1999;72:1117–24.
24. Mineur YS, Obayemi A, Wigestrand MB, Fote GM, Calarco CA, Li AM, et al. Cholinergic signaling in the hippocampus regulates social stress resilience and anxiety- and depression-like behavior. Proc Natl Acad Sci. 2013;110:3573–8.
25. Brunzell DH, McIntosh JM. Alpha7 nicotinic acetylcholine receptors modulate motivation to self-administer nicotine: implications for smoking and schizophrenia. Neuropsychopharmacology. 2012;37:1134–43.
26. Leonard S, Mexal S, Freedman R. Smoking, genetics and schizophrenia: evidence for self medication. J Dual Diag. 2007;3:43–59.
27. Kumari V, Postma P. Nicotine use in schizophrenia: the self medication hypotheses. Neurosci Biobehav Rev. 2005;29:1021–34.
28. Forchuk C, Norman R, Malla A, Martin ML, McLean T, Cheng S, et al. Schizophrenia and the motivation for smoking. Perspect Psychiatry Care. 2002;38:41–9.
29. Adler LE, Hoffer LD, Wiser A, Freedman R. Normalization of auditory physiology by cigarette smoking in schizophrenic patients. Am J Psychiatry. 1993;150:1856–61.
30. Cadenhead KS, Geyer MA, Swerdlow NR, Shafer K, Diaz M, Clementz BA, et al. Sensorimotor gating deficits in schizophrenic patients and their relatives. Biol Psychiatry. 1998;43:33.
31. Freedman R, Waldo M, Bickford-Wimer P, Nagamoto H. Elementary neuronal dysfunctions in schizophrenia. Schiz Res. 1991;4:233–43.
32. Clementz BA, Geyer MA, Braff DL. Multiple site evaluation of P50 suppression among schizophrenia and normal comparison subjects. Schiz Res. 1998;30:71–80.
33. Sherr JD, Myers C, Avila MT, Elliott A, Blaxton TA, Thaker GK. The effects of nicotine on specific eye tracking measures in schizophrenia. Biol Psychiatry. 2002;52:721–8.
34. Olincy A, Johnson LL, Ross RG. Differential effects of cigarette smoking on performance of a smooth pursuit and a saccadic eye movement task in schizophrenia. Psychiatry Res. 2003;117:223–36.
35. Tanabe J, Tregellas JR, Martin LF, Freedman R. Effects of nicotine on hippocampal and cingulate activity during smooth pursuit eye movement in schizophrenia. Biol Psychiatry. 2006;59:754–61.
36. Kumari V, Soni W, Sharma T. Influence of cigarette smoking on prepulse inhibition of the acoustic startle response in schizophrenia. Hum Psychopharmacol Clin Exp. 2001;16:321–6.
37. George TP, Vessicchio JC, Termine A, Sahady DM, Head CA, Pepper WT, et al. Effects of smoking abstinence on visuospatial working memory function in schizophrenia. Neuropsychopharmacolgy. 2002;26(1):75–85.
38. Sacco KA, Termine A, Seyal A, Dudas MM, Vessicchio JC, Krishnan-Sarin S, et al. Effects of cigarette smoking on spatial working memory and attentional deficits in schizophrenia - involvement of nicotinic receptor mechanisms. Arch Gen Psychiatry. 2005;62(6):649–59.
39. Jacobsen LK, D'Souza DC, Mencl WE, Pugh KR, Skudlarski P, Krystal JH. Nicotine effects on brain function and functional connectivity in schizophrenia. Biol Psychiatry. 2004;55(8):850–8.
40. Myers CS, Robles O, Kakoyannis AN, Sherr JD, Avila MT, Blaxton TA, et al. Nicotine improves delayed recognition in schizophrenic patients. Psychopharmacology. 2004; 174(3):334–40.
41. Dalack GW, Becks L, Hill E, Pomerleau OF, Meador-Woodruff JH. Nicotine withdrawal and psychiatric symptoms in cigarette smokers with schizophrenia. Neuropsychopharmacology. 1999;21(2):195–202.
42. Smith RC, Singh A, Infante M, Khandat A, Kloos A. Effects of cigarette smoking and nicotine nasal spray on psychiatric symptoms and cognition in schizophrenia. Neuropsychopharmacolgy. 2002;27(3):479–97.
43. Miyamoto S, Miyake N, Jarskog LF, Fleischhacker WW, Lieberman JA. Pharmacological treatment of schizophrenia: a critical review of the pharmacology and clinical effects of current and future therapeutic agents. Mol Psychiatry. 2012;17:1206–27.

44. Dani JA. Roles of dopamine signaling in nicotine addiction. Mol Psychiatry. 2003; 8(3):255–6.
45. Dickinson JA, Kew JNC, Wonnacott S. Presynaptic alpha 7- and beta 2-containing nicotinic acetylcholine receptors modulate excitatory amino acid release from rat prefrontal cortex nerve terminals via distinct cellular mechanisms. Mol Pharmacol. 2008;74(2):348–59.
46. Wonnacott S, Kaiser S, Mogg A, Soliakov L, Jones IW. Presynaptic nicotinic receptors modulating dopamine release in the rat striatum. Eur J Pharmacol. 2000;393(1–3):51–8.
47. Covernton POJ, Lester RAJ. Prolonged stimulation of presynaptic nicotinic acetylcholine receptors in the rat interpeduncular nucleus has differential effects on transmitter release. Int J Develop Neurosci. 2002;20(3–5):247–58.
48. Schilstrom B, Rawal N, Mameli-Engvall M, Nomikos GG, Svensson TH. Dual effects of nicotine on dopamine neurons mediated by different nicotinic receptor subtypes. Int J Neuropsychopharmacol. 2003;6(1):1–11.
49. Mexal S, Frank M, Berger R, Adams CE, Ross RG, Freedman R, et al. Differential modulation of gene expression in the NMDA postsynaptic density of schizophrenic and control smokers. Mol Brain Res. 2005;139(2):317–32.
50. Tidey JW, Rohsenow DJ, Kaplan GB, Swift RM, Adolfo AB. Effects of smoking abstinence, smoking cues and nicotine replacement in smokers with schizophrenia and controls. Nic Tob Res. 2008;10:1047–56.
51. Moss TG, Sacco KA, Alien RM, Weinberger AH, Vessicchio JC, George TP. Prefrontal cognitive dysfunction is associated with tobacco dependence treatment failure in smokers with schizophrenia. Drug Alc Depend. 2009;104:94–9.
52. Bierut LJ, Stitzel JA, Wang JC, Hinrichs AL, Grucza RA, Xuei XL, et al. Variants in nicotinic receptors and risk for nicotine dependence. Am J Psychiatry. 2008;165(9):1163–71.
53. Saccone NL, Schwantes-An TH, Wang JC, Grucza RA, Breslau N, Hatsukami D, et al. Multiple cholinergic nicotinic receptor genes affect nicotine dependence risk in African and European Americans. Genes Brain Behav. 2010;9(7):741–50.
54. Berrettini W, Yuan X, Tozzi F, Song K, Francks C, Chilcoat H, et al. Alpha-5/alpha-3 nicotinic receptor subunit alleles increase risk for heavy smoking. Mol Psychiatry. 2008; 13(4):368–73.
55. Jackson KJ, Fanous AH, Chen J, Kendler KS, Chen X. Variants in the 15q25 gene cluster are associated with risk for schizophrenia and bipolar disorder. Psychiatry Genet. 2013;23:20–8.
56. Chen LS, Xian H, Grucza RA, Saccone NL, Wang JC, Johnson EO, et al. Nicotine dependence and comorbid psychiatric disorders: examination of specific genetic variants in the CHRNA5-A3-B4 nicotinic receptor genes. Drug Alc Depend. 2012;123:S42–51.
57. Fallin MD, Lasseter VK, Wolyniec PS, McGrath JA, Nestadt G, Valle D, et al. Genomewide linkage scan for schizophrenia susceptibility loci among Ashkenazi Jewish families shows evidence of linkage on chromosome 10q22. Am J Hum Genet. 2003;73(3):601–11.
58. Freedman R, Leonard S, Olincy A, Kaufmann CA, Malaspina D, Cloninger CR, et al. Evidence for the multigenic inheritance of schizophrenia. Am J Med Genet. 2001;105(8):794–800.
59. Gejman PV, Sanders AR, Badner JA, Cao QH, Zhang J. Linkage analysis of schizophrenia to chromosome 15. Am J Med Genet. 2001;105(8):789–93.
60. Kaufmann CA, Suarez B, Malaspina D, Pepple J, Svrakic D, Markel PD, et al. NIMH Genetics Initiative Millennium Schizophrenia Consortium: linkage analysis of African-American pedigrees. Am J Med Genet. 1998;81(4):282–9.
61. Liu CM, Hwu HG, Lin MW, Ou-Yang WC, Lee SFC, Fann CSJ, et al. Suggestive evidence for linkage of schizophrenia to markers at chromosome 15q13-14 in Taiwanese families. Am J Med Genet. 2001;105(8):658–61.
62. Leonard S, Freedman R. Genetics of chromosome 15q13-q14 in schizophrenia. Biol Psychiatry. 2006;60(2):115–22.
63. Riley BP, Makoff AM, Magudi-Carter M, Jenkins TJ, Williamson R, Collier DA, et al. Haplotype transmission disequilibrium and evidence for linkage of the CHRNA7 gene region to schizophrenia in Southern African Bantu families. Am J Med Genet. 2000;96:196–201.

64. Tsuang DW, Skol AD, Faraone SV, Bingham S, Young KA, Prabhudesai S, et al. Examination of genetic linkage of chromosome 15 to schizophrenia in a large veterans affairs cooperative study sample. Am J Med Genet. 2001;105(8):662–8.
65. Turecki G, Grof P, Grof E, D'Souza V, Cavazzoni P, Duffy A, et al. A genome scan using a pharmacogenetic approach indicates a susceptibility locus for BD on 15q14. Biol Psychiatry. 2000;47:69S.
66. Edenberg HJ, Foroud T, Conneally PM, Sorbel JJ, Carr K, Crose C, et al. Initial genomic scan of the NIMH genetics initiative bipolar pedigrees: chromosomes 3, 5, 15, 16, 17, and 22. Am J Med Genet. 1997;74(3):238–46.
67. Stefansson H, Rujescu D, Cichon S, Pietilainen OPH, Ingason A, Steinberg S, et al. Large recurrent microdeletions associated with schizophrenia. Nature. 2008;455(7210):232–7.
68. Stone JL, O'Donovan MC, Gurling H, Kirov GK, Blackwood DHR, Corvin A, et al. Rare chromosomal deletions and duplications increase risk of schizophrenia. Nature. 2008;455(7210):237–41.
69. Masurel-Paulet A, Andrieux J, Callier P, Cuisset JM, Le Caignec C, Holder M, et al. Delineation of 15q13.3 microdeletions. Clin Genet. 2010;78:149–61.
70. Miller DT, Shen Y, Weiss LA, Korn J, Anselm I, Bridgemohan C, et al. Microdeletion/duplication at 15q13.2q13.3 among individuals with features of autism and other neuropsychiatric disorders. J Med Genet. 2009;46(4):242–8.
71. Mewborn SK, Miley NL, Fantes JA, Brown RL, Butler MG, Christian SL, et al. Breakpoint junction fragments in Prader-Willi and Angelman syndrome (PWS/AS) deletion patients reveal variable breakpoints within large duplicons. Am J Hum Genet. 2002;71(4):736.
72. Boer H, Holland A, Whittington J, Butler J, Webb T, Clarke D. Psychotic illness in people with Prader Willi syndrome due to chromosome 15 maternal uniparental disomy. Lancet. 2002;359(9301):135–6.
73. Vogels A, De Hert M, Descheemaeker MJ, Govers V, Devriendt K, Legius E, et al. Psychotic disorders in Prader-Willi syndrome. Am J Med Genet A. 2004;127A(3):238–43.
74. Young JW, Crawford N, Kelly JS, Kerr LE, Marston HM, Spratt C, et al. Impaired attention is central to the cognitive deficits observed in alpha 7 deficient mice. Eur Neuropsychopharmacol. 2007;17(2):145–55.
75. Young JW, Meves JM, Tarantino IS, Caldwell S, Geyer MA. Delayed procedural learning in alpha 7-nicotinic acetylcholine receptor knockout mice. Genes Brain Behav. 2011;10:720–33.
76. Stevens KE, Adams CE, Yonchek J, Zheng L, Collins AC. C3H alpha7 nicotinic receptor heterozygote mice as a new model of schizophrenia. Schiz Res. 2010;117(2–3):510.
77. De Luca V, Wong AHC, Muller DJ, Wong GWH, Tyndale RF, Kennedy JL. Evidence of association between smoking and alpha 7 nicotinic receptor subunit gene in schizophrenia patients. Neuropsychopharmacolgy. 2004;29(8):1522–6.
78. Faraone SV, Su J, Taylor L, Wilcox M. A novel permutation testing method implicates sixteen nicotinic acetylcholine receptor genes as risk factors for smoking in schizophrenia families. Hum Hered. 2004;57(2):59–68.
79. Breese CR, Marks MJ, Logel J, Adams CE, Sullivan B, Collins AC, et al. Effect of smoking history on [^{3}H]nicotine binding in human postmortem brain. J Pharm Exp Ther. 1997; 282:7–13.
80. Perry DC, Davila-Garcia MI, Stockmeier CA, Kellar KJ. Increased nicotinic receptors in brains from smokers: membrane binding and autoradiography studies. J Pharm Exp Ther. 1999;289(3):1545–52.
81. Benwell ME, Balfour DJ, Anderson JM. Evidence that tobacco smoking increases the density of (-)-[^{3}H]nicotine binding sites in human brain. J Neurochem. 1988;50(4):1243–7.
82. Benhammou K, Lee MJ, Strook M, Sullivan B, Logel J, Raschen K, et al. [^{3}H]Nicotine binding in peripheral blood cells of smokers is correlated with the number of cigarettes smoked per day. Neuropharmacology. 2000;39:2818–29.
83. Breese CR, Lee MJ, Adams CE, Sullivan B, Logel J, Gillen KM, et al. Abnormal regulation of high affinity nicotinic receptors in subjects with schizophrenia. Neuropsychopharmacolgy. 2000;23(4):351–64.

84. Freedman R, Hall M, Adler LE, Leonard S. Evidence in postmortem brain tissue for decreased numbers of hippocampal nicotinic receptors in schizophrenia. Biol Psychiatry. 1995; 38(1):22–33.
85. Guan ZZ, Zhang X, Blennow K, Nordberg A. Decreased protein level of nicotinic receptor alpha7 subunit in the frontal cortex from schizophrenic brain. Neuroreport. 1999;10(8): 1779–82.
86. Marutle A, Zhang X, Court J, Piggott M, Johnson M, Perry R, et al. Laminar distribution of nicotinic receptor subtypes in cortical regions in schizophrenia. J Chem Neuroanat. 2001;22(1–2):115–26.
87. Court J, Spurden D, Lloyd S, McKeith I, Ballard C, Cairns N, et al. Neuronal nicotinic receptors in dementia with Lewy bodies and schizophrenia: alpha-bungarotoxin and nicotine binding in the thalamus. J Neurochem. 1999;73(4):1590–7.
88. Gault J, Robinson M, Berger R, Drebing C, Logel J, Hopkins J, et al. Genomic organization and partial duplication of the human α7 neuronal nicotinic acetylcholine receptor gene. Genomics. 1998;52(2):173–85.
89. Riley B, Williamson M, Collier D, Wilkie H, Makoff A. A 3-Mb map of a large segmental duplication overlapping the alpha 7-nicotinic acetylcholine receptor gene (*CHRNA7*) at human 15q13-q14. Genomics. 2002;79(2):197–209.
90. Locke DP, Jiang Z, Pertz LM, Misceo D, Archidiacono N, Eichler EE. Molecular evolution of the human chromosome 15 pericentromeric region. Cytogen Genome Res. 2005; 108(1–3):73–82.
91. Leonard S, Gault J, Hopkins J, Logel J, Vianzon R, Short M, et al. Association of promoter variants in the alpha 7 nicotinic acetylcholine receptor subunit gene with an inhibitory deficit found in schizophrenia. Arch Gen Psychiatry. 2002;59(12):1085–96.
92. Nachman MW, Crowell SL. Estimate of the mutation rate per nucleotide in humans. Genetics. 2000;156:297–304.
93. Houy E, Raux G, Thibaut F, Belmont A, Demily C, Allio G, et al. The promoter-194 C polymorphism of the nicotinic alpha 7 receptor gene has a protective effect against the P50 sensory gating deficit. Mol Psychiatry. 2004;9(3):320–2.
94. Stephens SH, Franks A, Berger R, Palionyte M, Fingerlin TE, Wagner B, et al. Multiple genes in the 15q13-q14 chromosomal region are associated with schizophrenia. Psychiatric Genetics. 2012;22:1–14.
95. Tregellas JR, Tanabe J, Rojas DC, Shatti S, Olincy A, Johnson L, et al. Effects of an alpha 7-nicotinic agonist on default network activity in schizophrenia. Biol Psychiatry. 2010;69:7–11.
96. Kem WR, Mahnir VM, Prokai L, Papke RL, Cao XF, LeFrancois S, et al. Hydroxy metabolites of the Alzheimer's drug candidate 3-[(2,4-dimethoxy)benzylidene]-anabaseine dihydrochloride (GTS-21): their molecular properties, interactions with brain nicotinic receptors, and brain penetration. Mol Pharmacol. 2004;65(1):56–67.
97. Ross RG, Hunter SK, McCarthy L, Beuler J, Hutchison AK, Wagner BD, et al. Perinatal choline effects on neonatal pathophysiology related to later schizophrenia risk. Am J Psychiatry. 2013;170:290–8.
98. Gault J, Hopkins J, Berger R, Drebing C, Logel J, Walton K, et al. Comparison of polymorphisms in the α7 nicotinic receptor gene and its partial duplication in schizophrenic and control subjects. Am J Med Genet B. 2003;123(1):39–49.
99. Sinkus ML, Lee MJ, Gault J, Logel J, Short M, Freedman R, et al. A 2-base pair deletion polymorphism in the partial duplication of the alpha 7 nicotinic acetylcholine gene (*CHRFAM7A*) on chromosome 15q14 is associated with schizophrenia. Brain Res. 2009;1291:1–11.
100. Raux G, Bonnet-Brilhault F, Louchart S, Houy E, Gantier R, Levillain D, et al. The-2 bp deletion in exon 6 of the 'alpha 7-like' nicotinic receptor subunit gene is a risk factor for the P50 sensory gating deficit. Mol Psychiatry. 2002;7(9):1006–11.
101. Flomen RH, Shaikh M, Walshe M, Schulze K, Hall M-H, Picchioni M, et al. Association between the 2-bp deletion polymorphism in the duplicated version of the alpha7 nicotinic receptor gene and P50 gating. Eur J Hum Genet. 2013;21:76–81.

102. Flomen RH, Davies AF, Di Forti M, La Cascia C, Mackie-Ogilvie C, Murray R, et al. The copy number variant involving part of the alpha 7 nicotinic receptor gene contains a polymorphic inversion. Eur J Hum Genet. 2008;16(11):1364–71.
103. Villiger Y, Szanto I, Jaconi S, Blanchet C, Buisson B, Krause KH, et al. Expression of an alpha 7 duplicate nicotinic acetylcholine receptor-related protein in human leukocytes. J Neuroimmunol. 2002;126(1–2):86–98.
104. Araud T, Graw S, Berger R, Neveu E, Bertrand D, Leonard S. The duplicated alpha 7 nicotinic receptor gene *CHRFAM7A* is a dominant negative regulator of *CHRNA7* expression. Biochem Pharmacol. 2011;82:904–14.
105. de Lucas-Cerrillo AM, Maldifassi MC, Arnalich F, Renart J, Atienza G, Serantes R, et al. Function of partially duplicated human alpha 7 nicotinic receptor subunit *CHRFAM7A* Gene: potential implications for the cholinergic anti-inflammatory response. J Biol Chem. 2011; 286(1):594–606.
106. Benfante R, Antonini RA, De Pizzol M, Gotti C, Clementi F, Locati M, et al. Expression of the α7 nAChR subunit duplicate form (*CHRFAM7A*) is down-regulated in the monocytic cell line THP-1 on treatment with LPS. J Immunol. 2011;230:74–84.
107. Gault J, Logel J, Drebing C, Berger R, Hopkins J, Olincy A, et al. Mutation analysis of the α7 nicotinic acetylcholine receptor gene and its partial duplication in schizophrenia patients. Am J Hum Genet. 1999;65(Supp.):A271.
108. Canastar A, Logel J, Graw S, Finlay-Schulz JF, Osborne C, Palionyte M, et al. DNA methylation and tissue-specific transcription of the α7 nicotinic receptor gene (*CHRNA7*). J Mol Neurosci. 2012;47:389–400.
109. Finlay-Schulz JA, Canastar A, Short M, El Gazzar MA, Leonard S. Transcriptional repression of the α7 nicotinic acetylcholine receptor subunit gene (*CHRNA7*) by AP-2α. J Biol Chem. 2011;286:42123–32.
110. Stevens KE, Bullock AE, Collins AC. Chronic corticosterone treatment alters sensory gating in C3H mice. Pharmacol Biochem Behav. 2001;69(3–4):359–66.
111. Bullock AE, Clark AL, Grady SR, Robinson SF, Slobe BS, Marks MJ, et al. Neurosteroids modulate nicotinic receptor function in mouse striatal and thalamic synaptosomes. J Neurochem. 1997;68(6):2412–23.
112. Carrasco-Serrano C, Criado M. Glucocorticoid activation of the neuronal nicotinic acetylcholine receptor alpha 7 subunit gene: involvement of transcription factor Egr-1. FEBS Lett. 2004;566(1–3):247–50.
113. Stefansson H, Steinthorsdottir V, Thorgeirsson TE, Gulcher JR, Stefansson K. Neuregulin 1 and schizophrenia. Ann Med. 2004;36(1):62–71.
114. Harrison PJ, Law AJ. Neuregulin 1 and schizophrenia: genetics, gene expression, and neurobiology. Biol Psychiatry. 2006;60(2):132–40.
115. Petryshen TL, Middleton FA, Kirby A, Aldinger KA, Purcell S, Tahl AR, et al. Support for involvement of neuregulin 1 in schizophrenia pathophysiology. Mol Psychiatry. 2005;10(4):366–74.
116. Yang X, Kuo YH, Devay P, Yu CR, Role L. A cysteine-rich isoform of neuregulin controls the level of expression of neuronal nicotinic receptor channels during synaptogenesis. Neuron. 1998;20(2):255–70.
117. Chen YJJ, Johnson MA, Lieberman MD, Goodchild RE, Schobel S, Lewandowski N, et al. Type III neuregulin-1 is required for normal sensorimotor gating, memory-related behaviors, and corticostriatal circuit components. J Neurosci. 2008;28(27):6872–83.
118. Mathew SV, Law AJ, Lipska BK, Davila-Garcia MI, Zamora ED, Mitkus SN, et al. Alpha 7 nicotinic acetylcholine receptor mRNA expression and binding in postmortem human brain are associated with genetic variation in neuregulin 1. Hum Mol Genet. 2007;16(23):2921–32.
119. Hsu YN, Edwards SC, Wecker L. Nicotine enhances the cyclic AMP-dependent protein kinase- mediated phosphorylation of alpha 4 subunits of neuronal nicotinic receptors. J Neurochem. 1997;69(6):2427–31.
120. Moss SJ, McDonald BJ, Rudhard Y, Schoepfer R. Phosphorylation of the predicted major intracellular domains of the rat and chick neuronal nicotinic acetylcholine receptor alpha 7 subunit by cAMP-dependent protein kinase. Neuropharmacology. 1996;35(8):1023–8.

121. Drisdel RC, Manzana E, Green WN. The role of palmitoylation in functional expression of nicotinic alpha 7 receptors. J Neurosci. 2004;24(46):10502–10.
122. Alexander JK, Govind AP, Drisdel RC, Blanton MP, Vallejo Y, Lam TT, et al. Palmitoylation of nicotinic acetylcholine receptors. J Mol Neurosci. 2010;40:12–20.
123. Yao JK, Leonard S, Reddy RD. Membrane phospholipid abnormalities in postmortem brains from schizophrenic patients. Schiz Res. 2000;42(1):7–17.
124. Castelan F, Castillo M, Mulet J, Sala S, Sala F, del Toro ED, et al. Molecular characterization and localization of the RIC-3 protein, an effector of nicotinic acetylcholine receptor expression. J Neurochem. 2008;105(3):617–27.
125. Valles AS, Barrantes FJ. Chaperoning alpha 7 neuronal nicotinic acetylcholine receptors. Biochim Biophys Acta. 1818;2012:718–29.
126. Miwa JM, Stevens TR, King SL, Caldarone BJ, Ibanez-Tallon I, Cheng X, et al. The prototoxin lynx1 acts on nicotinic acetylcholine receptors to balance neuronal activity and survival in vivo. Neuron. 2006;51(5):587–600.
127. Morishita H, Miwa JM, Heintz N, Hensch TK. Lynx1, a cholinergic brake, limits plasticity in adult visual cortex. Science. 2010;330(6008):1238–40.
128. Lester HA, Xiao C, Srinivasan R, Son CD, Miwa J, Pantoja R, et al. Nicotine is a selective pharmacological chaperone of acetylcholine receptor number and stoichiometry: implications for drug discovery. AAPS J. 2009;11(1):167–77.
129. Yang Y, Paspalas CD, Jin LE, Picciotto MR, Arnsten AFT, Wang M. Nicotinic α7 receptors enhance NMDA cognitive circuits in dorsolateral prefrontal cortex. Proc Natl Acad Sci. 2013;110(29):12078–83.
130. Levy RB, Aoki C. alpha 7 nicotinic acetylcholine receptors occur at postsynaptic densities of AMPA receptor-positive and -negative excitatory synapses in rat sensory cortex. J Neurosci. 2002;22(12):5001–15.
131. Li S, Li ZX, Pei L, Le AD, Liu F. The α7nACh-NMDA rceptor complex is involved in cue-induced reinstatement of nicotine seeking. J Exp Med. 2012;209(12):2141–7.
132. Yasui DH, Scoles HA, Horike S, Meguro-Horike M, Dunaway KW, Schroeder DI, et al. 15q11.2-13.3 chromatin analysis reveals epigenetic regulation of CHRNA7 with deficiencies in Rett and autism brain. Hum Mol Genet. 2011;20(22):4311–23.
133. Green MF. What are the functional consequences of neurocognitive deficits in schizophrenia. Am J Psychiatry. 1996;153:321–30.
134. Mihailescu S, Drucker-Colin R. Nicotine, brain nicotinic receptors, and neuropsychiatric disorders. Arch Med Res. 2000;31(2):131–44.
135. D'Souza MS, Markou A. Schizophrenia and tobacco smoking comorbidity: nAChR agonists in the treatment of schizophrenia-associated cognitive deficits. Neuropharmacology. 2012;62:1371–80.
136. McEvoy JP, Lieberman JA, Stroup TS, Davis SM, Meltzer HY, Rosenheck RA, et al. Effectiveness of clozapine versus olanzapine, quetiapine, and risperidone in patients with chronic schizophrenia who did not respond to prior atypical antipsychotic treatment. Am J Psychiatry. 2006;163:600–10.
137. Lieberman JA, Stroup TS. The NIMH-CATIE schizophrenia study: what did we learn? Am J Psychiatry. 2011;168:770–5.
138. Barnes NM, Sharp T. A review of central 5HT receptors and their function. Neuropharmacology. 1999;38:1083–152.
139. Shirazi-Southall S, Rodriguez DE, Nomikos GG. Effects of typical and atypical antipsychotics and receptor selective compounds on acetylcholine efflux in the hippocampus of the rat. Neuropsychopharmacolgy. 2002;26(5):583–94.
140. McEvoy J, Freudenreich O, McGee M, Vanderzwaag C, Levin E, Rose J. Clozapine decreases smoking in patients with chronic schizophrenia. Biol Psychiatry. 1995;37(8):550–2.
141. Meltzer HY, McGurk SR. The effects of clozapine, risperidone, and olanzapine on cognitive function in schizophrenia. Schiz Bull. 1999;25(2):233–55.
142. Adler LE, Hoffer LJ, Griffith J, Waldo MC, Freedman R. Normalization by nicotine of deficient auditory sensory gating in the relatives of schizophrenics. Biol Psychiatry. 1992; 32:607–16.

143. Nagamoto HT, Adler LE, Hea RA, Griffith JM, McRae KA, Freedman R. Gating of auditory P50 in schizophrenics: unique effects of clozapine. Biol Psychiatry. 1996;40(3):181–8.
144. Koike K, Hashimoto K, Takai N, Shimizu E, Komatsu N, Watanabe H, et al. Tropisetron improves deficits in auditory P50 suppression in schizophrenia. Schiz Res. 2005;76(1):67–72.
145. Oncken C, Gonzales D, Nides M, Rennard S, Watsky E, Billing C, et al. Efficacy and safety of the novel selective nicotinic acetylcholine receptor partial agonist, varenicline, for smoking cessation. Arch Int Med. 2006;166:1571–7.
146. Smith RC, Lindenmayer J-P, Davis JM, Cornwell J, Noth K, Gupta S, et al. Cognitive and antismoking effects of varenicline in patients with schizophrenia or schizoaffective disorder. Schiz Res. 2009;110:149–55.
147. Williams JM, Anthenelli RM, Morris CD, Treadow J, Thompson JR, Yunis C, et al. A randomized, double-blind, placebo-controlled study evaluating the safety and efficacy of varenicline for smoking cessation in patients with schizophrenia or schizoaffective disorder. J Clin Psychiatry. 2012;73:654–60.
148. Freedman R. Exacerbation of schizophrenia by varenicline. Am J Psychiatry. 2007; 164(8):1269.
149. Liu ME, Tsai SJ, Yang ST. Varenicline-induced mixed mood and psychotic episode in a patient with schizophrenia. CNS Spect. 2009;14:346.
150. Olincy A, Harris JG, Johnson LL, Pender V, Kongs S, Allensworth D, et al. Proof-of-concept trial of an alpha 7 nicotinic agonist in schizophrenia. Arch Gen Psychiatry. 2006;63(6):630–8.
151. Olincy A, Freedman R, Buchanan RW, Harris JG, Gold JM, Johnson LL, et al. Initial phase 2 trial of a nicotinic agonist in schizophrenia. Am J Psychiatry. 2008;165(8):1040–7.
152. Stephens SH, Logel J, Barton A, Franks A, Schultz J, Short M, et al. Association of the 5′-upstream regulatory region of the α7 nicotinic acetylcholine receptor subunit gene (*CHRNA7*) with schizophrenia. Schiz Res. 2009;109:102–12.
153. Zeisel SH. Choline: critical role during fetal development and dietary requirements in adults. Ann Rev Nutr. 2006;26:229–50.
154. Ross RG, Stevens KE, Proctor WR, Leonard S, Kisley MA, Hunter SK, et al. Cholinergic mechanisms, early brain development, and risk for schizophrenia. J Child Psychol Psychiatry. 2010;51:535–49.
155. Hunter SK, Corral N, Ponicsan H, Ross RG. Reliability of P50 auditory sensory gating measures in infants during active sleep. Neuroreport. 2008;19(1):79–82.
156. Liu ZP, Neff RA, Berg DK. Sequential interplay of nicotinic and GABAergic signaling guides neuronal development. Science. 2006;314:1610–3.
157. Arion D, Lewis DA. Altered expression of regulators of the cortical chloride transporters NRCC1 and KCC2 in schizophrenia. Arch Gen Psychiatry. 2010;68:21–31.
158. Bertrand D, Bertrand S, Cassar S, Gubbins E, Li JH, Gopalakrishnan M. Positive allosteric modulation of the alpha 7 nicotinic acetylcholine receptor: ligand interactions with distinct binding sites and evidence for a prominent role of the M2-M3 segment. Mol Pharmacol. 2008;74(5):1407–16.

Chapter 21
Current and Future Trends in Drug Discovery and Development Related to Nicotinic Receptors

Stephen P. Arneric and Emanuele Sher

Abstract The potential of nicotinic acetylcholine receptors (nAChRs) as therapeutic targets has evolved over the last 50 years. First focused on peripheral, muscle and ganglionic, therapeutics, it evolved toward more CNS focus, mainly thanks to the identification and cloning of various CNS nAChRs. A minimalistic approach was initially taken to identify drugs that worked on recombinant versions of the two main CNS nAChRs: the homomeric α_7 and the heteromeric $\alpha_4\beta_2$. From the early 1990s to the present, selective $\alpha_4\beta_2$ agonists and modulators have been discovered. In the last decade highly selective α_7 agonists and positive allosteric modulators (PAMs) have emerged. Unfortunately, success in the clinic has been limited. Improved understanding of the molecular nature of drug–receptor interactions, molecules with optimized pharmaceutical properties, biomarkers that reflect pharmacodynamic and clinical benefit, and foundational understanding of the underlying disease physiopathology are all needed for delivering the new nicotinic drugs of the twenty-first century. We here review advances in some of these areas, summarizing the growing evidence that CNS nAChRs represent a family of receptors with significant roles in human diseases and a strong potential for delivering much needed therapies to patients with unmet medical needs.

Keywords Nicotinic acetylcholine receptors (nAChRs) • Drug discovery • Therapeutic potential • Smoking cessation • Pain relief • Attention deficit hyperactivity disorder (ADHD) • Alzheimer's disease (AD) • Cognition • Schizophrenia • Anthelmintics • Neuromuscular blockers

S.P. Arneric, Ph.D. (✉)
Pain and Migraine Drug Hunting Team, Neuroscience Discovery, Lilly Research Centre, Eli Lilly and Company, Indianapolis, IN 46285, USA
e-mail: arneric_stephen@Lilly.com

E. Sher, M.D., Ph.D.
Neuroscience Discovery, Lilly Research Centre, Eli Lilly and Company, Erl Wood Manor, Sunninghill Road, Windlesham, Surrey GU206PH, UK
e-mail: sher_emanuele@lilly.com

R.A.J. Lester (ed.), *Nicotinic Receptors*, The Receptors 26,
DOI 10.1007/978-1-4939-1167-7_21

1 Introduction

The potential of nicotinic acetylcholine receptors (nAChRs) as therapeutic targets has evolved over the last 50 years with interest waxing and waning as new insights have emerged to support or refute each emerging finding. A review on the drug discovery perspective of this rich field can be found covering advances from 1987 to 2007 [1].

The focus of this chapter is to give an update on both the successes and challenges that have faced the therapeutic field of nAChRs from 2008 until 2013, and to project where the future trends in discovery and development efforts will be directed over the next 5–10 years.

1.1 The nAChR Family of Receptors: Structure and Function

In the late 1980s, Heinemann and co-workers and Patrick and co-workers cloned and expressed nAChRs from a variety of species [2]. The nAChR family has as its basic motif a functional pentamer consisting of five transmembrane spanning subunits around a central pore. Typically, neuronal nAChRs are composed of combinations of α (α_2–α_6) and β (β_2–β_4) subunits, homomers of α subunits (α_7 and α_9), and α subunit heteromers (α_9 and α_{10}). Mammals do not appear to have the α_8 subunit identified in avian brain. Each subunit has four transmembrane (TM) segments, a long extracellular N-terminal domain, an intracellular loop between TMs 3 and 4, and a short C-terminal domain. A major advance in understanding the structure of nAChRs at less than 3 Å resolution was the identification and crystallization of an ACh binding protein (AChBP) of unknown physiological function secreted by snail glial cells [3]. For comprehensive reviews on the structure and function of nAChRs we refer the readers to [4–8].

Drug discovery efforts over the last 25 years have primarily focused on neuronal nAChRs containing the $\alpha_4\beta_2$ heteromer and the α_7 homomer. At the former construct, ACh binds in a small pocket between the α_4 and β_2 subunits and in the latter, the binding site is defined by the adjacent α_7 subunits [7]. $\alpha_4\beta_2$ heteromers, however, may be anticipated to exhibit different pharmacology based on the precise subunit ratio. For example, the $\alpha_4(3)\beta_2(2)$ heteromer would be anticipated to be different in terms of both the number and structure of binding sites to the $\alpha_4(2)\beta_2(3)$ heteromer [9]. In the last few years emerging data suggests that $\alpha_7\beta_2$ heteromers may also play a unique and important role in specific brain areas [10, 11]. Moreover, substitution of the alternative subunits (e.g., α_5) can further significantly alter the properties of the channel [12].

Other proteins not directly part of the nAChR pentamer could become relevant for drug discovery in the near future. Auxiliary subunits are known to regulate the trafficking, localization, and gating kinetics of various ion channels, including

AMPA receptors, calcium and potassium channels, etc. [13]. More importantly, they increasingly appear to allow for selective pharmacological targeting of ion channel subtypes within a particular family. In recent years, new transmembrane proteins that might be important for the expression, function and, possibly, pharmacology of nAChRs have also been described. Boulin and co-workers [14] proposed to call this new class of proteins "Nicotinic receptor associated proteins" or NRAPs. One example of these NRAPs is the CNS protein Lynx-1 [15], also referred to as a "prototoxin," based on the finding that it adopts the three-fingered toxin fold characteristic of α-neurotoxins, such as α-Bungarotoxin (α-Bgtx), which bind and block nAChRs. Both soluble and membrane-bound Lynx-1 have been shown to modulate nAChR function. In particular, Lynx-1 modulation of $\alpha_4\beta_2$ receptors includes the promotion of large conductance single-channel openings, an increase in ACh EC_{50}, an enhanced desensitization, and a slower recovery from desensitization. How these biophysical changes affect the physiology of nicotinic transmission in the brain is still unclear. However, since Lynx-1 co-localizes with $\alpha_4\beta_2$ and α_7 nAChRs in the somato-dendritic compartment of neurons in many brain areas (the cerebral cortex, thalamus, substantia nigra, and cerebellum, in particular), it is likely that important modulatory roles for Lynx-1 in the mammalian brain will be soon discovered.

SLURP1 is a protein similar to Lynx1, secreted by keratinocytes and involved in immune regulation. Mutation in SLURP1 has been shown to cause a skin disease called Mal de Meleda [16].

MOLO-1 was recently identified by Boulin et al. [14] as a positive regulator of the levamisole-sensitive acetylcholine receptors (L-AChRs) at the *Caenorhabditis elegans* neuromuscular junction. It will be very important to study if orthologs or homologues of MOLO-1 are found associated also with CNS nAChRs and, specifically, within the human brain.

2 Stoking the Clinical Development Engine

In 2008 there were 17 compounds disclosed to be in clinical development and varenicline [Chantix™/Champix™] had been approved for smoking cessation only 2 years prior. With many compounds progressing to enter Phase 2 development, there was remarkable optimism that the field would yield a number of promising new therapeutics. A 2009 Nature Review in Drug Discovery by Taly and co-workers [7] captures this early wave of development compounds and comprehensively highlights some of the key scientific breakthroughs at that time. In their summary they state it was a time when there was a ".... gap between the two rapidly progressing fields in nicotinic receptor research: the knowledge on the atomic structure, functional organization and conformational transitions of the nAChRs, and the development of nicotinic agents as novel therapeutics for nervous-system disorders by the pharmaceutical industry."

3 Where Are the Smoking Guns?

From 2009 to 2012 there were 24 novel nAChR compounds disclosed to be in clinical development and at least a similar number of compounds being pursued in preclinical development ([8], Tables 21.1 and 21.2). Reports of psychotic episodes related to Chantix™ (varenicline) had raised a specter of target-related side effect that many feared would follow all subsequent $\alpha_4\beta_2$ preferring therapeutics. This clearly had a negative impact on Chantix™ sales which declined from a global sales high of nearly $1 billion in 2007, to approximately $0.7 billion in 2008–2010 and is anticipated to level off at $0.5–0.6 billion from 2013 until its patent expiration in 2018. It was also a source of continued pharmaceutical development conversation when discussing the continued advancement of molecules in the clinic, despite the fact that nicotine patches and other related nicotine products had not had the same incidence of bizarre psychotic side effects. In Sect. 4 we discuss the compounds remaining in 2013 that may provide the clinical momentum required to propel this field of therapeutics.

In parallel, a number of key scientific advances were being made in understanding how channel stoichiometry could drive pharmacology and, importantly, how positive allosteric modulation (PAM) and negative allosteric modulation (NAM) of the two major subtypes ($\alpha_4\beta_2$- and α_7-preferring) could be achieved. For those looking for a more extensive review of the scientific advances and archiving of numerous clinical and preclinical compounds up to 2012 we recommend the review by Hurst and co-workers [8]. Below are some key highlights through 2013.

3.1 *New nAChR Stoichiometries Drive Novel Pharmacology*

It has been known for a while that the hetero-pentameric nAChRs, such as $\alpha_4\beta_2$, can be assembled in different stoichiometries, such as $\alpha_4(2)\beta_2(3)$ and $\alpha_4(3)\beta_2(2)$, which have high affinity and low affinity for acetylcholine, respectively [9]. What has been emerging more recently is the dramatic importance of these changes in stoichiometry in defining the properties and the pharmacology of the receptors.

We have previously shown that synthetic compounds such as 5-I-A-85380, and TC-2559 are more active at the high affinity (HA) $\alpha_4(2)\beta_2(3)$ stoichiometry than the low affinity (LA) $\alpha_4(3)\beta_2(2)$ stoichiometry, which were achieved by injection of different ratios of α_4 and β_2 in *Xenopus laevis* oocytes [17]. These findings were confirmed and extended by Carbone et al. [18] who confirmed this selective pharmacology but using concatenated α_4 and β_2 subunits. Sazetidine-A is a particularly intriguing example. It was suggested to act as a "silent desensitizer" and as a functional antagonist in vivo [19, 20]. However, only one stoichiometry of $\alpha_4\beta_2$ receptors was studied in these papers and with a low sensitivity assay. To better understand this potential "new mechanism" of action of sazetidine-A, we expressed alternative stoichiometries of $\alpha_4\beta_2$ nAChR in *Xenopus laevis* oocytes and investigated the agonist properties

Table 21.1 α4β2-Preferring nAChRs compounds: clinical development status (2008–2013)

Compound	Company	Clinical stage	Targeted indication (s)	Comments	References
TC-1734/AZD-3480 (Ispronicline)	Targacept	Ph2	AD/CIAS/ADHD	Partial agonist; CIAS and ADHD discontinued with Astra-Zeneca; Compound reverted to Targacept for 2nd Ph 2 AD trial (ongoing in the USA and Eastern EU)—the 1st with Astra-Zeneca was inconclusive since donepezil did show a positive response	[57, 58], http://www.targacept.com/therapeutic-pipeline/TC1734.cfm; http://finance.yahoo.com/news/targacept-completes-recruitment-phase-2b-125500675.html
TC-6683/AZD1446	Targacept/ Astra-Zeneca	Ph2	AD	[NCT01125683] The status of this compound is unclear	[88],biz.yahoo.com/e/130315/trgt10-k.html; http://www.targacept.com/therapeutic-pipeline/AZD1446.cfm
TC-5214 [(+)-S-mecamylamine]	Targacept	Up to Ph3; Ph2 ongoing	Overactive bladder	The asset was partnered with Astra-Zeneca; however, subsequent Ph2/3 studies conducted for monotherapy and augmentation therapy for Major Depressive Disorders (MDD) ultimately failed. The MDD agreement with A-Z terminated in May 2012 and reverted to Targacept. It has been repurposed for overactive bladder with a trail beginning in May 2013 [NCT01868516]	[59], biz.yahoo.com/e/130315/trgt10-k.html
ATG003	CoMentis	Ph2	Diabetic macular edema/ age-related macular degeneration/ wet AMD	Racemic mixture of mecamylamine for: ophthalmic solution for treatment of Diabetic Macular Edema [NCT00536692]—completed 2008; adjunct eye-drop treatment of Age-Related Macular Degeneration [NCT00607750]-completed 2010; Wet AMD [NCT00414206]—completed 2010; Discontinuation uncertain given that CoMentis and Anvyl announce in June 2013 the formation of Alpharmagen, a joint venture to explore α7 nicotinic modulators (agonists and PAMs)	http://www.prnewswire.com/news-releases/comentis-and-anvyl-announce-formation-of-alpharmagen-a-joint-venture-to-explore-nicotinic-modulators-211549081.html; http://www.anvylllc.com/Pipeline.htm
TC-6499	Targacept/ GSK	Up to Ph2	Neuropathic pain/IBS	Full agonist; Ph 1 Neuropathic Pain discontinued in March 2009 with GSK; IBS-constipation trial completed in 2010 by Targacept—ClinicalTrials.gov—NCT01149200. Discontinued	http://www.fiercebiotech.com/press-releases/targacept-provides-update-tc-6499-and-pain-program-glaxosmithkline-alliance

(continued)

Table 21.1 (continued)

Compound	Company	Clinical stage	Targeted indication (s)	Comments	References
Dianicline (SSR-591,813)	Sanofi-Aventis	Up to Ph2	Smoking cessation	Partial agonist; In a large (602 healthy smoker) RCT study was shown to be ineffective to increase abstinence; but effective in craving and withdrawal symptoms. Discontinued	[63, 88]
ABT-594 (Tebanicline)	Abbott	Up to Ph2	Neuropathic pain	Full agonist; Demonstrated efficacy in diabetic peripheral neuropathy comparable to gabapentin but with very high AE profile. Discontinued	[89, 90]
ABT-089 (Pozanicline)	Abbott	Up to Ph2	ADHD	Partial agonist; Ph2b studies in adults showed efficacy, whereas trials in children proved ineffective. Discontinued	[91]
ABT-894 (Sofinicline)	Abbott/ NeuroSearch	Up to Ph2	ADHD/ neuropathic pain	Full agonist; ABT-894 was effective in adult ADHD, but not in children—well tolerated; Was ineffective, but well tolerated in Diabetic Neuropathic pain—the same indication ABT-594 was effective, but with significant nausea and emesis; reverted to NeuroSearch—which restructured to became Aniona in 2012. Discontinued?	[92, 93]
ABT-560	Abbott/ NeuroSearch	Up to Ph1	Cognition	Partial agonist; Discontinued (presumed—no formal announcement made)	http://www.phrma.org/sites/default/files/pdf/alzheimers2012.pdf
Lobeline	–	Up to Ph3	Smoking cessation	Discontinued. In a large (750 healthy smoker) RCT study was shown to be ineffective when given in a sublingual formulation	[72]
Varenicline [Chantix™]	Pfizer	Marketed; Ph1 patch;	Smoking cessation; smoking cessation (patch)	Life cycle management (LCM) indications were being explored: AD and ADHD indications discontinued; ClinicalTrials.gov has an impressive 222 clinical trials registered that explore a very wide spectrum of clinical conditions	[63]

Table 21.2 α7-Preferring nAChRs compounds: clinical development status (2008–2013)

Compound	Company	Clinical stage	Target indication	Comments	References
ASM-024	Asmacure	Ph 2	Asthma; COPD	Asmacure has released data suggesting a dual nicotinic/muscarinic pharmacology; Ph 2 initiated in July 2013 with a Dry Powder Inhaled (DPI) formulation	http://www.asmacure.com/userfiles/ckeditor/bwt08012013(1).pdf
TC-6987	Targacept	Up to Ph 2	Asthma/diabetes/undisclosed	Partial α7 agonist Targacept previously conducted exploratory clinical studies of TC-6987 to treat "inflammatory disorders"—development in this area has discontinued [asthma-NCT01296087; diabetes NCT01293669]. Further development of TC-6987 in another indication remains under consideration	http://www.targacept.com/therapeutic-pipeline/TC-6987.cfm
TC-5619	Targacept	Ph 2	CIAS and negative symptoms; AD?	Partial α7 agonist ADHD discontinued—did not meet primary efficacy endpoint; CIAS and negative symptoms of schizophrenia ongoing; AD being reconsidered	http://www.targacept.com/therapeutic-pipeline/TC-5619.cfm
GTS-21	CoMentis [in June 2013 merged with Anvyl to form Alpharmagen]	Ph2	AD/ADHD/Endotoxemia/CIAS	Partial α7 agonist AD [NCT00414622] completed in 2007; ADHD [NCT00419445], completed in 2010; GTS-21 on Inflammation and End-organ Dysfunction During Human Endotoxemia [NCT00783068]—completed in 2010; CIAS ongoing [NCT01400477]	[94]
AQW051	Novartis	Ph 2	AD/Parkinson's disease/CIAS	Partial α7 agonist Mild AD [NCT00582855] terminated 2011;PD [NCT01474421] completed March 2013; CIAS [NCT01730768] recruiting patients in July 2013]	–
ABT-126	AbbVie	Ph 2	CIAS/AD	Partial α7 agonist; CIAS [NCT01095562] completed in 2011; Mild to moderate AD [NCT00948909] completed January 2013; Mild to moderate AD [NCT01690195 and NCT01676935] enrolling in July 2013 with completion in 2014; CIAS [NCT01655680] recruiting 2012–2013 with completion by mid-2014	–

(continued)

Table 21.2 (continued)

Compound	Company	Clinical stage	Target indication	Comments	References
ENV-6124	EnVivo Pharma	Ph 3	AD/CIAS	See details in text for trials reported in 2012; Ph 3 CIAS study recruiting [NCT01714661] to examine Adjunctive Pro-Cognitive Treatment in Schizophrenia Subjects on Chronic Stable Atypical Antipsychotic Therapy—completion planned for late 2014	[31]
ABT-107	Abbott/ NeuroSearch	Up to Ph2	AD/CIAS	Discontinued (Presumed)—no activity for several years	[95–97]
MEM63908/ R4996	Memory Pharma/ Roche	Up to Ph2	AD	Discontinued (Presumed)—no activity for several years	–
AZD-0328	Astra-Zeneca	Ph 2	AD/CIAS	Discontinued in 2010 [NCT00669903]—did not meet target product profile	–
SSR-180711	Sanofi-Aventis	Ph 2	AD	Discontinued	http://www.schizophreniaforum.org/res/drc/detail.asp?id=281
MEM2454/ R3457	Memory Pharma/ Roche	Ph 1	AD/CIAS	Discontinued (Presumed)—no activity for several years. (Presumed)—no activity for several years	
XY-4083	Xytis	Ph 1 ?	Cognition	Positive Allosteric Modulator (PAM); The first PAM *reporting plans to be in Ph1 in 2009*; Xytis never progressed this compound further	http://www.thefreelibrary.com/Xytis+to+Initiate+First-In-Class+Phase+I+Clinical+Trials+with+Alpha+7…-a0189168297
AVL-3288	Anvyl [in June 2013 merged with CoMentis to form Alpharmagen]	Ph 1	CIAS	Type I PAM; CIAS [NCT01851603]; Other potential indications based on animal models have been recently reviewed [36, 98]	http://www.anylllc.com/Pipeline.htm

of sazetidine-A on both HA $\alpha_4(2)\beta_2(3)$ and LA $\alpha_4(3)\beta_2(2)$ nAChRs [21]. We found that sazetidine-A was indeed a full agonist on HA $\alpha_4(2)\beta_2(3)$ nAChRs, and a very partial agonist on LA $\alpha_4(3)\beta_2(2)$ nAChRs, with similar potency at both subtypes. We therefore concluded that sazetidine-A has no "new mechanism" of action in terms of desensitization, but rather a very interesting selectivity, mainly in terms of efficacy, toward different stoichiometries of the "same" $\alpha_4\beta_2$ receptor.

Ussing et al. [22] have recently characterized some new compounds at both HA $\alpha_4(2)\beta_2(3)$ and LA $\alpha_4(3)\beta_2(2)$ receptors. Their results confirmed that it is possible, also by rational design, to selectively target individual $\alpha_4\beta_2$ stoichiometries. Importantly, recent studies combining biochemistry and the use of heterozygous α_4 and β_2 KO mice elegantly demonstrated that different $\alpha_4\beta_2$ stoichiometries exist also in vivo [23], suggesting that agonists "biased" to specific stoichiometries of the nAChR could exert profoundly different in vivo effects.

The $\alpha_4\beta_2$ receptor not only comes in HA and LA stoichiometries, but can also assemble in trimeric combinations, for example with the α_6 subunit. The trimeric $\alpha_4\alpha_6\beta_2$ receptors seem to be selectively expressed in the nigro-striatal part of the dopaminergic system, at variance to $\alpha_6\beta_2$ receptors (lacking α_4) that are preferentially expressed in the meso-limbic pathway [24]. New compounds that target preferentially α_6-containing nAChRs have recently been disclosed [25].

Similar studies as those mentioned above with $\alpha_4\beta_2$ need to be extended now also to α_7 nAChRs (the prototypical homomeric nAChRs) since it has been shown they can co-assemble with the β_2 subunit [10, 11, 26, 27]. The native stoichiometry of these newly described hetero-pentameric $\alpha_7\beta_2$ receptors is not known. Exciting early available data pointed to a preferential expression of $\alpha_7\beta_2$ in basal forebrain cholinergic neurons and hippocampal GABAergic interneurons, cell types that are both highly relevant to the physiopathology of neuropsychiatric diseases, and important therapeutic targets. It is possible that, like with $\alpha_4\beta_2$, different stoichiometries of $\alpha_7\beta_2$ will also be discovered, with potentially different localizations, function, and pharmacology.

New findings suggest that also the known "duplicated" α_7 chimeric subunit, coded by the *CHRFAM7A* gene, and devoid of the orthosteric acetylcholine binding site, could co-assemble with "normal" α_7, and exert a dominant-negative effect on its expression and/or function [28, 29]. This has been investigated in recombinant systems in vitro and we recently replicated these findings (Zwart and Sher, unpublished data). However, to the best of our knowledge, no evidence of functional heteromeric α_7/dup-α_7 nAChRs has been reported yet in native preparations or in human brain. This is a human-specific issue that warrants further investigation, especially considering that both genes are linked to neuropsychiatric conditions ([30] and references therein).

In general, the possible presence of $\alpha_7\beta_2$ and α_7/dup-α_7 nAChRs in the human brain needs to be further investigated, not only because of its biological relevance, but also because this might shed some light into the reasons why several "typical" α_7 agonists failed to show significant efficacy in recent clinical trials. More foundational science determining the native stoichiometry and corresponding pharmacology is required to pave the road toward clinical success.

3.2 Recently Described "Co-agonist" Effects of Novel Ligands

nAChRs need to bind at least two or three agonists (or acetylcholine) in order to be fully activated [7, 8]. Some novel selective α_7 agonists, exemplified by EVP-6124 [31], have been shown to *activate* α_7 nAChRs at micromolar concentrations, but also to potentiate the actions of low concentrations of acetylcholine at nanomolar concentrations to activate α_7 nAChRs. It has been suggested that this could be a mechanism common to several drugs, including nicotine and selective agonists. The hypothesis is that at low acetylcholine concentrations only one orthosteric binding site will be occupied and unrelated orthosteric agonists will be able to bind to the free orthosteric site and synergize with acetylcholine. There is also evidence that at these low concentrations agonists could induce α_7 nAChR upregulation [32]. We should refer to this mechanism as co-agonism or orthosteric potentiation, to keep it distinct from the allosteric potentiation described below.

3.3 New Allosteric Modulators

In addition to extensive studies aimed at characterizing new nAChR selective agonists and their novel mechanisms of action, a large body of work has been generated in the last few years regarding the identification and characterization of a diverse group of allosteric modulators of nAChRs [33]. Potentiation of nAChRs has been demonstrated with a few compounds including galanthamine, a cholinesterase inhibitor approved for the treatment of Alzheimer's disease, AD [34]. However, while the nAChR potentiating effects of galanthamine might contribute to its clinical efficacy, the degree of potentiation is mild, at best, and relatively nonspecific. More potent and more selective nAChR allosteric modulators have been recently described, and reviews detailing the emerging pharmacologic tools for this important area have recently been published by Mantione and co-workers [35], and Pandya and Yakel [36]. Below we highlight some of these advances.

3.3.1 Mixed nAChR PAMs

Novel (2-amino-5-keto)thiazole compounds, such as LSN1078733 and LSN2087101, were identified by Lilly in high-throughput screening assays and their activities thoroughly profiled on seven subtypes of nAChRs expressed in mammalian cells and *Xenopus* oocytes [37]. Electrophysiological recordings, calcium imaging, and radioligand binding experiments all showed that these ligands are unique nAChR positive allosteric modulators, i.e., they are specific for α_7 and $\alpha_4\beta_2$ nAChRs but devoid of activity at muscle α_1 or ganglionic $\alpha_3\beta_4$.

The effects of these compounds on native nAChRs were also investigated [38], a key step in verifying the functional relevance of the pharmacology. Nicotine is known to increase the spontaneous firing rate of dopamine neurons in the somatic regions of the ventral tegmental area (VTA) and to stimulate dopamine release in the dopaminergic terminal fields in the striatum. The application of LSN1078733 or

LSN2087101 significantly enhanced both nicotine-induced dopamine cell firing in VTA slices and nicotine-stimulated dopamine release in striatal slices. ACh-induced, α_7-mediated, GABA release in rat hippocampal cultures was also enhanced by LSN2087101. In contrast, this potentiator had no effect on nicotine-induced noradrenaline release from hippocampal slices, mainly driven by $\alpha_3\beta_4$ nAChRs, confirming, in native preparations, the interesting subtype selectivity previously demonstrated in recombinant systems.

A new series of pyrazole carbamates has been described by the Amgen group, which show selective potentiation of acetylcholine-induced current through the $\alpha_4\beta_2$ nAChR [39]. These compounds did not displace ^{3}H-cytisine binding to $\alpha_4\beta_2$ from rat cerebral cortical membranes, suggesting that they are not binding to the orthosteric site. No published molecular biology data are available to confirm the identity of the allosteric binding site(s) for these compounds.

The scientists at NeuroSearch, in collaboration with Abbott, have also disclosed a series of $\alpha_4\beta_2$ potentiators exemplified by NS-9283 [40]. Interestingly, these new PAMs show selectivity for the different stoichiometries of $\alpha_4\beta_2$, with high preference for the low-affinity, $\alpha_4(3)\beta_2(2)$ stoichiometry. However, they also potentiate other brain subtypes such $\alpha_2\beta_2$, $\alpha_2\beta_4$ and $\alpha_4\beta_4$ but not α_3-containing receptors. In vivo effects of these PAMs have been reported in rat pain models but only when combined with an agonist [41]. It will be important to understand whether these compounds potentiate not only the beneficial, but also the detrimental (i.e., seizure and sympathomimetic) effects of nicotine.

3.3.2 Selective α_7 PAMs

Highly selective α_7 PAMs have been identified by several groups in recent years. The very first ones were actually reported back in 2005 when both the Pfizer and the Lilly groups presented data on novel and highly selective molecules [42, 43]. The magnitude of potentiation with these new compounds was very marked in comparison to the few previously known allosteric potentiators for α_7 nAChRs, like ivermectin or 5-Hydroxyindole [33]. In particular, some of these compounds, also referred to as type-II compounds, have dramatic effects on the desensitization of α_7 receptors and on their single-channel properties [42–44]. These highly selective α_7 potentiators enhanced choline-induced, α_7-mediated, GABA release in rat hippocampal cultures. The effects were reversibly prevented by the selective α_7 antagonist methyllycaconitine, MLA. Over the years, other selective α_7 PAMs have been discovered by many groups [8, 33].

3.3.3 Selective α_7 and $\alpha_4\beta_2$ NAMs

The inhibitory effect of (R,S)-dehydronorketamine was characterized recently by Moaddel et al. [45]. They showed that the block of α_7 was voltage-independent and that the compound did not competitively displace selective α_7-nicotinic acetylcholine receptor orthosteric ligands, suggesting that (R,S)-dehydronorketamine could act as a negative allosteric modulator of the α_7 nicotinic acetylcholine receptor.

A reported example of a naturally occurring $\alpha_4\beta_2$ NAM is progesterone, which inhibits the response of $\alpha_4\beta_2$ nAChRs in an allosteric manner [33]. Henderson et al. [46] identified a novel allosteric site on human $\alpha_4\beta_2$ nAChRs using a series of computational and in vitro approaches. The authors characterized the allosteric site via site-directed mutagenesis. Three amino acids (Phe118, Glu60, and Thr58) on the β_2 subunit were shown to participate in the inhibitory properties of the selective antagonist KAB-18. SAR studies with KAB-18 analogues and various mutant $\alpha_4\beta_2$ nAChRs also provided information concerning how different physicalchemical features influence the inhibition of nAChRs through this allosteric site.

Menthol, a compound with analgesic properties, is another compound that might act as an $\alpha_4\beta_2$ NAM, albeit with relatively low potency [47]. The authors showed that menthol did not affect nicotine's EC_{50} value for currents through recombinant human $\alpha_4\beta_2$ nAChRs but caused a significant reduction in nicotine's efficacy, as well as selective effects at the single-channel level. Taken together, the findings of Hans et al. indicate that menthol acts as a NAM of nAChRs.

3.3.4 α_7 PAM Binding Site Characterization

From studies with a series of subunit chimeras, we have identified the transmembrane regions of α_7 as being critical in mediating the potentiation of agonist-evoked responses [48]. Furthermore, we have identified five transmembrane amino acids that, when mutated, significantly reduce potentiation of α_7 nAChRs. The amino acids we have identified are located within the α-helical transmembrane domains TM1 (S222 and A225), TM2 (M253), and TM4 (F455 and C459). Mutation of either A225 or M253 individually has particularly profound effects, reducing potentiation of EC_{20} concentrations of acetylcholine to a tenth of the level seen with wild-type α_7. Reference to homology models of the α_7 nAChR, based on the 4 Å structure of the *Torpedo* nAChR, indicates that the side chains of all five amino acids point toward an intra-subunit cavity located between the four α-helical transmembrane domains. Computer docking simulations predict that the allosteric compounds such as PNU-120596 (type II) and LSN-2087101 (type I) may bind within this intra-subunit cavity, much as neurosteroids and volatile anesthetics are thought to interact with GABA-A and glycine receptors. Our findings suggest that this is a conserved modulatory allosteric site within neurotransmitter-gated ion channels, and suggest that different classes of PAMs might bind to overlapping sites but induce quite different gating changes in α_7 receptors.

3.3.5 Ago-Allosteric Compounds

TQS (4-(1-napthyl)-3*a*,4,5,9*b*-tetrahydro-3*H*-cyclopenta[*c*]quinoline-8-sulphonamide) has been described previously as a type II α_7 nAChR PAM [49]. We recently showed [50] that 4BP-TQS(4-(4-bromophenyl)-3*a*,4,5,9*b*-tetrahydro-3*H*-cyclopenta[*c*]quinoline-8-sulphonamide), a compound that is similar in chemical structure to TQS, has potent, but atypical, "agonist" activity at α_7 nAChRs. We further demonstrated that these atypical α_7 agonists, or ago-allosteric compounds, bind to the same transmembrane allosteric site we identified before for more typical α_7 PAMs [48, 50], but they are able to activate α_7 nAChRs in the absence of agonist binding at the orthosteric site. This supports the idea that ligand-induced channel activation can be driven by both orthosteric and allosteric agonists, presumably by affecting allosteric transitions and stabilizing the open conformation of the receptor. With an eye on the therapeutic potential of these ago-allosteric α_7 compounds, we recently confirmed that this novel mechanism is not unique to recombinant receptors but it is active also in native neurons, including rat hippocampal cultures and human iPSCS-derived neurons [51].

3.4 *Areas of Interesting but Limited Advances*

Limited advances have occurred with the pharmacology of α_9/α_{10} nAChR agonists or antagonists. McIntosh and co-workers have identified non-peptide molecules that display efficacy in preclinical pain models that represent a small breakthrough in this area [52]. However, the physicochemical properties of these molecules will require significant advances before they can be further developed into clinical experimentation.

nAChR agonists have been known to modulate cellular proliferation. In 2010 Schedel and co-workers reported the expression of α_7 receptors on platelets [53], and in 2011 [54] they reported that MLA, used at concentrations known to antagonize α_7, blocks megakaryocyte differentiation. These findings suggest that α_7 agonists may therapeutically augment cell differentiation and offer a new approach to the treatment of thrombocytopenia. In 2013, Gahring and co-workers [55] provided evidence that a novel subpopulation of bone marrow cells containing α_7 nAChRs that include hematopoietic progenitor cells can re-populate an animal's inflammatory/immune system. The authors suggest that: "α_7 exhibits a pleiotropic role in the hematopoietic system that includes both the direct modulation of proinflammatory cell composition and later in the adult the role of modulating pro-inflammatory responses that would impact upon an individual's lifelong response to inflammation and infection."

Finally, Stegemann and co-workers demonstrated that tropisetron directly reduces collagen synthesis in human dermal fibroblasts via an α_7 nAChR-dependent, not 5-HT3 mechanism, and was effective in a mouse model of bleomycin-induced scleroderma [56]. If the antifibrogenic and antifibrotic effects of other α_7-preferring compounds can be replicated with similar effects, the unmet needs of treating fibrotic diseases such as scleroderma may have a safe and effective avenue.

4 Future Trends: Will the Phoenix Rise?

In the vernacular of venture capitalists "past performance does not predict future returns." For those vested in the field of CNS nicotinic drug discovery we hope this is true. Over the last decade there has been a tremendous investment both in the basic sciences around nAChRs as well as the commitment to find new medicines that would treat unmet medical needs. The approval of Varenicline in 2006 was viewed as just the first of several soon to be approved nicotinic therapeutics, as a dozen $\alpha_4\beta_2$- and a dozen α_7-preferring compounds (Tables 21.1 and 21.2) were in clinical development at the height of this endeavor (2009–2012).

Unfortunately, this development surge has waned significantly, primarily due to clinical efficacy/safety profiles that did not warrant continued development, such that in 2013 only 11 compounds have been confirmed to remain in development. Interestingly, the stock history of Targacept, one of the industry leaders in nAChR pharmacology and development, has been a remarkably reflective barometer of the drug discovery trends from 2008 to 2013 (see Fig. 21.1).

In 2013 the current focus of the clinical pipeline is on α_7-preferring mechanisms, while many of the $\alpha_4\beta_2$-preferring compounds have been discontinued (see Table 21.1). Three compounds to keep an eye on are ENV-6124 (EnVivo Pharmaceutics), ABT-126 (AbbVie), and TC-5619 (Targacept). Two of these three have shown promising Phase 2 data for treating domains of schizophrenia not currently addressed by available therapies such as negative symptoms and cognitive deficits. If the ongoing Phase 3 data remain supportive, then the field could expect a resurgence of interest with commensurate discovery and clinical investments. If not,

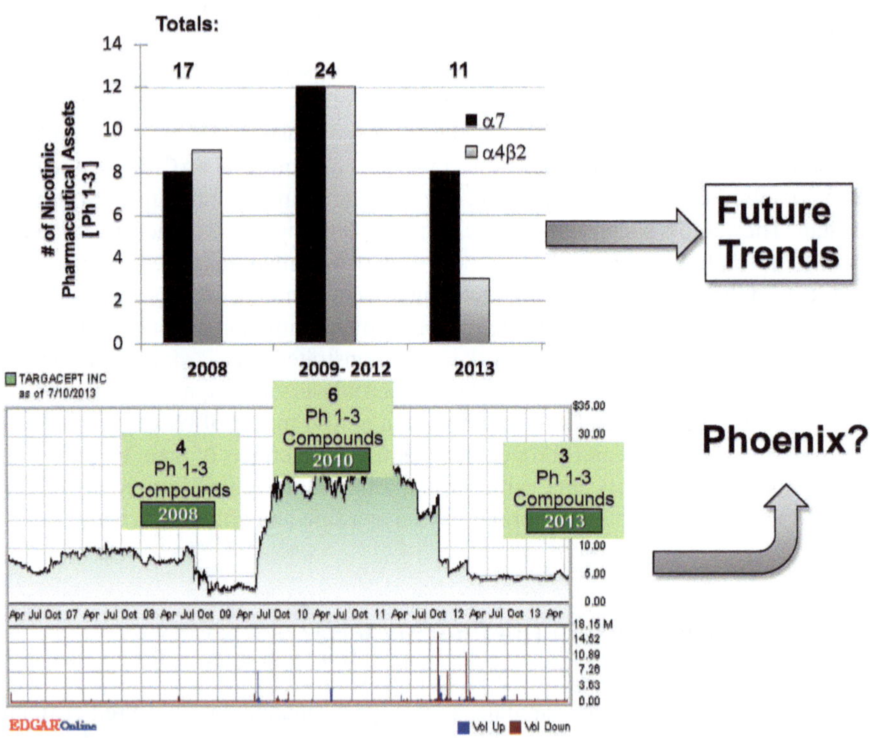

Fig. 21.1 Targacept stock: a barometer of the nAChR field (2008–2013)

and especially if ABT-126 and TC-5619 also do not meet their primary outcome objectives, then the fate observed with a majority of the $\alpha_4\beta_2$ approaches could possibly prevail for α_7-preferring nAChR treatments.

4.1 $\alpha_4\beta_2$-Preferring Clinical Compounds

We have chosen to summarize the status of those remaining compounds where the most information currently exists.

TC-1734: Also known as Ispronicline (and AZD3480) is a partial agonist at $\alpha_4\beta_2$, with excellent separation from α_7 (K_i = 11 nM $\alpha_4\beta_2$; >5,000 nM α_7) [8, 57, 58].

Targacept initiated a Phase 2b study evaluating the efficacy of a fixed dose of the full agonist TC-1734 head-to-head with donepezil, the marketed medication most often prescribed for Alzheimer's disease, in late 2011. The ongoing study is the second clinical trial of TC-1734 in Alzheimer's disease. The first was conducted by Astra-Zeneca and its outcome was inconclusive, as neither TC-1734 nor donepezil met the study's primary outcome measure. The ongoing study was designed to randomize approximately 300 patients to receive either a fixed dose of TC-1734 or donepezil daily over a 12-month period. The study was subject to a Special Protocol Assessment (SPA) agreement with the U.S. Food and Drug Administration as a potential registration trial. Results are not anticipated until 2014.

TC-5214: Is the (S)-(+)-enantiomer of mecamylamine, a pan nAChR antagonist (e.g., IC50s: $\alpha_4\beta_2$ = 2.5 μM; $\alpha_3\beta_4$ = 0.6 μM; α_7 = 6.9 μM). Targacept published preclinical work suggesting that this compound differentially interacted with the HA and LA states of $\alpha_4\beta_2$, an effect that was buttressed by the emerging data that compounds with similar properties had demonstrated activity in preclinical models of mood/depression [59, 60]. Targacept, together with Astra-Zeneca, mounted an aggressive clinical campaign to test this hypothesis in individuals with refractory depression. Reasons for the ultimate failure of these studies are unclear. Some have suggested it is related to the clinical recruitment sites chosen, the fact that other laboratories have had difficulty in reproducing the initial scientific findings on the HA/LA sites, or it simply, the hypothesis was proven incorrect for this patient population. Recent imaging results in patients with major depression [61] and bipolar disorder [62] showing that β_2*-nAChRs are reduced during depressive episodes may provide additional insight into the pharmacology required to achieve clinical benefit.

Notably, TC-5214 and Inversine™ are listed in 26 trials in ClinicalTrials.gov. A variety of industry studies have explored antagonism of $\alpha_4\beta_2$-like nAChRs across a spectrum of indications (see Table 21.1), while others are being conducted by NIDA to understand "Nicotinic Modulation of the Default Network of Resting Brain Function" [e.g., NCT01240616].

Targacept announced in September 2012 plans to pursue development of TC-5214 as a treatment for overactive bladder (OAB) (http://www.targacept.com/therapeutic-pipeline/TC-5214.cfm). They stated that they are planning to start a

Phase 2b clinical study in the second quarter of 2013. Given that (1) approximately 2,400 subjects in a different patient population demonstrated a well-established safety and tolerability profile for TC-5214; (2) additional preclinical studies have revealed physiological findings that they believe to be consistent with marketed treatments for OAB; and (3) TC-5214 is largely eliminated unchanged through the bladder, Targacept has taken a well-calculated decision to explore the low dose (0.5, 1.0, 2.0 mg, PO, BID) use of TC-5214 to treat OAB while minimizing unwanted systemic effects [NCT01868516]. These doses are ~10–25 % of those needed to lower blood pressure, or cause constipation, by racemic mecamylamine (Inversine™).

Varenicline (Chantix™/Champix™) is a potent partial agonist of $\alpha_4\beta_2$ and $\alpha_6/\alpha_4\beta_2$, and full agonist of α_7 [8, 63]. It is rather disappointing that more progress hasn't been made with this compound or the "$\alpha_4\beta_2$ approach" given the monumental clinical investment that Pfizer made, and the 222 clinical trials listed in ClinicalTrials.gov [http://clinicaltrials.gov/ct2/results?term=varenicline].

4.2 α_7-Preferring Clinical Compounds

The rationale of targeting α_7 nAChRs in cognition [64] and schizophrenia [65] has been recently reviewed.

EVP-6124: EVP-6124 is a partial α_7 agonist with high affinity for the receptor ($K_i = 4$ nM) and selectivity against all other nAChRs [8, 31]. While it also has affinity for the 5-HT3 receptor [IC50 ~ 10 nM], it is unclear whether this contributes to its favorable therapeutic profile. This compound has perhaps the most consistent development path of any other α_7 compound to date [see Table 21.2]. It has been exposed to ~1,700 people and has demonstrated reliable, linear oral pharmacokinetics (1–180 mg) that support once daily dosing. In 2009 EnVivo Pharmaceuticals announced an agreement with Mitsubishi Tanabe Pharma Corporation granting them exclusive rights to develop and commercialize EVP-6124 in Japan, Korea, Taiwan, Indonesia, and several other Asian markets.

In 2012 EnVivo released results from a Phase 2b trial of EVP-6124 for the treatment of AD. This double-blind, placebo-controlled study enrolled 409 patients with mild to moderate AD. Subjects, some of whom were receiving acetylcholinesterase inhibitor (AChEI) treatments (donepezil and rivastigmine) as well as some not on AChEIs treatment, received EVP-6124 0.3, 1.0, or 2.0 mg once daily or placebo for 24 weeks. The 2.0 mg dose met both primary endpoints with statistically significant positive effects on cognition ($p = 0.0189$) as measured by the Alzheimer's Disease Assessment Scale-Cognitive subscale-13 (ADAS-Cog-13) and clinical function ($p = 0.0253$) as measured by the Clinical Dementia Rating Scale Sum of Boxes (CDR-SB). The 2.0 mg dose also showed an improvement in cognition composite ($p = 0.0037$), memory composite ($p = 0.0088$), and executive function composite ($p = 0.0427$). EVP-6124 was well tolerated. The most frequent adverse events were mild to moderate gastrointestinal side effects in the 1.0 and 2.0 mg groups. Based on these data, EnVivo will continue to advance EVP-6124 for AD and is planning a Phase 3 trial for 2013.

EVP-6124 has also shown promising results in a randomized, double-blind, placebo-controlled, parallel, 12-week Phase 2b trial evaluating the safety and efficacy of two doses (0.3 and 1.0 mg per day) versus placebo in chronic schizophrenia patients on stable second-generation antipsychotic drugs (except clozapine). In addition to safety and tolerability, the trial's primary endpoint was overall cognition as measured by the CogState overall cognitive index. EVP-6124 had a clinically meaningful and statistically significant impact on patients' overall cognition as measured by the full CogState overall cognitive index, or "OCI" ($p=0.05$ for all patients treated with EVP-6124 versus placebo). Additionally, results from this Phase 2b trial demonstrated that patients treated with EVP-6124 (1.0 mg) showed clinically meaningful and statistically significant effects in key secondary endpoints: improvement in clinical function ($p=0.011$; as assessed by the Schizophrenia Cognition Rating Scale (SCoRS)) and reduction of the negative symptoms of schizophrenia ($p=0.028$; as measured by the Negative Symptom Scale of the Positive and Negative Symptoms Scale (PANSS)). Importantly, EVP-6124 was generally safe and well tolerated over the trial's 3-month dosing period. The most commonly reported adverse events were headache, nausea, and nasopharyngitis (all less than 4 %). There were no drug-related serious adverse events.

With these encouraging results EnVivo are working with leading clinicians and researchers to finalize the design of their Phase 3 Clinical Trials to maximize these learnings [http://www.envivopharma.com/news-item.php?id=32].

TC-5619: TC-5619 is a potent ($K_i=0.3$ nM) full agonist at α_7 nAChRs [8, 66]. Targacept has an ongoing Phase 2b study that is evaluating the compound as a treatment for negative symptoms and cognitive dysfunction in schizophrenia at sites in the USA and eastern Europe. This trial is a double-blind, placebo-controlled, parallel group study in schizophrenia patients with stable psychosis taking a fixed dose of an atypical antipsychotic. Following a 4-week screening period, patients will receive either one of two doses of TC-5619 (5 or 50 mg) or placebo, randomized in a 2:1:1 ratio (placebo:TC-5619 5 mg:TC-5619 50 mg). The primary outcome measure is change from baseline on the Scale for Assessment of Negative Symptoms (SANS). Key secondary outcome measures include the composite score on the CogState Schizophrenia Battery and the University of California, San Diego Performance-Based Skills Assessment, brief version [http://www.targacept.com/therapeutic-pipeline/TC-5619.cfm].

ABT-126: Limited published information exists on ABT-126 [64, 65]. Fourteen clinical trials with ABT-126 were registered in ClinicalTrials.gov as of August, 2013.

In July 2013 AbbVie reported results from a Phase 2a randomized, double-blind, placebo- and active-controlled, multi-center Phase 2 study in mild-to-moderate Alzheimer's dementia (Mini-Mental Status Examination [MMSE] score of 10–24, inclusive) [NCT00948909]. Subjects not currently receiving acetylcholinesterase inhibitors were randomized to ABT-126 (low or high dose QD), donepezil 10 mg QD, or placebo for 12 weeks. The relationship between ABT-126 plasma exposure and change in ADAS-Cog score/safety was evaluated as the doses selected were chosen based on efficacy seen in preclinical studies. The primary efficacy endpoint was the change from baseline to final evaluation in the 11-item Alzheimer's Disease Assessment Scale-cognitive subscale (ADAS-Cog) total score. The study randomized 274 patients with mean (SD) age of 73.9 (7.9) years. The mean (SD) baseline MMSE was 19.1 (3.8) and did not differ across treatment groups. High-dose ABT-126 showed a trend toward improvement on the 11-item ADAS-Cog (LS mean [SE] difference from placebo = −1.19 [0.90], $P = 0.095$); donepezil performed similarly (−1.43 [0.90], $P = 0.057$), demonstrating adequate assay sensitivity. Low-dose ABT-126 showed similar results to placebo. Exposure-response analyses suggested that higher ABT-126 exposures were associated with greater improvement on the ADAS-Cog ($P < 0.05$).

AbbVie concluded [Alzheimer's Association International Conference (AAIC) 07/14/2013—Abstract 01-06-03, Company Presentation] that there was evidence for an ABT-126 treatment effect on measures of cognition obtained in this proof-of-concept study. Improvement in the donepezil treatment group on the ADAS-Cog 11 item and ADAS-Cog 13 item indicates that the study design and conduct was adequate to demonstrate assay sensitivity. ABT-126 exhibited a dose–response and an exposure-response relationship for measure of cognition. ABT-126 exhibited a safety and tolerability profile that supports continued development. Overall, the efficacy and safety data support exploring a higher dose range in ongoing Phase 2b studies.

In 2012 Abbott initiated a Phase 2b randomized, double-blind, placebo- and active-controlled study to evaluate the efficacy and safety of ABT-126 in 410 patients with mild to moderate AD over 24 weeks [NCT01527916]. The primary outcome measure is the ADAS-cog; with secondary outcome measures of ADCS-ADL, Mini Mental Status Exam (MMSE); DEMentia Quality of Life (DEMQoL); Clinician Interview-Based Impression of Change (CIBIC-plus); Neuropsychiatric Inventory (NPI), and four others. Three doses of ABT-126 are to be compared with placebo and the active comparator, donepezil. AbbVie (previously Abbott) anticipates results in late 2013.

In parallel, two trials to examine cognitive disorders in schizophrenia [NCT01655680 and NCT01678755] have been recruiting during 2012–2013, with completion anticipated by mid-2014. These are randomized, double-blind, placebo-controlled, dose-ranging, parallel-group, Ph 2 studies of the safety and efficacy of ABT-126 in the treatment of cognitive deficits in schizophrenia. Eligible subjects will take study drug as an add-on treatment to their antipsychotic treatment regimen for 24 weeks. The primary outcome measure will be MCCB (MATRICS Consensus Cognitive Battery).

4.3 Other nAChR-Related Therapeutics

It is easy to forget that efforts are continuing to improve therapeutic product for neuromuscular transmission, smoking cessation, and anthelmintics. This section will very briefly highlight these three therapeutic spaces.

4.3.1 Neuromuscular Blockers

Incremental improvements in blockers of the neuromuscular junction for use during surgical procedures have continued over the last several decades with varying advantages/disadvantages being recently reviewed [67]. They can be grouped into three main categories: (1) Short acting (onset 0.5–2 min/duration 5–30 min) that includes Mivacurium (Mivacron™), Rapacuronium (Rapion™), Rocuronium (Zemeron™), and succinylcholine; (2) Intermediate acting (onset 2–5 min/duration 20–90 min) that includes Atracurium (Tracrium™), cisatracurium (Nimbex™), pancuronium (Pavulon™), and vecuronium (Norcuron™); and (3) Long acting (onset 3–6 min/duration 70–100 min) that includes doxacurium (Nuromax™), pipecurium (Arduan™), and tubocurarine.

Compounds that reverse the actions of neuromuscular blockers have also been formally advanced. For example, neostigmine methylsulfate (Bloxiverz™), the first FDA-approved version of neostigmine, was only recently approved in May, 2013, to reverse the effects of non-depolarizing neuromuscular blocking agents (www.Drugs.com, 06/03/13). Neostigmine is a cholinesterase inhibitor that does not affect cholinergic transmission in the CNS, but will increase synaptic acetylcholine levels at the peripheral neuromuscular junction. In contrast, Merck (via merger with Schering) has continued to have challenges in getting Sugammadex™ (originally Org-25969) approved in the USA. Sugammadex™, initially endorsed by an FDA Advisory Committee back in 2008, was to be the first and only selective relaxant binding agent (SRBA) given to anesthesiologists to rapidly and predictably reverse within minutes any depth of muscle relaxation induced by rocuronium and vecuronium [68]. Remarkably, Sugammadex™ is marketed in 40 countries other than the USA with more than five million vials of Sugammadex™ having been sold as of March 2013. This highlights some of the uncertain challenges moving products through different regulatory agencies.

4.3.2 Smoking Cessation

Marketed smoking cessation agents include bupropion (Zyban™, GSK), varenicline (Chantix™/Champix™, Pfizer), and nicotine patches (Niquitin CQ™, GSK) which all interact with the nAChRs via differing mechanisms and with different degrees of efficacy [69, 70]. In Eastern Europe cytisine has shown efficacy to treat smoking cessation [71], whereas lobeline was shown to be ineffective in a Ph3 study [72]. Cahill and co-workers recently summarized the field of nicotine partial agonists for smoking cessation [73].

Over the last 2 years numerous alternative approaches have stalled in development or have been discontinued. Phase 1 efforts that are stalled and are awaiting potential development partners include Nicotine MDTS (Acrux), ARD-1600 (Aradigm), QuitPak (Cary Pharmaceuticals), SEL-068 (Selecta), and Nicotine patch (NAL Pharmaceuticals). Phase 2 efforts that have been terminated or are assumed to be terminated include Niccine (Independent Pharma), TA-NIC (Celtic Pharma), and NIC002 (Cytos/Novartis). NicVax (Nabi Pharma/GSK) a long-term

Phase 3 effort failed to meet their primary endpoints resulting in GSK declining its licensing option.

Two efforts that remain in development are EVP-6124 (EnVivo's α_7-preferring agonist), a Phase 2 investigator initiated trial that is expected to finish in 2014, and PF-05402536/PF-06413367 (Pfizer), Phase 1 vaccines. Whether the promising aspects of $\alpha_6/\alpha_4\beta_2$ modulation [25, e.g., Targacept/GSK identify TC-5653 as dual $\alpha_6/\alpha_4\beta_2$ ligand] will have the opportunity to be explored further in the clinic may depend upon the success of the nicotinic field as a whole.

4.3.3 Anthelmintics

An area of drug discovery that has been grossly neglected is therapeutics that target poverty-related tropical diseases caused by intestinal nematodes: ascariasis (caused by *Ascaris lumbricoides*), trichuriasis (caused by *Trichuris trichiura* or whipworm), and hookworm disease (caused by *Necator americanus* and *Ancylostoma duodenale*). These parasites (Hookworms, Ascaris, and Trichuris or HAT) are amongst the most common human parasitic infections, with an estimated 0.6 billion people infected with hookworms, ~1 billion infected with Ascaris, and ~0.7 billion infected with Trichuris [74]. Not surprisingly, HAT infections have been reported to have an impact on human growth, nutrition, fitness, stature, metabolism, cognition, immunity, school, attendance/performance, earnings, and pregnancy [74–77]. Despite this impact, HAT infections remain one of the most prevalent and important infectious diseases in the world where few treatment options exist. The nAChR agonists levamisol and pyrantel belong to one of the two classes of anthelmintics approved by The World Health Organization (WHO) [78]. Tribendimidine is the most recent member of this class that has been entered into human trials in China [79]. Finding a safe and effective treatment not subject to treatment resistance would have tremendous global health and economic impact.

4.4 Critical Technical Advances

In the paragraphs below we discuss emerging preclinical science that, to our opinion, will help deliver the nicotinic drugs of the future. Two areas are related to a better understanding of the functional binding properties of the drugs to nAChRs, both in crystal structures and in the human living brain, respectively. A third one, which will also act as a bridge to the study of the human brain, is the use of iPSC-derived human neurons, from both normal human volunteers (NHVs) and patients.

4.4.1 Crystal Structures and Molecular Modeling

A critical advance to the crystallography of nAChRs was given by the identification, purification, and crystallization of a soluble AChBP from snails (see above and [3, 8]). Several homology models have been developed in recent years to understand drug binding sites on these extracellular receptor domains. More recently, Li et al. [80] developed an α_7-AChBP chimera and determined X-ray crystal structures of the resulting pentamer and its complex with the agonist epibatidine.

Because the ligand-binding site and flanking regions consist entirely of α_7 residues, the structures provide the highest resolution images that have so far been obtained of the AChR in regions that govern ligand recognition and the initial steps in signal transduction. Furthermore, the structures provide realistic templates for computational drug design, as well as bases for probing structure–function relationships of the physiologically and clinically important neuronal α_7 AChR.

4.4.2 Emerging nAChR-Related PET Traces

Successful imaging of nAChR with positron-emission tomography (PET) or single-photon emission computed tomography (SPECT) would dramatically enhance our ability to evaluate novel nicotinic drugs, on one hand, and to better understand the role of these receptors in physiology and in diseases. PET imaging of $\alpha_4\beta_2$ nAChR in human subjects is a current reality [81], and very recently has shown intriguing insights into the similarities in the direction of changes of β_2*-containing nAChRs during the depressive component of bipolar disorder [63] and depression [62], as well as the differences in dynamic changes observed during the euthymic stage of bipolar, suggesting that there can be rapid neuroplastic changes in nAChR circuits during the course of these mood swings. These data highlight that some patient populations are in a very dynamic, evolving process. Use of imaging tools such as this will be critical in understanding which nAChR therapy should be used, and when.

Unfortunately, we still lack human PET radioligands for α_7 nAChRs. Abbott Laboratories have recently described [^{11}C]A-833834 and [^{11}C]A-752274 as potential PET tracers for imaging α_7 nAChRs [82]. Also Rotering et al. have disclosed promising data with [^{18}F]NS14490 [83]. From a preclinical point of view, a new α_7-specific radioligand has been described that has the advantage of being an agonist, and looks promising at least for ex vivo binding assays [84].

4.4.3 First Data on nAChRs in iPSC-Derived Human Neurons

Nordberg and colleagues described human embryonic derived stem cells that upon differentiation into neurons expressed mRNA for some nicotinic receptors, α_3, α_4, and α_7 in particular [85]. However, no functional nAChR-mediated responses were found. Young et al. reported on a broad ion channel characterization of another embryonically derived human stem cell differentiated into neurons [86]. They also showed the presence of mRNA for some nAChR subunits but did not report any functional responses. We recently showed that an iPSC-derived human neuronal cell line with forebrain characteristics expresses functional α_7 nAChR [87] as well as mRNA related to other nAChR subunits (manuscript in preparation). These are exciting developments that will certainly help testing new drugs on native human nAChRs and help translation into the clinic.

5 Conclusions

Basic science is driving an exciting new wave of knowledge on the molecular structure of CNS nAChRs, the identification of drugs that affects more subtly the different subtypes of brain receptors, better understanding of CNS nAChRs dynamics/plasticity in human diseases and, importantly from a translational point of view, better biomarkers to drive rational clinical drug investigations. The promise of efficacy in negative symptoms and cognitive impairments associated with schizophrenia would re-invigorate commitment by the pharmaceutical industry to invest in nicotinic medicines across a number of maladies that literally affect billions of people.

References

1. Arneric SP, Holladay M, Williams M. Neuronal nicotinic receptors: a perspective on two decades of drug discovery research. Biochem Pharmacol. 2007;74:1092–101.
2. Heinemann S, Boulter J, Deneris E, Conolly J, Duvoisin R, Papke R, et al. The brain nicotinic acetylcholine receptor gene family. Prog in Brain Res. 1990;86:195–203.
3. Smit AB, Syed NI, Schaap D, van Minnen J, Klumperman J, Kits KS, et al. A glia-derived acetylcholine-binding protein that modulates synaptic transmission. Nature. 2001;411:261–8.
4. Karlin A. Emerging structure of the nicotinic acetylcholine receptors. Nat Rev Neurosci. 2002;3:102–14.
5. Dani JA, Bertrand D. Ann Rev Pharmacol Toxicol. 2007;47:699–729.
6. Gotti C, Moretti M, Gaimarri A, Zanardi A, Clementi F, Zoli M. Heterogeneity and complexity of native brain neuronal nicotinic receptors. Biochem Pharmacol. 2007;74:1102–11.
7. Taly A, Corringer P-J, Guedin D, Pierre Lestage P, Changeux J-P. Nicotinic receptors: allosteric transitions and therapeutic targets in the nervous system. Nat Rev Drug Discov. 2009;8:733–50.
8. Hurst R, Rollema H, Bertrand D. Nicotinic acetylcholine receptors: from basic science to therapeutics. Pharmacol Ther. 2013;137:22–54.
9. Zwart R, Vijverberg HP. Four pharmacologically distinct subtypes of alpha4beta2 nicotinic acetylcholine receptor expressed in Xenopus laevis oocytes. Mol Pharmacol. 1998;54:1124–31.
10. Liu Q, Huang Y, Xue F, Simard A, DeChon J, Li G, et al. A novel nicotinic acetylcholine receptor subtype in basal forebrain cholinergic neurons with high sensitivity to amyloid peptides. J Neurosci. 2009;29:918–29.
11. Liu Q, Huang Y, Shen J, Steffensen S, Wu J. Functional α7β2 nicotinic acetylcholine receptors expressed in hippocampal interneurons exhibit high sensitivity to pathological level of amyloid β peptides. BMC Neurosci. 2012;13:155–66.
12. Groot-Kormelink PJ, Boorman JP, Sivilotti LG. Formation of functional alpha3beta4alpha5 human neuronal nicotinic receptors in Xenopus oocytes: a reporter mutation approach. Br J Pharmacol. 2001;134:789–96.
13. Kato AS, Gill MB, Yu H, Nisenbaum E, Bredt DS. TARPs differentially decorate AMPA receptors to specify neuropharmacology. Trends Neurosci. 2010;33:2414–8.
14. Boulin T, Raptil G, Briseno-Roal L, Stigloher C, Richmond JE, Paoletti P, et al. Positive modulation of a Cys-loop acetylcholine receptor by an auxiliary transmembrane subunit. Nat Neurosci. 2012;15:1374–81.
15. Miwa JM, Ibanez-Tallon I, Crabtree GW, Sánchez R, Sali A, Role LW, et al. Lynx1, an endogenous toxin-like modulator of nicotinic acetylcholine receptors in the mammalian CNS. Neuron. 1999;23:105–14.

16. Chimienti F, Hogg RC, Plantard L, Lehmann C, Brakch N, Fischer J, et al. Identification of SLURP-1 as an epidermal neuromodulator explains the clinical phenotype of Mal de Meleda. Hum Mol Genet. 2003;12:3017–24.
17. Zwart R, Broad LM, Xi Q, Lee M, Moroni M, Bermudez I, Sher E. 5-I A-85380 and TC-2559 differentially activate heterologously expressed α4β2 nicotinic receptors. Eur J Pharmacol. 2006;539:10–7.
18. Carbone AL, Moroni M, Groot-Kormelink PJ, Bermudez I. Pentameric concatenated (alpha4)(2)(beta2)(3) and (alpha4)(3)(beta2)(2) nicotinic acetylcholine receptors: subunit arrangement determines functional expression. Br J Pharmacol. 2009;156:970–81.
19. Xiao Y, Fan H, Musachio JL, Wei ZL, Chellappan SK, Kozikowski AP, et al. Sazetidine-A, a novel ligand that desensitizes α4β2 nicotinic acetylcholine receptors without activating them. Mol Pharmacol. 2006;70:1454–60.
20. Cucchiaro G, Xiao Y, Gonzalez-Sulser A, Kellar KK. Analgesic effects of Sazetidine-A, a new nicotinic cholinergic drug. Anesthesiology. 2008;109:512–9.
21. Zwart R, Carbone AL, Moroni M, Bermudez I, Mogg AJ, Folly EA, et al. Sazetidine-A is a potent and selective agonist at native and recombinant $\alpha_4\beta_2$ nicotinic acetylcholine receptors. Mol Pharmacol. 2008;73:1838–43.
22. Ussing CA, Hansen CP, Petersen JG, Jensen AA, Rohde LAH, Ahring PK, et al. Synthesis, pharmacology, and biostructural characterization of novel α4β2 nicotinic acetylcholine receptor agonists. J Med Chem. 2013;56:940–51.
23. Gotti C, Moretti M, Meinerz NM, Clementi F, Gaimarri A, Collins AC, et al. Partial deletion of the nicotinic cholinergic receptor α_4 or β_2 subunit genes changes the acetylcholine sensitivity of receptor-mediated $^{86}Rb^+$ efflux in cortex and thalamus and alters relative expression of α4 and β2 subunits. Mol Pharmacol. 2008;73:1796–807.
24. Champtiaux N, Gotti C, Cordero-Erausquin M, David DJ, Przybylski C, Lena C, et al. Subunit composition of functional nicotinic receptors in dopaminergic neurons investigated with knock-out mice. J Neurosci. 2003;23:7820–9.
25. Breining SR, Bencherif M, Grady SR, Whiteaker P, Marks MJ, et al. Evaluation of structurally diverse neuronal nicotinic receptor ligands for selectivity at the α_6* subtype. Bioorg Med Chem Lett. 2009;19:4359–63.
26. Khiroug SS, Harkness PC, Lamb PW, Sudweeks SN, Khiroug L, Millar NS, et al. Rat nicotinic ACh receptor α_7 and β_2 subunits co-assemble to form functional heteromeric nicotinic receptor channels. J Physiol. 2002;540(2):425–34.
27. Murray TA, Bertrand D, Papke RL, George AA, Pantoja R, Srinivasan R, et al. $\alpha_7\beta_2$ nAChRs assemble and function, and are activated primarily via their α_7-α_7 interfaces. Mol Pharmacol. 2012;81:175–88.
28. Araud T, Graw S, Berger R, Lee M, Neveu E, Bertrand D, et al. The chimeric gene CHRFAM7A, a partial duplication of the CHRNA7 gene, is a dominant negative regulator of α_7* nAChR function. Biochem Pharmacol. 2011;82:904–14.
29. De Lucas-Cerrillo AM, Maldifasi MC, Arnalich F, Renart J, Atienza G, Serantes R, et al. Function of partially duplicated human α_7 nicotinic receptor subunit CHRFAM7A gene: potential implications for the cholinergic anti-inflammatory response. J Biol Chem. 2011;286:594–606.
30. Sinkus M, Lee MJ, Gault J, Logel J, Short M, Freedman R, et al. A 2-base pair deletion polymorphism in the partial duplication of the alpha7 nicotinic acetylcholine gene (CHRFAM7A) on chromosome 15q14 is associated with schizophrenia. Brain Res. 2009;1291:1–11.
31. Prickaerts J, van Goethem NP, Chesworth R, Shapiro G, Boess FG, Methfessel C, et al. EVP-6124, a novel and selective α_7 nicotinic acetylcholine receptor partial agonist, improves memory performance by potentiating the acetylcholine response of α_7 nicotinic acetylcholine receptors. Neuropharmacology. 2012;62:1099–110.
32. Werkheiser JL, Sydserff S, Hubbs SJ, Ding M, Eisman MS, Perry D, et al. Ultra-low exposure to alpha-7 nicotinic acetylcholine receptor partial agonists elicits an improvement in cognition that corresponds with an increase in alpha-7 receptor expression in rodents: implications for low dose clinical efficacy. Neuroscience. 2011;186:76–87.

33. Bertrand D, Gopalakrishnan M. Allosteric modulation of nicotinic acetylcholine receptors. Biochem Pharmacol. 2007;74:1155–63.
34. Santos MD, Alkondon M, Pereira EF, Aracava Y, Eisenberg HM, Maelicke A, et al. The nicotinic allosteric potentiating ligand galantamine facilitates synaptic transmission in the mammalian central nervous system. Mol Pharmacol. 2002;61:1222–34.
35. Mantione E, Micheloni S, Alcaino C, New K, Mazzaferro S, et al. Allosteric modulators of $\alpha_4\beta_2$ nicotinic acetylcholine receptors. Fut Med Chem. 2012;4:2217–30.
36. Pandya AA, Yakel JL. Effects of neuronal nicotinic acetylcholine receptor allosteric modulators in animal behavior studies. Biochem Pharmacol. 2013;86:1054. doi:10.1016/j.bcp.2013.05.018.
37. Broad LM, Zwart R, Pearson KH, Lee M, Wallace L, McPhie GI, et al. Identification and pharmacological profile of a new class of selective nicotinic acetylcholine receptor potentiators. J Pharmacol Exp Ther. 2006;318:1108–17.
38. De Filippi G, Mogg AJ, Phillips KG, Zwart R, Sher E, Chen Y. The subtype-selective nicotinic acetylcholine receptor positive allosteric potentiator 2087101 differentially facilitates neurotransmission in the brain. Eur J Pharmacol. 2010;643:218–24.
39. Springer SK, Woodin KS, Berry V, Boezio AA, Cao L, Clarkin K, et al. Synthesis and activity of substituted carbamates as potentiators of the $\alpha_4\beta_2$ nicotinic acetylcholine receptor. Bioorg Med Chem Lett. 2008;18:5643–7.
40. Grupe M, Jensen AA, Ahring PK, Christensen JK, Grunnet M. Unravelling the mechanism of action of NS9283, a positive allosteric modulator of $(\alpha_4)3(\beta_2)2$ nicotinic ACh receptors. Br J Pharmacol. 2013;168:2000–10.
41. Zhu CZ, Chin CL, Rustay NR, Zhong C, Mikusa J, Chandran P, et al. Potentiation of analgesic efficacy but not side effects: co-administration of an $\alpha_4\beta_2$ neuronal nicotinic acetylcholine receptor agonist and its positive allosteric modulator in experimental models of pain in rats. Biochem Pharmacol. 2011;82:967–76.
42. Hurst RS, Hajo's M, Raggenbass M, Wall TM, Higdon NR, Lawson JA, et al. A novel positive allosteric modulator of the α_7 neuronal nicotinic acetylcholine receptor: in vitro and in vivo characterization. J Neurosci. 2005;25:4396–405.
43. Sher E, Zwart R, De Filippi G, Baldwinson T, Pearson KH, McPhie GI, et al. Pharmacological and electrophysiological characterisation of a new class of selective alpha-7 nicotinic receptor potentiators. Society for Neurosci Abs. 2005; 951.10.
44. Palczynska MM, Jindrichova M, Gibb AJ, Millar NS. Activation of α_7 nicotinic receptors by orthosteric and allosteric agonists: influence on single-channel kinetics and conductance. Mol Pharmacol. 2012;82:910–7.
45. Moaddel R, Abdrakhmanova G, Kozak J, Jozwiak K, Toll L, Lucita Jimenez L, et al. Subanesthetic concentrationsof(R, S)-ketamine metabolites inhibit acetylcholine-evoked currents in α7 nicotinic acetylcholine receptors. Eur J Pharmacol. 2013;698:228–34.
46. Henderson BJ, González-Cestari TF, Yi B, Pavlovicz RE, Boyd RT, Li C, et al. Defining the putative inhibitory site for a selective negative allosteric modulator of human $\alpha_4\beta_2$ neuronal nicotinic receptors. ACS Chem Neurosci. 2012;3:682–92.
47. Hans M, Wilhelm M, Swandulla D. Menthol suppresses nicotinic acetylcholine receptor functioning in sensory neurons via allosteric modulation. Chem Senses. 2012;37:463–9.
48. Young GT, Zwart R, Walker AS, Sher E, Millar NS. Potentiation of α_7 nicotinic acetylcholine receptors via an allosteric transmembrane site. Proc Natl Acad Sci U S A. 2008;105:14686–91.
49. Grønlien JH, Håkerud M, Ween H, Thorin-Hagene K, Briggs CA, Gopalakrishnan M, et al. Distinct profiles of α7 nAChR positive allosteric modulation revealed by structurally diverse chemotypes. Mol Pharmacol. 2007;72:715–24.
50. Gill JK, Savolainena M, Young GT, Zwart R, Sher E, Millar NS. Agonist activation of α_7 nicotinic acetylcholine receptors via an allosteric transmembrane site. Proc Natl Acad Sci U S A. 2011;108:5867–72.
51. Gill JK, Chatzidaki A, Ursu D, Sher E, Millar NS. Contrasting properties of α_7-selective orthosteric and allosteric agonists examined on native nicotinic acetylcholine receptors. PLoS One. 2013;8(1):e55047. doi:10.1371/journal.pone.0055047.

52. Wala EP, Crooks PA, McIntosh JM, Holtman Jr JR. Novel small molecule a9-a10 nicotinic receptor antagonist prevents and reverses chemotherapy-evoked neuropathic pain in rats. Anesth Analg. 2012;115:713–20.
53. Schedel A, Thornton S, Schloss P, Klüter H, Bugert P. Human platelets express functional alpha7-nicotinic acetylcholine receptors. Arterioscler Thromb Vasc Biol. 2011;31:928–34.
54. Thornton S, Schedel A, Besenfelder S, Klüter H, Bugert P. Cholinergic drugs inhibit in vitro megakaryopoiesis via the alpha7-nicotinic acetylcholine receptor. Platelets. 2011;22:390–5.
55. Gahring LC, Enioutina EY, Myers EJ, Spangrude GJ, Efimova OV, Kelley TW, Tvrdik P, Capecchi MR, Rogers SW. Nicotinic receptor alpha7 expression identifies a novel hematopoietic progenitor lineage. PLoS One. 2013;8(3):e57481. doi:10.1371/journal.pone.0057481.
56. Stegemann A, Sindrilaru A, Eckes B, del Rey A, Heinick A, Schulte JS, et al. Tropisetron suppresses collagen synthesis in skin fibroblasts via α_7 nicotinic acetylcholine receptor and attenuates fibrosis in a scleroderma mouse model. Arthritis Rheum. 2013;65:792–804.
57. Gatto GJ, Bohme GA, Caldwell WS, Letchworth SR, Traina VM, Obinu MC, et al. TC-1734: an orally active neuronal nicotinic acetylcholine receptor modulator with antidepressant, neuroprotective and long-lasting cognitive effects. CNS Drug Rev. 2004;10:147–66.
58. Dunbar G, Kuchibhatla R, Lee G. A randomized double-blind study comparing 25 and 50 mg TC-1734 (AZD3480) with placebo, in older subjects with age-associated memory impairment. J Psychopharmacol. 2011;25:1020–9.
59. Lippiello PM, Beaver JS, Gatto GJ, James JW, Jordan KG, Traina VM, et al. TC-5214 (S-(+)-mecamylamine): a neuronal nicotinic receptor modulator with antidepressant activity. CNS Neurosci Ther. 2008;14:266–77.
60. Picciotto MR, Addy NA, Mineur YS, Brunzell DH. It is not "either/or": activation and desensitization of nicotinic acetylcholine receptors both contribute to behaviors related to nicotine addiction and mood. Prog Neurobiol. 2008;84:329–42.
61. Saricicek A, Esterlis I, Maloney KH, Mineur YS, Ruf BM, Muralidharan A, et al. Persistent β2*-nicotinic acetylcholinergic receptor dysfunction in major depressive disorder. Am J Psychiatry. 2012;169:851–9.
62. Hannestad JO, Cosgrove KP, Dellagioia NF, Perkins E, Bois F, Bhagwagar Z, et al. Changes in the cholinergic system between bipolar depression and euthymia as measured with [^{123}I]5IA single photon emission computed tomography. Biol Psychiatry. 2013;74:768. doi:10.1016/j.biopsych.2013.04.004. pii: S0006-3223(13)00335-1.
63. Rollema H, Shrikhande A, Ward KM, Tingley 3rd FD, Coe JW, O'Neill BT, et al. Pre-clinical properties of the alpha4beta2 nicotinic acetylcholine receptor partial agonists varenicline, cytisine and dianicline translate to clinical efficacy for nicotine dependence. Br J Pharmacol. 2010;160:334–45.
64. Wallace TL, Porter RH. Targeting the nicotinic alpha7 acetylcholine receptor to enhance cognition in disease. Biochem Pharmacol. 2011;82:891–903.
65. Wallace TL, Bertrand D. Alpha7 neuronal nicotinic receptors as a drug target in schizophrenia. Expert Opin Ther Targets. 2013;17:139–55.
66. Hauser TA, Kucinski A, Jordan KG, Gatto GJ, Wersinger SR, Hesse RA, et al. TC-5619: an alpha7 neuronal nicotinic receptor-selective agonist that demonstrates efficacy in animal models of the positive and negative symptoms and cognitive dysfunction of schizophrenia. Biochem Pharmacol. 2009;78:803–12.
67. Piriyapatsom A, Bittner EA, Hines J, Schmidt UH. Sedation and paralysis. Respir Care. 2013;58:1024–37.
68. Yang LP, Keam SJ. Sugammadex: a review of its use in anaesthetic practice. Drugs. 2009;69:919–42.
69. Wu P, Wilson K, Dimoulas P, Mills EJ. Effectiveness of smoking cessation therapies: a systematic review and meta-analysis. BMC Public Health. 2006;6:300.
70. Arias HR. Molecular interaction of bupropion with nicotinic acetylcholine receptors. J Pediat Biochem. 2010;1:185–97.
71. Hajek P, McRobbie H, Myers K. Efficacy of cytisine in helping smokers quit: systematic review and meta-analysis. Thorax. 2013;68:1037. doi:10.1136/thoraxjnl-2012-203035.

72. Glover ED, Rath JM, Sharma E, Glover PN, Laflin M, Tonnesen P, et al. A multicenter phase 3 trial of lobeline sulfate for smoking cessation. Am J Health Behav. 2010;34:101–9.
73. Cahill K, Stead LF, Lancaster T. Nicotine receptor partial agonists for smoking cessation. Cochrane Database Syst Rev. 2012; 4: CD006103. doi: 10.1002/14651858. CD006103.pub6
74. Bethony J, Brooker S, Albonico M, Geiger SM, Loukas A, et al. Soil transmitted helminth infections: ascariasis, trichuriasis, and hookworm. Lancet. 2006;367:1521–32.
75. Albonico M, Allen H, Chitsulo L, Engels D, Gabrielli AF, et al. Controlling soil-transmitted helminthiasis in pre-school-age children through preventive chemotherapy. PLoS Negl Trop Dis. 2008;2:e126.
76. Hall A, Hewitt G, Tuffrey V, de Silva N. A review and meta-analysis of the impact of intestinal worms on child growth and nutrition. Matern Child Nutr. 2008;1:118–236.
77. Hotez PJ, Brindley PJ, Bethony JM, King CH, Pearce EJ, et al. Helminth infections: the great neglected tropical diseases. J Clin Invest. 2008;118:1311–21.
78. Keiser J, Utzinger J. Efficacy of current drugs against soil-transmitted helminth infections: systematic review and meta-analysis. JAMA. 2008;299:1937–48.
79. Hu Y, Xiao S-H, Aroian RV. The new anthelmintic tribendimidine is an L-type (Levamisole and Pyrantel) nicotinic acetylcholine receptor agonist. PLoS Negl Trop Dis. 2009;3:e499. doi:10.1371/journal.pntd.0000499.
80. Li S-H, Huang S, Bren N, Noridomi K, Dellisanti CD, Sine SM, et al. Ligand-binding domain of an α_7-nicotinic receptor chimera and its complex with agonist. Nat Neurosci. 2011;14: 1253–60.
81. Brasic JR, Cascella N, Kumar A, Zhou Y, Hilton J, Raymont V, et al. Positron emission tomography experience with 2-[^{18}F]fluoro-3-(2(S)-azetidinylmethoxy)pyridine (2-[^{18}F]FA) in the living human brain of smokers with paranoid schizophrenia. Synapse. 2012;66:352–68.
82. Horti AG, Ravert HT, Gao Y, Holt DP, Bunnelle WH, Schrimpf MR. Synthesis and evaluation of new radioligands [^{11}C]A-833834 and [^{11}C]A-752274 for positron-emission tomography of α_7-nicotinic acetylcholine receptors. Nucl Med Biol. 2013;40:395–402.
83. Rötering S, Scheunemann M, Fischer S, Hiller A, Peters D, Deuther-Conrad W, et al. Radiosynthesis and first evaluation in mice of [^{18}F]NS14490 for molecular imaging of α_7 nicotinic acetylcholine receptors. Bioorg Med Chem. 2013;21:2635–42.
84. Maier DL, Hill G, Ding M, Tuke D, Einstein E, Gurley D, et al. Pre-clinical validation of a novel alpha-7 nicotinic receptor radiotracer, [^{3}H]AZ11637326: target localization, biodistribution and ligand occupancy in the rat brain. Neuropharmacology. 2011;61:161–71.
85. Nilbratt M, Porras O, Marutle A, Hovatta O, Nordberg A. Neurotrophic factors promote cholinergic differentiation in human embryonic stem cell-derived neurons. J Cell Mol Med. 2010;14:1476–84.
86. Young A, Machacek DW, Dhara SK, Nacleish PR, Benveniste M, Dodla MC, et al. Ion channels and ionotropic receptors in human embryonic stem cell derived neural progenitors. Neuroscience. 2011;192:793–805.
87. Mazurov AA, Miao L, Bhatti BS, Strachan JP, Akireddy S, Murthy S, et al. Discovery of 3-(5-chloro-2-furoyl)-3,7-diazabicyclo[3.3.0]octane (TC-6683, AZD1446), a novel highly selective $\alpha4\beta2$ nicotinic acetylcholine receptor agonist for the treatment of cognitive disorders. Med Chem. 2012;55:9181–94.
88. Tonstad S, Holme I, Tønnesen P. Dianicline, a novel $\alpha_4\beta_2$ nicotinic acetylcholine receptor partial agonist, for smoking cessation: a randomized placebo-controlled clinical trial. Nicotine Tob Res. 2011;13:1–6.
89. Rowbotham MC, Duan WR, Thomas J, Nothaft W, Backonja MM. A randomized, double-blind, placebo-controlled trial evaluating the efficacy and safety of ABT-594 in patients with diabetic peripheral neuropathic pain. Pain. 2009;146:245–52.
90. Dutta S, Hosmane BS, Awni WM. Population analyses of efficacy and safety of ABT-594 in subjects with diabetic peripheral neuropathic pain. AAPS J. 2012;14:168–75.
91. Apostol G, Abi-Saab W, Kratochvil CJ, Adler LA, Robieson WZ, Gault LM, et al. Efficacy and safety of the novel $\alpha_4\beta_2$ neuronal nicotinic receptor partial agonist ABT-089 in adults with attention-deficit/hyperactivity disorder: a randomized, double-blind, placebo-controlled crossover study. Psychopharmacology. 2012;219:715–25.

92. Bain EE, Robieson W, Pritchett Y, Garimella T, Abi-Saab W, Apostol G, et al. A randomized, double-blind, placebo-controlled phase 2 study of α4β2 agonist ABT-894 in adults with ADHD. Neuropsychopharmacology. 2013;38:405–13.
93. Rowbotham MC, Arslanian A, Nothaft W, Duan WR, Best AE, Pritchett Y, et al. Efficacy and safety of the α4β2 neuronal nicotinic receptor agonist ABT-894 in patients with diabetic peripheral neuropathic pain. Pain. 2012;153:862–8.
94. Freedman R, Olincy A, Buchanan RW, Harris JG, Gold JM, Johnson L, et al. Initial phase 2 trial of a nicotinic agonist in schizophrenia. Am J Psychiatry. 2008;165:1040–7.
95. Malysz J, Anderson DJ, Grønlien JH, Ji J, Bunnelle WH, Håkerud M, et al. In vitro pharmacological characterization of a novel selective alpha7 neuronal nicotinic acetylcholine receptor agonist ABT-107. J Pharmacol Exp Ther. 2010;334:863–74.
96. Bitner RS, Bunnelle WH, Decker MW, Drescher KU, Kohlhaas KL, Markosyan S, et al. In vivo pharmacological characterization of a novel selective alpha7 neuronal nicotinic acetylcholine receptor agonist ABT-107: preclinical considerations in Alzheimer's disease. J Pharmacol Exp Ther. 2010;334:875–86.
97. Radek RJ, Robb HM, Stevens KE, Gopalakrishnan M, Bitner RS. Effects of the novel α_7 nicotinic acetylcholine receptor agonist ABT-107 on sensory gating in DBA/2 mice: pharmacodynamic characterization. J Pharmacol Exp Ther. 2012;343:736–45.
98. AlSharari SD, et al. Functional role of alpha7 nicotinic receptor in chronic neuropathic and inflammatory pain: studies in transgenic mice. Biochem Pharmacol. 2013;86:1201. http://dx.doi.org/10.1016/j.bcp.2013.06.018.

MIX
Papier aus verantwortungsvollen Quellen
Paper from responsible sources
FSC® C105338

If you have any concerns about our products, you can contact us on
ProductSafety@springernature.com

In case Publisher is established outside the EU, the EU authorized representative is:
Springer Nature Customer Service Center GmbH
Europaplatz 3, 69115 Heidelberg, Germany

Printed by Libri Plureos GmbH
in Hamburg, Germany